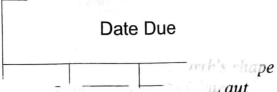

rth's shape

aut

This book investigates, through the problem of the earth's shape, part of the development of post-Newtonian mechanics by the Parisian scientific community during the first half of the eighteenth century. In the *Principia* Newton first raised the question of the earth's shape. John Greenberg shows how continental scholars outside France influenced efforts in Paris to solve the problem, and he also demonstrates that Parisian scholars, including Bouguer and Fontaine, did work that Alexis-Claude Clairaut used in developing his mature theory of the earth's shape. This evolution of Parisian mechanics proved not to be the replacement of a Cartesian paradigm by a Newtonian one, a replacement that might be expected from Thomas Kuhn's formulations about scientific revolutions, but a complex process instead involving many areas of research and contributions of different kinds from the entire scientific world. For example, Newtonian "normal" science does not take into account crucial developments in continental mathematics used to tackle the problem at issue.

Greenberg both explores the myriad of technical problems that underlie the historical development of part of post-Newtonian mechanics, which have been analyzed only rarely by Western scholars, and embeds his technical discussion in a framework that involves social and institutional history, politics, and biography. Instead of focusing exclusively on the historiographical problem, Greenberg shows as well that international scientific communication was as much a vital part of the scientific progress of individual nations during the first half of the eighteenth century as it is today.

THE PROBLEM OF
THE EARTH'S SHAPE FROM
NEWTON TO CLAIRAUT

*The rise of mathematical science in eighteenth-century
Paris and the fall of "normal" science*

JOHN L. GREENBERG

CAMBRIDGE
UNIVERSITY PRESS

CAMBRIDGE UNIVERSITY PRESS
Cambridge, New York, Melbourne, Madrid, Cape Town, Singapore,
São Paulo, Delhi, Dubai, Tokyo

Cambridge University Press
The Edinburgh Building, Cambridge CB2 8RU, UK

Published in the United States of America by Cambridge University Press, New York

www.cambridge.org
Information on this title: www.cambridge.org/9780521130998

First published 1995
This digitally printed version 2010

A catalogue record for this publication is available from the British Library

Library of Congress Cataloguing in Publication data
Greenberg, John Leonard, 1945–
The problem of the earth's shape from Newton to Clairaut : the rise of mathematical
science in eighteenth-century Paris and the fall of "normal" science/John L. Greenberg.
p.   cm.
Includes bibliographical references and index.
ISBN 0-521-38541-5
1. Earth – Figure – History. 2. Physical sciences – France – History – 18th century.
I. Title.
QB283.G74   1995
525′.1 – dc20   94-690
                                                                                      CIP

ISBN 978-0-521-38541-1 Hardback
ISBN 978-0-521-13099-8 Paperback

To my wife Maïté
without whose loving care, encouragement, and
help this book would never have been written

and

to my children Philippe and Sylvie

and

to the memory of my parents
Leonard S. Greenberg
(1913–1987)

and

Margaret O'Briant Greenberg
(1911–1992)

Contents

Illustrations

Preface

The story recounted in the following pages is a chapter in the history of physical sciences in France. During the period in which a large part of this story takes place, the physical sciences in France have usually been represented in ways that accord with the following scheme: the publication of Isaac Newton's *Principia* in 1687 rounds off a seventeenth-century European scientific revolution. Now, the seventeenth-century scientific revolution scarcely began with Newton, but an irreconcilable conflict between the Newtonian world system and the preceding "paradigm," the Cartesian world system, means that Newton's *Principia* was certainly a revolutionary work. Although "crucial experiments" need not enter into the genesis of a scientific revolution (Copernicus, for example, simply reinterpreted old data; it was not until much later that any new observations began to play a role in determining the course of the "Copernican Revolution"), they did function decisively in giving Newton the victory over Descartes. "Normal science" was the "mopping up" of Newton's system by the French, beginning in the 1730s and 40s. The story ends with the death of Descartes's last advocate. Cartesianism becomes a dead issue. Any reader of Kuhn (1962 or 1970) should recognize the jargon I have used here and have no trouble imagining the scheme. Moreover, the plan may well strike Kuhn's reader as thoroughly plausible. Thus before beginning my story, I offer the following words of forethought. The story may be viewed as simply depicting an inevitable mopping up of a section of Newton's *Principia* – the articulation of a "Newtonian paradigm," and its extension to cover anomalies, by means of "puzzle solving" within the "paradigm," to use Kuhn's language. But it might also involve considerations that go beyond the confines of Newtonianism alone. If so, the problem of what makes up normal science, which Kuhn contrasts with "revolutionary" science, is raised. In that case, deciding which of two interpretations of the story that follows makes the most sense requires dealing with current issues in the history and philosophy of science.

I shall not sketch here the conclusions that I have drawn from the following story. If I did I could risk reducing the story's verve. I have tried to write a narrative that keeps the reader wondering.

I will say, however, that problems of figures of equilibrium is a theme that reappears throughout the story. The story takes place in Paris during the first half of the eighteenth century. Today the investigation of problems of figures of equilibrium includes dynamical issues that were not dealt with during that time. The kinds of questions asked today involve solving functional equations and integral equations. Since the eighteenth century very powerful, highly ingenious methods have been developed to tackle these questions. Despite the invention of such mathematical tools, however, it is still not known to this day whether all figures of equilibrium of a rotating fluid mass that attracts in accordance with the universal inverse-square law of gravitation have been found. This gives an idea of just how difficult problems of equilibrium have become since the first half of the eighteenth century. The little concerning the technical aspects of problems of figures of equilibrium which appear in this story, including the advances that were made during the period that the story covers, could probably be summarized in a page or two. Even so, these few achievements were arrived at only with great effort, because the problems that were solved were still very difficult for the time. One of our aims is to understand how the problems were solved at all and the larger implications that such a success has.

Consequently, treating the questions that underlie the historiographic issue raised above necessitates that technical matters be examined in some depth and detail. However, it would be a mistake to think that the story that follows consists of nothing but the study of technical concerns. The technical discussion is embedded in a framework that involves social history, institutional history, politics, and biography that I interpret in ways that, to the best of my knowledge, cannot be found elsewhere. Remarks concerning social history, institutional history, politics, and biography are dispersed throughout the narrative, but there are also entire sections of the story that should in principle appeal to the historian who is not "internalist." The story's meaning goes beyond what could possibly emerge from a study confined solely to the technical side of the particular problems that make up the recurrent theme that I have already mentioned. If I manage to hold the reader's interest and attention, then I may succeed in transporting him to Paris as it existed during the first half of the eighteenth century.

In addition to answering this historiographic question, which involves current issues in the history and philosophy of science, the story has a moral that concerns the growth of science in a sense that differs from the one that is at issue in the historiographic problem. We shall find that international scientific communication was as much a vital part of the scientific progress of individual nations during the first half of the eighteenth century as it is today.

Research for this book was made possible by a University of Wisconsin travel grant in 1977, a National Science Foundation national needs postdoctoral fellowship in 1979–80, a Fulbright research scholarship in 1984–5, a National

Science Foundation scholar's award in 1984–5, the U.S.–France Exchange of Scientists in 1984–6, a visiting research associateship ("poste rouge") with the Centre National de la Recherche Scientifique in 1986, an American Council of Learned Societies fellowship in 1987–8, a National Science Foundation scholar's award in 1990–1, and the French Société de Secours des Amis des Sciences. I am grateful to the officials in charge of the Archives de l'Académie Royale des Sciences (Paris), Bibliothèque de l'Institut (Paris), Observatoire de Paris, Bibliothèque Nationale (Paris), Archives Nationales (Paris), Archives de l'Ecole Polytechnique (Palaiseau), Library of the Royal Society (London), and the Bernoulli Archives at the Universitätsbibliothek, Basel, for allowing me to examine their collections.

I thank Dan Siegel for accepting in 1979 an early version of this work as a Ph.D. "thesis." I profited greatly from René Taton's seminars at the Centre Alexandre Koyré in Paris in January and February of 1977, in 1979–81, and in 1984–5. I owe special thanks to Monsieur Taton who sponsored my National Science Foundation national needs postdoctoral fellowship in 1979–80 and my Fulbright research scholarship in 1984–5 and to Maurice Caveing who made me visiting research associate with the Centre National de la Recherche Scientifique in 1986.

I thank Bob Weinstock for helping me understand extremely difficult parts of Isaac Newton's *Principia* and for having saved me from making some serious errors pertaining to nomenclature. I also thank Clifford Truesdell for patiently answering my idiotic questions about elementary fluid mechanics. I thank John Pappas whose knowledge of the eighteenth-century French Enlightenment is vast and who taught me much about that Enlightenment. He helped me to portray the French men of science who appear in this book not simply as scientists but as *French* scientists. I thank Jim Casey for having saved me from making some great blunders concerning the application of the principle of solidification.

I also thank Dan Siegel, George Downs, and Bob Weinstock for providing the subsidy needed to publish this book.

I call attention to the fact that I only discovered David Beeson's *Maupertuis: An Intellectual Biography* while my own manuscript was being copyedited and that I only discovered Rob Iliffe's article of 1993 on Maupertuis, Jesper Lützen's article of 1994 on Liouville, and John Heilbron's *Weighing the Imponderables and Other Quantitative Science around 1800* while the page proofs were undergoing final correction.

This book was written in very difficult circumstances. Since October of 1983 I have been afflicted with spastic paraparesis, an insidiously progressive form of multiple sclerosis. I have been ill the whole time that I did much of the research and all of the writing of this book. I apologize for not having included a subject index. By the time that I received proofs I was too physically handicapped to handle the preparation of a subject index for such an intricate work. I am indebted to Benedicte Bilodeau at the Centre Alexandre Koyré, Claudine Pouret at the Archives de l'Académie Royale des Sciences, Claudine Billoux at the Archives de l'Ecole Polytechnique, Sally Grover at the Library of the Royal Society, and Fritz Nagel of the Bernoulli Edition for sending me material when I was no longer able to travel.

I also thank my faithful correspondents who keep constantly in touch by letter or by phone and who sometimes visit me. They have helped to keep me going. They are Clifford Truesdell, Paul Halmos, Bob Weinstock, Ivor Grattan-Guinness, Jim Cross, John Pappas, John Hirschfield, and George Downs.

I thank Helen Wheeler for cheerfully assisting in the production of this book. She had to put up with *a lot*.

I thank my son Philippe and daughter Sylvie whom I often neglected while working on the contents of this book, which means ever since they were born. Finally I thank Maïté, my wife and guardian angel for more than twenty-two years. She was a constant source of moral support in my struggle to write this book and is a constant source of moral support in my battle against my illness.

I

Isaac Newton's theory of a
flattened earth
(1687, 1713, 1726)

In the first edition of the *Principia* (1687), Isaac Newton reasoned that the earth cannot be a perfect sphere. Newton took the French astronomer Jean Richer's experiments with seconds pendulums in Cayenne, French Guyana, carried out in 1672–73, as the starting point that led him to conclude this. Richer found that a seconds pendulum had a shorter length in Cayenne than in Paris. To Newton this meant that the effective gravitational force per unit of mass, which I shall simply call the effective gravity hereafter, at the earth's surface had a smaller magnitude in the vicinity of the equator than in France. This Newton attributed in part to a decrease in the magnitude of the earth's centrifugal force of rotation with latitude. Newton assumed the earth initially to have been a fluid body, and he presupposed as well that in a fluid state the earth would be spherical, were it not for its rotation. He hypothesized that the centrifugal force of rotation caused the earth to flatten at its poles. As evidence to support his belief, Newton had John Flamsteed's and Gian-Domenico Cassini's telescopic observations of Jupiter's flattening at the poles.

I briefly sketch the version of Newton's theory of the earth's shape which appears in Book III of the third edition of the *Principia* (1726). By the 1730s, Parisian men of science were familiar with the third edition, and I shall be mainly concerned with scientific developments in Paris in the 1730s and 40s in the story that follows, although a considerable amount of background from earlier years must also be examined and discussed in connection with these developments, in order that they may be better understood. Since the values of the key parameters involved in the problem of the earth's shape do not change appreciably from one edition of the *Principia* to another, nothing is lost in restricting our attention to Newton's final presentation of his theory.[1]

Newton used the value of a degree of latitude between Amiens, France, and nearby Malvoisine, France, measured by the French astronomer Jean Picard in 1669–70, to determine the radius of an earth presumed to be homogeneous and spherical.[2] Knowing the earth's sidereal rate of diurnal rotation, Newton calculated the magnitude of the centrifugal force per unit of mass at the equator of the spherical earth. He also found the magnitude of the attraction, meaning the magnitude of the gravitational force per unit of mass, at the equator of the homogeneous spherical earth assumed to attract according to the universal inverse-square law in the following manner. Experiments made with seconds pendulums and with bodies falling freely in Paris furnished Newton with the magnitude of the component of effective gravity in Paris which is perpendicular to the earth's surface there. Knowing the latitude of Paris, Newton computed the magnitude of the component of centrifugal force per unit of mass in Paris perpendicular to the earth's surface there in terms of the magnitude of the centrifugal force per unit of mass at the equator of the spherical earth. Continuing to assume the earth to be spherical, Newton determined the magnitude of the attraction at the equator as follows: the magnitude of the attraction at the equator equals the magnitude of the attraction in Paris, which equals the magnitude of the component of effective gravity in Paris perpendicular to the earth's surface there plus the magnitude of the component of centrifugal force per unit of mass in Paris perpendicular to the earth's surface there. Newton then found the ratio of his calculated values of the magnitude of the centrifugal force per unit of mass and the magnitude of the attraction at the equator to be the fraction $\frac{1}{289}$.

(In 1690, Christiaan Huygens determined the same ratio to be this same fraction by hypothesizing that a central force of attraction, instead of Newton's universal inverse-square law of attraction, acts upon an earth assumed to be homogeneous and spherical, where Huygens supposed the center of force to be located at the center of the body that is attracted. In either case, the magnitude of the attraction along the surfaces of homogeneous *spherical* bodies is constant, while the variations of the magnitude of the effective gravity with latitude along the surfaces of Newton's and Huygens's spheres must be the same, since the two spheres have the same radii and rotate at the same rate. Hence Newton and Huygens arrived at the same fraction, $\frac{1}{289}$.)[3]

Newton then theorized that the earth was in reality a homogeneous body that attracts according to the universal inverse-square law and whose surface was an ellipsoid of revolution which is slightly flattened at its poles (see Figure 1.) He took the polar axis, which is the axis that the earth revolves around, to be the axis of symmetry of the ellipsoid of revolution. He postulated that the ratio of the magnitude of the centrifugal force per unit of mass at the equator to the magnitude of the attraction there remained $\frac{1}{289}$. Whatever the earth's actual shape, no one doubted that it differed only slightly from that of a sphere. Presumably then such a small deformation did not cause the ratio $\frac{1}{289}$ to change appreciably.[4]

To calculate the "ellipticity" $\varepsilon \equiv (E - P)/P$, a parameter that measures the degree of flattening or oblongness of a figure of revolution at its poles, where E is the equatorial radius of the figure and P is the polar radius of the figure, Newton assumed the earth to consist of homogeneous matter. He also supposed that two columns that join the center of the homogeneous body shaped like a flattened

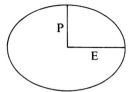

Figure 1. Meridian of a flattened ellipsoid of revolution.

ellipsoid of revolution to points at its surface and that lie along its polar and equatorial axes would balance. Expressed in terms of the calculations that Newton actually did, this assumption amounted to hypothesizing that the two columns weigh the same. Now, two columns that connect the center of a homogeneous solid body shaped like a flattened ellipsoid of revolution which revolves around its axis of symmetry (its polar axis) to points at its surface and which lie along the polar and equatorial axes of the body could certainly weigh the same. But what Newton really had in mind were two homogeneous fluid columns that he theorized would not displace each other. Two solid homogeneous columns could not displace each other, even if they did not weigh the same. Thus, as I mentioned, he treated the homogeneous rotating figure that he took to represent the earth to be a fluid figure, not a solid one. But in addition, in balancing two columns from the center to the surface of such a figure which lie along the polar and equatorial axes of the figure, Newton, in effect, not only treated the earth as a homogeneous rotating fluid figure, but he assumed as well that all of the columns from the center to the surface of the figure balance or weigh the same.

The polar axis of a flattened ellipsoid of revolution is its axis of symmetry. I note here for later purposes in Chapter 1, as well as in Chapters 4, 6, and 9, that in a homogeneous figure shaped like a flattened ellipsoid of revolution which attracts according to the universal inverse-square law and which revolves around its axis of symmetry, it is enough to consider columns from the center to the surface of such a figure which all lie in the same plane of a meridian at the surface of the figure without loss of generality. The reason is as follows. Suppose that two columns from the center to the surface of the figure do not lie in the same plane of a meridian at the surface of the figure. The point at the surface of the figure on either column must be located on some meridian at the surface of the figure, and the column in question of course lies in the plane of that meridian. But as a result of the symmetry of the figure of revolution, the fact that the axis of symmetry of the figure is also an axis of symmetry of the attraction produced by the figure, and the fact that the figure revolves around its axis of symmetry and not around some other line, so that the axis of symmetry of the figure is also an axis of symmetry of the centrifugal force per unit of mass, it follows that the other column can be revolved around the axis of symmetry of the figure into the plane of the meridian just defined without changing the length or the weight of this column.

The trouble with Newton's assumption that all of the columns from the center to the surface of a homogeneous figure shaped like a flattened ellipsoid of revolution which attracts according to the universal inverse-square law and which revolves around its axis of symmetry can balance or weigh the same is that he did *not* actually *verify* it. That is, he did *not* actually prove that all of the columns from the center to the surface of such a figure *can* balance or weigh the same. I shall return to this matter in Chapter 6.

Moreover, Newton in fact really assumed somewhat more. The polar axis of any figure of revolution is its axis of symmetry. Newton assumed that any homogeneous figure of revolution which revolves around its axis of symmetry and whose columns from center to surface all balance or weigh the same is a figure of relative equilibrium. As we shall see in Chapter 4, this need not be true in general. In Chapter 4 we shall discover that when homogeneous figures of revolution which revolve around their axes of symmetry are assumed to attract according to certain hypotheses of attraction, all of the columns from the centers to the surfaces of these figures can balance or weigh the same, yet these figures are not figures of relative equilibrium. Whether a homogeneous figure of revolution which revolves around its axis of symmetry and whose columns from center to surface all balance or weigh the same can be a figure of relative equilibrium or not depends upon the law according to which the figure attracts, as we shall observe in Chapter 4. A homogeneous figure shaped like an ellipsoid of revolution flattened at its poles which attracts according to the universal inverse-square law, which revolves around its axis of symmetry, and whose columns from center to surface all balance or weigh the same *does* turn out to be a possible figure of relative equilibrium, but we shall have to wait until Chapter 9 to see that this is true and to learn why it is true. I say a "possible" figure of relative equilibrium, because the ratio $\omega^2/2\pi\rho$ has a certain upper limit whose value is less than 1, where ω is the angular speed of rotation and ρ is the density of the figure. (In other words, for a given density ρ, the angular speed of rotation ω has an upper limit.) If the ratio $\omega^2/2\pi\rho$ exceeds this upper limit for a particular homogeneous figure shaped like an ellipsoid of revolution flattened at its poles which attracts in accordance with the universal inverse-square law and which revolves around its axis of symmetry (its polar axis), then such a figure cannot exist as a figure of equilibrium, even if the columns from the center of such a figure to its surface should all theoretically balance or weigh the same. None of the authors whose works I will discuss in this book took this upper limit into account. They did not realize that such an upper limit existed. The upper limit was first discovered only much later by Laplace.[5]

Now, Newton could also have assumed that the earth was originally a homogeneous fluid mass without having to assume that this mass existed in a state of equilibrium. However, by not assuming the homogeneous fluid mass to be in a state of equilibrium, Newton would have been left with no way of handling the problem of the earth's shape as he imagined it. But as we shall see in Chapters 4, 6, and 9, there are other ways to treat the problem of the earth's shape without having to assume the earth originally to have been a fluid mass, much less a figure of equilibrium. Specifically, we shall find that the earth can also be viewed instead as a solid, stratified figure that attracts according to the universal inverse-square law and that need not be a figure of equilibrium.

Finally, with regard to figures of equilibrium, among the homogeneous figures that attract according to the universal inverse-square law which Newton considered, he found ones that turn out to fulfill a particular condition that must be satisfied in order that homogeneous figures of equilibrium exist – namely, the principle of balanced columns, which means that all columns from the center of a homogeneous figure to its surface balance or weigh the same. (In Chapter 4 we shall see why the principle of balanced columns is a principle of equilibrium for homogeneous fluid figures of equilibrium, and in Chapter 9 we shall discover that the principle of balanced columns follows from another, more general principle of equilibrium for homogeneous fluid figures of equilibrium.) The part of the story told in these pages which has to do with figures of equilibrium mainly involves the discovery and use of various conditions that must be fulfilled in order that a fluid mass be a figure of equilibrium. Today we would call such conditions *necessary* conditions for figures of equilibrium to exist. The story does not deal at all with the problem of determining conditions that are *sufficient* for figures of equilibrium to exist.

The advantages of imagining the fluid figure that represented the earth to be shaped like an ellipsoid of revolution are evident. When the *Principia* first appeared, the tools of mathematical analysis had not yet been well developed. (Indeed, mathematical analysis only came into being at this time.) Newton could use Apollonian geometry as a guide in calculating attractions produced by homogeneous bodies shaped like flattened ellipsoids of revolution which attract according to the universal inverse-square law. He could then utilize these calculations to compute the weights of the two columns mentioned above, taking, of course, the effects of centrifugal force of rotation into account. In Proposition 91, Corollary 3 of Book I of the *Principia*, Newton stated that the magnitude of the attraction at points in the interior of a homogeneous body shaped like an ellipsoid of revolution which attracts in accordance with the universal inverse-square law varies directly as the distances of the points from the center of the body. If the body rotates as well and if the polar axis is its axis of symmetry and its axis of rotation, then the magnitude of the centrifugal force per unit of mass at points along an equatorial axis also varies directly as the distances of the points from the center of the body. (The magnitude of the centrifugal force per unit of mass equals zero at all points along the polar axis.) The two variations together make the magnitudes of the resultants of the effective gravity at points along the two axes vary directly as the distances of the points from the center of the body, too.[6]

Using this last result, Newton calculated the weights of two columns from the center of the body to its surface along its two axes. To be more specific, using Proposition 91, Corollary 2 of Book I of the *Principia*, Newton found the magnitude of the attraction at the equator of a homogeneous body shaped like a flattened ellipsoid of revolution which attracts according to the universal inverse-square law. Newton took $\frac{1}{289}$ to be the ratio of the magnitude of the centrifugal force per unit of mass to the magnitude of the attraction at the equator of the body when it revolves around its axis of symmetry (its polar axis), in which case $\frac{288}{289}$ is the ratio of the magnitude of the effective gravity to the magnitude of the attraction at the equator of the rotating body. This ratio multiplied by the magnitude of the attraction at the equator of the body equals the magnitude of the effective gravity at the equator of the body. Newton then utilized his result mentioned previously concerning the variation of the magnitude of the effective gravity along the body's equatorial axis to calculate the weight of a column that connects the equator of the body and its center. Similarly, after determining the magnitude of the attraction at a pole of a homogeneous body shaped like a flattened ellipsoid of revolution which attracts according to the universal inverse-square law, again by employing Proposition 91, Corollary 2 of Book I of the *Principia*, Newton used his result regarding the variation of the magnitude of the attraction in the body's interior to compute the weight of a column from a pole of the body to its center. If $\frac{1}{289}$ was taken to be the ratio of the magnitude of the centrifugal force per unit of mass to the magnitude of the attraction at the equator of the body when it revolves around its axis of symmetry, Newton discovered that the two columns weighed the same when the ellipticity ε of the body was equal to $\frac{1}{229}$.

Newton came to other conclusions as well. In so doing he made actual use of the assumption that *all* of the columns from the center to the surface of the figure that he had found balance or weigh the same. That is, Newton maintained that attraction and centrifugal force per unit of mass at points along any column from the center to the surface of a homogeneous figure shaped like a flattened ellipsoid of revolution which attracts in accordance with the universal inverse-square law and which revolves around its axis of symmetry determine how much that column weighs in exactly the same way that attraction and centrifugal force per unit of mass at points along two columns from the center of the figure to its surface which lie along its polar and equatorial axes determine how much those two columns weigh. Thus Newton concluded that the weight of any column from the center of a homogeneous figure shaped like a flattened ellipsoid of revolution which attracts according to the universal inverse-square law and which revolves around its axis of symmetry to a point at its surface was regulated in the same manner as the weights of two columns from the center of the figure to its surface which lie along its two axes. Applying this reasoning backward to an arbitrary column from the center of the figure to a point on its surface, and using the assumption that a short column weighs exactly the same as a long one, Newton deduced that the magnitude of the effective gravity at all points on the surface of his homogeneous figure shaped like a flattened ellipsoid of revolution which attracts according to the universal inverse-square law, which revolves around its axis of symmetry, and whose columns from center to surface were all assumed to balance or weigh the same varies inversely as the distances of the points from the center of the figure.[7] Consequently, he declared, although without furnishing the reader with any proof, that the increase in the magnitude of the effective gravity along the surface of the figure from its equator to one of its poles is nearly proportional to the square of the sine of the latitude and, moreover, that the increase in degrees of latitude with latitude is also nearly proportional to the square of the sine of the latitude.[8]

Newton used these findings to calculate theoretical values of the magnitude of the effective gravity and lengths of degrees of latitude at various latitudes along the earth's surface. He then compared his tables obtained in this way with the observations of the day made in Paris, Cayenne, Gorée (an island off the coast of Senegal, near Dakar), and the Caribbean. Newton noted that the magnitude of the component of effective gravity in Paris perpendicular to the earth's surface there exceeded the magnitudes of the components of the effective gravity in places at southern latitudes perpendicular to the earth's surface at those places, by which he meant that the lengths of seconds pendulums measured in Paris exceeded the lengths of seconds pendulums measured near the equator, by amounts greater than the ones that his theory required. Newton concluded that if these measurements could be trusted, then the earth was not homogeneous, but a little denser at its center than at its surface and a little flatter at its poles than was his homogeneous body shaped like a flattened ellipsoid of revolution.

(Newton, in fact, had greater faith in Richer's measurements than in other measurements made near the equator. He commented that if the observed difference between lengths of seconds pendulums in Paris and in Cayenne were reduced slightly to allow for lengthening of metallic cords in the Torrid Zone caused by the heat there, then the observed increase in the magnitude of the effective gravity with latitude modified in this way would agree closely with the increase in the magnitude of the effective gravity with latitude which his theory applied to the earth required, assuming that the earth is homogeneous.)

To affirm that Newton plainly demonstrated the various results and conclusions that his theory, outlined above, led him to would be to overstate the case. For Newton actually did no more than sketch his theory and its consequences himself. His theory differs considerably from what we would regard as a mathematical theory, and this fact probably explains in part why Derek Whiteside wrote so little in *The Mathematical Papers of Isaac Newton* about the theory. During the first fifty years that followed the publication of the *Principia*, Newton's theory of the earth's shape struck even the most reputable continental mathematicians of his time as incomprehensible.

I give some examples of the difficulties that the reader faced. As I mentioned earlier, Newton stated in Corollary 3 to Proposition 91 of Book I of the *Principia* that the magnitude of the attraction at points along any line segment from the center of a homogeneous body shaped like an ellipsoid of revolution which attracts according to the universal inverse-square law to a point on its surface varies directly as the distances of the points from the center of the body. But Newton's statement of this corollary is not precise, nor is his proof of what he did state complete.

In the first half of the proof, Newton arrived at the result that within the cavity of a homogeneous shell whose inner and outer surfaces are concentric geometrically similar ellipsoids of revolution having a common axis of symmetry and which attracts in accordance with the universal inverse-square law, the net attraction is zero. But his proof of this result has a gap: it depends upon the surprising equality of a certain pair of line segments, which Newton claimed to establish by invoking without proof a directly equivalent geometric statement (namely, that a certain other pair of line segments that lie along the same line have the same midpoint) that is not at all obvious.[9]

The second part of Newton's proof of Corollary 3 to Proposition 91 is based on Corollary 3 to Proposition 72 of Book I of the *Principia*. As it is stated by Newton, however, the meaning of this corollary is not immediately evident; and he chose not to present a proof of it – a proof which, had it been included, might have brought to full light the meaning of the corollary. Moreover, Proposition 72 is itself a difficult proposition, whose "meaning may remain uncertain even after several readings ... "; in its "excruciating terseness," the proof that Newton offers does little, unfortunately, to clarify the statement that it is supposed to prove.[10]

Finally, Newton never talked about the directions of attraction at points. But at each point there is the *total*, or *resultant of the*, attraction at the point, which has a certain well-determined direction. And then there are also *components* of the total attraction at the point in every *other* direction. To determine the weight of a column from the center of a figure to its surface one is interested in the magnitudes of the components of the attraction at points along the column which are directed toward the point at the surface. In fact, it can be shown, using methods patterned upon Newton's tersely stated arguments that make up his proof of Proposition 72 of Book I of the *Principia*, that the magnitudes of the *components* of attraction *in any fixed direction* at points along any line segment from the center of a homogeneous body shaped like an ellipsoid of revolution which attracts according to the universal inverse-square law to a point on its surface vary directly as the distances of the points from the center of the body. Furthermore, this result can be extended to cover the *total* attractions at points along any such line segment. That is, it can be demonstrated as well, again using Newton's proof of Proposition 72 of Book I of the *Principia* as a guide, that the *total* attractions at all points along a line segment from the center of a homogeneous figure shaped like an ellipsoid of revolution which attracts in accordance with the universal inverse-square law to a point on its surface *all have the same direction* and that their magnitudes vary directly as the distances of the points from the center of the figure. Indeed, it *follows immediately* from this last result that the magnitudes of the *components* of attraction *in any fixed direction* at points along any such line segment *must* vary directly as the distances of the points from the center of the figure.

Corollary 3 to Proposition 72 of Book I of the *Principia* can in fact be restated and proved in such a way that it gives, in conjunction with the result that Newton arrived at in the first half of his proof Corollary 3 to Proposition 91 of Book I of the *Principia*, all of these various results. It is possible that Newton realized this himself – that he could have stated this form of Corollary 3 to Proposition 72 of Book I of the *Principia* and simply did not. In order to do so, he would have had to introduce and explicate concepts like "magnitude of a force," "direction of a force," and "components of forces" distinguished from "total forces." Although he did not introduce such concepts, he may very well have understood them. But he did not bother to make such ideas clear for the benefit of his reader.[11]

And because Newton did not introduce and explain such concepts, one must greatly stretch the imagination to contend that Newton proved the different, specific results mentioned above, which require such concepts even to state them, in a way that readers of his time could understand these results and their proofs. He really did not even state the various results, much less establish them individually. As I observed, the different results and their proofs may have been clear to Newton. Newton possibly thought to himself a phrase that mathematicians today often use when teaching or giving talks: "It is obvious that ...." But what may have been obvious to Newton was almost never obvious to his contemporaries.

In practice, Newton used Corollary 3 to Proposition 91 of Book I of the *Principia* to infer that the magnitudes of the components of attraction at points along any line segment from the center of a homogeneous body shaped like an ellipsoid of revolution which attracts according to the universal inverse-square law to a point at its surface, *in the direction of that line segment,* vary directly as the distances of the points from the center of the body – although as I say, Newton in fact mentioned nothing in stating or proving this corollary which would lead one to conclude how the magnitudes of such *components* of attraction at points along the line segment actually vary. This particular result is what he used to reason about the way that the weights of columns from the center to points on the surface of a homogeneous figure shaped like an ellipsoid of revolution flattened at its poles which attracts in accordance with the universal inverse-square law and all of whose columns from center to surface are assumed to balance or weigh the same are regulated and to conclude that the magnitudes of the effective gravity at all points on the surface of such a figure vary inversely as the distances of the points from the center of the figure. As I indicated earlier, to arrive at this conclusion, Newton made direct use of the assumption that all of the columns from the center to the surface of the figure do balance or weigh the same.[12]

Moreover, Newton asserted that the ratio of the magnitude of the attraction at a pole to the magnitude of the attraction at the equator of a homogeneous body shaped like an ellipsoid of revolution which attracts according to the universal inverse-square law and whose polar radius is equal to $\frac{100}{101}$ of its equatorial radius (in other words, whose ellipticity is $\frac{1}{100}$) is $\frac{501}{500}$. As the basis for this conclusion, he cited Proposition 91, Corollary 2 of Book I of the *Principia*, mentioned above, which is a general, geometric theorem that can be used to determine the magnitudes of the attractions at points along the polar axes, in the exteriors, of homogeneous bodies that attract in accordance with the universal inverse-square law and whose surfaces are ellipsoids of revolution which are either flattened or elongated at their poles. (Here the attraction at a point on the polar axis of such a body in its exterior is the *total* attraction at the point produced by the body, which is directed along the polar axis because of symmetry.) But Newton stated the theorem without demonstrating it in detail.[13]

Then he maintained, again without going into details, that if this figure shaped like a flattened ellipsoid of revolution revolved around its axis of symmetry (its polar axis) and if columns from the center of the figure to its surface which lie along its polar and equatorial axes balanced or weighed the same, then the ratio of the magnitude of the centrifugal force per unit of mass to the magnitude of the attraction at the equator of this figure would be $\frac{4}{505}$.

Finally, he introduced, again without providing the reader with any demonstration, the equation

$$\frac{\delta}{1/100} = \frac{\phi}{4/505}, \tag{1.1}$$

from which he found $\frac{1}{229}$ to be the ellipticity δ of a homogeneous body shaped like a flattened ellipsoid of revolution which attracts according to the universal inverse-square law, which revolves around its axis of symmetry (its polar axis), and whose columns from center to surface which lie along the polar and equatorial axes of the body balance or weigh the same, when the ratio ϕ of the magnitude of the centrifugal force per unit of mass to the magnitude of the attraction at the equator of the body is the earth's value $\frac{1}{289}$. Newton did not explain where this equation came from or how he found it; he gave no account whatever of his procedure.

In fact, although Newton did not say so, this equation holds true *only* for a homogeneous figure shaped like a flattened ellipsoid of revolution which attracts in accordance with the universal inverse-square law, which revolves around its axis of symmetry (its polar axis), whose columns from center to surface all balance or weigh the same, and whose ellipticity is "infinitesimal," where an infinitesimal number ε is a nonzero number ε for which $\varepsilon^2 \ll \varepsilon$ is true. (To express the matter another way, equation (1.1) holds good only to terms of first order.) Newton must have known and understood this himself, considering the approximations that he had to have made in order to arrive at equation (1.1). Nevertheless, for some reason this did not stop him from applying equation (1.1) to Jupiter, as we shall see in Chapter 5, even though Newton knew from Flamsteed's and Gian-

Domenico Cassini's telescopic observations that Jupiter's ellipticity was *not* infinitesimal.

For reasons that we shall discover in Chapter 7, in 1729 the Paris Academician Pierre-Louis Moreau de Maupertuis became one of Johann I Bernoulli's students of mathematics. In a letter that he wrote to Bernoulli in March of 1731, Maupertuis admitted that he "did not understand Monsieur Newton's method at all" for finding the earth's shape. Maupertuis equated the weights of all columns that joined the center of a homogeneous figure of revolution which revolved around its axis of symmetry (its polar axis) and which was attracted by a force that varied with distance from a fixed center of force, which Maupertuis assumed to be located at the center of the figure of revolution, to points along a meridian at the surface of the figure. Huygens and Jakob Hermann had already done much the same years earlier for certain specific cases, in the *Discours de la cause de la pesanteur* (1690) and *Phoronomia* (1716), respectively.

[Because of what I explained earlier in Chapter 1, it was enough to consider columns from the center to the surface of such a homogeneous figure of revolution which all lie in the same plane of a meridian at the surface of the figure, because of the symmetry of the figure of revolution, the fact that the attraction produced by the center of force situated at the center of the figure is symmetric around the axis of symmetry of the figure, and the fact that the figure revolves around its axis of symmetry and not around some other line, so that the centrifugal force per unit of mass is symmetric around the axis of symmetry of the figure. These conditions meant that all of the columns from the center to the surface of such a homogeneous figure of revolution balanced or weighed the same, if all of the columns from the center to the surface of the figure which lie in the same plane of a meridian at the surface of the figure balanced or weighed the same.]

Maupertuis generalized the problem, and in every case the figure came out flattened at the poles. But like these earlier authors, Maupertuis determined a value for the earth's flattening at the poles which differed from Newton's value. Maupertuis simply could not follow Newton's arguments involving the universal inverse-square law of attraction. And because he could not, he did not understand why, for example, Newton thought that the earth was shaped like an ellipsoid of revolution.[14]

(In fact, we shall see in Chapter 6 that Newton himself never truly explained in the *Principia* why he believed the earth to be shaped like an ellipsoid of revolution and that it was left to people who could follow Newton's presentation of his theory of the earth's shape to justify much later what Newton failed to make clear in the *Principia*. In general, when a central force attracts, whose magnitude is proportional to r^n, where n can equal $0, \pm 1, \pm 2, \pm 3, \pm 4, \ldots$ and where r is the distance from the center of force, a homogeneous figure of revolution which revolves around its axis of symmetry (its polar axis), whose columns from center to surface all balance or weigh the same, and whose center is located at the center of force is not an ellipsoid unless $n = 1$. Huygens had solved the case $n = 0$ and Hermann the cases $n = 0$ and $n = 1$. Maupertuis examined the more general case where n is zero or any positive or negative integer.)

Bernoulli, left without peer in mathematics after Gottfried Wilhelm Leibniz died in 1716 and Newton died in 1727, replied that he could do no better than Maupertuis. The great mathematician from Basel told his French friend: "You say that you do not understand Monsieur Newton's method at all. I tried to understand it. I read and reread what he had to say concerning the subject, but, like you, I could not understand a thing. I do not know whether the fault lies with my impatience, resulting from my reaction to his references to things back in Book I, or whether I do not understand how he applied these [to his theory of the earth's shape in Book III]. In a word, all I found was obscurity and impenetrability." Since Bernoulli had found other places in the *Principia* where Newton, in difficulty, talked what Bernoulli called "gibberish" in order to try to dodge problems, Bernoulli doubted that Newton himself completely understood what he was doing.[15] Did Bernoulli simply mean to take another swipe at an old rival? What good did it do to ridicule Newton four years after Newton's death? What then did Bernoulli intend by his remarks? I think Bernoulli simply gives us evidence that the *Principia* was no easier to read two hundred fifty years ago than it is today.

Newton often solved problems in mechanics in the *Principia* using geometry that neither truly followed the style of the ancient Greeks nor was a kind that was employed on the Continent.[16] To be sure, surfaces generated by conic sections were still, by and large, the only ones that could be investigated in any detail in Newton's time.[17] Maupertuis knew that well enough. But did this explain why Newton thought that the earth was shaped like an ellipsoid of revolution? Maupertuis did not have the faintest idea. Sometimes Newton simply mentioned the basic elements that underlay certain conclusions that he had come to and then illustrated intermediate steps in his train of thought by means of numerical examples, as in the case of his "derivation" of the equation presented here as (1.1), rather than work out demonstrations in detail for the reader's benefit.

As I noted previously, having deduced that the magnitude of the effective gravity at all points on the surface of his homogeneous figure shaped like a flattened ellipsoid of revolution which attracts according to the universal inverse-square law, which revolves around its axis of symmetry, and whose columns from center to surface were all assumed to balance or weigh the same varied inversely as the distances of the points from the center of the figure, Newton stated without proof two "$\sin^2$ laws" that express the increase in effective gravity with latitude and the increase in degrees of latitude with latitude. If these two laws followed from the assertion that the magnitude of the effective gravity at all points on the surface of his figure varied inversely as the distances of the points from the center of the figure, Newton did not make clear how or why. (We shall see in Chapter 3 that Newton's failure to explain himself could have easily led to misinterpretations.) Sometimes Newton employed theorems whose meanings were not evident and whose proofs were so terse that they did not help shed light on what was even stated in the theorems, like Proposition 72 of Book I of the *Principia*. Sometimes he stated and utilized corollaries of theorems which he did not demonstrate at all and, as a result, whose meanings were ambiguous, like Corollary 3 to Proposition 72 of Book I of the *Principia*. And because of the ambiguity of this particular corollary, Corollary 3 to Proposition 91 of Book I of the *Principia*, which depends upon this other corollary, can be interpreted in various ways, all of which are true but different. At other times he uttered unsubstantiated, and sometimes ambiguous, conjectures, without providing the reader with the slightest hint that might have helped illuminate his reasoning. Occasionally he seemed to contradict himself. Once in a while, Newton even made errors.

The question is whether the *Principia*, had it been without gaps, inconsistencies, and errors, would have stimulated other people to do research. It appears that the *Principia* became an indispensible tool of research *because of* its "defects." Within an historical context, the work's value lies *in* its gaps, inconsistencies, and errors. Whether intended or not, the *Principia*'s density, obscurity, and various other drawbacks and inconveniences are advantages or strengths, not weaknesses. At least they seem to me to be what motivated other researchers to do a great deal of further work in some branches of mechanics. The *Principia*'s gaps, inconsistencies, and errors prompted other researchers to investigate problems that they might not have otherwise gotten around to dealing with as quickly. Without these gaps, inconsistencies, and errors this later research might not have started up as quickly. The *Principia* reads like anything but today's self-contained physics or mechanics textbook. Nor do such books teach people how to do research. Their purpose is to help them to understand and assimilate what is already known. Reading the *Principia* requires know-how. For readers with enough knowledge and ability to locate the border lines between demonstrated conclusions and true but undemonstrated assertions, and who could verify the undemonstrated ones, as well as determine where the boundaries between truth and error lay, the *Principia* served as a source of problems for research as exciting and as fascinating as Galileo's works had been. Some researchers who penetrated the *Principia* enough to discover its limits went on to expand Newton's ideas beyond his own work, and they extended the scope of Newton's world system in important ways by this means. The story that follows illustrates this. Had the *Principia* truly been a completely self-contained "mechanics course," consisting of nothing but straightforward demonstrations of whatever particular conclusions follow from its principles, without any gaps or limitations or conjectures or errors, would it have provoked mathematicians and, in this way, induced them to try to determine the range of Newton's world system and in so doing enable some of them to show that its range could be enlarged?

Let us consider for a moment, say, Frans Van Schooten Jr.'s *Geometria*, first published in 1649 and then republished in two volumes in 1659 and 1661. This work filled in the many gaps that Descartes had left in his *Géométrie* (1637). Although it advanced Dutch mathematics in the short term, it had an inhibiting effect in the long run, because "the contents were so comprehensive that people came to regard Van Schooten's work as the last word on the subject. As a result all efforts were directed toward interpreting and understanding the *Geometria* instead of using it as a stimulus for new mathematical activities."[18] I do not wish to argue that perspicuity is the quickest route to oblivion and that works more readable than the *Principia* have not been influential. The problem is simply to try to understand how and why a work as formidable as the *Principia* could have had such authority.

In discussing the problem of the earth's shape in the *Principia,* Newton began with a theory of that shape whose mathematics he did not explain very clearly. He then tried to patch up the "mathematical theory" by tinkering with certain details of the theory in a qualitative way in order to justify observations that did not accord with the theory. His guesswork included no mathematics (or, if it did, Newton at any rate provided no clues whatever that might have shed some light on any mathematics involved in his maneuvers). Newton simply appears to have followed his nose. The conjectures that he introduced to deal with anomalies seem to take the form of what we would today call "intuitions." At least, he did not explain or prove the conjectures, nor did he furnish any evidence that supported them. He did not demonstrate mathematically how the anomalous observations in question followed from these conjectures. This is true in the case of the earth. What appeared to be the loose ends and inconsistencies to researchers in Paris in the 1730s caused them to reexamine the whole problem of the earth's shape from a theoretical standpoint. My ultimate aim here is to investigate this reconsideration in detail. Like much of the diffusion of the *Principia* in the years after 1687, the assimilation of Newton's theory of the earth's shape takes two different directions: (1) Try to understand the theory, work out Newton's method, and add missing details; (2) Find errors in the theory, correct them, and advance beyond.

2

The state of the problem of the earth's shape in the 1720s: Stalemate

2.1. *Dortous de Mairan's phenomenological theory of an elongated earth (1720)*

In 1722, the Paris Academy of Sciences published Jacques Cassini's long account of the Academy's preceding fifty years of geodetic expeditions to determine the size of the earth from the length of a mean degree of latitude and the shape of the earth by means of observed variations in the lengths of degrees of latitude in France along the meridian through Paris. Originally undertaken with navigation and map making in mind, the project was suspended several times during the latter part of the reign of Louis XIV. Financial shortages brought on by the wars of that era caused the interruptions. The venture was finally completed in 1718, three years after Louis's death. Cassini concluded from his and his colleagues' measurements that the earth is nearly spherical, but not perfectly so. Instead, it was shaped like an ellipsoid of revolution slightly elongated at the poles.[1]

Newton and Huygens, on the contrary, had concluded earlier, using theoretical reasoning, that the earth is shaped like a figure of revolution slightly flattened at the poles. They started from the experiments with seconds pendulums which measured the variation of the magnitude of the effective gravity with latitude. However, differing hypotheses of attraction had led them to different values of the flattening. (Unlike Newton, who had assumed that the earth attracts according to the universal inverse-square law, Huygens assumed that attraction was a constant central force – that is, that the magnitude of attraction at all points was constant and that the attraction at a point was directed along the line segment that joins the point to a fixed center of force, where Huygens assumed the center of force to be situated at the earth's center.)

In a *mémoire* that he read at the Paris Academy in 1720, Jean-Jacques Dortous de Mairan contested the assumption that the two sets of terrestrial observations (the observed variations of degrees of latitude with latitude and the observed variation of the magnitude of the effective gravity with latitude) were incompatible.[2] In raising the issue, Mairan called attention to Huygens's theory but not to

Newton's. He exploited a calculation that Huygens had done. Huygens had determined that if Paris and Cayenne were situated on a homogeneous spherical body at whose surface the magnitude of attraction is constant, whose radius is such as to make one degree of latitude at its surface equal to Picard's degree of latitude, and which revolves around its polar axis so that the ratio of the magnitude of the centrifugal force per unit of mass to the magnitude of the attraction at its equator is the earth's value $\frac{1}{289}$, then the decrease in the magnitude of the effective gravity along the surface of this homogeneous spherical body from Paris to Cayenne, measured by the shortening of the length of a seconds pendulum, would be "a little less than what Monsieur Richer had [actually] found [it to be] at Cayenne." Here Mairan quoted Huygens, but Mairan added that Richer had in fact found that the length of a seconds pendulum had to be shortened in Cayenne by more than twice the amount that Huygens later determined from theory.[3]

(When the attraction at a point in or on the surface of a figure of revolution does not depend upon the density of the figure of revolution, which is true of a central force of attraction, for example, since the magnitude of the attraction at points produced by such a force varies only as a function of the distance of the points from the center of force, then it really makes no difference whether the figure of revolution is assumed to be homogeneous or not when determining its shape – that is, its ellipticity – or calculating the variation of the magnitude of the effective gravity with latitude along its surface when the figure of revolution revolves around its axis of symmetry. I shall discuss this point further in Chapter 9.)

Mairan thought that this difference proved that Huygens had not reasoned correctly. Mairan believed that if one approached the problem the right way, one ought to find that for a figure of revolution which revolves around its axis of symmetry so that the ratio of the magnitude of the centrifugal force per unit of mass to the magnitude of the attraction at its equator is the earth's value $\frac{1}{289}$, the decrease in the magnitude of the effective gravity along the surface of the figure from a point whose latitude is the same as the latitude of Paris to a point whose latitude is the same as the latitude of Cayenne should agree better with the observed decrease than Huygens's decrease did.

(We saw in Chapter 1 that Newton's homogeneous body shaped like an ellipsoid of revolution infinitesimally flattened at the poles which attracts according to the universal inverse-square law and which revolves around its axis of symmetry so that the ratio of the magnitude of the centrifugal force per unit of mass to the magnitude of the attraction at its equator is the earth's value $\frac{1}{289}$ also made the magnitude of the effective gravity decrease along its surface from a point whose latitude is the same as the latitude of Paris to southern latitudes by amounts less than the ones observed along the earth's surface, assuming that the measurements of the day of the magnitudes of effective gravity at different latitudes were acceptable, which meant neglecting the effects of climactic changes upon the lengths of seconds pendulums.)

And what did Mairan think the proper approach to be? Mairan insisted that *both* sets of terrestrial observations had to be accepted from the beginning and that the correct law that governs the earth's attraction was one that accorded with both sets. In particular this meant admitting or recognizing an elongated earth as a fact.

Mairan believed effective gravity to be perpendicular to the earth's surface at all points on its surface. This he concluded from observations that waters at the earth's surface which do not slope find their own levels. This would not occur, were there a component of effective gravity tangent to the earth's surface at some horizontal place on the surface. Such a component would not permit water at the place to come to rest, but would cause it to flow along the surface at that place instead. He gave other evidence based on observations as well for what he called "one of the most inviolable laws of nature."[4]

In fact, Huygens had first stated the law, often referred to as the principle of the plumb line, long before, in his *Discours de la cause de la pesanteur* (1690).[5] But Huygens did not actually apply the principle in practice. He utilized Newton's principle of balanced columns instead to determine the shapes of homogeneous bodies that were flattened, nearly spherical figures of revolution which revolved around their axes of symmetry. (In Chapter 5 we will learn why Huygens probably *could not have applied* the principle of the plumb line to find the shapes of homogeneous figures of revolution which revolved around their axes of symmetry. Moreover, Huygens tacitly assumed that the principles of the plumb line and balanced columns were equivalent. That is, for a given law of attraction and a given ratio of the magnitude of the centrifugal force per unit of mass to the magnitude of the attraction at the equator, he implicitly supposed that both principles determine the same homogeneous figure of revolution which revolves around its axis of symmetry. In so doing, Huygens essentially hypothesized that the two principles determine figures of equilibrium. In Chapter 4 we shall discover that the assumption that both principles determine the same homogeneous figure of revolution which revolves around its axis of symmetry is not necessarily true, and when it is not true, the different homogeneous figures of revolution determined by the two principles individually are not figures of equilibrium. We recall that I already mentioned in Chapter 1 that a homogeneous figure of revolution which revolves around its axis of symmetry and whose columns from center to surface all balance or weigh the same need not be a figure of equilibrium.)

Whereas earlier authors had derived equations of the meridians along the surfaces of bodies shaped like flattened figures of revolution which revolve around their axes of symmetry, Mairan did not attempt to determine equations of the meridians along the surfaces of bodies shaped like elongated figures of revolution which revolve around their axes of symmetry. He refrained from doing this because he felt that the paucity of sufficiently accurate terrestrial observations made it impossible to decide between various possible shapes. Thus, for example, he thought that Cassini's assumption that the elongated earth was shaped like an ellipsoid of revolution was an unnecessarily restricted hypothesis. (After all, Huygens's flattened nearly spherical figure of revolution which revolved around its axis of symmetry was not shaped like an ellipsoid of revolution.) Mairan showed that along a meridian at the surface of *any* elongated figure of revolution, degrees of latitude increase from pole to equator, provided, he stipulated, that a meridian contains no points of inflexion, where a trend can change continuously from increasing to decreasing, and no cusps, where a trend can change discontinuously. In other words, degrees of latitude increase from pole to equator along the surface of *any* decent elongated figure of revolution. (In the story told in this book any figure of revolution which represents a planet like the earth or Jupiter will always be symmetric with respect to the plane of its equator. This, for example, rules out pear-shaped figures.) Hence, in Mairan's view, the observed variations in lengths of degrees of latitude of the day did not alone suffice to specify a particular elongated shape. In using the theory of curvature of plane curves to establish this, Mairan had occasion to cite Huygens, the father of this theory, and Pierre Varignon, the French Jesuit mathematician who had translated Huygens's work in this subject into the language of the differential calculus.

Mairan's demonstration illustrated the nature of his paper. Mairan took terrestrial observations to be primary. Hypotheses – and in this particular case hypotheses of attraction – were of secondary importance to him. They served a purpose that was not his chief concern. He had no use at all for dissimilar hypotheses that observations could not discriminate. Mairan attempted to find out how the magnitude of effective gravity varies from pole to equator along the surfaces of rotating spherical bodies, along the surfaces of elongated figures of revolution of given shape which revolve around their axes of symmetry, and along the surfaces of flattened figures of revolution of given shape which revolve around their axes of symmetry. (As we shall discover in examining Mairan's research closely, it makes no difference whether these figures are assumed to be homogeneous or not.) He then hoped that the law according to which the earth attracts could be *inferred* from his study of such variations.

Mairan continued to make extensive use of the theory of curvature of plane curves in his essay, in endeavoring to treat the problem of the earth's shape phenomenologically. I give a brief résumé of the elements of the theory used by Mairan. Consider a figure of revolution elongated at its poles A and B (see Figure 2). Mairan assumed attraction to be perpendicular to the surfaces at all points on the surfaces of figures of revolution. Thus the directions of attraction at points along a meridian $ADBE$ at the surface of the elongated figure of revolution, which Mairan assumed are perpendicular to the meridian at those points, determine the evolute $GOHK$ of the meridian $ADBE$ at the surface of the elongated figure of revolution. (The radius of curvature RT of the meridian at R is the radius of the osculating circle of the meridian at R. This radius determines the point T on the evolute of the meridian which corresponds to R.) The "pointed"

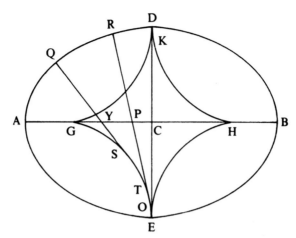

Figure 2. Evolute of a meridian of an elongated figure of revolution (Mairan 1720: Figure 1).

surface of revolution, produced by revolving the evolute $GOHK$ of the meridian $ADBE$ around the elongated figure's axis of symmetry AB, is the evolute of the surface of the elongated figure of revolution.[6]

Mairan now examined what happened when the elongated figure of revolution revolved around its axis of symmetry AB. Mairan first assumed attraction to be constant in magnitude at points on the surfaces of figures of revolution. He hypothesized in addition that the centrifugal force per unit of mass alone causes the magnitude of the effective gravity to vary along the surfaces of the figures when the figures revolve around their axes of symmetry.[7] (Thus there is no need to presuppose that figures be homogeneous here.)

He took the line segments DC, RP, QY, which are parts of the radii of curvature of a meridian $ADBE$ through points D, R, Q at the surface of a figure of revolution elongated at its poles A and B (see Figure 3) to represent the directions of attraction at D, R, Q. Thus he presumed that these directions at points D, R, Q at the surface of the figure were perpendicular to the surface at those points. Mairan then determined how the magnitudes of the components of the centrifugal force per unit of mass in these directions at points along the meridian vary with latitude when the elongated figure of revolution revolves around its axis of symmetry AB. Here I let $|XY|$ designate the length of the line segment XY. The magnitude of centrifugal force per unit of mass at a point R varies directly with the distance $|RN|$ of R from the axis of symmetry AB, the axis that the figure revolves around. Mairan showed that the magnitude of the component of the centrifugal force per unit of mass at R in the direction of RP increases from the pole A to the point D on the equator directly as $|RN|\sin((\pi/2) - \theta)$ (or $|RN|\cos\theta$), where θ is the latitude of point R $((\pi/2) - \theta = $ angle $RPA)$. Because he assumed the magnitude of the attraction to be constant at the surface of the figure, Mairan stated that it followed that when the figure revolves around its axis of symmetry AB, the

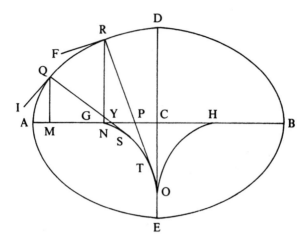

Figure 3. Radii of curvature of a meridian of an elongated figure of revolution (Mairan 1720: Figure 4).

magnitude of the effective gravity decreases from a pole to the equator along a meridian at the surface of the elongated figure of revolution.[8]

Assuming that the same conditions as above hold, Mairan next intended to prove that along the surfaces of a spherical figure and a figure of revolution elongated at its poles which revolve around their axes of symmetry at the same rate (meaning that the two figures rotate with the same angular speed), the ratio of the magnitude of the component of centrifugal force per unit of mass in the direction of the attraction at a point on the surface at a given latitude to the magnitude of the centrifugal force per unit of mass at the equator is smaller for the elongated figure of revolution than it is for the spherical figure. From this he deduced that the magnitudes of the components of centrifugal force per unit of mass in the directions of the attraction at points on the surfaces of the two figures increase more from a pole to the equator along the surface of the elongated figure of revolution than they do along the surface of the spherical figure. Since he assumed attraction to be constant in magnitude at all points on the surfaces of the figures, Mairan concluded that, accordingly, the magnitude of the effective gravity decreases more from a pole to the equator along the surface of the elongated figure of revolution than it does along the surface of the spherical figure.

Mairan reasoned as follows (see Figure 4). He showed that the results that he aimed to establish would not depend upon the relative sizes of the two figures. The magnitude of the component of the centrifugal force per unit of mass at point V on the meridian HDE at the surface of a spherical figure whose center is at C in the direction perpendicular to the surface of the spherical figure at V increases from the pole H to the point D on the equator directly as $|VZ|\sin((\pi/2) - \theta)$ (or

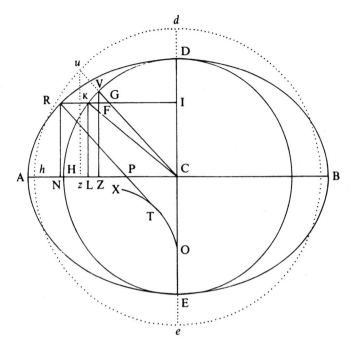

Figure 4. Meridians of an elongated figure of revolution and inscribed sphere
(Mairan 1720: Figure 5).

$|VZ|\cos\theta)$, where θ is the latitude of point $V((\pi/2) - \theta = $ angle VCH), when the spherical figure revolves around its diameter through HC. Now consider the point u on the meridian Ade at the surface of a second spherical figure whose center is also at C and which revolves around its diameter through hC at the same rate as the first spherical figure revolves around its diameter through HC, where point u has the same latitude as point V and where the meridians HDE and Ade at the surfaces of the two spherical figures lie in the same plane. Since triangles Czu and CZV are similar triangles, it follows that

$$\frac{|uz|}{|VZ|} = \frac{|uC|}{|VC|}. \tag{2.1.1}$$

At the same time

$$\frac{|uC|}{|VC|} = \frac{|dC|}{|DC|} \tag{2.1.2}$$

is true. From equalities (2.1.1) and (2.1.2) it follows that

$$\frac{|uz|}{|VZ|} = \frac{|dC|}{|DC|} \tag{2.1.3}$$

and, consequently, that

$$\frac{|uz|}{|dC|} = \frac{|VZ|}{|DC|}.$$ (2.1.4)

If both sides of equality (2.1.4) are multiplied by $\sin((\pi/2) - \theta)$, the equality

$$\frac{|uz|\sin((\pi/2) - \theta)}{|dC|} = \frac{|VZ|\sin((\pi/2) - \theta)}{|DC|}$$ (2.1.5)

results. But equality (2.1.5) means that the ratios of the magnitudes of the components of the centrifugal force per unit of mass perpendicular to the surfaces of the two spherical figures at two points V and u with the same latitude to the magnitudes of the centrifugal forces per unit of mass at the equators of the two spherical figures are equal when the two spherical figures revolve around their polar axes at the same rate. Hence the size of the spherical figure had no effect on the ratios that interested Mairan. A spherical figure of any size can be chosen. Mairan picked a spherical figure that had the same equatorial radius as the elongated figure of revolution. More specifically he chose the spherical figure inscribed in the elongated figure of revolution. We shall see why it was convenient to do so in a moment.

The angles that the line perpendicular to a meridian at the surface of the spherical figure at a point on the meridian at the surface of the spherical figure and that the line perpendicular to a meridian at the surface of the elongated figure of revolution at a point on the meridian at the surface of the elongated figure of revolution make with the common axis of symmetry $AHCB$ of the two figures are, respectively, the latitudes of the point on the meridian at the surface of the spherical figure and the point on the meridian at the surface of the elongated figure of revolution. Thus, for example, point R on the meridian $ADBE$ at the surface of the elongated figure of revolution has the same latitude as point V on the meridian HDE at the surface of the spherical figure, where the two meridians lie in the same plane, because RP, a line segment perpendicular to the surface of the elongated figure of revolution at R, is parallel to VC, a line segment perpendicular to the surface of the spherical figure at V. Consequently point R at the surface of the elongated figure of revolution corresponds to point V at the surface of the spherical figure. Now, the magnitudes of the centrifugal forces per unit of mass at the equators of the two figures are equal, since the equators of the two figures coincide and the two figures revolve around their common axis of symmetry $AHCB$ at the same rate. (Mairan chose the particular spherical figure the way that he did in order to have the magnitudes of the centrifugal forces per unit of mass at the equators of the two figures equal to each other.) Moreover, the latitudes of R and V are equal; hence the sines of the complements of the latitudes of R and V are equal as well. Therefore it was enough for Mairan to show that the magnitude of the centrifugal force per unit of mass at R is smaller than the magnitude of the centrifugal force per unit of mass at V. This meant demonstrating that $|RN| < |VZ|$, which is what Mairan did.[9]

Keeping the same assumptions, Mairan stated that along the surfaces of a spherical figure and a figure of revolution flattened at its poles which revolve around their axes of symmetry at the same rate, the ratio of the magnitude of the component of centrifugal force per unit of mass in the direction of the attraction at a point on the surface at a given latitude to the magnitude of the centrifugal force per unit of mass at the equator is smaller for the spherical figure than it is for the flattened figure of revolution. From this he inferred that the magnitudes of the components of the centrifugal force per unit of mass in the directions of the attraction at points on the surfaces of the two figures increase more from a pole to the equator along the surface of the spherical figure than they do along the surface of the flattened figure of revolution. Again because he assumed attraction to be constant in magnitude at all points on the surfaces of the figures, he concluded that, as a result, the magnitude of the effective gravity decreases more from a pole to the equator along the surface of the spherical figure than it does along the surface of the flattened figure of revolution.

Mairan argued as follows. He was again free to choose the size of the spherical figure arbitrarily. He picked a spherical figure that had the same equatorial radius as the flattened figure of revolution. More precisely he chose the spherical figure circumscribed about the flattened figure of revolution. Again we shall see why it was convenient to do so in a moment.

For the same reasons as the ones given previously, the point R on the meridian $ADBE$ at the surface of the flattened figure of revolution (see Figure 5) has the same latitude as the point V on the meridian ABS at the surface of the spherical figure, where the two meridians lie in the same plane. Here line segment $SECD$ lies along the common axis of symmetry of the two figures. Moreover the magnitudes of the centrifugal forces per unit of mass at the equators of the two figures are equal, since the equators of the two figures coincide and the two figures revolve around their common axis of symmetry at the same rate. (Again, Mairan chose the spherical figure the way that he did in order to have the magnitudes of the centrifugal forces per unit of mass at the equators of the two figures equal.) Hence it sufficed for Mairan to show that $|VZ| < |RN|$, since it follows from this inequality that the magnitude of the centrifugal force per unit of mass at V is smaller than the magnitude of the centrifugal force per unit of mass at R.[10]

Combining these findings together, Mairan concluded that along the surfaces of elongated and flattened figures of revolution which revolve around their axes of symmetry at the same rate, the magnitudes of the components of centrifugal force per unit of mass in the directions of the attraction at points on the surfaces of the two figures increase more from a pole to the equator along the surface of the elongated figure of revolution than they do along the surface of the flattened figure of revolution. Mairan attributed what he called a common belief in just

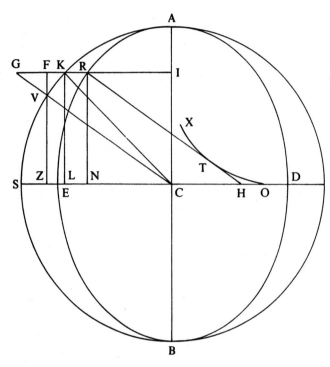

Figure 5. Meridians of a flattened figure of revolution and circumscribed sphere (Mairan 1720: Figure 7).

the opposite to a failure to compare the relevant components of centrifugal force per unit of mass at points with the same latitudes on the surfaces of the two rotating figures to the centrifugal forces per unit of mass at their equators. He suspected that people in the past had instead only compared the centrifugal forces per unit of mass at the equators of figures of revolution "with the same solidity"[11] which revolve around their axes of symmetry at the same rate. Here he presumably meant figures that have the same volume, tacitly assuming that all of his figures are homogeneous and have the same density. In such cases, he stated, the flattened figure of revolution has the largest equatorial radius; the elongated figure of revolution has the smallest equatorial radius; and the spherical figure has a radius somewhere between the equatorial radii of the other two figures. Thus the magnitude of the centrifugal force per unit of mass was greatest at the equator of the flattened figure of revolution and least at the equator of the elongated figure of revolution.

Thus far Mairan made clear his intent. According to observations, the magnitude of the effective gravity decreased less at the surface of Huygens's homogeneous rotating spherical body from a point whose latitude is the same as the latitude of Paris to a point whose latitude is the same as the latitude of Cayenne than it did at the earth's surface from Paris to Cayenne. In demonstrating that the magnitude of the effective gravity decreases more from pole to equator along the surfaces of homogeneous elongated figures of revolution than it does along the surfaces of homogeneous spherical figures or homogeneous flattened figures of revolution when the attraction is assumed to be perpendicular to the surfaces at all points on the surfaces of the three different kinds of figures, when the attraction is assumed to be constant in magnitude at all points on the surfaces of the three different kinds of figures, and when the three different kinds of figures revolve around their axes of symmetry at the same rate, Mairan hoped to show that the observed variation of the magnitude of the effective gravity with latitude along the earth's surface could be made to agree better with homogeneous elongated figures of revolution which revolve around their axes of symmetry than it did with Huygens's homogeneous rotating spherical figure or Huygens's homogeneous flattened figure of revolution which revolved around its axis of symmetry.

Mairan's arguments, however, appear peculiar when examined closely. By taking the distances of points from the axes of rotation in his diagrams as the measure of the magnitudes of the centrifugal forces per unit of mass at points on the surfaces of the three different kinds of figures that he compared, Mairan treated each of his diagrams as a rigid figure, whose parts all revolve around the axis of symmetry of the figure at the same rate, and he tacitly assumed that the three different kinds of figures compared all revolve around their axes of symmetry at the same rate too. But then it is too much to expect that the principle of the plumb line, which Mairan, we remember, called "one of the most inviolable laws of nature," will hold simultaneously at the surfaces of two or more such figures that revolve around their axes of symmetry at the same rate. Indeed, Mairan's analysis leads to no apparent connection between particular shapes of figures and rates at which figures revolve around their axes of symmetry through the principle of the plumb line. It is implausible that Mairan's figures, whose shapes are specified ahead of time, entail no restrictions on the rates at which the figures revolve around their axes of symmetry in order that the principle of the plumb line hold at all points on their surfaces. After all, both Newton and Huygens had linked specific values of the ellipticity of a homogeneous infinitesimally flattened figure of revolution which revolves around its axis of symmetry to the same specific rate at which the figure revolves around its axis of symmetry. (The shapes of Newton's and Huygens's flattened figures of revolution which revolved around their axes of symmetry and the values of the ellipticities of the two figures differed, because Newton's and Huygens's hypotheses of attraction differed.) But Mairan stipulated shapes of figures of revolution beforehand, and he said nothing about the rates at which these figures revolve around their axes of symmetry. From this we can conclude that Mairan went amiss.

This brings us to the fundamental problem. The directions of effective gravity at points on the surfaces of Mairan's figures of revolution when these figures revolve around their axes of symmetry are *not* the directions that Mairan thought – namely, the directions of attraction at the points. Moreover the principle of the plumb line does *not* hold at the surfaces of Mairan's figures when the figures revolve around their axes of symmetry. If attraction is constant in magnitude at the surface and perpendicular to the surface at all points on the surface of a figure of revolution, as Mairan assumed, then the principle of the plumb line cannot possibly hold at all points on the surface of the figure when the figure revolves around its axis of symmetry. The centrifugal force per unit of mass at each point on the surface of the figure combines with the attraction perpendicular to the surface of the figure at each point on the surface of the figure to produce a resultant of effective gravity at each point on the surface of the figure which in general is not perpendicular to the surface of the figure at the point. In order for effective gravity to be perpendicular to the surface of the figure of revolution at all points on the surface of the figure, the figure would have to change its shape when it revolves around its axis of symmetry. For example, when the ellipticity is zero, in which case the figure of revolution is a spherical figure, then if attraction is perpendicular to the surface of the spherical figure at all points on the surface of the figure when the figure does not move, it is clearly impossible for the principle of the plumb line to hold at all points on the surface of the spherical figure if the figure rotates, no matter what the rate of rotation.[12]

To get a clearer idea of what is at issue here, let us recall that Christiaan Huygens assumed attraction to be constant in magnitude at all points on the surface of the homogeneous infinitesimally flattened figure of revolution which he determined and whose shape he took to be the earth's shape. In fact, the principle of the plumb line holds all points on the surface of his figure. But Huygens assumed that attraction was a *central* force whose magnitude was constant *everywhere* and whose center of force was located at the center of the figure. In general, then, the magnitude of the attraction produced by this center of force was constant at all points on the surface of *any* figure, whatever the shape of the figure and wherever the figure was situated. Provided the center of force of attraction is located at the center of a figure of revolution, attraction at a point on the surface of this figure is directed toward the center of the figure. Attraction is therefore in general *not* perpendicular to the surface of a nonspherical figure of revolution at points on the surface of the figure when the center of force of attraction is situated at the center of the figure. (Clearly the direction of the attraction at a point on a meridian along the surface of a nonspherical figure of revolution does lie in the plane of that meridian when the center of force of attraction is located at the center of the figure.)

Consequently nonspherical figures of revolution attracted by centers of force located at the centers of the figures could exist which revolve around their axes of symmetry in ways (that is, at rates) that would make the resultant of effective gravity at each point on the surfaces of such figures, found by combining the attraction at each point on the surfaces of such figures and the centrifugal force per unit of mass at each point on the surfaces of such figures, be perpendicular to the surfaces of the figures at all points on the surfaces of the figures. The rate at which a particular nonspherical figure of revolution must revolve around its axis of symmetry in order for this to happen would depend upon the particular shape of the figure. In other words, the principle of the plumb line could hold at all points on the surfaces of these nonspherical figures of revolution, when these figures revolve around their axes of symmetry at certain rates, which depend upon the shapes of the particular figures.

(Huygens possibly confused Mairan. We remember that Huygens had calculated the variation of the magnitude of the effective gravity with latitude along the surface of a homogeneous rotating spherical body and then compared the variation that he computed with the one observed at the earth's surface. But the principle of the plumb line cannot hold at the surface of Huygens's spherical body if the spherical body rotates. Since Huygens's spherical body rotated "slowly," however, meaning that the ratio $\phi = \frac{1}{289}$ of the magnitude of the centrifugal force per unit of mass to the magnitude of the attraction at its equator was infinitesimal (that is, $(\frac{1}{289})^2 \ll \frac{1}{289}$), Huygens probably disregarded the effect of the rotation on the directions of effective gravity for the sake of his argument. In other words, he probably assumed that the direction of the attraction at a point on the surface of the spherical body, which is perpendicular to the surface of the figure at the point, and the direction of the effective gravity at the same point on the surface of the spherical body when the spherical body rotates are the same for all practical purposes when the spherical body rotates "slowly." But this could have conceivably caused Mairan to misinterpret the problem.)

What, in Mairan's opinion, had led Huygens to believe that the earth was a figure of revolution flattened at the poles? Like Newton, Huygens assumed that the earth was originally a fluid body that would have been spherical if it did not rotate. Mairan concluded that this assumption induced Huygens to think that the earth was flattened at the poles. Mairan did not doubt that a flattening effect of centrifugal forces of rotation existed. He understood that such forces would shorten the axis around which a figure of revolution revolves, if it were at all possible for the length of the axis of rotation of the figure to change. This Mairan acknowledged to be sound mechanics. Indeed, when the Dutch optician, microscopist, and self-styled physicist Nicolas Hartsoëker denied the effects of centrifugal force of rotation which Newton and Huygens had both maintained existed, Mairan outspokenly rejoined in the *Journal des Savants* for 1722, the same year that his Paris Academy *mémoire* was published. Mairan could not understand why Hartsoëker refused to accept Newton's and Huygens's conclusions concerning the effects of centrifugal force of rotation, since their conclusions "depend on the experiments made with pendulums" alone, and "not on geometrical reasoning," as Newton had shown in Proposition 2 of Book III of the *Principia*.[13] What Mairan questioned was not the flattening effect of centrifugal forces of rotation but, instead, the assumption that the earth initially would have been spherical if it

did not rotate. Were the earth, say, originally a homogeneous stationary elongated fluid figure of revolution in equilibrium, centrifugal forces produced by
revolving the figure around its axis of symmetry (its polar axis) would act to make
the body less elongated, but not necessarily flatten it, Mairan affirmed.

(At the place in his *mémoire* where Mairan stated that revolving a homogeneous immobile fluid figure of revolution in equilibrium around its axis of symmetry
would cause the shape of the figure to change, he understood that the centrifugal
forces of rotation would make the directions of effective gravity at points on the
surface of the figure change. Namely, he thought that the directions of effective
gravity at points on the surface of the originally motionless figure would change
into directions that are perpendicular to the surface of the new, rotating figure at
all points on the surface of this figure, because he believed that the homogeneous
rotating fluid figure which had a different shape than the figure at rest would also
be a figure of equilibrium. (We recall, however, that I mentioned at the beginning
of this chapter that in Chapter 4 we would find that the principle of the plumb can
hold at all points on the surface of a homogeneous figure of revolution which
revolves around its axis of symmetry and which is *not* a figure of equilibrium.) But
in analyzing how centrifugal forces of rotation cause the magnitude of the
effective gravity to vary from pole to equator along the surfaces of rotating
spheres, of elongated figures of revolution which revolve around their axes of
symmetry, and of flattened figures of revolution which revolve around their axes
of symmetry, Mairan somehow forgot about these changes in direction of
effective gravity due to rotation and failed to take them into account in his
investigation.)

Mairan clearly saw the implications of such assumptions. If the earth were in
the beginning a fluid figure of equilibrium which did not move, then the principle
of the plumb line held at its surface in that state. This meant that attraction was
perpendicular to the earth's surface at all points on its surface. But if this
motionless earth had an elongated shape as well, the two conditions together
implied that attraction could not be everywhere directed toward one point, the
earth's center, contrary to what would be the case if the stationary earth were
spherical. In other words, attraction could not be a central force, as Huygens had
assumed it to be, if the originally immobile earth were elongated. Consequently,
as we shall see shortly, Mairan had to exclude central-force laws as possible laws
that govern the earth's attraction in order to justify continuing his train of
thought.

Mairan sought a way to characterize the directions of attraction at points when
attraction is not a central force but the directions of attraction at points have an
axis of symmetry. When attraction is perpendicular to the surface at all points on
the surface of a figure of revolution which does not move, he envisioned curves in
the interior of the stationary figure which lie in the plane of one of the meridians
ADBE at the surface of the figure and which are perpendicular to that meridian
such that the direction of attraction at each point on one of the curves lies in the

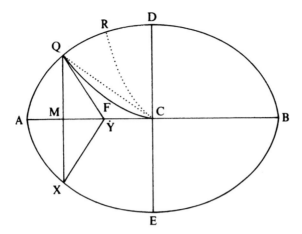

Figure 6. Meridian of a nonspherical figure of revolution and plane curve from the center of the figure to the meridian which are tangent to the directions of the attraction at points on the curve (Mairan 1720: Figure 8).

plane of that meridian and is tangent to that curve (see Figure 6). He specified general conditions on attraction which would make such curves go from the surface of the figure to its center, as in *QFC*, and he restricted his study to a consideration of these cases.[14]

In the event that the directions of attraction at all points in the plane of a meridian at the surface of the figure were perpendicular to that meridian, these curves would be line segments. These segments would be parts of the same lines as the radii of curvature of the meridian. But the radii of curvature of a meridian join points on the meridian to points on the evolute of the meridian. They do not join points on the meridian to the center of the figure unless the evolute of the meridian reduces to the point at the center of the figure, in which case the meridian must be a circle, the figure of revolution must be spherical, and the attraction must be a central force.[15]

To describe the directions of attraction within and upon the surface of a nonspherical figure of revolution which does not move, when attraction is perpendicular to the surface at all points on the surface of the figure and where the attraction is assumed to satisfy the conditions that ensure that the directions of attraction at all points lie in planes of the meridians at the surface of the figure and are tangent to curves which also lie in planes of the meridians at the surface of the figure and which meet at the center of the figure, Mairan utilized what he called "one of Mr. [Johann I] Bernoulli's famous [problems of] *Trajectories*."[16] Mairan alluded here to the problem of orthogonal trajectories. I shall have a great deal to say about this problem later, in Chapter 7 on integral calculus. For the moment I limit myself to Mairan's application of the problem.

Mairan imagined a motionless nonspherical figure of revolution to be made up of a nest of concentric similar surfaces of revolution, infinite in number, one inside another, which go from the surface of the nonspherical figure of revolution to its center. The nest of concentric similar surfaces of revolution is a family of surfaces, and Mairan assumed in addition that all of the surfaces in the family had a common axis of symmetry, namely, the axis of symmetry of the nonspherical figure of revolution. Each point within the nonspherical figure of revolution was situated on exactly one member of the family of surfaces of revolution, and Mairan assumed moreover that the direction of attraction at a point inside the nonspherical figure of revolution was perpendicular at the point to the particular surface of revolution in the family on which the point was located. The directions of attraction in the plane of a meridian at the surface of the nonspherical figure of revolution were therefore tangent to the curves that cut at right angles all of the meridians of the members of the family of surfaces of revolution which lie in that plane. The curves that cut a family of plane curves at right angles are called the orthogonal tragectories of the family. The problem of orthogonal trajectories is to find the curves that cut a given family of plane curves at right angles.

Mairan illustrated the problem by finding the plane curves that join points on a meridian at the surface of a figure shaped like an elongated ellipsoid of revolution to the center of the figure, such that the direction of attraction at each point on one of the plane curves lies in the plane of the meridian at the surface of the figure and is tangent to the curve at the point. It is the family of surfaces of revolution that defines the directions of attraction: the direction of attraction at each point on one of the surfaces of revolution in the family is perpendicular to the surface at the point.

For Mairan, part of the problem was to determine a family of concentric similar surfaces of revolution. He showed, for example, that the family of concentric surfaces of revolution could not consist of surfaces that all have the same evolute as the elongated ellipsoid of revolution which is the surface of the given figure (see Figure 7). Such surfaces of revolution are elongated ellipsoids of revolution which are everywhere parallel to (that is, equidistant from) each other and the elongated ellipsoid of revolution which is the surface of the given figure. In that case, a line segment that joins any point on a meridian ADBE of the elongated ellipsoid of revolution which is the surface of the given figure to the corresponding point on the evolute of that meridian (that is, any radius of curvature of the particular meridian ADBE of the elongated ellipsoid of revolution which is the surface of the given figure) would intersect all of the elongated ellipsoids of revolution in the family at right angles, since the members of the family are all involutes of the elongated ellipsoid of revolution which is the surface of the given figure. By the same token, however, these orthogonal trajectories do not go to the center of the given figure as required, since the given figure is not spherical.[17] (Because two elongated ellipsoids of revolution in the same family of involutes are equidistant from each other, the ratios of their major semiaxes to their minor semiaxes cannot be equal. Hence the elongated ellipsoids of revolu-

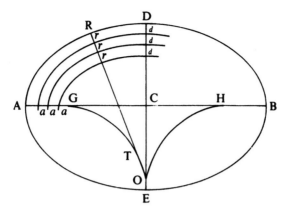

Figure 7. Radii of curvature of a meridian of an elongated ellipsoid of revolution and plane curves that all have the same evolute as the meridian of the elongated ellipsoid of revolution (Mairan 1720: Figure 9).

tion in the family do not have the same ellipticity. Consequently they are not geometrically similar – unless, of course, the surface of the given figure and its involutes are all spheres, in which case they all have the same ellipticity, namely zero.)

Therefore, the family of concentric similar surfaces of revolution must differ from the family just described if the trajectories everywhere perpendicular to its members are to meet at the center of the given figure. Mairan determined a family of surfaces of revolution where this is the case (see Figure 8): a family of concentric, geometrically similar elongated ellipsoids of revolution which are concentric with the given figure shaped like an elongated ellipsoid of revolution, which all have a common axis of symmetry, namely, the axis of symmetry *AB* of the given figure, and where the ratio of the major semiaxis to the minor semiaxis of an elongated ellipse that is a meridian of each member of the family is the same as the ratio of the major semiaxis to the minor semiaxis of an elongated ellipse that is a meridian *ADBE* of the elongated ellipsoid of revolution which is the surface of the given figure. (Hence the elongated ellipsoids of revolution in the family all have the same ellipticity, namely, the ellipticity of the elongated ellipsoid of revolution which is the surface of the given figure.) He determined the trajectories orthogonal to all of the members of a family composed of the kind of elongated ellipses just mentioned, where all of the elongated ellipses lie in the same plane of a meridian *ADBE* of the elongated ellipsoid of revolution which is the surface of the given figure. As I just said, the ratio of the major semiaxis to the minor semiaxis is the same for each member of the family of elongated ellipses. Let

$$k \equiv \sqrt{\frac{m}{p}} \qquad (2.1.6)$$

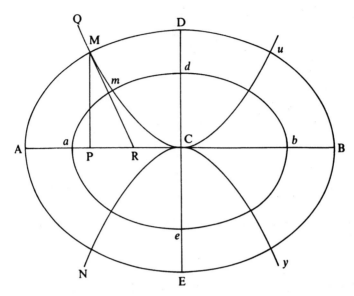

Figure 8. Meridians of a family of concentric, elongated ellipsoids of revolution which all have the same ellipticity, hence which are geometrically similar, and the parabolas orthogonal to the members of the family (Mairan 1720: Figure 10).

designate this constant ratio, where m and p are both greater than zero. Then the equation of the family of elongated ellipses can be written as

$$\frac{x^2}{k^2 a^2} + \frac{y^2}{a^2} = 1 \qquad (2.1.7)$$

where a is the parameter of the family of elongated ellipses. Mairan found the ordinary "differential" equation for plane curves orthogonal to all of the members of the family of elongated ellipses to be

$$\frac{p\,dy}{y} = \frac{m\,dx}{x}. \qquad (2.1.8)$$

(A "differential" equation is an equation in which differentials appear.)
The integral of equation (2.1.8) is a family of parabolas

$$y^p = nx^m, \qquad (2.1.9)$$

where the constant of integration n is the parameter of the family of parabolas and n is greater than zero. The vertices of all of the parabolas in the family are all situated at the center of the given figure. The particular kind of parabolas in the family (ordinary, cubical, semicubical, etc.) depends upon the ratio of the major semiaxis to the minor semiaxis of any one of the elongated ellipses in the given family, $(m/p)^{1/2}$, a ratio that is the same for all of the elongated ellipses in the family, as I have indicated. Thus we see that the particular kind of parabolas in the family of orthogonal trajectories depends upon the value of the constant ratio given in (2.1.6) determined by the elongated ellipses in the given family, and we see how the kind of parabolas depends upon this ratio. Mairan applied the whole exercise to the particular elongated ellipsoid of revolution which Jacques Cassini considered to be the earth's surface based on his and his colleagues' observations of variations of degrees of latitude with latitude in France along the meridian through Paris.[18]

Assuming the field of attraction to extend beyond the surface of a figure of revolution (or, as Mairan stated, to extend "to the limits of the terrestrial vortex"[19]), Mairan maintained that one could in principle do experiments above the surface of a stationary figure of revolution, using suspended chains, flexible tubes filled with liquid, or other means, to determine the curves in the plane of a meridian at the surface of the figure of revolution to which directions of attraction are tangent. Then, by inversion, a meridian at the surface of the figure of revolution could be determined, assuming that the figure does not move. Namely, the meridian at the surface of the figure of revolution which lies in the plane in question is just one of the trajectories orthogonal to these curves. Finally, the surface of the figure of revolution is then found by revolving the meridian around its axis of symmetry. Mairan illustrated the idea by presenting the preceding problem backward. That is, he stated the result that is just the reverse of the one that he had demonstrated: if the family of parabolas is given, the trajectories orthogonal to its members must be elongated ellipses.

Mairan acknowledged, however, that in practice observations would not permit the portions of these curves which reach to the limits of the field of attraction (that is, to the limits of the terrestrial vortex) from the surface of the figure of revolution to be differentiated from straight lines, in case the motionless figure is nearly spherical like the earth. In this event, the curves could not actually be found by doing experiments using the apparatus mentioned.

Moreover, despite his way of thinking, Mairan did not clearly define a problem with a unique solution. Any number of hypotheses of attraction, each specified by means of curves in the plane of a meridian at the surface of an immobile nonspherical figure of revolution which join the center of the figure to that meridian and which are perpendicular to that meridian, could be imagined so that they accord with a given set of values of attraction at the surface of the figure, where the directions of attraction at all points on the surface of the figure are assumed to be perpendicular to the surface of the figure at the points. Each hypothesis of attraction would give rise to a distinct family of plane curves. Likewise, the family of meridians of the surfaces of revolution in the family of surfaces of revolution within the given figure of revolution whose members are concentric with the given figure, lying in the same plane as the meridian in question at the surface of the given figure, all of whose members are also orthogonal to these plane curves, would depend upon the particular hypothesis of attraction, too. In other words, each hypothesis of attraction would determine a different family of surfaces of revolution whose members are concentric with the given figure.

(To be sure, there is only one family of surfaces whose members are concentric with and *geometrically similar* to the surface of a nonspherical figure of revolution. For example, in the case of the figure shaped like an elongated ellipsoid of revolution discussed earlier, the ratio of the major semiaxis to the minor semiaxis of one of the meridians at the surface of the figure uniquely determines the other elongated ellipsoids of revolution in the family. Because these elongated ellipsoids of revolution are all geometrically similar, the ratios of the major semiaxes to the minor semiaxes of their meridians must all be the same. Namely, they must all equal the ratio of the major semiaxis to the minor semiaxis of a meridian at the surface of the given figure. However, the stipulation that the members of the family of surfaces all be geometrically similar is *not needed* to ensure that the conditions that must be met, stated above, are all fulfilled.)

Consequently the problem of orthogonal trajectories, as Mairan presented it, not only had little practical use, but did not have much theoretical importance, either. The whole exercise served no real purpose in Mairan's *mémoire*. As I noted previously, it could in principle be used as a way of describing the directions of attraction when attraction is not a central force but the directions of attraction do have an axis of symmetry. The family of surfaces of revolution inside the given figure of revolution which does not move just turn out to be what were later called "level surfaces" of the attraction, meaning that the attraction at any point on such a surface is perpendicular to the surface at the point. But, as we shall see in Chapter 9, the notion of a "level surface" only emerged in the work of other people, and its utility only became evident in the work done by these other people, both for reasons that have nothing to do with Mairan's *mémoire*. In Mairan's *mémoire*, the problem of orthogonal trajectories remained an academic exercise of no use.

Then why did Mairan discuss the problem in his *mémoire*? He probably did so for no other reason than to call attention to some recent developments in mathematics. I shall study those developments, for reasons that do not relate to and have nothing to do with Mairan's *mémoire*, in Chapter 7. They ultimately turn out to have a very important part in the story told here.

Until now Mairan had postulated attraction to be constant in magnitude at the surfaces of his homogeneous stationary spherical figures and homogeneous stationary figures of revolution, as well as perpendicular to the surfaces of these figures at all points on their surfaces. He did so for two reasons. First, Huygens had previously assumed the same to be true for his homogeneous motionless spherical figure. Second, Mairan said that this hypothesis simplified the study of variations of the magnitude of the effective gravity at the surfaces of his figures when the figures revolved around their axes of symmetry. That is, it was only necessary to analyze the effects of centrifugal force per unit of mass at the surfaces of the figures, since attraction was constant in magnitude at the surfaces of the figures, so that there were no variations of the magnitude of the effective gravity which involved attraction to have to take into account – or so he mistakenly thought, having apparently failed to understand what caused the magnitude of the effective gravity to vary along the surface of Huygens's homogeneous infinitesimally flattened figure of revolution when the figure revolved around its axis of symmetry.

(As I mentioned earlier, Huygens had indeed assumed attraction to be constant in magnitude at the surface of his homogeneous infinitesimally flattened figure of revolution, but he had also supposed attraction to be a *central* force whose magnitude was constant everywhere and whose center of force was situated at the center of the figure. Hence the directions of attraction at points on the surface of Huygens's figure were not in general perpendicular to the surface of the figure at those points, so that the principle of the plumb line could hold at all points on the surface of the figure when the figure revolved around its axis of symmetry at a certain rate. Mairan, however, always took attraction at points on the surfaces of his figures of revolution to be perpendicular to the surfaces of the figures at those points, in which case it was impossible for the principle of the plumb line to hold at all points on the surfaces of the figures when the figures revolved around their axes of symmetry, no matter what their rates of rotation.)

Assuming attraction to be constant in magnitude at the surface of a homogeneous immobile elongated figure of revolution and to be perpendicular to the surface at all points on the surface of the figure as well, Mairan maintained that it followed that the magnitude of attraction would be the same at all points along a curve in the plane of a meridian at the surface of the figure which joins that meridian to the center of the figure and which is perpendicular to the family of surfaces of revolution inside the figure which characterize the hypothesis of attraction. In other words, the magnitude of attraction is constant along any one of the plane curves to which directions of attraction in the plane of a meridian at the surface of the figure are tangent.

This he deduced as follows. Since the given figure of revolution is elongated and the magnitude of attraction is constant at its surface, the magnitude of attraction at points on the surface of the figure is not a function of the distances of points on the surface of the figure from the center of the figure. Mairan concluded from this fact that there was no reason to think that the magnitude of attraction varied with distance from the center of the figure at points along a trajectory orthogonal to the members of the family of surfaces inside the figure. This opinion he expressed in the form of a rhetorical question. Thus supposing attraction to be constant in magnitude at the surface of the given elongated figure of revolution, Mairan inferred that the magnitude of attraction at points along a trajectory orthogonal to the members of the family of surfaces within the given motionless elongated figure of revolution was independent of the distances of points on the trajectory from the center of the figure. From this he concluded in addition that the magnitude of attraction at all points along all parts of a trajectory orthogonal to the members of the family of surfaces within the given immobile elongated figure of revolution, whether the parts of the orthogonal trajectory lie inside or outside the given figure, did not depend upon the distances of the points from the center of the figure.

Mairan did *not* deduce that attraction was a *central* force whose magnitude was constant and whose center of force was situated at the center of the given figure. Given the assumption that attraction is perpendicular to the surface at all points on the surface of a stationary elongated figure of revolution, attraction could not be a central force whose magnitude either varies as a function of the distance from the center of force or is constant, where in both cases the center of force is located at the center of the figure, because a *spherical* figure is the *only* figure at whose surface the directions of the attraction produced by a central force whose center of force is situated at the center of the figure are perpendicular to the surface of the figure at all points on the surface of the figure. (Indeed, I repeat, Mairan had to rule out central-force laws altogether as possible laws that govern the earth's attraction, as we shall see in a moment).

Mairan noted, however, that not only were such ideas "contrary to the received opinions of the day concerning attraction,"[20] by which he meant that his conclusion that attraction was constant in magnitude at points along the plane curves to which directions of attraction in the plane of a meridian at the surface of a figure of revolution are tangent did not accord with the ideas of the day regarding attraction, but that they were based as well on one particular hypothesis that had been easier for Huygens to justify making than he. That is, both Huygens and Mairan had assumed the magnitude of attraction to be constant at the surface of the figures that each had taken to represent the originally motionless earth. Mairan had supposed the earth to be an elongated figure of revolution in the beginning when it did not move. Having assumed attraction to be constant in magnitude at its surface, he reasoned that this meant that attraction had to be constant in magnitude everywhere within its range of action. Consequently, unless attraction was constant in magnitude everywhere within its range of action, Mairan believed it unreasonable to think that the magnitude of attraction did not vary at the surface of the motionless elongated figure of revolution which he took to represent the primeval earth. Huygens, on the other hand, had taken the initially immobile earth to be spherical. As a result, Mairan declared, Huygens did not have to suppose attraction to be constant in magnitude everywhere within its range of action, simply because he had assumed the magnitude of attraction to be constant at the surface of the spherical figure that he had taken to represent the originally stationary earth. Specifically, Mairan remarked that Huygens could have made attraction be constant in magnitude at the surface of the spherical figure without having to suppose that attraction is a central force whose magnitude is constant and whose center of force is located at the center of the figure, which is in fact what Huygens had assumed. Instead,

Mairan noted, Huygens could have just as easily hypothesized that attraction is a central force whose magnitude varies as a function of the distance to a center of force situated at the center of the spherical figure, and this would have led to the same result that Mairan had in mind. That is, in this case too, the magnitude of the attraction would be constant at the surfaces of all spherical bodies whose centers were located at the center of force.

(With Mairan's errors described earlier in mind, I note that Huygens may have taken the magnitude of attraction to be constant everywhere within its range of action for precisely the same reasons that Mairan assumed the magnitude of attraction to be constant at the surface of his immobile elongated figure of revolution. Namely, if attraction is not assumed to be constant in magnitude everywhere within its range of action, then it will not in general be constant in magnitude at the surfaces of *nonspherical* bodies like Mairan's elongated figure of revolution. Hence if Huygens had not assumed the magnitude of his central force of attraction to be constant everywhere, then it would not be constant in magnitude along the surfaces of nonspherical figures of revolution whose centers are situated at the center of force. In particular, it would not be constant in magnitude at the surface of his *homogeneous infinitesimally flattened figure of revolution which revolves around its axis of symmetry*. But this would be contrary to what Huygens had assumed! Mairan, however, apparently failed to observe this, which suggests, once again, that Mairan did not truly understand Huygens's theory of the shape of a homogeneous infinitesimally flattened figure or revolution which revolves around its axis of symmetry.)

But if contrary to what Mairan had assumed until now, attraction does not have a constant magnitude everywhere but is a $1/r$ central force instead, or, say, a $1/r^2$ central force, "as is commonly believed to be the case today,"[21] where r is the distance from the center of force, then Mairan said that his preceding arguments from which it followed that the magnitude of the effective gravity decreased more from a pole to the equator along the surface of an elongated figure of revolution than along the surfaces of a spherical figure or a flattened figure of revolution when the three figures revolved around their axes of symmetry at the same rate, arguments which were based only upon a greater increase in the magnitude of the centrifugal force per unit of mass from a pole to the equator along the surface of the elongated figure of revolution than along the surfaces of the spherical figure or the flattened figure of revolution when the three figures revolved around their axes of symmetry at the same rate, "lost all of their force."[22] Mairan meant by this that it did not follow from his prior reasoning alone that the variations of the magnitudes of the effective gravity with latitude along the surfaces of elongated figures of revolution which revolve around their axes of symmetry would necessarily accord better with the observed variation of the magnitude of the effective gravity with latitude along the earth's surface than did the variations of the magnitudes of the effective gravity with latitude along the surfaces of flattened figures of revolution which revolve around their axes of symmetry. (As we saw, however, the arguments that Mairan referred to here had no validity, anyway!) In the cases of $1/r$ or $1/r^2$ central forces of attraction, the magnitudes of the attraction and the centrifugal force per unit of mass increase together from a pole to the equator along the surface of an elongated figure of revolution which revolves around its axis of symmetry. The increase in the former could conceivably neutralize the increase in the latter, if not actually exceed it, so as to produce no decrease, or even a net increase, in the magnitude of the effective gravity from a pole to the equator along the surface of an elongated figure of revolution which revolves around its axis of symmetry instead of a decrease.

(By the same token, although Mairan did not mention the fact, the same $1/r$ or $1/r^2$ central forces of attraction would cause the magnitude of the attraction to diminish from a pole to the equator along the surface of a flattened figure of revolution which revolves around its axis of symmetry. This decrease could possibly combine with the increase in the magnitude of the centrifugal force per unit of mass from a pole to the equator along the surface of such a figure to decrease the magnitude of the effective gravity from a pole to the equator along the surface of the figure even more, resulting in a cumulative decrease in the magnitude of the effective gravity along the surface of the figure from the point whose latitude is the same as the latitude of Paris to the point whose latitude is the same as the latitude of Cayenne which was greater than the corresponding decrease determined by Huygens for a rotating spherical body. If the net decrease in the magnitude of the effective gravity along the surface of the flattened figure of revolution which revolves around its axis of symmetry, from the point whose latitude is the same as the latitude of Paris to the point whose latitude is the same as the latitude of Cayenne, did not exceed the observed decrease in the magnitude of the effective gravity from Paris to Cayenne, or if it overreached the observed decrease by an insignificant amount, then such a flattened figure of revolution would accord better with terrestrial observations than either Huygens's spherical figure or Huygens's flattened figure of revolution did when central forces whose magnitudes are *constant* and whose centers of force are situated at the centers of the two figures were assumed to attract the two figures. In that event, it would no longer serve any purpose to exploit the difference between Huygens's calculated decrease in the magnitude of the effective gravity and the observed decrease, which is what Mairan was attempting to do.)

To handle this problem, Mairan had to show, once and for all, that the force that the earth attracts in accordance with cannot be a central force. At the same time, he had to establish a measure of the magnitude of the attraction at different points when attraction is not a central force. (His discussion of the problem of orthogonal trajectories only bore upon the directions of the attraction.)

Mairan, as I have said, doubtlessly believed that the measurements of degrees of latitude made by the Paris Academy's astronomers were correct. Someone with Mairan's phenomenological or positivistic viewpoint would not have been inclined to question the accuracy of the work of colleagues whose primary business, making measurements, was the same as his own. From the variations in degrees of latitude with latitude which the astronomers found, they concluded that the earth was at present elongated. From this Mairan inferred that the earth must have been even more elongated before it began to rotate, since he supposed that the earth was originally a homogeneous stationary fluid figure of revolution in equilibrium, which then became less elongated when it began to revolve around its axis of symmetry. He also assumed that the principle of the plumb line held at the surface of the immobile earth, again because he presumed the initially motionless earth to have been a figure of equilibrium. Since attraction was consequently perpendicular to the surface of the stationary earth at all points on its surface, the directions of attraction at the earth's surface could not all have been aimed at one point in the beginning, namely, the point at the center of the earth, because the surface of the primeval earth was not a sphere according to Mairan. Mairan deduced from this argument that the kind of force that makes the earth attract could not be a central force.

Mairan admitted that the way that he established this conclusion perhaps seemed odd, but he insisted that the proper objective was to accept the terrestrial observations of the day and use them as clues for determining the earth's original shape and its real law of attraction. Thus he believed that nature's true laws are to be determined empirically. These laws are then used as constraints or controls or indicators or guides in inventing theories instead of proceeding vice versa (that is, first imagining hypotheses, then drawing conclusions from the hypotheses). From the earth's existing shape, inferred from the terrestrial observations of the day, Mairan thought that the earth's initial shape could be found and, as a result, that the "true directions of attraction" could be determined. As Mairan expressed the matter,

whatever shape the earth has at present, once the [plane] curve that is supposed to generate the earth's shape by revolving the curve around the earth's [polar] axis [in other words, one of the earth's meridians] has been determined by observations, the earth's original shape can be found by inversion (*par l'inverse*), and, consequently, the true directions of attraction can be determined, too. In a word, it is up to observations to show us how centrifugal forces changed the earth's shape [from what it was originally to what we observe it to be today] and not up to centrifugal forces to fix some shape that the earth must have today [assuming, by hypothesis, that it initially had a certain shape].[23]

Thus the right approach was to work backward from present to past and not forward from past to present. In short, the earth's original shape was a problem to be solved empirically. From what we have seen of Mairan's intentions and objectives thus far, we can safely conclude that his views and attitudes were firmly entrenched by that positivistic tradition in French scientific thought which goes back at least as far as Mersenne, Gassendi, Pascal, and Roberval in the midseventeenth century.[24] I shall say more about this tradition in a moment.

In fact, however, Mairan gave absolutely no indication how the earth's original shape was to be determined from its present shape "by inversion" and, accordingly, how the original directions of the earth's attraction were to be found. But this is consistent with his having failed to link specific shapes of rotating figures with specific rates of rotation. Newton and Huygens, we recall, had established such connections.

Besides, it is not at all clear what the "true directions of attraction" mentioned in the passage quoted even means. Presumably Mairan's "true directions" were the directions of attraction at points on the surface of the originally stationary earth. These directions could be determined, if the shape that the earth had in the beginning could be found, because Mairan assumed these directions to be perpendicular to the surface at all points on the surface of the originally motionless earth. But it is not necessarily the case that these directions are any more "true," meaning inherent, than other directions.

If the directions of attraction are independent of the shape of a body that attracts itself, then the directions of attraction do not change if rotating the body causes the shape of the body to change; what Mairan called the "true directions of attraction" can be thought of as having a sense; and these directions would be the ones that I just mentioned, even though as I said Mairan did not explain how the earth's initial shape and, accordingly, how these directions could be found in practice. For example, the directions of attraction are independent of the shapes of bodies that attract themselves according to central forces of attraction. However, if the earth attracted itself in accordance with a central force where the center of force is situated at the earth's center, then the originally stationary earth at whose surface attraction is perpendicular to the surface at all points on the surface would be shaped like a sphere. But a fluid spherical body would flatten once it began to rotate, whereas the Paris Academy's astronomers had found the earth to be elongated at its poles by making astronomical observations and geodetic measurements. And for this reason Mairan had to rule out central forces of attraction.

On the other hand, the directions of attraction can depend upon the shape of a body that attracts itself. This happens, for example, when a body attracts according to the universal inverse-square law. In that case the directions of attraction change if rotating the body causes the shape of the body to change. (I shall discuss this fact and its implications fully in Chapter 9.) These two different kinds of laws of attraction explain in part why Newton and Huygens arrived at homogeneous flattened figures of revolution which represented the rotating earth but whose shapes and infinitesimal ellipticities differed, even though both Newton and Huygens assumed that their figures revolved around their axes of symmetry at the same rate (that is, the ratios of the magnitudes of the centrifugal force per unit of mass to the magnitudes of the attraction at the equators of the two figures had the same value, $\frac{1}{289}$). But when the directions of attraction depend upon the shape of the body that attracts itself, the "true directions of attraction" is an expression with no intrinsic meaning.

Most important, contrary to what Mairan suggested in the passage quoted, it is not true that the earth's original shape and the original directions of attraction can be determined from the earth's present shape, deduced from observations, by purely empirical means.

Indeed, Mairan's paper includes two hidden hypotheses, his positivistic outlook notwithstanding. When Mairan adduced the evidence based on observations that the principle of the plumb line currently held at the earth's surface, he added that he would also explain later what he called "the a priori reason" why this principle proved true.[25] By this he must have meant a belief that the earth was not just initially elongated, but originally a stationary figure of equilibrium as well. At least, there seems to be no other way for him to have justified the supposition that the principle of the plumb line held at the surface of the motionless elongated earth, from which he inferred that the earth's attraction could not be a central force (although as I have already noted the principle of the plumb can hold at all points on the surface of a homogeneous figure of revolution which revolves around its axis of symmetry and which is *not* a figure of equilibrium). But Mairan's belief is an hypothesis. It obviously cannot be verified directly. I shall identify Mairan's second, concealed hypothesis shortly.

Having excluded hypothesized central forces as possible forces that determine the law that governs the earth's attraction, Mairan applied the theory of curvature of plane curves to determine how the magnitude of attraction must vary at the surfaces of figures of revolution. If attraction is everywhere perpendicular to the surface at all points on the surface of a stationary figure of revolution elongated at the poles, the direction of attraction at a point R on a meridian $ADBE$ at the surface of the figure lies along the radius of curvature RT of the meridian at the point R (see Figure 9). Mairan concluded that the magnitude of attraction at a point R on the surface of the elongated figure of revolution varies inversely as the product of the length $|RT|$ of the radius of curvature of the meridian at R and the distance $|RY|$ of R from the point on the axis of symmetry AB where this radius of curvature intersects the axis of symmetry AB.[26]

The radius of curvature RT of a meridian $ADBE$ at the surface of a motionless figure of revolution flattened at the poles at a point R on the meridian does not cross the axis of symmetry ED. Instead, the point T on the evolute of the meridian through R is on the same side of the axis of symmetry ED, in the plane of that meridian, as the point R (see Figure 10). In this case, if attraction is everywhere perpendicular to the surface at all points on the surface of this immobile figure of revolution flattened at the poles, Mairan declared that the magnitude of attraction at R varies inversely as the product of the length $|RT|$ of the radius of curvature at R of the meridian through R and the length $|RY|$ of the prolongation of this radius of curvature to the axis of symmetry ED.[27]

Mairan's demonstration of his law of attraction includes his second hidden hypothesis. "Accepting as fact, as the majority of modern physicists and astronomers do, that [the magnitude of the] attraction on bodies increases the closer they are to the center of attraction, inversely as the square of the distance to the center,"[28] Mairan explained $1/r^2$ central forces of attraction the same way that

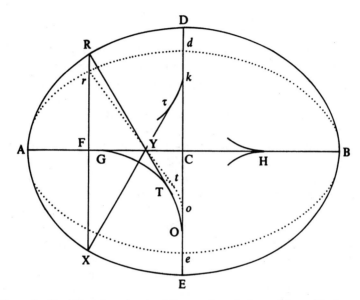

Figure 9. Meridian of an elongated figure of revolution and its radii of curvature (Mairan 1720: Figure 12).

one customarily accounted for the law of photometry (that is, the fact that the power of a candle varies inversely as the square of the distance to the candle). Namely, Mairan stated that the flux of the lines of attraction across spheres whose centers were situated at the center of force remains constant, so that the density of the lines of attraction varies inversely as the square of the distance from the center of force. Mairan used this kind of argument to deduce his own law of attraction, which differed from the laws of attraction for the central forces of attraction which he had to rule out.

He took the point T as a local center of force for the attraction at R, and he treated the portion of the meridian $ADBE$ through R at the surface of an elongated figure of revolution in the vicinity of R as approximately a circular arc with radius RT (see Figure 9). For lines of force in a plane, whose center of force is situated at T, the linear flux of lines of attraction across circles in the plane whose centers are located at T is constant. At the same time, the circumferences of these circles measure the linear flux across them. Since the circumference of the circle through R whose center is at T is $2\pi|RT|$, $2\pi|RT|$ is the measure of this constant flux. From this Mairan concluded that the linear density of the lines of attraction at R along the meridian through R varies inversely as the length $|RT|$ of the radius of curvature RT of this meridian at R.[29]

Now, the density of the lines of attraction at a point R on a meridian $ADBE$ at the surface of an elongated figure of revolution (that is, the intensity, or magnitude, of the attraction at R) is not linear or one dimensional but two dimensional

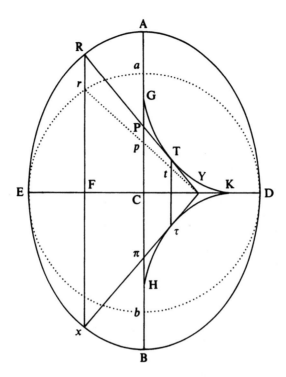

Figure 10. Meridian of a flattened figure of revolution and its radii of curvature (Mairan 1720: Figure 13).

instead (the flux of the lines across an infinitesimally small area surrounding *R* at the surface of the elongated figure of revolution). At a point on the equator of the elongated figure of revolution, such as *D*, the direction of attraction at *D* (that is, the direction of the radius of curvature *DO* at *D* of the meridian *ADBE* through *D* at the surface of the figure) is the same as the direction of the radius *DC* of the circular equator. The equator is perpendicular to the meridians, so that the radii of curvature of the meridians at the points where the meridians intersect the equator lie in the same plane as the radii of the equator. In fact, the radius *DC* of the equator lies along the radius of curvature *DO* of the meridian through *D* at *D*. Mairan took the point *C* as a second local center of force for the attraction at *D* – namely, one that determines the linear density of the lines of attraction at *D* along the equator, which is perpendicular at *D* to the meridian through *D* (see Figure 9). Using the same kind of arguments as those above, he concluded that the linear density of the lines of attraction at *D* along the equator varies inversely as the length |*DC*| of the radius *DC* of the circular equator. Thus the two-dimensional density of the lines of attraction at *D* varies inversely as the product |*DO*| × |*DC*|.[30]

More generally, for a point R on a meridian $ADBE$ at the surface of the elongated figure of revolution which is not situated at the equator, the circle at the surface of the elongated figure of revolution through R which is parallel to the equator is not perpendicular at R to the meridian through R. In this case, the lines along which the directions of attraction at such points R lie (in other words, the directions of the radii of curvature RT of the meridians through such points R at such points R) for all points R on the same circle parallel to the equator intersect at a point Y on the axis of symmetry AB (RYX revolved around AB generates a cone with apex at Y). Mairan took this point Y as the second local center of force for the lines of attraction at R – namely, one that determines the linear density of the lines of attraction at R along a circle through R whose radius is RY and which is perpendicular at R to the meridian $ADBE$ through R (see Figure 9). Again, reasoning the same way that he had above, Mairan concluded that the linear density of the lines of attraction at R along the circle through R whose radius is RY and which is perpendicular at R to the meridian $ADBE$ through R varies inversely as $|RY|$. The two-dimensional density of the lines of attraction at R is, once again, the product of the linear densities. Thus, Mairan concluded, attraction at a point R on a meridian $ADBE$ at the surface of an elongated figure of revolution varies inversely as the product $|RT| \times |RY|$.[31]

Applying analagous arguments to flattened figures of revolution (see Figure 10), Mairan derived the law of attraction at the surfaces of such figures which I stated earlier.[32]

We note that like the laws of attraction which central forces of attraction determine, Mairan's law of attraction does not depend upon the density of the figure, so that it really does not matter whether the figure is assumed to be homogeneous or not. Indeed, if the figure were spherical instead of an elongated or flattened figure of revolution, the evolute GTO of each meridian at the surface of the spherical figure would reduce to a point, the center C of the spherical figure, in which case $T = C$ and $RT = RC$ for radii of curvature at all points R on a meridian at the surface of the spherical figure. Moreover, the point Y that corresponds to the circle at the surface of the spherical figure through a point R which is parallel to the equator, where R is not located on the equator, is also equal to C, so that $RY = RC$. In addition, if the point D is on the equator, and DO is the radius of curvature at D of the meridian through D, then the radius of curvature of the meridian at D obviously equals the radius of the spherical figure. In other words, $O = C$, and $DO = DC$. These two results show that in case a figure is spherical, Mairan's law states that attraction at a point on the surface of the spherical figure varies inversely as the square of the radius of the spherical figure. In short, the attraction in this case just turns out to be a $1/r^2$ central force, where r stands for distance from the center of force.

This, of course, is a foregone conclusion; it follows inevitably from Mairan's argument. His law of attraction involves the second of the two hypotheses "hidden" in his paper – namely, if attraction at the surface of a spherical figure that does not move is perpendicular to the surface at all points on the surface of the figure, then its magnitude is proportional to $1/r^2$, where r is the radius of the spherical figure. In other words, the attraction is a central force. Moreover, the attraction is not just any central force. The magnitude of the attraction is not, say, proportional to some power of the distance r from the center of the spherical figure which differs from 2. The attraction is necessarily an inverse-square central force. (It is not certain that Mairan would have realized that this was a hypothesis. To Mairan it may have appeared to be as "factual" as the law of photometry. He was clearly trying to draw a parallel between his law of attraction and the law of photometry.)

Mairan then imagined a nonspherical figure of revolution to consist of a nest of surfaces geometrically similar to the surface of the figure of revolution, infinite in number, one inside another, which all have a common axis of symmetry, namely, the axis of symmetry of the nonspherical figure of revolution, and which go from the surface of the figure of revolution to its center. Each point within the figure of revolution is situated on exactly one of these surfaces, and according to Mairan the attraction at a point inside the figure of revolution was determined by his law of attraction applied to the particular surface that the point is located on and to the point in question situated on this surface.

(However, as I mentioned before, in connection with Mairan's application of the problem of orthogonal trajectories to specify the directions of attraction at points when attraction is not a central force, a figure of revolution can be thought of as composed of any number of different nests of surfaces, so that Mairan could not truly extend his law of attraction in a unique way to cover points in the interior of a figure of revolution. Moreover, we note that Mairan's law of attraction at points on the surface of a figure of revolution and his extension of this law to points inside the figure of revolution depends only on geometry (the curvatures of the surfaces in the nest of surfaces that make up the figure of revolution) and not on the density of the figure of revolution. Changing a homogeneous figure of revolution into a heterogeneous figure of revolution which has the same shape as the homogeneous figure of revolution does not change the law of attraction.)

Now, the lengths of the radii of curvature of a meridian at the surface of a figure of revolution elongated at the poles increase from a pole to the equator along the meridian. Moreover, the lengths of the portions of the radii of curvature of a meridian at the surface of a figure of revolution elongated at the poles which join points on the meridian to the points where the radii of curvature of the meridian intersect the axis of symmetry of the elongated figure of revolution also increase from a pole to the equator along the meridian. From these increases Mairan concluded that if the priciple of the plumb line holds at the surface of a stationary figure of revolution elongated at the poles, which Mairan assumed to be the case and which means that attraction at the surface of the elongated figure of revolution is perpendicular to the surface of the figure at all points on the surface, then the magnitude of the attraction at the surface of the elongated figure of revolution decreases from a pole to the equator.

At the same time, the lengths of the radii of curvature of a meridian at the surface of a figure of revolution flattened at the poles decrease from a pole to the equator along the meridian. Furthermore, the lengths of the prolongations of these radii of curvature to the axis of symmetry of the flattened figure of revolution decrease from a pole to the equator along the meridian, too. From these decreases Mairan concluded that if the principle of the plumb holds at the surface of a motionless figure of revolution flattened at the poles, which, again, Mairan presumed to be true and which means that attraction at the surface of the flattened figure of revolution is perpendicular to the surface of the figure at all points on the surface, then the magnitude of attraction at the surface of the flattened figure of revolution increases from a pole to the equator.

In short, Mairan concluded that his law of attraction strengthened arguments that supported elongated figures of revolution and weakened arguments that supported flattened figures of revolution, where accounting for the observed decrease in the magnitude of the effective gravity from the earth's north pole to its equator was concerned.

But centrifugal force per unit of mass also had to be taken into consideration. When attraction was assumed to have constant magnitude at the surfaces of figures of revolution and to be perpendicular to the surfaces at all points on the surfaces of such figures, Mairan thought that he had demonstrated that the magnitude of the component of centrifugal force per unit of mass in the direction of the effective gravity increased more from the poles to the equators along meridians at the surfaces of elongated figures of revolution which revolve around their axes of symmetry than along meridians at the surfaces of flattened figures of revolution which revolve around their axes of symmetry, when the two kinds of figures revolve around their axes of symmetry at the same rate. Mairan believed that this showed that the magnitude of the effective gravity decreased more from the poles to the equators along meridians at the surfaces of elongated figures of revolution which revolve around their axes of symmetry than along meridians at the surfaces of flattened figures of revolution which revolve around their axes of symmetry, when the two kinds of figures revolve around their axes of symmetry at the same rate and attraction was taken to have constant magnitude at their surfaces and to be perpendicular to their surfaces at all points on their surfaces.

(As we saw, however, he gave a worthless proof of this proposition, because the directions of effective gravity at points on the surfaces of his figures of revolution when these figures revolved around their axes of symmetry were not the directions that he thought, namely, the directions of attraction at the points, and because the principle of the plumb line failed to hold at the surfaces of his figures when these figures revolved around their axes of symmetry. Consequently the conclusion that he came to was equally useless and invalid.)

Now Mairan had to determine whether variations in the magnitudes of attraction from the poles to the equators along meridians at the surfaces of such figures changed these trends or not. What concerned Mairan was the following possibility: according to Mairan's law of attraction described earlier, the magnitude of attraction decreases from a pole to the equator along a meridian at the surface of an elongated figure of revolution. Assuming that the principle of the plumb line holds at all points at the surface of such a figure when the figure revolves around its axis of symmetry, in which case the effective gravity at the surface of the figure is perpendicular to the surface of the figure at all points on the surface, the decrease in the magnitudes of the components of attraction in the directions of effective gravity from a pole to the equator at points along a meridian at the surface of the figure would make the magnitudes of the components of centrifugal force per unit of mass in the directions of effective gravity increase less from a pole to the equator at points along a meridian at the surface of the figure than would be the case if the magnitude of attraction were constant at the surface of the figure and the effective gravity were perpendicular to the surface of the figure at all points on the surface when the figure revolved around its axis of symmetry. This followed, Mairan stated, because the magnitude of attraction decreases from a pole to the equator along a meridian at the surface of an elongated figure of revolution according to the law of attraction which he proposed, which required that the magnitudes of the components of centrifugal force per unit of mass needed to make the principle of the plumb line hold at the surface of the elongated figure of revolution when the figure revolves around its axis of symmetry be larger toward the poles than would be the case if the magnitude of the attraction were constant at the surface of the figure. Since the magnitudes of these components of centrifugal force per unit of mass near a pole are larger to start with, the magnitudes of these components of centrifugal force per unit of mass increase from a pole to the equator correspondingly less along a meridian at the surface of the elongated figure of revolution when the figure revolves around its axis of symmetry than would be the case if the magnitude of the attraction were constant at the surface of the figure. This reduced increase could conceivably cause the magnitude of the effective gravity to decrease less from a pole to the equator along a meridian at the surface of an elongated figure of revolution when the figure revolves around its axis of symmetry than would be the case if the magnitude of the attraction were constant at the surface of the figure.

To restate the argument, when Mairan assumed that the magnitude of the attraction was constant at the surface of an elongated figure of revolution, he found a certain increase in the magnitudes of the components of centrifugal force per unit of mass in the directions of effective gravity from a pole to the equator at points along a meridian at the surface of the figure which was greater than the increase in the magnitudes of the components of centrifugal force per unit of mass in the directions of effective gravity at points along meridians at the surfaces of spherical figures and flattened figures of revolution from their poles to their equators, when the three kinds of figures revolved around their axes of symmetry at the same rate. Now, when Mairan assumed that his new law of attraction held at the surface of an elongated figure of revolution, the magnitude of the attraction decreased along a meridian at the surface of the figure from a pole to its equator. Mairan considered the possibility that this decrease might reduce the increase in the magnitudes of the components of centrifugal force per unit of mass in the directions of the effective gravity at points along a meridian at the surface of the figure from a pole to its equator when the figure revolved around its axis of symmetry from what it had been when the magnitude of the attraction was assumed constant at the surface of the figure and that this reduced increase could possibly lessen the decrease in the magnitude of the effective gravity along a meridian at the surface of the figure from a pole to its equator when the figure revolved around its axis of symmetry from what it had been when the magnitude of the attraction was assumed constant at the surface of the figure and thereby cause the elongated figure of revolution to lose the advantage it had of having a decrease in the magnitude of the effective gravity along its meridians from a pole to its equator which accorded better with the decrease in the magnitude of the effective gravity observed from Paris to Cayenne than the decreases in the magnitude of the effective gravity along meridians at the surfaces of spherical figures and flattened figures of revolution from their poles to their equators did when the three kinds of figures revolved around their axes of symmetry at the same rate. Mairan thought that he had previously established this advantage when the magnitude of the attraction was assumed constant at the surfaces of the three kinds of figures. In other words, even though the magnitude of the attraction now decreased along a meridian at the surface of the elongated figure of revolution from a pole to its equator according to Mairan's proposed law of attraction, whose effect by itself would be to decrease the magnitude of the effective gravity along the meridian from a pole to the equator when the figure revolved around its axis of symmetry, Mairan considered the possibility that a reduced increase in the magnitudes of the components of centrifugal force per unit of mass in the directions of the effective gravity at points along the meridian from a pole to the equator when the figure revolved around its axis of symmetry from what it had been when the magnitude of the attraction was assumed constant at the surface of the figure could produce a net decrease in the magnitude of the effective gravity along the meridian from a pole to the equator when the figure revolved around its axis of symmetry which was less than what it had been when the magnitude of the attraction was assumed constant at the surface of the figure.

Without providing detailed arguments, Mairan simply asserted that although the magnitude of the attraction is not constant at the surfaces of nonspherical figures of revolution, according to his law of attraction, the effect of attraction at the earth's surface, assuming that the earth attracts in accordance with Mairan's law at its surface, nevertheless greatly exceeds the opposing effect that centrifugal force per unit of mass has at all points on the earth's surface. He cited as evidence for this claim the infinitesimal ratio $\frac{1}{289}$ of the magnitude of the centrifugal force per unit of mass to the magnitude of the attraction at the earth's equator. Consequently, as Mairan stated the matter, his law of attraction

always adds much more to [the magnitude of the] effective gravity, and to the lengths of seconds pendulums, in going toward the poles [along a meridian at the surface of an elongated figure of revolution when the figure revolves around its axis of symmetry] than this reduced increase in the magnitude of the centrifugal force per unit of mass [from a pole to the equator along a meridian at the surface of the elongated figure of revolution when the figure revolves around its axis of symmetry, produced by the increase in the magnitudes of the components of centrifugal force per unit of mass near a pole needed to make the principle of the plumb line hold at the surface of the figure when the figure revolves around its axis of symmetry] takes away.[33]

(I should mention that Mairan added to the confusion in his *mémoire* by continually employing the term "pesanteur" throughout it, which he sometimes used to designate what I have called the magnitude of the effective gravity, while at other times he used the term to mean what I have called the magnitude of the attraction. The reader had to determine from the context the particular meaning of the word each time it was used.)

Once more, however, Mairan erred, just as he did when he assumed the magnitude of the attraction to be constant at the surfaces of spherical figures and elongated and flattened figures of revolution. Again the directions of effective gravity at points on the surfaces of his figures of revolution when these figures revolve around their axes of symmetry are not the directions that he thought, namely, the directions of attraction at the points. Although the magnitude of attraction is no longer constant but varies instead at the surfaces of nonspherical figures of revolution, according to Mairan's law of attraction, the directions of attraction at the surfaces of nonspherical figures of revolution are still perpendicular to the surfaces of such figures at all points on their surfaces, according to this law of attraction, and Mairan mistakenly took these directions to be the directions of effective gravity at points on the surfaces of nonspherical figures of revolution when these figures revolved around their axes of symmetry.

Moreover, it is again impossible for effective gravity at the surface of a nonspherical figure of revolution to be perpendicular to the surface of the figure at all points on the surface of the figure when the figure revolves around its axis of symmetry, if Mairan's law of attraction holds at the surface of the figure. This is true because the attraction perpendicular to the surface of the figure at all points on the surface of the figure combines with the centrifugal force per unit of mass at points on the surface of the figure to produce a resultant of effective gravity which cannot in general be perpendicular to the surface of the figure at points on the surface of the figure. (The same is clearly true of spherical figures as well, at whose surfaces the magnitude of attraction is constant, according to Mairan's law of attraction. The directions of attraction at points on the surfaces of such figures are again perpendicular to the surfaces at these points, according to this law of attraction.) In other words, the principle of the plumb line again cannot hold at all points at the surface of an elongated figure of revolution when the figure revolves around its axis of symmetry, if Mairan's law of attraction holds at the surface of the figure, for exactly the same reasons that kept the principle of the plumb line from holding at all points on the surface of the figure when Mairan assumed the magnitude of attraction to be constant at the surface of the figure and when the figure revolved around its axis of symmetry.

Mairan's phenomenological (or, at least, supposedly phenomenological) study of attraction accorded with the viewpoint of many Paris Academicians of the time and the astronomers in particular. Indeed, the line of argument which Mairan developed in his paper probably exemplified the way that these people worked. Their outlook, as I mentioned earlier, can be traced back to the skepticism established in France in the seventeenth century by Mersenne, Gassendi, Pascal, Roberval, and others. Historians who depict Cartesianism as the dominant theme in French science of the later seventeenth and early eighteenth centuries, in their search for conflicts between Cartesian and Newtonian paradigms, ignore this current in French scientific thought.

Mairan's *mémoire*, for example, has been called Cartesian.[34] In some respects his essay illustrates the antithesis of Cartesianism. It is true that references to the terrestrial vortex do appear in his paper, and I mentioned one of them. In another place in his paper, Mairan briefly talked about what an attraction that is not a central force would require concerning the shape of this vortex which produces the attraction–namely, the vortex would have to have an irregular shape.[35] Here Mairan alluded to Huygens's vortex, which was supposed to be spherically shaped in order to cause attraction to act like a central force. But these remarks are incidental to the discussion in Mairan's *mémoire*; they are not central to the problem that interested Mairan and that he tried to deal with in the *mémoire*. Mairan was concerned in general with the fine details of phenomena. Neither Descartes nor his followers were ever particularly renowned for an interest in such things. A first-rate experimenter, even as a young scholar, Mairan duplicated successfully, perhaps for the first time in France, Newton's experiments on light and color, which had baffled an earlier generation of French experimenters.[36] He wrote his *mémoire* of 1720 with the interests of the experimenter in mind.

It seems likely that Mairan viewed his hypothesis of attraction as an ad hoc one needed to fit all of the terrestrial observations of the day. He did not doubt the observations themselves, because he trusted experimenters and observational astronomers. That is to say, he did not question "facts," because he had faith in researchers who gathered facts. (Whether his law of attraction is completely ad hoc is debatable, however, because of the two "hidden" hypotheses that I mentioned.)

Mairan ignored Jupiter's flattening at its poles, which Flamsteed, Gian-Domenico Cassini, and James Pound had seen in their telescopes. No one else in Paris at the time paid any attention to this observed flattening, either. Mairan conceivably could have justified disregarding it, insofar as any variation of the magnitude of the effective gravity with latitude at Jupiter's surface is unobservable. In 1732, the Paris Academy of Sciences Secrétaire Perpétuel Bernard le Bovier de Fontenelle declared in a good positivistic manner that nothing prevented the earth from contradicting Newton's and Huygens's theories of the shapes of celestial bodies, even if the flattened Jupiter accorded with these theories.[37] A unified theory of the shapes of planets was clearly the furthest thing from Fontenelle's mind. Phenomenological considerations of the kind that the Paris Academy had long considered important induced Mairan to write his Paris Academy *mémoire* of 1720, not Cartesianism.

Contrary to customary belief, the Cartesian–Newtonian disputes have little if anything to do with the real issues that underlay the controversy about the earth's shape. After all, Mairan's principal opponent in his paper of 1720, Christiaan Huygens, had himself arrived at a flattened figure of the earth from undeniably Cartesian hypotheses, just as Jakob Hermann did after him. Newtonian hypotheses of attraction are not the only hypotheses of attraction which require that the earth be flattened at its poles. In other words, vortices are not the issue. But historians have created much confusion regarding these matters up to the present day.

Why is Mairan's *mémoire* of 1720 thought to be Cartesian? In all probability no one has examined it carefully for the past two hundred fifty years. Contemporary studies very likely just repeat the hearsay that appears in older histories. Mairan was a member of the Malebranche Group; hence he was no adversary of Descartes. I shall go into the confusion concerning Mairan's *mémoire* again briefly in Chapter 3 and talk more about the Malebranche Group that Mairan belonged to in Chapters 3 and 9. A variety of factors probably explains why historians have neglected Mairan's *mémoire*. One stands out among the others, however. This involves Mairan's style of writing, which I shall say more about when I return to his *mémoire* in the next section of this chapter and touch upon another of his works in Chapter 5. Suffice it to say for the moment that plodding away at reading Mairan's *mémoire* of 1720 demands a great deal of patience and stamina.

2.2. J.T. Desaguliers's rejoinder (1725)

In the *Philosophical Transactions* of the Royal Society of London for 1725, John Theophilus Desaguliers, a Newtonian experimental physicist and Huguenot refugee who replaced Francis Hauksbee as Newton's chief experimenter when Hauksbee died in 1713, published three articles concerning the problem of the earth's shape. In the last of these articles he tried to refute Mairan's Paris

Academy *mémoire* of 1720.[38] He claimed to prove that a stationary homogeneous fluid figure of equilibrium must be shaped like a sphere–that a homogeneous elongated fluid mass initially at rest would transform itself by means of its attraction into a spherical fluid figure "for the same Reason that Drops of Mercury, of Water, and other Fluids, put on such a Figure."

Desaguliers, however, was no mathematician. His "demonstration" of this assertion was simplistic to say the least. He vaguely reasoned that such a change of shape was the only way to make all columns from the center of a motionless homogeneous fluid figure to the surface of the figure balance or weigh the same. Here he assumed that the homogeneous fluid mass attracts according to the universal inverse-square law.[39] But his argument does not prove what he assumed it did. His reasoning was typical of certain followers of Newton in Great Britain who tried to read the *Principia* or at least pretended to, who thought that they had understood the work, and who in fact understood very little if any of it at all. Desaguliers did not have the faintest idea what a conclusive proof of what he had tried to show would involve. Mathematics still had a long way to advance before it became possible to use it to demonstrate with certainty, toward the end of the eighteenth century, that a homogeneous fluid figure shaped like an elongated ellipsoid of revolution which attracts in accordance with the universal inverse-square law cannot be a figure of equilibrium. It turns out that such a figure can never be a figure of equilibrium because it can be shown mathematically to fail to fulfill what I stated in the preceding section of this chapter to be two necessary conditions that homogeneous figures of equilibrium must satisfy– namely, the principle of the plumb line and the principle of balanced columns. And it was not rigorously proven until the beginning of the twentieth century that a homogeneous immobile fluid figure of equilibrium which attracts according to the universal inverse-square law must be spherically shaped.[40]

(Moreover, as I mentioned in Chapter 1, even Newton did not really demonstrate in the *Principia* that a homogeneous figure shaped like a flattened ellipsoid of revolution which attracts according to the universal inverse-square law, whose ellipticity is infinitesimal, which revolves around its axis of symmetry, and whose columns from center to surface are all assumed to balance or weigh the same actually *is* a possible figure of equilibrium.)

Desaguliers also observed that Mairan did not explain why a homogeneous stationary fluid figure of revolution assumed originally to have been elongated at its poles should reduce its elongation, once it began to revolve around its axis of symmetry, just enough to give it a shape whose ellipticity is $-\frac{1}{96}$, which was the value of the ellipticity of Cassini's figure shaped like an elongated ellipsoid of revolution which was supposed to represent the earth. Here Desaguliers argued on better grounds. As I noted in Chapter 1, both Newton and Huygens had related the extent of the earth's flattening to the ratio of the magnitude of the centrifugal force per unit of mass at the earth's equator to the magnitude of the attraction there, a ratio whose value is $\frac{1}{289}$. This ratio depends upon the earth's rate of rotation, but Mairan did not specify the rates at which figures of revolution revolve around their axes of symmetry. Here Desaguliers, in effect, alluded to a defect in Mairan's analysis which I discussed in the preceding section of this chapter. Namely, Mairan, we recall, had made it appear as if there were no connection between the particular shapes of fluid figures of revolution and the rates at which the figures revolve around their axes of symmetry.

Desaguliers also recognized the other errors in Mairan's arguments which I pointed out in the preceding section of this chapter. In demonstrating that the magnitudes of the components of centrifugal force per unit of mass in the directions of the attraction increase more from a pole to the equator at points along a meridian at the surface of an elongated figure of revolution than they do along a meridian at the surface of a spherical figure, when attraction is assumed to be constant in magnitude at all points on the surfaces of the two figures and the two figures revolve around their axes of symmetry at the same rate, Mairan confused matters. He took the directions of the radii of curvature of meridians at points along meridians at the surfaces of the two figures to be the directions of attraction at these points. Mairan began with figures of revolution at rest and assumed the directions of attraction at points along meridians at the surfaces of the figures to lie along the radii of curvature of the meridians at these points. In other words, he assumed attraction to be perpendicular to the surfaces at all points on the surfaces of stationary figures of revolution, in accordance with the principle of the plumb line. He then found the magnitudes of the components of centrifugal force per unit of mass in these directions at points on the surfaces of the figures when the figures revolve around their axes of symmetry. Finally Mairan deduced the way that the magnitude of the effective gravity varies from the poles to the equators at the surfaces of the figures when the figures revolve around their axes of symmetry from the way that the magnitudes of these particular components of centrifugal force per unit of mass at points on the surfaces of the figures vary from the poles to the equators of the figures when the figures revolve around their axes of symmetry. This means that he took the directions of effective gravity at points on the surfaces of the figures when the figures revolved around their axes of symmetry to be the directions of attraction at these points.

But in doing so, Desaguliers noted, Mairan failed to take into account the changes in the directions of effective gravity at points on the surfaces of his figures which revolution of the figures around their axes of symmetry brings about. Mairan overlooked the fact that the centrifugal forces produced by rotation not only change the magnitudes of the effective gravity at points on the surfaces of the figures, but they cause the directions of the effective gravity at points on the surfaces of the figures to change as well. As a result of this oversight Mairan had not examined the correct components of centrifugal force per unit of mass at points on the surfaces of his figures when the figures revolved around their axes of symmetry. Accordingly, since he investigated the wrong components of centrifugal force per unit of mass at points on the surfaces of his figures when the figures revolved around their axes of symmetry, he had not found the way that the magnitude of the effective gravity varies from the poles to the equators at the surfaces of the figures when the figures revolved around their axes of symmetry, because he had not looked at the effective gravity at points on the surfaces of the figures but at components of the effective gravity at points on the surfaces of the figures in the wrong directions instead. Or to express the matter another way, he had chosen incorrectly the directions of the effective gravity at points on the surfaces of the figures when the figures revolved around their axes of symmetry.

Worse yet, the principle of the plumb line could not even hold at the surfaces of Mairan's figures when they revolved around their axes of symmetry, because the directions of the effective gravity at points on the surfaces of the figures are perpendicular to the surfaces of the figures at these points when the figures do not revolve around their axes of symmetry, since in this case the effective gravity just reduces to the attraction. Hence the directions of the effective gravity at points on the surfaces of the figures can no longer be perpendicular to the surfaces of the figures at these points when the figures do revolve around their axes of symmetry.

One might think that for a spherical figure that rotates or a figure of revolution which revolves around its axis of symmetry "slowly" enough (meaning a figure whose ratio ϕ of the magnitude of the centrifugal force per unit of mass to the magnitude of the attraction at its equator is infinitesimal–that is, $\phi^2 \ll \phi$), where the attraction is again assumed to have constant magnitude at the surfaces of figures and to be perpendicular to the surfaces of figures at all points on the surfaces of figures, the effective gravity at points on the surfaces of such figures ought to be almost perpendicular to the surfaces of the figures at these points (in other words, should almost be in the same directions as attraction at points on the surfaces of the figures), in which case Mairan's arguments could conceivably hold approximately.

(As I mentioned in the preceding section of this chapter, Huygens appears to have assumed that the two directions are the same in the case of his rotating spherical figure. Moreover, as I noted in Chapter 1, Newton stated that the magnitude of the effective gravity at points along a meridian at the surface of a homogeneous figure shaped like a flattened ellipsoid of revolution which attracts in accordance with the universal inverse-square law, whose ellipticity is infinitesimal, which revolves around its axis of symmetry, and whose columns from center to surface are all assumed to balance or weigh the same varies inversely as the distances of the points from the center of the figure, whereas in fact this variation only holds exactly for the magnitudes of the components of the effective gravity at the points along the meridian which are directed toward the center of the figure. However, since the ellipticity of Newton's figure is infinitesimal, which means that the figure is nearly spherical, the magnitudes of the resultants of the effective gravity at points along a meridian at the surface of the figure and the magnitudes of the components of the effective gravity at these same points which are directed toward the center of the figure are equal to terms of first order and cannot be distinguished for all practical purposes.[41]

As I also observed in Chapter 1, Newton stated as well that the magnitude of the effective gravity at points along a meridian at the surface of a homogeneous figure shaped like a flattened ellipsoid of revolution which attracts according to the universal inverse-square law, whose ellipticity is infinitesimal, which revolves around its axis of symmetry, and whose columns from center to surface are all assumed to balance or weigh the same varies directly as the squares of the sines of the latitudes of the points. Here again he tacitly supposed that the resultants of the effective gravity at points along a meridian at the surface of the figure are directed toward the center of the figure, when in actuality these resultants are perpendicular to the surface of the figure at these points, for reasons that I shall discuss in Chapter 6. As I also mentioned in Chapter 1, the figure does turn out to be a possible figure of equilibrium, for reasons that we shall discover in Chapter 9, in which case the principle of the plumb line *must* hold at all points at the surface of the figure when the figure truly is a figure of equilibrium, for the reasons that I stated in the preceding section of this chapter and at the beginning of this section of this chapter. Again, however, since the ellipticity of Newton's figure is infinitesimal, which means that the figure is almost spherical, the two different directions at a point on its surface are nearly the same. In other words, Newton's assumption and what is actually true cannot be distinguished for all practical purposes.[42]

Thus in stating both of these results, Newton himself in effect took the resultants of the effective gravity at points along a meridian at the surface of a homogeneous figure shaped like a flattened ellipsoid of revolution which attracts according to the universal inverse-square law, whose ellipticity is infinitesimal, which revolves around its axis of symmetry, and whose columns from center to surface are all assumed to balance or weigh the same to be directed toward the center of the figure.)

But Desaguliers maintained that in the case of Cassini's figure shaped like an elongated ellipsoid of revolution, the angle that a perpendicular to the surface at a point on the surface of the figure makes with the direction of the effective gravity at the point (in other words, the direction of the plumb line at the point) when attraction is assumed to be constant in magnitude at all points on the surface of the figure and to be perpendicular to the surface of the figure at all points on the surface of the figure, as Mairan had hypothesized in the demonstration in question, and when the figure revolves around its axis of symmetry so that the ratio of the magnitude of the centrifugal force per unit of mass to the magnitude of the attraction at the equator of the figure is the earth's value $\frac{1}{289}$ would be large enough to observe and to involve observable effects. In particular, the principle of the plumb line would be found not to hold at the surface of the figure.

(As I mentioned at the beginning of this section of this chapter, a homogeneous fluid figure shaped like an elongated ellipsoid of revolution which attracts in accordance with the universal inverse-square law can never be a figure of·

equilibrium because it fails to fulfill two necessary conditions that homogeneous figures of equilibrium must satisfy–the principle of the plumb line and the principle of balanced columns. But in Chapters 6 and 9 we shall discover that the principle of the plumb line *can* hold at the surfaces of *solid stratified* figures shaped like elongated ellipsoids of revolution which attract according to the universal inverse-square law, whose ellipticities are infinitesimal, and which revolve around their axes of symmetry. However we shall also see in Chapter 9 that if Cassini's figure is assumed to be such a figure, it still cannot represent the earth. Because if the principle of the plumb line did hold at its surface, as the principle of the plumb line is observed to hold at the surface of the earth, then assuming Cassini's figure to be stratified, to attract according to the universal inverse-square law, and to revolve around its axis of symmetry so that the ratio of the magnitude of the centrifugal force per unit of mass to the magnitude of the attraction at its equator is the earth's value $\frac{1}{289}$ would require that the decrease in the magnitude of the effective gravity from a pole to the equator along its surface be much greater than the observed decrease in the magnitude of the effective gravity from a pole to the equator along the earth's surface.)

After having demolished Mairan's arguments, by recognizing and pointing out errors that invalidated Mairan's analysis because the mistakes could not be corrected, Desaguliers went on unwittingly to help confuse the issues himself. Mairan believed that he had shown that the ratio of the magnitude of the centrifugal force per unit of mass at a given latitude to the magnitude of the centrifugal force per unit of mass at the equator was smaller at the surface of an elongated figure of revolution than it was at the surfaces of a spherical figure or a flattened figure of revolution, when all three figures revolved around their axes of symmetry at the same rate, so that the magnitude of the centrifugal force per unit of mass increased more from a pole to the equator along a meridian at the surface of the elongated figure of revolution than it did along the surface of the spherical figure or along the surface of the flattened figure of revolution when all three figures revolved around their axes of symmetry at the same rate. Desaguliers declared that the magnitude of the centrifugal force per unit of mass would, as Mairan had affirmed, indeed increase more from a pole to the equator and that the magnitude of the effective gravity would decrease more from a pole to the equator along the surface of an elongated figure of revolution than they would along the surface of a spherical figure when the two figures were "of the same Solidity," when the magnitude of the attraction had the same constant value at all points on the surfaces of the two figures and when the two figures revolved around their axes of symmetry at the same rate. Likewise, Desaguliers stated that the magnitude of the centrifugal force per unit of mass would, as Mairan had also maintained, increase more from a pole to the equator and that the magnitude of the effective gravity would decrease more from a pole to the equator along the surface of a spherical figure than they would along the surface of a flattened figure of revolution when the two figures were "of the same Solidity," when the magnitude of the attraction had the same constant value at all points on the surfaces of the two figures and when the two figures revolved around their axes of symmetry at the same rate.

It is odd that Desaguliers should now grant the truth of Mairan's results just stated, after having shown that the principle of the plumb line could not even hold at the surface of a rotating spherical figure or at the surface of an elongated figure of revolution which revolves around its axis of symmetry or a flattened figure of revolution which revolves around its axis of symmetry, when the magnitude of the attraction is assumed to be constant at all points on the surface of the figure and perpendicular to the surface of the figure at all points on the surface of the figure. The only way that Desaguliers could have come to such a conclusion is if he took the effective gravity at the surfaces of figures of revolution and the constant magnitude of the attraction at the surfaces of figures of revolution to mean different things than they meant to Mairan.

Desaguliers translated into English Mairan's demonstration that the magnitudes of the components of the centrifugal force per unit of mass in the directions of the attraction increase more from a pole to the equator along the surface of an elongated figure of revolution than they do along the surface of a spherical figure, when the magnitude of the attraction is constant at the surfaces of the two figures and perpendicular to the surfaces of the two figures at all points on the surfaces of the two figures and when the two figures revolve around their axes of symmetry at the same rate. Then he gave what he thought was an illustration of Mairan's results. In fact, what Desaguliers did is rather vague. Moreover, in trying to reproduce Mairan's results in his own way, Desaguliers made mistakes. I examine what Desaguliers did, and I point out some of the problems with his account of Mairan's findings.

As I just mentioned, Desaguliers concerned himself with elongated figures of revolution, spherical figures, and flattened figures of revolution "of the same Solidity." Since the figures were all homogeneous and were all implicitly assumed to have the same density, this means that Desaguliers studied figures that had the same volume. In this case the elongated figure of revolution has the smallest equatorial radius; the flattened figure of revolution has the largest equatorial radius; and the spherical figure has a radius that lies between the values of the radii of the other two figures. Then if the three figures revolve around their axes of symmetry at the same rate, the magnitude of the centrifugal force per unit of mass is smallest at the equator of the elongated figure of revolution; the magnitude of the centrifugal force per unit of mass is greatest at the equator of the flattened figure of revolution; and the magnitude of the centrifugal force per unit of mass at the equator of the spherical figure lies between the values of the magnitudes of the centrifugal force per unit of mass at the equators of the other two figures. Mairan let c stand for the magnitude of the centrifugal force per unit of mass at the equator of the elongated figure of revolution; he let $c + 1$ designate the magnitude of the centrifugal force per unit of mass at the equator of the spherical figure; and he let $c + 2$ symbolize the magnitude of the centrifugal force per unit of mass at the equator of the flattened figure of revolution.

Now, the magnitudes of the centrifugal forces per unit of mass at the equators of figures of revolution which revolve around their axes of symmetry at the same rate (that is, with the same angular speed) vary directly as the equatorial radii of the figures. (Mairan, we recall, made use of this fact to establish the way he thought that the magnitude of the effective gravity varies with latitude at the surfaces of elongated figures of revolution, spherical figures, and flattened figures of revolution when these figures revolve around their axes of symmetry at the same rate.) But Desaguliers expressions c, $c + 1$, and $c + 2$ for the magnitudes of the centrifugal forces per unit of mass at the equators of the three figures are not linearly related. If c is the magnitude of the centrifugal force per unit of mass at the equator of the elongated figure of revolution, the magnitudes of the centrifugal forces per unit of mass at the equators of the spherical figure and the flattened figure of revolution would have to have the form kc. In the case of the spherical figure the value of k is the ratio of the equatorial radius of the spherical figure to the equatorial radius of the elongated figure of revolution. In the case of the flattened figure of revolution the value of k is the ratio of the equatorial radius of the flattened figure of revolution to the equatorial radius of the elongated figure of revolution. In other words, the expressions c, $c + 1$, and $c + 2$ for the magnitudes of the centrifugal forces per unit of mass which Desaguliers stipulated at the equators of the three figures do not conform to the problem as Mairan had formulated it.

Desaguliers let $c + 2 - l$ designate the magnitude of the centrifugal force per unit of mass at a point on the surface of the flattened figure of revolution whose latitude is the same as the latitude of Paris. Here Desaguliers seems to have chosen l to conform with the principle of the plumb line. That is, Desaguliers assumed effective gravity to be perpendicular to the surface of the flattened figure of revolution at this particular point when the figure revolved around its axis of symmetry, and he let $c + 2 - l$ stand for the magnitude of the component of centrifugal force per unit of mass at the point in the direction of effective gravity at the point, for he chose l so that $c + 2$ "is diminished on Account of a shorter Co-sine of Latitude [that is, the cosine of the angle that the direction perpendicular to the surface of the figure at the point makes with the axis of rotation of the figure which is the axis of symmetry of the figure. Mairan also realized that the magnitude of the component of centrifugal force per unit of mass at the particular point in question perpendicular to the surface of the figure at the point was less than the magnitude of the total centrifugal force per unit of mass at the point whose direction was perpendicular to the axis of rotation of the figure which is the axis of symmetry of the figure], and likewise on Account of its Obliquity to the Line of tendency."[43] Desaguliers is referring to the oblique angle that the relevant component of centrifugal force per unit of mass at the point, which is perpendicular to the surface of the figure at the point, makes with the direction of the resultant of attraction at the point, a direction that cannot be perpendicular to the surface of the figure at the point if the principle of the plumb line is to hold at the point when the figure revolves around its axis of symmetry. This angle did not exist in Mairan's investigation, because Mairan took attraction to be perpendicular to the surface of a figure of revolution at points on the surface of the figure, which attraction cannot be if the principle of the plumb line is to hold at all points on the surface of the figure of revolution when the figure revolves around its axis of symmetry.

Now, earlier in his paper Desaguliers had used the phrase "the whole Attraction or Gravity of the Body."[44] In other words, he thought of gravity as synonymous with attraction in his paper. Desaguliers let g designate what he called the "uniform Gravity,"[45] by which he meant in this case the constant value of the magnitude of attraction at all points on the surfaces of the figures.

Then he let $g - (c + 2 - l)$ denote what he called the "Diminution of Gravity at *Paris.*"[46] Because of the form that this expression has, as well as the phrase that he used to characterize it, here he could only have meant the magnitude of the effective gravity at the point on the surface of the flattened figure of revolution whose latitude is the same as the latitude of Paris. Moreover, because he chose l to make the component of centrifugal force per unit of mass at the point whose magnitude is $c + 2 - l$ lie in the direction of effective gravity at the point, which is perpendicular to the surface of the figure at the point, the component of attraction at the point whose magnitude is g must also lie in the same direction, in order that the expression $g - (c + 2 - l)$ make sense. In other words, at a point on the surface of the flattened figure of revolution which is neither at a pole nor on the equator of the flattened figure of revolution, g must actually be the magnitude of the *component* of attraction at the point which is *perpendicular* to the surface of the flattened figure of revolution at the point. That is, g cannot be the magnitude of the resultant of the attraction at the point, because the resultant of the attraction at a point on the surface of the flattened figure of revolution which is neither at a pole nor on the equator of the flattened figure of revolution cannot be perpendicular to the surface of the figure at the point. For if it were, the principle of the plumb line could not hold at points on the surface of the flattened figure of revolution when the figure revolves around its axis of symmetry, which is precisely the problem that Desaguliers called attention to in Mairan's *mémoire.*

In short, Desaguliers evidently did not assume the constant magnitude of the attraction at all points on the surfaces of the three different figures of revolution to mean the same thing that it meant to Mairan. Instead Desaguliers took the constant value of the magnitude of the attraction at the surfaces of the three different figures of revolution, which he designated by g, and took his values of the magnitudes of centrifugal force per unit of mass at different latitudes at the surfaces of these three figures to be, respectively, the magnitudes of the *components* of the attraction and of the centrifugal force per unit of mass at points on the surfaces of the three figures which are *perpendicular* to the surfaces of the figures at the points, in cases where the resultant of the attraction and the total centrifugal force per unit of mass are assumed to be such as to make the resultant of the effective gravity be perpendicular to the surfaces of the three figures at all points on the surfaces of the three figures at the same time, when the three figures revolve around their axes of symmetry at the same rate, so that the principle of the plumb line holds at all points on the surfaces of the three rotating figures. (The principle of the plumb line holds at all points on the surface of a figure of revolution which revolves around its axis of symmetry when the component of the attraction tangent to the surface of the figure at a point on the surface of the figure and the component of centrifugal force per unit of mass tangent to the surface of the figure at the same point cancel each other at all points on the surface of the figure. This occurs when these particular components of attraction and centrifugal force per unit of mass at all points on the surface of the figure are oppositely directed and have equal magnitudes.) In order for this to happen, the resultants of attraction at points on the surfaces of the three different figures must be different for each figure. I shall return to this point below.

Desaguliers let $g - (c + 2)$ symbolize the "Diminution of Gravity ... at the Æquator,"[47] by which he could only have meant the magnitude of the effective gravity at a point on the equator of the flattened figure of revolution, because of the form that this expression has and the phrase that he used to characterize it. Here the centrifugal force per unit of mass and the attraction at a equator, whose magnitudes are $c + 2$ and g, respectively, are both evidently perpendicular to the surface of the flattened figure of revolution at the point on the equator. (In other words, in this case $c + 2$ and g are themselves the magnitudes of the total centrifugal force per unit of mass and of the resultant of attraction, respectively, at the point on the equator of the flattened figure of revolution.)

Hence l was the difference between the magnitude of the effective gravity at the point on the surface of the flattened figure of revolution whose latitude is the same as the latitude of Paris and the magnitude of the effective gravity at the point on the equator of the flattened figure of revolution. Here l stands for a positive number, because the magnitude of effective gravity decreases from the point on the surface of the flattened figure of revolution whose latitude is the same as the latitude of Paris to the point on the equator of the figure. (Or to express the matter another way, the magnitude of the centrifugal force per unit of mass increases from the point on the surface of the flattened figure of revolution whose latitude is the same as the latitude of Paris to the point on the equator of the figure.) The value of l depends upon the particular shape of the flattened figure of revolution. That is, the value of l depends upon the distance of the point on the surface of the flattened figure of revolution whose latitude is the same as the latitude of Paris from the axis of rotation of the flattened figure of revolution which is the axis of symmetry of the figure, the curvature at the point of the meridian at the surface of the flattened figure of revolution through the point, which determines the perpendicular to the meridian at the point in the plane of the meridian, and the angle that this perpendicular to the meridian at the point makes with the axis of rotation of the flattened figure of revolution which is the axis of symmetry of the figure.

Desaguliers let $(c + 1) - (l + m)$ stand for the magnitude of the component of centrifugal force per unit of mass at a point on the surface of the spherical figure whose latitude is the same as the latitude of Paris in the direction of effective gravity at the point, meaning perpendicular to the surface of the spherical figure at the point when the spherical figure revolved around its polar axis, since the principle of the plumb line held at the surface of the spherical figure when the spherical figure revolved around its polar axis.

Then he let $g - ((c + 1) - (l + m))$ designate the magnitude of the effective gravity at the point, where g must again be the magnitude of the component of attraction at the point which is perpendicular to the surface of the spherical figure at the point, in order that the expression $g - ((c + 1) - (l + m))$ make sense.

Then Desaguliers let $g - (c + 1)$ stand for the magnitude of the effective gravity at a point on the equator of the spherical figure, where the centrifugal force per unit of mass and the attraction at the equator, whose magnitudes are $c + 1$ and g, respectively, are both perpendicular to the surface of the spherical figure at the point on the equator. (In other words, in this case $c + 1$ and g are themselves the magnitudes of the total centrifugal force per unit of mass and of the resultant of attraction, respectively, at the point on the equator of the spherical figure.)

Hence the difference between the magnitude of effective gravity at the point on the surface of the spherical figure whose latitude is the same as the latitude of Paris and the magnitude of effective gravity at the point on the equator of the spherical figure was $l + m$, where m stands for a positive number.

Why did Desaguliers choose his symbols in such a way that the expression for this difference would have the form $l + m$ instead of, say, $l - m$? Here he must have *used* Mairan's result that the magnitude of effective gravity decreased more from a pole to the equator along the surface of the spherical figure than along the surface of the flattened figure of revolution when the two figures revolved around their axes of symmetry at the same rate. (Of course, Desaguliers's use of Mairan's result does not quite follow from Mairan's result. Desaguliers's effective gravity and Mairan's effective gravity are not the same. Desaguliers's effective gravity was perpendicular to the surfaces of figures of revolution at all points on the surfaces of the figures of revolution when the figures of revolution revolved around their axes of symmetry at the same rate; Mairan studied components of effective gravity which in general were not perpendicular to the surfaces of figures of revolution at points on the surfaces of the figures of revolution when the figures of revolution revolved around their axes of symmetry at the same rate.)

Just as Desaguliers specified no particular numerical values for c, he gave no specific numerical values for l or m, either. Any value of c depends upon the common rate at which the three figures of revolution revolve around their axes of symmetry and the equatorial radius of the elongated figure of revolution. Any value of l depends upon the particular shape of the flattened figure of revolution. That is, the value of l depends upon the distance of the point on the surface of the flattened figure of revolution whose latitude is the same as the latitude of Paris from the axis of rotation of the flattened figure of revolution which is the axis of symmetry of the figure, the curvature at the point of the meridian at the surface of the flattened figure of revolution through the point, which determines the perpendicular to the meridian at the point in the plane of the meridian, and the angle that this perpendicular to the meridian at the point makes with the axis of rotation of the flattened figure of revolution which is the axis of symmetry of the figure. The value of m depends upon the value of $l + m \equiv K$. Any value of $l + m \equiv K$ depends upon the distance of the point on the surface of the spherical figure whose latitude is the same as the latitude of Paris from the axis of rotation of the spherical figure (the polar axis) and the angle that the perpendicular at the point to the circular meridian at the surface of the spherical figure through the point in the plane of the meridian makes with the axis of rotation of the spherical figure. Hence any value of m depends upon the particular value of K as well as upon the particular value of l.

Next Desaguliers let $c - (l + m + n)$ symbolize the magnitude of the component of centrifugal force per unit of mass at a point on the surface of the elongated figure of revolution whose latitude is the same as the latitude of Paris in the direction of effective gravity at the point, meaning perpendicular to the surface of the elongated figure of revolution at the point when the elongated figure of revolution revolved around its axis of symmetry, since the principle of the plumb line held at the surface of the elongated figure of revolution when the elongated figure of revolution revolved around its axis of symmetry.

Then he let $g - (c - (l + m + n))$ designate the magnitude of the effective gravity at the point, where g must again be the magnitude of the component of attraction at the point which is perpendicular to the surface of the elongated figure of revolution at the point, in order that the expression $g - (c - (l + m + n))$ make sense.

He let $g - c$ stand for the magnitude of the effective gravity at a point on the equator of the elongated figure of revolution, where the centrifugal force per unit of mass and the attraction at the equator, whose magnitudes are c and g, respectively, are both perpendicular to the surface of the elongated figure of revolution at the point on the equator. (In other words, c and g are themselves the magnitudes of the total centrifugal force per unit of mass and of the resultant of attraction, respectively, at the point on the equator of the elongated figure of revolution in this case.)

Consequently the difference between the magnitude of effective gravity at the point on the surface of the elongated figure of revolution whose latitude is the same as the latitude of Paris and the magnitude of effective gravity at the point on the equator of the elongated figure of revolution was $l + m + n$, where n symbolized a positive number.

Why did Desaguliers choose his symbols in such a way that the expression for this difference would have the form $l + m + n$ instead of, say, $l + m - n$? Here Desaguliers again probably *utilized* Mairan's finding that the magnitude of effective gravity decreased more from a pole to the equator along the surface of the elongated figure of revolution than along the surface of the spherical figure when the two figures revolved around their axes of symmetry at the same rate. (Again, however, Desaguliers's use of Mairan's result does not truly follow from Mairan's result, because Desaguliers's effective gravity and Mairan's effective gravity are not the same. Desaguliers and Mairan were not talking about the same components of effective gravity at points on the surfaces of figures of revolution which revolve around their axes of symmetry at the same rate.)

As usual, Desaguliers specified no particular numerical values for n. Any value of n depends upon the particular shape of the elongated figure of revolution. That is, the value of n depends upon the value of $l + m + n \equiv L$. The value of L depends upon the distance of the point on the surface of the elongated figure of revolution whose latitude is the same as the latitude of Paris from the axis of rotation of the elongated figure of revolution which is the axis of symmetry of the figure, the curvature at the point of the meridian at the surface of the elongated figure of revolution through the point, which determines the perpendicular to the meridian at the point in the plane of the meridian, and the angle that this perpendicular to the meridian at the point makes with the axis of rotation of the elongated figure of revolution which is the axis of symmetry of the figure. Moreover, any value of n also depends upon the value of $l + m \equiv K$. Hence any value of n depends upon the particular value of K as well as upon the particular values of L.

Desaguliers observed that the ratios of the magnitudes of effective gravity at points on the surfaces of the flattened figure of revolution, the spherical figure, and the elongated figure of revolution whose latitudes are the same as the latitude of Paris to the effective gravity at points on the equators of the flattened figure of revolution, the spherical figure, and the elongated figure of revolution, when these ratios are expressed in terms of his particular representations of effective gravity at the various points, are, respectively,

$$\frac{g - (c + 2 - l)}{g - (c + 2)},$$

$$\frac{g - ((c + 1) - (l + m))}{g - (c + 1)}, \tag{2.2.1}$$

and

$$\frac{g - (c - (l + m + n))}{g - c}.$$

Desaguliers stated that it followed from what Mairan had shown that the third ratio in (2.2.1) was larger than the other two ratios in (2.2.1), in which case the magnitude of effective gravity decreased the most in going from a point on the surface of the elongated figure of revolution whose latitude is the same as the latitude of Paris to a point on the equator of the elongated figure of revolution. (Consequently, seconds pendulums had to be shortened the most in going from the point whose latitude is the same as the latitude of Paris to the equator along the surface of the elongated figure of revolution.[48])

Desaguliers could not *himself* have proved if this result was actually true following the method he used to present what Mairan had done, since he did not specify numerical values of any of the quantities that he used letters to represent. These letters can be given numerical values in arbitrary ways. But for given numerical values of c, l, m, and n, a numerical value of g can always be found which makes the third ratio in (2.2.1) not be the largest of the three ratios in (2.2.1) and which consequently makes the result just stated be false.[49]

This interpretation of Desaguliers's analysis of Mairan's result is the only one that enables us to make any sense out of Desaguliers's statement that the way that the magnitude of effective gravity varies with latitude which Mairan found to hold at the surfaces of the three different kinds of figures of revolution which revolve around their axes of symmetry at the same rate is correct when the magnitude of the attraction is assumed constant at the surfaces of the three different kinds of figures. For if Desaguliers had taken the constant g to mean the same thing that Mairan did, namely, the magnitude of the *resultant of* the attraction at each point on the surfaces of the three different kinds of figures, where the resultant of the attraction was, moreover, perpendicular to the surfaces at each point on the surfaces of the three different kinds of figures, then Desaguliers would have been making the same mistake as the one that he pointed out in Mairan's *mémoire*, because in that case the effective gravity cannot be perpendicular to the surfaces of the three different kinds of figures at points on their surfaces when the three different kinds of figures revolve around their axes of symmetry. In other words, the principle of the plumb line could not hold at the surfaces of the three different kinds of figures when the figures revolve around their axes of symmetry were this the case. (Desaguliers's failure to distinguish *resultants* of attraction from *components* of attraction can doubtlessly be attributed in part to Newton's having failed to make clear in the *Principia* the difference between the two.)

Desaguliers's examination of Mairan's results was entirely qualitative. All he did was try to illustrate Mairan's results; he didn't actually try to prove anything. We saw where he made use of the very results at issue. And if he utilized these results, that means he did not demonstrate them. Desaguliers gave no specific numerical values to any of the letters that he used to stand for the magnitudes of centrifugal forces per unit of mass and the magnitude of attraction. Nor did he indicate what the *resultants of* the attraction at points on the surfaces of each of the three different figures of revolution had to be in order to make the principle of the plumb line hold simultaneously at all points on the surfaces of the three different kinds of figures of revolution when these figures revolved around their axes of symmetry at the same rate and where the magnitudes of the components of attraction perpendicular to the surfaces at all points on the surfaces of the three different kinds of figures of revolution had the same constant value g. These resultants of attraction would have to be different for each of the three different kinds of figures of revolution. But Desaguliers did not specify the resultants of attraction in each of the three cases. For each of the three different kinds of figures of revolution these resultants of attraction would depend upon the common rate at which the three different kinds of figures of revolution revolve around their axes of symmetry. For the two different kinds of nonspherical figures of revolution the resultants of attraction would also depend upon the particular shapes of the two different kinds of figures. But Desaguliers did not specify the particular shapes of the two different kinds of nonspherical figures of revolution, nor did he specify the common rate at which the three different kinds of figures of revolution revolve around their axes of symmetry. Normally when the law that determines the resultants of attraction at all points is stipulated beforehand, the shapes of figures of revolution and the rates at which these figures revolve around their axes of symmetry are related to each other through the condition that the principle of the plumb line hold at all points on the surfaces of the figures. Like Mairan, Desaguliers established no connections between the shapes of figures of revolution and the rates at which these figures revolve around their axes of symmetry, but the explanation for Desaguliers having determined no such connections differs from the explanation for Mairan's having failed to establish such connections. Unlike Mairan, Desaguliers did not specify beforehand any laws that determined the resultants of attraction. What Desaguliers did do instead is stipulate that the magnitudes of the components of attraction perpendicular to the surfaces of figures of revolution at all points on the surfaces of the figures have the same constant value g.

Moreover, as I have already pointed out, no matter what value c has, c, $c + 1$, and $c + 2$ cannot stand for the magnitudes of the centrifugal forces per unit of mass at the equators of an elongated figure of revolution, a spherical figure, and a flattened figure of revolution, respectively, "of the same Solidity" when these three figures revolve around their axes of symmetry at the same rate.

Desaguliers then maintained that Mairan's conclusions failed to hold for elongated and flattened figures of revolution when the magnitude of the attraction is not constant at their surfaces but obeys the universal inverse-square law of attraction at their surfaces instead. This he judged to be the other major inadequacy of Mairan's paper. Earlier we saw why Mairan needed to exclude central forces as possible forces that determine the law that governs the earth's attraction. But, Desaguliers noted, Mairan did not deal at all with Newton's universal inverse-square law of attraction in his *mémoire*. And bodies that are not spherical which attract in accordance with such a law do not attract like central forces in their interiors or in their exteriors, but Mairan did not take this into account. Simply eliminating central forces consequently did not suffice. (Moreover, unlike the laws of attraction that Mairan considered, the magnitude of the attraction at a point produced by a body that attracts according to the universal inverse-square law *does* depend upon the density of the body. I shall discuss this point further in Chapter 9.)

Desaguliers tried to show that if the universal inverse-square law of attraction is assumed to hold, which is what his statement that "[the magnitude of] the Force of Gravity [which in this case should mean the effective gravity, not the attraction, and which consequently is inconsistent with Desaguliers's prior use of the term "Gravity," a point which I shall return to below], in different Places on the Earth's surface, is reciprocally as the Distance from the Center"[50] entails, then the variation of the magnitude of the effective gravity from a pole to the equator along a meridian at the surface of Cassini's homogeneous figure shaped like an elongated ellipsoid of revolution whose ellipticity is $-\frac{1}{96}$ and which revolves around its axis of symmetry so that the ratio of the magnitude of the centrifugal force per unit of mass to the magnitude of the attraction at its equator is the earth's value $\frac{1}{289}$ could never be made to agree with the observed variation of the magnitude of the effective gravity with latitude at the earth's surface–that is, with the experiments done with seconds pendulums. Seconds pendulums would have to be lengthened, not shortened, in going from a point on the surface of Cassini's figure whose latitude is the same as the latitude of Paris to a point on the equator of Cassini's figure. Desaguliers reached this conclusion in part using Newton's statement in the *Principia* that the magnitude of the effective gravity at points on the earth's surface varies inversely as the distances of the points from the center of the earth, assuming that the earth attracts according to the universal inverse-square law. This is the import of the passage just quoted. Newton had concluded that this variation in the magnitude of the effective gravity held at the earth's surface *from* the principle of balanced columns–that is, that all columns from the earth's center to its surface balance or weigh the same. We remember in Chapter 1 that Newton had discovered that the magnitude of the effective gravity at points along a meridian at the surface of a homogeneous figure shaped like a flattened ellipsoid of revolution which attracts according to the universal inverse-square law, whose ellipticity is infinitesimal, which revolves around its axis of symmetry, and whose columns from center to surface are all assumed to balance or weigh the same varies inversely as the distances of the points from the center of the figure. To Desaguliers this meant that the magnitude of the effective gravity must increase, not decrease, along a meridian at the surface of Cassini's homogeneous figure shaped like an infinitesimally elongated ellipsoid of revolution from a point whose latitude is the same as the latitude of Paris to a point on the equator.

But Newton had never stated that the principle of balanced columns could hold for homogeneous figures shaped like *elongated* ellipsoids of revolution which attract according to the universal inverse-square law, whose ellipticities are infinitesimal, and which revolve around their axes of symmetry.

(Moreover, as I mentioned in Chapter 1, Newton did not really even *prove* in the *Principia* that all columns from the center to the surface of a homogeneous figure shaped like a *flattened* ellipsoid of revolution which attracts according to the universal inverse-square law, whose ellipticity is infinitesimal, and which revolves around its axis of symmetry, actually *can* balance or weigh the same.)

Desaguliers's conclusion would indeed be true if the columns from the center to the surface of Cassini's homogeneous figure shaped like an elongated ellipsoid of revolution whose ellipticity is $-\frac{1}{96}$, which revolves around its axis of symmetry so that the ratio of the magnitude of the centrifugal force per unit of mass to the magnitude of the attraction at its equator is the earth's value $\frac{1}{289}$, and which is now assumed to attract in accordance with Newton's universal inverse-square law did all balance or weigh the same. However, we recall that I mentioned at the beginning of this section of this chapter that a homogeneous figure shaped like an elongated ellipsoid of revolution which attracts according to the universal inverse-square law fails to fulfill two necessary conditions that any homogeneous figure must satisfy in order to be a figure of equilibrium – namely, the principle of balanced columns and the principle of the plumb line.

What is more, Desaguliers mistakenly thought that he had proven the true statement that the columns from the center to the surface of a homogeneous elongated fluid figure of revolution which attracts according to the universal inverse-square law cannot all balance or weigh the same. Consequently based only on what Desaguliers believed he had demonstrated, it made no sense for him to apply the variation of the magnitude of the effective gravity at the earth's surface which Newton determined theoretically to a homogeneous figure shaped like an elongated ellipsoid of revolution which attracts in accordance with the universal inverse-square law, since that variation in the magnitude of the effective gravity depends upon the principle of balanced columns, unless perhaps Desaguliers thought that the columns from the center to the surface of a homogeneous *solid* figure shaped like an elongated ellipsoid of revolution which attracts according to the universal inverse-square law can all balance or weigh the same. However, if this could truly happen, then the weights of the columns would not change if the figure suddenly became fluid, assuming that the shape of the figure does not change. But then we would have a homogeneous elongated fluid figure of revolution which attracts in accordance with the universal inverse-square law and whose columns from center to surface all balance or weigh the same, which Desaguliers claimed he had shown to be impossible. Thus no matter what Desaguliers may have actually believed, he was contradicting himself.

(In Chapter 4 we will discover that if a homogeneous figure of revolution which revolves around its axis of symmetry is assumed to attract according to certain non-Newtonian hypotheses of attraction, then all of the columns from the center of the figure to its surface can balance or weigh the same even though the figure is *not* a figure of equilibrium. Now, such a figure can exist if it is solid. But the fact that all of the columns from the center of such a figure to its surface can balance or weigh the same even though the figure is not a figure of equilibrium does not mean that the magnitude of the effective gravity at points on the surface of such a figure necessarily varies inversely as the distances of the points from the center of the figure. Newton only showed that such a variation in the magnitude of the effective gravity follows if the columns from center to surface of a homogeneous figure shaped like a *flattened ellipsoid of revolution* which attracts *according to the universal inverse-square law* and whose ellipticity is infinitesimal all balance or weigh the same. He did not demonstrate this result to be true for any other hypothesis of attraction or for any other figure.)

Thus Desaguliers really had no basis for coming to the conclusions that he did about the way that the magnitude of the effective gravity varies with latitude at the surface of Cassini's homogeneous figure shaped like an elongated ellipsoid of revolution whose ellipticity is $-\frac{1}{96}$ and which revolves around its axis of symmetry so that the ratio of the magnitude of the centrifugal force per unit of mass to the magnitude of the attraction at its equator is the earth's value $\frac{1}{289}$, assuming that the figure attracts according to the universal inverse-square law, which is not what Mairan had hypothesized.

(As I mentioned at the beginning of this section of this chapter, we shall discover in Chapters 6 and 9 that the principle of the plumb line *can* hold at the surfaces of *solid stratified* figures shaped like elongated ellipsoids of revolution which attract according to the universal inverse-square law, whose ellipticities are infinitesimal, and which revolve around their axes of symmetry. However we shall also see in Chapter 9 that if Cassini's figure is assumed to be such a figure, it still cannot represent the earth. Because if the principle of the plumb line did hold at its surface, as the principle of the plumb line is observed to hold at the earth's surface, then assuming Cassini's figure to be stratified and to attract according to the universal inverse-square law would require that the decrease in the magnitude of the effective gravity from a pole to the equator along its surface be much greater than the observed decrease in the magnitude of the effective gravity from a pole to the equator along the earth's surface.)

Returning to Mairan's propositions concerning the way that the magnitude of the effective gravity varies from the poles to the equators along the surfaces of elongated figures of revolution, spherical figures, and flattened figures of revolution when the magnitude of the attraction was assumed constant at their surfaces and when the figures revolved around their axes of symmetry at the same rate, Desaguliers maintained that Mairan's results no longer held if the figures attracted according to the universal inverse square law instead.

In order to apply Newton's theory of the earth's shape appearing in the *Principia* to Mairan's homogeneous nonspherical figures of revolution, Desaguliers had to treat Mairan's homogeneous nonspherical figures of revolution as if they were all shaped like ellipsoids of revolution. Newton had only developed his theory of the earth's shape for figures shaped like ellipsoids of revolution. And there is evidence that Desaguliers did knowingly treat all of Mairan's homogeneous nonspherical figures of revolution as if they were shaped like ellipsoids of revolution.

Namely, in translating into English Mairan's demonstration that the magnitudes of the components of the centrifugal force per unit of mass in the directions of the attraction increase more from a pole to the equator along the surface of an elongated figure of revolution than along the surface of a spherical figure, when the magnitude of the attraction is constant at the surfaces of the two figures and perpendicular to the surfaces of the two figures at all points on their surfaces and when the two figures revolve around their axes of symmetry at the same rate, Desaguliers translated Mairan's phrase "Ayant décrit la courbe ovale quelconque *ADBE*, comme ci-dessus" as "Having describ'd an oval Curve of any Kind, as for Example, the Ellipse* *ADBE* abovementioned...." Desaguliers used the * to call the reader's attention to the figure in Mairan's *mémoire* which accompanied the passage that he translated, a figure that looks like an ellipse. Thus Desaguliers added the words "the Ellipse" in his translation of a passage in which Mairan did not mention elongated ellipses or any other specific examples of "oval [meaning elongated] curves."[51]

The trouble, however, is that Mairan's homogeneous nonspherical figures of revolution were *not* in general shaped like ellipsoids of revolution. We recall that Mairan thought that assuming elongated figures of revolution to be shaped like elongated ellipsoids of revolution was a hypothesis about the earth's shape which was too restricted. Consequently, no conclusions that Desaguliers came to concerning the way that the magnitude of the effective gravity varies with latitude at the surfaces of homogeneous figures shaped like ellipsoids of revolution which attract according to the universal inverse-square law and which revolve around their axes of symmetry, even if his conclusions were correct in these cases, need be true of Mairan's homogeneous nonspherical figures of revolution which were not shaped like ellipsoids of revolution, when these figures revolved around their axes of symmetry and were assumed to attract in accordance with the universal inverse-square law.

Assuming that figures attract according to the universal inverse-square law, Desaguliers expressed the magnitudes of the attraction at the equators of a homogeneous flattened figure of revolution, a homogeneous spherical figure, and a homogeneous elongated figure of revolution, where the three figures all have "the same Solidity,"[52] as $g - s$, g, and $g + s$, respectively, where s symbolizes a positive number, where we assume that the three figures all have the same density and therefore the same volume, and where we also suppose that the two nonspherical figures of revolution are shaped like ellipsoids of revolution for the reasons just stated. In other words, according to Desaguliers expressions for the magnitudes of the attractions at the equators of the three figures, the magnitude of the attraction should be smallest at the equator of a homogeneous figure shaped like a flattened ellipsoid of revolution; the magnitude of the attraction should be greatest at the equator of a homogeneous figure shaped like an elongated ellipsoid of revolution; and the magnitude of the attraction at the equator of the homogeneous spherical figure should lie halfway between the values of these other two magnitudes, where the three figures all have "the same Solidity."

In fact, if a homogeneous figure shaped like a flattened ellipsoid of revolution whose ellipticity is infinitesimal, a homogeneous spherical figure, and a homogeneous figure shaped like an elongated ellipsoid of revolution whose ellipticity is infinitesimal all have the same density and, moreover, all have "the same Solidity," meaning the same volume, to terms of first order, then it can be shown that the magnitude of the attraction at the equator of the flattened figure of revolution does indeed have the smallest value, that the magnitude of the attraction at the equator of the elongated figure of revolution does have the largest value, and that the magnitude of the attraction at the equator of the spherical figure does lie between the values of the other two magnitudes. However, it also turns out that the value of the magnitude of attraction at the equator of the spherical figure is not exactly the *average* of the values of the other two magnitudes, contrary to what Desaguliers's three expressions $g - s$, g, and $g + s$ imply. (On the other hand, the value of the magnitude of the attraction at the equator of the spherical figure and the average of the values of the other two magnitudes are *almost* but not *exactly* the same to terms of first order.[53])

Using the expressions $g - s$, g, and $g + s$, together with the (incorrect) expressions $c + 2$, $c + 1$, and c for the magnitudes of the centrifugal forces per unit of mass at the equators of the flattened figure of revolution, the spherical figure, and the elongated figure of revolution "of the same Solidity" which revolve around their axes of symmetry at the same rate, which I explained earlier, Desaguliers found that the magnitudes of the effective gravity at the equators of the flattened figure of revolution, the spherical figure, and the elongated figure of revolution "of the same Solidity" which revolve around their axes of symmetry at the same rate, where we assume the two nonspherical figures of revolution to be shaped like ellipsoids of revolution, have the forms $g - s - (c + 2)$, $g - (c + 1)$, and $g + s - c$, respectively. It is easy to show that

$$g - s - (c + 2) < g - (c + 1) < g + s - c \qquad (2.2.2)$$

is always true, whatever the values of g, c, and s.[54] From (2.2.2) Desaguliers concluded that seconds pendulums will be longest at the equator of the elongated figure of revolution, where again we have assumed the two nonspherical figures of revolution to be shaped like ellipsoids of revolution.

In fact Desaguliers conclusion is irrelevant. We recall that when the magnitudes of the components of attraction perpendicular to the surfaces at all points on the surfaces of the three different kinds of homogeneous figures of revolution had the same constant value g, where the two nonspherical figures of revolution were not assumed to be necessarily shaped like ellipsoids of revolution, where the three figures had "the same Solidity" and the same density, meaning the same volume, where the three figures revolved around their axes of symmetry at the same rate, and where the principle of the plumb line was assumed to hold at the same time at all points on the surfaces of the three figures, then the magnitudes of the effective gravity at the equators of the flattened figure of revolution, the spherical figure, and the elongated figure of revolution had the forms $g - (c + 2)$, $g - (c + 1)$, and $g - c$, respectively. But

$$g - (c + 2) < g - (c + 1) < g - c \qquad (2.2.3)$$

always holds, whatever the values of g and c. From (2.2.3) it follows that seconds pendulums were longest at the equator of the elongated figure of revolution, too, when the magnitudes of the components of attraction perpendicular to the surfaces at all points on the surfaces of the three different kinds of homogeneous figures of revolution had the same constant value g. But this is to be expected, since the magnitude of the centrifugal force per unit of mass is least at the equator of the elongated figure of revolution, since this figure has the shortest equatorial radius, when the three figures have "the same Solidity" and the same density, meaning the same volume, and revolve around their axes of symmetry at the same rate. It is not the *lengths* of seconds pendulums *at the equators of the three figures* that matters. It is rather the *variation* in the lengths of seconds pendulums with latitude along the surface of *each figure individually* which is important. Desaguliers's observation in fact just illustrates the mistake that Mairan had warned against making, when Mairan spoke of drawing conclusions *only* from the lengths of seconds pendulums *at the equators* of flattened figures of revolution, spherical figures, and elongated figures of revolution "of the same Solidity" and the same density, meaning the same volume, which revolve around their axes of symmetry at the same rate.

Desaguliers now changed the meanings of g and s. Assuming that the homogeneous elongated figure of revolution, which he implicitly assumed to be shaped like an elongated ellipsoid of revolution, attracts according to the universal inverse-square law, Desaguliers let g stand for the "Gravity at *Paris*,"[55] which in this case meant the magnitude of the component of attraction at a point on the surface of the elongated figure of revolution whose latitude is the same as the latitude of Paris which is perpendicular to the surface of the figure at the point, and he let s designate the amount by which the magnitude of the attraction at the equator of this figure exceeds the magnitude of the component of attraction at the point on the surface of the figure whose latitude is the same as the latitude of Paris which is perpendicular to the surface of the figure at the point. In other words, he let $g + s$ symbolize the "Gravity at the Æquator,"[56] meaning the magnitude of the attraction at the equator of the figure, where s stands for a positive number.

(Once again, however, Desaguliers's argument includes a gap: Desaguliers said that the magnitude of the attraction at the equator of the figure was greater than the magnitude of the component of attraction at the point on the surface of the figure whose latitude is the same as the latitude of Paris which is perpendicular to the surface of the figure at the point because the equator was closer to the center of the figure than the point on the surface of the figure whose latitude is the same as the latitude of Paris. But Desaguliers did not show if the magnitude of the attraction at the equator of the figure truly does exceed the magnitude of the component of attraction at a point on the surface of the figure whose latitude is the same as the latitude of Paris which is perpendicular to the surface of the figure at the point. To do so he would have had to be able to compute the magnitudes of the attraction at points on the surface of a homogeneous figure shaped like an elongated ellipsoid of revolution which attracts in accordance with the universal inverse-square law when the points are not located at the equator. But not even Newton calculated the magnitudes of the attraction at points on the surface of a homogeneous figure shaped like an elongated ellipsoid of revolution which attracts according to the universal inverse-square law when the points were not situated at a pole or on the equator of the figure.)

Desaguliers again let c symbolize the magnitude of the centrifugal force per unit of mass at the equator of the elongated figure of revolution, and he let $l + m + n$ stand for the difference between the magnitude of the centrifugal force per unit of mass at the equator of the figure and the magnitude of the component of centrifugal force per unit of mass at the point on the surface of the figure whose latitude is the same as the latitude of Paris which is perpendicular to the surface of the figure at the point. (We note that the expression $l + m + n$ is the same as the expression that Desaguliers used to stand for the difference between the magnitude of the effective gravity at a point on the surface of an elongated figure of revolution whose latitude is the same as the latitude of Paris and the magnitude of the effective gravity at a point on the equator of the elongated figure of revolution when he assumed that the magnitudes of the components of attraction perpendicular to the surfaces at all points on the surfaces of the three different kinds of homogeneous figures of revolution had the same constant value g, where the two nonspherical figures of revolution were not assumed to be necessarily shaped like ellipsoids of revolution, where the three figures had "the same Solidity" and the same density, meaning the same volume, where the three figures revolved around their axes of symmetry at the same rate, and where the principle of the plumb line was assumed to hold at the same time at all points on the surfaces of the three figures.)

Then $g + s - c$ represented the "diminish'd Gravity... at the Æquator,"[57] meaning the magnitude of the effective gravity at the equator of the elongated figure of revolution, which is perpendicular to the surface of the figure at the equator, and $g - (c - (l + m + n)) = g + l + m + n - c$ stood for the "diminish'd Gravity... at Paris,"[58] meaning the magnitude of the effective gravity at the point on the surface of the figure whose latitude is the same as the latitude of Paris, which is perpendicular to the surface of the figure at the point.

Desaguliers now simply stated without explanation that "by making all possible allowances, in favour of Monsieur Mairan's hypotheses,"[59] it was impossible ever to make $s \leqslant l + m + n$. Thus he claimed that $g + l + m + n - c < g + s - c$ was always true. In other words, he maintained that the magnitude of the effective gravity at the equator of the elongated figure of revolution was always greater than the magnitude of the effective gravity at the point on the surface of the elongated figure of revolution whose latitude is the same as the latitude of Paris. Consequently, seconds pendulums would always have to be longer at the equator of the elongated figure of revolution than at the point on the surface of the elongated figure of revolution whose latitude is the same as the latitude of Paris, contrary to experience.

But Desaguliers proved no such claim. He did no calculations. He never gave numerical values to the letters that he used to designate the various quantities involved in the problem. Consider alone the question of determining values of s when s has the second of its two meanings above. In this case s is the s in the expression $g + s$ for the magnitude of the attraction at the equator of the homogeneous figure shaped like an elongated ellipsoid of revolution which attracts according to the universal inverse-square law, where g represents the magnitude of the component of attraction at a point on the surface of the homogeneous figure shaped like an elongated ellipsoid of revolution whose latitude is the same as the latitude of Paris which is perpendicular to the surface of the figure at the point and where s symbolizes a positive number. Any particular value of s depends upon the particular shape of the homogeneous figure shaped like an elongated ellipsoid of revolution which attracts according to the universal inverse-square law. But as I have already noted above, even Newton only calculated values of the magnitude of attraction at points on the surface of a homogeneous figure shaped like an elongated ellipsoid of revolution which attracts according to the universal inverse-square law when the points were located at a pole or on the equator of the figure.

Why did Desaguliers take s to stand for a positive number? He probably used Newton's statement that the magnitude of the *effective gravity* at points on the earth's surface varies inversely as the distances of the points from the earth's center to conclude that the magnitude of the *attraction* at points on the earth's surface varied inversely as the distances of the points from the earth's center too. As I have already mentioned, Desaguliers said that the magnitude of the attraction at the equator of the homogeneous figure shaped like an elongated ellipsoid of revolution which attracts in accordance with the universal inverse-square law was greater than the magnitude of the component of attraction at a point on the surface of the figure whose latitude is the same as the latitude of Paris which is perpendicular to the surface of the figure at the point because the equator was closer to the center of the figure than the point on the surface of the figure whose latitude is the same as the latitude of Paris was, and he probably used Newton's statement as the basis for making such an assertion. What evidence exists for thinking that Desaguliers did this? As I suggested, Desaguliers appears to have used the term "Gravity" inconsistently. In stating the variation of the magnitude of the effective gravity at the earth's surface which Newton determined theoretically, Desaguliers, we recall, employed the expression "the Force of Gravity." However, as I also indicated, Desaguliers first utilized the term "Gravity" to mean "Attraction." Thus he seems to have used the term in contradictory ways. Be that as it may, I also noted that Newton only stated the result in question for homogeneous figures shaped like *flattened* ellipsoids of revolution which attract according to the universal inverse-square law, *not elongated ones*, and, moreover, the result at issue only concerned the variation of the magnitude of the *effective gravity* at points on the surface of such a figure, not the variation of the magnitude of *attraction* at the points. Indeed, in Chapter 6 we shall see that the magnitudes of the attraction at points on the surface of a homogeneous figure shaped like a flattened ellipsoid of revolution which attracts according to the universal inverse-square law *cannot* vary inversely as the distances of the points from the center of the figure.

As I mentioned, Desaguliers's conclusion would indeed be true if the columns from the center to the surface of a homogeneous figure shaped like an infinitesimally elongated ellipsoid of revolution which attracts according to the universal inverse-square law all balanced or weighed the same. However, as I also noted, such a figure fails to fulfill two necessary conditions that any homogeneous figure must satisfy in order to be a figure of equilibrium. These two conditions are precisely the principle of balanced columns and the principle of the plumb line.

To summarize why Desaguliers's conclusions about the way the magnitude of the effective gravity varies with latitude at the surface of a homogeneous figure shaped like an elongated ellipsoid of revolution which attracts according to the universal inverse-square law and which revolves around its axis of symmetry have no foundations and therefore are not valid:

1. Newton had *only* stated the result in question which Desaguliers tried to apply, which Newton had inferred from the principle of balanced columns, for homogeneous figures shaped like *flattened* ellipsoids of revolution which attract according to the universal inverse-square law and which revolve around their axes of symmetry, *not* for homogeneous figures shaped like *elongated* ellipsoids of revolution which attract according to the universal inverse-square law and which revolve around their axes of symmetry. As I have indicated above, homogeneous figures shaped like elongated ellipsoids of revolution which attract according to the universal inverse-square law cannot even satisfy the principle of balanced columns;

2. the result at issue concerned the way the magnitude of the *effective gravity* varied with latitude at the surface of such a figure, *not the attraction*;

3. Mairan's nonspherical figures of revolution were *not* in general shaped like ellipsoids, and when his figures were not, Newton's theory could not be brought to bear upon the figures in any manner.

In short, if Desaguliers found real flaws in Mairan's attempt to reconcile all of the terrestrial observations of the day, he did not defend Newton's theory of the earth's shape very convincingly or successfully.

Furthermore, Desaguliers "derived" all of his conclusions here assuming that the universal inverse-square law of attraction holds, which is not a law of attraction that Mairan considered. Therefore in a sense Desaguliers talked past Mairan here, even though Desaguliers only did so because Mairan had failed to deal with Newton's law of attraction in his Paris Academy mémoire. (We shall observe throughout much of the remainder of the story told here that trying to confirm this law in the terrestrial realm was not an easy task at all.)

Desaguliers also attacked Mairan's ad hoc law of nonconstant attraction at the surfaces of figures. He claimed to prove that a homogeneous elongated fluid figure of revolution initially at rest which attracts according to Mairan's proposed law would transform itself into a spherical fluid figure. Here Desaguliers explicitly stated that Mairan's elongated fluid figure of revolution was shaped like an elongated ellipsoid of revolution,[60] which further supports my hypothesis that Desaguliers treated all of Mairan's elongated figures of revolution as if they were shaped like elongated ellipsoids of revolution. As he did in the case of the universal inverse-square law of attraction, Mairan argued that this change of shape was necessary to make all of the columns from the center of a stationary homogeneous figure to its surface balance or weigh the same. Moreover, he maintained that the elongated fluid mass would change into a spherical fluid mass faster if it attracted in accordance with Mairan's proposed law of attraction than it would if it attracted according to the universal inverse-square law.[61] To be sure, if the motionless figure is spherical, Mairan's ad hoc law of attraction reduces, as we have seen, to a law determined by an inverse-square central force, where the center of force is situated at the center of the figure, in which case all columns from the center of the spherical figure to its surface do balance or weigh the same. Once again, however, Desaguliers hardly presented a rigorous, persuasive proof that a spherical figure is the only stationary homogeneous figure for which this is true. His reasoning here is no more convincing than his argument was for the case that the homogeneous fluid mass attracts according to the universal inverse-square law.

In addition, taking centrifugal force of rotation into account, Desaguliers found that a seconds pendulum in Paris would have to be shortened at the equator nearly an inch, or about five times the amount of shortening actually observed, if the earth attracted according to Mairan's ad hoc law.[62]

But Desaguliers did not attack Mairan's ad hoc law of attraction where it was most vulnerable. Namely, just like Mairan's hypothesis of attraction whose magnitude is constant at the surfaces of figures, the ad hoc law of attraction would not permit the principle of the plumb line to hold at the surface of a figure of revolution when the figure revolved around its axis of symmetry. This is where the law was really deficient. It would be overly generous to call Desaguliers's paper a successful rejoinder to Mairan's Paris Academy mémoire.[63]

Had Mairan not read the theory of the earth's shape in either the first or second editions of the *Principia* when he wrote the original version of his Paris Academy *mémoire* of 1720? Does that explain why he failed to discuss Newton's theory? The published version of Mairan's *mémoire* does include one reference to Newton, but it is a rather incidental one. In a note in a margin of one of the pages of his printed *mémoire*, Mairan cited the version of Newton's theory that appears in the second edition of the *Principia*, in connection with the earlier doubts of some astronomers that lengths of seconds pendulums shorten at all as latitudes decrease, as well as in connection with the debates about the question whether climactic changes alter the lengths of seconds pendulums enough to invalidate experiments made with them at different latitudes and consequently invalidate the conclusions about changes in the magnitude of the effective gravity with latitude which were drawn from these experiments. Mairan cited Newton in order to do away with any lingering doubts about both matters. However, it was not really Newton's theory that Mairan called attention to. Instead Mairan mentioned what Newton had to say about the trustworthiness of the experiments done with seconds pendulums upon which the theories of Newton and Huygens and the analyses of Mairan were all based.[64] The fact that the reference to Newton appears in the margin of one of the pages of Mairan's printed *mémoire* could signify that Mairan first looked at Newton's theory only a short time before his *mémoire* was published, in 1722. On the other hand, the location of the reference in the margin may indicate nothing more than the secondary nature of Mairan's comments and may not help determine when Mairan first came across Newton's theory.

When the French astronomer Joseph-Nicolas Delisle, who had already defended Newton's *Principia* for several years when Mairan read his *mémoire* to the Paris Academy, summarized that version of the *mémoire* in a letter of August 1720 to his friend and colleague in astronomy Jacques-Eugène d'Allonville, Chevalier de Louville, who was himself an early French defender of the *Principia*, Louville replied: "I cannot understand anything in Monsieur de Mairan's *mémoire*. He must not have seized the thought of Messieurs Newton and Huygens."[65] If Mairan did not understand Newton's theory, whenever he finally came upon it for the first time, then he was no worse off than either Johann I Bernoulli or Maupertuis! Considering what he omitted in his *mémoire*, it is hard to see how in 1722 Mairan could have possibly understood something as basic to Newton's theory of attraction as the facts that a homogeneous spherical body that attracts according to the universal inverse-square law acts like a central force, both in its exterior and its interior, while a homogeneous nonspherical mass that attracts in accordance with the same law does not behave like a central force, either in its exterior or its interior.[66]

Mairan wrote a tedious, wordy *mémoire*. In fact, its style is customary of Mairan and may explain in part why no one has carefully studied the *mémoire* recently. Maupertuis did not find that Mairan treated the problem of the earth's shape any more satisfactorily than Newton had. Neither resolved the problem to suit Maupertuis. Nor did Mairan's arguments convince Johann I Bernoulli, either.[67]

Desaguliers's rejoinder to Mairan's Paris Academy *mémoire* seems to have failed to stimulate men of science in Paris to reconsider the problem of the earth's shape. At least, they give no sign of having paid any real attention to the details in any of Desaguliers's three articles published in the *Philosophical Transactions* for 1725 in the years that followed, perhaps because they were written in English, a language of which most Frenchmen had a very poor command. Mairan certainly did not read Desaguliers's articles. In a letter to Colin MacLaurin, dated January 1743, in which he thanked the Scottish mathematician for having sent him a copy of his *A Treatise of Fluxions*, published in 1742, Mairan lamented his inability to read English.[68] Maupertuis did not know English, either,[69] but during his stay of twelve weeks in London in the summer of 1728, he would have learned the substance of Desaguliers's articles in conversations with British scientists who spoke French or Latin and perhaps in conversations with (Huguenot descendant) Desaguliers himself. Maupertuis referred both to Mairan's *mémoire* and to Desaguliers's objections having to do with Mairan's statements that figures of equilibrium could be elongated, in his *Discours sur les différentes figures des astres*, published in November 1732, and then again in a Paris Academy *mémoire* of 1733 on geodesy, but Maupertuis did so both times only in passing.[70] He did no more than state Mairan's and Desaguliers's opposing positions, only to put the disagreement aside immediately. Now, the conflict could have aroused the interest that Maupertuis had during the years 1729–30 in the problem of homogeneous figures of revolution whose columns from center to surface all balance or weigh the same,[71] which ultimately led him to publish his *Discours* in 1732 as well as an article on such figures in the *Philosophical Transactions* for the same year. However, by 1733 Maupertuis had put this problem aside and had turned instead to geodesy as the means of ultimately resolving the controversy over the earth's shape. Scientists in Paris also appear to have taken no note of what Desaguliers had to say about the faults in the measurements of degrees of latitude in France along the meridian through Paris made by astronomers in Paris before 1719. Desaguliers's observations can be found in the first of his three articles published in the *Philosophical Transactions* for 1725. In this particular article, Desaguliers meant to rebut the conclusions appearing in Jacques Cassini's *De la grandeur et de la figure de la terre* (1722), the work that recounted the fifty years of the Paris Academy's expeditions that ended in 1718.[72] But when Maupertuis revived the geodetic approach for determining the earth's shape, in the 1730s, for reasons not unrelated to those defects in the earlier measurements which Desaguliers had already pointed out in the *Philosophical Transactions* for 1725, Maupertuis only did so under the influence of entirely independent sources.[73] These I shall discuss in the next chapter.

Some writers would have us believe that Desaguliers's rejoinder to Mairan's *mémoire* immediately brought about a pitched battle between scientists in Paris and English scientists, waged along nationalistic and ideological lines, centered on the problem of the earth's shape, whose solution would determine the victor.[74] Given the kind of ignorance and inattention that I have just described, it is very hard to imagine how the two papers in question could possibly have given rise to such a battle. Indeed, it is easy to doubt that such a battle ever took place. But the papers by Mairan and Desaguliers involve matters and raise issues that are frequently misunderstood by historians.

3

The revival of geodesy in Paris (1733–1735)

It is well known that beginning in 1732, the *Principia's* natural philosophy penetrated the Paris Academy of Sciences. This came about as a result of a drive originally led by Maupertuis. I have argued elsewhere that although this entry of Newton's world system occurred at the same time as a revival in Paris in the 1730s of the geodetic approach to determine the earth's shape, the simultaneity of the two events was wholly fortuitous.[1] Briefly here is what happened.

In 1733, Maupertuis touched off a revival of geodesy in Paris as a means of finding the earth's shape. He acted under the influence of two sources. In the spring of 1733 the Paris Academy announced that the Paduan scholar Giovanni Poleni had won the Academy's prize for that year. Around the same time Poleni presented to the Academy a copy of a booklet that he had published in 1729. In this work Poleni proposed, among other things, to determine the earth's shape by means of measuring degrees of longitude instead of the usual degrees of latitude. Maupertuis adopted Poleni's idea and made it the focus of a paper that he presented to the Paris Academy in June of 1733.[2]

Early in 1733, the Dutch *Journal Historique de la République des Lettres* also published a long, illuminating review of Poleni's tract. In it the author, who was probably Elie de Joncourt,[3] who founded the Dutch journal together with W. J. s'Gravesande among others, extended Poleni's approach. He introduced two equations, which do not appear in Poleni's pamphlet, for finding the ratio of the earth's axes (equivalently, the earth's ellipticity) assuming that the earth is shaped like an ellipsoid of revolution. One equation could be solved for the ratio of the earth's axes, if the radii of two different circles at the earth's surface parallel to the earth's equator whose latitudes can be arbitrarily chosen (equivalently, the lengths of degrees of longitude at any two, given latitudes at the earth's surface) were specified. The other equation could be solved for the ratio of the earth's axes, if the lengths of a degree of latitude and a degree of longitude at any arbitrarily chosen latitude at the earth's surface were given. Maupertuis more than likely read this review of Poleni's booklet, because in his paper in June of 1733 he provided mathematical derivations of these two equations. He did this without acknowledging where or from whom he had gotten the equations. However it can hardly be accidental that his equations were precisely the same as those that had appeared earlier in the review of Poleni's tract published in the Dutch journal.

But historians have readily confused the issues. The *Principia's* theory of the earth's shape no doubt helped cause Maupertuis to develop an interest in determining homogeneous figures of revolution which revolved around their axes of symmetry and whose columns from center to surface were all assumed to balance or weigh the same. However, Maupertuis showed by balancing columns that connect all points along a meridian at the surface of a homogeneous figure of revolution to the center of the figure, in his *Discours* (1732) and in an abridged, Latin version of this work which appeared in print slightly earlier in the *Philosophical Transactions* for 1732, where the figure revolved around its axis of symmetry, that *all* hypotheses of central forces of attraction whose magnitudes vary like r^n, where r is the distance from a center of force situated at the center of the figure of revolution, make *all* homogeneous figures of revolution come out flattened, not just Huygens's hypothesis ($n = 0$) and Hermann's hypotheses ($n = 0, n = 1$). (The figures of revolution are shaped like ellipsoids of revolution only when $n = 1$.[4]) In other words, not only Newton's universal inverse-square law of attraction but *all* of the leading hypotheses of attraction of the day together with the principle of balanced columns applied to homogeneous figures of revolution which revolved around their axes of symmetry required that such figures of revolution be flattened. Such results, however, contradicted the findings of Jacques Cassini and his colleagues, who deduced from measured variations of degrees of latitude with latitude in France along the meridian through Paris that the earth was shaped like an elongated ellipsoid of revolution. Maupertuis urged a renewed effort to determine the "facts of the matter," rather than continue to dwell upon competing, speculative theories.[5] By this he meant that only careful measurements would conclusively settle the matter. Thus did Newton's French advocate accept the Paris Academy's positivism as valid. From 1733 to 1735, Paris Academicians busied themselves with perfecting practical geographical and astronomical methods for determining the earth's shape.

In a Paris Academy *mémoire* of 1735, Maupertuis derived the equation

$$2A - 2\epsilon = 3(m^2 - 1)AS^2, \tag{3.1}$$

which holds for a body shaped like an ellipsoid of revolution whose ellipticity is infinitesimal (see Figure 11). Here *PapA* is a meridian at the surface of the body, line segment *EK* is perpendicular to the surface of the body at the point *E* on the meridian *PapA*, angle *AKE* is the latitude of the point *E*, A = the length of a degree of latitude at the equator, ϵ = the length of a degree of latitude at the latitude whose sine is S, m = the polar radius *CP* of the body, the axis of symmetry of the body is its polar axis *Pp*, and the equatorial radius *CA* of the body is normalized to equal 1. Maupertuis noted that if the body is flattened at its poles, in which case $\epsilon > A$ when $S > 0$, and $1 > m$, then equation (3.1) can be rewritten as

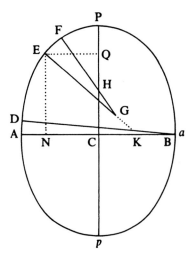

Figure 11. Meridian of an ellipsoid of revolution whose ellipticity is infinitesimal (Maupertuis 1735: figure on p. 99).

$$\epsilon - A = \tfrac{3}{2}(1 - m^2)AS^2. \tag{3.1'}$$

He rewrote equation (3.1) this way in order to show that the increase in degrees of latitude from the equator to a pole along the surface of such a body shaped like a flattened ellipsoid of revolution varies directly as the square of the sine of the latitude.

We observed in Chapter 1 that Newton had stated a result like this in the *Principia.* But Newton did not publish an explicit equation, much less provide the reader with a demonstration of the result. In deriving equation (3.1) Maupertuis made use of quantities whose orders were the same as the order of the ellipticity of the ellipsoid of revolution. But he discarded terms of second order and higher orders in these quantities to arrive at this equation. In other words, he showed that equation (3.1) is true *only* for bodies shaped like ellipsoids of revolution whose ellipticities are *infinitesimal.* Newton had not made clear that a result like (3.1') held *only* for bodies shaped like flattened ellipsoids of revolution whose ellipticities were *infinitesimal.* Moreover, Maupertuis proved that equation (3.1) holds true for *all* bodies shaped like ellipsoids of revolution whether they be flattened or elongated, provided their ellipticities are infinitesimal. In addition, Maupertuis's derivation of equation (3.1) does not involve any assumptions about the density of the body. Nor did Maupertuis make any assumptions about attraction in deriving equation (3.1). The body need not attract at all, or it can be assumed to attract according to any arbitrary hypothesis. In particular, equation (3.1') holds good for *any* homogeneous or heterogeneous, attracting or nonattracting body shaped like a flattened ellipsoid of revolution whose ellipticity is infinitesimal. Newton had not even mentioned these facts, much less demonstrated them. He stated in words the result that Maupertuis concluded to be expressed by equation (3.1′) while in the midst of outlining his theory of a homogeneous figure shaped like a flattened ellipsoid of revolution which attracts according to the universal inverse-square law, whose ellipticity is infinitesimal, which revolves around its axis of symmetry, and whose columns from center to surface are all assumed to balance or weigh the same.

Indeed, Newton stated the result at the same time that he stated that the increase in the magnitude of the effective gravity from the equator to a pole along the surface of such a homogeneous figure varies directly as the square of the sine of the latitude. This juxtaposition of the two "sin²" laws could have easily misled readers of the *Principia*, since equation (3.1′) has nothing whatever to do with Newton's theory of such a homogeneous figure. That is, equation (3.1′) turns out to be totally independent of Newton's theory of the earth's shape. Indeed, as I say, the derivation of equation (3.1′) does not depend in any way upon assumptions about the density of the figure or upon assumptions about how the figure attracts. All that is involved is the geometry of the figure.

Maupertuis purposely introduced equation (3.1) to help justify his conviction that it was necessary to measure two degrees of latitude at the earth's surface at latitudes as far apart as possible–namely, near the earth's equator and near one of the earth's poles–in order to determine the earth's ellipticity δ by direct measurement with the greatest possible accuracy. Here of course he assumed the earth to be shaped like an ellipsoid of revolution in order to be able to make use of equation (3.1). He contended that these two degrees of latitude would differ by the largest amount. In particular, he maintained that their difference would be so large that errors made in measuring the degrees of latitude would not nullify using equation (3.1) to determine the earth's ellipticity.[6] (In other words, the errors would not be of the same order as the difference between the two degrees of latitude, but of higher order instead.) When A has been found by measurement and ϵ has also been determined by measurement at a latitude as far north as possible whose sine is S, then equation (3.1) can be used to find m:

$$m = \sqrt{\frac{2}{3}\frac{(A-\epsilon)}{AS^2} + 1}. \qquad (3.1'')$$

But as a result of the normalization mentioned, m is the ratio of the earth's polar axis to the earth's equatorial axis. Consequently m can be used to calculate the earth's ellipticity δ, since $\delta \equiv (1 - m)/m$.

In fact, equation (3.1) can be simplified a good deal, and then δ can be found easily using the reduced equation. Since $\delta \equiv (1 - m)/m$, it follows that

$$m = \frac{1}{1+\delta}. \qquad (3.2)$$

If both sides of (3.2) are squared and then 1 is subtracted from both sides of the equation that results,

$$m^2 - 1 = \frac{-\delta^2 - 2\delta}{(1 + \delta)^2}$$
(3.3)

follows. But the right-hand side of (3.3) equals -2δ to first order in δ. Now, equation (3.1) can be rewritten as

$$A = \frac{2\epsilon}{2 - 3(m^2 - 1)S^2}.$$
(3.1''')

If -2δ is substituted for $m^2 - 1$ in (3.1'''), the approximation

$$A \simeq \frac{2\epsilon}{2 - 3(-2\delta)S^2} = \frac{\epsilon}{1 + 3\delta S^2}$$
(3.4)

which holds to first order in δ results.
   Then if (3.4) is solved for δ,

$$\delta \simeq \frac{\dfrac{\epsilon}{A} - 1}{3S^2}$$
(3.5)

results. At a latitude near a pole, $S < 1$ and $S \simeq 1$. If the body shaped like an ellipsoid of revolution is flattened and nearly spherical and ϵ is the length of a degree of latitude at this latitude, then

$$\frac{\epsilon}{A} > 1 \quad \text{and} \quad \frac{\epsilon}{A} \simeq 1. \quad \text{Hence} \quad 0 < \frac{\epsilon}{A} - 1 \quad \text{and} \quad 0 \simeq \frac{\epsilon}{A} - 1$$

in which case the right-hand side of (3.5) is indeed a positive number that is infinitesimal.
   The Paris Academy had already planned an expedition to Peru (what is now Equador). The Academicians involved set sail in 1735, and the expedition endured until 1744. In his *mémoire* of 1735, Maupertuis adduced evidence that a second Paris Academy expedition, to Lapland, was needed. Such an expedition got underway in 1736, and it lasted until 1737. During the period 1733–35, Paris Academicians pretty much left aside the problem of determining the earth's shape theoretically.
   Even today these geodetic expeditions are sometimes depicted as occasions for carrying out "crucial experiments" to decide between Cartesian and Newtonian world systems.[7] I have already said in the preceding chapter that Mairan's Paris Academy *mémoire* of 1720 has been interpreted as an exercise in Cartesian

physics, which together with Desaguliers's rejoinder of 1725 started an ideological war between scientists in Paris and English scientists.[8] Such an interpretation of Mairan's *mémoire* seems to me to miss what Mairan really aimed to do in that work, as I have also already argued. It is true that Mairan, a faithful supporter of Malebranche, hoped to reconcile Newton's world system with Descartes's, just as other physicists in the Malebranche Group, to which Mairan belonged, hoped to do.[9] Hence Mairan did not oppose Cartesianism. On the other hand, it is hard to see how an individual who possessed six editions of Newton's *Opticks* and seven editions of Newton's *Principia* could possibly be interpreted as Newton's adversary.[10] Perhaps Mairan was merely a bibliophile who knew the monetary value of rare, acclaimed works. Be that as it may, what motivated Mairan's *mémoire* of 1720 was his desire to harmonize theory with the observations that phenomenologically oriented Paris Academicians considered to be fundamental. Generally speaking, Paris Academicians were less likely to challenge the accuracy of measurements than they were to dispute theories that were at bottom conjectural. When the Paris Academicians were finally forced to recognize overlooked sources of errors in making observations and measurements, they did so. Indeed, the revival of geodesy in Paris in the 1730s turns out to be just such a case.[11] Maupertuis's campaign to renew efforts in Paris in the 1730s to use geodesy to find the earth's shape embodied and manifested the principal concerns of his colleagues. After having done some work on the earth's shape treated as a theoretical problem, Maupertuis finally acknowledged that measurements, not hypotheses, would ultimately have to decide the question. Measurements would have to remain the final arbiter.

Maupertuis and his contemporaries did not all use the Cartesian–Newtonian philosophical disputes to muddle scientific issues of the time the way that we do today. In 1737, the Swiss *savant* Samuel Koenig told his friend Maupertuis that a Cartesian model could easily be adapted to conform to the flattened earth inferred from the measurements of degrees of latitude made in France and in Lapland.[12] Indeed, as I mentioned earlier, Huygens and Hermann applied Cartesian principles (attraction toward one fixed center of force) to the theoretical question, which led them to flattened figures of revolution. The problem was that their flattened, nearly spherical figures of revolution were subsequently found not to be flat enough to accord with the extent of the earth's flattening deduced from the new measurements of degrees of latitude. But the assumptions that Huygens and Hermann made and the means that they used to determine their figures were certainly more Cartesian in spirit than the discussion that appears in Mairan's *mémoire*. And as Koenig pointed out, Huygens's and Hermann's assumptions could be modified so that flatter figures would result. (Koenig had specifically in mind here replacing a hypothesized fixed center of force situated at the earth's center by a solid nucleus at the earth's center which consists of a collection of fixed centers of force, each of which attracts like a central force, which Descartes had also hypothesized.)

Are the mathematical demonstrations in Maupertuis's *Discours*, all of which start from Cartesian hypotheses (central forces), only meant to fill space? Or were they perhaps intended to divert attention, in order to make less apparent a preference for Newton, whom it is not clear that Maupertuis truly understood *at all* at this point? Historians who use a conflict between Cartesian and Newtonian paradigms to portray the development of eighteenth-century physical science fail to ask such questions, much less deal with the problems of answering them. Concerning themselves primarily with the philosophical aspects of the *Discours*, they rarely, if ever, study the work's mathematical details. Many simply rely on the popularized version of the "conflict" described by Voltaire, whose witticisms display above all Voltaire's naïveté and lack of any real understanding of the scientific problems.

In addition to Cartesianism, nationalism has diverted the attention of historians from the true issues. This has only helped to jumble these issues even more. During the last thirty years of Louis XIV's reign, a period of almost continual warfare between France and its neighbors, chauvinism doubtlessly reached a peak. Contrary to some wishful thinking of historians from the allied countries that won World War II, which these historians reveal in books and articles that they wrote during the years after the war, in which they suggest that the sciences were never at war during Louis XIV's reign, Louis's wars nevertheless harmed the sciences of the period, certainly in France. The wars interrupted what had been a rather steady flow of communication between French and British and between French and German scientists. Before the wars, international exchange had served a purpose that was of vital importance. This exchange had caused European science to advance. Once the wars began, French men of science, and French mathematicians in particular, suffered greatly from the isolation that the wars entailed. French mathematicians fell behind.[13] The delayed reception of Newton's *Opticks* and *Principia* in Paris, as well as the facts that Swiss, and even Italian, mathematicians rapidly outdistanced Parisian mathematicians in developing and improving the Leibnizian calculus, must be viewed with Louis XIV's wars as a background.

While traces of nationalism may have remained during the thirty years of peace that followed the signing of the Treaty of Utrecht in 1713–a peace no doubt furthered by the death of the militant Louis XIV in 1715–there is also no question that previously blocked lines of communication reopened. The British and French scientific communities became more and more aware of each other's acts and words. Scientists of the two nationalities again paid more attention to each other, even if they did not always agree. It is scarcely accidental that Newton's *Opticks* and *Principia* both found general acceptance in Paris during this period, although both works date back to the years during the wars.

In the postwar years, Huguenot refugees in Holland and in England, who had been forced to flee France following the Revocation of the Edict of Nantes in 1685, took important steps to pave the way, through publications in French, for foreigners to exert influence in France. We know that the Huguenot journalists in Holland inspired the Frenchmen who led the French Enlightenment. This was the way that the journalists avenged themselves on the French for the Revocation of the Edict of Nantes. The Huguenot Pierre Coste's French translation of Newton's *Opticks*, published in Amsterdam in 1720, is an outstanding example of the effort by the refugees. Coste also translated into French John Locke's *Essay Concerning Human Understanding*. Leibniz profited greatly from this translation (which is not to say that the German rationalist agreed with the English empiricist's philosophy).[14] Yet another notable example of the effort by the refugees was the Huguenot Pierre des Maizeaux's *Recueil de diverses pièces sur la philosophie... par Messieurs Leibniz, Clark, Newton, et autres auteurs célèbres*, also published in Amsterdam in 1720. The philosophical issues that Newton's world system raised were brought to the attention of the French less by perusal of the *Principia*, "that intractable book," than by reading philosophers who perceived how deeply the issues went. Maizeaux's discussion of the philosophical questions in the French language made these questions accessible to a French audience.[15]

Why does confusion arise? In March 1731, Maupertuis used the expression "dispute between the English and the French," in describing to Johann I Bernoulli what induced Mairan to write his *mémoire* of 1720. Did Maupertuis mean to imply that an ideological war, waged along nationalistic lines, had begun? Was he talking about more than just the plain fact that Newton's conclusions conflicted with Cassini's conclusions? It is hard to see how. After the expedition to Lapland, certain Paris Academicians complained that Maupertuis and his co-workers confirmed the shape of the earth that "an Englishman [Newton] and a Dutchman [Huygens]" had determined.[16] But this grievance illustrates nothing more than French chauvinism directed toward France's traditional enemies. Such remarks do not raise Cartesianism as an issue in the slightest way. Indeed, as I have already said, Huygens followed Cartesian principles in determining the shape that he found for the earth. In 1738, Maupertuis published anonymously his "disinterested" account of the controversy over the earth's shape. He wrote it with insincerity to support Cassini's findings, but the attentive reader could not have failed to realize where the author's true sentiments lie, because of all of the sarcasm and obvious chicanery that fill the entire work. In his book the French mathematician tried to persuade the reader that nationalism had little to do with the real issues.[17] Maupertuis noted that two Englishmen, Thomas Childrey and Thomas Burnet, had both contended that the earth was elongated at its poles. Moreover, Childrey did so long before Newton and Huygens maintained that the earth was flattened at its poles.[18] The second of Desaguliers's three essays in the *Philosophical Transactions* for 1725 consists almost entirely of quotations from John Keill's refutation of Burnet's theory, which Keill wrote in 1698.[19] If Huygens and Hermann, both of whom reasoned that the earth was flattened at the poles, were influenced by anyone, it was by Descartes. In working out their theories of a flattened earth, Newton and Huygens started with the experiments of another Frenchman, Richer.[20] The reasons why Childrey, Burnet, and the astronomer from Strasbourg Samuel Eisenschmidt believed the earth to be elongated at its poles had little, if anything to do with Cartesian physics.[21] What annoyed the arrogant, impatient Maupertuis, who composed the satirical work, was that he knew that he had the facts right. Whether the French liked it or not, the issues regarding the earth's shape cut across national lines. But older Paris Academicians acted hostilely when confronted with such observations–although it is not always clear whether the evidence itself or Maupertuis's pranks instead repelled them.

If Maupertuis's underhanded doings paid off in the long run, they did him little good in the short term. His behavior did not stand him in good stead with his older colleagues in the Paris Academy. Two of Maupertuis's partisans, Voltaire and D'Alembert, likened their ally to a martyr. Voltaire viewed Maupertuis as France's Galileo in the battle to promote Newtonianism in Paris.[22] Such portrayals idealize Maupertuis's behavior during this struggle. In fact, his conduct was hardly exemplary. He resorted to tactics as reprehensible as his "disinterested" account of the controversy over the earth's shape mentioned previously. In the 1730s, Maupertuis acquired a reputation for acting as a gadfly in his campaign to promote Newtonianism in Paris. In a review of the second edition of Maupertuis's *Discours* published in the *Journal des Savants* in 1742, the staid writers for the sober journal confessed that they had never reviewed the first edition of the work. They recalled that the first edition aroused too many passions concerning aspects of Newton's world system which were hardly known, much less understood, in France at the time.[23] As an organ of the Paris Academy, the journal thus made clear what had been the Academy's position with regard to Maupertuis's *Discours* when it first appeared in 1732. Even Maupertuis's close friend Laurent Angliviel de La Beaumelle found Maupertuis's behavior during the 1730s difficult to defend in his biography of Maupertuis. In his crusade for Newton's world system, Maupertuis enlisted the younger Paris Academicians, who were easier to influence because they had not yet formed immutable opinions and, consequently, were more prepared and more willing to accept Newton's ideas in natural philosophy than the older Paris Academicians were. Some of the younger Academicians had already been inclined to do so even before they entered the Academy. The group led by Maupertuis badgered, intimidated, cajoled, coerced, and ridiculed the Cartesians in the Academy. La Beaumelle could not depict Maupertuis's trickery in praiseworthy terms.[24] In 1734, Maupertuis traced the idea of attraction back to Kepler and to two Frenchmen, Pascal and Roberval.[25] This might be interpreted as simply another attempt by Maupertuis to denationalize the issues. However, the skeptical Secrétaire Perpétuel of the Paris Academy Fontenelle had his doubts. He retorted: Did Maupertuis want to "claim a glory for his fatherland, or justify the English at the expense of the French?"[26] A member of the old school in the

Academy, Fontenelle wondered if all Maupertuis had tried to do was to pull the wool over his countrymen's eyes. In Fontenelle's opinion, all of the advantages in Maupertuis's supposedly "neutral" account of the Cartesian and Newtonian world systems in the *Discours* went to "the English natural philosopher."[27] The bravery attributed to Maupertuis by Voltaire and D'Alembert scarcely does justice to Maupertuis's opponents in the Paris Academy. On the other hand, we shall discover in Chapter 5 that Maupertuis paradoxically acted in a way that was absolutely necessary at the time!

As La Beaumelle tells the story, the opposition of the members of the old school in the Paris Academy to the conclusions that the members of the expedition of Lapland drew from their measurements was not so much directed toward Newton as it was toward Maupertuis. The findings reached during a mere nine months of work in the arctic region threatened to invalidate the results based upon astronomical observations and geodetic measurements painstakingly accumulated during five expeditions in France carried out over a period of fifty years by the leading observers among the astronomers in Paris.[28] Theory had little, if anything to do with the quarrel; the conclusions that the members of the Lapland expedition came to, which were founded only on the geographic measurements and astronomical observations made in Lapland, created a situation that now threw doubt upon the reliability of just such measurements and observations made earlier in France by the men who had been regarded as the best observers among the astronomers in Paris. The results of the Lapland expedition now called in question the very activity that had been assumed all along to be the strong point of these individuals. It is difficult to imagine that the older astronomers in Paris did not fear that their reputations were at stake.

In short, irrelevant Cartesian backlash, trumped up nationalistic pride, bruised egos, and threatened reputations all contributed to misunderstanding. They shed no light whatever on the scientific issues involved in the revival of geodesy in Paris in the 1730s.[29]

4

Pierre Bouguer and the theory of homogeneous figures of equilibrium (1734)

In the spring of 1734, while the revival of geodesy in Paris was in full swing, the Paris Academician Pierre Bouguer read a paper to the Paris Academy on the problem of homogeneous figures of equilibrium.[1] Preceding generations of Paris Academy mathematicians had spent considerable time translating into the infinitesimal calculus work in mathematics and in mechanics done earlier by mathematicians of other nationalities. Pierre Varignon, for example, had translated the work of Huygens (the theory of the curvature of plane curves) and Newton (the motions of orbiting bodies which are governed by central forces) into the differential calculus. As we shall see in the next chapter, Bouguer's contemporary Maupertuis tried to translate Newton's theory of the earth's shape into continental mathematical language, without total success. Bouguer on the contrary did not translate work in mathematics and in mechanics done earlier by other mathematicians into the infinitesimal calculus. Instead he used the infinitesimal calculus as a tool to discover and prove fundamentally new results in mechanics which could not have easily been found any other way. In 1733, in a Paris Academy *mémoire* on minimizing the impulse on a body caused by fluid shock, Bouguer utilized for the first time that anyone ever had in Paris the inversion of differentiation and integration (now called "differentiation under the integral sign").[2] Bouguer belonged to a new generation of mathematicians and mathematical scientists in the Paris Academy of Sciences. Although his work in mathematics perhaps does not equal that of some of the other members of the group, whom we shall meet in Chapters 6 and 7,[3] it does nevertheless illustrate the progress made in the integral calculus in Paris in the 1730s, after three decades of failure of the integral calculus to advance in the hands of Paris Academy mathematicians. I shall discuss this failure and its causes as well as the subsequent improvement and the reasons for it in some detail in Chapter 7.[4]

It is only with Bouguer's paper of 1734 that we can truly begin to talk about figures of equilibrium. Why? Until Bouguer wrote his paper, people had simply assumed that in order to demonstrate that a homogeneous figure of revolution which revolved around its axis of symmetry was a figure of equilibrium, one simply had to show that the figure satisfied either one of two principles (the principle of balanced columns or the principle of the plumb line) and nothing more. In his paper of 1734, Bouguer investigated whether the principles of balanced columns and the plumb line were equivalent. For it seemed to Bouguer that people had also long tacitly assumed that the two principles required that essentially the same conditions be fulfilled, hence that the two principles were equivalent, and that consequently if a homogeneous figure of revolution which revolved around its axis of symmetry satisfied one principle, then it necessarily satisfied the other. In other words, people thought that both principles necessarily determined the same homogeneous figure of revolution which revolved around its axis of symmetry and that this homogeneous figure of revolution was a figure of equilibrium. Consequently people believed that it sufficed to check that such a homogeneous figure of revolution satisifed either one of the two principles in order to conclude that it was a figure of equilibrium. (In practice, Newton, Huygens, Hermann, and Maupertuis had all balanced columns in trying to find what they assumed were such figures.)

Bouguer discovered that the two principles in fact were *not* interrelated. One principle only applied to the surface of a homogeneous figure of revolution which revolved around its axis of symmetry; the other principle solely concerned the interior of such a figure. In carrying out his research, Bouguer examined a class of hypotheses of attraction more general than central forces of attraction.

While Bouguer's hypotheses of attraction generalized central forces of attraction, they differed considerably from Mairan's ad hoc law of attraction, which, we recall, in effect included inverse-square central forces of attraction as a special case. Unlike Mairan, Bouguer did not use evolutes of meridians at the surfaces of homogeneous figures of revolution to specify directions of attraction at points on the surfaces of such figures. For one thing, in Mairan's investigation homogeneous figures of revolution were given beforehand. But in Bouguer's theory homogeneous figures of revolution were not given at the start; they had to be found by solving problems. For another, the directions of attraction at points on the surfaces of Mairan's homogeneous figures of revolution were always perpendicular to the surfaces of the figures of revolution at these points. But in Bouguer's theory the directions of attraction at points on the surfaces of homogeneous figures of revolution were not necessarily perpendicular to the surfaces of the homogeneous figures of revolution at these points. This is because the homogeneous figures of revolution revolved around their axes of symmetry, and Bouguer sometimes assumed that the principle of the plumb line held at the surfaces of homogeneous figures of revolution when the homogeneous figures of revolution revolved around their axes of symmetry, in which case the directions of attraction at points on the surfaces of the homogeneous figures of revolution cannot be perpendicular to the surfaces of the homogeneous figures of revolution at these points.

Thus, for example, Bouguer did not use, as Mairan had done, the problem of orthogonal trajectories to characterize the directions of attraction when attraction is not a central force. Bouguer did not mistake, as Mairan had done, the directions of *attraction* at points on the surfaces of homogeneous nonrotating figures of revolution, which *are perpendicular* to the surfaces of the homogeneous figures of revolution at these points when the principle of the plumb line holds at the surfaces of such figures of revolution, for the directions of *effective gravity* at points on the surfaces of homogeneous figures of revolution which revolve around their axes of symmetry, which *are not perpendicular* to the surfaces of the homogeneous figures of revolution at these points when the directions of attraction at these points are perpendicular to the surfaces of the homogeneous figures of revolution at these points.

Assuming conditions of symmetry which assured that all homogeneous figures considered were figures of revolution, where the figures revolve around their axes of symmetry AB, Bouguer stipulated that if $AKBL$ is a meridian at the surface of a homogeneous figure of revolution, then the lines that indicate the directions of attraction at all points on the meridian $AKBL$ are perpendicular to a given oval curve $ADBE$ in the plane of the meridian $AKBL$, called a directrix, which is concentric with the meridian $AKBL$. The common center of the two closed curves $AKBL$ and $ADBE$ is the center C of the homogeneous figure of revolution.[5] More generally, Bouguer stipulated that each of these lines also specifies the direction of the attraction at each point on it. (To express the matter another way, the plane curves in the plane of the meridian $AKBL$ to which the lines in the plane of the meridian $AKBL$ which indicate the directions of attraction at points in the plane of the meridian $AKBL$ are tangent are themselves all straight lines that are perpendicular to the directrix $ADBE$.) For example, in Figure 12 line segments MG and BG in the plane of the meridian $AKBL$ are both perpendicular to the directrix $ADBE$. Hence the attraction at all points along the line segments MG and BG, respectively, must be directed toward G. In case the given oval curve $ADBE$ is a circle, attraction at all points in the plane of the meridian $AKBL$ is everywhere directed toward a single point, the center of the circle $ADBE$, which is located at the center C of the homogeneous figure of revolution. In the event that the directrix $ADBE$ is not a circle, the lines in the plane of the meridian $AKBL$ which specify the directions of attraction at points in the plane of the meridian $AKBL$ do not all meet at one point, the center C of the homogeneous figure of revolution. Bouguer also allowed the magnitude of attraction at points in the plane of the meridian $AKBL$ to vary, both as a function of the distance of a point situated on a particular line in the plane of the meridian $AKBL$ which indicates a particular direction of attraction in the plane of the meridian $AKBL$ from the point where that line intersects the axis of symmetry AB of the homogeneous figure of revolution, as well as a function that varies with the directions of these lines in the plane of the meridian $AKBL$. For example, even when the directrix

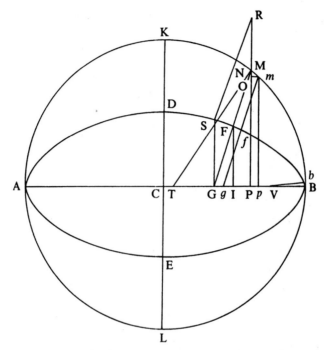

Figure 12. Meridian of a homogeneous figure of revolution and the corresponding directrix (Bouguer 1734: Figure 1).

$ADBE$ was a circle, Bouguer considered cases where the magnitude of attraction increased (or decreased) toward the polar axis AB or toward an equatorial axis KL of the homogeneous figure of revolution at points in the plane of the meridian $AKBL$ a constant distance from the center C of the homogeneous figure of revolution.

What induced Bouguer to look more deeply into the principles of balanced columns and the plumb line and to question their equivalence? Mairan's Paris Academy *mémoire* of 1720 could have had an influence. In that *mémoire* Mairan had refused to accept the idea that two sets of terrestrial observations (variations in degrees of latitude with latitude and variations of the magnitude of the effective gravity with latitude measured using seconds pendulums) necessarily contradicted each other. Perhaps Mairan's *mémoire*, despite its errors, suggested to Bouguer that one should not expect that the principles of the plumb line and balanced columns necessarily amount to the same thing either. In order to show that the two principles are, in fact, not equivalent, Bouguer had to abandon the classic hypotheses of central forces of attraction, as we shall see in a moment.

Other considerations make a connection between Mairan's Paris Academy *mémoire* of 1720 and Bouguer's Paris Academy *mémoire* of 1734 even more plausible. In a letter he wrote to Johann I Bernoulli in 1731, a few days after Bouguer had entered the Paris Academy, Maupertuis explained to the mathematician from Basel that Mairan had been Bouguer's original and chief backer.[6] Bouguer and his patron had already mutually influenced each other on several occasions during the 1720s. In 1721 Mairan took into consideration Bouguer's discussions of the problem of measuring the tonnage or capacity of ships in his own work on that subject. In 1721 Mairan also read a *mémoire* to the Paris Academy in which he propounded the problem of determining the intensity that sunlight loses in traversing the atmosphere, as well as suggested a way of solving the problem. As a result of Mairan's paper Bouguer became interested in problems of measuring light, which eventually led him to write a highly original treatise on the gradation of light (that is, light's different intensities) in 1729. In 1724 a treatise by Bouguer dealing with the problem of putting masts on ships affected the way that Mairan handled the same problem. Bouguer's treatise so thoroughly impressed Mairan that in 1725 Mairan proposed the best way of putting masts on ships as the subject of the Paris Academy's contest for 1727. Bouguer submitted his treatise and won the prize.[7] Mairan was one of the judges of the contest for 1727. Bouguer went on to win two more Paris Academy prizes, in 1729 and 1731. Mairan was one of the judges of both of those contests too.

In his paper of 1734 on figures of equilibrium, Bouguer first looked for homogeneous figures of revolution which accorded with his generalization of Newton's principle of balanced columns: if G is any point on the axis of symmetry AB of a homogeneous figure of revolution which revolves around its axis of symmetry and if M is any point on a meridian $AKBL$ at the surface of the homogeneous figure of revolution, then the column from M to G and the column from B to G which lies along the axis of symmetry AB of the homogeneous figure of revolution must balance or weigh the same. This means that any two columns which connect two points on a meridian $AKBL$ at the surface of the homogeneous figure of revolution which revolves around its axis of symmetry AB to the same point G on the axis of symmetry must balance or weigh the same. As a result of the symmetry of the homogeneous figure of revolution, the fact that the axis of symmetry AB of the homogeneous figure of revolution is also an axis of symmetry of Bouguer's hypotheses of attraction, and the fact that the homogeneous figure of revolution revolves around its axis of symmetry AB and not around some other line, so that the axis of symmetry AB of the homogeneous figure of revolution is also an axis of symmetry of the centrifugal force per unit of mass, mean that *any* two columns that join two points on the surface of the homogeneous figure of revolution to the same point G on the axis of symmetry AB of the homogeneous figure of revolution must balance or weigh the same, for essentially the same reasons that I explained in Chapter 1 concerning Newton's figures. In other words, in this case *all* columns from a point G on the axis of symmetry AB of the homogeneous figure of revolution (the polar axis) to the surface of the figure must balance or weigh the same. Bouguer called the weight of a column whose length is infinitesimal the "effort" of effective gravity along the column. The weight of a column whose length is finite was then the sum of the "efforts" along the columns of infinitesimal length which make up the column whose length is finite.[8] Bouguer assumed the density of the homogeneous figure of revolution to equal 1. Taking into account centrifugal force of rotation, which has no effect at points along the axis of symmetry AB of the homogeneous figure of revolution, Bouguer equated the weights, expressed as integrals, of the column from M to G and the column from B to G. (Both columns happen to be perpendicular to the directrix $ADBE$ in the plane of the meridian $AKBL$, which means that the attraction at each point on one of the columns is directed along that column.) By this means he obtained an equation. (The equality of the weights of the column from M to G and the column from B to G becomes an equation by equating the weights of columns from M to G and from B to G for all points M on the meridian $AKBL$ at the surface of the homogeneous figure of revolution. Doing so requires use of the assumption that all columns from points M to G balance or weigh the same.) To construct his equation Bouguer utilized various curvilinear coordinates defined in terms of the lines in the plane of a meridian $AKBL$ at the surface of the homogeneous figure of revolution which specify the directions of attraction at points in the plane of the meridian $AKBL$, the directrix $ADBE$ that determines these lines, the segments of the axis of symmetry AB of the homogeneous figure of revolution which these lines intercept, the lines in the plane of the meridian $AKBL$ which indicate the directions of centrifugal force per unit of mass at points in the plane of the meridian $AKBL$, and the lines in the plane of the meridian $AKBL$ which specify the directions of effective gravity at points in the plane of the meridian $AKBL$. In short, he let the geometry of the problem as he conceived the problem fix the coordinates that he employed. At the time this was still consider-ed to be the most natural manner to attempt to treat problems in mechanics. Bouguer also assumed the variation of the magnitude of attraction with direction at points in the plane of the meridian $AKBL$ to depend upon his chosen variables in a way that led to an equation for the principle of balanced columns whose variables were separated. (Specifically, Bouguer's expression for the magnitude of attraction has the form of a product of a function ξ that depends only on the direction of a line in the plane of the meridian $AKBL$ which specifies a direction of attraction in the plane of the meridian $AKBL$ (more precisely, a function of the sine of the angle that such a line makes with the axis of symmetry AB of the homogeneous figure of revolution) and a function p that depends only on the distance of points situated on such a line from the point where the line intercepts the axis of symmetry AB of the homogeneous figure of revolution. He did this because such an equation could then be solved. Once the equation was solved, a meridian $AKBL$ at the surface of a homogeneous figure of revolution whose density equals 1, which revolves around its axis of symmetry AB, and which is consistent with the generalized principle of balanced columns could be construc-

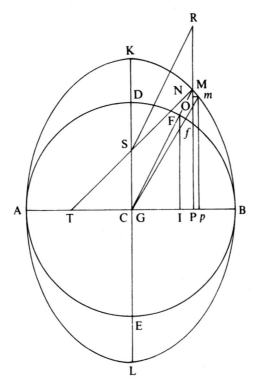

Figure 13. Application of the principle of the plumb line along a meridian of a homogeneous figure of revolution (Bouguer 1734: Figure 2).

ted using the solution of the equation, after certain variables were eliminated by expressing them in terms of the other variables.

Bouguer next sought homogeneous figures of revolution at whose surfaces the principle of the plumb line holds. For the laws of attraction described here earlier, Bouguer succeeded in reducing the application of the principle of the plumb line at the surface of a homogeneous figure of revolution which revolves around its axis of symmetry to a problem in solving ordinary "differential" equations. (A "differential" equation is an equation in which differentials appear.) This was a notable advance for the time in the use of the integral calculus.

To summarize Bouguer's arguments: Let M be a point on a meridian $AKBL$ at the surface of a homogeneous figure of revolution (see Figure 13). Bouguer again assumed that the density of the homogeneous figure of revolution equals 1. Let the direction of attraction at M lie in the direction of the line segment MG, where line segment MG is perpendicular to the directrix $ADBE$ in the plane of the meridian $AKBL$; let the direction of centrifugal force per unit of mass at M lie in the direction of the line segment MR in the plane of the meridian $AKBL$; and let the direction of the resultant of effective gravity at M lie in the direction of the line segment MT in the plane of the meridian $AKBL$. Suppose that $|MG|$ designates the magnitude of attraction at M and that $|MR|$ stands for the magnitude of centrifugal force per unit of mass at M. Then the resultant of the effective gravity at M is directed toward S, and $|MS|$ is the magnitude of the resultant of the effective gravity at M, where line segment MS is the diagonal of the parallelogram $RMGS$ in the plane of the meridian $AKBL$. Then, since $|MR| = |GS|$, equality (4.1) follows.

$$\frac{\text{Magnitude of Attraction at } M}{\text{Magnitude of Centrifugal Force per Unit of Mass at } M}$$

$$= \frac{|MG|}{|MR|} = \frac{|MG|}{|GS|}. \tag{4.1}$$

By means of equality (4.1), Bouguer represented geometrically the ratio of the magnitude of the attraction at M to the magnitude of centrifugal force per unit of mass at M.[9] Now, suppose that line segment MS is perpendicular to the surface of the homogeneous figure of revolution at M (that is, that the principle of the plumb line holds at point M on the meridian $AKBL$ at the surface of the homogeneous figure of revolution). Bouguer took m to be a point on the meridian $AKBL$ at the surface of the homogeneous figure of revolution which is infinitesimally near M, and, by the similarity of finite and infinitesimal triangles, which followed from having assumed MS to be orthogonal to the surface of the figure of revolution at M, he concluded that

$$\frac{|MG|}{|GS|} = \frac{|MO|}{|MN|}, \tag{4.2}$$

where O and N are determined in the following way. Line segments mO and mN in the plane of the meridian $AKBL$ are perpendicular to line segments MP and MG in the plane of the meridian $AKBL$, respectively.[10] Now, point M is a point chosen arbitrarily on the meridian $AKBL$ at the surface of the homogeneous figure of revolution. Bouguer transformed geometrical relation (4.2) into a first-order ordinary "differential" equation of a meridian $AKBL$ at the surface of the homogeneous figure of revolution.[11] Employing the same variables that he had used earlier in balancing columns and assuming the variation of the magnitude of attraction with direction at points in the plane of the meridian $AKBL$ to depend on these variables in the same way that he had supposed before, Bouguer arrived at a first-order ordinary "differential" equation of first degree whose variables were separated, hence an equation that again could be solved for a meridian $AKBL$ at the surface of a homogeneous figure of revolution whose density equals 1, which revolves around its axis of symmetry AB, and which accords with the principle of the plumb line. As Bouguer expressed the matter, he had reduced the problem of applying the principle of the plumb line at the surface of a homogeneous figure of revolution to the "inverse method of tangents," in the language of the time[12] – which is to say, to integrating in finite terms first-order ordinary "differential" equations of first degree or else to reducing such equations to quadratures. (In Chapter 7 I shall discuss the separation of variables in first-order ordinary "differential" equations of first degree.)

Bouguer next consolidated his two equations, which he had found by applying the principles of balanced columns and the plumb line separately to a homogeneous figure of revolution whose density equals 1 and which revolves around its axis of symmetry AB, in order to determine requirements for both principles to hold simultaneously. When homogeneous figures of revolution whose densities equal 1 and whose shapes are determined individually by the two equations in question revolve around their axes of symmetry at the same rate (that is, at the same angular speed), the terms involving magnitudes of centrifugal forces per unit of mass which appear in the second equation and in the equation that results when the first equation is differentiated are the same. Hence these terms can be eliminated from these two equations. The equation that results after these terms are eliminated from these two equations is an equation that expresses conditions on the law of attraction which must be fulfilled in order that the two principles hold at the same time for a homogeneous figure of revolution whose density equals 1 and which revolves around its axis of symmetry AB.

I briefly summarize Bouguer's principal conclusions. Bouguer found that if the attraction was taken to be "absolutely constant," by which he meant that the magnitude of the attraction was constant at points along the particular line in the plane of the meridian $AKBL$ which specifies any particular direction of the attraction in the plane of the meridian $AKBL$, as well as independent of the directions of these lines in the plane of the meridian $AKBL$, then "the curve $ADBE$, which determines the directions of primitive weight [that is, the attraction], can only be a circle, as in Fig. 2 [Figure 13], and that all of these directions · meet in one and the same point, which will be the center of the circle."[13] Bouguer added: "It is thus shown that as soon as primitive weight [in other words, the attraction] is entirely constant, there must be only one, unique point of tendency, or central point, in order that either one of our two principles necessarily entails that the other principle holds as well."[14]

In fact, Bouguer's analysis was not totally accurate. Later his conclusion would be shown to be faulty. Namely, it turns out that when the magnitude of the attraction is constant at points along the particular line in the plane of a meridian $AKBL$ at the surface of a homogeneous figure of revolution which indicates any particular direction of the attraction in the plane of the meridian $AKBL$ as well as independent of the directions of these lines in the plane of the meridian $AKBL$, then the lines in the plane of the meridian $AKBL$ which specify the directions of the attraction at points in the plane of the meridian $AKBL$ need *not* all meet at one point, the center C of the homogeneous figure of revolution, in order that both the principles of balanced columns and the plumb line hold at the same time. Nevertheless, as we shall see in Chapter 9, it was Bouguer's work specifically which helped lead Alexis-Claude Clairaut to the right result later. That is, Clairaut found the correct result in the course of rectifying what he perceived to be Bouguer's error.

When the lines in the plane of the meridian $AKBL$ which indicate the directions of the attraction at points in the plane of the meridian $AKBL$ do not all meet at one point, the center C of the homogeneous figure of revolution (in other words, when the directrix $ADBE$ in the plane of the meridian $AKBL$ is not a circle), Bouguer found that, in general, the magnitude of the attraction at points in the plane of the meridian $AKBL$ had to vary with the directions of the lines in the plane of the meridian $AKBL$ which specify the directions of the attraction at points in the plane of the meridian $AKBL$, in order that both the principles of balanced columns and the plumb line hold simultaneously. He investigated cases where the magnitude of the attraction at points along the particular line in the plane of a meridian $AKBL$ at the surface of a homogeneous figure of revolution which indicates any particular direction of the attraction in the plane of the meridian $AKBL$ varies directly as y^m, where y stands for the distance of a point on the line from the point where the line and the axis of symmetry AB of the homogeneous figure of revolution intersect (in short, varies with distance the same way that the magnitude of attraction at points varies with the distances of points from a center of force in the case of r^m central forces, where r designates the distance of a point from a center of force). He determined that the only time that the magnitude of attraction must not vary with the directions of the lines in the plane of the meridian $AKBL$ which specify the directions of the attraction at points in the plane of the meridian $AKBL$ when the directrix $ADBE$ in the plane of the meridian $AKBL$ is not a circle and the principles of balanced columns and the plumb line both hold at the same time is when $m = -1$ – that is, when the magnitude of the attraction varies inversely as the distance y of a point on the particular line in the plane of a meridian $AKBL$ at the surface of the figure of revolution which specifies any particular direction of attraction in the plane of the meridian $KBLA$ from the point where the line and the axis of symmetry AB of the homogeneous figure of revolution intersect.

Bouguer then turned to hypotheses of central forces – that is, the lines in the plane of a meridian $AKBL$ at the surface of a homogeneous figure of revolution which indicate the directions of the attraction at points in the plane of the meridian $AKBL$ all meet at one point, the center C of the homogeneous figure of revolution. In order that the principles of balanced columns and the plumb line both hold at the same time for such hypotheses, he found that, in general, just the opposite of the situation just described had to be true. That is, the magnitude of attraction had to vary with the directions of the lines in the plane of a meridian $AKBL$ at the surface of a homogeneous figure of revolution which specify the directions of attraction at points in the plane of the meridian $AKBL$ in general, in

order that the principles of balanced columns and the plumb line both hold
simultaneously in cases where the lines in the plane of the meridian $AKBL$ which
indicate the directions of attraction at points in the plane of the meridian $AKBL$
do *not* all meet at one point, the center C of the homogeneous figure of revolution.
But the magnitude of attraction cannot vary with those lines' directions, if the two
principles are to hold at the same time when the lines in the plane of the meridian
$AKBL$ which indicate the directions of attraction at points in the plane of the
meridian $AKBL$ *do* all meet at one point, the center C of the homogeneous figure
of revolution.

As luck would have it, Bouguer suggested, Huygens and Hermann – as well as
Maupertuis, in the *Discours* of 1732 – had assumed that central forces of attraction act whose magnitudes at points in any plane that includes the center of force
did not depend upon the angles that the lines in that plane which specify the
directions of attraction at points in that plane, which necessarily pass through the
center of force, make with one of these lines chosen as an axis of reference,
independently of the particular plane that includes the center of force, hence
whose magnitudes at points did not vary with direction, but depended solely on
the distances of points from the center of force. Otherwise, had Huygens,
Hermann, and Maupertuis hypothesized that central forces of attraction act
whose magnitudes at points depended upon such angles instead, the principles of
balanced columns and the plumb line applied separately would have resulted in
different figures of revolution, contrary to a supposition tacitly made that the two
principles always lead to the same figure of revolution. The figures of revolution
found in each case would not have been figures of equilibrium, which was again
contrary to implicit assumption.

Once again, however, Bouguer did not perceive certain subtleties, which
caused him to err in using his analysis to come to his conclusions. Later Clairaut
would show that the principle of balanced columns *can* be made to accord with
the principle of the plumb line for a homogeneous figure of revolution attracted
by a force that is directed at all points toward the same fixed center of force
located at the center of the homogeneous figure of revolution but whose
magnitudes at points in a plane that includes the center of force depend upon the
angles just mentioned as well. Instead Clairaut found that the failure of *other*,
independent principles of equilibrium to hold in the presence of such forces rules
out figures of equilibrium. Nevertheless, just as in the preceding problem, it was
Bouguer's work that served as the catalyst for Clairaut's later work, and
Bouguer's mistakes once again ultimately helped Clairaut discover the correct
result. This, too, we shall find out more about in Chapter 9.

Bouguer next determined the homogeneous figures of revolution which each of
the principles of the plumb line and balanced columns applied separately led to
when the two homogeneous figures of revolution revolved around their axes of
symmetry AB at the same rate. (That is, Bouguer solved the problem when the
two homogeneous figures of revolution revolve around their axes of symmetry

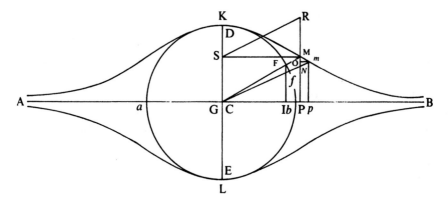

Figure 14. Meridian of a homogeneous conchoid of revolution (Bouguer 1734: Figure 3).

AB (their polar axes) with the same angular speed, which means that the magnitude of the centrifugal force per unit of mass at the points which make up the two homogeneous figures of revolution varies directly as the distances of the points from the axes of symmetry AB of the homogeneous figures of revolution, so that, in particular, the magnitudes of the centrifugal force per unit of mass at two such points each of which revolves around one of the two axes of symmetry AB of the two homogeneous figures of revolution at the same distances from these axes have the same value.) In each of the two cases attraction was assumed to be directed at all points toward the center C of the homogeneous figure of revolution, and the magnitude of attraction at a point in the plane of a meridian AKBL at the surface of the homogeneous figure of revolution was assumed to vary directly as the product of a power m of the distance of the point from the center C of the homogeneous figure of revolution and a power n of the sine of the angle that the particular line in the plane of the meridian AKBL which indicates the particular direction of attraction at the point in the plane of the meridian AKBL and which passes through the center C of the homogeneous figure of revolution makes with the axis of symmetry AB of the homogeneous figure of revolution, which in fact is itself just one of the lines in the plane of the meridian AKBL which specify the directions of attraction at points in the plane of the meridian AKBL and which pass through the center C of the homogeneous figure of revolution. He wanted to find out how different the two homogeneous figures of revolution could become.

   Thus on the one hand he showed that if the principle of balanced columns were applied, a homogeneous figure (Figure 14) shaped like an "infinitely long spheroid" of revolution, generated by revolving the conchoid AKB around its asymptote AB, where the circle aKbL whose center is at C is the directrix in the

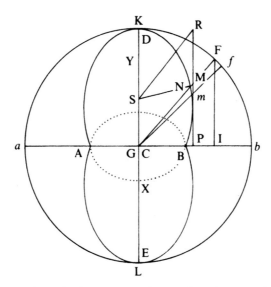

Figure 15. Meridian of a homogeneous figure of revolution made up of two identical ellipsoids of revolution which intersect (Bouguer 1734: Figure 4).

plane of the meridian *AKBL* at the surface of the homogeneous figure of revolution, resulted in case both exponents *m* and *n* equal 1.[15]

On the other hand, he demonstrated that the principle of the plumb line determined a "finite" homogeneous figure of revolution (Figure 15), generated by revolving around the axis *AB* the closed plane curve *AKBL* formed by unifying two identical ellipses *AKB* and *ALB* which lie in the same plane and which intersect, where the center *C* of the closed plane curve formed this way is situated at a common focus of each ellipse and where the circle *aKbL* whose center is at *C* is the directrix in the plane of the meridian *AKBL* at the surface of the homogeneous figure of revolution, when the conditions were the same as those in the preceding example (that is, both exponents *m* and *n* equal 1).[16]

Bouguer noted that if for each of the two homogeneous figures of revolution the magnitude *g* of attraction at a point on a line in the plane of the meridian *AKBL* which indicates a direction of attraction in the plane of the meridian *AKBL*, whose distance from the point *G* where the line intersects the axis of symmetry *AB* of the homogeneous figure of revolution equals *a*, is made to become increasingly larger than the magnitude *f* of the centrifugal force per unit of mass at points whose distance from *AB* equals *a* (in other words, if the ratio *f/g* is made to approach zero), the lengths of the two axes *KL* and *AB* in the second example approach each other and almost become equal, while the length of the axis *AB* remains infinitely long compared with the axis *KL* in the first example. In short, Bouguer showed that when this particular hypothesis of attraction was assumed to hold, the two principles applied separately led to homogeneous figures of revolution which revolved around their axes of symmetry *AB* at the same rate whose shapes were extremely different. (Today we would say that Bouguer showed in essence that while the principles of balanced columns and the plumb line are both necessary conditions that must be fulfilled in order that homogeneous figures be figures of equilibrium, neither principle is a sufficient condition for homogeneous figures of equilibrium to exist.)

Bouguer then tried to describe what he thought would happen if such hypotheses underlie the true law that governs attraction. He had shown that a homogeneous fluid figure of revolution which attracts according to such a law could not be a figure of equilibrium, because both the principles of balanced columns and the plumb line must hold simultaneously in order that a homogeneous fluid figure of revolution be a figure of equilibrium, which the two principles cannot do if the homogeneous fluid figure of revolution attracts in accordance with such a law. But Bouguer also believed that such a law of attraction would not permit either principle alone to hold true either. That is, he maintained that each principle would interfere with a permanent formation of a homogeneous fluid figure of revolution which accorded with the other principle, so that

neither of the two laws [the principles of balanced columns and the plumb line] could be observed to hold, because each would continually prevent the other from holding, while each, like a mechanical or physical cause, would constantly try to rule alone. Therefore the planet cannot have any determinate shape, but instead it alternately takes different shapes, Figures 3 and 4 [Figures 14 and 15, the two examples described immediately above] representing extremes, where all of the fluid parts do not simply agitate but exist in a state of perpetual disruption.[17]

Here Bouguer clearly struggled to differentiate a state of disequilibrium from a mere oscillation or fluctuation of homogeneous fluid inside and outside the bounding surface of some well-determined homogeneous figure of equilibrium. (He did not employ the term "stability" to try to differentiate an oscillating mass of homogeneous fluid which resembles or approximates a mass of homogeneous fluid in equilibrium which has a definite shape from the phenomenon that he endeavored to describe here. Concepts like stable equilibrium, unstable equilibrium, disequilibrium, and so forth were not yet clearly understood and the various terms for these concepts had not yet been coined. I shall return to this problem in Chapter 9. In fact, clarification in a precise way of the notion of stability, which involved among other things defining various degrees of stability, did not really take place until the nineteenth century.)

Bouguer remarked that in case the earth's attraction were of this kind, the perpetual disruption that the attraction would cause would, in principle, have an effect upon the waters at the earth's surface and upon the atmosphere, as well as upon fluids in capillary tubes. He realized, however, that it would take a considerable variation of the magnitude of attraction with the directions of the lines in the plane of a meridian at the earth's surface which pass through a hypothetical center of force situated at the earth's center and which specify the directions of attraction at points along the meridian to produce effects that could be observed at the earth's surface. Still, he doubted that the magnitude of a force of attraction whose lines that indicate the directions of attraction at points in the plane of a meridian at the earth's surface all meet at the earth's center could be exactly the same toward the poles as toward the equator at points along the meridian, if the force of attraction were produced by the centrifugal forces of vortices. Since the slightest variation in magnitude of a central force of attraction with direction at points in the plane of a meridian at the surface of a homogeneous figure of revolution would suffice to cause the principles of balanced columns and the plumb line to counteract each other and consequently produce disequilib-rium of the homogeneous figure of revolution, Bouguer thought that an observed presence or absence of equilibrium of fluids at the earth's surface – that is, the success or failure of fluids at the earth's surface to find their own levels – could be used to check such hypotheses of the earth's attraction. (Here, of course, the earth must be assumed to be a homogeneous figure of revolution, and the waters at the earth's surface must be assumed to have the same density as the earth.) Thus did Bouguer imagine how vortex theories of the earth's attraction could be tested, namely, by observing the condition of fluids at the earth's surface (the success or failure of the principle of the plumb line to hold at the earth's surface). Bouguer did not believe that laws of attraction deduced from vortex theories could permit fluids at the earth's surface to remain in a state of equilibrium. Once again, he isolated the principle of the plumb line, this time as a means of testing vortex theories of attraction, at the surface of the solid earth.

Such a "test," however, involves a kind of fallacy, which Bouguer overlooked or did not notice. At least he did not call the reader's attention to it. A perpetual disruption of fluids at the earth's surface caused by central forces of attraction whose magnitudes also depend upon the angles that the lines in a plane of a meridian at the earth's surface which pass through a hypothetical center of force situated at the earth's center and which specify the directions of attraction at points in the plane of that meridian make with one of these lines chosen as an axis of reference follows, in Bouguer's argument, as a consequence of assuming that *the whole earth together with the waters at the earth's surface make up a homogeneous fluid figure of revolution*. In other words, in order to apply Bouguer's test, the earth cannot itself be assumed to be a solid mass, and the fluids at the earth's surface must be assumed to have the same density as the homogeneous fluid earth. The trouble is that as Bouguer himself showed, the principle of the plumb line *alone* can hold at the surfaces of *certain homogeneous figures of revolution* which revolve around their axes of symmetry, when central forces of attraction act whose magnitudes at points in the planes of meridians at the surfaces of these homogeneous figures of revolution depend upon the directions of the lines in the planes of these meridians which specify the directions of attraction at points in the planes of these meridians, where the lines pass through hypothetical centers of force situated at the centers of these homogeneous figures of revolution.

In fact, Bouguer's paper contains all of the basics needed to resolve an apparent flaw in his test. For as we have seen, Bouguer was, in effect, also the first person to determine a homogeneous figure of revolution which revolves around its axis of symmetry, which is not a figure of equilibrium, but at whose outer surface the principle of the plump line holds. Since he doubted that the magnitude of the force of attraction resulting from hypothesized vortices could be independent of the angles that the lines in a plane of a meridian $AKBL$ at the surface of a homogeneous figure of revolution which indicate the directions of attraction at points in the plane of the meridian $AKBL$ make with one of these lines chosen as an axis of reference, even if the lines that specify the directions of attraction in the plane of the meridian $AKBL$ all meet at a single point, namely, the center of the homogeneous figure of revolution, the principle of the plumb line would probably not hold at the earth's surface, in case such hypotheses did truly underlie the law that governs the earth's attraction. For in the one case of this kind which he did solve explicitly, the homogeneous figure of revolution at whose outer surface the principle of the plumb line held while the principle of balanced columns failed to hold in its interior turned out to be one formed by unifying two homogeneous bodies that intersect and that are shaped like identical ellipsoids of revolution. The shape of this figure was one that differed too much from any shape that the earth could reasonably be expected to have. Hence if the principle of the plumb line were observed to hold at the earth's surface, that would disprove such hypotheses of attraction.

It is true that Bouguer only developed his ideas, worked out the problems that these involved, and came to conclusions for particular cases of hypotheses of attraction which under *no* circumstances would allow homogeneous figures of equilibrium to exist, namely, central forces of attraction whose magnitudes at points not only depend on the distances of points from the center of force but also depend on the angles that the lines in a plane which meet at the center of force and which specify the directions of attraction at points in the plane make with one of these lines chosen as an axis of reference. He did not, for example, try to find homogeneous figures of revolution which revolve around their axes of symmetry and at whose outer surfaces the principle of the plumb line holds but the principle of balanced columns does not hold in their interiors, in cases where the homogeneous figures of revolution attract in accordance with a law that *would* nevertheless permit both principles to hold simultaneously for *other* homogeneous figures of revolution *shaped differently*, which attract according to the same law, and which revolve around their axes of symmetry. (As I mentioned above, Clairaut did discover later, thanks to mistakes that Bouguer made, that the two principles can hold at the same time for a homogeneous figure of revolution, yet the figure fails to be a figure of equilibrium. In other words, as we would express the matter today, Clairaut eventually found that the principles of balanced

columns and the plumb line together are not a sufficient condition for homogeneous figures of equilibrium to exist.) Nor would such figures have illustrated the
particular point that Bouguer was trying to make in his paper, so it is not
surprising that he did not consider such figures that attract according to such
laws. (Moreover, as we shall see in Chapter 9, not only do laws of attraction
determined by fixed centers of force, where the magnitude of attraction at points
depends only upon the distances of the points from the fixed centers of force,
permit both principles to hold at the same time for certain homogeneous figures
of revolution when the centers of forces are situated at the centers of the figures of
revolution and when the figures of revolution revolve around their axes of
symmetry, as Bouguer had shown, but if *one* of the two principles holds for a
homogeneous figure of revolution which revolves around its axis of symmetry,
which is the only kind of rotating figure that Bouguer investigated, then if the
homogeneous figure of revolution is attracted by such a central force, it turns out
that the *other* principle must also hold as a result. In other words, *either both*
principles hold or else *neither* one holds for such homogeneous figures of
revolution, when the figures of revolution are attracted by such central forces.)

In any case, Bouguer was still the first person to find a homogeneous figure of
revolution which revolves around its axis of symmetry and which attracts in a
way that allows the principle of the plumb line to hold at its surface yet the figure
is not a figure of equilibrium. In so doing, he isolated the principle of the plumb
line in a manner that helped to shed a whole new light on that principle. Later this
particular elucidation of the principle would lead Clairaut to put the principle to
use in other ways than simply as a condition that a homogeneous figure of
revolution which revolves around its axis of symmetry must necessarily satisfy in
order that the figure be a figure of equilibrium. For the principle can be applied
just as well at the surfaces of stratified figures of revolution which revolve around
their axes of symmetry, which happen to be solid, and which are not necessarily
figures of equilibrium, but which attract in accordance with laws that would
nevertheless permit homogeneous figures of revolution which revolve around
their axes of symmetry to exist which allow both the principles of balanced
columns and the plumb line to hold at the same time. In Chapter 9 we shall find
that Clairaut utilized the principle in this context and learn why he did so. (Here it
is important to note that Mairan did *not* use the principle of balanced columns to
do any such thing.)

In his paper, which was published in the Paris Academy's *Mémoires* for 1734,
Bouguer did not deal with the following question at all: Do homogeneous figures
of revolution which revolve around their axes of symmetry and which attract
according to the universal inverse-square law exist which would permit both the
principles of balanced columns and the plumb line to hold simultaneously? In his
paper Bouguer did not study what happens when homogeneous figures which
revolve around their axes of symmetry attract in accordance with this law. In
Chapter 6 we shall find out that homogeneous figures of revolution which attract

according to the universal inverse-square law *do* exist which allow both the principles of balanced columns and the plumb line to hold at the same time.

But Bouguer's investigation of non-Newtonian hypotheses of attraction led Bouguer to make some fundamental discoveries in connection with homogeneous figures of equilibrium. Whether he had actually converted to Newtonianism by 1734 or not remains a question open to debate. The early 1730s seem most likely to have been a period of transition for him during which he changed over from Cartesianism to Newtonianism.[18] His initial sponsor and ally Mairan remained a Malebranchist, which meant, among other things, an eclectic who tried to reconcile the Newtonian world system with the Cartesian world system. We shall discover in the next chapter that Bouguer did not belong to Maupertuis's band of junior Paris Academicians who campaigned openly on behalf of Newtonianism during the 1730s. If Bouguer continued to lean toward Cartesian physics in 1734, or at least had not completely abandoned it by that time, his work of that year would demonstrate, in this event, that after the *Principia* was published, Newtonians alone did not advance science. Newton's followers did not acquire a monopoly in the domain of scientific progress. A Cartesian could still have insight, as well as original ideas that would prove in the long run to be more fruitful than those of Newton's many unknowing apes, zealots, and devotees.

5

Maupertuis: On the theory of the earth's shape (1734)

5.1. Harassing Bouguer

Bouguer's paper on the problem of homogeneous figures of equilibrium caused Maupertuis to write and read before the Paris Academy a paper of his own a few months later on the same problem.[1] Maupertuis, we recall, had ceased to investigate theoretically the problem of the earth's shape in order to focus upon geodesy. In this chapter I look into the question why Maupertuis momentarily reconsidered the theoretical problem. Maupertuis had a twofold purpose: to deal with the theoretical problem, but also to take Bouguer to task. This was not the first time that Maupertuis reacted to Bouguer's treatment of a mathematical or scientific problem. I shall say more about that in a moment.

Another incident in 1734 may have also helped induce Maupertuis to return for a short time to the theoretical problem of the earth's shape from geodesy. In May of 1734 the Paris Academy announced that Johann I Bernoulli and his son Daniel shared the Paris Academy prize for that year. The subject of the contest had been the causes of the inclinations of the planetary orbits to the plane of the ecliptic. In the essay that he submitted to the Academy for the prize the older Bernoulli applied principles of vortex theory to the problem. According to Johann I Bernoulli's vortex theory of the inclinations of the planetary orbits to the plane of the ecliptic, the earth should be elongated at the poles. For the *first* time, someone claimed to establish a connection between Cartesian physics and an earth whose shape is elongated. From September to November of 1734, Maupertuis and his young associate Alexis-Claude Clairaut visited Johann I Bernoulli in Basel, probably in order to discuss with him the problem of the earth's shape. Bernoulli's prize-winning essay likely induced Maupertuis and Clairaut to make the trip.[2] Less than a month before leaving Paris for Basel, Maupertuis read his paper to the Paris Academy on the problem of homogeneous figures of equilibrium.

At the beginning of the version of his paper published in the Paris Academy's *Mémoires* for 1734, Maupertuis recapitulated in five pages the version of Bouguer's nineteen-page paper printed in the same volume. After stating that he would investigate Bouguer's problem "in a way that differs from Mr. Bouguer's,"[3]

Maupertuis did little more than fiddle inconsequentially with that paper's essentials, merely abridging Bouguer's discussion. He redid Bouguer's calculations, adding nothing substantively new. He did express in different coordinates Bouguer's two equations for the principles of balanced columns and the plumb line. Maupertuis discarded altogether Bouguer's oval curve $ADBE$, which Bouguer had used in his paper to determine the directions of attraction at points in a plane of a meridian $AKBL$ at the surface of a homogeneous figure of revolution and which Bouguer had also used to fix some of the variables that he utilized to solve the problems that he treated in his paper. Instead, Maupertuis expressed in rectangular coordinates Bouguer's condition for the principle of the plumb line to hold. In fact, such coordinates were not commonly employed at this time to solve problems in mechanics, and Maupertuis's application of them here may indeed be one of the earliest uses of them in this context.[4] Maupertuis rederived all of Bouguer's principal results, including the ones that would later be shown to be incorrect. He concluded the introductory section of his paper by remarking that he judged Bouguer's hypotheses of attraction expressed in terms of lines that specify the directions of attraction at points in the plane of a meridian at the surface of a homogeneous figure of revolution and which do not all meet at the same point, the center of the homogeneous figure of revolution, but intersect the axis of symmetry of the homogeneous figure of revolution at different points instead, where the magnitude of attraction at points on any one of the lines depends upon the direction of the line as well as upon the distances of the points from the point where the line intersects the axis of symmetry, to require the addition of hypotheses that he considered to be unnatural "geometric fictions."[5]

Maupertuis clearly disliked Bouguer. Although Maupertuis entered the Paris Academy before Bouguer did, Bouguer was more than seven months older than Maupertuis (Bouguer was born on 10 February 1698; Maupertuis on 28 September 1698). Bouguer was a former child prodigy from Le Croisic, in Maupertuis's native Brittany. As I mentioned in Chapter 4, Bouguer won three Paris Academy prizes before joining the Paris Academy in 1731. Bouguer's having received so much attention from the Academy and having been commended so often by the members of that society may have annoyed his vain, jealous fellow Breton from St. Malo.[6] Perhaps Maupertuis thought that Bouguer had been pampered too much and was spoiled. Perhaps the inferior mathematician simply envied the precociousness that Bouguer had displayed as a youth. Bouguer's patron Mairan had himself won three Bordeaux Academy prizes before entering the Paris Academy. In an unprecedented action, Mairan joined the Paris Academy directly as associate geometer in December 1718. Thus he bypassed the assistant membership that newly elected members of the Academy were customarily designated. Maupertuis was not as lucky; he entered the Academy as assistant geometer in December 1723. Bouguer duplicated Mairan's triumph. Bouguer was also admitted to the Paris Academy as associate geometer in September 1731, replacing, of all people, Maupertuis, who had been promoted from associate geometer to pensioner geometer. And, as I mentioned in Chapter 4, Maupertuis explained to

Johann I Bernoulli, on the occasion of Bouguer's election to the Paris Academy, that the new Academician was Mairan's man. Seeing Bouguer take his place as associate member of the Academy must have irritated Maupertuis in his moment of glory and reduced the joy of his own advancement.

(Mairan and Bouguer may have entered the Paris Academy as associate members because of their ages. Mairan was forty years old when he joined the Academy, and Bouguer was thirty-three years old when he entered. Hence both were well past the minimum age of twenty years required for membership when they were admitted to the Academy. Maupertuis was twenty-five years old when he joined the Academy. Mairan's and Bouguer's ages should be compared with Maupertuis's in trying to determine the Paris Academy's policies concerning the standing of newly admitted members.[7] Furthermore, Bouguer's reputation as an erstwhile child prodigy who had won three Paris Academy prizes and Mairan's having won three Bordeaux Academy prizes himself are factors that probably should be taken into account in trying to explain why Mairan and Bouguer were admitted to the Paris Academy as associate members.)

Bouguer began his career as a Paris Academician as a nonresident member of the Academy. In 1714, at the age of fifteen, he replaced his father, a reputable hydrographer who had recently died, as professor of hydrography in Le Croisic. He remained in that position until 1730, when he moved to Le Havre in order to teach hydrography there. He stayed in Le Havre until 1735, when he left for Peru with other members of the Paris Academy in order to measure a degree of latitude in the vicinity of the equator.

Maupertuis severely criticized the writings of Mairan and his associate Bouguer. Let us recall that I mentioned in Chapter 2 that reading Mairan's Paris Academy *mémoire* of 1720 requires a great deal of patience and endurance. Maupertuis thought Mairan's treatise on the aurora borealis, published by the Paris Academy in 1732, "extremely long."[8] In relating to Johann I Bernoulli how the printing of the fat volume caused the publication of the Paris Academy's *Mémoires* for 1731 to be delayed, Maupertuis snickered caustically and scornfully: "It is a marvel, the fertility of a man who writes a work as long as Newton's *Principia* on such a phenomenon [the aurora borealis]."[9] In his anonymously published, "impartial" review of the problem of the earth's shape, published in 1738, which I talked about in Chapter 3, Maupertuis, with the same inimitable wit that appears throughout this work, let Mairan have the last word on the subject of the earth's shape. Contrary to the findings of the members of the expedition to Lapland, the earth was elongated at the poles, as Mairan had argued, Maupertuis declared at the end of the book![10] For a long time Maupertuis denied having written the "impartial" review. He even did so to his good friend Johann II Bernoulli, Johann I Bernoulli's third son. Maupertuis appears to have persuaded the younger Bernoulli to publish in the *Journal helvétique* for July 1740 an anonymous letter written by Maupertuis to him. Maupertuis also convinced Johann II Bernoulli to publish in the *Journal helvétique* for September 1740 extracts of a letter from Mairan to the older Bernoulli, in which Mairan praised the anonymously published, "impartial" review. One can only imagine Mairan's embarrassment and anger when he finally discovered the author's true identity! Maupertuis also thought Bouguer's works "too long" and written "à la Mairan."[11]

Maupertuis aimed to make comprehensible what seemed incomprehensible, and he simply could not tolerate writing that seemed to do just the opposite. Accordingly he had absolutely no patience with works whose lengths, in his opinion, far exceeded their substance, and he made clear his dissatisfaction with such works publicly in his own printed work. But he did this without naming particular authors, for reasons that I shall explain in the next section of this chapter. Thus, for example, in his "Balistique arithmétic," published in the Paris Academy's *Mémoires* for 1731, Maupertuis "dared," using the differential calculus, to reduce the contents of the most voluminous treatises on ballistics, including a long *mémoire* on the subject written by the Malebranchist Guisnée in the early part of the century, which Maupertuis did not refer to specifically in his printed *mémoire*, to a page[12] – the page on motions of projectiles in a vacuum when gravity is uniform, which is now a standard part of the chapter on mechanics in any general physics text and which includes the problem of determining the angle at which to launch a projectile in order to produce the longest range of the projectile treated as a maximum–minimum problem in the differential calculus.

Worse yet, Maupertuis at times resorted to some of the most underhanded tactics to ridicule what he considered to be verbose writing. In a Paris Academy *mémoire* of 1732 on "curves of pursuit," Bouguer took fourteen pages to reduce the second-order ordinary "differential" equation for these curves to a first-order ordinary "differential" equation and then to determine the integral of the latter equation. He used geometric arguments to introduce numerous changes of variables along the way. He introduced integrating factors and performed many algebraic manipulations as well. For some reason, Bouguer wanted to avoid using logarithmic expressions, even when he recognized their practibility or convenience.[13] In a paper on "curves of pursuit," which he read before the Academy a week after Bouguer had read the original version of his published *mémoire* before the Academy, Maupertuis in one stroke solved the second-order ordinary "differential" equation directly for the curves in question using logarithmic integration. The solution of Bouguer's problem which appears in the printed version of Maupertuis's paper, which is also published in the Paris Academy's *Mémoires* for 1732, is expressed in a fifteen-line paragraph! For good measure, Maupertuis proposed a generalization of Bouguer's problem, as well as what he thought was its solution, on the second page, which is the last page, of his published *mémoire*![14] There is no mistaking Maupertuis's intent here; he clearly meant to outdo the one-time child prodigy.

Afterward, Maupertuis continued to show hostility toward Bouguer and tried to make him appear like a fool in public and in private. Not only did he reduce Bouguer's paper of 1734 on homogeneous figures of equilibrium to its bare essence, but he contradicted Bouguer's assertion that Huygens had actually applied both the principle of the plumb line and the principle of balanced columns in determining the earth's shape. Huygens in fact did not employ the principle of the plumb line to tackle the problem in question. He only used the principle of balanced columns to find the earth's shape. Huygens doubtlessly could not have utilized the principle of the plumb line for this purpose for reasons that Maupertuis explained. Namely, Maupertuis maintained that Huygens could not have found homogeneous figures of revolution which revolve around their axes of symmetry, which are attracted by fixed centers of force situated at the centers of the homogeneous figures of revolution, where the magnitude of the attraction is constant at all points, and at whose surfaces the principle of the plumb line holds, because in order to do so Huygens "would have needed to use the inverse method of tangents," by which Maupertuis meant solving ordinary "differential" equtions, "which was little known at the time," meaning little known in 1690, when Huygens published his *Discours de la cause de la pesanteur*, in which Huygens treated the problem of the earth's shape, as I mentioned in Chapter 1.[15] (Nor would Huygens ever truly become an advocate of, much less an adept in, the infinitesimal calculus in the years between 1690 and 1695, the year that he died.) Perhaps Maupertuis deliberately attempted to exploit Bouguer's error. That is, perhaps Maupertuis tried to make it appear as if Bouguer had not read Huygens's work very carefully, or, if he had, then Bouguer had not understood it in that case.

Be that as it may, the slip that Bouguer made here was really rather insignificant. Besides, Maupertuis did not mention the fact that Hermann, who was reasonably skilled at solving ordinary "differential" equations for his time, had indeed applied both the principles of balanced columns and the plumb line in order to determine the earth's shape, just as Bouguer claimed Hermann had done. Bouguer's promoter Mairan had already pointed out Hermann's use of the principle of the plumb line to find the earth's shape, too, in his Paris Academy *mémoire* of 1720. If Maupertuis had ever taken the trouble to struggle with the details of that distasteful paper, he would have known that Mairan mentioned in the paper that by utilizing "integral calculus" Hermann had found the same homogeneous figures shaped like infinitesimally flattened figures of revolution which Huygens had determined. Mairan reminded the reader that Hermann did this in reaction to some of Bernard Nieuwentijt's criticisms of the infinitesimal calculus. Nieuwentijt was one of Leibniz's leading Dutch opponents. Hermann then employed synthetic geometry to demonstrate the results that he had arrived at by using the integral calculus, meaning the inverse method of tangents in this case. He made use of the demonstration to construct solutions, and he published both the geometric demonstration and the constructions in his *Phoronomia*, published in 1716.[16]

During the early part of the eighteenth century, European mathematicians made progress in the differential calculus faster on the whole than they did in the integral calculus. I have argued elsewhere that during the first three decades of the eighteenth century, research in the integral calculus done by mathematicians in Paris lagged considerably behind research in the integral calculus done by mathematicians in other European countries, and, in particular, by the members of the Basel School, to which two generations of Bernoullis, Hermann, and Leonhard Euler belonged.[17] I shall discuss some of the specific difficulties that slowed mathematicians in Paris down in Chapter 7. The lowly state of the art in Paris began to improve in the 1730s, for reasons that I have also discussed briefly elsewhere[18] and will go into again in more detail in Chapter 7. Bouguer's application in 1734 of the principle of the plumb line at the surfaces of homogeneous figures of revolution which revolve around their axes of symmetry and which attract according to different laws of attraction was not only an achievement in the use of the integral calculus but, as it turns out, illustrates the retarded progress in the integral calculus and its use in Paris as well.

Even when Maupertuis recognized Bouguer's merit he abused Bouguer. Maupertuis knew very well that Bouguer's paper of 1734 was extraordinarily good[19] but he did not want to admit that at this time. Maupertuis also appropriated, without acknowledgment, ideas appearing in the essay that Bouguer submitted to the Paris Academy in 1731 for the Academy prize of 1732. The subject of the contest of 1732 was the causes of the inclinations of the planetary orbits to the plane of the ecliptic. When Bouguer submitted his essay to the Paris Academy in 1731, he had not yet been elected to the Academy; therefore he was still eligible to compete for Academy prizes. Since the Academy did not deem any of the essays that it received to be worthy of a prize, the contest of 1732 was held over until 1734, and the value of the prize was doubled. As I mentioned above, Johann I Bernoulli and his son Daniel won the contest of 1734 and shared the prize. None of the essays submitted to the Academy for the prize of 1732 were published before 1734. Since Bouguer had become a member of the Paris Academy in the interim, he was no longer eligible to compete for the Academy prize of 1734. Nevertheless, his essay of 1732 received honorable mention in the contest of 1734 and was published. But Maupertuis had served as one of the judges of the contest of 1732, and he exploited some of the ideas appearing in Bouguer's essay of 1732, which remained unpublished in 1732, in his own *Discours*, which was published in 1732, to attack the long-presumed epistemological certitude of Cartesian action by contact. Using Bouguer's ideas, Maupertuis succeeded in undermining this philosophical position.[20] By this means Maupertuis paved the way for the reception in Paris of Newtonian attraction at a distance. But Maupertuis did not acknowledge Bouguer's contribution in the slightest way.

Bouguer should be contrasted with Clairaut, who was a child prodigy when he entered the Paris Academy in 1731, the same year that Bouguer did, at the age of eighteen. Clairaut presented his first paper to the Academy when he was twelve years old. He joined the Academy at the age of eighteen after the requirement of minimum age was suspended for the first and only time in the Academy's history.

Clairaut got along well with Maupertuis. Clairaut joined Maupertuis's drive to promote mathematics and the *Principia* at the Academy. Whereas in the beginning Clairaut indulged Maupertuis, his senior by almost fifteen years, Bouguer never flattered, kowtowed to, or curried favor with the mathematician with the inflated and fragile ego, who is usually considered to have been Newton's leading advocate in the Paris Academy in the 1730s. But Bouguer's behavior toward Maupertuis should hardly startle or astonish anyone, in view of what I have said above. (Moreover, as I have also mentioned, Bouguer was more than seven months older than Maupertuis, not fifteen years younger, like Clairaut. Based on age alone there was little reason for Bouguer to try to humor Maupertuis.) Years later, Charles-Marie de La Condamine informed the Yugoslavian Jesuit Roger-Joseph Boscovich, who resided in Italy but who visited Paris where he made numerous friends, that Bouguer detested Maupertuis.[21]

Nor did Bouguer appear to care for at least some of Maupertuis's followers and supporters. La Condamine, three years younger than Maupertuis, one of Maupertuis's closest friends and a participant in the Newtonian campaign, headed the Academy's expedition to Peru of 1735–44 together with Bouguer and Louis Godin. During the expedition, Bouguer repeated some astronomical observations that he had already taken. As a result he discovered that he had made a significant error in his first set of observations. He also found the source of the error as well as means of correcting the mistake. Bouguer either requested or demanded that he take all of the credit for locating the error, for finding the method that enabled him to discover the error and for inventing the method that he used to rectify the mistake. La Condamine did not accept Bouguer's wishes.[22] Bouguer and La Condamine became mortal enemies. They wrote separate, independent accounts of the expedition to Peru for the Academy concerning what had been accomplished and by whom.

Whether a possible uncontrolled *amour propre* of Bouguer's own caused Bouguer some of his troubles is a little hard to ascertain. Bouguer believed, probably rightly, that his superior abilities in mathematics and in physics, and in optics in particular where taking astronomical observations during the expedition to Peru was involved, made a great deal of difference during the expedition to Peru. Because of his talents and skills Bouguer had an important part in assuring the success of the venture and in determining its outcome. Naturally Bouguer felt that he deserved a great deal of credit as a result. But La Condamine thwarted Bouguer's desires, because La Condamine happened to be better known outside scientific circles than Bouguer was. Accordingly the French public concluded that La Condamine made the principal contributions to the expedition, and Bouguer believed that he received much less credit than he was entitled to.[23]

Having returned to Paris from Peru, the resentful Bouguer attacked the French mathematician Alexandre Savérien. In 1745, Savérien published a new theory of the rigging of ships. In the work he deviated from the principles appearing in

Bouguer's treatise that won the Paris Academy prize of 1727 on the best way of putting masts on ships. Savérien chose instead to use some calculations of Johann I Bernoulli, who subsequently praised Savérien's work. Savérien's approach to the problem of rigging ships not only displeased Bouguer; Savérien's theory exasperated and infuriated him. Bouguer sharply criticized Savérien and his theory.[24] Whether Bouguer's complaints truly illustrate vanity or simply a rightful bitterness with regard to the expedition to Peru which Bouguer could not hide and which made him sour is difficult to determine with certainty.[25]

5.2. Politics at the Paris Academy: "La magouille"

If he treated Bouguer shabbily and abominably, and if he used disreputable, cruel tactics in general at the Paris Academy, Maupertuis may have paradoxically had little choice but to resort to such unscrupulous methods. For Paris Academicians did not tolerate any frank, open, printed criticism of their work. Academicians treated each other as severely as they dealt with outsiders. All traces of dissension and signs of discord or controversy among Academicians were almost always eliminated, carefully and systematically, from work to be published. The suppression of candid opinions before publication not only applied to the Paris Academy's *Mémoires*, but all other French journals and reviews were also expected to censor articles that they intended to publish which involved the work of Paris Academicians.

Maupertuis's concealed or disguised mockery of Bouguer's paper on "curves of pursuit" I have already mentioned. In the version of his paper on the "Principe du Repos" which he read before the Paris Academy in February of 1740, Maupertuis claimed to reduce eighty-seven pages of earlier work by the French Jesuit mathematician and Paris Academician Pierre Varignon, another diffuse, prolix, confusing writer of an earlier period, to two lines! In the version of Maupertuis's paper published in the Paris Academy's *Mémoires*, the two lines of text are not presented as a simplification of what Varignon had done. Indeed, no reference to Varignon's work appears at all in the printed *mémoire*![26] In July of 1742, the Academy ruled that a paper by Maupertuis on determining the tonnage of ships would only be published in the Academy's *Mémoires* if the author added to the manuscript kind words about the work of earlier writers who had tried to handle the problem.[27] In a paper he read to the Academy in late April of 1731 on the separation of variables in first-order ordinary "differential" equations of first degree, Maupertuis said that he had succeeded in separating the variables in the same "differential" equations as the ones whose variables John Craige had already separated in his *De Calculo fluentium* (1718) using what Maupertuis regarded as a "rather confused" method. In the version of Maupertuis's paper published in the Paris Academy's *Mémoires*, the phrase "rather confused" which Maupertuis had used to describe Craige's method does not appear, however.

The phrase was obviously deleted from the original version of the paper by Maupertuis or by the Academy's censors.[28]

Nor was Maupertuis by any means the only target. In a paper that the Chevalier de Louville read to the Paris Academy in June 1722, in which the author claimed to refute an earlier paper read by Joseph Saurin, Louville accused Saurin of inattention. But all such remarks, and any other references to the controversy between the two Academicians, including a section entitled: "Refutation du mémoire de Mr. Saurin," were left out of the version of Louville's paper published in the Paris Academy's *Mémoires*.[29]

Nor could the Academy's Secrétaire Perpétuel Fontenelle allow Varignon's last paper, which the Jesuit mathematician submitted to the Academy before his death in 1722 and in which he replied to an attack that an Italian clergyman had previously made against him, to remain unchanged before publication in the Academy's *Mémoires*. In "deleting everything that the *mémoire* included which was personal and polemical, Monsieur Fontenelle contented himself with communicating to the public what he found to be useful and instructive."[30]

To cite yet another example, in a paper that he read before the Paris Academy in June of 1734, Clairaut solved mathematics problems that resembled precursors of the problems of "trajectories" worked out by Leibniz, Jakob Bernoulli, and the Marquis de L'Hôpital in *Acta Eruditorum* for 1697. These mathematicians responded to a challenge issued by Johann I Bernoulli in *Acta Eruditorum* for 1696 and 1697. (I mentioned one of the problems, the problem of "orthogonal trajectories," in Chapter 2.) Following the custom of the time, the mathematicians only presented solutions. Normally they did not reveal their methods. There was one exception, however. Jakob Bernoulli applied a method of undetermined coefficients and exponents to solve a certain problem, which Clairaut found unsatisfactory, because the method was "indirect."[31] In a paper that he read to the Academy in January of 1735, Alexis Fontaine remarked that he had informed Clairaut afterward that not only had Newton also solved the same problem, but that Newton had even sketched both the correct approach to the problem and the correct way to solve it as well. Fontaine observed that Newton had done both of these things in the *Philosophical Transactions* and in *Acta Eruditorum* for 1697, although Clairaut had failed to mention Newton at all in the paper that he had read before the Academy in June of 1734.[32] In the version of this paper which was published in the Paris Academy's *Mémoires* for 1734, Clairaut pointed out that Newton had both resolved the problem in question and provided clues leading to the correct method for solving it. Clairaut did so without referring to the fact that Fontaine had brought this oversight to his notice.[33] Furthermore, all references to Clairaut's inattention were eliminated from the final version of Fontaine's paper, which was also published in the Paris Academy's *Mémoires* for 1734.[34] I shall have occasion in Chapter 7 to return to these events concerning Clairaut and Fontaine again and to the particular circumstances involved.

The history of the Paris Academy of this period abounds in such examples. For instance, during his famous debate with Clairaut in 1747–48, the Comte de Buffon took issue with Clairaut for attempting to modify the universal inverse-square law of attraction, which Clairaut did in order to try to account for the anomalous motion of the moon's apogee. Buffon read a paper to the Academy in which he made some remarks that he cautioned Clairaut not to take personally. This apology, which appears as the last sentence of the version of the paper which Buffon read to the Academy, was stricken from the version of Buffon's paper published in the Paris Academy's *Mémoires*.[35]

When criticisms of Paris Academicians did find their way into print, commotion invariably broke out. At the beginning of the eighteenth century, the Academy warned Varignon and his adversary at that moment, the algebraist Michel Rolle, to limit themselves to the issues in the controversy then going on in Paris about the Leibnizian calculus and to refrain from making purely personal, offensive, irrelevant remarks.[36] While the papers read by Varignon and Rolle regarding their dispute were recorded in the Paris Academy's unpublished proceedings (*Registres des Procès-Verbaux*), most were never published in the Academy's *Mémoires* for the years 1701–4. Instead, a fight between Rolle and Saurin, who was not yet a member of the Paris Academy at this time, but who was a mathematics editor of the *Journal des Savants* and another of the earliest French defenders of the Leibnizian calculus, ensued from the controversy. It took place in the pages of the *Journal des Savants* for the years 1702–5. The quarrel was a terrible one. The antagonists peppered each other with hostile, malicious remarks; they wrangled bitterly; and their mutual insults descended to the personal level. The viciousness of this feud, more than any other single factor, probably influenced the Academy's future policy toward dissension within its ranks, which included individuals who were not yet members of the Academy but who were destined to join the Academy soon. For example, the feud was more than likely what caused Fontenelle beginning in 1728 to limit the diffusion in the Paris Academy's *Mémoires* of certain aspects of the *vis–viva* controversy which appeared in the Academy's unpublished proceedings.[37] This controversy had unofficially been going on already for several years. In 1706, an Academy jury finally ruled that if Rolle and Saurin were to continue their dispute that began in the *Journal des Savants*, they would do so in a civilized way.[38] The members of the Academy avoided a direct decision in favor of an outsider by electing Saurin an assistant geometer in March 1707. Two months later Saurin was promoted to fill a pensioner geometer's position in the Academy which had been vacated.[39] Saurin's astonishingly rapid advance from assistant member of the Academy to pensioner in the Academy, without even having to be named an associate member of the Academy first, probably can be attributed to his success in having brought the worst of the debate about the Leibnizian calculus at the Academy to a close. (Saurin was also already fifty-one years old when he entered the Academy, which may have helped the members of the Academy justify promoting Saurin quickly to pensioner and thereby ultimately enabled them to put through this rather unusual action without the Crown's opposition.[40])

In the late 1720s, Paris Academy mathematicians vilified Louis-Bertrand Castel, science editor of the Jesuit *Journal de Trévoux*, because of the journalistic liberties that he took in his reviews of their work which he published in that journal. They did not tolerate the merely lukewarm support of their work which he occasionally gave. The mathematicians reacted hostilely and tried to silence Castel.[41] In Chapter 7 on integral calculus I shall have occasion to turn to some of Castel's criticisms of the work done by Paris Academy mathematicians.

At the beginning of the eighteenth century, Antoine Parent, a prolific Paris Academician, failed to see many of his writings get printed in the Academy's *Mémoires*. Consequently, he published his *Recherches de mathématiques et de phisique* in several volumes on his own instead. In 1705, when Guillaume Amontons stumbled upon critical references to his work and to that of nine other Paris Academicians in one of the volumes, he demanded that Parent publicly retract what he had written. The Academy reprimanded Parent and insisted that he publicly withdraw the offending opinions that he had written, too.[42] Parent wrote too candidly and freely in his own publications about the work of his colleagues in the Academy. As a result, he alienated them. In fact many were less capable than he. In this particular case the offender paid a high price. Parent's behavior, together with the lack of interest in his more applied research which the mathematicians, the specialists in mechanics, and the physicists in the Academy showed, not to mention the complete lack of understanding of this work by the engineers for whom it was intended, combined to banish Parent from the annals of science and cause him to sink into oblivion. He was nevertheless a truly original thinker where mechanics and the strength of materials are concerned.[43] Yet he did not even advance beyond his initial position of assistant member of the Academy. His failure to be promoted has been attributed to, among other things, his aggressive, tactless, critical, and uncompromising candor in dealing with his colleagues.[44] L'Hôpital, who was five years older than Parent and the best of the senior mathematicians in Paris at the beginning of the eighteenth century, certainly recognized Parent's potential. In 1701, when Leibniz asked Fontenelle for a young collaborator to help him deepen his new binary system of arithmetic, L'Hôpital recommended Parent, who was willing to work for Leibniz in return for a salaried position in the new Berlin Academy. (By this time Parent probably already foresaw the troubles that he would have in Paris.) But the collaboration never came about. Although he was president of the Berlin Academy on paper, Leibniz could not promise any remuneration for the work that the junior Paris Academician might do. Leibniz's own powers as president of the Berlin Academy were limited, and finances at the time were available only for essentials.[45] Moreover, L'Hôpital died prematurely, in 1704. Consequently he was unable to do Parent any further good.

I have cited these various examples in order to justify conjecturing that possibly nothing short of Maupertuis's use of chicanery and deceit in Paris would have permitted Maupertuis to criticize before the public eye certain old ideas, an act necessary to make the advantages of new ideas stand out, and by this means pave the way for receiving and accepting the new ideas. Maupertuis's machinations and stratagems served a vital purpose in breaking through the Paris Academy's reactionary politics – its policies of inflexibility, inertia, and secrecy. In other words, as intolerable as Maupertuis's behavior toward others may seem now, a less brazen, less impudent, less arrogant, less presumptuous, less conniving individual with less temerity might not have succeeded under the circumstances. Maupertuis's "reprehensible" conduct was exactly what was required in Paris at the time!

Indeed, we know that intellectuals *of all kinds* had to resort to dishonest tactics in France at this time–for example, philosophers. Thus Benedict Spinoza's French champions were obliged to defend their hero by writing anti-Spinozist treatises in which Spinoza's ideas were disguised.[46] These authors hoped that the censors who reviewed their works and passed judgment would be as obtuse as, say, Mairan, who, we recall, had praised the anonymously published "disinterested" account of the controversy about the earth's shape, whose author was none other than Mairan's adversary Maupertuis!

I cite just one instance of the stiff price that a man of science could pay if he kept quiet in Paris. As I mentioned in Chapter 2, the astronomer Joseph-Nicolas Delisle was one of the first Frenchmen to support Newton's *Principia*. He advocated the *Principia* as a tool in mathematical astronomy as early as the period 1717–25.[47] But he did this in letters to colleagues, confidants, and patrons; he did not publicly campaign. He never did succeed in convincing the powers that be. Delisle also recognized almost immediately that the conclusions about the earth's shape which Jacques Cassini drew in 1718 from the measurements of degrees of latitude which his father, he, and some of their colleagues in the Paris Academy had made had their limitations. But Delisle did not make his doubts public at the time. The idea of measuring degrees of longitude also crossed Delisle's mind at this time, but, again, he only communicated this idea in letters to an individual whom he felt that he could trust, the king's representative at the Paris Academy, Abbé Jean-Paul Bignon. But nothing resulted from any of Delisle's efforts, and in 1725 Delisle left the Paris Academy for the new academy in St. Petersburg and for what he hoped would be greener pastures. As I mentioned in Chapter 3, Maupertuis revived geodesy in Paris in 1733 under the influence of Giovanni Poleni. At this time the same kinds of reservations about Cassini's measurements and conclusions which had occurred to Delisle but which Delisle had not made public came to light in Paris thanks to the review of Poleni's work published in the Dutch *Journal Historique de la République des Lettres* early in 1733. In fact, the idea of measuring degrees of longitude entered Delisle's mind long before it entered Poleni's.[48] Delisle tried to assert his priority in the matter but without success.[49] He complained to no avail that he had been robbed of a novel idea in geodesy. But he did not accuse his friend and correspondent Poleni of theft or reproach him for having published the idea first. Instead Delisle blamed Jacques Cassini and Giacomo Filippo Maraldi, Delisle's senior colleagues in Paris. They held the power in the Paris Academy when Delisle's idea came to his mind, and Delisle guessed correctly that they would never have accepted its scientific merits. Indeed, Cassini and Maraldi found out about Delisle's idea from the Abbé Bignon, Delisle's "confidant," and the two astronomers in fact did not approve of it.[50] Delisle would have had to resort to considerable campaigning and maneuvering behind the scenes at the time that his idea suggested itself to him in order to have produced an outcome that differed from the way things actually turned out, and he evidently felt that he was not in a position to do this in Paris, or else he was simply not aggressive enough or plucky enough or courageous enough to try. Delisle, like Maupertuis, was ambitious and often acted in his own self-interest. Unlike Maupertuis, however, Delisle either lacked the courage or the nerve or else he did not have the temerity or the imprudence to behave the way that Maupertuis did at the Paris Academy.

One wonders what would have happened to Maupertuis, had he, too, kept his mouth shut, instead of exasperating the older Paris Academicians.

5.3. Newton's theory of the earth's shape first "expounded" in French

If Bouguer in Maupertuis's opinion had talked at too great a length instead of concisely about the problem of homogeneous figures of equilibrium in his Paris Academy *mémoire* of 1734, Newton by contrast with Bouguer had not written enough in the *Principia* about his theory of the earth's shape to suit Maupertuis. In the last section of his own Paris Academy *mémoire* of 1734, Maupertuis tried to transform the words and geometry that made up Newton's theory of the earth's shape into the continental infinitesimal calculus. In effect, he attempted in 1734 to do for Newton's theory what he had done in the *Discours* in 1732 for Huygens's and Hermann's theories of the earth's shape. He made an effort to clarify Newton's theory of the earth's shape by expanding into fifteen pages what Newton had compressed into ten pages. The result was the first exposition in the French language of the theory of the earth's shape appearing in the *Principia*, third edition.[51]

In fact, Maupertuis had already begun earlier to try to deal with the problem of converting Newton's laws of attraction into a form that continental men of science could understand. This he did in a paper that he read to the Paris Academy in 1732–33. In the paper he employed integral calculus to demonstrate some of Newton's laws of attraction for spheres and nonspherical surfaces and homogeneous bodies shaped like spheres which attract according to the universal inverse-square law. Specifically, Maupertuis translated into continental mathematical language the geometrical theorems appearing in the second edition of the *Principia*, pp. 173–211 (Sections XII, XIII, and XIV of Book I).[52] Thus, for example, using algebra and the integral calculus, Maupertuis demonstrated Propositions 70–74 of Book I of the *Principia*–namely, the propositions in which Newton showed that a homogeneous spherical body that attracts according to the universal inverse-square law in fact attracts like a $1/r^2$ central force in its exterior and like an r central force in its interior. At this time, however, Maupertuis did not treat Newton's theorems concerning attractions at the poles and equators of homogeneous figures shaped like ellipsoids of revolution which attract in accordance with the universal inverse-square law. These theorems he made an attempt to handle in his Paris Academy *mémoire* of 1734 instead.

For a man who is generally thought to have been among the most devout supporters of Newton at this time, Maupertuis did some odd things. In the Paris Academy *mémoire* of 1734, he showed how to calculate the ratio of the magnitude of the centrifugal force per unit of mass to the magnitude of the attraction at the equator of an earth assumed to be shaped like a sphere and to revolve around its polar axis at a constant rate (that is, at a constant angular speed), without using experiments made with seconds pendulums or with bodies falling freely as Newton and Huygens had done. Maupertuis utilized instead the period of the moon's revolution around the earth, where he assumed that the moon revolved around the earth at a constant speed in a circle whose center was situated at the center of the earth, the radius of the lunar orbit relative to the earth's radius (that is, the ratio of the radius of the moon's orbit to the earth's radius), and the period of the earth's revolution around its polar axis. He also used the fact that the magnitude of the centripetal acceleration of a celestial body moving in a circle at a constant speed varies inversely as the square of the radius of the circle. This fact is a consequence of Kepler's third law of motion together with Newton's expression for the magnitude of the centripetal acceleration of a body moving in a circle at a constant speed (which was the same as Huygens's expression for the magnitude of the centrifugal force per unit of mass of a body moving in a circle at a constant speed). The centripetal acceleration of such a body is directed toward the center of the circle that is the body's path of motion, and Maupertuis implicitly assumed that the centripetal acceleration of the revolving body was produced by the attraction due to a center of force located at the center of the circle that the body moves along, where the magnitudes of the attraction at points caused by this center of force varied inversely as the squares of the distances of the points from this center of force. In other words, he assumed that the magnitude of the attraction due to the center of force varies with the distance from the center of force the same way that the magnitudes of the centripetal accelerations of bodies that revolve around the center of force in circles at constant speeds vary with the distances of the bodies from the center of force. This is a natural assumption to make, if it is hypothesized that the central force of attraction causes the centripetal acceleration of the body revolving in a circle whose center is situated at the center of force. Maupertuis thought his own method of calculating the ratio of the magnitude of the centrifugal force per unit of mass to the magnitude of the attraction at the equator of a spherical earth that revolves around its polar axis at a constant rate to be more accurate than other methods, because there was less of a chance for errors to be made.

He generalized the procedure to calculate the ratio of the magnitude of the centrifugal force per unit of mass to the magnitude of the attraction at the equator of a spherical celestial body assumed to be revolving around its polar axis at a constant rate–upon whose surfaces, of course, no experiments with seconds pendulums or with freely falling bodies could be carried out, anyway. He did this using the celestial body's radius, its observed period of revolution around its polar axis, the observed period of a satellite assumed to revolve around the celestial body at a constant speed in a circle whose center is located at the center of the celestial body, and the radius of the satellite's orbit relative to the radius of the celestial body around which it revolves (that is, the ratio of the radius of the satellite's orbit to the radius of the celestial body around which it revolves). He also treated the satellite as if it were attracted by a fixed center of force situated at the center of the celestial body around which it revolves, where the magnitudes of the attraction at points produced by this center of force varied inversely as the squares of the distances of the points from this center of force. He did this for the same reasons as he did for the earth and moon.[53]

(What Maupertuis failed to recognize, or, at least neglected to mention, is that Newton made use of the same kinds of data and the same relations among the data with different ends in view–namely, Newton used them to compute the ratio of Jupiter's density to the earth's density, in Proposition 8 of Book III of the third edition of the *Principia*.[54])

Applying his calculations to Jupiter, Maupertuis determined the ratio of the magnitude of the centrifugal force per unit of mass to the magnitude of the attraction at Jupiter's equator to be $1/(7.48)$. Taking this to be the value of the ratio of the magnitude of the centrifugal force per unit of mass to the magnitude of the attraction at the equator of a figure in his theory of homogeneous figures of revolution which revolve around their axes of symmetry which are attracted by single, fixed centers of force situated at the centers of the homogeneous figures of revolution, where the magnitudes of the attraction at points produced by a center of force vary solely with the distances of the points from the center of force, and where the columns from the center of such a homogeneous figure of revolution to its surface are all assumed to balance or weigh the same, which he had first presented in his article in the *Philosophical Transactions* of 1732 and again in his *Discours* published in 1732, Maupertuis found that when the magnitudes of the attraction at points were assumed to vary inversely as the squares of the distances of the points from the center of force, the value of Jupiter's ellipticity came out to be $1/(14.96)$, which he said was a value that was very close to the value $\frac{1}{15}$ of the extent of Jupiter's flattening found by Gian-Domenico Cassini using telescopes in 1691 and confirmed by Philippe de La Hire. (The two values 1/(14.96) and $\frac{1}{15}$ are in fact equal to terms of first order. This is true because

$$\frac{1}{14.96} = \frac{1}{15 - 0.04} = \frac{1}{15}\left(\frac{1}{1 - ((0.04)/15)}\right),$$

which equals $\frac{1}{15}(1 + \frac{1}{375})$ to terms of first order, because $\frac{1}{375}$ is infinitesimal, $((\frac{1}{375})^2 \ll \frac{1}{375})$.

Does $\frac{1}{15} \times \frac{1}{375} = \frac{1}{a} \times \frac{1}{b}$ where $\frac{1}{a}$ and $\frac{1}{b}$ are both infinitesimal? The answer is yes, because $\frac{1}{15} \times \frac{1}{375} = \frac{1}{5625} \simeq \frac{1}{70} \times \frac{1}{80}$, and $\frac{1}{70}$ and $\frac{1}{80}$ are both infinitesimal $((\frac{1}{70})^2 = \frac{1}{4900} \ll \frac{1}{70}, (\frac{1}{80})^2 = \frac{1}{6400} \ll \frac{1}{80})$. Hence $\frac{1}{15} \times \frac{1}{375}$ is a term of second order and can be discarded. Consequently $\frac{1}{14.96} \simeq \frac{1}{15}(1 + \frac{1}{375})$ equals $\frac{1}{15}$ to terms of first order.) Maupertuis remarked as well that his value $\frac{1}{14.96}$ of Jupiter's ellipticity was also closer to the value $1/(12\frac{11}{48})$ of Jupiter's ellipticity found by the English astronomer James Pound using telescopes in 1719 than was the value $1/(9\frac{1}{3})$ of Jupiter's ellipticity which Newton determined by assuming that Jupiter is a homogeneous figure shaped like a flattened ellipsoid of revolution which attracts according to the universal inverse-square law, which revolves around its axis of symmetry, and whose columns from center to surface are all assumed to balance or weigh the same. Maupertuis mentioned that the difference between the values of Jupiter's ellipticity found using telescopes and the value of Jupiter's ellipticity which Newton had determined theoretically had led Newton to conclude that Jupiter was not homogeneous, but denser toward the plane of its equator than toward its poles, and that, according to Newton, because of this variation of density the planet was not as flat as it would be if it were homogeneous, so that Jupiter had an ellipticity that was closer to the observed values than to the value that Newton had found theoretically.

(One reason why the ellipticity of Jupiter required by Newton's theory of homogeneous figures shaped like flattened ellipsoids of revolution which attract according to the universal inverse-square law, which revolve around their axes of symmetry, and where the columns from the center of such a homogeneous figure of revolution to its surface are all assumed to balance or weigh the same differed so much from the values of Jupiter's ellipticity determined by means of telescopic observations is that Jupiter's ellipticity is observed *not to be infinitesimal.* In other words, incorrect assumptions about Jupiter's density are not the only possible way to explain the difference between the values of Jupiter's ellipticity found by means of telescopic observations and Newton's theoretically determined value of Jupiter's ellipticity. I shall return to this problem shortly.)

Did Maupertuis doubt that the telescopic observations which indicated the extent of Jupiter's flattening were reliable? Did he think them too crude to test either hypothesized central forces of attraction or Newton's universal inverse-square law of attraction? Maupertuis had, after all, finally decided that observations, not theory, would reveal the earth's true shape. It is true that in the abridged, Latin version of the *Discours*, published in the *Philosophical Transactions* for 1732, he referred to all of his derivations of shapes of bodies from hypothesized central forces of attraction as "more mathematical than physical."[55] In the *Discours*, he stated that his determinations of shapes were not "exact," but he did not suggest that central forces lacked physical foundations.[56] And why did he recapitulate in his Paris Academy *mémoire* of 1734 his entire theory of homogeneous figures of revolution which revolve around their axes of symmetry, which are attracted by fixed centers of force located at the centers of the homogeneous figures of revolution, where the magnitudes of attraction at points produced by a center of force vary solely with the distances of the points from the center of force, and where the columns from the center of such a homogeneous figure of revolution to its surface are all assumed to balance or weigh the same, which he had already presented in the *Philosophical Transactions* for 1732 and again in his *Discours* published in 1732, if he truly believed such a theory to be imaginary? Did he only mean to fill up space? But Maupertuis was supposed to be the perfectionist when it came to writing briefly! Was the stickler for concise writing simply not playing the game according to his own rules? Moreover, in the paper in the *Philosophical Transactions*, which he wrote for a British audience, he did not mention at all the facts that the degree of Jupiter's flattening required by the theory of inverse-square central forces accorded with the observed amounts of Jupiter's flattening, while the value of Jupiter's ellipticity which Newton had determined theoretically did not agree with the observed amounts of Jupiter's flattening. Did Maupertuis deliberately adjust his remarks to suit particular audiences, in an attempt to cover all ground? Was he simply hedging his bets? A complex individual whose intentions are not always easy to ascertain, Maupertuis provides us with no simple answers. One scholar's reference almost two decades ago to Maupertuis as "not a very good Newtonian" seems just, however.[57]

In applying Newton's theory to Jupiter, Maupertuis began with the equation

$$\frac{X}{1/229} = \frac{\Omega/\Pi}{F/P}, \tag{5.3.1}$$

where X = Jupiter's ellipticity, $\frac{1}{229}$ = the earth's ellipticity, F = the magnitude of the centrifugal force per unit of mass at the earth's equator, P = the magnitude of the attraction at the earth's equator (so that $F/P = \frac{1}{289}$), Ω = the magnitude of the centrifugal force per unit of mass at Jupiter's equator, and Π = the magnitude of the attraction of Jupiter's equator. Now, Newton did not exhibit such an equation, but in fact equation (5.3.1) is just an analogue of Newton's equation

$$\frac{\delta}{1/100} = \frac{\phi}{4/505} \tag{1.1}$$

which I discussed in Chapter 1 and which Newton used to determine the value of the earth's ellipticity to be $\delta = \frac{1}{229}$ when the value of the ratio ϕ of the magnitude of the centrifugal force per unit of mass to the magnitude of the attraction at the equator of the homogeneous figure shaped like a flattened ellipsoid of revolution which attracts according to the universal inverse-square law, which revolves around its axis of symmetry, and whose columns from center to surface are all assumed to balance or weigh the same was taken to be the value of the ratio at the earth's equator, which is $\frac{1}{289}$. Although Newton did not say so in the *Principia*, equation (1.1) only holds good to terms of first order. In other words, it only holds when δ and ϕ are infinitesimal (that is, when $\delta^2 \ll \delta$ and $\phi^2 \ll \phi$). Like Newton's equation (1.1), Maupertuis's equation (5.3.1) is only valid to terms of first order. Moreover, the two equations (1.1) and (5.3.1) are equivalent to terms of first order. But in treating Jupiter, Newton displayed instead of equation (5.3.1) an equation for Jupiter's ellipticity X expressed in terms of the following quantities: the square of the ratio of the period of Jupiter's revolution around its axis of symmetry (its polar axis) to the period of the earth's revolution around its axis of symmetry (its polar axis), the ratio of Jupiter's density to the earth's density, and his value of the earth's ellipticity, which is $\frac{1}{229}$. Using this equation Newton found $X = 1/(9\frac{1}{3})$ to be the value of Jupiter's ellipticity.[58] Maupertuis showed how to convert his equation (5.3.1.) for Jupiter's ellipticity into Newton's equation for Jupiter's ellipticity. (In fact, it is a foregone conclusion that this could be done, because of what I said earlier about the way that Newton calculated the ratio of Jupiter's density to the earth's density in Proposition 8 of Book III of the third edition of the *Principia*.)

Once again, Maupertuis mentioned the large difference between Newton's theoretical value $1/(9\frac{1}{3})$ of Jupiter's ellipticity and the values of Jupiter's ellipticity that Cassini and Pound had determined using their telescopes. Maupertuis also restated that Newton had attempted to patch up his theory of Jupiter's shape in order to make theory and observations agree. Again, we remember that it was Maupertuis who complained about the way others failed to write succinctly. In his opinion they either dragged on or else they rambled and got sidetracked. And again the question arises: why then did Maupertuis write more than was necessary? Why did he need to repeat that Newton's theory required that results hold which did not accord with the actual telescopic observations? Did he do this in order not to let the reader forget that central forces worked better in this particular case?

Did Maupertuis believe that Newton had shown that his theory held true for homogeneous figures shaped like flattened ellipsoids of revolution which attract according to the universal inverse-square law, which revolve around their axes of symmetry, where the columns from the center of such a homogeneous figure of revolution to its surface are all assumed to balance or weigh the same, and whose ellipticities are *finite*? Maupertuis evidently did not think so. He concluded his exposition of Newton's theory by remarking that the shape determined "is only an approximation; nor does Mr. Newton try to make it appear to be an exact one. The closer [the shapes of] celestial bodies approach ellipsoids, and the ellipsoids

approach spheres, the smaller the errors."[59] Thus Maupertuis understood
Newton to have developed a theory of homogeneous figures shaped like flattened
ellipsoids of revolution which attract in accordance with the universal inverse-
square law, which revolve around their axes of symmetry, and where the columns
from the center of such a homogeneous figure of revolution to its surface are all
assumed to balance or weigh the same which he *only* showed to be true when the
ellipticities of the kinds of homogeneous figures of revolution in question are
infinitesimal.

But Maupertuis also declared that the equation

$$\frac{a}{b} = \frac{(ab) - f}{[ab]}$$
(5.3.2)

holds good for *all* homogeneous figures shaped like flattened ellipsoids of
revolution which attract according to the universal inverse-square law, which
revolve around their axes of symmetry, and where the columns from the center of
such a homogeneous figure of revolution to its surface are all assumed to balance
or weigh the same, whether the ellipticities of these homogeneous figures of
revolution are *infinitesimal or finite*, where a/b is the ratio of the polar radius to the
equatorial radius of a homogeneous figure shaped like a flattened ellipsoid of
revolution which attracts according to the universal inverse-square law, which
revolves around its axis of symmetry, and whose columns from center to surface
are all assumed to balance or weigh the same, $f=$ the magnitude of the centrifugal
force per unit of mass at the equator of the figure, $(ab)=$ the magnitude of the
attraction at the equator of the figure, and $[ab]=$ the magnitude of the attraction
at one of the two poles of the figure. He maintained that equation (5.3.2) could be
used to determine the shapes of the kinds of homogeneous figures of revolution in
question exactly, meaning that the ellipticities of the figures could be found when
the ellipticities were finite, "provided the difficult problem of computing attractions at the equators of [homogeneous figures shaped like] flattened ellipsoids of
revolution whose ellipticities are finite [and which attract in accordance with the
universal inverse-square law] could be solved."[60] In fact, equation (5.3.2) follows
as a simple consequence of equating the weights of columns from the center to the
surface which lie along the polar and equatorial axes of a homogeneous figure
shaped like a flattened ellipsoid of revolution whose ellipticity is finite, which
attracts according to the universal inverse-square law, and which revolves
around its axis of symmetry.

The trouble is that Newton often called figures shaped like ellipsoids of
revolution "spheroids," a term that does not by itself specify whether the
ellipticities of such figures are infinitesimal or finite. (Newton used the term
"spheroid" to stand for the figure itself, not the surface of the figure as we do
now.[61]) In fact, Newton did demonstrate that Corollaries 2 and 3 to Proposition
91 of Book I of the *Principia*, which express geometrically two characteristics or

properties of the attraction produced by homogeneous figures shaped like ellipsoids of revolution which attract in accordance with the universal inverse-square law and which I discussed in Chapter 1, *do* prove true for homogeneous figures shaped like ellipsoids of revolution whose ellipticities are *finite*. But did *the whole of* Newton's theory of homogeneous figures shaped like flattened ellipsoids of revolution which attract in accordance with the universal inverse-square law, which revolve around their axes of symmetry, and where the columns from the center of such a homogeneous figure of revolution to its surface are all assumed to balance or weigh the same, hold true for the homogeneous figures of revolution in question when the ellipticities of these figures are *finite*, too? Maupertuis's equation (5.3.2) reveals that Maupertuis simply assumed that at least parts of Newton's theory held good when the ellipticities of the homogeneous figures of revolution at issue are finite. Newton, in Maupertuis's opinion, had simply not demonstrated this to be the case. (As we shall see in a moment, Maupertuis could not have believed that Newton's "$\sin^2$ law" of the variation of the increase in the magnitude of effective gravity with latitude at points on the surfaces of homogeneous figures shaped like flattened ellipsoids of revolution which attract according to the universal inverse-square law, which revolve around their axes of symmetry, and where the columns from the center of such a homogeneous figure of revolution to its surface are all assumed to balance or weigh the same held when the ellipticities of the figures are finite.) Maupertuis gave no indication that he realized that equation (5.3.2) might be a problem itself. That is, he did not consider the possibility that homogeneous figures shaped like flattened ellipsoids of revolution which attract in accordance with the universal inverse-square law, which revolve around their axes of symmetry, where the columns from the center of such a homogeneous figure of revolution to its surface all balance or weigh the same, and whose ellipticities are *finite might not exist.*

[But Maupertuis alone should not be criticized for having failed to do this. For Newton was not blameless in this matter. He could be worse than simply obscure. After all, he applied his equation

$$\frac{\delta}{1/100} = \frac{\phi}{4/505},$$
(1.1)

which *only* holds for homogeneous figures shaped like flattened ellipsoids of revolution which attract according to the universal inverse-square law, which revolve around their axes of symmetry, where the columns from the center of such a homogeneous figure of revolution to its surface are all assumed to balance or weigh the same, and whose ellipticities δ and ratios ϕ of the magnitudes of the centrifugal force per unit of mass to the magnitudes of the attraction at their equators are both *infinitesimal*, to Jupiter. But by the time that Newton wrote the *Principia* there was sufficient telescopic evidence to prove that Jupiter's ellipticity is *finite, not infinitesimal.* Moreover, as I have already mentioned in Chapters 1 and 2, we shall discover in Chapter 6 that Newton never really even demonstrated that homogeneous figures shaped like flattened ellipsoids of revolution which attract in accordance with the universal inverse-square law, which revolve around their axes of symmetry, where the columns from the center of such a homogeneous figure of revolution to its surface all balance or weigh the same, and whose ellipticities are infinitesimal *can exist.*]

In reality, Maupertuis made no headway at all with the difficult parts of Newton's theory of the earth's shape. Maupertuis elaborated the theory using continental mathematical language when he could, and he simply repeated the English scientist's words and geometry when he could not. Maupertuis's study of attractions of homogeneous figures shaped like ellipsoids of revolution which attract in accordance with the universal inverse-square law, in which Maupertuis employed algebra and integral calculus, really turned out to be more an academic discourse in elementary analysis than research whose results included tools that proved to be useful in solving problems. In discussing attractions of homogeneous figures shaped like ellipsoids of revolution which attract according to the universal inverse-square law in his Paris Academy *mémoire* of 1734, Maupertuis did little more than formally derive expressions, which have the form of integrals, for the magnitudes of the attraction at the poles of such figures. Specifically, using algebra and the integral calculus, he determined an expression for the magnitude of the attraction at a pole of a homogeneous figure shaped like a flattened ellipsoid of revolution which attracts according to the universal inverse-square law, which involved the quadrature of the circle (which in today's terms means inverse trigonometric functions), and he found an expression for the magnitude of the attraction at a pole of a homogeneous figure shaped like an elongated ellipsoid of revolution which attracts according to the universal inverse-square law, which involved the quadrature of the hyperbola (hence ultimately logarithmic integrals).

But Maupertuis did not discuss the real problem where the earth's shape is concerned; how to evaluate these integrals in order to obtain numerical values. Thus he did not use his expressions to explain the numerical examples that Newton had introduced in presenting his theory of the earth's shape but whose underlying calculations Newton had not made clear. Newton had used these calculations to find a value of the earth's ellipticity equal to $\frac{1}{229}$. For example, Maupertuis did not use his expressions to verify Newton's assertions, which I mentioned in Chapter 1, that if the ratio of the polar radius to the equatorial radius of a homogeneous figure shaped like a flattened ellipsoid of revolution which attracts according to the universal inverse-square law is $\frac{100}{101}$, then the ratio of the magnitude of the attraction at a pole to the magnitude of the attraction at the equator of this homogeneous figure of revolution equals $\frac{501}{500}$, and that if, moreover, the columns from the center of this homogeneous figure of revolution to its surface which lie along the polar and equatorial axes of the figure balance or weigh the same, then the ratio of the magnitude of the centrifugal force per unit of mass to the magnitude of the attraction at the equator of the figure is $\frac{4}{505}$.

Maupertuis evidently dodged the problem of determining numerical values of his formal expressions for the magnitudes of the attraction at the poles of homogeneous figures shaped like ellipsoids of revolution which attract according to the universal inverse-square law. Such a problem could not be easily handled at the time, because nonalgebraic integrals appeared in the expressions. Maupertuis only affirmed that Newton determined the ratio $\frac{501}{500}$ by doing "similar calculations,"[62] meaning calculations similar to his own. Here Maupertuis surely bluffed, for he did not actually carry out any integrations, much less clarify what calculations Newton had done. Maupertuis simply repeated Newton's rather incomplete arguments, developing or expanding the parts that were already more or less evident or straightforward. Furthermore, Maupertuis derived formal expressions for the magnitudes of the attraction at the poles of homogeneous figures shaped like ellipsoids of revolution whose ellipticities are *finite* and which attract according to the universal inverse-square law. But Maupertuis did not try to show, say, that by making suitable first-order approximations in his expressions for the magnitudes of attraction, which hold when the ellipticities are infinitesimal, these expressions would simplify to ones whose numerical values could actually be computed and, in particular, to ones that could be used to produce the particular numerical value of the ratio of the magnitude of the attraction at a pole to the magnitude of the attraction at the equator which Newton had presented, namely $\frac{501}{500}$. Perhaps Maupertuis did not try to reduce his expressions for the magnitudes of the attraction at the poles of homogeneous figures shaped like ellipsoids of revolution which attract according to the universal inverse-square law, in cases where the ellipticities are infinitesimal, to expressions that could be made to give numerical values, by making approximations that hold good to first order, because he could not see how to do so.

Newton had claimed to show that at all points on the surface of a homogeneous figure shaped like a flattened ellipsoid of revolution which attracts in accordance with the universal inverse-square law, which revolves around its axis of symmetry, and whose columns from center to surface are all assumed to balance or weigh the same, the magnitude of the effective gravity at points on the surface of the homogeneous figure of revolution varies inversely as the distances of the points from the center of the figure. Newton stated that it followed from this result that the increase in the magnitude of the effective gravity along the surface of the homogeneous figure of revolution from its equator to its north pole must vary directly as the square of the sine of the latitude, but he did not furnish the reader with any demonstration of this assertion. Nor did Newton indicate whether the ellipticities of the homogeneous figures of revolution in question had to be infinitesimal in order for this "sin² law" to hold or whether "sin² law" held as well for homogeneous figures of revolution whose ellipticities are finite.

Maupertuis tried to demonstrate that for homogeneous figures shaped like flattened ellipsoids of revolution which attract according to the universal inverse-square law, which revolve around their axes of symmetry, where the columns

from the center of such a homogeneous figure of revolution to its surface are all
assumed to balance or weigh the same, and whose ellipticities are infinitesimal,
the increase in the magnitudes of the components of the effective gravity at points
on the surface of such a homogeneous figure of revolution which are directed
toward the center of the figure does vary directly as the squares of the sines of the
latitudes of the points, but he botched up the calculations. The errors that he
made suggest that he may have been incapable of carrying out computations
involving all but the simplest first-order approximations. (At the same time,
however, both Maupertuis's faulty proof as well as Maupertuis's faulty proof
corrected make use of the hypothesis that the homogeneous figures of revolution
in question are nearly spherical. This assumption shows that the "sin² law" of the
variation of the increase in the magnitudes of these components of effective
gravity with latitude at points on the surfaces of the homogeneous figures of
revolution at issue *only* holds when the ellipticities of the figures are *infinitesimal.*
This law *cannot* hold when the ellipticities of the figures are *finite.*[63] Consequently
it is hard to see how Maupertuis could have possibly believed that this part of
Newton's theory held true when the ellipticities of such homogeneous figures of
revolution are finite. As I noted in Chapter 3, Maupertuis did make first-order
approximations to arrive at equation (3.1), but in fact this turns out to be a very
easy exercise.)

Nevertheless, as I mentioned above, all the blame for having failed to differentiate homogeneous figures shaped like flattened ellipsoids of revolution whose
ellipticities are infinitesimal from homogeneous figures shaped like flattened
ellipsoids of revolution whose ellipticities are finite should not be laid upon
Maupertuis, because Newton himself paved the way for treating the two kinds of
figures as if they did not differ in any way.

Maupertuis simply displayed Newton's critical equation

$$\frac{\delta}{1/100} = \frac{\phi}{4/505} \tag{1.1}$$

which is valid, and only valid, for homogeneous figures shaped like flattened
ellipsoids of revolution which attract according to the universal inverse-square
law, which revolve around their axes of symmetry, where the columns from the
center of such a homogeneous figure of revolution to its surface are all assumed to
balance or weigh the same, and whose ellipticities are infinitesimal. Newton had
used equation (1.1) to determine a value of the earth's ellipticity δ equal to $\frac{1}{229}$ by
letting the ratio ϕ of the magnitude of the centrifugal force per unit of mass to the
magnitude of the attraction at the equator of a homogeneous figure shaped like a
flattened ellipsoid of revolution which attracts according to the universal inverse-
square law, which revolves around its axis of symmetry, and whose columns from
center to surface are all assumed to balance or weigh the same equal the value of
the ratio at the earth's equator, which is $\frac{1}{289}$. Maupertuis did not try at all to

explain equation (1.1). He undoubtedly did not understand where the equation came from or how Newton had discovered it.

[Maupertuis does appear to have derived the equation

$$\frac{X}{1/229} = \frac{\Omega/\Pi}{F/P} \tag{5.3.1}$$

for Jupiter's ellipticity X from Newton's equation (1.1), where $\frac{1}{229}$ = the earth's ellipticity, F/P = the ratio of the magnitude of the centrifugal force per unit of mass to the magnitude of the attraction at the earth's equator = $\frac{1}{289}$, and Ω/Π = the ratio of the magnitude of the centrifugal force per unit of mass to the magnitude of the attraction at Jupiter's equator, by himself. This suggests that he did conclude correctly from reading Newton that the quotient gotten by dividing the ellipticity ρ by the ratio θ of the magnitude of the centrifugal force per unit of mass to the magnitude of the attraction at the equator is the *same* for *all* homogeneous figures of revolution which Newton's theory covered.

That is, the quotient ρ/θ is the same for all homogeneous figures of revolution in question, provided ρ and θ are infinitesimal. According to equation $\frac{\delta}{1/100} = \frac{\phi}{4/505}$, (1.1) the constant value of this quotient is $\frac{1/100}{4/505}$.

But this value of the quotient can be simplified to $\frac{5}{4}$ as follows. Equation (1.1) can be rewritten as

$$\delta = \frac{1/100}{4/505}\phi. \tag{5.3.3}$$

Now, equation (5.3.3) can be rewritten as

$$\delta = \frac{505}{400}\phi = \left(\frac{5}{4} + \frac{1}{80}\right)\phi = \frac{5}{4}\phi + \frac{1}{80}\phi \tag{5.3.3'}$$

But $\frac{1}{80}$ is infinitesimal, since

$$\left(\frac{1}{80}\right)^2 = \frac{1}{6400} \ll \frac{1}{80}.$$

Consequently, since ϕ and $\frac{1}{80}$ are both infinitesimal, the term $\frac{1}{80}\phi$ on the right-hand side of equation (5.3.3') is a term of second order and can be discarded, in which case equation (5.3.3') becomes

$$\delta = \frac{5}{4}\phi. \tag{5.3.4}$$

Equation (5.3.4) holds to terms of first order when δ and ϕ are infinitesimal, so that equation (1.1) reduces to (5.3.4.)

Whether Maupertuis truly understood *why* this is true is another matter. It seems improbable. After all, Maupertuis should not have applied equation (5.3.1) to Jupiter, because as I mentioned above, Jupiter's ellipticity is not infinitesimal. But then Newton's application to Jupiter of his own equation for Jupiter's ellipticity, an equation that is also based on equation (1.1), remained unfounded even in his *own* work, for the same reason. It seems more likely that Maupertuis just blindly followed Newton.]

Finally, Maupertuis gave no sign in 1734 of understanding any better why Newton thought the earth to be shaped like an ellipsoid of revolution than he did when he wrote Johann I Bernoulli in March of 1731. In the three-year interim Maupertuis does not appear to have made much progress in his attempt to understand Newton's theory of the earth's shape, if he made any progress at all.

Maupertuis in effect treated Newton's theory of the shapes of the earth and other planetary bodies, which appears in the *Principia*, as a complete theory. He did not find any faults or weaknesses or detect any loose ends or inconsistencies. If Newton's theory were truly a finished product, the story of its diffusion in Paris might have ended here. But the story does not.

6

Alexis-Claude Clairaut's first theories of the earth's shape

6.1. Newton's theory completed (1737)

Early in 1737 the Paris Academy expedition to Lapland to measure a degree of latitude was still going on. At this time Anders Celsius, the Swedish astronomer and foreign member of the expedition, went from Lapland to London carrying a paper that Clairaut had written in Latin which dealt with Newton's theory of the earth's shape. The paper was read before the Royal Society late in March of 1737 and was subsequently published in the volume of the *Philosophical Transactions* for 1737–8.[1] In his covering letter to Royal Society Secretary Cromwell Mortimer, Clairaut contended that Newton's choice of the ellipsoid of revolution for the earth's shape was "without basis" in the *Principia*. Clairaut believed that Newton no doubt had reasons for choosing the shape that he did. However, Newton did not explain these in the *Principia*, Clairaut maintained. Clairaut noted that if Newton had chosen a different shape at the start, this could have led him to a different ratio of the earth's polar axis to the earth's equatorial axis.[2]

In Chapter 1 we saw that Newton used Proposition 91, Corollary 3 of Book I of the *Principia* to conlcude that the magnitudes of the components of the attraction at points along any line segment from the center to the surface of a homogeneous figure shaped like an ellipsoid of revolution which attracts according to the universal inverse-square law, where the relevant components of the attraction at points along such a line segment are the ones that have the same direction as that line segment, vary directly as the distances of the points from the center of the figure. If the figure revolves around its axis of symmetry, the magnitudes of the centrifugal force per unit of mass at points along an equatorial axis of the figure also vary directly as the distances of the points from the center of the figure. In Proposition 19, Book III of the *Principia*, Newton combined these two variations to calculate the magnitudes of the effective gravity at points along two columns from the center to the surface of a homogeneous figure shaped like a flattened ellipsoid of revolution which attracts according to the universal inverse-square law and which revolves around its axis of symmetry, where the two

Alexis-Claude Clairaut (engraved by Delafosse in the style of Carmontel, Bibliothèque Nationale, Paris).

columns lay along the polar and equatorial axes of the figure, respectively. Since the columns lay along the polar and equatorial axes of the figure, Newton in effect calculated the magnitudes of the total or the resultants of the attraction at points along the polar axis, where the magnitude of the centrifugal force per unit of mass is zero, and the magnitudes of the total or the resultants of the effective gravity at points along the equatorial axis. He used these computations to determine the weights of these two columns, which he then equated.

Why did Newton find the weights of these two columns and then equate these weights? In Proposition 20, Book III of the *Principia*, Newton contended that *all* of the columns from the center to the surface of a homogeneous fluid figure shaped like a flattened ellipsoid of revolution which attracts according to the universal inverse-square law and which revolves around its axis of symmetry at a certain rate (that is, at a certain angular speed) balance or weigh the same, where the ellipticity of the figure depends upon the rate at which the figure revolves around its axis of symmetry. He equated the weights of the two columns from the center of the homogeneous fluid figure of revolution in question to the surface of the figure which lie along the polar and equatorial axes of the figure in order to determine the ellipticity of the figure. But he believed that the weight of any column from the center of this figure to the surface of the figure is determined through the attraction and the centrifugal force per unit of mass at points along the column in exactly the same way that the weights of the two columns from the center of the figure to the surface of the figure which lie along the polar and equatorial axes of the figure are determined through the attraction and the centrifugal force per unit of mass at points along these two columns. Thus Newton thought that the weight of any column from the center of the figure to the surface of the figure could be found in principle the same way that he found the weights of the two columns from the center of the figure to the surface of the figure which lie along the polar and equatorial axes of the figure. Reversing in his mind the calculations that would have to be done to find the weight of an arbitrary column from the center of the figure to the surface of the figure and noting that a short column weighs exactly the same as a long one since all of the columns from the center of the figure to the surface of the figure balance or weigh the same, Newton concluded that the magnitude of the effective gravity at all points on the surface of the figure varies inversely as the distances of the points from the center of the figure.

But Newton never actually *demonstrated* in the *Principia* that if two columns from the center to the surface of a homogeneous fluid figure shaped like a flattened ellipsoid of revolution which attracts in accordance with the universal inverse-square law and which revolves around its axis of symmetry balance or weigh the same, where the two columns lie along the polar and equatorial axes of the figure, then *all* of the columns from the center to the surface of the figure balance or weigh the same – or, equivalently, that the magnitudes of the components of the effective gravity at *all* points on the surface of the homogeneous fluid figure shaped like a flattened ellipsoid of revolution which attracts according to the universal inverse-square law and which revolves around its axis of symmetry, where the relevant component of the effective gravity at each point on the surface of the figure is the one that points to the center of the figure, *do* vary inversely as the distances of the points from the center of the figure.[3] Instead he just *assumed* that all of the columns from the center of such a figure to the surface of the figure would balance or weigh the same if the two columns from the center of the figure to the surface of the figure which lie along the polar and equatorial axes of the figure balance or weigh the same, and then he determined the ellipticity of the figure by equating the weights of the two columns from the center of the figure to the surface of the figure which lie along the polar and equatorial axes of the figure.

In his paper of 1737 in the *Philosophical Transactions*, Clairaut filled in this gap. In so doing, he penetrated Newton's theory of the shape of a homogeneous fluid figure of revolution which attracts in accordance with the universal inverse-square law, which revolves around its axis of symmetry, and whose columns from center to surface are all assumed to balance or weigh the same to a degree that no Frenchman, Maupertuis included, had succeeded in doing before him. Clairaut demonstrated that the columns from the center to the surface of a homogeneous fluid figure shaped like a flattened ellipsoid of revolution which attracts according to the universal inverse-square law and which revolves around its axis of symmetry "slowly," meaning that the ratio ϕ of the magnitude of the centrifugal force per unit of mass to the magnitude of the attraction at the equator of the figure is infinitesimal ($\phi^2 \ll \phi$), hence whose ellipticity δ is infinitesimal as well ($\delta^2 \ll \delta$), *can* all balance or weigh the same. (In actuality, Clairaut showed that if the two columns from the center of the figure to the surface of the figure which lie along the polar and equatorial axes of the figure balance or weigh the same, then all of the columns from the center of the figure to the surface of the figure *which lie in the same plane as the polar and equatorial axes of the figure* will balance or weigh the same. However, as I indicated in Chapter 1, it is enough to show that this much is true without any loss of generality. For if all of the columns from the center of the figure to the surface of the figure which lie in the same plane of a meridian at the surface of the figure balance or weigh the same, it follows immediately that all of the columns from the center of the figure to the surface of the figure will balance or weigh the same, for reasons that I explained in Chapter 1.)

To prove that this is true, the *Principia* did not suffice. In computing the weights of two columns from the center to the surface of a homogeneous figure shaped like a flattened ellipsoid of revolution which attracts according to the universal inverse-square law and which revolves around its axis of symmetry, where one column connects the center of the figure to one of the poles of the figure and the other column joins the center of the figure to a point on the equator of the figure, Newton actually had to calculate the values of the magnitudes of the attraction at only two points, both of which are situated at the surface of the figure, one on the polar axis of the figure and the other on an equatorial axis of the figure, using Proposition 91, Corollary 2 of Book I of the *Principia*. After computing the magnitudes of the attraction at a pole of the figure and at a point on the equator of the figure, Newton could use his geometrical theorem (Proposition 91, Corollary 3 of Book I of the *Principia*) to determine ratios that gave the values of the magnitudes of the attraction at points elsewhere on the two axes of the figure, which Newton needed in order to calculate the weights of the two columns from the center of the figure to the surface of the figure which lie along the two axes of the figure. But to fill in the missing gap, Clairaut needed to know the magnitudes of the relevant components of effective gravity at *all* points on the surface of the homogeneous figure shaped like a flattened ellipsoid of revolution which attracts according to the universal inverse-square law and which revolves around its axis of symmetry – namely, those components of effective gravity at the points which are directed toward the center of the figure. Maupertuis's *mémoire* of 1734 could not help Clairaut deal with this problem; Maupertuis had simply reproduced Newton. A demonstration that all of the columns from the center to the surface of a homogeneous fluid figure shaped like a flattened ellipsoid of revolution which attracts according to the universal inverse-square law and which revolves around its axis of symmetry can all balance or weigh the same required the use of more mathematics than Maupertuis could muster or create.

In the paper on the laws of attraction which he read before the Paris Academy in 1732–33, Maupertuis did put forward the plausible hypothesis, which appears in a passage left out of the version of the paper published in the Paris Academy's *Mémoires*, that nonalgebraic integrals sometimes involving complicated logarithmic integrals, which arose in applying analysis (that is, the integral calculus) to study homogeneous spherical figures that attract in accordance with universal laws of attraction which have the form $f(r) = r^n$, where $f(r)$ is the magnitude of the attraction between two points that attract each other and where r is the distance between the two points, when n has certain values not equal to -2 (specifically, $n = -1, -3$, or -5), could have induced Newton to formulate his theory of attractions geometrically instead of analytically. (Maupertuis also briefly mentioned Newton's follower Roger Cotes's achievements pertaining to this particular problem in integral calculus. Here Maupertuis probably alluded to Cotes's posthumously published *Harmonia mensurarum* (1722).[4]).

Again let $|XY|$ stand for the length of line segment XY. Let $AEaeA$ designate a meridian at the surface of Newton's homogeneous figure shaped like a flattened ellipsoid of revolution which attracts according to the universal inverse-square law and which revolves around its axis of symmetry (see Figure 16). A meridian at the surface of such a figure is a flattened ellipse. According to Clairaut, the problem was to show that the magnitude of the particular component of attraction at any point N on the flattened ellipse $AEaeA$ which is a meridian at the surface of Newton's homogeneous rotating figure of revolution which is directed

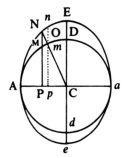

Figure 16. Clairaut's figure used for determining the magnitude of the attraction at a pole of a homogeneous flattened ellipsoid of revolution which attracts according to the universal inverse-square law and whose ellipticity is infinitesimal (Clairaut 1809: Figure 12).

toward the center C of the figure minus the magnitude of the particular component of centrifugal force per unit of mass at N which is again directed toward the center C of the figure varies inversely as the length of the radius CN of the figure from the center C of the figure to N (in other words, the difference between the two magnitudes is directly proportional to $1/|CN|$).[5] If the calculations that Newton had mentally reversed, which I alluded to above, are now inverted, a demonstration of the result stated in the preceding sentence would also prove that all of the columns from the center of the figure to the surface of the figure which lie in the plane of the flattened ellipse $AEaeA$ which is a meridian at the surface of the figure balance or weigh the same.

Calculating the exact values of the magnitudes of the attractions at arbitrary points on the surface of a homogeneous figure shaped like an ellipsoid of revolution which attracts according to the universal inverse-square law was still a difficult problem in integral calculus at this time. Everyone agreed, however, that the earth's shape and a sphere did not differ much, whatever shape the earth might have. Assuming that the earth is shaped like a nearly spherical ellipsoid of revolution, the earth's ellipticity δ, whatever its value, was consequently infinitesimal, which meant that δ is a nonzero number for which $\delta^2 \ll \delta$ is true. The controversy centered upon the sign of δ (+ if the earth is flattened at the poles, − if the earth is elongated at the poles). Consequently a theory accurate to terms of first order in δ seemed adequate to Clairaut. Such a theory allowed Clairaut to make approximations, by neglecting the squares and higher powers of the ellipticity δ, as well as the squares and higher powers of all other quantities of the same order as δ, such as the ratio ϕ of the magnitude of the centrifugal force per unit of mass to the magnitude of the attraction at the equator of the figure. Clairaut aimed to simplify the expressions for the magnitudes of the attraction at points on the surfaces of the homogeneous figures of revolution in question, which in general involved complicated integrals. Whereas these expressions almost never could be evaluated, the simplified expressions arrived at by making approximations could be at least be evaluated.

In doing parts of his calculations Clairaut made errors; other parts of his calculations Clairaut explained incorrectly. First I shall sketch Clairaut's argument. Then I shall point out parts of the calculations and the intermediate results that Clairaut arrived at which appear troublesome. Finally, I will show that Clairaut's ultimate conclusions can be saved. Some portions of the calculations which seem problematic at first sight are perfectly valid. Clairaut simply explained them inaccurately, perhaps because he was careless. But as we shall see, in reaching his final results, Clairaut either did doctor his calculations to suit his needs, or else he unknowingly made errors that cancelled each other. For this reason I examine Clairaut's calculations in more detail than would be necessary if Clairaut had made no mistakes.

Clairaut began by computing the approximate value of the magnitude of the attraction that a nearly spherical homogeneous figure shaped like an ellipsoid of revolution flattened at its poles which attracts according to the universal inverse-square law and whose meridians are flattened ellipses $AEaeA$ exerts at one of its poles A. He took the uniform density of the homogeneous figure of revolution to be 1. Clairaut said that he used Proposition 91, Corollary 2 of Book I of the *Principia*, a geometric theorem, to guide him in doing the calculation of the approximate value of the magnitude of the attraction at the pole A of the figure analytically.[6] He divided the computation into two parts. First he calculated the approximate value of the magnitude of the attraction at the pole A produced by the homogeneous figure which attracts according to the universal inverse-square law, whose density is 1, and which is composed of the two meniscus caps formed by revolving $AEaDA$ around the axis of symmetry Aa of the homogeneous figure shaped like an ellipsoid of revolution flattened at its poles which attracts according to the universal inverse-square law, whose density equals 1, and whose meridians are flattened ellipses $AEaeA$. Then he added to the expression that he found for this magnitude the magnitude of the attraction at A produced by the homogeneous spherical figure which attracts according to the universal inverse-square law, whose density is 1, and whose radius is the polar radius AC of the homogeneous figure shaped like a flattened ellipsoid of revolution which attracts according to the universal inverse-square law, whose density equals 1, and whose meridians are flattened ellipses $AEaeA$. (The circle $ADadA$ is the meridian at the surface of the homogeneous spherical figure which is situated in the plane of the flattened ellipse $AEaeA$ which is a meridian at the surface of the homogeneous figure shaped like a flattened ellipsoid of revolution. See Figure 16.)

Clairaut let N be a point on the flattened ellipse $AEaeA$ which is a meridian at the surface of the homogeneous figure shaped like flattened ellipsoid of revolution which attracts according to the universal inverse-square law and whose density equals 1. He let the polar radius of the flattened ellipsoid of revolution $AC = r$, and he let $AP = u$, where P is the point where the line through N perpendicular to the axis of symmetry Aa of the flattened ellipsoid of revolution intersects the axis of symmetry Aa. The ellipticity a of the flattened ellipsoid of revolution

$$\frac{CE - CA}{CA} = \frac{CE - r}{r},$$

which is a positive number, since $CA < CE$. It follows from the definition of the ellipticity a that $r + DE = CD + DE = CE = r + ar$. Hence $DE = ar$. The ellipticity a is infinitesimal ($a^2 \ll a$), since the homogeneous figure shaped like a flattened ellipsoid of revolution which attracts according to the universal inverse-square law and whose density equals 1 is assumed to be nearly spherical. Let $PN = x$. Then the equation of the flattened ellipse $AEaeA$ expressed in terms of the quantities x, r, u, and a is

$$\frac{x^2}{(r + ar)^2} + \frac{(r - u)^2}{r^2} = 1. \tag{6.1.1}$$

Hence

$$x = (r + ar)\sqrt{\frac{r^2 - (r - u)^2}{r^2}} = (1 + a)\sqrt{2ru - u^2}. \tag{6.1.2}$$

Moreover, $PM^2 + (r - u)^2 = CM^2 = r^2$, where M is the point where the line segment PN intersects the circle $ADadA$ which is the meridian at the surface of the homogeneous spherical figure situated in the plane of the flattened ellipse $AEaeA$. Hence

$$PM = \sqrt{r^2 - (r - u)^2} = \sqrt{2ru - u^2}. \tag{6.1.3}$$

It follows that

$$NM = PN - PM = x - PM = a\sqrt{2ru - u^2}. \tag{6.1.4}$$

Furthermore, $u^2 + PM^2 = AP^2 + PM^2 = AM^2$, or $u^2 + (2ru - u^2) = AM^2$, so that

$$AM = \sqrt{2ru}. \tag{6.1.5}$$

Clairaut stated that the volume of the ring whose thickness is du formed by revolving $NnmM$ around the axis of symmetry Aa of the flattened ellipsoid of revolution, where n is a point on the flattened ellipse $AEaeA$ which is very near N, p is the point where the line through n perpendicular to the axis of symmetry Aa of the flattened ellipsoid of revolution intersects the axis of symmetry Aa, in which case p is very near P, and m is the point where the line segment pn intersects the circle $ADadA$ which is the meridian at the surface of the homogeneous spherical figure located in the plane of the flattened ellipse $AEaeA$, in which case m is very near M, is

$$\frac{ac}{r}(2ru - u^2)du, \tag{6.1.6}$$

where $c \equiv$ the circumference of the circle $ADadA$ divided by the radius AC of the circle $ADadA = 2\pi$.

In order to calculate the approximate value of the magnitude of the attraction at the pole A produced by the homogeneous figure which attracts according to the universal inverse-square law, whose density is 1, and which is composed of the two meniscus caps formed by revolving $AEaDA$ around the axis of symmetry Aa of the flattened ellipsoid of revolution, Clairaut now computed the magnitude of the attraction at pole A due to a particle at M, which is one of the particles that make up the figure composed of the two meniscus caps. Because the axis of symmetry Aa of the flattened ellipsoid of revolution is also the axis of symmetry of the figure composed of the two meniscus caps, the line that specifies the direction of the resultant of the attraction at the pole A produced by the figure composed of the two meniscus caps has the same direction as the axis of symmetry Aa. In other words, the axis of symmetry Aa is part of this line. The components of attraction at the pole A in all other directions produced by the particles that make up the figure composed of the two meniscus caps mutually cancel. Consequently in order to calculate the attraction that the particle at M contributes to the attraction at the pole A produced by this figure, it is enough to calculate the magnitude of the component of attraction at pole A produced by a particle at M whose direction is specified by the line that includes the axis of symmetry Aa. The magnitude of the total attraction at A due to a particle at M equals $1/AM^2$. The magnitude of the component of the attraction at A produced by a particle at M which has the same direction as the axis of symmetry Aa equals $1/AM^2$ multiplied by the cosine of the angle PAM, which equals

$$\frac{1}{AM^2} \times \frac{AP}{AM} = \frac{1}{2ru} \times \frac{u}{\sqrt{2ru}}. \tag{6.1.7}$$

Clairaut next assumed that all of the particles that make up line segment NM attract the pole A as if they were all situated at M, because NM is so small by comparison with the polar radius $AC = r$ of the flattened ellipsoid of revolution. Indeed, $NM/r \leqslant DE/r =$ the ellipticity a, which is infinitesimal. (It can be shown that making this assumption amounts to making an approximation that holds to terms of first order in the ellipticity a.) Clearly the same supposition applies equally to all of the particles that make up the annulus formed by revolving NM around the axis of symmetry Aa. Furthermore, Clairaut presumed the same hypothesis to be true of all of the particles that make up the ring whose thickness is du formed by revolving $NnmM$ around the axis of symmetry Aa. Having made these assumptions, Clairaut multiplied (6.1.7) by (6.1.6), from which it followed that

$$\frac{1}{2ru} \times \frac{u}{\sqrt{2ru}} \times \frac{ac}{r}(2ru - u^2)\,du = \frac{ac}{2r^2\sqrt{2r}}(2r\sqrt{u} - u\sqrt{u})\,du \qquad (6.1.8)$$

was the approximate value of the magnitude of the attraction at the pole A produced by this ring.

Clairaut then integrated expression (6.1.8) from $u = 0$ to $u = 2r$, and he determined the integral to equal

$$\tfrac{8}{15}ac. \qquad (6.1.9)$$

Thus according to Clairaut, (6.1.9) expressed the approximate value of the magnitude of the attraction at the pole A produced by the figure composed of the two meniscus caps whose densities equal 1.

Clairaut next affirmed that the magnitude of the attraction at the pole A produced by the homogeneous spherical figure whose density is 1 and whose radius is the polar radius $AC = r$ of the flattened ellipsoid of revolution is

$$\tfrac{2}{3}c. \qquad (6.1.10)$$

Adding the two expressions (6.1.9) and (6.1.10) together, Clairaut found

$$\tfrac{2}{3}c + \tfrac{8}{15}ac \qquad (6.1.11)$$

to be the expression for the approximate value of the magnitude of the attraction at one of the poles A of a homogeneous figure shaped like an ellipsoid of revolution flattened at its poles which attracts according to the universal inverse-square law, whose polar radius $AC = r$, whose ellipticity a is positive and infinitesimal, whose density is 1, and whose meridians are flattened ellipses $AEaeA$.

Clairaut stated as a corollary that if the homogeneous figure which attracts in accordance with the universal inverse-square law, whose polar radius $AC = r$, and whose density is 1 is shaped instead like an ellipsoid of revolution elongated at its poles whose ellipticity is infinitesimal, in which case the infinitesimal ellipticity of the ellipsoid of revolution is negative, hence it has the form $-a$ where a is a positive number, then

$$\tfrac{2}{3}c - \tfrac{8}{15}ac \qquad (6.1.12)$$

will express the approximate value of the magnitude of the attraction produced by this homogeneous figure of revolution at one of its poles A.

Clairaut now considered again the attraction produced by the homogeneous figure shaped like a flattened ellipsoid of revolution which attracts according to the universal inverse-square law, whose polar radius $AC = r$, whose ellipticity a is positive and infinitesimal, whose density is 1, and whose meridians are flattened ellipses $AEaeA$. He calculated the approximate value of the magnitude of the particular component of the attraction at a point N on the flattened ellipse $AEaeA$ which is a meridian at the surface of this figure which points to the center C of the figure, where N can be any point on the meridian $AEaeA$ at the surface of the figure. Using new notation, he let $AC = a$, $CE = b$, and $CN = r$. Clairaut demonstrated a theorem to show that the approximate value of the magnitude of this component of attraction at N was the same as the approximate value of the magnitude of the attraction at N produced by a homogeneous figure shaped like an ellipsoid of revolution which attracts according to the universal inverse-square law, one of whose poles is at N, whose ellipticity is infinitesimal, whose density is 1, and whose axis of symmetry is the diameter of the homogeneous figure shaped like a flattened ellipsoid of revolution which attracts according to the universal inverse-square law, whose polar radius $AC = a$ in Clairaut's new notation, whose ellipticity a in Clairaut's old notation is positive and infinitesimal, whose density is 1, and whose meridians are flattened ellipses $AEaeA$ which traverse this figure from N through C. The polar radius of the homogeneous figure shaped like an ellipsoid of revolution which attracts according to the universal inverse-square law, one of whose poles is at N, whose ellipticity is infinitesimal, and whose density is 1 is consequently the line segment $CN = r$ in Clairaut's new notation. Moreover, according to Clairaut, the radius along the longer of the two axes of this figure is its polar radius $CN = r$ in Clairaut's new notation, and the radius along the shorter of the two axes of the figure is equal to $b\sqrt{a/r}$ in Clairaut's new notation. It follows from the last two statements that the homogeneous figure shaped like an ellipsoid of revolution which attracts according to the universal inverse-square law, one of whose poles is at N, whose ellipticity is infinitesimal, and whose density is 1 is elongated at its poles. Thus Clairaut reduced the problem of calculating the approximate value of the magnitude of the component of the attraction at N directed toward C produced by the homogeneous figure shaped like a flattened ellipsoid of revolution which attracts according to the universal inverse-square law, whose polar radius $AC = a$ in Clairaut's new notation, whose ellipticity a in Clairaut's old notation is positive and infinitesimal, whose density is 1, and whose meridians are flattened ellipses $AEaeA$ to the problem of calculating the approximate value of the magnitude of the attraction at a pole of a homogeneous figure shaped like an ellipsoid of revolution which attracts according to the universal inverse-square law, whose ellipticity is infinitesimal, and whose density is 1. In other words, Clairaut reduced the new problem to the problem that he had just solved.

[In fact, Clairaut stated that he was determining the approximate value of the magnitude of the *total* or the *resultant of* the attraction at N, not the magnitude of the *component* of the attraction at N which points to C, produced by the homogeneous figure shaped like a flattened ellipsoid of revolution which attracts according to the universal inverse-square law, whose polar radius is $AC = a$ in Clairaut's new notation, whose ellipticity a in Clairaut's old notation is positive and infinitesimal, whose density is 1, and whose meridians are flattened ellipses

AEaeA. But it is the magnitude of the component of the attraction at *N* produced by the homogeneous figure shaped like a flattened ellipsoid of revolution which attracts according to the universal inverse-square law, whose polar radius *AC* = *a* in Clairaut's new notation, whose ellipticity *a* in Clairaut's old notation is positive and infinitesimal, whose density is 1, and whose meridians are flattened ellipses *AEaeA* which is directed toward the center *C* of this figure which Clairaut needed to know in order to determine whether columns from the center to the surface of this figure all balance (that is, all weigh the same) or not.

Now, as I just mentioned, the polar axis, which is the axis of symmetry, of the homogeneous figure shaped like an ellipsoid of revolution which attracts according to the universal inverse-square law, one of whose poles is at *N*, whose ellipticity is infinitesimal, and whose density is 1, passes through the center *C* of the homogeneous figure shaped like a flattened ellipsoid of revolution which attracts according to the universal inverse-square law, whose polar radius *AC* = *a* in Clairaut's new notation, whose ellipticity *a* in Clairaut's old notation is positive and infinitesimal, whose density is 1, and whose meridians are flattened ellipses *AEaeA*. Indeed, the line segment *CN* = *r* in Clairaut's new notation is just the polar radius of the first figure, as I also just noted. But the polar axis of this figure indicates the direction of the total or the resultant of the attraction at *N* produced by this figure. Hence the direction of the total or the resultant of the attraction at *N* produced by the homogeneous figure shaped like an ellipsoid of revolution which attracts according to the universal inverse-square law, one of whose poles is at *N*, whose ellipticity is infinitesimal, whose density is 1, and whose polar radius is the line segment *CN* = *r* in Clairaut's new notation is the same as the direction of the component of the attraction at *N* directed toward *C* produced by the homogeneous figure shaped like a flattened ellipsoid of revolution which attracts according to the universal inverse-square law, whose polar radius *AC* = *a* in Clairaut's new notation, whose ellipticity *a* in Clairaut's old notation is positive and infinitesimal, whose density is 1, and whose meridians are flattened ellipses *AEaeA*. In short, the two attractions have the same directions.

In effect, Clairaut *implicitly equated* the magnitude of the total or the resultant of the attraction at *N* produced by the homogeneous figure shaped like an ellipsoid of revolution which attracts according to the universal inverse-square law, one of whose poles is at *N*, whose ellipticity is infinitesimal, whose density is 1, and whose polar radius is the line segment *CN* = *r* in Clairaut's new notation and the magnitude of the component of the attraction at *N* which points to *C* produced by the homogeneous figure shaped like a flattened ellipsoid of revolution which attracts according to the universal inverse-square law, whose polar radius is *AC* = *a* in Clairaut's new notation, whose ellipticity *a* in Clairaut's old notation is positive and infinitesimal, whose density is 1, and whose meridians are flattened ellipses *AEaeA*. But how did Clairaut justify equating the magnitudes of the two attractions?

The ellipticity of the homogeneous figure shaped like an ellipsoid of revolution which attracts according to the universal inverse-square law, one of whose poles is at N, whose density is 1, and whose polar radius is the line segment $CN = r$ in Clairaut's new notation is infinitesimal, and Clairaut was only calculating a value of the magnitude of the total or the resultant of the attraction at N produced by this figure which held true to terms of first order. At the same time, the ellipticity of the homogeneous figure shaped like a flattened ellipsoid of revolution which attracts according to the universal inverse-square law, whose polar radius $AC = a$ in Clairaut's new notation, whose density is 1, and whose meridians are flattened ellipses $AEaeA$, which is the figure that we started with, is also infinitesimal. Presumably then, the magnitude of the total or the resultant of the attraction at N produced by the homogeneous figure shaped like an ellipsoid of revolution which attracts according to the universal inverse-square law, one of whose poles is at N, whose ellipticity is infinitesimal, whose density is 1, and whose polar radius is the line segment $CN = r$ in Clairaut's new notation and the magnitude of the component of the attraction at N directed toward C produced by the homogeneous figure shaped like a flattened ellipsoid of revolution which attracts according to the universal inverse-square law, whose polar radius $AC = a$ in Clairaut's new notation, whose ellipticity a in Clairaut's old notation is positive and infinitesimal, whose density is 1, and whose meridians are flattened ellipses $AEaeA$ differ by terms of second order and higher orders at most. That is, the magnitudes of the two attractions at N are equal to terms of first order. At least, Clairaut tacitly assumed this to be the case, although he did not actually prove this to be true.]

To calculate the approximate value of the magnitude of the attraction at N produced by the homogeneous figure shaped like an ellipsoid of revolution which attracts according to the universal inverse-square law, one of whose poles is at N, whose ellipticity is infinitesimal, whose density is 1, and whose polar radius is the line segment $CN = r$ in Clairaut's new notation, Clairaut said that the expression

$$\tfrac{2}{3}c - \tfrac{8}{15}ac \qquad (6.1.12)$$

in his old notation must be used. According to Clairaut, to find this value the expression

$$\frac{r - b\sqrt{a/r}}{r} = 1 - \frac{b}{r}\sqrt{\frac{a}{r}} \qquad (6.1.13)$$

in his new notation must be substituted for a in (6.1.12). Since expression (6.1.12) was, according to Clairaut, only supposed to be used when the infinitesimal ellipticity of a homogeneous figure shaped like an ellipsoid of revolution which attracts according to the universal inverse-square law, whose polar radius $AC = r$ in Clairaut's old notation, and whose density equals 1 is negative (specifically, when the ellipticity equals $-a$ in Clairaut's old notation, where a is a positive number), the implication here is that expression (6.1.13) is always a positive number. This in fact is consistent with Clairaut's statement above that the radius along the longer of the two axes of the homogeneous figure shaped like an ellipsoid of revolution which attracts according to the universal inverse-square law, one of whose poles is at N, whose polar radius is the line segment $CN = r$ in Clairaut's new notation, whose ellipticity is infinitesimal, and whose density is 1 is its polar radius $CN = r$ in Clairaut's new notation and that the radius along the shorter of the two axes of this figure is equal to $b\sqrt{a/r}$ in Clairaut's new notation. Indeed, according to this statement, the ellipticity of the figure is the negative number

$$\frac{b\sqrt{a/r} - r}{r}, \tag{6.1.13*}$$

which means that the figure is elongated at its poles. Moreover, expression (6.1.13*) equals $-$ expression (6.1.13), which means that expression (6.1.13) is a positive number.

To find the result of substituting expression (6.1.13) for a in expression (6.1.12), Clairaut first let

$$c = pr \tag{6.1.14}$$

in the expression

$$\tfrac{2}{3}c - \tfrac{8}{15}ac. \tag{6.1.12}$$

The expression that results is

$$\tfrac{2}{3}pr - \tfrac{8}{15}apr. \tag{6.1.12'}$$

He then simplified the expression

$$1 - \frac{b}{r}\sqrt{\frac{a}{r}} \tag{6.1.13}$$

by neglecting all terms of second order and higher orders in the infinitesimal ellipticity of his homogeneous figure shaped like a flattened ellipsoid of revolution which attracts according to the universal inverse-square law, whose polar radius $AC = a$ in Clairaut's new notation, whose density is 1, and whose meridians are flattened ellipses $AEaeA$, as well as all terms of second order and higher orders in quantities that have the same order as this ellipticity. Namely, he let m stand for the positive number that is the infinitesimal ellipticity of his homogeneous figure shaped like a flattened ellipsoid of revolution which attracts according to the universal inverse-square law, whose polar radius $AC = a$ in Clairaut's new notation, whose density is 1, and whose meridians are flattened ellipses $AEaeA$. Since $AC = a$ and $CE = b$ in Clairaut's new notation, it follows that $m \equiv (b - a)/a$ in Clairaut's new notation, in which case $b = a + ma$. Clairaut let $r = a + na$ in his new notation, in which case

$$n = \frac{r - a}{a} \leqslant \frac{b - a}{a} = m,$$

since $a \leqslant r \leqslant b$, so that n is also infinitesimal. He then substituted these expressions for b and r into (6.1.13). After expanding in a power series the expression that resulted from these substitutions and then discarding all terms n^2, m^2, and mn of second order, as well as all terms of higher order, in the power series, he found

$$\tfrac{3}{2}n - m \qquad (6.1.13')$$

to be expression (6.1.13) reduced to a form which held true to terms of first order. He then substituted expression (6.1.13') for a in the expression

$$\tfrac{2}{3}pr - \tfrac{8}{15}apr. \qquad (6.1.12')$$

Consequently expression (6.1.12') became expression

$$\tfrac{2}{3}pr - \tfrac{8}{15}(\tfrac{3}{2}n - m)pr = \tfrac{2}{3}pr - \tfrac{4}{5}prn + \tfrac{8}{15}prm. \qquad (6.1.12'')$$

Substituting $r = a + na$ for r in expression (6.1.12'') and again neglecting all terms of second order, Clairaut finally arrived at

$$\tfrac{2}{3}pa - \tfrac{2}{15}pan + \tfrac{8}{15}pam \qquad (6.1.12''')$$

as the expression in his new notation for the approximate value of the magnitude of the component of the attraction at N which is directed toward C produced by the homogeneous figure shaped like a flattened ellipsoid of revolution which attracts according to the universal inverse-square law, whose polar radius $AC = a$ in Clairaut's new notation, whose ellipticity m in Clairaut's new notation is positive and infinitesimal, whose density is 1, and whose meridians are flattened ellipses $AEaeA$. He cited as special cases $n = 0$, in which case $r = a + na = a$, and $N =$ the pole A of his homogeneous figure shaped like a flattened ellipsoid of revolution, where the magnitude of the attraction is therefore

$$\tfrac{2}{3}pa + \tfrac{8}{15}pam, \qquad (6.1.15)$$

and $n = m$, in which case $r = a + na = a + ma = b$, and N is a point E on the equator of his homogeneous figure shaped like a flattened ellipsoid of revolution, where the magnitude of the attraction is accordingly

$$\tfrac{2}{3}pa + \tfrac{6}{15}pam. \qquad (6.1.16)$$

I shall now examine Clairaut's assumptions and results more closely.

1. Clairaut found the expression

$$\tfrac{8}{15}ac \qquad (6.1.9)$$

in his old notation to be that part of the approximate value of the magnitude of the attraction at a pole A of a homogeneous figure shaped like a flattened ellipsoid of revolution which attracts according to the universal inverse-square law, whose polar radius $AC = r$ in Clairaut's old notation, whose ellipticity a in Clairaut's old notation is positive and infinitesimal, whose density is 1, and whose meridians are flattened ellipses $AEaeA$ produced by the figure composed of the two meniscus caps. Expression (6.1.9) depends only on the ellipticity a in Clairaut's old notation of the flattened ellipsoid of revolution. It does not depend on the polar radius $AC = r$ of the flattened ellipsoid of revolution in Clairaut's old notation. It seems improbable that the attraction at the pole A produced by the figure composed of the two meniscus caps does not depend at all on the polar radius of the flattened ellipsoid of revolution.

2. Since c stands for 2π in Clairaut's notation, the magnitude of the attraction at the surface of a homogeneous spherical figure which attracts in accordance with the universal inverse-square law, whose density is 1, and whose radius is r equals

$$\tfrac{4}{3}\pi r = \tfrac{2}{3}(2\pi r) = \tfrac{2}{3}cr$$

By adding the expression

$$\tfrac{2}{3}c \tag{6.1.10}$$

to the expression (6.1.9) in order to arrive at expression (6.1.11), Clairaut must have tacitly assumed that *the polar radius $AC = r$ of the flattened ellipsoid of revolution in his old notation equals 1*. Yet he claimed that

$$\tfrac{2}{3}c + \tfrac{8}{15}ac \tag{6.1.11}$$

in his old notation expressed the approximate value of magnitude of the attraction at a pole A of a homogeneous figure shaped like a flattened ellipsoid of revolution which attracts according to the universal inverse-square law, whose polar radius $AC = r$ in Clairaut's old notation, whose ellipticity a in Clairaut's old notation is positive and infinitesimal, whose density is 1, and whose meridians are flattened ellipses $AEaeA$. But expression (6.1.11) does not depend upon r, and it seems unlikely that the attraction produced by the figure in question at one of its poles A does not depend upon its polar radius.

3. Clairaut stated that if a homogeneous figure which attracts in accordance with the universal inverse-square law, whose polar radius $AC = r$ in Clairaut's old notation, and whose density is 1 is shaped like an ellipsoid of revolution which is elongated at its poles and whose ellipticity is infinitesimal, in which case the infinitesimal ellipticity of the figure will be a negative number and therefore will have the form $-a$ in Clairaut's old notation, where a is a positive number, then the approximate value of the magnitude of the attraction at a pole A of this figure will be

$$\tfrac{2}{3}c - \tfrac{8}{15}ac \tag{6.1.12}$$

in his old notation. But to arrive at expression (6.1.12), Clairaut had to have tacitly supposed that the polar radius $AC = r$ of the elongated ellipsoid of revolution in his old notation equals 1. The proof that this is true can be seen to follow from the derivation of expression (6.1.12) which I give below. Yet Clairaut contended that (6.1.12) expressed the magnitude of the attraction at a pole A of a homogeneous figure shaped like an elongated ellipsoid of revolution which attracts according to the universal inverse-square law, whose polar radius $AC = r$ in Clairaut's old notation, whose infinitesimal ellipticity is a negative number and therefore has the form $-a$ in Clairaut's old notation, where a is a positive number, and whose density is 1. But expression (6.1.12) does not depend upon r, and, again, it seems improbable that the attraction produced by the figure in question at one of its poles A does not depend upon its polar radius.

4. Clairaut introduced a homogeneous figure shaped like an ellipsoid of revolution which attracts according to the universal inverse-square law, whose ellipticity is infinitesimal, one of whose poles is at N, whose density is 1, whose polar radius is the line segment $CN = r$ in Clairaut's new notation, and which attracts N with an approximate magnitude whose value is the same as the value of the approximate magnitude of the component of the attraction at N which points to C produced by the original homogeneous figure shaped like an ellipsoid of revolution flattened at its poles which attracts according to the universal inverse-square law, whose polar radius $AC = a$ in Clairaut's new notation, whose ellipticity a in Clairaut's old notation is positive and infinitesimal, whose density is 1, and whose meridians are flattened ellipses $AEaeA$. Clairaut maintained that the radius of the homogeneous figure shaped like an ellipsoid of revolution, one of whose poles is at N, along the longer of the two axes of this figure is the polar radius $CN = r$ of the figure in his new notation and that the radius of the figure along the shorter of the two axes of the figure is equal to $b\sqrt{a/r}$ in his new notation. To express the matter another way, Clairaut's statement means in particular that the homogeneous figure shaped like an ellipsoid of revolution, one of whose poles is at N, is always elongated at its poles, as I noted above. But these statements do not always hold true. For example, the first statement is true at the equator, where $N = E$, hence $r = CN = CE = b$ in Clairaut's new notation, and

$$b\sqrt{\frac{a}{r}} = b\sqrt{\frac{a}{b}} < b,$$

because $a < b$ and therefore $a/b < 1$. However, at the pole A, $N = A$, therefore $r = CN = CA = a$ in Clairaut's new notation and

$$b\sqrt{\frac{a}{r}} = b\sqrt{\frac{a}{a}} = b.$$

But a is less than b. In other words, $r < b\sqrt{a/r}$ in this particular case. Likewise, the second statement is not always true, because if $N = A$, then the corresponding ellipsoid of revolution is just the ellipsoid of revolution which is the outer surface of the original homogeneous figure shaped like an ellipsoid of revolution which attracts according to the universal inverse-square law, whose polar radius $AC = a$ in Clairaut's new notation, and whose density is 1 which we started with, which is flattened at its poles. (Its ellipticity m in Clairaut's new notation is positive and infinitesimal, and its meridians are flattened ellipses $AEaeA$.)

5. Clairaut says that we must apply the expression

$$\tfrac{2}{3}c - \tfrac{8}{15}ac \tag{6.1.12}$$

in his old notation to the homogeneous figure shaped like an ellipsoid of revolution which attracts according to the universal inverse-square law, whose ellipticity is infinitesimal, one of whose poles is at N, whose density is 1, and whose polar radius is the line segment $CN = r$ in Clairaut's new notation in order to determine the approximate value of the magnitude of the attraction at N produced by this homogeneous figure. But he stated that expression (6.1.12) is only to be applied to a homogeneous figure shaped like an elongated ellipsoid of revolution whose ellipticity is infinitesimal (in other words, a homogeneous figure shaped like an ellipsoid of revolution whose infinitesimal ellipticity is a negative number and therefore has the form $-a$ in Clairaut's old notation, where a is a positive number). This confirms that the ellipsoid of revolution, one of whose poles is at N, is supposed to be elongated at its poles. Indeed, as I also mentioned previously, the requirement that expression (6.1.12) always be used entails the same things as do Clairaut's statements that appear in 4, the preceding paragraph. But as I also indicated in paragraph 4, the ellipsoid of revolution cannot be elongated at its poles when $N = A$, because the corresponding ellipsoid of revolution is just the ellipsoid of revolution which is the outer surface of the original homogeneous figure shaped like an ellipsoid of revolution which attracts according to the universal inverse-square law, whose polar radius $AC = a$ in Clairaut's new notation, and whose density is 1 which we started with, which is flattened at its poles. (Its ellipticity m in Clairaut's new notation is positive and infinitesimal, and its meridians are flattened ellipses $AEaeA$.) If the expression (6.1.13) in Clairaut's new notation were always a positive number, it would in fact be the correct number to substitute for a in the expression (6.1.12) in Clairaut's old notation. But if $N = A$, in which case $r = a$, then (6.1.13) equals $(a - b)/a$, which is a negative number since $a < b$. It equals $-(b - a)/a = -m$, where m in Clairaut's new notation stands for the positive number that is the infinitesimal ellipticity of the ellipsoid of revolution which is the outer surface of the original homogeneous figure shaped like an ellipsoid of revolution flattened at its poles which attracts according to the universal inverse-square law, whose polar radius $AC = a$ in Clairaut's new notation, whose density is 1, and whose meridians are flattened ellipses $AEaeA$ which we began with.

Now, if we substitute the negative number $(6.1.13) = -m$ for a in the expression

$$\tfrac{2}{3}c - \tfrac{8}{15}ac \tag{6.1.12}$$

in Clairaut's old notation, which ordinarily we are not supposed to do according to Clairaut, because the a in the expression (6.1.12) stands for a positive number (the absolute value a of the negative, infinitesimal ellipticity $-a$, in Clairaut's old notation, of a homogeneous figure shaped like an ellipsoid of revolution infinitesimally elongated at its poles which attracts according to the universal inverse-square law, whose polar radius $AC = r$ in Clairaut's old notation, and whose density is 1), the expression that results is

$$\tfrac{2}{3}c + \tfrac{8}{15}mc, \qquad (6.1.11')$$

which in fact in Clairaut's new notation is just the approximate value of the magnitude of the attraction at A produced by the original homogeneous figure shaped like an ellipsoid of revolution flattened at its poles which attracts according to the universal inverse-square law, whose polar radius $AC = a$ in Clairaut's new notation, whose ellipticity m in Clairaut's new notation is positive and infinitesimal, whose density is 1, whose meridians are flattened ellipses $AEaeA$, and one of whose poles is at A. That is, (6.1.11') is the same as (6.1.11) in Clairaut's old notation. In other words, if expression (6.1.13) is substituted for a in expression (6.1.12) when $N = A$, the correct expression for the approximate value of the magnitude of the attraction at A produced by the original homogeneous figure shaped like an ellipsoid of revolution flattened at its poles which attracts according to the universal inverse-square law, whose polar radius $AC = a$ in Clairaut's new notation, whose ellipticity m in Clairaut's new notation is positive and infinitesimal, whose density is 1, whose meridians are flattened ellipses $AEaeA$, and one of whose poles is at A results, even though expression (6.1.13) is not a negative number in this case but is a positive number instead, nor is the homogeneous figure shaped like an ellipsoid of revolution which attracts in accordance with the universal inverse-square law, whose ellipticity is infinitesimal, one of whose poles is at A, and whose density equals 1 elongated at its poles but is flattened at its poles instead.

But, I repeat, the trouble is that, according to Clairaut, the expression

$$\tfrac{2}{3}c - \tfrac{8}{15}ac \qquad (6.1.12)$$

is only supposed to stand for the approximate value of the magnitude of the attraction at a pole of a homogeneous figure shaped like an ellipsoid of revolution elongated at its poles which attracts in accordance with the universal inverse-square law, whose polar radius $AC = r$ in Clairaut's old notation, whose infinitesimal ellipticity is a negative number and therefore has the form $-a$ in Clairaut's old notation, where a is a positive number, and whose density is 1. Expression (6.1.12) is not supposed to be applied to homogeneous figures shaped like ellipsoids of revolution flattened at their poles which attract according to the universal inverse-square law, whose ellipticities are positive and infinitesimal, and whose densities equal 1. In fact, if $N = E$, in which case $r = b$ in Clairaut's new notation, then the expression

$$\frac{r - b\sqrt{a/r}}{r} = \frac{b - b\sqrt{a/b}}{b}$$

in Clairaut's new notation is a positive number, because $a/b < 1$, and it *can* therefore be substituted for a in (6.1.12) in Clairaut's old notation. It does appear that the homogeneous figure shaped like an ellipsoid of revolution which attracts according to the universal inverse-square law, whose ellipticity is infinitesimal, whose density equals 1, and one of whose poles is at $N = E$ is indeed shaped like an elongated ellipsoid of revolution whose radius along its longer axis is its polar radius $CE = b$ and whose infinitesimal ellipticity is the negative number

$$-\left(\frac{b - b\sqrt{a/b}}{b}\right) = \frac{b\sqrt{a/b} - b}{b}.$$

6. Even if it made sense to apply (6.1.12) in Clairaut's old notation in the particular way that Clairaut said to use it at all points N, which does not appear to be the case (in particular, as I explained in numbered paragraph 5, it does not make sense to apply expression (6.1.12) at $N = A$, if we acknowledge the way that Clairaut says to use expression (6.1.12) to be correct), expression (6.1.12) still could only really be utilized when the polar radius of the homogeneous figure shaped like an ellipsoid of revolution which attracts according to the universal inverse-square law, whose ellipticity is infinitesimal, one of whose poles is at N, whose density is 1, and whose polar radius is the line segment $CN = r$ in Clairaut's new notation equals 1, as I indicated in paragraph 3, although Clairaut failed to perceive this fact, as I also noted in paragraph 3. But Clairaut provided no argument that assures that the polar radius of his homogeneous figure shaped like an ellipsoid of revolution which attracts according to the universal inverse-square law, whose ellipticity is infinitesimal, one of whose poles is at a particular point N, whose density is 1, and whose polar radius is the line segment $CN = r$ in Clairaut's new notation necessarily equals 1, even if the polar radius of the original homogeneous figure shaped like an ellipsoid of revolution flattened at its poles which attracts according to the universal inverse-square law and at whose surface the point N is situated, whose polar radius $AC = a$ in Clairaut's new notation, whose ellipticity m in Clairaut's new notation is positive and infinitesimal, whose meridians are flattened ellipses $AEaeA$, and whose density is 1 which we started with does equal 1.

7. Clairaut treated p in his expression

$$\tfrac{2}{3}pa - \tfrac{2}{15}pan + \tfrac{8}{15}pam \qquad (6.1.12''')$$

for the approximate value of the magnitude of the attraction at points N at the surface of his homogeneous figure shaped like a flattened ellipsoid of revolution which attracts according to the universal inverse-square law, whose polar radius $AC = a$ in Clairaut's new notation, whose ellipticity m in Clairaut's new notation is positive and infinitesimal, whose density is 1, and whose meridians are flattened ellipses $AEaeA$ as if p were constant. But if p were constant, then pr would vary with r. But pr does not vary with r; it is equal to the constant c. That is, Clairaut let

$$c = pr \qquad (6.1.14)$$

in the course of doing his calculation, in which case $p = c/r$.

Difficulties 1, 2, 3, 6, and 7 can be resolved. To show this, I shall calculate myself the volume of the ring whose thickness is du formed by revolving $NnmM$ around the axis of symmetry Aa. This volume is equal to the area of the annulus formed by revolving NM around the axis of symmetry Aa multiplied by the thickness du of the ring (see Figure 16). The area of the annulus is the difference between the areas of the outer and inner circles that bound the annulus. Expressed in terms of Clairaut's old notation for the various parameters, this area equals

$$\pi PN^2 - \pi PM^2 = \pi(PN^2 - PM^2)$$

$$= \pi((1+a)^2(2ru - u^2)) - \pi(2ru - u^2) = \pi((1+a)^2 - 1)(2ru - u^2)$$

$$= \pi(2a + a^2)(2ru - u^2), \qquad (6.1.17)$$

which equals

$$2\pi a(2ru - u^2) \qquad (6.1.17')$$

after the term in expression (6.1.17) which includes a^2 as a factor, which is a term of second order in the infinitesimal ellipticity a, where a is a positive number, is discarded. The following argument also leads to the same conclusion. Since $NM = a\sqrt{2ru - u^2}$ is a very small number, the annulus formed by revolving NM around the axis of symmetry Aa has to terms of first order in the infinitesimal ellipticity a an area equal to the circumference (that is, the length) of the circle that is the inner boundary of the annulus multiplied by the width of the annulus. But $2\pi PN = 2\pi\sqrt{2ru - u^2}$ is the circumference of the inner boundary of the annulus and $NM = a\sqrt{2ru - u^2}$ is the width of the annulus. The product of these two numbers is

$$2\pi a(2ru - u^2), \qquad (6.1.17')$$

which is the same as the preceding expression for the area of the annulus. Then

$$2\pi a(2ru - u^2)\,du \qquad (6.1.18)$$

expresses the volume of the ring whose thickness is du formed by revolving $NnmM$ around the axis of symmetry Aa. Since Clairaut let c stand for 2π, expression (6.1.18) is written as

$$ac(2ru - u^2)\,du \qquad (6.1.18')$$

in Clairaut's old notation. But Clairaut had stated that

$$\frac{ac}{r}(2ru - u^2)\,du \qquad (6.1.6)$$

expresses in his old notation the volume of the ring whose thickness is du formed by revolving $NnmM$ around the axis of symmetry Aa. Thus we see that the r in the denominator of the factor $(ac)/r$ in expression (6.1.6) does not belong there. Its presence cancelled out a factor r that should appear in the numerator of the correct expression for the attraction at the pole A produced by the ring whose thickness is du formed by revolving $NnmM$ around the axis of symmetry Aa. If Clairaut had integrated from $u = 0$ to $u = 2r$ the correct expression for the attraction at the pole A produced by this ring, he would have found

$$\tfrac{8}{15} acr \tag{6.1.19}$$

to be the correct expression in his old notation for the approximate value of the magnitude of the attraction at the pole A produced by the homogeneous figure composed of the two meniscus caps which attract according to the universal inverse-square law and whose densities equal 1. Instead Clairaut integrated from $u = 0$ to $u = 2r$ the incorrect expression (6.1.8) for the attraction at the pole A produced by the ring whose thickness is du formed by revolving $NnmM$ around the axis of symmetry Aa, and he found as a result the erroneous expression

$$\tfrac{8}{15} ac \tag{6.1.9}$$

in his old notation for the approximate value of the magnitude of the attraction at the pole A produced by the homogeneous figure composed of the two meniscus caps which attract according to the universal inverse-square law and whose densities equal 1. Thus we see that the approximate value of the magnitude of the attraction at the pole A produced by the homogeneous figure composed of the two meniscus caps which attract according to the universal inverse-square law and whose densities equal 1 *does* indeed depend on the polar radius $AC = r$ in Clairaut's old notation of the homogeneous figure shaped like a flattened ellipsoid of revolution which attracts according to the universal inverse-square law, whose ellipticity a in Clairaut's old notation is positive and infinitesimal, whose density is 1, and whose meridians are flattened ellipses $AEaeA$. This clears up the difficulty in paragraph 1.

The magnitude of the attraction at the surface of a homogeneous spherical figure which attracts according to the universal inverse-square law, whose density is 1, and whose radius equals r is $\tfrac{4}{3}\pi r = \tfrac{2}{3}(2\pi r)$. Again, since Clairaut let c stand for 2π, $\tfrac{2}{3}(2\pi r)$ is written as

$$\tfrac{2}{3} cr \tag{6.1.20}$$

in Clairaut's notation.

If we add expressions (6.1.19) and (6.1.20) together, we find

$$\tfrac{2}{3} cr + \tfrac{8}{15} acr \tag{6.1.21}$$

to be the correct expression in Clairaut's old notation for the approximate value of the magnitude of the attraction at a pole A of a homogeneous figure shaped like an ellipsoid of revolution flattened at its poles which attracts according to the universal inverse-square law, whose polar radius $AC = r$ in Clairaut's old notation, whose ellipticity a in Clairaut's old notation is positive and infinitesimal, whose density is 1, and whose meridians are flattened ellipses $AEaeA$. Instead Clairaut added expressions (6.1.9) and (6.1.10) together, and he found as a result the false expression

$$\tfrac{2}{3}c + \tfrac{8}{15}ac \qquad (6.1.11)$$

in his old notation for the approximate value of the magnitude of the attraction at a pole A of a homogeneous figure shaped like an ellipsoid of revolution flattened at its poles which attracts according to the universal inverse-square law, whose polar radius $AC = r$ in Clairaut's old notation, whose ellipticity a in Clairaut's old notation is positive and infinitesimal, whose density equals 1, and whose meridians are flattened ellipses $AEaeA$. Thus we see that the approximate value of the magnitude of the attraction at a pole A of a homogeneous figure shaped like an ellipsoid of revolution flattened at its poles which attracts according to the universal inverse-square law, whose polar radius $AC = r$ in Clairaut's old notation, whose ellipticity a in Clairaut's old notation is positive and infinitesimal, whose density is 1, and whose meridians are flattened ellipses $AEaeA$ *does* indeed depend upon the polar radius of the figure. This resolves difficulty 2.

In setting

$$c = pr \qquad (6.1.14)$$

in his expression

$$\tfrac{2}{3}c - \tfrac{8}{15}ac \qquad (6.1.12)$$

in his old notation for the approximate value of the magnitude of the attraction at a pole A of a homogeneous figure shaped like an ellipsoid of revolution elongated at its poles which attracts according to the universal inverse-square law, whose polar radius $AC = r$ in Clairaut's old notation, whose infinitesimal ellipticity is a negative number and therefore has the form $- a$ in Clairaut's old notation, where a is a positive number, and whose density equals 1, an expression that he called a "corollary" to (6.1.11), in order to arrive at the expression

$$\tfrac{2}{3}pr - \tfrac{8}{15}apr, \qquad (6.1.12')$$

Clairaut either fudged his calculation, realizing that he had made an error somewhere earlier, or else he simply lost track of what he was doing and unknowingly made two errors, the first one being (6.1.11) and treating p as a constant, which it is not, since $p = c/r$, being the second one. Be that as it may, Clairaut needed an r in each of the two terms in the expression (6.1.12')! If Clairaut had derived the correct expression

$$\tfrac{2}{3}cr + \tfrac{8}{15}acr \qquad (6.1.21)$$

instead of the false expression

$$\tfrac{2}{3}c + \tfrac{8}{15}ac, \tag{6.1.11}$$

he would not have needed to let c equal pr, a step that Clairaut neither explained nor justified. Nor could he have done so! Letting

$$c = pr \tag{6.1.14}$$

serves *no other purpose* than to get back the r that Clairaut lost in arriving at the false expression

$$\tfrac{8}{15}ac \tag{6.1.9}$$

for the approximate value of the magnitude of the attraction at a pole A of a homogeneous figure shaped like a flattened ellipsoid of revolution which attracts according to the universal inverse-square law, whose polar radius $AC = r$ in Clairaut's old notation, whose ellipticity a in Clairaut's old notation is positive and infinitesimal, whose density is 1, and whose meridians are flattened ellipses $AEaeA$ produced by the homogeneous figure composed of the two meniscus caps which attract according to the universal inverse-square law and whose densities equal 1. We recall that the correct expression for the approximate value of the magnitude of this attraction is

$$\tfrac{8}{15}acr. \tag{6.1.19}$$

Furthermore, it is easy to see by comparing expressions that c is the constant that should appear in place of p in (6.1.12'), (6.1.12''), (6.1.12'''), (6.1.15), and (6.1.16). This resolves difficulty 7.

Moreover, if Clairaut had derived expression (6.1.21), he would not have had to resort to the doubtful tactic of employing an expression

$$\tfrac{2}{3}c - \tfrac{8}{15}ac \tag{6.1.12}$$

that, as I observed in difficulty 3, only applies to a homogeneous figure shaped like an ellipsoid of revolution elongated at its poles which attracts according to the universal inverse-square law, whose infinitesimal ellipticity is a negative number and therefore has the form $-a$ in Clairaut's old notation, where a is a positive number, whose density equals 1, and whose polar radius equals 1 in order to compute the approximate values of the magnitudes of the attractions at the poles of homogeneous figures shaped like ellipsoids of revolution elongated at their poles which attract in accordance with the universal inverse-square law, whose infinitesimal ellipticities are negative numbers and therefore have the form $-a$ in Clairaut's old notation, where a stands for positive numbers, and whose densities equal 1, but whose polar radii do *not* necessarily equal 1. This is the difficulty that I mentioned as 6. If Clairaut had calculated correctly, he would have found the right expression

$$\tfrac{2}{3}cr - \tfrac{8}{15}acr, \tag{6.1.22}$$

in his old notation for the approximate value of the magnitude of the attraction at a pole A of a homogeneous figure shaped like an ellipsoid of revolution elongated at its poles which attracts according to the universal inverse-square law, whose polar radius $AC = r$ in Clairaut's old notation, whose infinitesimal ellipticity is a negative number and therefore has the form $-a$ in Clairaut's old notation, where a is a positive number, and whose density equals 1. Thus we see that the approximate value of the attraction at a pole A of the figure in question *does* indeed depend upon its polar radius. This clears up difficulty 6 and helps clear up difficulty 3. I note that the people who printed Clairaut's article did not introduce the errors. The two fundamental mistakes ((6.1.11) and treating p as a constant, which it is not, since $p = c/r$) are present in Clairaut's manuscript written in Latin.[7]

The following also helps resolve difficulty 3. Clairaut's calculations, including the mistakes that he made, require that the r in (6.1.22) in Clairaut's old notation for the approximate value of the magnitude of the attraction at a pole A of a homogeneous figure shaped like an ellipsoid of revolution elongated at its poles which attracts in accordance with the universal inverse-square law, whose infinitesimal ellipticity is a negative number and therefore has the form $-a$ in Clairaut's old notation, where a is a positive number, and whose density equals 1 be the *polar* radius of the elongated ellipsoid of revolution, *not* the *equatorial* radius of the elongated ellipsoid of revolution. But in stating that (6.1.12) is a "corollary" of his calculation of the approximate value of the magnitude of the attraction at a pole of a homogeneous figure shaped like an ellipsoid of revolution flattened at its poles which attracts according to the universal inverse-square law, whose polar radius $AC = r$ in Clairaut's old notation, whose ellipticity a in Clairaut's old notation is positive and infinitesimal, whose density equals 1, and whose meridians are flattened ellipses $AEaeA$ – that is, in claiming that (6.1.12) is a "corollary" of (6.1.11) – Clairaut did not explain why the radius of the homogeneous figure shaped like an ellipsoid of revolution elongated at its poles used to arrive at expression (6.1.12) must be the polar radius $AC = r$ of this figure in Clairaut's old notation, since Clairaut did not even hint why his "corollary" (6.1.12) should follow as a consequence of his "theorem" (6.1.11), much less show in detail how (6.1.12) follows from (6.1.11). (Actually, his "corollary" (6.1.12) holds only for the case when $r = 1$, because of the missing r in his "theorem" (6.1.11).) Therefore I furnish the explanation that Clairaut neglected to provide.

First, however, I prove the "corollary" that Clairaut should have stated. Let us consider Figure 17. The elongated ellipse $PNEP$ is a meridian at the surface of a homogeneous figure shaped like an ellipsoid of revolution elongated at its poles which attracts according to the universal inverse-square law, one of whose poles is at P, whose polar radius $OP = r + ar$ in Clairaut's old notation, whose equatorial radius $OE = r$ in Clairaut's old notation, whose ellipticity is negative and infinitesimal, and whose density equals 1. The ellipticity of the elongated

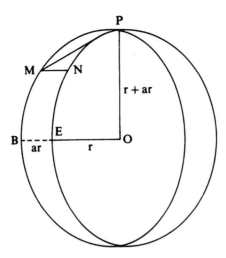

Figure 17. Figure used for determining the magnitude of the attraction at a pole of a homogeneous elongated ellipsoid of revolution which attracts according to the universal inverse-square law and whose ellipticity is infinitesimal.

ellipsoid of revolution

$$\equiv \frac{EO - OP}{OP} = \frac{r - (r + ar)}{r + ar} = \frac{-ar}{r + ar} = -a\left(\frac{1}{1 + a}\right).$$

But $a^2 \ll a$, because the positive number a is infinitesimal. Hence to terms of first order in a, the ellipticity of the elongated ellipsoid of revolution

$$= -a\left(\frac{1}{1 + a}\right) \simeq -a(1 - a) \simeq -a.$$

Now, if we assume that all of the particles that make up MN attract the pole P with the same force as the particle at M (in other words, attract the pole P as if all of the particles that make up MN were located at M), which is the same kind of hypothesis that Clairaut made in calculating the approximate value of the magnitude of the attraction at a pole A of a homogeneous figure shaped like an ellipsoid of revolution flattened at its poles which attracts according to the universal inverse-square law, whose ellipticity a in Clairaut's old notation is positive and infinitesimal, whose density equals 1, and whose meridians are flattened ellipses $AEaeA$, then one could guess from the expression

$$\tfrac{8}{15}acr \tag{6.1.19}$$

in Clairaut's old notation for the approximate value of the magnitude of the attraction at a pole A of a homogeneous figure shaped like an ellipsoid of revolution flattened at its poles which attracts in accordance with the universal inverse-square law, whose polar radius $AC = r$ in Clairaut's old notation, whose ellipticity a in Clairaut's old notation is positive and infinitesimal, whose density equals 1, and whose meridians are flattened ellipses $AEaeA$ produced by the homogeneous figure composed of the two meniscus caps which attract according to the universal inverse-square law and whose densities equal 1 formed by revolving $AEaDA$ around the axis of symmetry Aa in his diagram, which Clairaut would have found had he not mistakenly included an r in the denominator of his expression

$$\frac{ac}{r}(2ru - u^2)\,du \qquad\qquad (6.1.6)$$

in his old notation for the volume of the ring whose thickness is du formed by revolving $NnmN$ around the axis of symmetry Aa in his diagram, that the expression for the approximate value of the magnitude of the attraction at the pole P produced by the homogeneous figure composed of the two meniscus caps which attract in accordance with the universal inverse-square law and whose densities equal 1 which surround the homogeneous figure shaped like an ellipsoid of revolution elongated at its poles which attracts according to the universal inverse-square law, whose polar radius $OP = r + ar$ in Clairaut's old notation, whose infinitesimal ellipticity is a negative number and therefore has the form $- a$ in Clairaut's old notation, where a is a positive number, and whose density is 1 in Figure 17 and which together with this homogeneous figure shaped like an ellipsoid of revolution elongated at its poles make up a homogeneous spherical figure which attracts in accordance with the universal inverse-square law, whose density is 1, whose radius $OP = r + ar$ in Clairaut's old notation, and whose meridian $PMBP$ is a circle situated in the plane of the elongated ellipse $PNEP$ which is a meridian at the surface of the homogeneous figure shaped like an ellipsoid of revolution elongated at its poles ought to be

$$\tfrac{8}{15}ac\,(r + ar) \qquad\qquad (6.1.23)$$

in Clairaut's old notation. In fact this analogy with (6.1.19) can be easily verified to be true by integration. Moreover, the expression for the magnitude of the attraction at the pole P produced by the homogeneous spherical figure which attracts according to the universal inverse-square law, whose radius $OP = r + ar$ in Clairaut's old notation, whose density equals 1, and whose meridian $PMBP$ is a circle situated in the plane of the elongated ellipse $PNEP$ which is a meridian at the surface of the homogeneous figure shaped like an ellipsoid of revolution elongated at its poles is

$$\tfrac{2}{3}c\,(r + ar) \qquad\qquad (6.1.24)$$

in Clairaut's old notation. But then the expression for the approximate value of the magnitude of the attraction at the pole P produced by the homogeneous figure shaped like an ellipsoid of revolution elongated at its poles which attracts according to the universal inverse-square law, whose polar radius $OP = r + ar$ in Clairaut's old notation, whose infinitesimal ellipticity is a negative number and therefore has the form $-a$ in Clairaut's old notation, where a is a positive number, and whose density equals 1 can be determined by subtracting the expression (6.1.23) for the approximate value of the magnitude of the attraction at the pole P produced by the homogeneous figure composed of the two meniscus caps which attract according to the universal inverse-square law and whose densities equal 1 from the expression (6.1.24) for the magnitude of the attraction at the pole P produced by the homogeneous spherical figure which attracts according to the universal inverse-square law and whose density is 1. But the difference between these two expressions is just

$$\tfrac{2}{3}c(r + ar) - \tfrac{8}{15}ac(r + ar) \qquad (6.1.25)$$

in Clairaut's old notation. Expression (6.1.25) can be rewritten as

$$\tfrac{2}{3}cr + \tfrac{2}{3}acr - \tfrac{8}{15}ac(1 + a)r, \qquad (6.1.25')$$

and expression (6.1.25') is the same as expression

$$\tfrac{2}{3}cr + \tfrac{2}{3}acr - \tfrac{8}{15}acr \qquad (6.1.25'')$$

to terms of first order in a. But expression (6.1.25'') reduces to

$$\tfrac{2}{3}cr + \tfrac{2}{15}acr \qquad (6.1.25''')$$

in Clairaut's old notation. We shall see in Chapter 8 that expression (6.1.25''') is the same as the expression that Daniel Bernoulli arrived at in 1740 using an entirely different method of computing the approximate value of the magnitude of the attraction at a pole P of a homogeneous figure shaped like an ellipsoid of revolution elongated at its poles which attracts according to the universal inverse-square law, whose polar radius $OP = r + ar$ in Clairaut's old notation, whose infinitesimal ellipticity is a negative number and therefore has the form $-a$ in Clairaut's old notation, where a is a positive number, and whose density is 1. The fact that Bernoulli found expression (6.1.25'''), which is valid to terms of first order in a, using a method for doing the calculation which differed from Clairaut's method confirms that expression

$$\tfrac{2}{3}c(r + ar) - \tfrac{8}{15}ac(r + ar) \qquad (6.1.25)$$

in Clairaut's old notation is indeed correct to terms of first order in a. Finally, we note that the radius in the expression (6.1.25) must be the polar radius $OP = r + ar$ of the elongated ellipsoid of revolution in Clairaut's old notation. The equatorial radius $OE = r$ of the elongated ellipsoid of revolution in Clairaut's old notation cannot be substituted for the polar radius $OP = r + ar$ of the elongated ellipsoid of revolution in expression (6.1.25), because (6.1.25) can also be rewritten as

$$\tfrac{2}{3}c(1+a)r - \tfrac{8}{15}ac(1+a)r, \tag{6.1.26}$$

and expression (6.1.26) is the same as expression

$$\tfrac{2}{3}c(1+a)r - \tfrac{8}{15}acr \tag{6.1.26'}$$

to terms of first order in a. But the first term $\tfrac{2}{3}c(1+a)r$ in expression (6.1.26′) does *not* equal $\tfrac{2}{3}cr$ to terms of first order in a, because $\tfrac{2}{3}car$ in the first term $\tfrac{2}{3}c(1+a)r$ in expression (6.1.26′) is itself a term of first order in a, not a term of higher order in a, hence $(2/3)\,car$ cannot be discarded. Hence

$$\tfrac{2}{3}cr - \tfrac{8}{15}acr \tag{6.1.22}$$

is *not* an expression for the approximate value of the magnitude of the attraction at a pole P of a homogeneous figure shaped like an ellipsoid of revolution elongated at its poles which attracts according to the universal inverse-square law, whose infinitesimal ellipticity is a negative number and therefore has the form $-a$ in Clairaut's old notation, where a is a positive number, whose density equals 1, and whose equatorial radius $OE = r$, because expression (6.1.22) is not valid to terms of first order in a. This explains why the r in expression (6.1.22) *must* be the *polar* radius of the elongated ellipsoid of revolution in Clairaut's old notation, not the equatorial radius of the elongated ellipsoid of revolution in Clairaut's old notation. This clears up the rest of difficulty 3.

The difficulties mentioned in paragraphs 4 and 5 can also be resolved:

i. If N is a point at the surface of a homogeneous figure shaped like an ellipsoid of revolution flattened at the poles which attracts according to the universal inverse-square law, whose polar radius $AC = r$ in Clairaut's old notation, whose ellipticity a in Clairaut's old notation is positive and infinitesimal, whose density equals 1, and whose meridians are flattened ellipses $AEaeA$, where $CN = r$ and $r < b\sqrt{\tfrac{a}{r}}$ in Clairaut's new notation holds (thus, for example, when $N = A$), then the homogeneous figure shaped like an ellipsoid of revolution which attracts according to the universal inverse-square law, whose ellipticity is infinitesimal, one of whose poles is at N, whose density is 1, and whose polar radius is the line segment $CN = r$ in Clairaut's new notation is in fact shaped like an ellipsoid of revolution flattened at its poles whose infinitesimal ellipticity is the positive number

$$\frac{b\sqrt{a/r}-r}{r}$$

in Clairaut's new notation. But in this case

$$\frac{r-b\sqrt{a/r}}{r} \tag{6.1.13}$$

is the negative number

$$-\left(\frac{b\sqrt{a/r}-r}{r}\right).$$ (6.1.27)

If we substitute expression (6.1.27) for a in the expression (6.1.22) in Clairaut's old notation, which ordinarily we should not do, because according to Clairaut (6.1.22) should be an expression that *only* applies to a homogeneous figure shaped like an ellipsoid of revolution *elongated* at its poles which attracts according to the universal inverse-square law, whose polar radius $AC = r$ in Clairaut's old notation, whose infinitesimal ellipticity is a negative number and therefore has the form $-a$ in Clairaut's old notation, where a is a positive number, and whose density equals 1, the expression that results is

$$\tfrac{2}{3}cr + \tfrac{8}{15}\left(\frac{b\sqrt{a/r}-r}{r}\right)cr,$$ (6.1.21′)

which is precisely the expression for the approximate value of the magnitude of the attraction at a pole of a homogeneous figure shaped like an ellipsoid of revolution flattened at its poles which attracts according to the universal inverse-square law, whose density is 1, whose polar radius equals r, and whose infinitesimal ellipticity is the positive number

$$\frac{b\sqrt{a/r}-r}{r}.$$

ii. If N is a point at the surface of a homogeneous figure shaped like an ellipsoid of revolution flattened at the poles which attracts according to the universal inverse-square law, whose polar radius $AC = r$ in Clairaut's old notation, whose ellipticity a in Clairaut's old notation is positive and infinitesimal, whose density equals 1, and whose meridians are flattened ellipses $AEaeA$, where $CN = r$ and

$$b\sqrt{a/r} < r$$

in Clairaut's new notation holds (thus, for example, when $N = E$), then the homogeneous figure shaped like an ellipsoid of revolution which attracts according to the universal inverse-square law, whose ellipticity is infinitesimal, one of whose poles is at N, whose density equals 1, and whose polar radius is the line segment $CN = r$ in Clairaut's new notation is in fact shaped like an ellipsoid of revolution elongated at its poles whose infinitesimal ellipticity is the negative number

$$\frac{b\sqrt{a/r}-r}{r} = -\left(\frac{r-b\sqrt{a/r}}{r}\right)$$

in Clairaut's new notation. In this case we can substitute the positive number

$$\frac{r-b\sqrt{a/r}}{r}$$ (6.1.13)

for a in the expression

$$\tfrac{2}{3} cr - \tfrac{8}{15} acr \qquad (6.1.22)$$

in Clairaut's old notation, and the expression that results is

$$\tfrac{2}{3} cr - \tfrac{8}{15} \left(\frac{r - b\sqrt{a/r}}{r} \right) r, \qquad (6.1.22')$$

which is just the expression for the approximate value of the magnitude of the attraction at a pole of a homogeneous figure shaped like an ellipsoid of revolution elongated at its poles which attracts according to the universal inverse-square law, whose density is 1, whose polar radius equals r, and whose infinitesimal ellipticity is the negative number

$$\frac{b\sqrt{a/r} - r}{r},$$

whose absolute value is (6.1.13).

Therefore we conclude from i and ii that Clairaut did not really do anything wrong where difficulties 4 and 5 are concerned. He simply explained what he was doing incorrectly. The false statements that he sometimes made regarding 4 and 5 do not affect his calculations. That is to say, the inaccuracy of his statements do not nullify these calculations. I shall summarize how difficulties 4 and 5 are resolved, using the expressions that Clairaut should have derived but did not. To make a long story short, the homogeneous figure shaped like an ellipsoid of revolution which attracts according to the universal inverse-square law, whose ellipticity is infinitesimal, one of whose poles is at N, whose density is 1, and whose polar radius is the line segment $CN = r$ in Clairaut's new notation is not always elongated at its poles, contrary to what Clairaut's statements imply. The figure can be flattened at its poles instead (for example, when $N = A$). Equivalently, the expression

$$\frac{r - b\sqrt{a/r}}{r} \qquad (6.1.13)$$

in Clairaut's new notation which corresponds to the figure in question need not be a positive number, again contrary to what Clairaut's statements imply. Expression (6.1.13) can be a negative number, namely, when the figure in question is flattened at its poles, not elongated at its poles. Expression (6.1.13) is always equal to $- M$, where M is the ellipticity of the figure in question. If the figure is elongated at its poles, expression (6.1.13) is a positive number and M is a negative number. If the figure in question is flattened at its poles, expression (6.1.13) is a negative number and M is a positive number. But in *either* case, expression

$$\tfrac{2}{3} cr - \tfrac{8}{15} acr \qquad (6.1.22)$$

in Clairaut's old notation can be used. In case the figure in question is elongated at its poles (for example, when $N = E$), its ellipticity $M = -$ expression (6.1.13), where expression (6.1.13) is a positive number, and Clairaut says that we are to substitute expression (6.1.13) for a in expression (6.1.22), in accordance with his "corollary" for determining the approximate value of the magnitude of the attraction at a pole of a homogeneous figure shaped like an ellipsoid of revolution elongated at its poles which attracts according to the universal inverse-square law, whose polar radius $AC = r$ in Clairaut's old notation, whose infinitesimal ellipticity is a negative number and therefore has the form $-a$ in Clairaut's old notation, where a is a positive number, and whose density is 1. However, if the figure in question is flattened at its poles (for example, when $N = A$), in which case its ellipticity $M = -$ expression (6.1.13), where expression (6.1.13) is a negative number, then if we substitute expression (6.1.13) for a in expression (6.1.22), even though, according to Clairaut, expression (6.1.22) was only meant to apply to a homogeneous figure shaped like an elongated ellipsoid of revolution which attracts according to the universal inverse-square law, whose infinitesimal ellipticity is a negative number and therefore has the form $-a$ in Clairaut's old notation, where a is a positive number, and whose density equals 1, an expression

$$\tfrac{2}{3} cr + \tfrac{8}{15} Mcr \qquad (6.1.21'')$$

results, which is just the correct expression for the approximate value of the magnitude of the attraction at a pole of a homogeneous figure shaped like a flattened ellipsoid of revolution which attracts according to the universal inverse-square law, whose polar radius $AC = r$ in Clairaut's old notation, whose ellipticity M is positive and infinitesimal, whose density is 1, and whose meridians are flattened ellipses $AEaeA$. Clairaut's own calculations involve substitutions of this second kind, although Clairaut talked as if they did not. Clairaut did not explain what he was doing correctly. However, even though Clairaut made false statements about what he was doing, these statements do not translate into erroneous calculations.

The rest of Clairaut's calculations, which consist mostly of first-order approximations, hold true. We have only to replace p by c in his expressions (6.1.12'), (6.1.12''), and (6.1.12''') to find

$$\tfrac{2}{3} ca - \tfrac{2}{15} can + \tfrac{8}{15} cam \qquad (6.1.12'''')$$

as the general expression in Clairaut's new notation for the approximate value of the magnitude of the component of attraction at a point N on the surface of a homogeneous figure shaped like an ellipsoid of revolution flattened at its poles which attracts according to the universal inverse-square law, whose center is at C, whose ellipticity m in Clairaut's new notation is positive and infinitesimal, whose density is 1, and whose meridians are flattened ellipses $AEaeA$ which is directed toward C, where $AC = a$ and $CN = r = a + na$ define n in Clairaut's new notation.

The special cases of (6.1.12‴) which Clairaut cited become

$$\tfrac{2}{3} ca + \tfrac{8}{15} cam, \tag{6.1.15′}$$

which expresses the approximate value of the magnitude of the attraction at a pole of the figure (where $N = A$, hence $r = CN = AC = a$ and $n = 0$), and

$$\tfrac{2}{3} ca + \tfrac{6}{15} cam, \tag{6.1.16′}$$

which expresses the approximate value of the magnitude of the attraction at a point on the equator of the figure (where $N = E$, hence $n = m$, since $a + ma = b = CE = CN = r = a + na$ in Clairaut's new notation).

Clairaut now stated that if the axes of the homogeneous fluid figure shaped like an ellipsoid of revolution flattened at its poles which attracts according to the universal inverse-square law, whose density is 1, whose meridians are flattened ellipses $AEaeA$, and which revolves around its axis of symmetry "differ by a very small quantity, which for greater perspicuity I shall call infinitely small," which means that the ellipticity of the figure is infinitesimal, and if the "gravity" of the column CE from the center C of the figure to a point E on the equator of the figure and the "gravity" of the column AC from the center C of the figure to a pole A of the figure be the same, meaning that the two columns balance or weigh the same, or, equivalently, by the "Newtonian Principles," if the equality

$$\frac{\text{Effective Gravity at } E}{\text{Attraction at } A} = \frac{|CA|}{|CE|} \tag{6.1.28}$$

holds, then it necessarily follows that the equality

$$\frac{\text{Effective Gravity } (N, CN)}{\text{Attraction at } A} = \frac{|CA|}{|CN|} \tag{6.1.29}$$

holds too at all points N on the surface of the homogeneous fluid figure, where Effective Gravity at $E \equiv$ The Magnitude of the Effective Gravity at E, Effective Gravity $(N, CN) \equiv$ The Magnitude of the Component of the Effective Gravity at N Which is Directed Toward C, Attraction at $A \equiv$ The Magnitude of the Attraction at A, and where $|XY|$ again designates the length of line segment XY. In other words, to terms of first order in the ellipticity of the figure and all other quantities of the same order as the ellipticity of the figure, the component of the effective gravity at a point on the surface of the figure which is directed toward the center of the figure varies inversely as the distance of the point from the center of the figure (that is, to terms of first order in the ellipticity of the figure and all other quantities of the same order as the ellipticity of the figure, the magnitudes of these components of the effective gravity at points on the surface of the figure vary inversely as the lengths of the radii of the figure at these points). To express the matter another way, applying the "Newtonian Principles" in reverse, if the column AC that lies along the polar axis of the figure from the center C of the figure to the surface of the figure and the column CE that lies along an equatorial axis of the figure from the center C of the figure to the surface of the figure weigh the same to terms of first order in the ellipticity of the figure and all other quantities of the same order as the ellipticity of the figure, then *all* of the columns CN from the center C of the figure to points N on the surface of the figure will weigh the same as the column AC and the column CE to terms of first order in the ellipticity of the figure and all other quantities of the same order as the ellipticity of the figure.[8] (As we shall now see, Clairaut really only showed to terms of first order that equality (6.1.29) follows from equality (6.1.28) for points N on a flattened ellipse $AEaeA$ which is a meridian at the surface of the figure in the plane of the polar axis and the equatorial axis. Again, however, as I indicated in Chapter 1, it is enough to show that this much is true without any loss of generality. For if all of the columns from the center C of the figure to points N on the surface of the figure *which lie in the plane of a meridian at the surface of the figure determined by the columns AC and CE* balance or weigh the same, it follows immediately that *all* of the columns from the center C of the figure to points N on the surface of the figure will balance or weigh the same, for reasons that I explained in Chapter 1.)

To prove that this is true, Clairaut inserted in the appropriate places in equality (6.1.28) the expression in his new notation which he found for the approximate value of the magnitude of the attraction at the equator of a homogeneous figure shaped like an ellipsoid of revolution flattened at its poles which attracts according to the universal inverse-square law, whose polar radius $AC = a$ in Clairaut's new notation, whose center is at C, whose ellipticity m in Clairaut's new notation is positive and infinitesimal, whose density is 1, and whose meridians are flattened ellipses $AEaeA$ (expression

$$\tfrac{2}{3}\,pa + \tfrac{6}{15}\,pam \tag{6.1.16}$$

in Clairaut's new notation, which is valid to terms of first order, where p, of course, mistakenly appears in place of c), a letter f that stood for the magnitude of the centrifugal force per unit of mass at the equator of this figure of revolution when the figure revolves around its axis of symmetry Aa so that the equality (6.1.28) holds, whose expression he had yet to determine, and the expression in his new notation which he found for the approximate value of the magnitude of the attraction at a pole of a homogeneous figure shaped like an ellipsoid of revolution flattened at its poles which attracts according to the universal inverse-square law, whose polar radius $AC = a$ in Clairaut's new notation, whose center is at C, whose ellipticity m in Clairaut's new notation is positive and infinitesimal, whose density is 1, and whose meridians are flattened ellipses $AEaeA$ (expression

$$\tfrac{2}{3}\,pa + \tfrac{8}{15}\,pam \tag{6.1.15}$$

in Clairaut's new notation, which is valid to terms of first order, where p again mistakenly appears in place of c). Equality (6.1.28) thus became an equality that only held good to terms of first order, because the approximate values of the magnitudes of the attraction which Clairaut inserted into this equality were only valid to terms of first order. Since $|CA| = a$, m is the infinitesimal ellipticity of the figure, which is a positive number, and $|CE| = a + ma$ in Clairaut's new notation, Clairaut wrote the quotient

$$\frac{|CA|}{|CE|} \quad as \quad \frac{a}{a + ma} = \frac{1}{1 + m}$$

in his new notation. He reduced the expression $1/(1 + m)$ to $1 - m$ by neglecting terms in m^2 and higher powers of m. He then treated equality (6.1.28), after replacing the quotient $|CA|/|CE|$ on the right-hand side of this equality by $1 - m$, as an equation for f which held true to terms of first order. In solving this equation for f, Clairaut had to discard a term that included m^2 as a factor, which is a term of second order. By this means he determined the expression for the magnitude f of the centrifugal force per unit of mass at the equator of the figure when equality (6.1.28) held good to terms of first order to be

$$f = \tfrac{8}{15} \, pam \qquad (6.1.30)$$

in his new notation, where p again mistakenly appears in place of c.

Taking (6.1.30) to express the magnitude f of the centrifugal force per unit of mass at the equator of the figure, Clairaut next calculated to terms of first order what the magnitude of the component of centrifugal force per unit of mass at N directed toward C would have to be. In fact, he actually computed the magnitude of this component at the corresponding point M on the surface of the homogeneous spherical figure which attracts according to the universal inverse-square law, whose density is 1, whose center is also at C, whose radius equals the polar radius $AC = a$ in Clairaut's new notation of the homogeneous fluid figure shaped like an ellipsoid of revolution flattened at its poles which attracts according to the universal inverse-square law, whose center is at C, whose ellipticity m in Clairaut's new notation is positive and infinitesimal, whose density is 1, and which revolves around its axis of symmetry Aa so that equality (6.1.28) holds, whose meridians are circles $ADadA$ which lie in the planes of the flattened ellipses $AEaeA$ which are meridians at the surface of the homogeneous fluid figure shaped like an infinitesimally flattened ellipsoid of revolution which attracts according to the universal inverse-square law and whose density is 1, whose polar axis coincides with the axis of symmetry Aa of this figure, and which revolves around this common polar axis of the two figures at the same rate (that is, at the same angular speed) as the other figure (see Figure 16). It sufficed to calculate the magnitude of this component of centrifugal force per unit of mass instead of the magnitude of the other component of centrifugal force per unit of mass because the values of the magnitudes of the two components are equal to terms of first order (that is, they differ at most by quantities of second order and higher orders). (The point M on the meridian $ADadA$ at the surface of the homogeneous spherical figure which corresponds to a point N on the meridian $AEaeA$ at the surface of the homogeneous figure shaped like an infinitesimally flattened ellipsoid of revolution is the point where the line through N perpendicular to the common axis of rotation Aa of the two figures intersects the surface of the homogeneous spherical figure.[9] Clairaut found the expression for the magnitude of this component of centrifugal force per unit of mass at N to terms of first order to be

$$\tfrac{8}{15} \, pan \qquad (6.1.31)$$

in his new notation, where $|CN| = a + na$ in his new notation and p again mistakenly appears in place of c. Clairaut then subtracted $\tfrac{8}{15} \, pan$ from the expression

$$\tfrac{2}{3} \, pa - \tfrac{2}{15} \, pan + \tfrac{8}{15} \, pam \qquad (6.1.12''')$$

in his new notation for the approximate value of the magnitude of the component of attraction at N which points to C, which is valid to terms of first order, where p again mistakenly appears in place of c. Clairaut thus arrived at the expression

$$\tfrac{2}{3} \, pa - \tfrac{2}{3} \, pan + \tfrac{8}{15} \, pam \qquad (6.1.32)$$

in his new notation, where p again mistakenly appears in place of c, for the approximate value of the magnitude of the component of effective gravity at N directed toward C when equality (6.1.28) holds – in other words, when columns from the center C of the homogeneous figure in question to the surface of the figure which lie along the polar axis of the figure and along one of the equatorial axes of the figure balance or weigh the same. Expression (6.1.32) only holds good to terms of first order, which means that the ellipticity of the figure must be infinitesimal.

Now that Clairaut had determined to terms of first order the magnitude of the component of effective gravity at N which points to C when equality (6.1.28) holds, his problem was to show that the magnitude of this component of effective gravity at N varied inversely as the distance of N from the center C of the figure. In other words, he had to demonstrate that equality (6.1.28) entailed equality (6.1.29), hence that *all* columns from the center C of the homogeneous figure in question to points N on a flattened ellipse $AEaeA$ which is a meridian at the surface of the figure balance or weigh the same when equality (6.1.28) holds. To prove that this is true, Clairaut solved the following problem in a manner that did not depend on the preceding calculations: what does the expression for the magnitude of the component of the effective gravity at N directed toward C which accords with the equality

$$\frac{\text{Effective Gravity}\,(N, CN)}{\text{Attraction at } A} = \frac{|CA|}{|CN|} \qquad (6.1.29)$$

to terms of first order look like? Clairaut wrote the quotient $|CA|/|CN|$ as

$$\frac{a}{a + na} = \frac{1}{1 + n}$$

in his new notation, which he reduced to $1 - n$ by neglecting terms in n^2 and higher powers of n. (The quantity n is infinitesimal, because $a + na = |CN| \leqslant |CE| = a + ma$ means that $n \leqslant m$, and the ellipticity m is positive and infinitesimal.) He substituted $1 - n$ for the quotient on the right-hand side of equality (6.1.29), and he inserted as well into equality (6.1.29) the expression in his new notation which he found for the approximate value of the magnitude of the attraction at a pole of a homogeneous figure shaped like an ellipsoid of revolution flattened at its poles which attracts according to the universal inverse-square law, whose polar radius $AC = a$ in Clairaut's new notation, whose center is at C, whose ellipticity m in Clairaut's new notation is positive and infinitesimal, whose density is 1, and whose meridians are flattened ellipses $AEaeA$ (expression

$$\tfrac{2}{3}pa + \tfrac{8}{15}pam \qquad (6.1.15)$$

in Clairaut's new notation, which is valid to terms of first order, where p again mistakenly appears in place of c). Clairaut then treated equality (6.1.29) as an equation for the approximate value of the magnitude of the component of effective gravity at N which points to C which held true to terms of first order. In solving the equation, Clairaut had to neglect a term that included nm as a factor, which is a term of second order. As a result of solving the equation, Clairaut found exactly the same expression

$$\tfrac{2}{3}pa - \tfrac{2}{3}pan + \tfrac{8}{15}pam \qquad (6.1.32)$$

for the approximate value of the magnitude of the component of effective gravity at N directed toward C which his previous calculations involving equality (6.1.28) had led to. In other words, to terms of first order, equality (6.1.28) does indeed entail equality (6.1.29). (We note that although c should have appeared everywhere that p does, the presence of p instead of c does not nullify the conclusions that Clairaut came to, since these conclusions only depend upon Clairaut's having found in two different ways the same expression (6.1.32) for the approximate value of the magnitude of the component of effective gravity at N which points to C. The expression for the approximate value of the magnitude of this component which Clairaut would have found, had he not made the two mistakes that fortunately cancelled each other in his earlier calculations ((6.1.11) and treating p as a constant, which it is not, since $p = c/r$), is of course just

$$\tfrac{2}{3}ca - \tfrac{2}{3}can + \tfrac{8}{15}cam.) \qquad (6.1.32')$$

For reasons that I shall explain in a moment, Clairaut's particular methods of approximation do not appear as a central issue in the story that I tell here, which is why I do not discuss them in more detail. I only mention some general features. Sometimes when Clairaut employed analysis, the order of an approximation that he made is evident—for example, when he neglected terms of second order and higher orders in binomial expansions or in infinite series more generally. At other times, however, especially when he used geometry, he did not make the orders of his approximations clear. For example, he would make an assumption that simplified a geometric argument, which amounted to making an approximation, without converting what he had done into analysis. If he actually verified analytically that such an approximation held good to terms of first order, he did not display the evidence in his paper. I give an example of one such geometric approximation that he made–in fact, the most "trivial" one among his numerous approximations. I mentioned it previously. In calculating the magnitude of the attraction at the pole A of a homogeneous figure shaped like an ellipsoid of revolution flattened at its poles which attracts according to the universal inverse-square law, whose polar radius $AC = r$ in Clairaut's old notation, whose ellipticity a in Clairaut's old notation is positive and infinitesimal, whose meridians are flattened ellipses $AEaeA$, and whose density is 1 produced by the two meniscus caps formed by revolving $AEaDA$ around the axis of symmetry Aa (see Figure 16), Clairaut maintained: "Now, because of the smallness of NM, we may account all the particles of matter contained in that space as equally attracting the corpuscle at A."[10] It can indeed be shown using analysis that this approximation is valid to terms of first order in the ellipticity a,[11] but Clairaut did no such thing in his paper. Clairaut made similar kinds of approximations all the way along in his paper. Sometimes he made the approximations tacitly. For example, as I observed before, he implicitly equated the magnitude of the total or the resultant of the attraction at N produced by the homogeneous figure shaped like an ellipsoid of revolution which attracts according to the universal inverse-square law, one of whose poles is at N, whose ellipticity is infinitesimal, whose density is 1, and whose polar radius is the line segment $CN = r$ in Clairaut's new notation and the magnitude of the component of the attraction at N directed toward C produced by the homogeneous figure shaped like a flattened ellipsoid of revolution which attracts according to the universal inverse-square law, whose polar radius is $AC = a$ in Clairaut's new notation, whose ellipticity a in Clairaut's old notation is positive and infinitesimal, whose density is 1, and whose meridians are flattened ellipses $AEaeA$. But Clairaut did not even state what he was doing here, much less demonstrate that the two magnitudes of attraction do equal each other to terms of first order. Even when he utilized infinite series, in which case the terms of first order and the neglected terms of second order and higher orders appear explicitly, he did not keep track of the accumulated error made in truncating infinite series. Clairaut combined analysis and geometry in a way that included simplified, approximate geometric arguments, truncated infinite series, and so forth, but Clairaut did not estimate the order of the total error that resulted from making such approximations. But the order of the total accumulated error is obviously of some concern in approximate calculations that involve a quantity that is itself infinitesimal, namely, the ellipticity. Clairaut's arguments and the calculations that followed from them would have ended in a meaningless exercise, had the total error in fact been of the same order as the infinitesimal ellipticity of the homogeneous figure shaped like an ellipsoid of revolution flattened at its poles. I shall return to this problem in a moment. As we shall see in Chapters 8 and 9, Clairaut would eventually have to take this drawback of his calculations into consideration.

Clairaut did *not* show that equality (6.1.28) entails equality (6.1.29) for a homogeneous figure shaped like a flattened ellipsoid of revolution which attracts according to the universal inverse-square law, which revolves around its axis of symmetry (its polar axis), and whose ellipticity is *finite*. Nor did he suggest that such a result was even true in this case.

Maupertuis, we recall, had stated at the end of his Paris Academy *mémoire* of 1734 that equality (6.1.28) (that is, Maupertuis's balancing two columns from the center to the surface of a homogeneous figure shaped like an ellipsoid of revolution flattened at its poles which attracts according to the universal inverse-square law and which revolves around its axis of symmetry, where the two columns lie along the polar axis of the figure and along one of the equatorial axes of the figure, respectively, which resulted in the equality

$$\frac{a}{b} = \frac{(ab) - f}{[ab]}$$

(5.3.2)

that Maupertuis wrote at the end of his *mémoire*, which is the same as equality (6.1.28)) holds for a homogeneous figure shaped like an ellipsoid of revolution flattened at its poles which attracts according to the universal inverse-square law, which revolves around its axis of symmetry, whose ellipticity is finite, and whose columns from center to surface all balance or weigh the same. But Maupertuis simply *assumed* here that a homogeneous figure shaped like an ellipsoid of revolution flattened at its poles which attracts according to the universal inverse-square law, which revolves around its axis of symmetry, whose ellipticity is finite, and whose columns from center to surface all balance or weigh the same could exist. Namely, he supposed that such a figure existed when equality (6.1.28) held for the figure.

But Clairaut did not contend in his paper of 1737 that equality (6.1.28) was useful for determining whether or not a homogeneous figure shaped like an ellipsoid of revolution flattened at its poles which attracts according to the universal inverse-square law, which revolves around its axis of symmetry, whose ellipticity is finite, and whose columns from center to surface all balance or weigh the same could even exist. For Clairaut did not show in his paper of 1737 that such a figure could exist, much less demonstrate that such a figure necessarily exists. Moreover, two columns from the center to the surface of a homogeneous figure of revolution which attracts according to the universal inverse-square law, which revolves around its axis of symmetry, whose ellipticity is infinitesimal or finite, and which is *not* shaped like an ellipsoid of revolution flattened at its poles, where the two columns lie along the polar axis of the figure and along one of the equatorial axes of the figure, respectively, could *also* balance or weigh the same. But this condition alone does not assure that *all* of the columns from the center of the figure to the surface of the figure will balance or weigh the same. (In fact, if two columns from the center of such a figure to the surface of such a figure which lie along the polar axis of the figure and along one of the equatorial axes of the figure, respectively, do balance or weigh the same, the columns from the center of the figure to the surface of the figure will *not* in general all balance or weigh the same. It has been known since the nineteenth century, however, that rotating homogeneous figures which attract according to the universal inverse-square law, whose ellipticities are infinitesimal or finite, and which are not shaped like ellipsoids of revolution flattened at their poles do exist whose columns from their centers to their surfaces do all balance. Moreover, some of these figures are not even figures of revolution–for example, figures shaped like the Jacobi ellipsoids.)

Recall that Newton equated the weights of two columns from the center to the surface of the homogeneous figure shaped like an ellipsoid of revolution flattened at its poles which attracts according to the universal inverse-square law, which revolves around its axis of symmetry, and which he chose to represent the earth, where the two columns lie along the polar axis of the figure and along one of the equatorial axes of the figure, respectively, for a given ratio ϕ of the magnitude of the centrifugal force per unit of mass to the magnitude of the attraction at the equator of the figure, whose value he took to equal the value $\frac{1}{289}$ of the ratio at the earth's equator. But it does not follow immediately from the equality of the weights of these two columns that *all* of the columns from the center of this figure to the surface of the figure will balance or weigh the same, which is the reason why Clairaut wrote his paper of 1737. Indeed, as I just mentioned, balancing two columns from the center to the surface of a homogeneous figure of revolution which attracts according to the universal inverse-square law, which revolves around its axis of symmetry, whose ellipticity is infinitesimal or finite, and which is not shaped like an ellipsoid of revolution flattened at its poles, where the two columns lie along the polar axis of the figure and along one of the equatorial axes of the figure, respectively, does not mean that all of the columns from the center of the figure to the surface of the figure will necessarily balance or weigh the same. In general the columns from the center of such a figure to the surface of the figure will not all balance or weigh the same.

The equality (5.3.2) that Maupertuis wrote at the end of his Paris Academy *mémoire* of 1734 reveals that Maupertuis could not distinguish what Newton had demonstrated in his theory of the earth's shape from what Newton had not. Maupertuis had completely failed to realize that Newton had *not* shown that a homogeneous figure shaped like an ellipsoid of revolution flattened at its poles which attracts according to the universal inverse-square law and which revolves around its axis of symmetry *is* a figure whose columns from center to surface *can* all balance or weigh the same. Newton did not even do this for such a figure whose ellipticity is infinitesimal, much less for one whose ellipticity is finite. In other words, Maupertuis had completely missed a crucial gap in Newton's theory, namely, the absence of any proof that a homogeneous figure shaped like an ellipsoid of revolution flattened at its poles which attracts according to the universal inverse-square law and which revolves around its axis of symmetry is a figure whose columns from center to surface can all balance or weigh the same. We recall that Maupertuis had told Johann I Bernoulli in 1731 that he did not understand why Newton chose an ellipsoid of revolution as the earth's shape. It is safe to say that in 1734, Newton's leading "promoter" in the Paris Academy did not really understand Newton's choice any better, either!

I mentioned the disadvantages of Clairaut's approximations above because they would eventually trouble him, for reasons that I shall explain in detail in Chapters 8 and 9. For the moment it is enough to mention that in 1740, the Edinburgh mathematician Colin MacLaurin produced a rigorously exact, geometrical theory of homogeneous figures shaped like ellipsoids of revolution which attract according to the universal inverse-square law. MacLaurin made no approximations. Consequently his theory held for figures shaped like ellipsoids of revolution whose ellipticities were finite as well as for figures shaped like ellipsoids of revolution whose ellipticities were infinitesimal. In the most elementary part of his theory, MacLaurin showed how to determine the exact values of the magnitudes of the attraction at arbitrary points on the surface of a homogeneous figure shaped like an ellipsoid of revolution which attracts in accordance with the universal inverse-square law and whose ellipticity is finite. (Essentially he was able to reduce the problem to finding the exact values of the magnitudes of the attraction at the poles and at the equators of such figures.) He utilized the results of this part of his theory to demonstrate, for the first time ever, that a homogeneous figure shaped like a flattened ellipsoid of revolution which attracts in accordance with the universal inverse-square law, whose ellipticity is finite, which revolves around its axis of symmetry, and whose columns from center to surface all balance or weigh the same *can* indeed exist – or, to express the matter another way, that equality (6.1.28) *does* entail equality (6.1.29) when equality (6.1.28) holds for a homogeneous figure shaped like a flattened ellipsoid of revolution which attracts according to the universal inverse-square law, which revolves around its axis of symmetry, and whose ellipticity is finite. I shall sketch in Chapter 8 and 9 MacLaurin's method of finding the exact values of the magnitudes of the attraction at arbitrary points on the surface of a homogeneous figure shaped like an ellipsoid of revolution which attracts in accordance with the universal inverse-square law and whose ellipticity is finite, and at the same time I shall also outline how he used the values of the magnitudes of the attraction which he determined to show that equality (6.1.28) does entail equality (6.1.29) when equality (6.1.28) holds for a homogeneous figure shaped like a flattened ellipsoid of revolution which attracts according to the universal inverse-square law, which revolves around its axis of symmetry, and whose ellipticity is finite. For the moment it suffices to say that in proving the result stated in the second part of the preceding sentence, MacLaurin established the meaningfulness of the equality (5.3.2) that Maupertuis wrote at the end of his Paris Academy *mémoire* of 1734. Maupertuis did not even realize at the time that equality (5.3.2) was then a problem in itself.

(Of course, when the ellipticities are finite, Newton's equation

$$\frac{\delta}{1/100} = \frac{\phi}{4/505} \tag{1.1}$$

no longer holds, since equation (1.1) is only valid to terms of first order. That is, equation (1.1) only holds good when the ellipticity δ and the ratio ϕ of the magnitude of the centrifugal force per unit of mass to the magnitude of the attraction at the equator of a homogeneous figure shaped like a flattened ellipsoid of revolution which attracts according to the universal inverse-square law, which revolves around its axis of symmetry, and whose columns from center to surface all balance or weigh the same are infinitesimal.[12])

The geometry that MacLaurin employed to arrive at his findings maintained an advantage, in this one instance at least, over analysis for some time to come. Not until the 1770s did mathematicians succeed in applying analysis with the same degree of rigor to establish the results that MacLaurin had found thirty years earlier concerning attractions produced by homogeneous figures shaped like ellipsoids of revolution which attract in accordance with the universal inverse-square law and whose ellipticities are finite. Then it took analysts yet another decade to advance beyond MacLaurin's work. For this reason I have not gone into the details of Clairaut's approximate calculations of the magnitudes of the attractions produced by homogeneous figures shaped like nearly spherical ellipsoids of revolution which attract according to the universal inverse-square law. Nor do I intend to dwell upon such details in examining Clairaut's further work on the problem of the earth's shape, because mathematicians did not resolve the particular difficulty at issue (determining attractions produced by homogeneous or stratified figures shaped like ellipsoids of revolution which attract in accordance with the universal inverse-square law) during the period that I cover in the story that I tell here. They only did so years later.[13] We shall discover in Chapters 7 and 9 that mathematics is of major importance in this story, whose development it is essential to understand in order to follow the story, but the mathematics involved is not the mathematics connected with the problem of finding the attractions produced by homogeneous or stratified figures shaped like ellipsoids of revolution which attract according to the universal inverse-square law.

Clairaut emphasized in his paper of 1737 that the findings that he presented in that paper only held true approximately. He pointed out that the same results could hold just as well for other homogeneous nearly spherical figures of revolution which attract in accordance with the universal inverse-square law, which revolve around their axes of symmetry, and whose ellipticities, whose magnitudes of the attraction which they produce, and whose ratios of the magnitudes of the centrifugal force per unit of mass to the magnitudes of the attraction at their equators are the same to first order as those of homogeneous figures of revolution shaped like flattened ellipsoids of revolution which attract according to the universal inverse-square law, which revolve around their axes of symmetry, whose ellipticities are infinitesimal, and whose columns from center to surface all balance. Namely, the columns from the centers to the surfaces of these other homogeneous nearly spherical figures of revolution would also be found to balance, after terms of second order and higher orders in the ellipticities of these other homogeneous figures of revolution as well as all terms of second order and higher orders in quantities that have the same order as these ellipticities have been discarded in the calculations involved in determining the weights of columns from the centers of these other homogeneous figures of revolution to the surfaces of these figures.

Thus Clairaut observed that one could have just as easily inferred from the demonstration in his paper of 1737 that a homogeneous figure of revolution obtained by slightly altering or deforming a homogeneous figure shaped like a flattened ellipsoid of revolution which attracts according to the universal inverse-square law, whose ellipticity is infinitesimal, which revolves around its axis of symmetry, and whose columns from center to surface all balance or weigh the same is also a homogeneous figure of revolution which attracts according to the universal inverse-square law, whose ellipticity is infinitesimal, which revolves around its axis of symmetry, and whose columns from center to surface will all balance or weigh the same, because Clairaut only showed that the conditions necessary and sufficient for all of the columns from the center to the surface of a homogeneous figure shaped like a flattened ellipsoid of revolution which attracts according to the universal inverse-square law, whose ellipticity is infinitesimal, and which revolves around its axis of symmetry to balance or weigh the same hold after terms of second order and higher orders in the ellipiticity of such a homogeneous figure of revolution as well as all terms of second order and higher orders in quantities that have the same order as the ellipticity of the figure are neglected in the calculations involved in determining the weights of the columns from the center to the surface of the figure.

In reaching his conclusions, Clairaut incidentally furnished all of the basics and essentials needed to understand the origins of Newton's equation

$$\frac{\delta}{1/100} = \frac{\phi}{4/505}, \tag{1.1}$$

where δ = the infinitesimal ellipticity of a homogeneous figure shaped like an ellipsoid of revolution flattened at its poles which attracts according to the universal inverse-square law, which revolves around its axis of symmetry, and whose columns from center to surface all balance or weigh the same and ϕ = the ratio of the magnitude of the centrifugal force per unit of mass to the magnitude of the attraction at the equator of this figure. That is, Clairaut provided all of the particular approximate calculations, valid to first order, needed to produce an equation that is indistinguishable from Newton's equation (1.1) to terms of first order (in other words, the two equations differ by terms of second order and higher orders at most). The equation in fact is equation (5.3.4). Clairaut, however, did not derive and display equation (5.3.4) in his paper of 1737, for he was not so much concerned in this paper with equations as he was with the whole basis that underlay Newton's theory of the earth's shape. These foundations, which Newton had failed to explain, are what Clairaut questioned at this time.

Because $\left(\frac{1}{230}\right)^2 \ll \frac{1}{230}$, it seemed good enough to Clairaut to apply the first-order theory to the earth. In other words, for a figure whose ellipticity δ was as small as Newton's value $\frac{1}{230}$ for the earth's ellipticity, no one could doubt that Clairaut's arguments sufficed (disregarding the problem of estimating errors in making approximations).[14] Hence Newton had indeed chosen the correct shape for the earth, treating the earth as a homogeneous figure of revolution which attracts according to the universal inverse-square law, whose ellipticity is infinitesimal, which revolves around its axis of symmetry, and whose columns from center to surface all balance or weigh the same, although he had not made clear in the *Principia* his reasons for picking the particular shape that he did (an ellipsoid of revolution flattened at its poles). Newtons's theory lacked well-explained mathematical proofs. Barring the two mistakes that he made, which were not conceptual errors but mechanical errors of a technical sort, which luckily cancelled each other (or else he made one of the mistakes wittingly in order to neutralize an error that he realized that he had made somewhere earlier in his calculations), Clairaut provided exactly the kinds of mathematical arguments that Newton's theory required to be intelligible but that Newton had left out. Clairaut added the missing details needed to supply Newton's theory of the earth's shape with foundations that the theory did not seem to Clairaut to have. He was the first Frenchman to understand Newton's theory of the earth's shape.

6.2. First advances beyond Newton's theory (1738)

More than a year after returning to Paris from Lapland in August of 1737, Clairaut sent a second paper on the theory of the earth's shape, which he wrote in French this time, to the Royal Society.[15] Because he had been preoccupied with his paper on the phenomenon of stellar aberration discovered by James Bradley in 1728, which he read before the Paris Academy in December of 1737 and in February of 1738,[16] Clairaut delayed completing the new paper on the theory of the earth's shape. (In his paper on stellar aberration, which was published in the Paris Academy's *Mémoires* for 1737, Clairaut described the rules that the phenomenon obeyed, as Bradley had stated them, and he showed how these rules could be derived. The members of the expedition to Lapland had to take stellar aberration into account in making astronomical observations.[17]) The Reverend

John Colson, Lucasian Professor of Mathematics at Cambridge University, translated Clairaut's new paper on the theory of the earth's shape into English. Like Clairaut's paper of 1737 on the theory of the earth's shape, the translation of the new paper on the theory of the earth's shape was also published in the volume of the *Philosophical Transactions* for 1737–38.[18]

In this second paper Clairaut first recalled the objective of his first paper. Newton did "not acquaint us why" he chose an ellipsoid of revolution as the earth's shape; nor can "we perceive how he had satisfied himself in this particular," Colson wrote. Some other shape chosen instead could have led to a different value for the ratio of the earth's axis.[19] Newton's failure to explain why he chose the shape that he did, which Clairaut regarded as a deficiency of Newton's theory, had prompted Clairaut to write the previous paper.

After comparing their measurement of a degree of latitude in Lapland with Picard's measurement of a degree of latitude in France, the members of the expedition to Lapland concluded that the earth was flattened at the poles. Moreover, they calculated the value of the earth's ellipticity to be $\frac{1}{177}$, which made the earth flatter than Newton's homogeneous figure shaped like an ellipsoid of revolution flattened at the poles which attracts according to the universal inverse-square law, which revolves around its axis of symmetry, whose ellipticity is $\frac{1}{230}$, and whose columns from center to surface all balance or weigh the same. And the experiments done earlier with seconds pendulums at different latitudes seemed to indicate that the increase in the magnitude of the effective gravity with latitude along the earth's surface was greater than the one that Newton's theory of a homogeneous figure shaped like an ellipsoid of revolution flattened at the poles which attracts in accordance with the universal inverse-square law, which revolves around its axis of symmetry, and whose columns from its center to its surface all balance or weigh the same required when applied to the earth. We remember from Chapter 1 that Newton had allowed for the possibility of a greater observed increase in the magnitude of the effective gravity with latitude along the earth's surface than the one that his theory required when applied to the earth. Making allowance for a greater increase in the magnitude of the effective gravity with latitude along the earth's surface than the increase in the magnitude of the effective gravity with latitude along the surface of the particular homogeneous figure shaped like an ellipsoid of revolution flattened at the poles which attracts according to the universal inverse-square law, which revolves around its axis of symmetry, and whose columns from center to surface all balance or weigh the same which represented the earth in his theory, by totally disregarding any effects that climactic changes might have on the lengths of the metallic cords of seconds pendulums, Newton declared, without explaining his reasoning, that the earth would have to be flatter than the particular homogeneous figure that represented the earth in his theory and also be denser toward its center than toward its surface (Article XXIV and following). And the conclusions that the members of the Lapland expedition drew from their measurements did indeed appear to confirm the first of Newton's two statements.

Newton had also concluded from telescopic observations that although Jupiter was flattened at the poles, its ellipticity was less than the ellipticity that his theory required when applied to that planet. The fact that Jupiter's observed ellipticity did not follow as a consequence of his theory had led Newton to assert, on page 416 of the third edition of the *Principia*, that Jupiter was more dense in the plane of its equator than toward its poles, because, as Colson translated Clairaut, "its Moisture is more dried up by the Heat of the Sun" there.[20]

But the ways that Newton rationalized why various observations differed from the results that followed from his theory thoroughly perplexed Clairaut. It seemed peculiar and unnatural to Clairaut to attribute to the sun an effect upon far away Jupiter which, if real, should have that much greater an effect upon the much closer earth. Yet Newton did not call attention to any such effect of the sun upon the earth. Nor did Clairaut understand why Newton had not tried to ascribe "a like Cause," in Colson's words, to explain the anomalies involving the earth and Jupiter – in particular, a difference in densities at the centers and at the surfaces of both the earth and Jupiter to account for these anomalies.[21] Here Clairaut imagined the unified theory of the shapes of planets which was the furthest thing from the minds of Fontenelle and the other positivists in the Paris Academy. We recall that they had not considered such a theory to be necessary, even if they had believed that such a theory could at least exist in principle, which does not appear to have been the case.

As I suggested in Chapter 1, Newton tried to patch up what had begun in the *Principia* as a mathematical theory by tinkering with certain details of the theory in a qualitative way in order to justify observations that did not accord with the theory. In the process of doing this, he made claims that appear to be little more than "intuitions." At least, he did not explain or prove the claims, nor did he furnish any evidence that supported them. He did not demonstrate mathematically how the anomalous observations in question followed from these claims. This is true in the case of the earth. Moreover, he also introduced ideas to "explain" other observations that did not accord with his theory. But he did not show mathematically how the anomalous observations in question followed from these ideas, either. This is true in the case of Jupiter. Moreover, the ideas that Newton introduced to deal with Jupiter seemed to Clairaut to be inconsistent with the claims that Newton made about the earth. If Newton had mathematical bases for his various claims and ideas, he nevertheless withheld them when he published the *Principia*. Could Newton's claims and ideas be furnished with mathematically sound foundations? Clairaut intended to try. In his paper of 1738, he revealed discoveries that he made while investigating Newton's seemingly inconsistent claims and ideas.

Specifically, Clairaut determined the ellipticity of a nearly spherical figure shaped like an ellipsoid of revolution flattened at its poles which was composed of infinitely many concentric ellipsoids of revolution flattened at their poles, all of which had the same infinitesimal ellipticity α and all of which had a common axis of symmetry (the axis of symmetry of the figure, which is the polar axis of the figure). He supposed that the figure attracted in accordance with the universal inverse-square law, that the figure revolved around its axis of symmetry, and that the principle of the plumb line held at all points on the surface of the figure. Clairaut assumed the density of each ellipsoid of revolution to be uniform, but he allowed the density to vary from ellipsoid to ellipsoid. With the densities of the ellipsoids of revolution as well as the infinitesimal ratio of the magnitude of the centrifugal force per unit of mass to the magnitude of the effective gravity at the equator of the figure both given at the start, Clairaut found the infinitesimal ellipticity α of that particular figure made up of such ellipsoids of revolution at whose surface the principle of the plumb line held. For the same reasons as those that he gave in his paper of 1737, he neglected terms that included α^n as factors when he did his calculations, where $n \geqslant 2$, as well as all other terms that included as factors quantities of the same orders as α^n, such as R^n, for $n \geqslant 2$, where R is the ratio of the magnitude of the centrifugal force per unit of mass to the magnitude of the effective gravity at the equator of the figure. Once again Clairaut sometimes made clear what terms he was discarding in doing his calculations. This was true when he employed analysis. On the other hand, he also neglected terms implicitly without clarifying what these terms looked like. This he did when he simplified the geometry of a nearly spherical stratified figure shaped like an ellipsoid of revolution flattened at its poles which attracts according to the universal inverse-square law, whose ellipticity is infinitesimal, which revolves around its axis of symmetry, and at whose surface the principle of the plumb line holds.

The methods, procedures, and calculations that appear in the paper of 1738 strongly resemble those that appear in the paper of the year before. In 1737, Clairaut had to calculate the magnitudes of certain components of the attraction and of the effective gravity at arbitrary points N on the surface of a homogeneous figure shaped like an ellipsoid of revolution flattened at its poles which attracts according to the universal inverse-square law, which revolves around its axis of symmetry, and whose ellipticity is infinitesimal, where the points N were situated on a meridian at the surface of the figure, in order to determine how the figure should revolve around its axis of symmetry in order to make the magnitudes of the relevant components of the effective gravity at the points N on the meridian at the surface of the figure vary inversely as the distances of the points from the center of the figure (namely, the components of the effective gravity at points N on a meridian at the surface of the figure which are directed toward the center C of the figure). In the paper of 1738, he had to do similar kinds of computations, in order to apply the principle of the plumb line at arbitrary points N on the surface of the stratified figure of revolution when the figure revolves around its axis of symmetry. The calculations of course were more complex and required more work than the ones that Clairaut did in the paper of 1737. Again, however, because of the symmetry of the stratified figure of revolution, the fact that the axis of symmetry of the stratified figure is also an axis of symmetry of the attraction produced by the stratified figure, and the fact that the stratified figure revolves around its axis of symmetry and not around some other line, so that the axis of symmetry of the stratified figure is also an axis of symmetry of the centrifugal force per unit of mass, it sufficed to do the calculations at points N located on a flattened ellipse *EbeBE* which is a meridian at the surface of the nearly spherical stratified figure shaped like an ellipsoid of revolution flattened at its poles which attracts according to the universal inverse-square law, whose ellipticity α is infinitesimal, and which revolves around its axis of symmetry.

Clairaut imagined the magnitude of the attraction at a point N on a flattened ellipse *EbeBE* which is a meridian at the surface of his nearly spherical stratified figure shaped like a flattened ellipsoid of revolution which attracts according to the universal inverse-square law produced by one of the flattened ellipsoids of revolution inside the figure to be the difference between the magnitudes of the attractions at N produced by two concentric homogeneous figures shaped like flattened ellipsoids of revolution which attract according to the universal inverse-square law, whose surfaces are infinitesimally close to each other, whose densities are the same, and whose ellipticities are infinitesimal and equal (to α), where N is situated outside both figures. Consequently, he needed to determine the magnitudes of the attractions produced by homogeneous figures shaped like flattened ellipsoids of revolution which attract according to the universal inverse-square law and whose ellipticities are infinitesimal at points located outside the figures.

Moreover, in order to apply the principle of the plumb line at a point N on a flattened ellipse *EbeBE* which is a meridian at the surface of his stratified figure when the figure revolved around its axis of symmetry, Clairaut had to find the magnitudes and the directions of several components of the attraction at N produced by the figure and several components of the effective gravity at N when the figure revolved around its axis of symmetry and not just the magnitudes of the one component of the attraction and the one component of the effective gravity which were enough for Clairaut to consider in the paper of 1737 (namely, the magnitudes of the components of the attraction and of the effective gravity at N which are directed toward the center C of a homogeneous figure shaped like a flattened ellipsoid of revolution which attracts according to the universal inverse-square law, whose ellipticity is infinitesimal, which revolves around its axis of symmetry, and whose density is 1, in order to determine how the figure should revolve around its axis of symmetry in order to make the columns from the center of the figure to the surface of the figure all balance or weigh the same).

To express the densities of an infinite number of concentric, individually homogeneous ellipsoids of revolution flattened at their poles, all of which have the same infinitesimal ellipticity α and all of which have a common axis of symmetry (the axis of symmetry of the stratified figure that the ellipsoids of revolution make up, which is the polar axis of the figure), Clairaut let the density D be an arbitrary, given function of distance r along the polar axis of the stratified figure, in which case r was, in effect, the polar radius of an ellipsoid of revolution, measured along the polar axis of the stratified figure. According to Clairaut, an arbitrary function $D(r)$ of r which expressed the densities of the ellipsoids of revolution could be compounded of monomials

$$D(r) = fr^p + gr^q + hr^s + ir^t + \text{etc.} \qquad (6.2.1)$$

"by the Property of Series...." (The densities of the individually homogeneous ellipsoids of revolution are non-negative, which means that $0 \leqslant D(r)$ for all $0 \leqslant r \leqslant$ the length of the polar axis of the stratified figure. Here f, g, h, i, etc. are the coefficients in and p, q, s, t, etc. are the exponents in the infinite series for $D(r)$ in (6.2.1). In practice Clairaut seems to have assumed that all of the exponents in the infinite series for $D(r)$ in (6.2.1) were greater than or equal to zero.) Clairaut actually only kept two terms

$$D(r) = fr^p + gr^q \qquad (6.2.2)$$

of such infinite series for $D(r)$ in (6.2.1) in his calculations, because he claimed that additional terms of such series would not change the results that he found or the conclusions that he arrived at. He said that he would explain the reasons for this as he went along.[22]

In fact, Clairaut's theory holds as well for a nearly spherical stratified figure shaped like a flattened ellipsoid of revolution which attracts according to the universal inverse-square law, which revolves around its axis of symmetry, at whose surface the principle of the plumb line holds, and which consists of a *finite* number of strata each of which has *finite* thickness and each of whose inner and outer surfaces are ellipsoids of revolution flattened at their poles, where all of the ellipsoids of revolution which bound the strata are concentric, all of these ellipsoids of revolution have a common axis of symmetry (the axis of symmetry of the stratified figure, which is the polar axis of the figure), and the ellipticities of the ellipsoids of revolution are all infinitesimal and equal (to α).[23]

Clairaut also called attention to the fact that, once again, the results that he found would hold just as well for *any* nearly spherical stratified figure of revolution which attracts according to the universal inverse-square law, whose ellipticity is infinitesimal, which revolves around its axis of symmetry, at whose surface the principle of the plumb line holds, and which leads to the same calculations, after terms of second order and higher orders have been neglected. In other words, because Clairaut's calculations were approximate, they neither proved that the results that followed from them held exactly for a nearly spherical stratified figure shaped like a flattened ellipsoid of revolution which attracts according to the universal inverse-square law, which revolves around its axis of symmetry, at whose surface the principle of the plumb line is assumed to hold, and which is composed of infinitely many concentric, individually homogeneous ellipsoids of revolution flattened at their poles, all of which have the same infinitesimal ellipticity α and all of which have a common axis of symmetry (the axis of symmetry of the stratified figure, which is the polar axis of the figure) nor that the same results could not be found to hold to terms of first order for some other nearly spherical stratified figure of revolution which attracts according to the universal inverse-square law, which revolves around its axis of symmetry, at whose surface the principle of the plumb line holds, and which consists of an infinite number of concentric, individually homogeneous strata that are surfaces of revolution, all of which have a common axis of symmetry (the axis of symmetry of the stratified figure of revolution, which is the polar axis of the figure), and whose shapes differ slightly from flattened ellipsoids of revolution whose ellipticities are all infinitesimal and equal (to α).

Clairaut began the technical part of his paper of 1738 by determining the approximate value of the magnitude of the attraction produced by a homogeneous figure shaped like a flattened ellipsoid of revolution which attracts according to the universal inverse-square law, whose ellipticity is infinitesimal, and whose density is 1 at any point outside the figure and situated on the axis of symmetry of the figure. Clairaut found this value in much the same way that he had computed in 1737 the approximate value of the magnitude of the attraction at a pole of a homogeneous figure shaped like an ellipsoid of revolution flattened at its poles which attracts according to the universal inverse-square law, whose ellipticity is infinitesimal, and whose density equals 1. In effect, the new calculation just generalizes the one of the year before.

[In computing the approximate values of the magnitudes of the attraction at points along the axis of symmetry in the exterior of a homogeneous figure shaped like an ellipsoid of revolution flattened at its poles which attracts according to the universal inverse-square law, whose ellipticity is infinitesimal, whose density equals 1, and whose polar radius equals r, Clairaut found an expression for the magnitude of the attraction which reduces to the correct expression (6.1.21) when the point along the axis of symmetry of the figure in the exterior of the figure is chosen to be located at the pole of the figure. In his paper of 1738 Clairaut explained in less detail how he determined the approximate values of the magnitudes of the attraction at points along the axis of symmetry of the figure situated outside the figure than he had explained in his paper of 1737 how he found the approximate value of the magnitude of the attraction at the pole of the figure. That is, he did not reproduce step by step the chain of arguments and computations which led to his expression for the magnitudes of the attraction at points along the axis of symmetry of the figure in the exterior of the figure. It is easy to see that the method that he sketched is essentially the same as the one that he used in his paper of 1737 to find the approximate value of the magnitude of the attraction at the pole of the figure. But one thing is certain. In doing the calculations in 1738 which led to the more general expression, Clairaut evidently did not make the error that he did in his paper of 1737 in deriving the false expression (6.1.11) for the approximate value of the magnitude of the attraction at the pole of the figure, because, as I just mentioned, the expression in his paper of 1738 reduces to the correct expression (6.1.21) when the point along the axis of symmetry of the figure outside the figure is taken to be situated at the pole of the

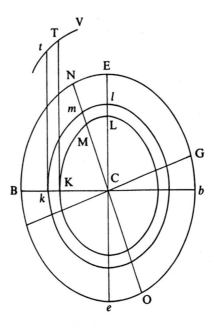

Figure 18. Meridian of a stratified figure whose pole is at B and whose polar radius is CB composed of an infinite number of concentric, individually homogeneous flattened ellipsoids of revolution all of which have a common axis of symmetry, whose ellipticities are all infinitesimal and equal, and which attract according to the universal inverse-square law (Clairaut 1738b: Figure 2).

figure. I shall give more evidence below that in writing his paper of 1738 Clairaut found and corrected the mistake that he had made in his paper of 1737.]

Clairaut needed this generalization in order to solve a problem that did not arise in his paper of 1737. In the technical part of his paper of 1738 Clairaut next determined the approximate value of the magnitude of the attraction at a pole B of a nearly spherical stratified figure shaped like an ellipsoid of revolution flattened at its poles which attracts according to the universal inverse-square law, whose polar radius $CB = e$, and which consists of an infinite number of concentric, individually homogeneous flattened ellipsoids of revolution all of which have a common axis of symmetry (the axis of symmetry of the stratified figure, which is its polar axis) and whose ellipticities are all infinitesimal and equal (to α) (see Figure 18). The densities of the flattened ellipsoids of revolution which are the strata of the figure vary in the manner that I described above (that is, according to expression (6.2.1)). In order to do this calculation, Clairaut interpreted the magnitude of the attraction produced at the pole B of the stratified figure by a single flattened ellipsoid of revolution whose ellipticity α is infinitesimal, whose polar radius $CK = r$, where $0 \leqslant r \leqslant e = CB$, and whose density is 1–that is, by a single stratum of the stratified figure, whose density is taken to be equal to 1–as the difference between the magnitudes of the attractions produced at the pole B of the stratified figure by two concentric homogeneous figures shaped like flattened ellipsoids of revolution which attract according to the universal inverse-square law, whose densities equal 1, which have a common axis of symmetry (the axis of symmetry of the stratified figure, which is the polar axis of the figure), whose ellipticities are infinitesimal and are equal (to α), whose polar radii $CK = r$, where $0 \leqslant r \leqslant e = CB$, are nearly the same, and whose surfaces are infinitesimally close to each other. But this difference is just the differential of the magnitude of the attraction produced at the pole B of the stratified figure by a homogeneous figure shaped like a flattened ellipsoid of revolution which attracts in accordance with the universal inverse-square law, whose ellipticity α is infinitesimal, whose polar radius $CK = r$, where $0 \leqslant r \leqslant e = CB$, and whose density is 1. In his manuscript in French, however, Clairaut substituted the British term "fluxion" for the continental term "differential," and he employed the notation for fluxions in his calculations instead of the notation for differentials.[24] (Clearly, Clairaut deliberately chose to use terms and notation that he knew would appeal to a British audience and that the British would understand, in an effort to please the British.) Now, the pole B of the stratified figure is not the pole of any of the flattened ellipsoids of revolution which are the strata that make up the stratified figure, except the flattened ellipsoid of revolution which is the surface of the stratified figure. Or to express the matter another way, the pole B of the stratified figure lies in the exterior along the polar axis of each homogeneous figure shaped like a flattened ellipsoid of revolution which attracts in accordance with the universal inverse-square law, whose ellipticity α is infinitesimal, whose polar radius $CK = r$, where $0 \leqslant r \leqslant e = CB$, and whose density equals 1, except the homogeneous figure shaped like a flattened ellipsoid of revolution which attracts in accordance with the universal inverse-square law, whose ellipticity α is infinitesimal, whose density is 1, and whose polar radius $CK = r = e = CB$ and consequently whose pole is at B. This explains why Clairaut had to generalize the calculation that he did in 1737 of the approximate value of the magnitude of the attraction at a pole of a homogeneous figure shaped like a flattened ellipsoid of revolution which attracts according to the universal inverse-square law, whose ellipticity is infinitesimal, and whose density equals 1 to include the approximate values of the magnitudes of the attraction at all points in the exterior along the axis of symmetry of such a homogeneous figure. He needed the generalization in order to determine the approximate value of the magnitude of the attraction at the pole B of the stratified figure produced by all of the flattened ellipsoids of revolution of which the stratified figure is composed.

In order to compute this value, Clairaut found the approximate value of the magnitude of the attraction at the pole B of the stratified figure produced by a single flattened ellipsoid of revolution whose polar radius $CK = r$, where $0 \leqslant r \leqslant e = CB$, and whose density is 1, and he multiplied that value by the real density of that flattened ellipsoid of revolution. He did this for each of the flattened ellipsoids of revolution which make up the stratified figure. Finally, he found the sum of the approximate values of the magnitudes of the attractions at the pole B of the stratified figure produced by all of the flattened ellipsoids of revolution which the stratified figure is composed of. In other words, he determined the integral of the approximate values of the magnitudes of these attractions in order to find the approximate value of the magnitude of the total or the resultant of the attraction at the pole B of the stratified figure which is produced by the stratified figure. Once again, however, Clairaut substituted the British term "fluent" for the continental term "integral" in his manuscript in French.[25]

[When the ellipsoids of revolution are all taken to have the same density equal to 1, which means that $p = q = 0$ and $f + g = 1$ in expression (6.2.2) for the density $D(r)$ of the stratified figure, so that the stratified figure becomes a homogeneous figure, then the expression that Clairaut arrived at for the approximate value of the magnitude of the attraction at a pole B of the nearly spherical stratified figure shaped like an ellipsoid of revolution flattened at its poles which attracts according to the universal inverse-square law, whose polar radius $CB = e$, and whose ellipticity α is infinitesimal reduces, after an appropriate change of notation, to the expression

$$\tfrac{2}{3}ca + \tfrac{8}{15}cam \qquad (6.1.15')$$

for the approximate value of the magnitude of the attraction at a pole A of a homogeneous figure shaped like an ellipsoid of revolution flattened at its poles which attracts according to the universal inverse-square law, whose polar radius $AC = a$ in Clairaut's new notation in his paper of 1737, whose equatorial radius $CE = b$ in Clairaut's new notation in his paper of 1737, and whose ellipticity $m \equiv \frac{b-a}{a}$ in Clairaut's new notation in his paper of 1737 is infinitesimal, where $c = 2\pi$. Hence we have more evidence that in writing his paper of 1738 Clairaut did not make the mistake that had led him in his paper of 1737 to the false expression (6.1.11) for the approximate value of the magnitude of the attraction at the pole of such a homogeneous figure.]

Clairaut next stated that the approximate value of the magnitude of the attraction at any point N on a flattened ellipse $EbeBE$ which is a meridian at the surface of the nearly spherical stratified figure shaped like an ellipsoid of revolution flattened at its poles which attracts according to the universal inverse-square law, whose polar radius $CB = e$, and whose ellipticity α is infinitesimal was the same as the approximate value of the magnitude of the attraction at N produced by a certain other nearly spherical stratified figure shaped like an ellipsoid of revolution which attracts according to the universal inverse-square law, whose pole is at N, whose polar axis is NO and whose polar radius is CN in Figure 19, and which is composed of infinitely many concentric, individually homogeneous ellipsoids of revolution, all of which have a common axis of symmetry (the axis of symmetry of the stratified figure, which is the polar axis of

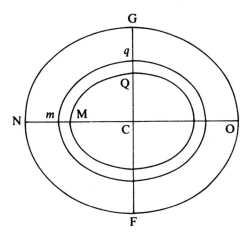

Figure 19. Meridian of a stratified figure whose pole is at N and whose polar radius is CN (see Figure 18) composed of an infinite number of concentric, individually homogeneous ellipsoids of revolution all of which have a common axis of symmetry, whose ellipticities are all infinitesimal and equal, and which attract according to the universal inverse-square law (Clairaut 1738b: Figure 3).

the figure), whose ellipticities are all infinitesimal and all equal, and which have the same densities as the flattened ellipsoids of revolution which make up the nearly spherical stratified figure shaped like an ellipsoid of revolution flattened at its poles which attracts according to the universal inverse-square law, whose polar radius $CB = e$, whose ellipticity α is infinitesimal, and whose meridians are flattened ellipses $EbeBE$. (Again, as a result of the various symmetries, it is enough to consider points N on any one of these flattened ellipses $EbeBE$.) Clairaut used the new result to determine the approximate value of the magnitude of the attraction produced by the stratified figure shaped like an ellipsoid of revolution flattened at its poles whose polar radius $CB = e$ and whose ellipticity α is infinitesimal at any point N on a flattened ellipse $EbeBE$ which is a meridian at the surface of this stratified figure.[26]

[For the same kinds of reasons that I mentioned in the preceding section of this chapter, what Clairaut really approximated here was the magnitude of the *component* of the attraction at N produced by the stratified figure shaped like a flattened ellipsoid of revolution whose ellipticity α is infinitesimal and whose meridians are flattened ellipses $EbeBE$ which is directed toward the center C of the stratified figure, not the magnitude of the *total* or the *resultant of* the attraction at N produced by this stratified figure. Again in effect Clairaut *implicitly equated* the magnitude of the attraction at N produced by the nearly spherical stratified figure shaped like an ellipsoid of revolution which attracts according to the universal inverse-square law, whose pole is at N, whose polar axis is NO, and whose polar radius is CN and the magnitude of the *component* of the attraction at N produced by the stratified figure shaped like a flattened ellipsoid of revolution whose ellipticity α is infinitesimal and whose meridians are flattened ellipses $EbeBE$ which is directed toward the center C of the stratified figure. He did this because the ellipticities of all of the flattened ellipsoids of revolution which this figure is composed of are all infinitesimal (they are all equal to the infinitesimal number α), and at the same time the ellipticities of the ellipsoids of revolution which make up the stratified figure whose pole is at N, whose polar axis is NO, and whose polar radius is CN are all infinitesimal too (they all have the same infinitesimal value). Thus the values of these two magnitudes of attractions at N presumably differ by terms of second order and higher order at most. That is, the two different values of the magnitudes of attractions at N are equal to terms of first order. Once again, however, just as he neglected to do at the corresponding place in his paper of 1737, Clairaut did not state precisely what he was doing here in his paper of 1738, equating the magnitudes of two attractions, much less actually demonstrate that the values of the magnitudes of the two attractions at N do equal each other to terms of first order.]

To establish the result that he stated, Clairaut used basically the same kind of arguments that he had employed in his paper of 1737 to show that an analogous result held for a point N on a flattened ellipse $AEaeA$ which is a meridian at the surface of a homogeneous figure shaped like a flattened ellipsoid of revolution which attracts according to the universal inverse-square law, whose ellipticity is infinitesimal, and whose density equals 1, which the new result again simply generalizes. That is, the particular result of 1737 in question is in fact just a special case of the new result. (We recall, however, that Clairaut described what he was doing in this part of his paper of 1737 rather confusedly. What he did was not wrong, but he explained himself incorrectly.) By this means, Clairaut reduced the problem of determining the approximate value of the magnitude of the attraction at any point N on a flattened ellipse $EbeBE$ which is a meridian at the surface of the flattened stratified figure whose polar radius $CB = e$ and whose ellipticity α is infinitesimal to the preceding problem of finding the approximate value of the magnitude of the attraction at a pole of a stratified figure shaped like an ellipsoid of revolution which attracts according to the universal inverse-square law and which consists of an infinite number of concentric, individually homogeneous ellipsoids of revolution, all of which have a common axis of symmetry (the axis of symmetry of the figure, which is the polar axis of the figure), and whose ellipticities are all infinitesimal and all equal.

[If we think of a homogeneous figure as a special case of a stratified figure whose individual strata all have the same density, we realize that in stating the result above, Clairaut must have also determined the approximate value of the magnitude of the attraction at a pole of a nearly spherical stratified figure shaped like an ellipsoid of revolution *elongated* at the poles which attracts according to the universal inverse-square law and which is made up of an infinite number of concentric, individually homogeneous ellipsoids of revolution elongated at the poles, all of which have a common axis of symmetry (the axis of symmetry of the stratified figure, which is the polar axis of the figure), and whose ellipticities are all infinitesimal and all equal. This is true because we remember that in the case of a homogeneous figure shaped like a flattened ellipsoid of revolution which attracts according to the universal inverse-square law, whose ellipticity is infinitesimal, and whose density is 1, which Clairaut treated in his paper of 1737, the homogeneous figure shaped like an ellipsoid of revolution which attracts according to the universal inverse-square law, whose ellipticity is infinitesimal, whose density is 1, whose pole is at a point N on a flattened ellipse $AEaeA$ which is a meridian at the surface of the flattened homogeneous figure, whose polar radius is CN where C is the center of the flattened homogeneous figure, and where the approximate value of the magnitude of the attraction at N produced by the figure equals the approximate value of the magnitude of the component of the attraction at N which points to C produced by the flattened homogeneous figure can be elongated at its poles. Whether this homogeneous figure whose pole is at N and whose polar radius is CN is flattened at its poles or elongated at its poles depends upon the particular point N on the surface of the flattened homogeneous figure. This we observed in the preceding section of this chapter to be true in the case of a homogeneous figure shaped like a flattened ellipsoid of revolution which attracts according to the universal inverse-square law, whose ellipticity is infinitesimal, and whose density equals 1, which Clairaut treated in his paper of 1737, although I just alluded to the fact that Clairaut neglected to point this truth out in that paper, as we also saw in the preceding section of this chapter. Indeed, he spoke instead as if the homogeneous figures whose poles are at N and whose polar radii are CN are always elongated at their poles. Perhaps Clairaut's failure to explain himself clearly in this part of his paper of 1737 accounts in part for the fact that in his paper of 1738 Clairaut did not present the approximate value of the magnitude of the attraction at a pole of a stratified figure shaped like an ellipsoid of revolution elongated at the poles which attracts according to the universal inverse-square law and whose ellipticity is infinitesimal as a separate, isolated result, even though he had stated the result in the special case – that is, the result for homogeneous figures shaped like ellipsoids of revolution elongated at the poles which attract according to the universal inverse-square law, whose ellipticities are infinitesimal, and whose densities equal 1 – separately in his paper of 1737. Nevertheless, in determining the approximate value of the magnitude of the attraction at a point N on a flattened ellipse $EbeBE$ which is a meridian at the surface of his flattened stratified figure whose polar radius $CB = e$ and whose ellipticity α is infinitesimal, Clairaut had to have made use of the result for stratified figures shaped like elongated ellipsoids of revolution. But this fact only shows up in the actual calculations that Clairaut did in his paper of 1738 to determine the approximate value of the magnitude of the attraction at N. He did not state the result for elongated stratified figures as a separate theorem in his paper of 1738.

When the flattened ellipsoids of revolution are all taken to have the same density equal to 1, which means that $p = q = 0$ and $f + g = 1$ in expression (6.2.2) for the density $D(r)$ of the stratified figure shaped like an ellipsoid of revolution flattened at its poles which attracts according to the universal inverse-square law, whose polar radius $CB = e$, and whose ellipticity α is infinitesimal, so that the stratified figure becomes a homogeneous figure, then the general expression that Clairaut found for the approximate value of the magnitude of the component of the attraction at a point N on a flattened ellipse $EbeBE$ which is a meridian at the surface of the stratified figure which is directed toward the center C of the figure reduces, after an appropriate change of notation, to the expression

$$\tfrac{2}{3}ca - \tfrac{2}{15}can + \tfrac{8}{15}cam \qquad (6.1.12'''')$$

for the approximate value of the magnitude of the component of the attraction at a point N on a flattened ellipse $AEaeA$ at the surface of a homogeneous figure shaped like an ellipsoid of revolution flattened at its poles which attracts according to the universal inverse-square law, whose polar radius $AC = a$ in Clairaut's new notation in his paper of 1737, whose equatorial radius $CE = b$ in Clairaut's new notation in his paper of 1737, whose ellipticity

$$m \equiv \frac{b-a}{a}$$

in Clairaut's new notation in his paper of 1737 is infinitesimal, and where $CN = r$ and

$$n \equiv \frac{r-a}{a}$$

in Clairaut's new notation in his paper of 1737, which points to the center C of the figure, where $c = 2\pi$. If we recall how expression (6.1.12'''') was derived in the preceding section of this chapter, we see that Clairaut must have made use of the result for stratified figures shaped like elongated ellipsoids of revolution in order to arrive at his general expression for the approximate value of the magnitude of the component of the attraction at a point N on a flattened ellipse $EbeBE$ which is a meridian at the surface of the stratified figure shaped like an ellipsoid of revolution flattened at its poles which attracts according to the universal inverse-square law, whose polar radius $CB = e$, and whose ellipticity α is infinitesimal which is directed toward the center C of the figure. Moreover, the fact that this general expression reduces, after an appropriate change of notation, to expression (6.1.12'''') when the flattened figure is homogeneous is more evidence that in writing his paper of 1738, Clairaut found and corrected the mistake that had led him in his paper of 1737 to the false expression (6.1.11) for the approximate value of the magnitude of the attraction at a pole of a homogeneous

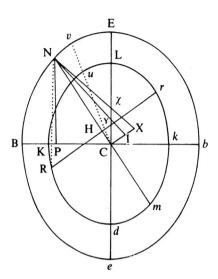

Figure 20. *Line segments in various directions at point N in Figure 18 in the plane of the meridian (Clairaut 1738b: Figure 5).*

figure shaped like an ellipsoid of revolution flattened at its poles which attracts according to the universal inverse-square law and whose ellipticity is infinitesimal.]

To apply the principle of the plumb line at the surface of the nearly spherical stratified figure shaped like an ellipsoid of revolution flattened at its poles which attracts according to the universal inverse-square law, whose polar radius $CB = e$, whose ellipticity α is infinitesimal, and whose meridians are flattened ellipses *EbeBE* when the figure revolved around its axis of symmetry, Clairaut also had to calculate the approximate values of the magnitudes of various components of the attraction in various directions at points N located on a flattened ellipse *EbeBE* which is a meridian at the surface of the stratified figure (see Figure 20).

For instance, he determined the approximate value of the magnitude of the component of the attraction produced by a homogeneous figure shaped like an ellipsoid of revolution flattened at its poles which attracts according to the universal inverse-square law, whose ellipticity α is infinitesimal, whose polar radius $CK = r$, where $0 \leqslant r \leqslant e = CB \equiv$ the polar radius of the stratified figure, and whose density is 1 at a point N situated on a flattened ellipse *EbeBE* which is a meridian at the surface of the stratified figure whose direction is indicated by the line segment CX which lies in the plane of the flattened ellipse *EbeBE* and which is perpendicular to CN, where C is the center of the stratified figure.[27] He then found the differential, which he again called the fluxion, of the expression for this value, which represented the approximate value of the magnitude of the component of the attraction at N whose direction is specified by the line segment CX produced by an ellipsoid of revolution flattened at its poles which attracts according to the universal inverse-square law, whose ellipticity α is infinitesimal, whose polar radius $CK = r$, where $0 \leqslant r \leqslant e = CB$, and whose density equals 1. He then multiplied this fluxion by the density of the ellipsoid of revolution flattened at the poles which attracts according to the universal inverse-square law and which is located inside the stratified figure, whose ellipticity α is infinitesimal, and whose polar radius $CK = r$, where $0 \leqslant r \leqslant e = CB$, which gave him the approximate value of the magnitude of the component of the attraction at N whose direction is indicated by the line segment CX produced by the ellipsoid of revolution flattened at the poles which attracts according to the universal inverse-square law and which is situated within the stratified figure, whose ellipticity α is infinitesimal, and whose polar axis $CK = r$, where $0 \leqslant r \leqslant e = CB$. Clairaut then found the sum of all of the approximate values of the magnitudes of the components of the attraction at N whose directions are specified by the line segment CX produced by the individual flattened ellipsoids of revolution which make up the flattened stratified figure whose polar radius $CB = e$ and all of which have the same infinitesimal ellipticity α. In other words, he determined the fluent of these values of the magnitudes of these components of the attraction at N. In this way, Clairaut found the approximate value of the magnitude of the component of the attraction at a point N whose direction is indicated by the line segment CX which lies in the plane of the flattened ellipse $EbeBE$ and which is perpendicular to CN at the point N on the flattened ellipse $EbeBE$ which is a meridian at the surface of the stratified figure shaped like an ellipsoid of revolution flattened at its poles which attracts according to the universal inverse-square law, whose polar radius $CB = e$, where C is the center of the stratified figure, and which is composed of infinitely many concentric, individually homogeneous flattened ellipsoids of revolution, all of which have a common axis of symmetry (the axis of symmetry of the stratified figure, which is the polar axis of the figure) and all of which have the same infinitesimal ellipticity α, which is produced by this stratified figure. Just as he did in 1737, Clairaut simplified geometric arguments which led to results that held good approximately instead of results that held true exactly. Once again, the orders to which the approximations that he made in the course of arriving at his results hold are sometimes concealed.[28]

Having already found the approximate value of the magnitude of the attraction produced by the stratified figure at a point N on a flattened ellipse $EbeBE$ which is a meridian at the surface of the figure, Clairaut then determined the direction of the attraction at N. (In finding this direction Clairaut mentioned that in fact what he had really determined previously in his paper was the approximate value of the magnitude of the *component* of the attraction at N which points to the center C of the stratified figure, not the approximate value of the magnitude of the *total* or the *resultant of* the attraction at N. By finding a fourth proportional for the approximate value of the magnitude of the component of the attraction at N

which is directed toward the center C of the stratified figure, the approximate value of the magnitude of the component of the attraction at N whose direction is indicated by the line segment CX which lies in the plane of the flattened ellipse $EbeBE$ and which is perpendicular to CN at N, and the length of CN, Clairaut determined the direction of the total or the resultant of the attraction at N, although Clairaut did not explain his reasoning here very clearly.[29])

Colson then stated that Clairaut supposed that his nearly spherical stratified figure shaped like an ellipsoid of revolution flattened at its poles which attracts according to the universal inverse-square law, whose polar radius $CB = e$, whose ellipticity α is infinitesimal, and whose meridians are flattened ellipses $EbeBE$ revolved around its axis of symmetry Bb, which is the polar axis of the figure, and "arrived at its permanent State."[30] Then the principle of the plumb line had to hold at all points on the surface of the stratified figure, otherwise "without this Condition there could be no Æquilibrium," Colson translated Clairaut as saying.[31] That is, Clairaut assumed his nearly spherical stratified figure shaped like an ellipsoid of revolution flattened at its poles which attracts according to the universal inverse-square law, whose polar radius $CB = e$, whose ellipticity α is infinitesimal, whose meridians are flattened ellipses $EbeBE$, and which revolves around its axis of symmetry Bb to be a fluid figure of equilibrium.

In effect, Clairaut applied the following form of the principle of the plumb line at points N on the surface of the stratified figure:

$$\frac{|CX|}{|CN|} = \frac{\text{Effective Gravity}(N, CX)}{\text{Effective Gravity}(N, CN)}, \qquad (6.2.3)$$

where Effective Gravity $(N, CX) \equiv$ the magnitude of the component of the effective gravity at N whose direction is indicated by the line segment CX, and Effective Gravity $(N, CN) \equiv$ the magnitude of the component of the effective gravity at N which is directed toward C, where C is the center of the stratified figure (see Figure 20). Here again $|AB|$ stands for the length of the line segment AB. If the flattened ellipse $EbeBE$ is the particular meridian at the surface of the stratified figure which N is situated on, then the point X along the line through the center C of the stratified figure which lies in the plane of the flattened ellipse $EbeBE$ and which is perpendicular to CN at C is determined in the following way. The line segment NX is perpendicular to the surface of the stratified figure at N. In other words, the line through the center C of the stratified figure which lies in the plane of the flattened ellipse $EbeBE$ and which is perpendicular to CN at C and the line perpendicular to the surface of the stratified figure at N intersect at X. It is easy to show why equality (6.2.3) expresses the condition that the total or the resultant of the effective gravity at all points N on a flattened ellipse $EbeBE$ which is a meridian at the surface of the stratified figure be perpendicular to the surface of the figure at all of these points.[32]

In fact, Clairaut did not display equality (6.2.3). Instead he immediately wrote a simpler form of this equality, which he evidently found by taking equality (6.2.3) and discarding all terms of second order and higher orders in the infinitesimal ellipticity α as well as all terms of second order and higher orders in all other quantities whose orders are the same as that of the infinitesimal ellipticity α. Specifically, he took as the denominator of the right-hand side of equality (6.2.3) only the magnitude of the component of the attraction at N which points to C. That is, he disregarded the magnitude of the component of the centrifugal force per unit of mass at N which is directed away from C. Furthermore, he took as the value of the magnitude of this component of the attraction at N what is in effect the value of the magnitude of the total or the resultant of the attraction at N produced by a stratified spherical figure whose radius $CB = e$, whose strata are spheres that have the same densities as the flattened ellipsoids of revolution which make up the stratified figure in question, and upon whose surface N is located. Again, Clairaut left out the details in making the approximations that he did, but, knowing that he sought a first-order theory, it is not difficult to reconstruct his arguments.[33]

Clairaut then made one additional simplification of the geometry of the nearly spherical stratified figure shaped like an ellipsoid of revolution flattened at its poles which attracts according to the universal inverse-square law, whose polar radius $CB = e$, whose ellipticity α is infinitesimal, and whose meridians are flattened ellipses $EbeBE$, which again amounted to making an approximation that is valid to terms of first order but which is concealed. This resulted in an equation that is a relation between the centrifugal force per unit of mass ϕ at the equator of the stratified figure and the ellipticity α of the stratified figure, expressed in terms of the densities of the flattened ellipsoids of revolution which make up the stratified figure, which exists when the principle of the plumb line holds at all points on the surface of the stratified figure. More precisely, Clairaut showed this equation to be independent of the particular point N on a flattened ellipse $EbeBE$ which is a meridian at the surface of the stratified figure to terms of first order in the ellipticity α of the figure. In this way Clairaut in effect established that a stratified figure of revolution which attracts according to the universal inverse-square law, which revolves around its axis of symmetry, at whose surface the principle of the plumb line holds, and which consists of an infinite number of concentric, individually homogeneous ellipsoids of revolution flattened at the poles, all of which have the same ellipticity α, all of which have a common axis of symmetry (the axis of symmetry of the stratified figure, which is the polar axis of the figure), and whose densities are given by expression (6.2.2) actually does exist to terms of first order in α. This means that such a figure exists when α is infinitesimal. The value of the magnitude of the centrifugal force per unit of mass ϕ at the equator of such a figure which the figure must have in order to exist is determined by the relation between α and ϕ in question, which depends upon the coefficients and exponents, f, g, p, and q in the expression (6.2.2) for the density $D(r)$ of the stratified figure.[34]

The coefficients and exponents f, g, p, and q in the expression (6.2.2) for the density $D(r)$ of the stratified figure can be chosen to make the density of the figure equal to 1 or any other positive number (namely, $p = q = 0$ and $f + g = a$ positive number), so that Clairaut's paper of 1738 implicitly includes one of the first proofs that homogeneous figures shaped like ellipsoids of revolution flattened at the poles which attract according to the universal inverse-square law, whose ellipticities are infinitesimal, which revolve around their axes of symmetry, and at whose surfaces the principle of the plumb line holds exist. (In fact the Scottish mathematician James Stirling had already treated this same problem in a paper that he published in the volume of the *Philosophical Transactions* for 1735. But Stirling gave a verbal account of the problem in that paper; he did not really demonstrate the result mathematically. I shall have occasion to refer to Stirling's paper again in Chapter 8.)

The results for this special case together with the results in Clairaut's paper of 1737 can easily be used to show that if the principle of the plumb line holds at all points on the surface of a homogeneous figure shaped like an ellipsoid of revolution flattened at the poles which attracts according to the universal inverse-square law, whose ellipticity is infinitesimal, and which revolves around its axis of symmetry, then the columns from the center of the figure to the surface of the figure will all balance or weigh the same, too. In other words, homogeneous figures of revolution which attract according to the universal inverse-square law, whose ellipticities are infinitesimal, and which revolve around their axes of symmetry exist which accord with both the principles of balanced columns and the plumb line at the same time.[35] This fact did not follow from the statements that Bouguer demonstrated and the conclusions that he drew from the statements and their demonstrations which appear in Bouguer's Paris Academy *mémoire* of 1734, for Bouguer did not investigate the universal inverse-square law of attraction in connection with the problems that he raised in that *mémoire*. Nor did Clairaut give at this time the slightest hint why homogeneous figures shaped like flattened ellipsoids of revolution which attract according to the universal inverse-square law, whose ellipticities are infinitesimal, and which revolve around their axes of symmetry exist which satisfy both the principles of balanced columns and the plumb line simultaneously. As we shall see in a moment, in his paper of 1738 he simply assumed that each one of the two principles was consistent with the other when figures attract in accordance with the universal inverse-square law.

Hermann's and Bouguer's applications of the principle of the plumb line involved solving problems in the "inverse method of tangents" (that is, solving ordinary "differential" equations), whereas Clairaut's application of the principle of the plumb line did not require that such problems be solved. Clairaut did not attack the same kind of problem which Hermann and Bouguer had tackled. The latter two individuals specified beforehand the magnitudes and the directions of the attraction everywhere. Their problem was to determine figures of revolution which revolved around their axes of symmetry at whose surfaces the principle of the plumb line held when such conditions on attraction were stipulated in advance. Unlike Hermann and Bouguer, Clairaut tackled a problem in which the total or the resultant of the attraction cannot be designated ahead of time, because the total or the resultant of the attraction depends upon the very shape of the figure sought. In other words, when figures attract according to the universal inverse-square law, the total or the resultant of the attraction depends upon the shapes of the figures that attract. Clairaut presupposed at the start the basic shapes of the figures that he sought, which Hermann and Bouguer did not do. Namely, Clairaut assumed that the figures were shaped like flattened ellipsoids of revolution. The only unknown parameters concerning the shapes of the figures that he sought were the ellipticities of the figures. In other words, Clairaut hypothesized from the beginning that he knew everything about the shapes of the figures that he looked for except their ellipticities, which is all that he had to find in applying the principle of the plumb line. As a result, ordinary "differential" equations that had to be solved did not arise in Clairaut's problem, whereas Bouguer's application of the principle of the plumb line led to ordinary "differential" equations which had to be solved.

On the one hand, Clairaut concluded from his various equations that when the principle of the plumb line held at the surface of his nearly spherical stratified figure shaped like an ellipsoid of revolution flattened at its poles which attracts according to the universal inverse-square law, whose polar radius $CB = e$, whose ellipticity α is infinitesimal, whose meridians are flattened ellipses $EbeBE$, and which revolves around its axis of symmetry, the increase in the magnitude of the effective gravity with latitude at points on the surface of the figure from the equator of the figure to a pole of the figure varied directly as the squares of the sines of the latitudes of the points. That is, Clairaut found the increase in the magnitude of the effective gravity with latitude along the surface of his stratified figure from the equator of the figure to a pole of the figure when the principle of the plumb line holds at the surface of the figure to vary with latitude exactly the same way that Newton maintained that the increase in magnitude of the effective gravity with latitude at points on the surface of a homogeneous figure shaped like an ellipsoid of revolution flattened at its poles which attracts according to the universal inverse-square law, which revolves around its axis of symmetry, and whose columns from center to surface all balance or weigh the same varied with latitude along the surface of the figure from the equator of the figure to a pole of the figure.

[In fact, if we follow Clairaut's calculations closely, we discover that Clairaut did not prove this result for the *total* or the *resultant* of the effective gravity at points N on the surface of his stratified figure, which are perpendicular to the surface of the figure at points N on the surface of the figure by the principle of the plumb line, but for the *components* of the effective gravity at points N on the surface of the figure which are directed toward the center C of the figure instead.

That this statement is true follows in part from the remarks that I made which had to do with Clairaut's calculations of the magnitudes of the attraction at points N on the surface of the figure. However, because the ellipticities of all of the flattened ellipsoids of revolution which make up the stratified figure are all infinitesimal (they are all equal to the infinitesimal number α), the value of the magnitude of the total or the resultant of the effective gravity at a point N on the surface of the stratified figure and the value of the magnitude of the component of the effective gravity at the point N which points to the center C of the stratified figure are presumably equal to terms of first order.]

On the other hand, Clairaut also showed that the magnitudes of the effective gravity at points N on the surface of his stratified figure did *not* vary inversely as the distances of the points from the center C of the stratified figure. Newton, we recall, had claimed that the magnitudes of the effective gravity at points on the surface of a homogeneous figure shaped like an ellipsoid of revolution flattened at its poles which attracts according to the universal inverse-square law, which revolves around its axis of symmetry, and whose columns from center to surface all balance or weigh the same did vary inversely as the distances of the points from the center of the homogeneous figure. We recall that in fact this variation of the magnitudes of the effective gravity at points on the surface of the homogeneous figure in question is equivalent to having the columns from the center of the homogeneous figure to the surface of the homogeneous figure all balance or weigh the same and that in his paper of 1737 Clairaut had demonstrated that Newton was indeed correct when the ellipticity of the homogeneous figure was infinitesimal.

(Once again, however, Clairaut was really talking about the magnitudes of the *components* of the effective gravity at points N on the surface of his stratified figure which are directed toward the center C of the stratified figure, not about the magnitudes of the *total* or the *resultant of* the effective gravity at points N on the surface of his stratified figure. But for the reasons mentioned previously, the values of the two magnitudes of the attraction at a point N on the surface of the stratified figure are the same for all practical purposes. That is, they are equal to terms of first order.)

Writing the ratio of the magnitude of the centrifugal force per unit of mass to the magnitude of the effective gravity at the equator of his nearly spherical stratified figure shaped like an ellipsoid of revolution flattened at its poles which attracts according to the universal inverse-square law, whose polar radius $CB = e$, whose ellipticity α is infinitesimal, whose meridians are flattened ellipses $EbeBE$, which revolves around its axis of symmetry, and at whose surface the principle of the plumb line holds as $1/m$, Clairaut derived from his prior equations a new equation that exists when the principle of the plumb line holds at the surface of the stratified figure. This new equation related the ellipticity α of the stratified figure to $1/m$ through the coefficients and exponents f, g, p, and q in the expression (6.2.2) for the density $D(r)$ of the stratified figure.[36]

If the exponents p and q in the expression (6.2.2) for the density $D(r)$ of the stratified figure are chosen to make the figure homogeneous (that is, if $p = q = 0$ and $f + g = a$ positive number), this new equation just reduces to the equation that Clairaut could have written in his paper of 1737 but did not, namely, equation (5.3.4). We remember that we saw in Chapter 5 that equation (5.3.4) cannot be distinguished to terms of first order from the equation

$$\frac{\delta}{1/100} = \frac{\phi}{4/505} \tag{1.1}$$

that Newton had used to determine the ellipticity δ of a homogeneous figure shaped like an ellipsoid of revolution flattened at its poles which attracts in accordance with the universal inverse-square law, which revolves around its axis of symmetry, and whose columns from center to surface all balance or weigh the same, where ϕ stood for the ratio of the magnitude of the centrifugal force per unit of mass to the magnitude of the attraction at the equator of the figure.[37] We recall that in his paper of 1737 Clairaut answered the question that had induced him to write that paper: Why had Newton chosen a homogeneous figure of revolution which attracts according to the universal inverse-square law, which revolves around its axis of symmetry, and whose columns from center to surface all balance or weigh the same to be shaped like an ellipsoid of revolution flattened at the poles? The fact that the equation (5.3.4) that Clairaut could have written in his paper of 1737 but did not and the one that his equation of 1738 relating α to $1/m$ through the coefficients and exponents f, g, p, and q in the expression (6.2.2) for the density $D(r)$ of the stratified figure reduces to when the stratified figure is assumed to be homogeneous (in other words, all of the flattened ellipsoids of revolution which make up the stratified figure are assumed to have the same density) are identical is further proof that if the principle of the plumb line holds at all points on the surface of a homogeneous figure shaped like an ellipsoid of revolution flattened at the poles which attracts according to the universal inverse-square law, whose ellipticity is infinitesimal, and which revolves around its axis of symmetry, then the columns from the center of the figure to the surface of the figure will all balance or weigh the same, too. In other words, we have more evidence that homogeneous figures of revolution which attract according to the universal inverse-square law, whose ellipticities are infinitesimal, and which revolve around their axes of symmetry exist which accord with both the principles of balanced columns and the plumb line at the same time.

Clairaut's new equation, however, could not be used alone to find out how the quantity $1/m$ determined the ellipticity α when the figure shaped like a flattened ellipsoid of revolution which attracts according to the universal inverse-square law, whose ellipticity α is infinitesimal, which revolves around its axis of symmetry, and at whose surface the principle of the plumb line holds was stratified and consequently heterogeneous, because the equation includes unobservable quantities. That is, in order to use the new equation to work out how the quantity $1/m$

determined the ellipticity α when the figure shaped like a flattened ellipsoid of revolution was stratified, the coefficients and exponents f, g, p, and q in the expression (6.2.2) for the density $D(r)$ of the stratified figure have to be stipulated beforehand. But the variation of the earth's density in its interior cannot be determined, hence the coefficients and exponents f, g, p, and q in the expression (6.2.2) for the density $D(r)$ of the stratified figure are consequently unknown for the earth, assuming the earth to be made up of the kind of strata that Clairaut specified.

Clairaut sought an equation that related α to observable parameters alone. He found what he wanted in the following equation: when $1/m$ is infinitesimal $((1/m)^2 \ll 1/m)$, Clairaut discovered that for any nearly spherical stratified figure shaped like a flattened ellipsoid of revolution which attracts in accordance with the universal inverse-square law, whose ellipticity α is infinitesimal, which revolves around its axis of symmetry, at whose surface the principle of the plumb line holds, and which is composed of infinitely many concentric, individually homogeneous flattened ellipsoids of revolution all of which have a common axis of symmetry (the axis of symmetry of the stratified figure, which is the polar axis of the figure) and whose ellipticities are all infinitesimal and all equal (to α), the equation

$$\frac{P' - P}{P} = \frac{10}{4m} - \alpha \tag{6.2.4}$$

is always true, whatever values the coefficients and exponents f, g, p, and q in the expression (6.2.2) for the density $D(r)$ of the stratified figure may have, where P' designates the magnitude of the attraction at a pole of the stratified figure shaped like a flattened ellipsoid of revolution, P stands for the magnitude of the effective gravity at the equator of the stratified figure shaped like a flattened ellipsoid of revolution, and $1/m \equiv$ the infinitesimal ratio of the magnitude of the centrifugal force per unit of mass to the magnitude of the effective gravity at the equator of the stratified figure shaped like a flattened ellipsoid of revolution. In other words, equation (6.2.4) is independent of the particular values of the coefficients and exponents f, g, p, and q in the expression (6.2.2) for the density $D(r)$ of the stratified figure.

Clairaut did not explain very intelligibly how he found equation (6.2.4). He restated that the increase in the magnitude of the effective gravity with latitude along the surface of his stratified figure shaped like a flattened ellipsoid of revolution which attracts according to the universal inverse-square law, whose ellipticity α is infinitesimal, which revolves around its axis of symmetry, and at whose surface the principle of the plumb line holds varied directly as the square of the sine of the latitude. In other words, he found that the increase in the magnitude of the effective gravity with latitude along the surface of his stratified figure shaped like a flattened ellipsoid of revolution which attracts in accordance with the universal inverse-square law, whose ellipticity α is infinitesimal, which revolves around its axis of symmetry, and at whose surface the principle of the plumb line holds varies with latitude exactly the same way that Newton claimed that the increase in the magnitude of the effective gravity with latitude along the surface of a homogeneous figure shaped like a flattened ellipsoid of revolution which attracts according to the universal inverse-square law, which revolves around its axis of symmetry, and whose columns from center to surface all balance or weigh the same varies with latitude. But when the principle of the plumb line holds at all points on the surface of the stratified figure shaped like a flattened ellipsoid of revolution, the values of the magnitudes of the effective gravity at points at just two different latitudes on the surface of the figure are enough to specify such a "sin² law" of the increase in the magnitude of the effective gravity with latitude along the surface of this figure completely and uniquely, Clairaut affirmed.

[Clairaut must have had in mind the following idea when he said this. The increase in the magnitude of the effective gravity with latitude at the surface of his stratified figure shaped like a flattened ellipsoid of revolution varies according to a "sin² law" that has the form

$$G - P = k \sin^2 \theta \qquad (6.2.5)$$

when the principle of the plumb line holds at all points on the surface of the figure, where G stands for the magnitude of the effective gravity at a point on the surface of the figure whose latitude is θ, P again designates the magnitude of the effective gravity at the equator of the figure, and k is a constant of proportionality. If P is measured at the equator of the figure and G_o is measured at a point on the surface of the figure whose latitude is $\theta_o \neq 0$, then $G_o - P$ can be found. But $G_o - P$ and $\sin^2\theta_o$ specify the constant of proportionality k in equation (6.2.5), because $\theta_o \neq 0$ means that $\sin^2\theta_o \neq 0$, hence

$$k = \frac{G_o - P}{\sin^2 \theta_o}. \qquad (6.2.6)$$

Since P and k are both constant and therefore do not change values, equation (6.2.5) uniquely determines G as a function of θ for all values of the latitude θ.]

Hence the two values of the magnitudes of the effective gravity at points on the surface of the stratified figure shaped like a flattened ellipsoid of revolution at two different latitudes fixes the values of the magnitudes of the effective gravity at points on the surface of the figure at all other latitudes as well, when the principle of the plumb line holds at all points on the surface of the figure. In other words, the values of the magnitudes of the effective gravity at just two different latitudes at the surface of the stratified figure shaped like a flattened ellipsoid of revolution uniquely determine the values of the magnitudes of the effective gravity at all other latitudes at the surface of the figure, too, when the principle of the plumb line holds at all points on the surface of the figure.

Clairaut concluded from this that only one other equation involving the coefficients and exponents f, g, p, and q in the expression (6.2.2) for the densities of the flattened ellipsoids of revolution which make up the stratified figure shaped like a flattened ellipsoid of revolution could possibly exist when the principle of the plumb line holds at all points on the surface of the stratified figure – namely, the equation

$$\frac{P' - P}{P} = \cdots,$$
(6.2.7)

where P' again designates the magnitude of the attraction at a pole of the stratified figure and P again stands for the magnitude of the effective gravity at the equator of the figure. The left-hand side of equation (6.2.7) is calculated using observed values of P' and P, both of which can be found in principle by measurement, and where the right-hand side of equation (6.2.7) is formed by replacing P' and P by the particular expressions for these quantities which Clairaut had determined theoretically when the principle of the plumb line holds at all points on the surface of the stratified figure. [In fact the theoretical expression for P' does not really depend upon the condition that the principle of the plumb line hold at all points on the surface of the stratified figure, because the centrifugal force per unit of mass is always zero along the axis of symmetry of the stratified figure, whatever rate the figure revolves around this axis. The expression for P' is simply the expression for the approximate value of the magnitude of the attraction at a pole of the stratified figure.] Equation (6.2.7) relates the ellipticity α of the stratified figure shaped like a flattened ellipsoid of revolution to the magnitudes of the effective gravity at points on the surface of the figure at two different latitudes through the coefficients and exponents f, g, p, and q in the expression (6.2.2) for the density $D(r)$ of the stratified figure, when the principle of the plumb line holds at all points on the surface of the figure. [In fact, the ratio on the left-hand side of equation (6.2.7) is just the relative increase in the magnitude of the effective gravity with latitude along the surface of the figure from the equator of the figure to one of the poles of the figure, when the principle of the plumb line holds at all points on the surface of the figure.]

Equation (6.2.7) and the equation that related α to $1/m$ through the coefficients and exponents f, g, p, and q in the expression (6.2.2) for the density $D(r)$ of the stratified figure shaped like a flattened ellipsoid of revolution when the principle of the plumb line holds at all points on the surface of the figure together form two equations for α, $1/m$, and $(P' - P)/P$ which exist when the principle of the plumb line holds at all points on the surface of the figure, where the three quantities α, $1/m$, and $(P' - P)/P$ are interrelated through the coefficients and exponents f, g, p, and q in the expression (6.2.2) for the density $D(r)$ of the stratified figure which appear in the two equations.

Now, the coefficients and exponents f, g, p, and q in the expression (6.2.2) for the density $D(r)$ of the stratified figure determine the total or the resultant of the attraction produced by any nearly spherical stratified figure shaped like an ellipsoid of revolution flattened at its poles which attracts according to the universal inverse-square law, whose polar radius $CB = e$, whose ellipticity α is infinitesimal, whose meridians are flattened ellipses $EbeBE$, and which is made up of infinitely many concentric, individually homogeneous flattened ellipsoids of revolution all of which have a common axis of symmetry (the axis of symmetry of the stratified figure, which is the polar axis of the figure), whose densities are expressed by $D(r)$ in (6.2.2), and whose ellipticities are all infinitesimal and all equal (to α). That is, the coefficients and exponents f, g, p, and q in the expression (6.2.2) for the density $D(r)$ of such a stratified figure determine how the figure attracts. In particular, the coefficients and exponents f, g, p, and q in the expression (6.2.2) for the density $D(r)$ of such a stratified figure determine the attraction at each point on the surface of the stratified figure. As a result the coefficients and exponents f, g, p, and q in the expression (6.2.2) for the density $D(r)$ of such a stratified figure will determine how the magnitude of the attraction varies with latitude at the surface of the stratified figure too. Then the coefficients and exponents f, g, p, and q in the expression (6.2.2) for the density $D(r)$ of such a stratified figure together with the value of $1/m$ which makes the principle of the plumb line hold at all points on the surface of the stratified figure when the figure revolves around its axis of symmetry will determine the effective gravity at each point on the surface of the stratified figure when the principle of the plumb line holds at all points on the surface of the stratified figure. Consequently the coefficients and exponents f, g, p, and q in the expression (6.2.2) for the density $D(r)$ of the stratified figure and $1/m$ will determine as well how the increase in the magnitude of the effective gravity must vary with latitude at the surface of the stratified figure when the principle of the plumb line holds at all points on the surface of the stratified figure.

But if we let $\theta_o = \pi/2$ and $G_o = P'$ in equation (6.2.6), then it follows from equation (6.2.5) that P and P', both of which can be found in principle using observed values of P' and P, uniquely determine how the increase in the magnitude of the effective gravity will vary with latitude at the surface of the stratified figure when the figure revolves around its axis of symmetry such that the principle of the plumb line holds at all points on the surface of the figure. Namely, if we let $\theta_o = \pi/2$ and $G_o = P'$ in equation (6.2.6), then equation (6.2.5) becomes

$$G - P = (P' - P)\sin^2\theta. \qquad (6.2.5')$$

Thus the magnitude of the effective gravity G at a point on the surface of the stratified figure whose latitude is θ must satisfy equation (6.2.5'). In other words, G must vary with θ according to equation (6.2.5'). Consequently, if equation (6.2.5') is given beforehand and is stipulated as a condition that must be fulfilled, it restricts the values that f, g, p, q, and $1/m$ can have if the principle of the plumb line is to hold at all points on the surface of the stratified figure.

But equation (6.2.5′) does not alone *uniquely* determine the values that f, g, p, q, and $1/m$ can have in order that the principle of the plumb line hold at all points on the surface of the stratified figure, for it is also easy to show that the converse of what I just stated is not true. For one thing, specifying the attraction at all points on the surface of a stratified figure shaped like an ellipsoid of revolution flattened at the poles which attracts according to the universal inverse-square law, whose ellipticity α is given, and which is composed of an infinite number of concentric, individually homogeneous flattened ellipsoids of revolution, all of which have a common axis of symmetry (the axis of symmetry of the stratified figure, which is the polar axis of the figure), and whose ellipticities are all equal (to α) is not enough by itself to fix uniquely the densities of the flattened ellipsoids of revolution which make up the figure. Perhaps the simplest way to realize this is to consider a stratified figure shaped like an ellipsoid of revolution which attracts according to the universal inverse-square law and whose ellipticity α is zero (in other words, a spherical figure).[38] Furthermore, that equation (6.2.5′) does not alone uniquely determine the values that f, g, p, q, and $1/m$ can have in order that the principle of the plumb line hold at all points on the surface of the stratified figure shaped like a flattened ellipsoid of revolution which attracts in accordance with the universal inverse-square law and whose ellipticity α is infinitesimal, when the figure revolves around its axis of symmetry, can be seen by examining the form that the equation

$$\frac{P' - P}{P} = \cdots ,$$ (6.2.7)

which exists when the principle of the plumb line holds at all points on the surface of the stratified figure, has.[39]

However, even if specifying the attraction at all points on the surface of a nearly spherical stratified figure shaped like an ellipsoid of revolution flattened at its poles which attracts according to the universal inverse-square law, whose polar radius $CB = e$, whose infinitesimal ellipticity α is given, whose meridians are flattened ellipses *EbeBE*, and which is made up of infinitely many concentric, individually homogeneous flattened ellipsoids of revolution all of which have a common axis of symmetry (the axis of symmetry of the stratified figure, which is the polar axis of the figure) and whose ellipticities are all infinitesimal and all equal (to α) and, consequently, fixing the way that the magnitude of the attraction varies with latitude at the surface of this stratified figure does not by itself uniquely determine the coefficients and exponents f, g, p, and q in the expression (6.2.2) for the density $D(r)$ of the stratified figure, and even if equation (6.2.5′) does not alone uniquely determine the values that f, g, p, q, and $1/m$ can have in order that the principle of the plumb line hold at all points on the surface of the stratified figure shaped like a flattened ellipsoid of revolution which attracts in accordance with the universal inverse-square law and whose ellipticity α is infinitesimal, when the figure revolves around its axis of symmetry, Clairaut showed that *all* of the coefficients and exponents f, g, p, and q in the expression (6.2.2) for the density $D(r)$ of the stratified figure can nevertheless be *simultaneously eliminated* from the *two* equations for α, $1/m$ and $(P' - P)/P$ which exist when the principle of the plumb line holds at all points on the surface of the stratified figure. That is, it does not matter that $(P' - P)/P$, $1/m$, and α do not uniquely determine the coefficients and exponents f, g, p, and q in the expression (6.2.2) for the density $D(r)$ of the stratified figure when the principle of the plumb line holds at all points on the surface of the stratified figure. For the coefficients and exponents f, g, p, and q in the expression (6.2.2) for the density $D(r)$ of the stratified figure can be simultaneously eliminated from the *two* equations for α, $1/m$, and $(P' - P)/P$. When this is done the single equation

$$\frac{P' - P}{P} = \frac{10}{4m} - \alpha, \tag{6.2.4}$$

results which is good when the principle of the plumb line holds at the surface of the stratified figure, *whatever* values the coefficients and exponents f, g, p, and q in the expression (6.2.2) for the density $D(r)$ of the stratified figure have. More precisely, Clairaut showed that this elimination can be carried out when $(1/m)^2 \ll 1/m$. In other words, when terms of second order and higher orders in $1/m$ are discarded, all of the coefficients and exponents f, g, p, and q in the expression (6.2.2) for the density $D(r)$ of the stratified figure can be simultaneously eliminated from the two equations for α, $1/m$, and $(P' - P)/P$ which exist when the principle of the plumb line holds at all points on the surface of the stratified figure, and by this means the single equation

$$\frac{P' - P}{P} = \frac{10}{4m} - \alpha, \tag{6.2.4}$$

which is independent of the coefficients and exponents f, g, p, and q in the expression (6.2.2) for the density $D(r)$ of the stratified figure and which is true when the principle of the plumb line holds at all points on the surface of the stratified figure, is established.[40]

Clairaut now restated Newton's assertions that appear on page 430 of the third edition of the *Principia*: experiments with seconds pendulums done at southern latitudes indicated that the increase in the magnitude of the effective gravity with latitude at the earth's surface was actually greater than the increase deduced from the Table in Proposition 20 of Book III which Newton had constructed for the earth, assuming the earth to be homogeneous. Newton concluded from this difference in theoretical values and measured values of the increase that the earth was denser toward its center than toward its surface and also flatter than the homogeneous figure shaped like a flattened ellipsoid of revolution which he had taken to represent the earth. Colson stated that from Clairaut's own "foregoing Theory, we may easily perceive, that if the Density of the Earth diminishes from the Centre toward the Superficies, the Dimunition of Gravity from the Pole towards the Equator will be greater than according to Sir Isaac's Table." Clairaut did not really make clear at this time exactly how this assertion followed from his theory, but he would do so later, as we shall see in Chapter 9. In any case, in 1738, the particular statement that Newton had made without any explanation, proof, or supporting evidence in the *Principia* about a greater density toward the earth's center than toward the earth's surface appeared to be correct to Clairaut.[41]

But equation (6.2.4) caused trouble. If the ratio $1/m$ of the magnitude of the centrifugal force per unit of mass to the magnitude of the effective gravity at the equator of the stratified figure has the earth's value $\frac{1}{288}$, the equation for the ellipticity α of the stratified figure becomes

$$\frac{P' - P}{P} = \frac{1}{115} - \alpha. \qquad (6.2.4')$$

Now, for the homogeneous figure shaped like an ellipsoid of revolution flattened at the poles which attracts according to the universal inverse-square law, which revolves around its axis of symmetry, and whose columns from center to surface all balance or weigh the same which Newton took as his model of the earth, $(P' - P)/P = $ the value $\frac{1}{230}$ of the ellipticity of this homogeneous figure.[42] But it followed from measurements of the increase of the magnitude of the effective gravity with latitude along the earth's surface that $(P' - P)/P$ was greater than $\frac{1}{230}$. Thus the increase of the magnitude of the effective gravity with latitude along the earth's surface was observed to be greater than the increase that Newton's model of the earth required, which invalidated Newton's homogeneous model of the earth. Newton had declared that if the earth were denser toward its center than toward its surface, this would make $(P' - P)/P$ be greater than $\frac{1}{230}$, as observed. Clairaut found that Newton's statement about a greater density toward the earth's center than toward the earth's surface accorded with the results that followed from the theory that he presented in his paper of 1738. However, in order for $(P' - P)/P$ to be greater than $\frac{1}{230}$, Clairaut's equation (6.2.4') required that α be less than $\frac{1}{230}$. In other words, if $(P' - P)/P$ is greater than $\frac{1}{230}$, then according to the theory in Clairaut's paper of 1738, if the earth is a nearly spherical stratified figure shaped like an ellipsoid of revolution flattened at the poles which attracts according to the universal inverse-square law, whose ellipticity α is infinitesimal, which revolves around its axis of symmetry, at whose surface the principle of the plumb line holds, and which is made up of infinitely many concentric, individually homogeneous ellipsoids of revolution flattened at the poles, all of which have a common axis of symmetry (the axis of symmetry of the stratified figure, which is the polar axis of the figure), and whose ellipticities are all infinitesimal and all equal (to α), then the earth should be less flat than the homogeneous figure shaped like a flattened ellipsoid of revolution which attracts according to the universal inverse-square law, which revolves around its axis of symmetry, whose ratio $1/m$ of the magnitude of the centrifugal force per unit of mass to the magnitude of the effective gravity at the equator is the same as the value of this ratio at the equator of the stratified figure (specifically, $1/m = \frac{1}{288}$ in this case), and whose columns from center to surface all balance or weigh the same. For the ellipticity of this homogeneous figure equals $\frac{1}{230}$, whereas the ellipticity α of the stratified figure is, according to equation (6.2.4′), less than $\frac{1}{230}$.

Clairaut allowed for the possibility that Newton had imagined the density to decrease from the center of the earth to the surface of the earth in some manner that differed from the ways that the density could be made to decrease from the center to the surface of a nearly spherical stratified figure shaped like an ellipsoid of revolution flattened at the poles which attracts according to the universal inverse-square law and whose ellipticity is infinitesimal, when the figure consists of an infinite number of concentric, individually homogeneous ellipsoids of revolution flattened at the poles, all of which have a common axis of symmetry (the axis of symmetry of the stratified figure, which is the polar axis of the figure), and whose ellipticities are all infinitesimal and all equal to each other. (The fact that the ellipticities of the flattened ellipsoids of revolution are all the same means that these flattened ellipsoids of revolution are all geometrically similar.) For Newton had not explained in what way he thought that the earth's density decreased from its center to its surface, and it was possible that the decrease could occur in a way that would make the earth flatter than the homogeneous figure in his theory.

Clairaut believed that he at least understood what caused Newton's disciple David Gregory, a rather feeble mathematician, to make a mistake in his commentary on Newton's statement about the earth being flatter than the homogeneous figure in Newton's theory. Clairaut discovered Gregory's error while reading Gregory's *Elements of Astronomy* [*Astronomia ... elementa*] (1702), Book III, Section 8, Proposition 52, Scholium. On the page before this Scholium Gregory took Newton's proposition that the magnitudes of the effective gravity at points on the surface of the figure vary inversely as the distances of the points from the center of the figure, when the figure attracts according to the universal inverse-square law, as the basis for a method of determining the earth's shape (that is, finding the ratio of the earth's polar axis to its equatorial axis) by measuring the magnitude of the effective gravity *alone*, using seconds pendulums, at two different latitudes.

[Thus Gregory implied that it was not necessary to measure, say, degrees of latitude at two different latitudes in order to determine by direct measurement the earth's ellipticity or, equivalently, the ratio of the earth's polar axis to its equatorial axis. We saw in Chapter 3 that in 1735 Maupertuis derived equation (3.1) that expressed the ellipticity of a homogeneous figure shaped like an ellipsoid of revolution flattened at the poles whose ellipticity is infinitesimal in terms of degrees of latitude at the equator of the figure and at another, arbitrarily chosen latitude at the surface of the figure. We recall that equation (3.1) confirmed

Newton's undemonstrated assertion in the *Principia* that the increase in degrees of latitude with latitude along the surface of a homogeneous figure shaped like an ellipsoid of revolution flattened at the poles varies directly as the square of the sine of the latitude and that the Paris Academy used equation (3.1) as the basis and the grounds for sending expeditions to Peru and to Lapland to measure degrees of latitude in those two countries. What Newton had failed to make clear in the *Principia* was that the ellipticity of the homogeneous figure shaped like a flattened ellipsoid of revolution had to be infinitesimal in order that the increase in degrees of latitude with latitude along the surface of the figure vary directly as the square of the sine of the latitude, but that the columns from the center of the figure to the surface of the figure did not have to balance or weigh the same, nor did the figure even have to attract according to the universal inverse-square law. Indeed, the figure did not have to attract at all for that matter, since attraction had no part in determining this "sin^2 law." Moreover, Maupertuis' derivation of equation (3.1) did not involve the density of the figure but solely its shape. Hence equation (3.1) could be applied just as easily to a stratified figure shaped like an ellipsoid of revolution. Thus it also provided a way of finding the ellipticity of a stratified figure shaped like a flattened ellipsoid of revolution whose ellipticity is infinitesimal by measuring degrees of latitude at the equator of the stratified figure and, say, at one of the poles of the figure.]

The easiest way to imagine what Gregory must have had in mind is to consider a pole of a figure and the equator of the figure. Suppose that the magnitude of the attraction at a pole of the figure is measured to be P' and that the magnitude of the effective gravity at the equator of the figure is measured to be P. Let r designate the polar radius of the figure and let R stand for the equatorial radius of the figure. If the magnitude of the effective gravity at a point on the surface of the figure varies inversely as the distance of the point from the center of the figure, it follows that

$$\frac{P'}{P} = \frac{R}{r}. \tag{6.2.8}$$

Then P' and P can be used to calculate P'/P, which, according to equality (6.2.8), equals the ratio R/r of the axes of the figure. [$(R - r)/r = (R/r) - 1$ is the ellipticity of the figure.] Clairaut suspected that Gregory had used Newton's proposition to infer, in the Scholium on the next page of his treatise, that the increase in the magnitude of the effective gravity with latitude which was observed to be greater along the earth's surface than the increase of effective gravity with latitude along the surface of the homogeneous figure shaped like a flattened ellipsoid of revolution which attracts according to the universal inverse-square law, which revolves around its axis of symmetry, and whose columns from center to surface all balance or weigh the same, which Newton had taken to represent the earth, meant that the earth was flatter than this homogeneous figure, because it follows from equality (6.2.8) that

$$\frac{P' - P}{P} = \frac{P'}{P} - 1 = \frac{R}{r} - 1 = \frac{R - r}{r}. \tag{6.2.8'}$$

Hence if $(P' - P)/P$ is found to be larger for the earth than for the homogeneous figure, then according to equality (6.2.8') the ellipticity $(R - r)/r$ must also be larger for the earth than for the homogeneous figure.

It is easy to see that this would be true, if the magnitude of the effective gravity at points on the surface of the figure did in fact vary inversely as the distances of the points from the center of the figure. Now, we recall that Newton had stated this proposition to be true for a homogeneous figure shaped like an ellipsoid of revolution flattened at the poles which attracts according to the universal inverse-square law, which revolves around its axis of symmetry, and whose columns from its center to its surface all balance or weigh the same without actually demonstrating the truth of the proposition for such a figure. And we remember that in his paper of 1737 Clairaut showed that this proposition did hold for a nearly spherical homogeneous figure shaped like an ellipsoid of revolution flattened at the poles which attracts in accordance with the universal inverse-square law, whose ellipticity is infinitesimal, which revolves around its axis of symmetry, and whose columns from its center to its surface all balance or weigh the same. But in his paper of 1738 Clairaut demonstrated that the proposition did *not* hold for a nearly spherical stratified figure shaped like an ellipsoid of revolution flattened at its poles which attracts according to the universal inverse-square law, whose polar radius $CB = e$, whose ellipticity α is infinitesimal, whose meridians are flattened ellipses $EbeBE$, and which is composed of an infinite number of concentric, individually homogeneous flattened ellipsoids of revolution, all of which have a common axis of symmetry (the axis of symmetry of the stratified figure, which is the polar axis of the figure), whose ellipticities are all infinitesimal and all equal (to α), and whose densities are not all the same, when the figure revolves around its axis of symmetry such that the principle of the plumb line holds at all points on the surface of the figure. [And as a result, equation (6.2.4) turns out to be true for such a figure, *not* equality (6.2.8').]

Clairaut tried to illustrate Gregory's error by example. In his *Discours* of 1732, Maupertuis had suggested that central forces of attraction whose magnitudes at points vary inversely as the squares of the distances of the points from the centers of force might be interpreted as a limiting case of the universal inverse-square law of attraction. Namely, when matter not situated at the center of a figure that attracts according to the universal inverse-square law is very rarefied when compared with the matter located at the center of the figure, the mutual attractions of matter not at the center of the figure can be neglected and only the attraction of matter toward the center of the figure need be considered.

[Maupertuis's observation, incidentally, might account for his having spent so much time dwelling upon "fictitious" central forces of attraction in his paper in the *Philosophical Transactions* for 1732, in his *Discours* (1732), and in his Paris Academy *mémoire* of 1734.[43] Nevertheless, while Maupertuis may have shown how to reduce inverse-square central forces of attraction conceptually to the universal inverse-square law of attraction, we must not overlook the fact that in the particular case that I discussed in Chapter 5 where Maupertuis determined the effects of inverse-square central forces of attraction to accord better with the phenomenon than the universal inverse-square law did, namely, in arriving at a value of Jupiter's ellipticity using inverse-square central forces of attraction which was closer to the values of Jupiter's ellipticity determined by late-seventeenth-century and early-eighteenth-century astronomers than the value that Newton found, the attraction produced by a homogeneous figure shaped like a flattened ellipsoid of revolution which attracts according to the universal inverse-square law *cannot* be reduced to the attraction produced by an inverse-square central force, *unless* one makes the implausible, although admittedly irrefutable, hypothesis that Jupiter is much more dense at its center than it is everywhere else. Moreover, we shall see in Chapter 9 that Maupertuis's discovery that hypothesizing that an inverse-square central force attracts Jupiter, where the center of force is assumed to be situated at Jupiter's center, led to a value of Jupiter's ellipticity which was very close to the values of Jupiter's ellipticity determined by late-seventeenth-century and early-eighteenth-century astronomers depended critically upon the period of Jupiter's fourth satellite, the satellite farthest from Jupiter's surface among the four satellites visible to the eye at the time using telescopes, which Maupertuis used in his calculations. But it was not easy for late-seventeenth-century and early-eighteenth-century astronomers to measure this period with precision, and a slight change in the value of this period leads to values of Jupiter's ellipticity which are not so close to the values of Jupiter's ellipticity determined by late-seventeenth-century and early-eighteenth-century astronomers, assuming that an inverse-square central force attracts Jupiter, where the center of force is assumed to be located at Jupiter's center.]

Clairaut examined this limiting case. Here the density decreases rapidly from the center of the figure to the surface of the figure (in fact, the density does so discontinuously). But when attraction acts like a central force whose magnitudes at points vary inversely as the squares of the distances of the points from the center of force, and if the ratio of the magnitude of the centrifugal force per unit of mass to the magnitude of the effective gravity at the equator of the figure has the earth's value $\frac{1}{288}$, then the ellipticity of the figure is $\frac{1}{576}$, Clairaut observed, which is a value less than $\frac{1}{230}$ and not greater than $\frac{1}{230}$ as Newton had claimed should be the case. At the same time, the value of the ratio $(P' - P)/P$ was estimated, based on observations, to be about $\frac{1}{144}$ for the earth, which is a value greater than $\frac{1}{230}$.[44] It made no difference that the results for homogeneous figures of revolution which revolve around their axes of symmetry and which are assumed to be attracted by central forces situated at the centers of the figures had been previously arrived at for the most part by balancing columns from the centers of the figures to the surfaces of the figures and not by applying the principle of the plumb line at the surfaces of the figures. Bouguer had demonstrated that if a homogeneous figure of revolution is attracted by a central force of attraction whose center of force is located at the center of the figure and whose magnitudes do not depend upon the angles that the directions of the attraction at points make with some axis of reference, and if the figure revolves around its axis of symmetry at a certain rate (that is, at a certain angular speed) such that columns from the center of the figure to the surface of the figure all balance or weigh the same, then the principle of the plumb line will hold at all points on the surface of the figure too, and conversely.[45]

Clairaut now turned to the principles that underlay his calculations. To determine the infinitesimal ellipticity α of his nearly spherical stratified figure shaped like an ellipsoid of revolution flattened at its poles which attracts according to the universal inverse-square law, whose polar radius $CB = e$, whose meridians are flattened ellipses $EbeBE$, which revolves around its axis of symmetry, and which is made up of infinitely many concentric, individually homogeneous flattened ellipsoids of revolution, all of which have a common axis of symmetry (the axis of symmetry of the stratified figure, which is the polar axis of the figure), and whose ellipticities are all infinitesimal and all equal (to α), he said that he had employed the principle of the plumb line at the surface of the figure instead of balancing columns from the center of the figure to the surface of the figure for two reasons. First, he claimed that applying the principle of the plumb line led to simpler calculations. Second, since the earth is actually solid, the principle of the plumb line seemed to Clairaut to be the more evidential, indisputable principle – an "indispensibly necessary" one, Colson stated in translating Clairaut.[46] Colson said that Clairaut thought the principle of the plumb line to be "of absolute necessity." Otherwise, if there were a component of the effective gravity tangent to the earth's surface at a place that does not slope, water at the earth's surface, no matter how small the amount, would fail to find its own level at that place and consequently would not remain at rest at the place, contrary to observation.[47] Evidently Clairaut, like Mairan, did not doubt that the principle of the plumb held at the earth's surface. When Leonhard Euler later informed Clairaut of the experiments in surveying carried out by Henri Kuhn, professor of mathematics in Danzig, which Kuhn took as evidence that a state of equilibrium of the waters at the earth's surface did not exist, Clairaut scoffed; he rejected Kuhn's claim; and he dismissed Kuhn's work in general as worthless. (I shall have occasion to return to Kuhn and his work again in Chapter 9.)

Colson stated: "we shall assume no other Ratio for that of the two Axes, than that of the Spheroid, which results from a Coincidence of these two Principles."[48] In other words, Clairaut said that he would only study figures that permitted both the principles of balanced columns and the plumb line to hold. To Clairaut this meant hypothesizing that his figures were figures of equilibrium, which in fact he had already done at the beginning of his paper. He said that he made such an

assumption because Newton and everyone else who had treated the problem of
the earth's shape until then had assumed that the earth had originally been in a
fluid state. (We recall, however, that I mentioned in Chapter 1 that the principles
of balanced columns and the plumb line are only necessary conditions that figures
must satisfy in order to be figures of equilibrium. In Chapter 9 we shall see that
the two principles together are not a condition that is sufficient for figures of
equilibrium to exist.)

In his Paris Academy *mémoire* of 1734, Bouguer had not investigated the
universal inverse-square law of attraction. He had not shown that homogeneous
figures of revolution which attract according to the universal inverse-square law
and which revolve around their axes of symmetry do exist which allow both
principles to hold simultaneously. But as I have already mentioned, Clairaut's
discoveries of 1737 and 1738 implicitly reveal (that is, Clairaut's demonstrations
of the results that he arrived at in his papers of 1737 and 1738 include implicit
proofs) that *homogeneous* figures shaped like flattened ellipsoids of revolution
which attract according to the universal inverse-square law, whose ellipticities
are infinitesimal, and which revolve around their axes of symmetry exist which
accord with both principles at the same time. Clairaut could not have failed to see
this. But he did not *prove* at the end of his paper of 1738 that *stratified* figures of
revolution which attract in accordance with the universal inverse-square law and
which revolve around their axes of symmetry exist which permit both principles
to hold simultaneously. He just *assumed* at this time that such figures existed.

In examining the principle of balanced columns at the end of his paper of 1738,
Clairaut first calculated to terms of first order the weight due to attraction alone
of a column CN from the center C to an arbitrary point N at the surface of a
nearly spherical stratified figure shaped like an ellipsoid of revolution flattened at
the poles which attracts according to the universal inverse-square law, whose
polar radius $CB = e$, whose ellipticity α is infinitesimal, whose meridians are
flattened ellipses $EbeBE$, and which is composed of an infinite number of
concentric, individually homogeneous flattened ellipsoids of revolution, all of
which have a common axis of symmetry (the axis of symmetry of the stratified
figure, which is the polar axis of the figure), whose ellipticities are all infinitesimal
and all equal (to α), and whose densities vary as

$$D(r) = fr^p + gr^q. \tag{6.2.2}$$

He displayed the general expression for this weight.

[In order to determine the weight of a column from the center C of the stratified
figure to a point N on the surface of the stratified figure due only to attraction,
Clairaut had to find the values of the magnitudes of the components of the
attraction which are directed toward C at points M *inside* the stratified figure.
Since the ellipsoids of revolution which make up the stratified figure are all
geometrically similar (that is, they all have the same ellipticity α) and all have a
common axis of symmetry (the axis of symmetry of the stratified figure, which is

the polar axis of the figure), Clairaut only had to determine the value of the magnitude of the component of the attraction which is directed toward C at such a point M produced by the stratified figure shaped like an ellipsoid of revolution flattened at its poles whose ellipticity is α and upon whose surface M lies. The reason is that Clairaut could apply Corollary 3 to Proposition 91 of Book I of the *Principia,* which I talked about in Chapter 1, to the problem, from which it followed that the value of the magnitude of the total or the resultant of the attraction at M produced by the figure made up of those ellipsoids of revolution which surround M, which form a shell upon whose inner surface M lies, is zero. Hence the value of the magnitude of the component of the attraction at M which is directed toward C produced by this shell upon whose inner surface M is situated is also zero.]

Clairaut then assumed that the stratified figure revolved around its axis of symmetry such that the principle of the plumb line held at all points on the surface of the figure; he equated the weights of a column CB from the center C of the figure to one of the poles B of the figure and a column CN from the center C of the figure to an arbitrary point N on the surface of the figure, where both columns CB and CN lie in the plane of a meridian at the surface of the figure; and he exhibited the equation that resulted, which held good to terms of first order. (Again, it suffices to consider such columns that lie in the plane of a meridian at the surface of the stratified figure because of all of the various symmetries.) To do this he calculated the difference in weights of columns CB and CN due to attraction alone. Then he set this difference equal to the magnitude of the sum of the components of the centrifugal force at points M along the column CN which are directed away from C when the principle of the plumb line holds at all points on the surface of the stratified figure. This sum he found as follows. The expression (6.2.2) for the densities of the ellipsoids of revolution which make up the stratified figure gives the density of the column CN at each point M along the column CN. Each point M along the column CN is the midpoint of an infinitesimally long segment of the column CN. The infinitesimal length of each of these segments is just the differential of the length of the column CN. (Here the infinitesimally long segments of the column CN are assumed to be shaped like circular cylinders whose cross sections have an area that is $\ll$ the lengths of these segments of the column CN, so that these segments of the column CN are one-dimensional. In other words, the volumes of these segments of the column CN reduce to the lengths of the segments.) Then the magnitude of the sum of the components of the centrifugal force at points M along the column CN which are directed away from C when the principle of the plumb line holds at all points on the surface of the stratified figure is just the integral along CN of the product of the density of the column CN at each point M along the column CN and the differential of the length of the column CN and the magnitude of the component of the centrifugal force per unit of mass at each point M along the column CN which is directed away from C when the principle of the plumb line holds at all points on the surface of the stratified figure. Again, however, Clairaut used the terms fluxions and fluents in place of differentials and integrals, and he employed the notation for fluxions instead of the notation for differentials in doing these calculations. As I say, in doing these calculations Clairaut produced an equation that held true to terms of first order. In other words, in doing the calculations he neglected terms that included α^n, $n \geqslant 2$, as factors, as well as all other terms that included quantities of the same orders as α^n, $n \geqslant 2$, as factors. [To find the weight of column CN due to attraction alone, the magnitude of the component of the centrifugal force per unit of mass at each point M along the column CN which is directed away from C when the principle of the plumb line holds at all points on the surface of the stratified figure is simply replaced in the calculation just described by the magnitude of the component of the attraction at each point M along the column CN which is directed toward C.]

Clairaut noted that the equation that he arrived at in this way required that conditions on the coefficients and exponents f, g, p, q, etc. in the expression (6.2.1) for the density $D(r)$ of the stratified figure be fulfilled in order that both of the principles of balanced columns and the plumb line hold simultaneously. Specifically, all but one of the coefficients and exponents f, g, p, q, etc. in the expression (6.2.1) for the density $D(r)$ of the stratified figure could be chosen arbitrarily. The equation that Clairaut arrived at by equating the weights of columns CB and CN related the remaining coefficient or exponent to the other coefficients and exponents. (The remaining coefficient or exponent could be expressed in principle if not in practice in terms of the other coefficients and exponents through the equation.) In short, Clairaut claimed to show that certain relations among the coefficients and exponents f, g, p, q, etc. in the expression (6.2.1) for the density $D(r)$ of the stratified figure must exist, in order that both of the principles of balanced columns and the plumb line hold at the same time.

Clairaut believed that when these conditions on the coefficients and exponents f, g, p, q, etc. in the expression (6.2.1) for the density $D(r)$ of the stratified figure were not satisfied, the figure determined using the principle of the plumb line could not be a figure of equilibrium. That is to say, in equating the weights of columns CB and CN from the center to the surface which lie in the plane of a meridian at the surface of his nearly spherical stratified figure shaped like an ellipsoid of revolution flattened at the poles which attracts according to the universal inverse-square law, whose polar radius $CB = e$, whose ellipticity α is infinitesimal, whose meridians are flattened ellipses $EbeBE$, which revolves around its axis of symmetry, at whose surface the principle of the plumb line holds, and which is composed of an infinite number of concentric, individually homogeneous flattened ellipsoids of revolution, all of which have a common axis of symmetry (the axis of symmetry of the stratified figure, which is the polar axis of the figure), and whose ellipticities are all infinitesimal and all equal (to α), and then determining the relations among the coefficients and exponents f, g, p, q, etc. in the expression (6.2.1) for the density $D(r)$ of the stratified figure which had to be fulfilled in order that columns CB and CN have the same weight, Clairaut in effect treated a stratified figure that met these conditions as a figure of equilibrium. At this time he thought that the principles of balanced columns and the plumb line together were a condition that is sufficient for figures of equilibrium to exist. (However, as I just mentioned, the principles of balanced columns and the plumb line are only necessary conditions that figures must satisfy in order to be figures of equilibrium. It turns out that the two principles together are not a condition that is sufficient for figures of equilibrium to exist, as we shall see in Chapter 9.)

Why did Clairaut go through all of this? What did the additional hypothesis that the principle of balanced columns hold for his stratified figures add to his investigation? After all, the variation of the earth's density in its interior cannot be determined, hence the coefficients and exponents f, g, p, q, etc. in the expression (6.2.1) for the density $D(r)$ of the stratified figure are unknown for the earth, assuming the earth to be made up of the kind of strata that Clairaut specified. Clairaut did not just hypothesize that the earth was a figure of equilibrium for the sake of the argument. He truly thought, as people had before him, like Newton and Huygens, that the earth is a figure of equilibrium. On the other hand, Colson interpreted Clairaut as saying that many people thought the usual grounds for believing the earth to be a figure of equilibrium to be "of small Force."[49] These people judged the hypothesis that the earth that is now solid was in a fluid state when it was created to be little more than speculation that could not be verified directly. Perhaps the earth never existed in such a fluid state. Clairaut believed that tangible evidence could be found which supported the traditional assumption that the earth was originally a figure of equilibrium. He personally thought his own argument more convincing – one that had "greater Weight," Colson translated Clairaut as saying.[50] This explains why Clairaut investigated what conditions equating the weights of columns CB and CN from the center to the surface of his stratified figure shaped like a flattened ellipsoid of revolution which attracts according to the universal inverse-square law, whose ellipticity α is infinitesimal, which revolves around its axis of symmetry, and at whose surface the principle of the plumb line holds might entail. He concluded that in order that all columns from the center to the surface of this figure balance or weigh the same when the principle of the plumb line holds at all points on the surface of the figure, relations among the coefficients and exponents f, g, p, q, etc. in the expression (6.2.1) for the density $D(r)$ of the figure had to exist, which the principle of the plumb line alone did not require.

To show why he believed the earth to be a figure of equilibrium, Clairaut reasoned as follows. He imagined that the deep waters at the earth's surface are connected by underground channels filled with water and that the whole system of oceans, seas, and underground channels of water remained in a state of equilibrium. This assumption involves more than the equilibrium of waters *at the earth's surface*, which exists because the principle of the plumb line holds at the earth's surface. It did not follow from Bouguer's Paris Academy *mémoire* of 1734 that the principle of the plumb line *alone* was enough to assure the equilibrium of such a system. What in Clairaut's opinion was needed to ensure that a state of equilibrium of such a system endured? Colson translated Clairaut as follows: "Considering the Earth as it is at present, and without carrying our Thoughts so far back as to its Formation, if the Ocean, which is now upon its Surface, has any considerable Depth, and if its Parts preserve a Communication with each other, from Region to Region, by subterraneous Canals; it [the ocean] can only keep an Equilibrium by this Means, because its Superficies is the same as it would have, were the whole a Fluid." Colson's English translation of Clairaut's statement is difficult to interpret. The passage is not written in straightforward language, and what it says is not unambiguous. Nor did Clairaut express himself more clearly in his manuscript written in French.[51] What Clairaut seems to have wanted to say is that if the earth's actual shape were not the same as the one that it would have if the whole earth were a mass of fluid in equilibrium, then a state of equilibrium of the system of oceans and seas at the earth's surface and underground channels filled with water which join the oceans and seas would not persist, contrary to what was observed to be true. At least, Clairaut could have meant this, given the context in which he made his statement (namely, his equating the weights of columns CB and CN from the center to the surface of his stratified figure shaped like a flattened ellipsoid of revolution at whose surface the principle of the plumb line holds). In his Paris Academy *mémoire* of 1734, Bouguer made the principle of the plumb line applied at the earth's surface not appear at first sight to be enough by itself to ensure that such a system would remain in a state of equilibrium. Clairaut tried to find grounds for treating the earth as a fluid figure of equilibrium besides the usual argument that the earth was a fluid mass in a state of equilibrium at the time of its formation, which was a claim that was impossible to verify directly. He argued instead that such a condition was necessary for the earth's system of oceans, seas, and underground channels filled with water which join the oceans and seas, whose existence Clairaut evidently did not doubt, to be in a state of equilibrium *now*. How Clairaut came to what seems to be the conclusion in Colson's translation of the passage that I quoted above that the earth is a figure of equilibrium is not entirely clear, but it seems likely, based on additional thoughts that Clairaut had, which I will turn to in a moment, that the "principle of solidification" was somehow involved. According to this principle, solidifying all of the fluid in a mass of fluid in equilibrium except the fluid in an arbitrary channel leaves the fluid in the channel in equilibrium.

Clairaut's reflections induced him to do another calculation different from the one that we just examined. Clairaut, we recall, did that calculation in order to determine the weights of columns from the center to the surface which lie in the plane of a meridian at the surface of his nearly spherical stratified figure shaped like an ellipsoid of revolution flattened at the poles which attracts according to the universal inverse-square law, whose polar radius $CB = e$, whose ellipticity α is infinitesimal, whose meridians are flattened ellipses $EbeBE$, which revolves around its axis of symmetry, at whose surface the principle of the plumb line holds, and which is made up of infinitely many concentric, individually homogeneous flattened ellipsoids of revolution, all of which have a common axis of symmetry (the axis of symmetry of the stratified figure, which is the polar axis of the figure), and whose ellipticities are all infinitesimal and all equal (to α), where the densities of the different columns from the center to the surface of the figure varied the same way as the densities of the individually homogeneous strata that the columns traversed. Clairaut now found instead the weights of rectilinear channels from the center to the surface which lie in the plane of a meridian at the surface of his nearly spherical stratified figure shaped like an ellipsoid of revolution flattened at the poles which attracts according to the universal inverse-square law, whose polar radius $CB = e$, whose ellipticity α is infinitesimal, whose meridians are flattened ellipses $EbeBE$, which revolves around its axis of symmetry, at whose surface the principle of the plumb line holds, and which is composed of an infinite number of concentric, individually homogeneous flattened ellipsoids of revolution, all of which have a common axis of symmetry (the axis of symmetry of the stratified figure, which is the polar axis of the figure), and whose ellipticities are all infinitesimal and all equal (to α), where the rectilinear channels were all filled with homogeneous fluid of the same density. Here he had in mind the idea of simulating oceans and seas at the surface of the earth which are in equilibrium and which are connected by underground channels that are also filled with water.

As a result of the new calculation he discovered that the straight channel CB filled with water from the center C to a pole B of the stratified figure in question and a straight channel CN also filled with water from the center C to an arbitrary point N on the surface of this figure balanced or weighed the same, where both straight channels CB and CN lie in the plane of a meridian at the surface of the figure, when the principle of the plumb line held at all points on the surface of the figure. In other words, he found the system made up of waters at the surface of the stratified figure in question and the two underground channels of water in the interior of this figure which join the waters at the surface of the figure to be in a state of equilibrium, when the principle of the plumb line held at all points on the surface of the figure. Clairaut noted in addition that whereas the prior calculation entailed that relations hold among the coefficients and exponents f, g, p, q, etc. in the expression (6.2.1) for the density $D(r)$ of the stratified figure in question, in order that columns from the center to the surface of the figure which lie in the plane of a meridian at the surface of the figure all balance or weigh the same, when the principle of the plumb line held at all points on the surface of the figure, the present calculation did not require that any such conditions involving the coefficients and exponents f, g, p, q, etc. in the expression (6.2.1) for the density $D(r)$ of this figure be fulfilled, in order that rectilinear channels from the center to the surface of the figure which lie in the plane of a meridian at the surface of the figure and which are all filled with homogeneous fluid of equal density all balance or weigh the same, when the principle of the plumb line held at all points on the surface of the figure. In other words, as long as the principle of the plumb line held at all points on the surface of the stratified figure in question, rectilinear channels from the center to the surface of the figure which lie in the plane of a meridian at the surface of the figure and which are all filled with homogeneous fluid of equal density all turned out to balance or weigh the same, however the densities of the individually homogeneous strata that make up the figure varied from stratum to stratum.

What purpose did this calculation serve? We noted above that the principle of the plumb line applied at the surface of the stratified figure in question should *not be enough by itself* to assure that a system composed of oceans and seas at the surface of this figure and underground channels filled with water inside this figure which connect the oceans and seas exists in a state of equilibrium. And yet Clairaut found that when the principle of the plumb line held at all points on the surface of this figure, straight channels from the center to the surface of the figure which lie in the plane of a meridian at the surface of the figure and which are all filled with homogeneous fluid of the same density all balanced or weighed the same. In other words, a system made up of homogeneous fluid at the surface of the stratified figure in question together with straight channels from the center to the surface of this figure which lie in the plane of a meridian at the surface of the figure and which are filled with homogeneous fluid, where all of the homogeneous fluid in the system has the same density, would be in a state of equilibrium, when the principle of the plumb line holds at all points on the surface of the figure. That is to say, the principle of the plumb line *alone did appear to be enough* to ensure that this state of equilibrium of the system exists. From the way that Clairaut spoke, we can only conclude that he inferred that in order for a state of equilibrium of such a system to exist, the stratified figure in question at whose surface the principle of the plumb line holds must be a figure of equilibrium.

Clairaut also reasoned that the result that followed from this calculation could be generalized. Viewing a pair of rectilinear channels from the center to the surface of the stratified figure in question which lie in the plane of a meridian at the surface of this figure as a single channel in the interior of the figure which joins two points situated on the meridian, Clairaut remarked that homogeneous fluid that fills *any* channel within the figure which lies in the plane of a meridian at the surface of the figure and which connects two points located on the meridian should be in a state of equilibrium, whatever shape the plane channel has, when the principle of the plumb line holds at all points on the surface of the figure.

To prove that this is true, Clairaut hypothesized that "independently of the Attraction of any Matter," the magnitude of the effective gravity at any point N on the surface of a nearly spherical stratified figure shaped like an ellipsoid of revolution flattened at the poles which attracts according to the universal inverse-square law, whose polar radius $CB = e$, whose ellipticity α is infinitesimal, whose meridians are flattened ellipses $EbeBE$, and which is composed of an infinite number of concentric, individually homogeneous flattened ellipsoids of revolution, all of which have a common axis of symmetry (the axis of symmetry of the stratified figure, which is the polar axis of the figure), and whose ellipticities are all infinitesimal and all equal (to α) is proportional to the magnitude of the attraction at the point N when the stratified figure revolves around its axis of symmetry such that the principle of the plumb line holds at all points on its surface.[52] (In fact what Clairaut really hypothesized here is that the approximate value of the magnitude of the component of the effective gravity at N which is directed toward the center C of the stratified figure is proportional to the approximate value of the magnitude of the component of attraction at N which points to the center C of the stratified figure.)

[What Clairaut took as the basis for this assumption is not clear. Indeed, the hypothesis appears to be false. To see this, let us assume that the strata that make up the figure all have the same density equal to 1, so that the stratified figure reduces to a homogeneous figure whose density is 1. Since Clairaut intended to prove that the stratified figure is a figure of equilibrium, we suppose that the homogeneous figure is a figure of equilibrium, in which case the columns from the center to the surface of the figure all balance or weigh the same. But Clairaut had shown in his paper of 1737 that the approximate value of the magnitude of the component of effective gravity at a point N on the surface of a homogeneous figure shaped like an ellipsoid of revolution flattened at the poles which attracts according to the universal inverse-square law, whose ellipticity is infinitesimal, which revolves around its axis of symmetry, and whose columns from center to surface all balance or weigh the same which is directed toward the center C of the figure varies inversely as the distance of N from C (in other words, varies directly as $1/|CN|$). (Here we recall that Clairaut proved what Newton had stated in the *Principia* but had not demonstrated in that work.)

Now, if the approximate value of the magnitude of the component of the effective gravity at N which points to the center C of this homogeneous figure is proportional to the approximate value of the magnitude of the component of the attraction at N which is directed toward the center C of the figure when the figure revolves around its axis of symmetry such that the principle of the plumb line holds at all points on its surface, which is Clairaut's hypothesis, this means that the approximate value of the magnitude of the component of the attraction at N which points to the center C of the figure also varies inversely as the distance of N from C (in other words, varies directly as $1/|CN|$). This follows from the fact that, as we have already observed, Clairaut's demonstrations of the results that he arrived at in his papers of 1737 and 1738 implicitly include the proof that if the principle of the plumb line holds at all points on the surface of a homogeneous figure shaped like a flattened ellipsoid of revolution which attracts according to the universal inverse-square law, whose ellipticity is infinitesimal, and which revolves around its axis of symmetry, then the columns from the center to the surface of the figure all balance or weigh the same too.

Now

$$\tfrac{2}{3}ca - \tfrac{2}{15}can + \tfrac{8}{15}cam \qquad (6.1.12'''')$$

is the general expression for the approximate value of the magnitude of the component of the attraction at a point N on the surface of a homogeneous figure shaped like an infinitesimally flattened ellipsoid of revolution which attracts according to the universal inverse square law and whose density is 1 which is directed toward the center C of the figure, where the polar radius of the figure $AC = a$ in Clairaut's new notation in his paper of 1737, the equatorial radius of the figure $CE = b$ in Clairaut's new notation in his paper of 1737, $CN = r$ in Clairaut's new notation in his paper of 1737, $m \equiv (b - a)/a \equiv$ the infinitesimal ellipticity of the figure in Clairaut's new notation in his paper of 1737, and $n \equiv (r - a)/a$ in Clairaut's new notation in his paper of 1737, where $c = 2\pi$. But if the approximate value of the magnitude of the component of attraction at N which is directed toward the center C of the figure varies inversely as the distance of N from C (in other words, varies directly as $1/|CN|$), then this means that

$$\tfrac{2}{3}ca - \tfrac{2}{15}can + \tfrac{8}{15}cam = \frac{k}{r} = \frac{k}{(a + na)} \qquad (6.2.9)$$

for some constant number k. But equality (6.2.9) means that

$$k = (\tfrac{2}{3}ca - \tfrac{2}{15}can + \tfrac{8}{15}cam)(a + na). \qquad (6.2.9')$$

Hence the expression on the right hand side of equality (6.2.9') must be constant. But n varies from 0 (when $r = a$) to m (when $r = b$). Thus in order for the expression on the right-hand side of equality (6.2.9') to be constant, this expression has to be independent of n. However, if we multiply one of the two factors in the expression on the right-hand side of equality (6.2.9') by the other factor in the expression on the right-hand side of equality (6.2.9'), we find that the product can be written as

$$\tfrac{2}{3}ca^2 - \tfrac{2}{15}ca^2n + \tfrac{8}{15}ca^2m + \tfrac{2}{3}ca^2n - \tfrac{2}{15}ca^2n^2 + \tfrac{8}{15}ca^2mn. \qquad (6.2.9'')$$

But the terms in expression (6.2.9'') which include n and n^2 do not mutually cancel each other. Hence expression (6.2.9'') is not independent of n, which means that the expression on the right-hand side of equality (6.2.9') is not constant, which means that no such constant number k on the left-hand side of equality (6.2.9') exists, which means that equality (6.2.9) does not hold. In fact, in Chapter 2 I suggested that thinking that the magnitudes of the attraction at points on the earth's surface vary inversely as the distances of the points from the earth's center, when the earth is taken to be the kind of homogeneous figure in question, was one of the errors that Desaguliers made in his paper in the *Philosophical Transactions* for 1725 in which he tried to refute Mairan's Paris Academy *mémoire* of 1720.]

Clairaut then claimed that it followed from this hypothesis that if a stationary homogeneous spherical fluid figure which attracts according to the universal inverse-square law revolves around an axis through its center, then the new homogeneous fluid figure produced by revolving the motionless homogeneous spherical fluid figure around this axis would have exactly the same form as the stratified figure and, moreover, would be in a state of equilibrium, where the axis that this homogeneous fluid figure revolves around is its axis of symmetry. Moreover, Clairaut implicitly assumed that the two figures are the same size (that is, that the polar radii of the two figures are equal).

[Again, what Clairaut took as his foundations for reaching this conclusion is unclear. Nor did Clairaut even state precisely the conditions that the result allegedly follows from. Clairaut did not specify at what rate (that is, at what angular speed) the stationary homogeneous spherical fluid figure must rotate in order that it will have the same shape as the stratified figure. He did not say whether the motionless homogeneous spherical fluid figure should rotate at the same rate as the stratified figure, which revolves around its axis of symmetry at the rate which allows the principle of the plumb line to hold at all points on its surface, or whether the stationary homogeneous spherical fluid figure should rotate instead at the rate that makes the ratio of the magnitude of the centrifugal force per unit of mass to the magnitude of the effective gravity at the equator of the new homogeneous fluid figure produced by rotation equal the value of this ratio at the equator of the stratified figure when the stratified figure revolves around its axis of symmetry so that the principle of the plumb line holds at all points on its surface, or whether the motionless homogeneous spherical fluid figure should rotate at yet some other rate.

If a stratified figure and a homogeneous figure are both shaped like flattened ellipsoids of revolution which attract according to the universal inverse-square law, whose ellipticities α and δ, respectively, are both infinitesimal, and which revolve around their axes of symmetry so that, respectively, the principle of the plumb line holds at all points on the surface of the stratified figure and the columns from the center to the surface of the homogeneous figure all balance or weigh the same, then if the ratios of the magnitudes of the centrifugal force per unit of mass to the magnitudes of the effective gravity at the equators of the two figures happen to be equal, this means that the $1/m$ in equation

$$\frac{P'-P}{P} = \frac{10}{4m} - \alpha \qquad (6.2.4)$$

equals the ϕ in equation

$$\delta = \frac{5}{4}\phi. \qquad (5.3.4)$$

In this case equation (6.2.4) and equation (5.3.4) can be combined, and when this is done the equation

$$\frac{P'-P}{P} = 2\delta - \alpha \qquad (6.2.10)$$

results. But equation (6.2.10) can be rewritten as

$$\frac{P' - P}{P} - \delta = \delta - \alpha. \tag{6.2.10'}$$

Then according to equation (6.2.10'), in order for δ to equal α, the stratified figure must attract and revolve around its axis of symmetry so as to make $(P' - P)/P = \delta$. We saw above, however, that in the case of the earth, Clairaut found that $(P' - P)/P > \delta$, from which it followed using equation (6.2.10') that $\delta > \alpha$, where

$$\delta = \frac{1}{230} \quad \text{and} \quad \phi = \frac{1}{m} = \frac{1}{288}.$$

In short, the conditions here will not make δ equal to α.]

Clairaut then applied what amounts to one form of the "principle of solidification," which is attributed to Simon Stevin, to this homogeneous fluid figure of equilibrium shaped like the stratified figure, in order to obtain a state of equilibrium of the fluid within a channel in the interior of the homogeneous fluid figure of equilibrium which lies in the plane of a meridian at the surface of this figure, which joins two points located on the meridian, and whose shape can be chosen arbitrarily.[53] He then declared that homogeneous fluid whose density is the same as the density of the homogeneous fluid figure of equilibrium and which fills a channel inside the stratified figure shaped like a flattened ellipsoid of revolution which lies in the plane of a meridian at the surface of this figure, which connects two points situated on the meridian, and whose form is identical to the form of the plane channel within the homogeneous fluid figure of equilibrium shaped like the stratified figure would also be in a state of equilibrium, "provided that the Space, which this Canal possesses in the [stratified] Globe, be not of so large an Extent, as to change the [resultant] Law of Attraction," according to Colson,[54] which suggests that Clairaut believed that the effective gravity at corresponding points inside and on the surfaces of the two figures is the same, hence that the effect of the effective gravity of one of the two figures upon a plane channel inside this figure which lies in the plane of a meridian at the surface of the figure and which joins two points on this meridian is the same as the effect of the effective gravity of the other of the two figures upon a plane channel within that figure which lies in the plane of a meridian at the surface of the figure and which connects two points on this meridian, when the two plane channels are identically shaped and are filled with homogeneous fluid of equal density (namely, the density of the homogeneous fluid figure of equilibrium), provided the channels not be so wide that the homogeneous fluid in the channel inside the stratified figure causes the total or the resultant of the attraction, hence the effective gravity, of the stratified figure to change.

Clairaut's argument above is obscure, not to say incomprehensible. Some of the conclusions that Clairaut reached are correct, but the reasoning that he used to arrive at these conclusions is often wrong, while other conclusions that he came to are simply untrue. As we shall see in Chapter 9, Clairaut would later contend that his nearly spherical stratified figure shaped like an ellipsoid of revolution flattened at the poles which attracts according to the universal inverse-square law, whose polar radius $CB = e$, whose ellipticity α is infinitesimal, whose meridians are flattened ellipses $EbeBE$, which revolves around its axis of symmetry, at whose surface the principle of the plumb line holds, and which is composed of an infinite number of concentric, individually homogeneous flattened ellipsoids of revolution, all of which have a common axis of symmetry (the axis of symmetry of the stratified figure, which is the polar axis of the figure), whose ellipticities are all infinitesimal and all equal (to α), and whose densities are not all the same *cannot* be a figure of equilibrium, so that it made no sense for him to equate in his paper of 1738 the weights of columns from the center of this stratified figure to the surface of the figure which lie in the plane of a meridian at the surface of the figure in order to determine relations that must hold among the coefficients and exponents f, g, p, q, etc. in the expression (6.2.1) for the density $D(r)$ of the figure in order for the figure to be a figure of equilibrium.

(Considering only what Clairaut thought he had discovered in 1738, without introducing any additional ideas, it seems paradoxical at first sight, not to say inconsistent, first to affirm that in order for the stratified figure shaped like a flattened ellipsoid of revolution to be a figure of equilibrium, certain relations among the coefficients and exponents f, g, p, q, etc. in expression (6.2.1) for the density $D(r)$ of the stratified figure must hold, and then to declare almost immediately afterwards that in order for the stratified figure shaped like a flattened ellipsoid of revolution to be a figure of equilibrium, apparently no such relations among the coefficients and exponents f, g, p, q, etc. in expression (6.2.1) for the density $D(r)$ of the stratified figure need exist. After all, the conditions that the first calculation entail are conditions that must hold in order for all *stratified* columns from the center of the figure to its surface to balance or weigh the same, and the principle of balanced *stratified* columns is a necessary condition that must hold in order for *stratified* figures of equilibrium to exist. However, we must not forget that in the first of Clairaut's two calculations discussed immediately above, the density of a column from the center of the stratified figure to the surface of the figure varied the same way as the densities of the individually homogeneous strata that the column traversed. That is, the column was *heterogeneous*. Its density *depended upon* the coefficients and exponents f, g, p, q, etc. in expression (6.2.1) for the density $D(r)$ of the stratified figure. Indeed, the density of the column varied itself as $D(r)$ in expression (6.2.1). On the other hand, in the second of Clairaut's two calculations, the fluid in a straight channel from the center of the stratified figure to the surface of the figure was *homogeneous*, and the density of the homogeneous fluid was the same in all such channels. In a moment we shall see why this second calculation did not require that any relations among the coefficients and exponents f, g, p, q, etc. in expression (6.2.1) for the density $D(r)$ of the stratified figure hold, and this will account for what at first sight look like contradictory results that follow from the two calculations.)

Clairaut's "intuition" that the system of waters at the earth's surface and the underground channels of water which join the waters at the earth's surface exist in a state of equilibrium can be substantiated. After all, Clairaut did calculate directly the weights of rectilinear channels all of which are filled with homogeneous fluid of the same density from the center to the surface and which lie in the plane of a meridian at the surface of his nearly spherical stratified figure shaped like an ellipsoid of revolution flattened at the poles which attracts according to the universal inverse-square law, whose polar radius $CB = e$, whose ellipticity α is infinitesimal, whose meridians are flattened ellipses $EbeBE$, which revolves around its axis of symmetry, and which is composed of an infinite number of concentric, individually homogeneous flattened ellipsoids of revolution, all of which have a common axis of symmetry (the axis of symmetry of the stratified figure, which is the polar axis of the figure), whose ellipticities are all infinitesimal and all equal (to α), and he found that the rectilinear channels of fluid all turned out to have the same weight when the principle of the plumb line held at all points on the surface of the figure. And the principle of the plumb line is observed to hold at all places on the earth's surface which do not slope. But this result does not mean that the stratified figure is a figure of equilibrium. The result follows instead from the principle of the plumb line applied at the surface of the stratified figure together with the fact that the total or the resultant of the attraction produced by a homogeneous or heterogeneous figure that attracts according to the universal inverse-square law always produces total efforts equal to zero around closed channels filled with homogeneous fluid, whether the figure is a figure of equilibrium or not, in which case a homogeneous or heterogeneous figure that attracts in accordance with the universal inverse-square law never causes homogeneous fluid in closed channels that lie within the interior of the figure to circulate around the closed channels. A homogeneous or heterogeneous figure can satisfy both the principles of balanced columns and the plumb line and *still not* be a figure of equilibrium. Only later, as we shall see in Chapter 9, did Clairaut discover that the absence of circulation of homogeneous fluid in closed channels that lie within the interior of figures is a condition for figures of equilibrium to exist which is independent of the principles of balanced columns and the plumb line and that the universal inverse-square law of attraction permits this condition to hold. But like the principles of balanced columns and the plumb line, this new condition too is only a *necessary* condition for figures of equilibrium to exist. It is not a condition *sufficient* for figures of equilibrium to exist. In Chapter 9 we shall learn why the total or the resultant of the attraction produced by a homogeneous or heterogeneous mass that attracts in accordance with the universal inverse-square law *never* at any instant produces a total effort not equal to zero around a closed channel filled with homogeneous fluid which lies within the interior of the mass, whether the mass is in a state of equilibrium at that instant or not. Consequently, as long as the principle of the plumb line holds at all points on the surface of a homogeneous or heterogeneous figure that attracts according to the universal inverse-square law, then homogeneous fluid that fills channels that lie in the interior of the figure and in the plane of a meridian at the surface of the figure and that connect two points on this meridian will remain in a state of equilibrium, whether the figure is a figure of equilibrium or not. (I mentioned in Chapter 1 that the problem of *sufficient* conditions for equilibrium goes beyond the limits of the story told in these pages, and I explained why it does so.) Since Clairaut succeeded in solving for a shape (that is, an ellipticity) by applying the principle of the plumb line at the surface of a nearly spherical stratified figure shaped like an ellipsoid of revolution flattened at the poles which attracts according to the universal inverse-square law, whose polar radius $CB = e$, whose ellipticity α is infinitesimal, whose meridians are flattened ellipses $EbeBE$, which revolves around its axis of symmetry, and which is made up of infinitely many concentric, individually homogeneous flattened ellipsoids of revolution, all of which have a common axis of symmetry (the axis of symmetry of the stratified figure, which is the polar axis of the figure), and whose ellipticities are all infinitesimal and all equal (to α), when the coefficients and exponents f, g, p, q, etc. in expression (6.2.1) for the density $D(r)$ of the figure are chosen arbitrarily [in other words, however $D(r)$ varies with r in expression (6.2.1)], his equating the weights of straight channels from the center of this stratified figure to the surface of the figure all of which are filled with homogeneous fluid of equal density and which lie in the plane of a meridian at the surface of the figure did not require that any relations among the coefficients and exponents f, g, p, q, etc. in expression (6.2.1) for the density $D(r)$ of the figure hold. In 1738 Clairaut saw that this is true, but he did not understand at this time why it is true.

[In Chapter 9 we shall see that if the channels of fluid that lie within the interior of a stratified figure that attracts according to the universal inverse-square law are so thin that changing the density of the fluid in the channels does not change the resultant of the attraction, then if the total efforts around closed channels of *homogeneous* fluid that lie within the interior of the stratified figure are equal to zero and if the principle of the plumb line holds at the surface of the figure, then the principle of balanced *stratified* columns also holds. Thus we shall see once again that the conditions that Clairaut assumed in doing his second calculation do not entail that any specific relations must hold among the coefficients and exponents f, g, p, q, etc. in expression (6.2.1) for the density $D(r)$ of the figure in order for all of the columns from the center of the figure to its surface to balance or weigh the same.]

If the nearly spherical stratified figure shaped like an ellipsoid of revolution flattened at the poles which attracts according to the universal inverse-square law, whose. polar radius $CB = e$, whose ellipticity α is infinitesimal, whose meridians are flattened ellipses $EbeBE$, which revolves around its axis of symmetry, and which is composed of an infinite number of concentric, individually homogeneous flattened ellipsoids of revolution, all of which have a common axis of symmetry (the axis of symmetry of the stratified figure, which is the polar axis of the figure), whose ellipticities are all infinitesimal and all equal (to α) is assumed to be *solid* instead of fluid, then Clairaut's calculations in his paper of 1738 which only involve the hypothesis that the principle of the plumb line holds at all points on the surface of the figure *are* valid, however. In particular, equation

$$\frac{P' - P}{P} = \frac{10}{4m} - \alpha \qquad (6.2.4)$$

holds in this case. We recall that Clairaut did allow for the possibility that Newton had imagined heterogeneous figures shaped like ellipsoids of revolution flattened at the poles which attract according to the universal inverse-square law and which were composed of concentric, individually homogeneous strata that were shaped differently from Clairaut's flattened ellipsoids of revolution, all of which have a common axis of symmetry (the axis of symmetry of the stratified figure, which is the polar axis of the figure) and whose ellipticities are all infinitesimal and all equal (to α). Clairaut thought that such a difference might account for the fact that his findings concerning the earth's shape did not accord with Newton's claims about the way that the earth is shaped. This meant that Clairaut could try to apply the principle of the plumb line at the surfaces of solid stratified figures shaped like ellipsoids of revolution flattened at the poles which attract according to the universal inverse-square law, whose ellipticities are infinitesimal, which revolve around their axes of symmetry, and whose strata have forms that differ from the ones that appear in his paper of 1738 to see if this would lead to results that agreed with Newton's claims about the earth's shape. In fact we shall see in Chapter 9 that Clairaut attempted later to do just this.

In his Paris Academy *mémoire* of 1734, which I examined in Chapter 4, Bouguer in effect had paved the way for such an application of the principle of the plumb line. (We shall see in Chapter 9 that Bouguer's *mémoire* definitely influenced Clairaut's future work on the problem of the earth's shape.) We recall that Bouguer had made it crystal clear that the principle of the plumb line only involved conditions at the surfaces of figures and not the states of the interiors of figures. He had proven, among other things, that it was possible to determine homogeneous figures of revolution which revolve around their axes of symmetry and at whose surfaces the principle of the plumb line holds but whose columns from the center of a figure to the surface of a figure do not all balance or weigh the same. In particular he demonstrated this to be true in situations where the principles of balanced columns and the plumb line could never both hold at the same time for any homogeneous figure of revolution, whatever its shape, because of the kinds of forces of attraction which he assumed to act in the cases in question. While establishing these results he actually found and exhibited a certain homogeneous figure of revolution which attracted according to a law that differed from the universal inverse-square law, which revolved around its axis of symmetry, and at whose surface the principle of the plumb line held but whose columns from center to surface did not all balance or weigh the same. Hence the figure in question could not possibly be a figure of equilibrium. And because such a figure cannot be a figure of equilibrium, it serves little if any purpose to assume such a figure to be in a fluid state. (In Chapter 1 I explained why it made no sense at this time to hypothesize that a figure is in a fluid state in case the figure is not a figure of equilibrium.) Conversely, a solid figure need not be assumed to be a figure of equilibrium, since, indeed, it may not be a figure of equilibrium.

[Mairan's Paris Academy *mémoire* of 1720, which I discussed in Chapter 2, by contrast with Bouguer's Paris Academy *mémoire* of 1734 was unimportant, and it undoubtedly did not influence Clairaut's work on the problem of the earth's shape. For one thing, we recall that Mairan did not actually determine figures. For another, although he had intended to apply the principle of the plumb line at the surfaces of homogeneous solid figures of revolution whose shapes were *given* beforehand and which revolved around their axes of symmetry, in fact he did not. The principle of the plumb line could not hold at the surfaces of his homogeneous solid figures of revolution when the figures revolved around their axes of symmetry, as Desaguliers pointed out.]

But it is safe to say that Clairaut did not understand stratified fluid figures of equilibrium when he wrote his paper of 1738. Bouguer had not dealt with the topic at all in his Paris Academy *mémoire* of 1734, and Clairaut probably contented himself in his paper of 1738 with putting forward some ideas that he had not thought about very deeply. Moreover, even if Clairaut had wanted to investigate stratified fluid figures of equilibrium more carefully in 1738, it seems unlikely that he could have done so at that time for lack of suitable mathematical tools. The mathematics that would prove to be useful to the task only began to appear later in Paris–in fact, in the months just after Clairaut sent his paper of 1738 to the Royal Society. In his classic history of the problem of the earth's shape, Isaac Todhunter took for granted that this mathematics existed at this time.[55] That was a mistake. The relevant mathematics only came into being during the period that Clairaut worked on the theory of the earth's shape. I shall now turn to the story of this new mathematics. The need to deepen the theory of stratified fluid figures of equilibrium I will take up in Chapters 8 and 9.

7

Interlude I: Integral calculus
(1690–1741)

7.1. *Leibniz, the Bernoullis, and the birth of the partial differential*
calculus (1690–1721)

7.1.1. Leibniz and Johann I Bernoulli (1690–1721). In the beginning a geometric
entity, an individual plane curve, was central to the very concept of differentiation
in Gottfried Wilhelm Leibniz's single-operator differential calculus. This created
a certain difficulty. Namely, differentials of quantities defined at points situated
on plane curves could only be taken *along* individual plane curves, for reasons
that accorded with the geometry of plane curves of the time. However, not long
after Leibniz invented the differential calculus, he imagined problems that seemed
to require differentiation of quantities defined at neighboring points located on
neighboring plane curves in a family of plane curves that all lie in the same plane
between or *across* the neighboring plane curves in the family. Thus Leibniz had to
broaden his original notion of differentiation along plane curves to include what
might be called "differentiation from curve to curve."

For example, in the early 1690s, Leibniz extended the idea of differention *along*
plane curves to differentiation *from one* plane curve *to another* plane curve located
nearby in the same plane, or differentiation from curve to curve, in finding the
"envelopes" of one-parameter families of plane curves, where all of the plane
curves in such a family lie in the same plane. To solve this problem, he converted
the parameter of a one-parameter family of plane curves that all lie in the same
plane, whose *constant* values determined the individual plane curves in the family,
which was the *only* way that the parameter of a one-parameter family of plane
curves that all lie in the same plane had been meaningfully interpreted until then,
into a *variable*. He differentiated with respect to the plane curves' spatial variables
and the parameter *alike*. In doing so, he utilized what amounts to the partial
differential calculus.

The major part of the early developments in the partial differential calculus
arose in connection with the problems issued by Leibniz and the Bernoulli
brothers Jakob and Johann I in the later 1690s to challenge the European
mathematicians of the era to solve problems that the Bernoullis believed could
only be handled by means of the infinitesimal calculus.[1] The integral calculus

developed more slowly than the differential calculus did. Difficulties with the former threatened to block the very formulation of some of these challenges, much less permit them to be solved.

In 1697, Leibniz hit upon the fact that differentiation and integration can be inverted. This result can be written as

$$d_a \int_{x_0}^{x} p(x,a)\,dx = \int_{x_0}^{x} d_a p(x,a)\,dx \qquad (7.1.1.1)$$

in the modernized notation of Engelsman (1984: 44). Leibniz's discovery helped surmount some of the obstacles that the integral calculus presented. Here d_a designates the partial differential operator with respect to a. It can be applied to an expression $p = p(x,a)$ in two finite variables x and a, to a one-parameter family of plane curves

$$y = f(x,a) \equiv \int_{x_0}^{x} p(x,a)\,dx \qquad (7.1.1.2)$$

whose parameter is a, and so forth. The parameter a functions as a finite variable; the technique (7.1.1.1) illustrates differentiation from curve to curve.

Johann I Bernoulli applied Leibniz's discovery immediately to the problem I mentioned in Chapter 2 of finding "orthogonal trajectories": given a one-parameter family of plane curves that lie in the same plane, find the plane curves that intersect all of the members of the family at right angles. Leibniz had introduced a method for dealing with the problem which was limited, with a few exceptions, to families of algebraic plane curves. Using this method, Leibniz treated the parameter a as a constant, with the intention of trying to eliminate it from the problem. Leibniz had established that if c is the slope of the curve S at a point on a curve S in a given family of plane curves that lie in the same plane, where c is neither 0 nor ∞ (in other words, the tangent to the curve S at the point is neither horizontal nor vertical), and if b is the slope at that same point of the plane curve orthogonal to S at the point and which lies in the same plane as the given family, then $cb = -1$. In particular, if

$$y = \int_{x_0}^{x} p(x,a)\,dx \qquad (7.1.1.2)$$

[or $dy = P(x,a)dx$] represents the given one-parameter family of plane curves, then $dx = -p(x,a)\,dy$ is consequently an equation for the plane curves orthogonal to all of the members of the family. (This condition or rule eventually came to be called the "Canon Hermannii," because Jakob Hermann published it, for the first time ever, years later.[2]) If (7.1.1.2) can be simplified to a finite equation $F(x,y,a) = 0$ that, moreover, can be solved for a explicitly, meaning that a can be expressed in terms of x and y: $a = a(x,y)$,[3] then Leibniz could reduce the problem of finding the orthogonal trajectories of the one-parameter family of plane curves (7.1.1.2) to solving the first-order ordinary "differential" equation $dx = -p(x, a(x, y))\, dy$ in the two finite variables x and y. The method failed, however, whenever the right-hand side of (7.1.1.2) was transcendental in such a way that made it impossible to solve (7.1.1.2) explicitly for the parameter a. [In fact, we have known since the nineteenth century that the same problem would arise if (7.1.1.2) were an algebraic equation of degree $\geqslant 5$ in the parameter a. Such an equation cannot be solved for a by radicals.] With this difficulty in mind, Bernoulli derived the "variable parameter equation"

$$(1 + p^2)dx + pq\,da = 0 \tag{7.1.1.3}$$

for the one-parameter family of plane curves (7.1.1.2), where $q \equiv f_a$ is the differential coefficient in the partial differential d_a of

$$f(x, a) \equiv \int_{x_0}^{x} p(x, a)\, dx \tag{7.1.1.4}$$

(in other words, $d_a f = q\,da$).[4] If Leibniz's result (7.1.1.1) is applied to $q \equiv f_a$, the variable-parameter equation (7.1.1.3) becomes

$$(1 + p^2)dx + \left\{ p \int_{x_0}^{x} p_a(x, a)dx \right\} da = 0 \tag{7.1.1.3'}$$

(where $d_a p = p_a da$). Equation (7.1.1.3') is a first-order ordinary "differential" equation in x and a. It is, in effect, an alternate first-order ordinary "differential" equation for the plane curves orthogonal to all of the members of the one-parameter family of plane curves (7.1.1.2). Equation (7.1.1.3') results in part from the unique values of the parameter a which an orthogonal trajectory makes correspond to different values of the abscissa x:

$$a = a(x).$$

If the variable-parameter equation (7.1.1.3') could be integrated in a way that makes this correspondence $a = a(x)$ explicit (that is, that makes it possible to express a in terms of x), then the problem of finding the orthogonal trajectories of the one-parameter family of plane curves (7.1.1.2) could be reduced to a quadrature:

$$y = \int_{x_0}^{x} p(x, a(x))\, dx, \tag{7.1.1.5}$$

where the constant of integration in the integral $a = a(x)$ of the variable-parameter equation (7.1.1.3') acts as the parameter of the family of orthogonal trajectories. However, unless

$$q = \int_{x_0}^{x} p_a(x, a)\, dx, \tag{7.1.1.6}$$

which appears inside the variable-parameter equation (7.1.1.3′), could be expressed algebraically in terms of x and a, the variable-parameter equation (7.1.1.3′) could not be integrated in the way just mentioned, because methods of solving first-order ordinary "differential" equations whose coefficients are transcendental had not yet been discovered. Bernoulli's method came to a dead end in such cases.

Moreover, Johann's astute and profound brother Jakob doubted Johann's whole approach to problems of orthogonal trajectories. He did not feel that Johann's method went to the heart of the matter. In an article on brachistochrones published in the issue of *Acta Eruditorum* for May 1697, in which he first publicly challenged other European mathematicians to find orthogonal trajectories, Johann judged the problem to be easy when the given families of plane curves were algebraic. He had convinced himself that families of transcendental plane curves were the ones that introduced the obstacles to solving problems of orthogonal trajectories, and he believed that the method of variable parameters (that is, differentiation from curve to curve) could ultimately be used to overcome the difficulties. But Jakob took issue with his brother on what was "easy." He was skeptical that Johann knew methods that truly provided the key to solving hard problems of orthogonal trajectories in general. In reply to his brother's assertions, Jakob gave an example to illustrate his point, in an article published in the issue of *Acta Eruditorum* for May 1698. In this article, Jakob exhibited a certain family of algebraic plane curves. Finding the plane curves orthogonal to all of the members of this family reduced in this particular case to solving a first-order ordinary "differential" equation of second degree when Leibniz's method, which was the relevant method in this case, was applied. The trouble was that this ordinary "differential" equation could not be integrated at this time. Johann had obviously failed to get at the crux of the matter. He missed the essence of problems of orthogonal trajectories–at least, when these problems are considered in full generality.

Johann apparently took his brother Jakob's criticisms seriously. In his own article on orthogonal trajectories published in the issue of *Acta Eruditorum* for October 1698, Johann solved geometrically the problem that his brother had proposed. Johann employed a method that was based neither on Leibniz's method nor on the variable parameter equation.[5] Research on the problem of orthogonal trajectories temporarily came to an end. The dormant period lasted fifteen years. Up to this point, the discoveries connected with differentiation from curve to curve (reversing the order of differentiation and integration, the variable-parameter equation) remained unpublished. Those mathematicians who participated in the late-seventeenth-century challenges published solutions alone. They did not publish the methods that they had used to arrive at their solutions. These methods they kept secret. The absence of proofs excited interest in mathematicians in much the same way and for much the same reasons as the obscurity of some of the demonstrations that appear in Newton's *Principia* did.

Late in 1715, shortly before Leibniz died in 1716, Leibniz and Johann I Bernoulli publicly revived problems of orthogonal trajectories by openly challenging British mathematicians to solve such problems. Leibniz and Bernoulli hoped to confound and outwit Newton's partisans who had behaved reprehensibly during the controversy over the invention of the infinitesimal calculus. That controversy had worsened during the fifteen-year period that European mathematicians ceased to do research on problems of orthogonal trajectories. Leibniz and Bernoulli felt that their understanding of the obstacles to solving such problems, which they had acquired during the 1690s and which had led them to invent the still unpublished method of variable parameters, gave them a definite advantage.

A pitched battle between continental and British mathematicians ensued from the challenge, which led to other challenges and counter challenges to solve difficult problems, including problems of orthogonal trajectories. Johann I Bernoulli and Brook Taylor were the principal contestants in this fight, but Newtonians John Keill, Henry Pemberton, James Stirling, and John Machin, not to mention the aged Newton himself, participated as well. Nikolaus I Bernoulli, Nikolaus II Bernoulli, and Hermann, three other members of the Basel School, also represented the other side. The controversy broadened. Around the time that Leibniz and Johann I Bernoulli issued their second challenge to British mathematicians, which I shall discuss shortly, Bernoulli accused Taylor of having plagiarized continental mathematicians in Taylor's recently published *Methodus incrementorum directa et inversa* (1715).[6] The continental mathematicians published the work that they did in connection with the various challenges and charges in *Acta Eruditorum* for the years 1716–21, and the British mathematicians published their work in the *Philosophical Transactions* for the same period.

Leibniz and Johann I Bernoulli proposed the problem of finding the orthogonal trajectories of the generalized brachistochrone cycloids, a one-parameter family of transcendental plane curves, as the ultimate problem to test all of the methods of solving problems of orthogonal trajectories. Taylor published a construction in the *Philosophical Transactions* for 1717, and Jakob Hermann published one in *Acta Eruditorum* for 1717. In *Acta Eruditorum* for 1718, Johann's son Nikolaus II published two of his father's constructions. None of these solutions involved the use of the partial differential calculus.

Why did Johann I Bernoulli not utilize the partial differential calculus at this time to solve this particular problem of orthogonal trajectories? In effect, he simply had made no more progress in solving the variable-parameter equation (7.1.1.3), which he had introduced in the 1690s, beyond what he had achieved at that time.[7] He still could not integrate the equation in the manner discussed above when the equation had transcendental coefficients. In the case of the generalized brachistochrone cycloids, the coefficients (specifically, q) of the corresponding variable parameter equation are transcendental. In 1716, Johann confided to Leibniz his failure to find methods of integrating the variable-parameter equation for families of transcendental plane curves (7.1.1.2) in general.[8] He admitted that he had made no more headway in studying families of transcendental plane curves than Leibniz had made in studying families of algebraic plane curves. Namely, finding the orthogonal trajectories in special cases was as far as either had gotten. Neither had been able to deal with the problem of orthogonal trajectories in complete generality. The warning that Jakob, now deceased, had given Johann eighteen years earlier returned to haunt him.

7.1.2. Nikolaus I Bernoulli (1719–21). Unpublished documents abound with evidence that Johann's nephew Nikolaus I Bernoulli made the next great advances in the partial differential calculus after the initial achievements of the 1690s. This Nikolaus did during the challenges of the late 1710s. He made explicit for the first time ever, although not as clearly in published work as in unpublished work, the equality

$$d_y d_a S = d_a d_y S \qquad (7.1.2.1)$$

of mixed, second-order partial differentials of an expression $S = S(y, a)$ in two finite variables y and a. His uncle Johann also tacitly employed this result, around the same time, to demonstrate that differentiation and integration could be inverted. But Johann did not call attention to the result in doing so, because he had failed to realize its significance.[9]

Nikolaus used equality (7.1.2.1) to solve restricted and general "completion problems," which he introduced to try to deal with the expression

$$q = \int_{x_0}^{x} p_a(x, a) \, dx, \qquad (7.1.1.6)$$

which was almost always unmanageable when it was transcendental. The restricted completion problem is to find $q = q(y, a)$ such that

$$dx = p(y, a) \, dy + q(y, a) \, da \qquad (7.1.2.2)$$

is a total "differential" equation, given an equation

$$d_y x = p(y, a) \, dy \qquad (7.1.2.3)$$

for a one-parameter family of plane curves that lie in the same plane and whose parameter is a. In the more general completion problem, an equation

$$d_y x = p(x, y, a) \, dy \qquad (7.1.2.4)$$

for a one-parameter family of plane curves that lie in the same plane and whose parameter is a is given and the problem is to find $q = q(x, y, a)$ in the total "differential" equation

$$dx = p(x, y, a) \, dy + q(x, y, a) \, da \qquad (7.1.2.5)$$

for the family of curves.

What purpose did the completion problem serve? The coefficient $q(y, a)$ in the total "differential" equation (7.1.2.2) which solves the restricted completion problem given an equation (7.1.2.3) for a one-parameter family of plane curves that lie in the same plane and whose parameter is a is the same as the coefficient $q(y, a)$ in the variable-parameter equation (7.1.1.3) for the plane curves orthogonal to all of the members of the given one-parameter family of plane curves that lie in the same plane and whose parameter is a. [Note: If we compare the equations here with the equations in the preceding section of this chapter, we see that x and y have been interchanged. They have simply traded places. I have followed Engelsman (1984: 99–105, 110–111) in transposing the variables here.]

Nikolaus I Bernoulli's other great accomplishment was to apply his solution of the restricted completion problem, a solution that amounts to showing that from the equality of mixed, second-order partial differentials (7.1.2.1) it follows that differentiation and integration can be reversed when conditions hold which are more general than the ones that Leibniz stated,[10] to integrate by means of an integrating factor the variable-parameter equation

$$\frac{1 + p^2}{p} dy + q da = 0 \qquad (7.1.2.6)$$

for the plane curves orthogonal to the generalized brachistochrone cycloids, whose coefficient q is transcendental.[11] [Equation (7.1.1.3) becomes equation (7.1.2.6) when both sides of equation (7.1.1.3) are divided by p when $p \neq 0$.] Nikolaus was the *only* mathematician at this time to solve this particular problem of orthogonal trajectories by integrating the corresponding variable-parameter equation. However, in the solution of the problem which he published in *Acta Eruditorum* for 1719, Johann's nephew only presented rules for constructing the orthogonal trajectories. These rules concealed the variable-parameter equation and its integration.[12] Had Nikolaus revealed his methods for determining the rules, his solution would have struck European mathematicians as a whole as utterly spectacular. Nikolaus's uncle Johann, we recall, had been unable to integrate the relevant variable-parameter equation, which forced him to construct the orthogonal trajectories of the generalized brachistochrone cycloids by other means instead.[13]

7.2. *The uphill struggle in Paris to master the integral calculus (1703–30)*

By the time that these advances in solving problems that involved the integral calculus were made, Parisian mathematicians had fallen far behind in the integral calculus, despite the presence of a long-established academy of sciences in Paris without equal in Germany, Switzerland, or Italy. The failure of Parisian mathematicians to keep pace in the integral calculus with mathematicians in Germany, Switzerland, or Italy probably occurred in part for reasons that I have discussed elsewhere. These include the wars between France and her neighbors, which, until the Treaty of Utrecht in 1713, virtually isolated French *savants*. In particular, the wars made the issues of *Acta Eruditorum* published in Germany at the end of the seventeenth century, which included much of the work of the 1690s on the infinitesimal calculus done by Leibniz and the Bernoulli brothers Johann I and Jakob, all but impossible to obtain in France during the wars. Beginning around 1720 the rate of exchange between French and German currency then became unfavorable to the French. As a consequence the *Acta* subsequently became too expensive for the French to buy regularly. Finally, the contact between Parisian mathematicians and Johann I Bernoulli, the only surviving pioneer of the continental calculus after the deaths of Jakob Bernoulli in 1705 and Leibniz in 1716, gradually ceased. This loss of communication resulted from the deaths, one by one, of Johann's principal French correspondents among the mathematicians: the Marquis de L'Hôpital's death in 1704, Pierre-Rémond de Montmort's death in 1719, and Pierre Varignon's death in 1722.[14] I shall return to these events and to the problems that they caused Parisian mathematicians later in this chapter, when I discuss the rebuilding of mathematics in Paris.

After L'Hôpital died, only two Parisian mathematicians, Pierre-Rémond de Montmort and François Nicole, made a serious, concerted effort to follow the progress made elsewhere in Europe in the integral calculus. (In 1711, Christophe-Bernard de Bragelogne hoped to tackle advanced research in integral calculus, but he abandoned the subject.) The Paris Academy of Sciences published a paper by Johann I Bernoulli on integration by partial fractions in its *Mémoires* for 1702.[15] The paper induced Montmort to write to Bernoulli in 1703. After waiting a year and a half and still having received no reply, Montmort wrote Bernoulli again concerning his difficulties with the integral calculus. This time Montmort cited L'Hôpital, who had died in the meantime, as a reference, in order to show Bernoulli that he was serious. He tried to coax Bernoulli to reply, which Bernoulli soon did. Thus began a regular correspondence between the two mathematicians. Montmort also personally financed the reproduction of Newton's *De quadratura curvarum* (1704), thereby making this work available to the French as early as 1707, despite the war going on between France and England.[16] Later Montmort began to correspond regularly with Nikolaus I Bernoulli as well.[17] Montmort's correspondences with both Bernoullis lasted until Montmort's death in 1719. The French mathematician profited greatly from these correspondences. He learned much about the integral calculus from the two Swiss mathematicians. Nicole, who was Montmort's protégé and collaborator at the beginning of the eighteenth century, also tried, like his patron, to keep abreast of new developments in the integral calculus.

The following example illustrates the activity of the two French mathematicians at the beginning of the eighteenth century. In 1703, Montmort, who was then twenty-four years old, tried to tackle problems in the integral calculus. These included a difficult one that L'Hôpital had proposed in the *Journal des Savants* for 1692: to rectify (which means to find the length of) "deBeaune's curve."

(Finding deBeaune's curve required that a certain ordinary "differential" equation be solved. As I have already mentioned in Chapters 4–6, solving such equations was called "the inverse method of tangents.") At the same time, Montmort attempted to rectify the cissoid, as did Nicole, who was then only nineteen years old. Montmort communicated his two rectifications, together with Nicole's rectification of the cissoid, to the *Journal des Savants*, where the solutions were published in 1703–Montmort's anonymously.[18]

While reminiscing about these juvenile, immature efforts in a letter to Nikolaus I Bernoulli in 1712, Montmort acknowledged that he had been rather young when he attacked the problem of rectifying deBeaune's curve, that only five or six mathematicians had known any integral calculus at the time, that the integral calculus had developed and progressed considerably in the interim, and that he had learned much since 1703 from having corresponded with Johann I Bernoulli and from having read *Acta Eruditorum*.[19]

Johann I Bernoulli had a singularly interesting and unusual part in the reception and diffusion in Paris of problems of finding orthogonal trajectories. Bernoulli had proposed a certain problem of orthogonal trajectories in the issue of *Acta Eruditorum* for May 1698, and he published solutions of this problem in the issue of *Acta Eruditorum* for October 1698. But one of the printed solutions was wrong, and his Paris Academy *mémoire* of 1702 on integration by partial fractions provided him with an occasion to correct the error.

In this *mémoire* Bernoulli first restated the problem of orthogonal trajectories in question: find the plane curves that intersect at right angles all of the members of a one-parameter family of parabolas that lie in the same plane, where the parabolas in the family all have a common axis, and where the value of the parameter of each parabola in the family equals the distance of the vertex of the parabola to the same fixed point on the common axis. Next he presented in the *mémoire* two different first-order ordinary "differential" equations each of whose solutions solved the problem. He added that the results were not limited to ordinary parabolas but included parabolas of any degree (cubical, semicubical, and so forth). With the errors corrected, the first of the two first-order ordinary "differential" equations could be integrated by means of partial fractions, which explains why Bernoulli introduced the particular problem of orthogonal trajectories at issue into his Paris Academy *mémoire* of 1702. But Bernoulli only published the two first-order ordinary "differential" equations whose solutions each solved the problem of orthogonal trajectories in question; he did not reveal in the *mémoire* how he found these equations.

Now, the issue of *Acta Eruditorum* for October 1698 is in fact the same issue of the *Acta* in which Bernoulli solved the problem of orthogonal trajectories which his brother Jakob had proposed in the issue of *Acta Eruditorum* for May 1698. We recall that Johann could not solve this problem using either Leibniz's method or the variable-parameter equation. The problem that Jakob had formulated is the following one: find the plane curves that intersect at right angles all of the members of a one-parameter family of parabolas that lie in the same plane, where the parabolas in the family all have a common axis, and where the length of the latus rectum of each parabola in the family equals the distance of the vertex of the parabola to a given fixed point.[20] Although each of the two Bernoulli brothers undoubtedly submitted his problem to *Acta Eruditorum* without any knowledge of the problem submitted by the other, the two problems, both of which were published in the issue of *Acta Eruditorum* for May 1698, look remarkably like variations of each other.

As I say, in his Paris Academy *mémoire* of 1702, Johann I Bernoulli did not divulge the methods that he used to arrive at the two first-order ordinary "differential" equations which he published in the *mémoire* and each of whose solutions was a solution to the problem of orthogonal trajectories which he had proposed in the issue of *Acta Eruditorum* for May 1698. Bernoulli did, however, introduce in the *mémoire* "a variable parameter ... of any parabola" which he designated by x ("en prenant x pour le variable paramètre d'une Parabole quelconque"), and the first of his two first-order ordinary "differential" equations which solve the problem include differentials with respect to the parameter x. Specifically, the first of his two first-order ordinary differential equations has the form $dx/x = F(z)\,dz$, where y stands for the ordinate of a plane curve that solves the problem, and where $y = xz$. [The second of Bernoulli's two first-order ordinary differential equations has the form $dy/y = G(z)\,dz$, where y now symbolizes the abscissa of a plane curve that solves the problem, and where the variable parameter $x = yz$.[21]] But Bernoulli did not reveal the procedures and techniques that he used to derive his two first-order ordinary "differential" equations. He omitted all details of whatever analysis he employed to obtain these equations, and oblique references to "variable parameters" without further clarification could scarcely have provided the inexperienced, untrained, or uninformed reader with any useful clues.

Late in 1714 and in the spring of 1715, Nicole read a paper before the Paris Academy on the problem of finding the plane curves that intersect at a specified constant angle all of the members of a given family of plane curves that lie in the same plane, of which the problem of orthogonal trajectories is a special case. In this paper Nicole derived, among other things, the two first-order ordinary "differential" equations that appear in Bernoulli's Paris Academy *mémoire* of 1702 on integration by partial fractions, each of which solved the problem of orthogonal trajectories at issue in that *mémoire*. The published version of Nicole's paper appears in the Paris Academy's *Mémoires* for 1715.[22] In the process of solving the first of the two first-order ordinary "differential" equations, Nicole differentiated with respect to the parameter x. (His method of solving this equation did not involve the use of the still unpublished variable-parameter equation, however.)

What prompted Nicole to work on Bernoulli's problem of orthogonal trajectories at just this particular moment, some twelve years after Varignon read the original version of Bernoulli's paper on integration by partial fractions before the

Paris Academy in December of 1702? This provocative question does not have an immediately obvious answer. Leibniz and Bernoulli did not openly challenge the British mathematicians to solve problems of orthogonal trajectories until late 1715.[23] Brook Taylor visited France in June of 1715, at which time he met Montmort and probably Nicole as well. At this time Taylor gave Montmort a copy of his recently published *Methods incrementorum directa et inversa* (1715). Taylor also presented a copy of his book to the Paris Academy, which the Academy gave to Varignon to review.[24] In July of 1715 Taylor sent Nicole two copies of his new book, which Taylor authorized Nicole to distribute.[25] Taylor's treatise subsequently influenced Nicole's work on finite differences.[26] On 31 March 1716 Nikolaus I Bernoulli wrote Montmort a letter in which he said that his uncle Johann and Leibniz had challenged the British mathematicians to find the plane curves orthogonal to all of the members of a one-parameter family of coplanar hyperbolas that all have the same transverse axis (hence the same center) and the same vertices and that his uncle and Leibniz would soon propose the problem of finding the plane curves orthogonal to the generalized brachistochrone cycloids as the final test.[27] Montmort answered the letter at the end of April 1716. He told Nikolaus that he had known about the challenge "for a long time" (here Montmort was presumably alluding specifically to the challenge to find the plane curves orthogonal to all of the members of a one-parameter family of coplanar hyperbolas that all have the same transverse axis–hence the same center–and the same vertices), that Nicole had solved it "immediately" (here again Montmort must have been alluding specifically to the problem of finding the plane curves orthogonal to all of the members of a one-parameter family of coplanar hyperbolas that all have the same transverse axis (hence the same center) and the same vertices, which in fact, as we shall see in a moment, Nicole had already solved at least six months before the challenge was ever issued), and he added that Taylor had written to say that he had resolved the problem "in all possible generality."[28] Montmort possibly learned about the challenge from Taylor, since Johann I Bernoulli did not write to Montmort in 1716. On 31 March 1716, thus before he had received Nikolaus I Bernoulli's letter to him written that same day, Montmort wrote Taylor to say that a few days earlier he had read a "copy of a very curious letter that Monsieur Leibniz wrote to the Abbé de Conty." Montmort advised Taylor to ask to be shown the letter if he had not already seen it.[29] The letter from Leibniz to the Abbé Antonio Conti concerned Leibniz's and Johann I Bernoulli's challenge to the British mathematicians to solve problems of orthogonal trajectories. In December of 1715 Conti had transmitted Leibniz's original wording of the challenge to the British mathematicians. Near the middle of April of 1716, Montmort wrote Taylor again, this time to say that he had written to Germany in order to propose to Leibniz a new problem of orthogonal trajectories which Taylor had formulated and had sent to Montmort.[30] (Taylor in fact did not define this problem well. He introduced a certain second-order ordinary "differential" equation to characterize the members of the given family of plane curves that all lie in the same plane. But the integral of such an equation includes two parameters or constants of integration, so that the equation does not determine a unique one-parameter family of plane curves that all lie in the same plane.) But all of these events occurred *after* Nicole had read to the Paris Academy his paper on the problem of finding the plane curves that intersect at a fixed angle all of the members of a given family of plane curves that lie in the same plane. Nicole only met Taylor after he had read his paper. Taylor presumably only became interested in problems of orthogonal trajectories himself at the end of 1715, when Leibniz and Johann I Bernoulli publicly challenged the British mathematicians to solve them. But Nicole had already read his paper to the Paris Academy by then. Indeed, he had finished reading it six months earlier. Nicole had doubtlessly known about Bernoulli's Paris Academy *mémoire* of 1702 on integration by partial fractions for a long time. After all, it had caused Montmort, Nicole's sponsor and collaborator at the beginning of the eighteenth century, to write Bernoulli the first time, in 1703, and the two French mathematicians worked closely together when Montmort wrote the letter.

Perhaps the most intriguing fact of all is that in his paper of 1714–15 on the problem of finding the plane curves that intersect at a fixed angle all of the members of a given family of plane curves that lie in the same plane, Nicole stated and solved the first problem of orthogonal trajectories which Leibniz originally challenged the British mathematicians six months later to solve. Leibniz communicated the problem to the British mathematicians in a letter dated 25 November 1715 which he sent to Conti, who was then in England. It is the first of the two problems of orthogonal trajectories which Nikolaus I Bernoulli talked about in his letter to Montmort dated 31 March 1716: find the plane curves orthogonal to all of the members of a one-parameter family of coplanar hyperbolas that all have the same transverse axis (hence the same center) and the same vertices. This problem turned out to be too simple. The British mathematicians easily solved it. Leibniz had in fact phrased the problem ambiguously in the letter to Conti. Leibniz had only meant the one-parameter family of hyperbolas to serve as an example that illustrated the general problem. Both he and Johann I Bernoulli intended that the British mathematicians deal with the problem in complete generality, not simply solve the special case that Leibniz had worded. This misunderstanding led Bernoulli to propose in March of 1716 the problem of finding the plane curves orthogonal to the generalized brachistochrone cycloids as the ultimate test, which Leibniz transmitted to Conti in a letter dated 3 April 1716. But exactly what circumstances induced Montmort's one-time client Nicole to take up the problem of finding the plane curves orthogonal to all of the members of a one-parameter family of coplanar hyperbolas that all have the same transverse axis (hence the same center) and the same vertices, as well as the particular problem of orthogonal trajectories that Bernoulli stated in his Paris Academy *mémoire* of 1702 on integration by partial fractions, whose solutions

Bernoulli published in that *mémoire* without any demonstrations, one year before such problems of orthogonal trajectories were publicly revived, reamin a mystery.

When the contest began late in 1715, the two main opponents were, as I mentioned, Johann I Bernoulli and Taylor. Now, Montmort happened to be a friend and confidant of both Bernoulli and Taylor. As a result, Montmort found himself mediating between the two competitors. While probability theory had become his preferred specialty by this time, there is no question that as a result of his correspondence with Johann I Bernoulli, Taylor,[31] and others who participated in the competition, like Nikolaus I Bernoulli, who worked behind the scenes, Montmort became the best informed Frenchman where matters that involved the use of the integral calculus to solve problems in mathematics which lay at the forefront of research in mathematics were concerned. But Montmort also had to promise to keep what he learned a secret. Nor did he lose the confidence of the contestants, all of whom trusted him. He took his knowledge with him to his grave in 1719.

Not even Nicole profited from his former patron's privileged position. In the spring of 1718, Nicole read a paper before the Paris Academy on the problem of finding the plane curves orthogonal to the generalized brachistochrone cycloids.[32] Leibniz and Johann I Bernoulli had formulated the challenge in two parts. First, they characterized geometrically the cycloids in question, and then they asked that the cycloids be represented analytically. Second, the plane curves orthogonal to all of the members of this family of cycloids had to be determined. Nicole found the correct analytic expression for the generalized brachistochrone cycloids as a one-parameter family of plane curves. In attempting to determine the plane curves orthogonal to all of these cycloids, Nicole treated the parameter a of the family of cycloids as a constant, which he tried to eliminate from the pair of equations[33]

$$dy = p(x, a) \, dx \tag{7.2.1}$$

$$dx = -p(x, a) \, dy. \tag{7.2.2}$$

Here I do not display the specific equations for this particular problem. I exhibit only the general form that equations have in problems of orthogonal trajectories. For families of transcendental plane curves that lie in the same plane, such as the generalized brachistochrone cycloids, equation (7.2.1) could not in general be solved explicitly for the parameter a. This is the same difficulty that had limited Leibniz's method of finding orthogonal trajectories. By means of assorted operations and procedures, Nicole overcame this obstacle. In particular, he multiplied both sides of equation (7.2.1) by a factor that changed the expression on the right-hand side of that equation into an expression with a finite, algebraic integral. Then he integrated the new equation that resulted from this multiplication. This integration led to an equation that Nicole *could* solve explicitly for the parameter a. Since he could already solve equation (7.2.2) explicitly for the parameter a, Nicole eliminated the parameter a from the problem. He did so, however, at the expense of introducing an integro-differential equation for the orthogonal trajectories. When he differentiated this equation, he arrived at a second-order ordinary "differential" equation for the orthogonal trajectories, instead of a first-order ordinary "differential" equation, which was to be preferred. Nicole managed, again by trial and error (for example, by introducing integrating factors once more), to reduce the second-order ordinary "differential" equation to an ordinary "differential" equation of first order whose variables were separated. Nicole had thus apparently reduced the whole problem of determining the orthogonal trajectories of the generalized brachistochrone cycloids to quadratures.

In fact, Nicole failed to solve the problem. Montmort communicated Nicole's solution of the problem and Nicole's method for attacking the problem to Johann I Bernoulli, but Bernoulli found Nicole's final expression for the solution of the problem to be wrong. The fact that it gave the correct result in the one particular case that Nicole tested was purely accidental, Bernoulli noted. Bernoulli showed Montmort where Nicole had erred.[34] In reply Montmort regretted the tendency of his spirited, skillful colleague to make mistakes.[35] Bernoulli nervertheless claimed to be impressed by Nicole's proficiency in the integral calculus. In particular, he liked what he called "the ingenious way" that Nicole reduced his second-order ordinary "differential" equation to an ordinary "differential" equation of first order.[36] (Such compliments from Bernoulli should be looked at with suspicion or skepticism, however. Bernoulli showered praise upon the work of such mediocre mathematicians as his friend Maupertuis, as we shall soon see, and upon mathematicians who caused him no troubles. They could not challenge the authority of the mathematician from Basel. On the other hand, Bernoulli was far less charitable to such first-rate, potentially dangerous competitors as his brother Jakob while Jakob was still alive, Taylor, and even his own son Daniel.)

Late in 1725 and early in 1726, in a paper tht he read before the Paris Academy, Nicole once again discussed the problem of finding the plane curves orthogonal to the generalized brachistochrone cycloids.[37] In the published version of the paper, which appears in the Paris Academy's *Mémoires* for 1725, Nicole began as he had in 1718 by deriving the analytic expression for the family of cycloids. In trying to determine the plane curves orthogonal to all of these cycloids, he again treated the parameter a of the family of cycloids as a constant, and he solved equation (7.2.2) for the parameter a in the same way as he had done in 1718. In tackling the part of the problem which remained, however, he proceeded differently than he had seven years earlier. He expanded the integrand of the integral that is the right-hand side of equation

$$y = \int_{x_0}^{x} p(x, a)\, dx, \tag{7.1.1.2}$$

which is an equation equivalent to equation (7.2.1), in an infinite series, and then he integrated this series term by term. He then substituted into the new series that resulted from this term-by-term integration the expression that he had found for the parameter a by solving equation (7.2.2) for a. By manipulating the resulting equation (specifically, by multiplying both sides of the equation, whose right-hand side is an infinite series, by a cleverly chosen factor) and then differentiating the new equation that followed from this operation, which involved differentiating term by term the infinite series that is the right-hand side of the new equation, Nicole arrived at an equation whose right-hand side was an infinite series whose terms all cancelled each other except the first term. This equation consequently reduced to a second-order ordinary "differential" equation for the orthogonal trajectories. By means of more manipulating (multiplying both sides of the equation by a judiciously chosen factor), Nicole reduced this equation to a first-order ordinary "differential" equation whose variables were not separated. Before dealing with this equation, Nicole reexpressed the integral that is the right-hand side of equation (7.1.1.2) as an infinite series that was different from the infinite series that he had originally derived to represent the integral that is the right-hand side of this equation. Substituting into this new infinite series the expression that he had determined for the parameter a by solving equation (7.2.2) for a, multiplying both sides of equation (7.1.1.2) expressed this way by a smartly chosen factor, and then differentiating the equation that resulted, which again involved differentiating term by term the infinite series that is the right-hand side of the equation, Nicole arrived at another equation whose right-hand side was an infinite series whose terms did not cancel each other but whose sum he recognized. This equation accordingly reduced to another second-order ordinary "differential" equation, which Nicole then reduced further to the same first-order ordinary "differential" equation whose variables were not separated which he had already derived. Nicole ultimately succeeded in transforming this equation into a first-order ordinary "differential" equation whose variables were separated. By this means he reduced the problem of finding the orthogonal trajectories of the generalized brachistochrone cycloids to quadratures. This time Nicole did arrive at the correct expression for the plane curves orthogonal to the generalized brachistochrone cycloids.

Nicole's sporadic work on problems of orthogonal trajectories during the decade 1715–25 appears to have attracted little attention in Paris at the time. Mairan evidently knew something about such problems. As I mentioned in Chapter 2, in his Paris Academy *mémoire* of 1720 Mairan referred to "Monsieur [Johann I] Bernoulli's famous *Trajectories.*" These Mairan illustrated, as I also noted in Chapter 2, by formulating and solving the easy problem of finding the plane curves orthogonal to all of the members of a family of concentric coplanar geometrically similar elongated ellipses whose major axes all lie along the same line, therefore whose minor axes all lie along the same line too.

In his Paris Academy *mémoire* of 1715 Nicole had, in effect, already solved part of this problem by inversion. Namely, Nicole did not begin that *mémoire* by attacking the problem of orthogonal trajectories which Johann I Bernoulli had presented in his Paris Academy *mémoire* of 1702 on integration by partial fractions. Nicole first determined instead in his *mémoire* of 1715 the plane curves orthogonal to all of the members of a one-parameter family of coplanar circles with a single point in common and with centers all lying along the same axis, the plane curves orthogonal to all of the members of a one-parameter family of coplanar ellipses with a common axis and a common vertex (*sommet*), the plane curves orthogonal to all of the members of a one-parameter family of coplanar hyperbolas with the same transverse axis (hence the same center) and the same vertices, which is the original problem of orthogonal trajectories which Leibniz later chose to illustrate the kinds of problems that he and Johann I Bernoulli challenged the British mathematicians to solve, and finally the plane curves orthogonal to all of the members of a one-parameter family of coplanar ordinary parabolas with the same axis and vertex. These problems Nicole handled by using what amounts to Leibniz's method for finding the plane curves orthogonal to a one-parameter family of coplanar algebraic curves. In the case of a family of coplanar ordinary parabolas

$$x = \frac{1}{z} y^2 \qquad (7.2.3)$$

whose parameter is z, Nicole found by eliminating the parameter z that the equation for the plane curves orthogonal to all of the parabolas is

$$\frac{x^2}{a^2} + \frac{y^2}{2a^2} = 1. \qquad (7.2.4)$$

In other words, the family of plane curves orthogonal to all of the parabolas is a one-parameter family of concentric coplanar geometrically similar elongated ellipses whose parameter is a, whose common center is situated at the common vertex of the parabolas, and whose major axes all lie along the same line, the y-axis, hence whose minor axes all lie along the same line as well, the x-axis. The ratio of the major semiaxis to the minor semiaxis of each ellipse in the family has the same value, namely, $\sqrt{2}/1$. Hence the ellipses in the family all have the same ellipticity.[38] This problem of orthogonal trajectories and Nicole's solution of it when viewed in reverse includes part of the solution of the problem of orthogonal trajectories which Mairan formulated in his Paris Academy *mémoire* of 1720. Namely, if the finite variables x and y are interchanged in equation (7.2.3), then equation

$$y^p = nx^m \qquad (2.1.9)$$

results, where $p = 1$, $m = 2$, and $n = 1/z$. But if $p = 1$ and $m = 2$ in (2.1.6), then $k = \sqrt{2}$ in (2.1.6) and in equation (2.1.7), in which case equations (2.1.7) and (7.2.4) are the same, when the finite variables x and y in equation (7.2.4) are transposed.

[Inverting Nicole's solution does not, however, solve the general problem that Mairan formulated in his Paris Academy *mémoire* of 1720, namely, the problem of finding the plane curves orthogonal to all of the members of a family of concentric coplanar geometrically similar elongated ellipses whose major axes all lie along the same line, therefore whose minor axes all lie along the same line too. The reason is that, as we saw in Chapter 2, Mairan found that some families of concentric coplanar geometrically similar elongated ellipses, where the major axes of all of the members of such a family all lie along the same line, consequently the minor axes of all of the members of such a family all lie along the same line as well, have as orthogonal trajectories families of coplanar parabolas that differ from ordinary parabolas. For example, they may be families of cubical parabolas, or families of semicubical parabolas, or families of still other kinds of parabolas. In other words

$$k \equiv \sqrt{\frac{m}{p}} \tag{2.1.6}$$

is only equal to $\sqrt{2}$ in equation

$$\frac{x^2}{k^2 a^2} + \frac{y^2}{a^2} = 1 \tag{2.1.7}$$

when $p = 1$ and $m = 2$ in equation (2.1.9), in which case equation (2.1.9) expresses a family of coplanar ordinary parabolas whose parameter is n. Thus when k is not equal to $\sqrt{2}$ in equation (2.1.7) for a family of concentric coplanar geometrically similar elongated ellipses whose major axes all lie along the same line, therefore whose minor axes all lie along the same line too, and whose parameter is a, the family of plane curves whose members are orthogonal to all of the members of this family and whose parameter is n, which is expressed by equation (2.1.9), consists of coplanar parabolas that are not ordinary parabolas. Nor, I should mention, is the one-parameter family of coplanar parabolas expressed by equation (2.1.9) the same as the one-parameter family of coplanar general parabolas which Johann I Bernoulli defined in his Paris Academy *mémoire* of 1702 on integration by partial fractions and which Bernoulli presented in that *mémoire* as the one-parameter family of plane curves that all lie in the same plane whose orthogonal trajectories were to be found.]

But Nicole's work on problems of orthogonal trajectories during the decade 1715–25 appears to have had very little, if any influence on mathematicians in Paris during this period. Why? For one thing, Nicole's solution of 1725 to the problem of finding the plane curves orthogonal to the generalized brachistochrone cycloids could hardly have struck the members of the European community of mathematicians in general as a great achievement. In addition to all of the other solutions to the same problem which had already been published between 1717 and 1719, Johann I Bernoulli published the methods that he used to find his solutions of 1718 to the problem in his son Nikolaus II's articles that appeared in *Acta Eruditorum* for 1720. The older Bernoulli had in fact written all of the mathematics for the article that was published under his son's name.[39] If Nicole had not seen the solutions of 1717 in time to make use of them in his own work on the same problem in 1718, it seems likely that by 1725 he knew in advance what the solution should look like.[40]

Indeed, the Jesuit Louis-Bertrand Castel insinuated as much, in his review of the Paris Academy's *Mémoires* for 1725 which was published in the *Journal de Trévoux*. Castel's commentary was one of several that he wrote which infuriated certain mathematicians who belonged to the Paris Academy and which rankled in their minds. Castel's candor incurred him the hatred of some of the Academy's mathematicians, who, incensed at his remarks, would have liked to be able to silence him. In Castel's opinion, Nicole had merely furnished another solution of a problem that had already been "successfully resolved several years earlier" by Castel's champions, "the members of the Royal Society," (which undoubtedly meant Taylor). Castel failed to mention the continental mathematicians' solutions of the problem which were published in *Acta Eruditorum*. Whether the fact that Castel was an anglophile explains why he did not call attention to them, or whether the relevant volumes of the *Acta* had simply failed to reach him because they were too expensive and consequently scarce is hard to say. In any case, there certainly was nothing particularly novel or original about Nicole's solution of 1725 to the problem of finding the plane curves orthogonal to the generalized brachistochrone cycloids.

The second volume of Charles-René Reyneau's textbook entitled *Analyse demontrée*, published in two volumes in 1708, included chapters on the integral calculus. In some of these chapters Reyneau discussed "the inverse method of tangents" (in other words, integrating in finite terms first-order ordinary "differential" equations of first degree when this can be done or else reducing such equations to quadratures). In 1728 Clairaut corrected errors that he found in Reyneau's textbook, and in 1739 Jean LeRond D'Alembert corrected other mistakes that remained in the second edition of the work, published in 1736. Castel found volume 2 of Reyneau's textbook to be prolix and full of obstacles and stumbling blocks (*embarras*), which made the volume practically incomprehensible and unreadable in Castel's opinion. Castel thought that it was better to read instead various writings by Wallis, Newton, Bernoulli, L'Hôpital, Edmund Stone, and Gregory of Saint Vincent.[41] In other words, Castel evidently found that Reyneau's textbook did not help make the works of some of Europe's better known creative mathematicians easier to understand! I shall have more to say about Castel and about Edmund Stone's book on the integral calculus later in this chapter.

It seems doubtful that the literature of orthogonal trajectories published in the volumes of the *Acta* for the years 1716–21 could have had much influence on mathematicians in Paris during much of the 1720s. For one thing the volumes were not readily available in Paris. I mentioned the reasons for this previously: a devaluation of French currency relative to German currency, which made the cost of German publications in France prohibitive.

Finally, discoveries favor the prepared mind. But during the 1720s, Parisian mathematicians had barely begun to close the gap that separated the level of their mathematical work from that of the members of the Basel School. The unstable relations between the two groups of mathematicians during this period did nothing to help Parisian mathematicians make progress in solving problems that involved the integral calculus. Before 1725 even mathematicians in Italy apparently handled problems of orthogonal trajectories better than Parisian mathematicians did. Until 1729, an insignificant minority of mathematicians in Paris, none of whom were truly men of top rank, struggled to catch up with mathematicians elsewhere on the Continent who worked on the integral calculus, but to no avail. Montmort personally found the situation in Paris to be shameful. In a letter to Taylor in 1718 he lamented: "It is disgraceful for France that we have no one capable of entering the lists with the English and the [Swiss] Germans."[42]

7.3. The revival of mathematics in Paris in the 1730s

7.3.1. Maupertuis and Johann I Bernoulli. Before I treat the questions of how and why mathematics was revived in Paris in the 1730s, I must first describe as clearly as possible the state of mathematics in Paris during the first three decades of the eighteenth century in order to do away with any doubts that might remain about whether what occurred in Paris in the 1730s was truly a rebirth of mathematics in that city or not.

Montmort probably made the understatement of his life when he told Johann I Bernoulli in 1709 that "the war has caused the sciences to languish a little in France...."[43] Nor did the situation rapidly improve in Paris where mathematics is concerned after the Treaty of Utrecht was signed in 1713. As I mentioned earlier, Montmort's death in 1719 followed by Varignon's death in 1722 ushered in a period that saw Johann I Bernoulli communicate very little, if at all, with the mathematicians who were left in the Paris Academy. Throughout most of the 1720s Bernoulli corresponded chiefly with Fontenelle and Mairan.

During these years Fontenelle evidently understood next to nothing about the issues that involved the use of the integral calculus to solve problems in mathematics which lay at the frontier of research in mathematics. Bernoulli had to advise him of a rather serious error that he made in his "Eloge de Monsieur Newton," published in the Paris Academy's *Histoire* for 1727. In reply to Leibniz's challenge of late 1715 to find the plane curves orthogonal to all of the members of a one-parameter family of coplanar hyperbolas that all have the same transverse axis (hence the same center) and the same vertices, Newton anonymously published in the *Philosophical Transactions* early in 1716 a solution to Leibniz's challenge as well as a sketch of what he called a method for solving problems in general of this nature. In his "Eloge der Monsieur Newton," Fontenelle praised Newton for having solved Leibniz's challenge. But Bernoulli told Fontenelle that Newton's so-called "general method" for solving such problems was totally unsubstantial. Bernoulli called the method that Newton had outlined "pure gibberish," "unintelligible jargon," and a "delusion."[44] In truth the method that Newton proposed for solving problems of orthogonal trajectories in general was entirely inadequate, which reveals that in 1727 Fontenelle still did not really have a very good grasp of the difficulties in solving problems of orthogonal trajectories in general. Now, in 1717 Bernoulli had explained to Montmort why Newton's general method was insufficient. Bernoulli called Montmort's attention to the fact that Newton never did try to use his general method to solve Bernoulli's challenge of March–April 1716 to find the plane curves orthogonal to the generalized brachistochrone cycloids.[45] This shows just how ignorant Fontenelle, the Paris Academy's Secrétaire Perpétuel since 1697, remained in 1727 of things that Montmort, now deceased, had known a decade earlier but had promised the mathematicians who had participated in the challenges to keep to himself. If Fontenelle ever knew before he wrote his "Eloge de Monsieur Newton" that other British mathematicians had also easily solved Leibniz's challenge of late 1715, he apparently forgot that fact.

Long before he wrote Newton's obituary, Fontenelle had already misjudged the true value of the scientific work done by a certain individual in the Paris Academy who made use of the integral calculus in his work, but this time Fontenelle did so by underestimating the importance of the work in question instead of overestimating its worth. In 1715, Saulmon, an unassuming member of the early-eighteenth-century Paris Academy, gave a remarkable demonstration, using the integral calculus, of the form that the free surface of a rotating cylindrical vortex has.[46] The Malebranchist Guisnée deemed that Saulmon's *mémoire* "was on a par with" ("allait de pair avec") the articles published by Leibniz and the Bernoullis.[47] Yet Saulmon disappeared inconspicuously from the scene. His death did not even occasion the customary "éloge" written by Fontenelle which was normally published in the Paris Academy's *Histoire*. Experts outside France, however, did not neglect Saulmon's work, although it took more than twenty years for the significance of the work to be recognized. Daniel Bernoulli duly acknowledged the merit of Saulman's experimental studies of cyclindrical vortices, which were published in the Paris Academy's *Mémoires* for 1716, in his *Hydrodynamica* (1738), some fourteen years after Saulmon died.[48]

Fontenelle apparently admitted himself that he had made mistakes in some of his articles published in the Academy's *Histoire*.[49] Years later Lagrange called Fontenelle's articles on mathematics and mechanics which were published in the Academy's *Histoire* "unintelligible gibberish." Lagrange added that in "wanting and trying to put things within reach of the layman, he [Fontenelle] often wrote articles in an obscure style which *savants* find incomprehensible."[50]

Fontenelle's *Eléments de la géométrie de l'infini*, published in 1727, which was Fontenelle's one serious attempt to do mathematics, was a book that some mathematicians "doubted was the work of a mathematician."[51]

Now let us consider Mairan. We have already seen in Chapter 2 how badly Mairan botched his Paris Academy *mémoire* of 1720. In Johann I Bernoulli's opinion, Mairan epitomized the complacence of the majority of French mathematicians to "scratch the surface" ("effleurer la superficie") of problems and questions in mathematics and mechanics, as evidenced by Mairan's disagreements with Bernoulli about the proper measure of the force of a body's motion.[52] These disagreements reached their peak when the *vis–viva* controversy intensified during the 1720s. Bernoulli's influence in Paris probably reached its low point when Bernoulli failed to win either of the Paris Academy's contests of 1724 or 1726, on the motions of hard bodies and elastic bodies, respectively, after collision. I suggest that it is not surprising that these defeats occurred within two years of the death of Bernoulli's last correspondent among the mathematicians in the Paris Academy, Pierre Varignon, who died in 1722.

What prompted Bernoulli to compete in the Paris Academy's annual contests, which were inaugurated in 1720 and were only open to people who did not belong to the Academy? Bernoulli simply could not believe that the Academy awarded the prize in its first contest to as mediocre a mathematician as Jean-Pierre de Crousaz of Lausanne, Switzerland.[53] In 1721 Crousaz published his *Commentaire sur l'analise des infiniments petits* (Paris: Montalant, 1721), in which he showed that he completely misunderstood the nature of extreme values and of their expression in the differential calculus. Bernoulli damned the book by faint praise, and Saurin criticized it severely.[54] The Paris Academy's standards evidently did leave a lot to be desired!

In Chapter 5 I mentioned a dispute between Saurin and the Chevalier de Louville which occurred in 1722 and which the Academy did its best to try to hide. The origins of the disagreement go back to a paper that Saurin wrote in 1720 in which Saurin proved in a manner "much too long and confused" to suit Bernoulli, by employing what Bernoulli regarded as "complicated calculations," something that seemed obvious to Bernoulli, namely, that the isochronous path of a real pendulum traversing a resistant medium like air cannot be a cycloid. The controversy in 1722 involved what seemed to be a paradox that Antoine Parent had raised earlier. Huygens had demonstrated that the cycloid is the path of isochronous motion in a vacuum. But Galileo had shown earlier that the unresisted motions of bodies that fall to the lowest point on a circle held in a vertical position, where the bodies descend along the chords that join the points on the circle to the lowest point on the circle, take place in equal times. Hence arose the question: In a vacuum should an infinitesimal portion of the path of a simple pendulum which begins at the lowest point on the pendulum's circular path be approximated by an infinitesimal arc of a cycloid or by an infinitesimal chord that subtends an infinitesimal arc of the pendulum's circular path? The findings of Galileo and Huygens seemed at first sight to conflict. Bernoulli noted that Saurin and Louville had both found Parent's error and had successfully refuted Parent's objections to Huygens's theory of the pendulum. But Bernoulli also mentioned that this had already been done years earlier. Moreover, while Saurin and Louville both correctly showed that the cycloid is the isochronous curve in a vacuum, whether the paths that are traversed along the cycloid have finite or infinitesimal lengths, Bernoulli found their demonstrations to be "very long," when in fact there were several "very short" ways to prove the same thing. From this example it seems clear that in Bernoulli's opinion, mathematicians in Paris in the 1720s handled the infinitesimal calculus clumsily, awkwardly, and cumbersomely and perhaps sometimes even misapplied it when using it as a tool to solve mathematical problems.[55]

In the late 1720s Bernoulli's fortune in Paris began to undergo a change for the better, and the pace of research in the infinitesimal calculus began to quicken and its quality began to improve there, largely as a result of the interest, zeal, and personal ambitions of Maupertuis. A one-time musketeer and soldier of fortune, Maupertuis had learned all that he could from Nicole, who had been his most important teacher of mathematics until then,[56] and from the other mathematicians in the Paris Academy. Unsatisfied with the way that these mathematicians utilized the infinitesimal calculus, Maupertuis went to Basel in 1729 in search of Johann I Bernoulli, who was left without peer in mathematics after Newton died in 1727. Bernoulli remained the only pioneer of the infinitesimal calculus who was still alive. To gain access to the notoriously cantankerous Swiss mathematician, Maupertuis graciously put up with having to matriculate at the University of Basel, which was a step that he needed to take in order to get to see Bernoulli. Maupertuis's action was unprecedented for a full-fledged Paris Academician. Most of the other members of the Academy no doubt would have shuddered at the very thought of doing such a thing. They would have regarded Maupertuis's way of proceeding as beneath their dignity. Urbane, cosmopolitan, and witty, Maupertuis captivated his Swiss host. No French mathematician, past or future, ever devoted himself more to Bernoulli than Maupertuis did. Maupertuis visited Bernoulli at his home in Basel at least twice. A difference of thirty years in their ages undoubtedly helped them establish their good relations. Bernoulli found Maupertuis to be so utterly fascinating and so thoroughly delightful that he confessed to Mairan that he "was charmed by his [Maupertuis's] conversation" and furthermore that he "made new discoveries in mathematics" which perhaps he "would never have done without Maupertuis's presence."[57] The union of Europe's leading mathematician and a French mathematician with a rather modest aptitude for mathematics paved the way for the subsequent developments in mathematics in Paris.[58] By the late 1730s what we now recognize, based on statistics, to be the highest level of interest in mathematics at the Paris Academy ever attained during the ancien régime was reached.[59] Thanks to Maupertius in large part and to Bernoulli through him, the integral calculus in particular advanced as it never had before in Paris. I have discussed these developments elsewhere,[60] but it is worth repeating and further elaborating some of them as they relate to the story told in these pages.

Although no great mathematician himself, Maupertuis did have charisma, in addition to the audacity that I mentioned in Chapters 3 and 5, which attracted younger mathematicians, even those who were more capable than he. Maupertuis was not afraid to take command. (He did, after all, become president of the Berlin Academy in 1746.) He possessed the attributes of a leader who inspired courage in others to break away from the Paris Academy's routines. Thus he, instead of someone else his age, like Bouguer for instance, organized and took charge of the campaign to promote Newtonianism at the Paris Academy. Maupertuis knew how to spur younger Academicians to do mathematics and once begun to persevere at doing mathematics, which was an ability that mathematicians earlier in the century, like Varignon, totally lacked. Maupertuis evidently knew how to scheme and manipulate, as we saw in Chapter 3. If he acted in part in his own self-interest, which seems likely, the fact remains that French people as a whole largely benefited from his behavior nevertheless, as we saw in Chapter 5. He had faults and qualities in amounts that were about equal.

Maupertuis was probably no better a mathematician than Varignon had been. In fact, Varignon may have been a better mathematician than Maupertuis. But Varignon had not known how to survey problems and to condense them and to state them clearly. Maupertuis, as we shall observe shortly, did. As a result Maupertuis gave other mathematicians in Paris, especially those more talented than he, a sense of direction and problems to focus their attention on. It was just such a person who could see the forest for the trees which the mathematics community in Paris did not have at the beginning of the eighteenth century. Varignon was long winded when he wrote. He always produced prolix, excessively detailed Paris Academy *mémoires* in which he invariably talked about nothing but extremely technical matters, which included much formalism without depth and which often involved generalization for its own sake.[61] Using the differential calculus, Varignon did show during the first decade of the eighteenth century how to treat problems in the motions of orbiting bodies which are governed by central forces, a part of mechanics which Newton had originated in the *Principia*. (On the other hand, Varignon never showed facility in doing integral calculus.[62]) However, Varignon either did not have the capacity or else he did not have the interest to stand back from the jumble of technical details involved in the way he dealt with a particular problem in mathematics or mechanics in order to clarify the larger, underlying issues so as to make them known to a general audience.[63] Despite the apparently genuine aptitude for mathematics which René-Antoine Ferchault de Réaumur showed in his earliest Paris Academy *mémoires*, written toward the end of the first decade of the eighteenth century, in which he attacked problems in the new infinitesimal calculus, Réaumur soon abandoned mathematics and switched to the life sciences.[64] Réaumur was originally Varignon's assistant in the Paris Academy, but Varignon could not hold onto Réaumur and keep him from giving up mathematics. Once again Varignon should be contrasted with Maupertuis. Maupertuis had a good eye for worthwhile mathematical problems. He knew how to pick problems in mathematics and describe the central aspects of them in a way that would attract others, not turn them away. He was able to describe the essential characteristics of mathematical problems succinctly in simple terms, without having to introduce a welter of technical detail which would only have served to obscure the primary issues involved. We shall see examples of this shortly as well as later in this chapter. As I just mentioned, it was not at all evident that the leading mathematicians in the Paris Academy earlier in the century, like Varignon, had the same capability or desire to publicize mathematics in order to promote it.

(In fact, Maupertuis himself showed an early interest in the life sciences, which grew in the 1740s to such a degree that he is perhaps best known today for his work that treats these sciences. In one of the first papers that he read to the Paris Academy, which was published in the Academy's *Mémoires* for 1727, Maupertuis studied salamanders. In another, published in the Academy's *Mémoires* for 1731, he treated scorpions. In other words, just because an Academician was interested in the life sciences, the Academy did not necessarily discourage him from liking and doing mathematics and vice versa.)

A second factor supported the revival of mathematics in Paris which began when Maupertuis met Johann I Bernoulli in 1729. The St. Petersburg Academy, founded in 1725, published the first two volumes of its *Commentarii*, for the years 1726 and 1727, in 1728 and 1729, respectively. These volumes included much high-level mathematics done by Bernoulli and by the members of the new academy (Johann's sons Nikolaus II and Daniel, Hermann, Euler, and Christian Goldbach, for example). In addition to the literature of mathematics published in *Acta Eruditorum*, some of which went all the way back to the end of the seventeenth century, which was ignored altogether in Paris once L'Hôpital, Montmort, and Varignon had died and which Bernoulli helped the Parisians to rediscover when he met Maupertuis, the first two volumes published by the new academy in St. Petersburg conveniently provided further material for Maupertuis and Bernoulli to discuss, and these two volumes helped kindle the common interests of the two mathematicians. The effect was to help spark in Paris Academicians a new interest in mathematics. Maupertuis said that he was particularly impressed by the first volume of the *Commentarii*, because this volume revealed that "the gentlemen in the Petersburg Academy work a lot on the integral calculus."[65]

In the first paper that he wrote since he began collaborating with Johann I Bernoulli, which he read before the Paris Academy late in 1730, Maupertuis took up a problem that Leibniz had first proposed in 1687: to find the trajectory of a falling body that approaches the horizon such that segments of the trajectory whose lengths are equal are traversed in equal times. At the end of the seventeenth century, Jakob Bernoulli, Johann I Bernoulli, and Varignon had treated the case where the body descends in a vacuum. Maupertuis examined the case where the body falls through a medium that resists the body's motion directly as the square of the body's speed. After deriving the equation of motion of a body that falls through such a medium, Maupertuis arrived at a first integral of this equation by applying a method that involved introducing an integrating factor and integrating by parts.[66]

Around the same time, Maupertuis discovered that he could use the same technique to separate the variables in first-order ordinary "differential" equations of first degree in two finite variables which had certain forms. Maupertuis made his procedure the basis of a paper on the separation of variables which he read before the Paris Academy in the spring of 1731. The printed version of the paper appears in the Paris Academy's *Mémoires* for 1731.[67] Johann I Bernoulli, who excelled at separating variables in first-order ordinary "differential" equations of first degree in two finite variables, called Maupertuis's method "ingenious" and "excellent."[68] As I suggested earlier in this chapter, however, Bernoulli was more inclined to praise inferior mathematicians who were his friends than mathematicians whom he considered to be his rivals. In the version of the paper that he read to the Paris Academy, Maupertuis called attention to the fact, which he stressed, that his method did "cover a type of equation that contains an infinity of special cases."[69] These special cases included first-order ordinary "differential" equations of first degree in two finite variables whose variables had already been separated by other mathematicians using one method or another. In truth, the paper may be one of the best, if not the best, mathematical papers that Maupertuis ever wrote. It is the paper that I mentioned in Chapter 5 as the one in which Maupertuis originally made disparaging remarks about John Craige, which the Academy obliged Maupertuis to eliminate before the paper could be published. Maupertuis had simply said that he found his own method of separating variables to be better and less complicated than the one that appears in Craige's *De Calculo fluentium*. The Academy did permit Maupertuis to publish his discovery that Craige had erred. Namely, Craige believed that he had separated the variables in first-order ordinary "differential" equations of first degree of a certain kind in two finite variables when in reality he had not. Maupertuis noted that neither his method nor Craige's method would separate the variables in equations of the particular type in question.

In fact, Maupertuis had not really discovered anything new. Thanks to his correspondences with Johann I Bernoulli, Nikolaus I Bernoulli, and Brook Taylor, Montmort learned of Craige's error in 1718, the very year that Craige's *De Calculo fluentium* came out. Moreover, the Bernoullis also told Montmort at that time that they believed that it was impossible in general to separate the variables in equations of the type whose variables Craige mistakenly thought that he had separated. Thus Maupertuis, like Fontenelle, remained ignorant in the 1720s of matters concerning the integral calculus which Montmort had been well aware of years before as a result of having corresponded with the two Bernoullis and with Taylor during the 1710s. Late in July of 1731 Maupertuis asked Johann I Bernoulli by letter if he knew about Craige's attempt in *De Calculo fluentium* to separate the variables in equations of the particular kind in question. Maupertuis told Bernoulli in the same letter that his own method of separating variables would not separate the variables in equations of the type at issue and that he thought that Craige's method would not separate the variables in equations of this kind, either, contrary to what Craige thought that he had achieved. Bernoulli replied in mid-August of 1731 that Craige had indeed made a serious error; Bernoulli pointed out Craige's mistake to Maupertuis; Bernoulli said that he doubted that the variables in equations of the type in question could in general be separated by any means; and Bernoulli added that he had informed Montmort about all of this in a letter that he wrote to Montmort shortly before Montmort died.[70]

Paradoxically the most important part of Maupertuis's paper on the separation of variables was never published. What Maupertuis really wanted to emphasize in that paper appears in the introduction to the paper, which for one reason or another was omitted from the version of the paper which was published in the Paris Academy's *Mémoires* for 1731. The version of the paper which he read to the Academy, which was copied into the Academy's unpublished proceedings,[71] is the version that is significant. In its introduction Maupertuis stressed that the separation of variables was a severely limited way of trying to handle first-order ordinary "differential" equations of first degree in two finite variables. Maupertuis regretted that the separation of variables, which Leibniz and Johann I Bernoulli had pioneered at the end of the seventeenth century, had since become the principal, not to say the sole objective of the integral calculus, and Maupertuis protested. The goal of this customary, standard approach to treating first-order ordinary "differential" equations of first degree in two finite variables, say, x and y, is to separate the equation into two parts

$$f(y)\,dy = g(x)\,dx, \tag{7.3.1.1}$$

where one part is on one side of the equality sign and the other part is on the other side of the equality sign and where each of the two parts consists of terms made up of expressions that include only one of the two finite variables x and y, together with the differential of that particular variable, by making changes of variables if necessary, by multiplying by integrating factors if need be, and so forth. Such a separation produces an equation that can be expressed as an equality of two integrals

$$\int_{y_0}^{y} f(y)\,dy = \int_{x_0}^{x} g(x)\,dx + C, \tag{7.3.1.2}$$

where each of the two integrals on the left-hand side and right-hand side of equation (7.3.1.2), respectively, is an integral in just one of the two finite variables

x and y, in which case each of the two integrals can be treated individually, and where C is a constant of integration. The problem of dealing with the first-order ordinary "differential" equation of first degree in two finite variables x and y is thereby reduced to the problem of integrating two separate quantities $f(y)$ and $g(x)$ each of which is expressed in terms of a single finite variable. In the event that both expressions $f(y)$ and $g(x)$ in (7.3.1.1) can be integrated in finite terms, the first-order ordinary "differential" equation of first degree in two finite variables x and y is transformed into a finite equation

$$F(y) = G(x) + C, \qquad (7.3.1.3)$$

which can also be written instead as

$$H(x, y, C) = 0. \qquad (7.3.1.4)$$

Equations (7.3.1.3) or (7.3.1.4) are solutions of the first-order ordinary "differential" equation of first degree in two finite variables x and y.

[A finite equation like (7.3.1.3) or (7.3.1.4) which looks like it expresses a one-parameter family of transcendental plane curves may in fact actually represent a one-parameter family of algebraic plane curves. For example, the first-order ordinary "differential" equation of first degree

$$\frac{p}{y} dy = \frac{m}{x} dx \qquad (2.1.8)$$

whose two finite variables x and y are separated, where $m > 0$ and $p > 0$, can be integrated in finite terms. An integral of equation (2.1.8) is

$$p \ln y = m \ln x + M, \qquad (7.3.1.5)$$

where M is another constant of integration. Now, $\ln y$ and $\ln x$ individually are transcendental expressions. But equation (7.3.1.5) can be simplified as follows. If we let $M \equiv \ln n$, where $n > 0$, then equation (7.3.1.5) becomes

$$\ln y^p = \ln x^m + \ln n = \ln (nx^m). \qquad (7.3.1.6)$$

But it follows from equation (7.3.1.6) that

$$y^p = nx^m, \qquad (2.1.9)$$

and equation (2.1.9) expresses a one-parameter family of algebraic curves (specifically, a one-parameter family of parabolas, where the kind of parabolas in the family depends upon the particular values of $m > 0$ and $p > 0$) whose parameter is n, where $n > 0$. If the finite equations (7.3.1.3) or (7.3.1.4) can be solved for one of the finite variables x and y in terms of the other, say y in terms of x, then the solutions of the first-order ordinary "differential" equation of first degree in two finite variables x and y can be written as

$$y = T(x, K) \qquad (7.3.1.7)$$

where K is an alternate constant of integration. For example, when $x > 0$, equation (2.1.9) can always be solved for y in terms of x when $n > 0$, whatever the values of $m > 0$ and $p > 0$. Namely,

$$y = n^{\frac{1}{p}} x^{\frac{m}{p}}. \tag{7.3.1.8}$$

Hence

$$T(x, K) \equiv K x^{\frac{m}{p}} \tag{7.3.1.9}$$

in (7.3.1.7) in this particular case, where $K \equiv n^{\frac{1}{p}}$.]

In case one or both of the expressions $f(y)$ and $g(x)$ in (7.3.1.1) cannot be integrated in finite terms, the problem of dealing with the first-order ordinary "differential" equation of first degree in two finite variables x and y has at least been reduced "to quadratures."

Whether both of the expressions $f(y)$ and $g(x)$ in (7.3.1.1) are integrable in finite terms or whether one or both of these expressions are not integrable in finite terms, in either case one kind of problem in the integral calculus (trying to treat first-order ordinary "differential" equations of first degree in two finite variables) is reduced to a problem in the integral calculus of a more elementary type.

To reduce an expression in a single finite variable to quadratures essentially meant to reduce the expression to an integral or to a sum of integrals that could not be (or was/were not known if it/they could be) represented as finite sums of elementary functions. In 1731 the known elementary functions were algebraic expressions as well as such transcendental expressions as circular functions (which soon came to mean the trigonometric functions), logarithms (equivalently, the areas under segments of an equilateral hyperbola) and logarithmic functions, and exponentials (which are the inverses of logarithms) and exponential functions. It was not known and could not be shown at this time that the rectifications (meaning the arc lengths) of segments of the ellipse and the hyperbola were themselves distinct elementary transcendents that could not be reduced to finite sums of expressions like those just mentioned. Some seventeenth-century mathematicians, including the great Scottish mathematician James Gregory, suspected that these rectifications were irreducible and treated them as such. That is, these mathematicians tried to express other integrals in terms of these rectifications.

Maupertuis emphasized how much current methods in the integral calculus left to be desired. If a first-order ordinary "differential" equation of first degree in two finite variables was not one that by "happy accident" could be integrated in finite terms and hence solved, or not one whose variables could be separated so that the "differential" equation could be transformed into an equality of two expressions that could at least be reduced to quadratures, even if the "differential" equation could not be integrated in finite terms, then one typically had no other choice but to abandon the equation entirely or else search for solutions of the equation which have the form of infinite series. Maupertuis acknowledged that

Newton's method of infinite series was the only "absolutely general method that the integral calculus has," but that "the solutions gotten by using it did not have the elegance of the solutions found by integration or by quadratures." Consequently Maupertuis believed that infinite series should only be employed "as the last resort in the hopeless cases."[72]

In fact Maupertuis told Johann I Bernoulli that he found infinite series "very tricky" and that his mind could not get accustomed to them.[73] Here Maupertuis revealed further his limitations as a mathematician which we saw in Chapter 5, where we found him unable to handle all but the simplest first-order approximations in connection with the problem of the earth's shape.

In the introduction of the version of his paper on the separation of variables which he read before the Paris Academy in the spring of 1731, Maupertuis gave an example that illustrated the difficulties that troubled him. Abraham DeMoivre, another French Huguenot like Desaguliers, who sought and took refuge in England, had proposed a problem to Maupertuis while Maupertuis visited England in 1728. The problem gave rise to a first-order ordinary "differential" equation of first degree in two finite variables where the two variables and their differentials in the terms that made up the equation were mixed together to such a degree that "one could spend a lot of time, or perhaps end up wasting a lot of time, if one tried to separate the variables" in the equation, Maupertuis observed. Maupertuis added that "one encountered too many examples of the same kind," which suggested to him "how useful it would be to perfect the integral calculus" and also convinced him "how far one still had to go to achieve this." What the integral calculus lacked according to Maupertuis was not methods for integrating *every* expression that included differentials. Maupertuis was sure that expressions that included differentials and that could not be integrated did exist. The problem instead was that one was satisfied if he succeeded in separating the variables in a first-order ordinary "differential" equation of first degree in two finite variables, for in that case it was easy to see whether the "differential" equation could be integrated in finite terms or, if not, at least be reduced to quadratures. As a result, Maupertuis affirmed, "the separation of variables is today the principal object of the integral calculus," but he declared that "one can scarcely hope to turn the separation of variables into an absolutely general method" for treating first-order ordinary "differential" equations of first degree in two finite variables.

Nor could one hope, Maupertuis believed, to convert what he judged to be "the best integration [or reduction to quadratures] yet" into an absolutely general method for dealing with this problem. Here he referred to Johann I Bernoulli's method for handling first-order ordinary "differential" equations of first degree in two finite variables, published in the first volume of the St. Petersburg *Commentarii*.[74] Maupertuis noted that the method applied "to all [first-order ordinary] 'differential' equations [of first degree in two finite variables] in which the sum of the exponents of the [two finite] variables in each term of the equation is the same for all terms [of the equation]." But he concluded that as a result of this condition that had to hold in order to be able to make use of the method, "one sees...that this fine method is still very limited."[75]

In the introduction to his paper of 1731 on the separation of variables, Maupertuis rightly minimized and played down the importance of what he had accomplished in the technical part of his paper, which, as I say, is the only part of his paper which was published in the Paris Academy's *Mémoires* for 1731. Namely, he duly emphasized that the use of the separation of variables would never suffice to deal with first-order ordinary "differential" equations of first degree in two finite variables. He declared that the technique of separating variables could never be used in general to integrate in finite terms and therefore solve, or at least to reduce to quadratures, first-order ordinary "differential" equations of first degree in two finite variables. Maupertuis eloquently stressed the importance of not limiting research in the integral calculus to the usual, routine, established objective of trying to separate the variables in first-order ordinary "differential" equations of first degree in two finite variables, even though that aim served as the very subject of the part of his paper which was published.

In some sense the controversy about the use of infinite series in the integral calculus was the remains or a vestige of the argument between Newton and Leibniz about who discovered the infinitesimal calculus. One issue in the dispute had been the part that infinite series played in that discovery. According to Leibniz and Johann I Bernoulli, Newton's achievements were "simply" in the realm of infinite series. Leibniz and Bernoulli did not consider Newton's work on infinite series to have been a part of the discovery of the infinitesimal calculus. They contended that the progress that Newton made in the study of infinite series had nothing to do with the development of the infinitesimal calculus. At the same time, Newton maintained that Leibniz's method of infinitesimals was not "universal" without infinite series. As far as Newton was concerned, his work on infinite series and his discovery of the infinitesimal calculus were intrinsically linked.

Leibniz dreamed of reducing the integral calculus to algorithms like those he had found for the differential calculus. His desire and hope to be able to do this was part of his grand scheme to produce a "universal characteristic," which he never did achieve.[76] In Leibniz's opinion the *method* for arriving at results mattered more than the *results* themselves. Leibniz had a much more abstract view of the infinitesimal calculus than Newton did. Newton had a standpoint that was more circumspect and concrete than Leibniz's. Newton hit upon the infinitesimal calculus while solving problems of interpolation and quadrature by using infinite series. But infinite series did not appeal to Leibniz, because their use could not be reduced to algorithms. Infinite series appeared to him to be devices to be utilized ad hoc in the integral calculus when all else failed. Leibniz believed that the true goal of the integral calculus was to try to reduce problems in the integral calculus to integrals of quantities expressed in terms of single finite variables which could be represented as finite sums of algebraic expressions, elementary transcendents like the circular functions, exponential functions, and logarithmic functions, and quadratures that mathematicians could not reduce further and for that reason led some mathematicians to believe that these quadratures very likely expressed other elementary transcendents.

Moreover, the bases for algebraic expressions, elementary transcendental expressions like the circular functions, exponential functions, and logarithmic functions, and the seemingly irreducible quadratures originally lay in the geometry of plane curves, just as the foundations of Leibniz's infinitesimal calculus itself did. In viewing infinite series as "inelegant," continental mathematicians may have reacted to the multitude of infinite series which British mathematicians produced during the latter part of the seventeenth century, as well as to the wave of uncritical ideas that accompanied the introduction of this abundance of infinite series. For one thing, only numerical considerations counted. Toward the end of the seventeenth century British mathematicians no longer sought geometrical foundations for infinite series. But infinite series had always had such foundations earlier in the seventeenth century.[77]

In fact, however, the very same thing happened to the continental infinitesimal calculus during the eighteenth century. As the years wore on, more and more transcendental quantities were formally expressed (like e^x and log x), and the continental calculus also became more and more a matter of juggling formulas involving algebraic and transcendental expressions, while the geometry that underlay the expressions was forgotten about and disregarded.[78] Thus the algebraic and transcendental expressions employed by mathematicians on the Continent also gradually took on a life of their own which was independent of geometry.

Continental mathematicians had other reasons to be dissatisfied with integrations that involved the use of infinite series. Infinite series cannot be used to help classify the plane curves that are solutions of first-order ordinary "differential" equations of first degree in two finite variables. In particular, in the event that such an equation can be integrated in finite terms, infinite series cannot be used to show how to decompose the integral into finite sums of algebraic expressions, elementary transcendents like the circular functions, exponential functions, and logarithmic functions. Or in case such an equation cannot be integrated in finite terms but can at least be reduced to quadratures, infinite series cannot be used to show how to express the equation in terms of finite sums of the expressions just mentioned and the quadratures that mathematicians could not simplify further which caused some mathematicians to think that these quadratures probably expressed other elementary transcendents. But classifying curves had become a problem of considerable interest to European mathematicians by this time.

For example, unless you recognize that the expression on the right-hand side of the equality

$$y = 1 + x + x^2 + x^3 + x^4 + \cdots + x^n + \cdots \tag{7.3.1.10}$$

has the sum $1/(1-x)$ when $-1 < x < 1$, you would not know that equality (7.3.1.10) represents a segment of an algebraic plane curve $y = \frac{1}{1-x}$, or $y - xy - 1 = 0$, when $-1 < x < 1$. Similarly, unless you realize that the expression on the right-hand side of the equality

$$y = 1 + \frac{x}{1!} + \frac{x^2}{2!} + \frac{x^3}{3!} + \frac{x^4}{4!} + \cdots + \frac{x^n}{n!} + \cdots \tag{7.3.1.11}$$

has the sum e^x for all finite values of x, you would not know that equality (7.3.1.11) represents the transcendental plane curve $y = e^x$. And if you do not recognize that the infinite series on the right-hand side of (7.3.1.11) has the sum e^x for all finite values of x, it is consequently unlikely that you would discover using infinite series that the integral of equation

$$\frac{p}{y} dy = \frac{m}{x} dx \tag{2.1.8}$$

is the one-parameter family of algebraic plane curves

$$y^p = nx^m \tag{2.1.9}$$

(specifically, a one-parameter family of parabolas, where the kind of parabolas in the family depends upon the particular values of $m > 0$ and $p > 0$) whose parameter is n, where $n > 0$. Furthermore, more than one series in x can represent an expression $\phi(x)$ in x. If one series is infinite, another series could terminate (which means that all terms in the series are zero after a certain term), in which case the equality $y = \phi(x)$ must represent an algebraic plane curve. Perhaps the most astonishing example of all is the following one. It seems highly unlikely that integration by means of infinite series would have enabled the Italian Count Giulio Carlo de' Toschi di Fagnano to discover in 1717 that even though the integral of

$$\frac{dx}{\sqrt{1 - x^4}}, \tag{7.3.1.12}$$

which expresses the arc length of a segment of a lemniscate, seems to be irreducible [indeed, the integral of (7.3.1.12) is a simple elliptic integral, which defines a transcendental function], in which case the first-order ordinary "differential" equation of first degree

$$\frac{dy}{\sqrt{1 - y^4}} = \frac{dx}{\sqrt{1 - x^4}} \tag{7.3.1.13}$$

in the two finite variables x and y appears at best to be reducible to quadratures

$$\int_{y_0}^{y} \frac{dy}{\sqrt{1-y^4}} = \int_{x_0}^{x} \frac{dx}{\sqrt{1-x^4}} + C, \qquad (7.3.1.14)$$

nevertheless equation (7.3.1.13) has an integral that is the algebraic plane curve

$$y = \sqrt{\frac{1-x^2}{1+x^2}} \qquad (7.3.1.15)$$

and

$$y = -\sqrt{\frac{1-x^2}{1+x^2}} \qquad (7.3.1.15')$$

[which may be expressed as

$$x^2 + y^2 + x^2 y^2 = 1 \qquad (7.3.1.16)]$$

when $-1 \leqslant x \leqslant 1$ and $-1 \leqslant y \leqslant 1$.

Today we know that Maupertuis's "hopeless" cases far outnumber the ones that have "elegant" solutions in closed form, so that the "last resort" turns out to be the rule instead of the exception. But it took a hundred years of further development of mathematical analysis to realize this. During the nineteenth century it was proven that not all first-order ordinary "differential" equations of first degree in two finite variables can be reduced to quadratures, much less be integrated in finite terms. (One such equation, the Riccati equation, has an important part in Chapter 9.) These equations do have solutions when they cannot be reduced to quadratures, but these solutions are distinct, irreducible transcendents that cannot be expressed as finite sums of, say, the elementary functions that were known in Maupertuis's day and the rectifications of the ellipse and the hyperbola. The solutions cannot be written in closed form; they can only be expressed as infinite series. But Maupertuis's viewpoint was the same as that of many of the continental mathematicians of his time who were better mathematicians than he was. This viewpoint included Maupertuis's opinion that just because infinite series were convenient, available devices, this did not mean that they should be misused and abused in trying to handle first-order ordinary "differential" equations of first degree in two finite variables.

In a paper on tractrices which he read before the Paris Academy in March of 1736, but which was never published in the Academy's *Mémoires*, Maupertuis returned to the general problem that he had raised in his paper of 1731 on the separation of variables. He commented further on Johann I Bernoulli's "method for integrating [in finite terms or reducing to quadratures first-order ordinary] "differential" equations [of first degree in two finite variables] without separating the variables," which was published in volume 1 of the St. Petersburg Academy's *Commentarii*. While Maupertuis declared that there was "nothing finer nor more ingenious than this method," he added that it could only be applied "to [first-order ordinary] "differential" equations [of first degree in two finite variables] which have a certain form, and this form is that in each term [of the equation] the variables have the same number of dimensions." He meant by this that if in a term of the equation the quantities that are not differentials in that term of the equation are multiplied together, where each of these quantities is one of the two finite variables raised to some power, and if this is done for each term of the equation, then the sum of the powers in each term of the equation is the same for all terms of the equation. Maupertuis had already mentioned this fact in his paper of 1731 on the separation of variables. But now he pointed out further that in this case the terms in such an equation are homogeneous in their variables, in which case the variables in the equation can always be separated. Thus he concluded that Bernoulli's "method only applies to [first-order ordinary] "differential" equations [of first degree in two finite variables] whose variables can be separated [in any case]."[79] (Later in this chapter we shall see how the variables in homogeneous first-order ordinary "differential" equations of first degree in two finite variables can always be separated.)

In his paper of 1736 on tractrices Maupertuis did not really add anything new that concerned the general problem of how to handle first-order ordinary "differential" equations of first degree in two finite variables. He just more or less repeated what he had already said about this problem in the introduction to his paper of 1731 on the separation of variables. This suggests that Parisian mathematicians had made little, if any, progress in treating the problem that he had raised in 1731, if they had made any effort at all in the five-year interim to tackle the problem, and Maupertuis thought that the time had come for Parisian mathematicians to begin to deal seriously with the problem. In his Paris Academy *mémoire* of 1734, for example, which I discussed in Chapter 4, Bouguer had defined the variables in such a way that these variables in the first-order ordinary "differential" equation of first degree which expressed his application of the principle of the plumb line in terms of these particular variables could be separated. Thus Bouguer had not come to terms with Maupertuis's problem. In his paper of 1736 on tractrices Maupertuis called his audience's attention to the fact that there still were many people who found infinite series to be adequate because they were good enough for all practical purposes. Maupertuis thought that such an attitude might have reduced the incentive of mathematicians in the Paris Academy to take up his challenge, and in his paper of 1736 on tractrices he condemned people for having such an attitude toward infinite series. Maupertuis urged his colleagues in the Paris Academy who were mathematicians to take up the problem that he had raised five years earlier.

As we shall see later in this chapter, Maupertuis's appeals did not continue to be disregarded or to go unheeded. Eventually they were answered.

7.3.2. Louis-Bertrand Castel. I mentioned in Chapter 5 and earlier in this chapter that as science editor of the *Journal de Trévoux*, which was published by the

Jesuits, the Jesuit Louis-Bertrand Castel incurred the hatred of quite a few Paris Academy mathematicians during the 1720s because he wrote reviews of their work in the Jesuit journal which were sometimes unenthusiastic or indifferent.

From 1720 until his death in 1757, Castel also belonged to the faculty of the Jesuit Collège de Louis-le-Grand in Paris. At this school he taught physics, mathematics, specialized courses in the infinitesimal calculus, and mechanics, which included statics and dynamics as well as the physics of light and sound, pyrotechny, clock-making, and civil and military architecture. In 1729 Castel was named professor of differential and integral calculus. Later he became professor of mathematics. For awhile he was also designated "executive of the chamber of physicists."[80]

Pedagogy interested Castel a great deal. He established a reputation as a good teacher very quickly. He did not do mathematical research. He was inclined to view institutions like the Paris Academy of Sciences as self-serving. He considered the Academy's members, whom he thought acted high and mighty, to be bookish, pedantic, and stubbornly opinionated.

Although he was probably one of the most outspoken French opponents of Newtonian science during the second quarter of the eighteenth century,[81] Castel was nevertheless a zealous supporter of British mathematics.[82] Early in his career Castel developed a tremendous admiration for Newton as a mathematician, even as he resisted and fought against Newton's natural philosophy. In 1720 someone loaned Castel a copy of Newton's *Principia* for two months. Feeling that this was not enough time to master the book, Castel resolved to copy it long-hand in order to be able to study the work at his leisure.[83]

In the 1720s Castel led a movement to humanize and popularize mathematics. He pioneered "mathematics without tears." He advocated a less formal, more intuitive approach to the teaching of mathematics.[84] Thus Castel waged a campaign for making mathematics less abstract and more concrete at least a decade before research mathematicians attempted to introduce exploratory devices into the teaching of mathematics in an effort to motivate the subject better.[85]

In 1728 Castel published his *Mathématique universelle, abrégée à l'usage et à la portée de tout le monde*, in which he defended Newton's method of infinite series as the basic tool of the integral calculus. Accordingly he maintained that the use of infinite series with undetermined coefficients "is even rather necessary, when it is a matter of solving an equation that includes various differences [that is, differentials], such as the equation

$$dx + ydx - dy = 0."^{86} \tag{7.3.2.1}$$

Continental mathematicians must have made merciless fun of such a claim applied to such a first-order ordinary "differential" equation of first degree in the two finite variables x and y, for the variables in this particular equation can easily be separated, and then the equation that results when the variables are separated can be integrated without further ado. Namely, equation (7.3.2.1) can be rewritten as

$$(1 + y)dx = dy, \tag{7.3.2.2}$$

which is a first-order ordinary "differential" equation of first degree in the two finite variables x and y whose variables can be readily separated. When the variables in equation (7.3.2.2) are separated, the equation

$$dx = \frac{1}{1 + y}dy \tag{7.3.2.3}$$

results. But equation (7.3.2.3) can be integrated when $-1 < y$, and its integral when $-1 < y$ is

$$x = \int_{y_0}^{y} \frac{1}{1 + y}dy + C = \ln(1 + y) + M = \ln(1 + y) + \ln n$$

$$= \ln(n(1 + y)), \tag{7.3.2.4}$$

where $C, M,$ and n are constants of integration and $n > 0$. When $-1 < y$ equation (7.3.2.4) can also be written instead as

$$n(1 + y) = e^x, \tag{7.3.2.5}$$

or as

$$y = \frac{e^x - n}{n}. \tag{7.3.2.6}$$

Castel's book was well received in England. Mathematicians in London found it "marvelous and extraordinary." As a result Castel was elected to the Royal Society in 1730. At the same time, Paris Academy mathematicians severely criticized the work.[87] Saurin ruthlessly and mockingly panned the book and ridiculed it unmercifully in his pamphlet entitled *Lettre... sur le Traité de mathématiques du P. C. et les extraits qu'il a faits dans les Journaux de Trévoux... des mémoires de l'Académie des Sciences,* which was published in 1730. Castel made some errors in his book. Saurin found them, and he did not fail to call the reader's attention to them. Saurin also used his booklet as an occasion to strike back at Castel for Castel's having, in Saurin's opinion, senselessly "attacked" Mairan and Nicole, among others, in his reviews in the *Journal de Trévoux.*[88]

But Castel very often spoke the truth about the work of Paris Academy mathematicians. We saw in Chapter 2 what Mairan's mathematics was like. Earlier in this chapter we read Castel's remarks on Nicole's solution of 1725 to the problem of finding the plane curves orthogonal to the generalized brachistochrone cycloids, and what Castel said about Nicole's solution was correct. We also recall that Castel commented on the second volume of the *Analyse démontrée,* published in two volumes in 1708 by Charles-René Reyneau, an Oratorian mathematician, a follower of Malebranche, and a member of the Paris Academy. Castel said that the reader of volume 2 of Reyneau's textbook would have to confront many obstacles and stumbling blocks ("embarras"), which was also true. In the second volume of his work Reyneau discussed the integral calculus, including "the inverse method of tangents" (in other words, solving ordinary "differential" equations). As I mentioned previously in this chapter, Castel thought that it was better to read instead various writings by Wallis, Newton, Bernoulli, L'Hôpital, Edmund Stone, and Gregory of Saint Vincent.[89] In other words, Castel evidently found that Reyneau's textbook did not help make the works of some of Europe's better known creative mathematicians easier to understand! In addition to the personal glory involved, Castel very likely viewed his election to the Royal Society as a way of getting even with certain members of the Paris Academy who had attacked him.[90]

In 1730 Edmund Stone published *The Method of Fluxions, Both Direct and Inverse*, a self-styled sequel, written by an Englishman, to L'Hôpital's famous textbook on the differential calculus entitled *L'Analyse des infiniments petits* (1696). In his review of Stone's book which was published in the *Journal de Trévoux* for January 1732, Castel contended that Newton had demonstrated that, contrary to Leibniz's way of thinking, solving problems in the integral calculus required the use of infinite series. That is, Castel maintained that problems in the integral calculus could not be solved in general without the summation of infinite series.[91] A couple of years later Castel wrote a long introduction to Rondet's French translation of Stone's book which was published in 1735.[92] In his introduction to the translation Castel basically attempted to trace historically the origins of the differential and integral calculus to the applications of infinite series by the Flemish mathematician Gregory of Saint Vincent. The writers of the review of the French translation of Stone's book which appeared in the *Journal des Savants* for June 1735 did not really review the translation. Except for one short, concluding paragraph, the review was a shrewd attack on Castel's introduction to the translation instead.[93] As I mentioned in Chapter 3, the *Journal des Savants* was an organ of the Paris Academy.

Late in March of 1736 Mairan sent Johann I Bernoulli a copy of Rondet's French translation of Stone's book.[94] Bernoulli harshly criticized both the book and its introduction. He dismissed them as the works of semilearned individuals. Stone made mistakes in his textbook, but the fact that both Stone and Castel stressed integration by means of infinite series is what really exasperated Bernoulli the most and made him angry. Bernoulli considered such an emphasis upon the use of infinite series in order to integrate to be utter nonsense. Just as Castel had done in his *Mathématique universelle*, Stone and Castel took a device that *sometimes* had to be used in order to integrate and made it appear to be a device that *always* had to be used in order to integrate. But Bernoulli discovered case after case where Stone employed infinite series in order to integrate, when in fact infinite series were not needed at all to solve the problems in these instances. Bernoulli even found signs that Stone mistook purely algebraic problems for problems in the differential and integral calculus.[95] Bernoulli in fact was more than just castigating Stone's book and Castel's introduction to the book; he was implicitly condemning the use of infinite series to integrate by the whole British school of mathematicians. Newton had been one of the first members of the school to do this.

Maupertuis did not mention Stone and Castel by name in the paper on tractrices which he read before the Paris Academy in March of 1736, in which he reproached people for regarding infinite series to be satisfactory because they were good enough for all practical purposes, but Maupertuis could very well have had Stone and Castel in mind at the time, since the French edition of Stone's book had only come out the year before.

By 1730 Bernoulli had already decided that Castel's knowledge of mathematics was "muddled, ill-digested, and purely historical." He judged Castel's *Mathématique universelle* to be suitable for "young scatterbrains" who were as "mentally deficient" as Castel.[96] Castel had angered Bernoulli by suggesting that there was something "contradictory" or "paradoxical" about one of the problems that Bernoulli discussed in his paper on isoperimetrical problems which was published in the Paris Academy's *Mémoires* for 1718 and which Castel reviewed in the *Journal de Trévoux* for June 1722. It had been known since the end of the seventeenth century that the cycloid was the brachistochrone, or the path of quickest descent, in a vacuum. That is, among all paths in a vertical plane which join the same two points in the plane, where the two points are neither situated on a horizontal line nor on a vertical line in the plane, the particular path along which a body descends the most rapidly in a vacuum from the higher point to the lower point is a segment of a cycloid. In his review of Bernoulli's Paris Academy *mémoire* Castel declared that in allegedly finding the path of quickest descent in a vacuum among all paths in a vertical plane *which have the same length* and which connect the same two points in the plane, where the two points are neither located on a horizontal line nor on a vertical line in the plane, Bernoulli implied that a segment of a cycloid was *not* the path of quickest descent in a vacuum among *all* paths in a vertical plane which join the two points in the plane.

Castel misunderstood what it meant to find the path of quickest descent in a vacuum among *all paths in a vertical plane which have the same, given length and which connect the same two points in the plane*, which is an isoperimetrical problem, as opposed to finding the path of quickest descent in a vacuum among *all paths in the vertical plane which join the two points in the plane* instead. He declared that "according to Monsieur Bernoulli, the cycloid is the curve of quickest descent from one point to another point, when in fact in the new problem [the isoperimetrical problem] it is still only a question of the quickest descent from one point to another point." How did Castel reach such a conclusion? He said that since a segment of a cycloid is the path of quickest descent in a vacuum among all segments of curves in a vertical plane which connect the same two points in the plane, where the two points are neither situated on a horizontal line nor on a vertical line in the plane, it followed that the segment of the cycloid is also necessarily the path of quickest descent in a vacuum among the members of a family of segments of curves which consists of some but not all segments of curves in a vertical plane which join the two points in the plane. As Castel expressed the matter, the condition that the segments of curves be isoperimetrical (in other words, that the segments of curves have the same, given length) and that the path of quickest descent among such segments of curves be sought "without changing anything in the nature of the problem, only appears at the very most to limit the problem and to render it less general." Consequently, according to Castel, "if the cycloid is the curve of quickest descent among curves of all kinds, that seems to be all the more reason that it should be the curve of quickest descent among curves of one kind." In other words, Castel interpreted the condition that the segments of curves be isoperimetrical as a restriction that simply reduced the number of segments of curves involved in the problem of finding the path of quickest descent in a vacuum among *all segments of curves in a vertical plane which connect the same two points in the plane*, where the two points are neither located on a horizontal line nor on a vertical line in the plane.

Now, the segment of the cycloid *would* indeed still be the path of quickest descent among all segments of curves which satisfy this condition, *provided* the segment of the cycloid *does* satisfy the condition. But Castel went wrong in analyzing the problem where this aspect of it is concerned. Why? Castel seems to have paid more attention to Fontenelle's summary of and commentary on Bernoulli's Paris Academy *mémoire*, which appears in the Paris Academy's *Histoire* for 1718 and parts of which Castel quoted in his review of Bernoulli's *mémoire*, than he did to Bernoulli's *mémoire* itself. This may explain in part why Castel got badly confused. (In the next section of this chapter we shall discover that, despite the hostility that members of the Paris Academy, particularly the mathematicians, showed to Castel, Fontenelle's writings in general greatly interested Castel, and we shall learn why they did so.) In his commentary Fontenelle stated that it seemed to be a "rather strange paradox" that a certain length could be chosen which is the length of an infinite number of paths in a vertical plane which join the same two points in the plane and yet this length not be the length of a segment of any cycloid which connects the two points in the plane. The implication here is that the solution of Bernoulli's isoperimetrical problem is not a cycloid. Fontenelle also noted that the given length could be chosen to be precisely the length of a segment of a cycloid in a vertical plane which joins two points in the plane. Fontenelle's second remark led Castel to imagine all other segments of curves in the vertical plane which connect the two points in the plane and which have the same length as the segment of this cycloid which joins the two points in the plane. Castel now evidently got all mixed up. Fontenelle's second comment apparently caused Castel to become thoroughly confused and to go astray completely. Combining Fontenelle's two observations, Castel some-how fallaciously deduced that it followed from Bernoulli's derivation of the path of quickest descent in a vacuum among all paths in a vertical plane which have the same length and which connect the same two points in the plane, where the two points are neither situated on a horizontal line nor on a vertical line in the plane, that a segment of a cycloid was not the path of quickest descent in a vacuum among all paths in a vertical plane which join the two points in the plane, hence the "paradox."[97]

What Castel overlooked or failed to understand is that in the case of the particular isoperimetrical problem in question, one chooses the given length of the segments of curves in a vertical plane which connect the same two points in the plane to be a length that *differs* from the length of the segment of the cycloid in the vertical plane which in a vacuum is the brachistochrone that joins the two points in the plane. Fontenelle's second remark above is thoroughly misleading in this regard. It is totally irrelevent to the isoperimetrical problem that Bernoulli tackled and solved. (Of course the path of quickest descent in a vacuum among *all* paths in a vertical plane which connect the same two points in the plane, where the two points are neither located on a horizontal line nor on a vertical line in the plane, is a fortiori the path of quickest descent in a vacuum among all paths in the vertical plane which join the two points in the plane and *which have the same length* as the path of quickest descent among all paths in the vertical plane which connect the two points in the plane.) Castel clearly misinterpreted the nature of isoperimetrical problems, and he may have done so with Fontenelle's help. Be that as it may, as far as Bernoulli was concerned, only an imbecile, which Castel was as far as Bernoulli was concerned, could make such a blunder, by confusing the problem of the brachistochrone with the particular isoperimetrical problem at issue.

At the beginning of his *Mathématique universelle* Castel wrote: "Geometry has advanced truths, objects that are scarcely developed, and points of view which are only like fleeting thoughts. Why hide that fact? Geometry has paradoxes, what appear to be contradictions, conclusions drawn within systems, and conclusions that are more like concessions, factional opinions, conjectures, and even para-logisms. And why not? For I think that it is men who made geometry and that geometry is only what men have made."[98] Further along in the same book Castel said:

In observing the success of their techniques for calculating and methods of analysis in the hands of a certain number of creative minds like Descartes, Newton, Leibniz, etc., modern geometers believed that nothing more than analysis and calculation was needed to make the most beautiful discoveries in geometry.... It is genius and not calculation that leads to the discoveries, or at the very most genius aided by calculation. Calculation is the instrument and not the craftsman. It is the mind that calculates, not the routine of calculation, which leads to these discoveries.... One must no doubt calculate to invent, but it is even more necessary to think, reason, combine, generalize, reduce, unite, disunite, evaluate, compensate for, etc., and that in a general and intelligible manner rather than by means of narrow trial and error with symbols and characters.[99]

Such was the outlook on mathematics of Bernoulli's "numskull," who was also the individual whom the author of *L'Esprit des lois*, published in 1748, in which Montesquieu introduced the concept of the separation of executive, legislative, and judicial powers, entrusted the task of educating his eight-year-old son Jean-Baptiste.[100] In Castel's estimation, doing real mathematics required imagination and its use. Castel stressed what it takes to do original work in mathematics, as opposed to the routine calculation that is often a sign of imitation. Someone interested in trying to do new work in mathematics at the time would have done well to read Castel and follow the advice implied in what he said. Castel's counsel was the sort that a technician lost in his calculations, like Varignon, say, could never have given.

In the next section of this chapter we shall see that Castel's words did not go to waste.

7.3.3. First French advances in the partial differential calculus I: Alexis Fontaine des Bertins (1732–1739). At approximately the same time that Maupertuis solved the problem of finding the path that a body follows as it falls through a medium that resists the body's motion as the square of the body's speed, where the body traverses segments of the path of equal lengths in equal amounts of time as it approaches the horizon, Johann I Bernoulli tackled and solved a considerably more difficult problem: finding the tautochrones, meaning the isochronous paths, of a body that falls through a medium that resists the body's motion as the square of the body's speed. Bernoulli wrote a paper in Latin on the problem, which Maupertuis translated into French and read before the Paris Academy early in 1731. The French translation was published in the Paris Academy's *Mémoires* for 1730. Bernoulli had won the Paris Academy prize of 1730. The subject of the contest had been the causes of the elliptical orbits of the planets and the rotations of their apsides. Bernoulli's victory meant that he had been rehabilitated in the Paris Academy. The publication of Maupertuis's French translation of his paper on tautochrones in the Academy's *Mémoires* further marked his rally.

In his Paris Academy *mémoire* of 1730 on finding tautochrones, Bernoulli began with the same equation of motion which appeared in Maupertuis's Paris Academy *mémoire* of 1730 on finding the path described in the first sentence of the paragraph above: the equation of motion for a body that falls through a medium that resists the body's motion as the square of the body's speed. Bernoulli then determined a first integral of this equation of motion using a technique that differed from the one that Maupertuis had employed in his Paris Academy *mémoire* of 1730 to find a first integral of this equation. Bernoulli's method involved making a change of variables and the use of an integrating factor.[101]

In 1732 and in 1734, respectively, Alexis Fontaine des Bertins, whom Maupertuis described to Bernoulli as a "very clever geometer and a very good friend" ("fort habile en géométrie et fort de mes amis"),[102] solved problems of finding brachistochrones, which, we recall, means finding the paths of quickest descent, and finding tautochrones. I have already discussed Fontaine's solutions of these problems elsewhere,[103] but I shall examine them again in greater detail here, because the mathematics that his solutions involve is precisely the mathematics that turns out to enter opportunely into later developments in the story about the problem of the earth's shape told in these pages.

To solve problems of finding brachistochrones and tautochrones, Fontaine introduced what he called his "fluxio-differential method." This calculus involved the use of two independent, first-order Leibnizian differential operators, "·" and "d", which could act upon quantities defined in terms of finite and/or infinitesimal variables in accord with the usual rules that applied to a Leibnizian differential operator:

$$\overline{A+B} = \dot{A} + \dot{B}, \quad \overline{AB} = \dot{A}B + \dot{B}A, \quad \left(\frac{\dot{A}}{B}\right) = \frac{B\dot{A} - A\dot{B}}{B^2},$$

$$d(A+B) = dA + dB, \quad d(AB) = AdB + BdA, \quad d\left(\frac{A}{B}\right) = \frac{BdA - AdB}{B^2}. \quad (7.3.3.1)$$

Quantities like A and B upon which "·" and "d" act in (7.3.3.1) could be defined in terms of finite independent variables and/or finite dependent variables and/or infinitesimal variables.[104] Moreover, Fontaine assumed that the two operations "·" and "d" commuted; that is,

$$d\overline{(\cdots)} = \overline{d(\cdots)}. \quad (7.3.3.2)$$

Why did Fontaine introduce a second, independent differential operator at all? As we shall observe shortly, Fontaine viewed problems of finding brachistochrones and tautochrones as problems that required differentiation of quantities defined at neighboring points located on neighboring plane curves in a family of plane curves that all lie in the same plane *between* or *across* the neighboring plane curves in the family. We recall at the beginning of this chapter that I called this kind of differentiation "differentiation from curve to curve." As I also mentioned at the beginning of this chapter, the difficulty was that a geometric entity, an individual plane curve, was central to the very concept of differentiation in the Leibnizian, single-operator differential calculus–at least, as Leibniz originally conceived the idea of differentiation. Differentials of quantities defined at points situated on plane curves could only be taken *along* individual plane curves, for reasons that accorded with the geometry of plane curves of the time.[105] Leibniz and Johann I Bernoulli in effect generalized the notion of differentiation when they introduced differentiation with respect to a parameter in a family of plane curves in a single parameter which all lie in the same plane. As we saw earlier in

Figure 21. A family of plane curves that all lie in the same plane.

this chapter, this led them to invent at the end of the seventeenth century a rudimentary form of the partial differential calculus. Fontaine instead introduced a second, independent differential operation as a way of tackling problems in which the use of differentiation of quantities defined at points located on plane curves taken along individual plane curves would not alone suffice to solve the problems. Although Leibniz and Johann I Bernoulli created the partial differential calculus specifically in order to deal with problems that seemed to require differentiation from curve to curve, neither the problems of finding brachistochrones nor finding tautochrones were included among the particular problems that Leibniz and Bernoulli considered.[106] As we shall see, this fact may help explain why Fontaine introduced a method that differed from theirs in order to solve problems that also appeared to him to necessitate differentiation from curve to curve.

Let us imagine a family of plane curves that all lie in the same plane (Figure 21). Let F be a quantity defined at points situated on the plane curves in the family. F, for example, may be expressed in terms of the coordinates through which the plane curves in the family are defined. Then $\dot{F}$ stands for an "infinitesimal" difference in F at two neighboring points located on the *same* plane curve in the family, so that in this case the infinitesimal difference is taken *along* the plane curve in the family, while dF symbolizes an infinitesimal difference in F at two neighboring points situated on two *different*, neighboring plane curves in the family, so that in this case the infinitesimal difference is taken *between* or *across* the two neighboring plane curves in the family. The foremost thing to keep in mind here is that a certain view of *plane curves as individual, distinct entities* determines the particular way that Fontaine formalized his calculus; *coordinates* and other *variables* only serve a secondary purpose in constructing the formalism. Fontaine proceeded the way that he did in accordance with the geometry that originally underlay the Leibnizian differential calculus.

Thus the dot "·" indicates differentiation of a quantity defined at points located on the plane curves in the family *along* a plane curve in the family, and the "d" signifies differentiation of a quantity defined at points situated on the plane curves in the family *from* a plane curve in the family to *another* plane curve in the family which is located nearby. In other words, "·"-differentiation of a quantity defined at points located on a plane curve in the family means differentiation of the quantity *along* that plane curve in the family, and "*d*"-differentiation of a quantity defined at points situated on the plane curves in the family means differentiation of the quantity *between* or *across* neighboring plane curves in the family. Just as difficulties with the integral calculus interfered with analytically formulating and solving problems of finding the plane curves that intersect at right angles all of the members of a one-parameter family of plane curves that all lie in the same plane (problems of orthogonal trajectories), problems of finding the plane curves that intercept segments of the members of a one-parameter family of plane curves that all lie in the same plane and which bound equal areas, and problems of finding the plane curves that intercept segments of the members of a one-parameter family of plane curves that all lie in the same plane and which have the same arc lengths, they also did the same for problems of finding brachistochrones and finding tautochrones. To circumvent the obstacles, Fontaine applied the operator "*d*" in his "·", "*d*" calculus to carry out a kind of "differentiation from curve to curve." Many years later, Lagrange rightly termed Fontaine's method of 1732–34 "a type of calculus of variations [une espèce de calcul de variation]."[107]

In 1732, the year that he determined the brachistochrone in a vacuum, Fontaine did not yet belong to the Paris Academy of Sciences. (He joined the Academy in May of 1733.) Consequently his solution of 1732 to the problem of finding the brachistochrone in a vacuum does not appear either in the unpublished proceedings of the Paris Academy (*les Registres des Procès-Verbaux*) or in the Academy's *Mémoires*. Fontaine only published a solution to the problem of finding the brachistochrone in a vacuum for the first time at the beginning of his complete works, published in 1764. He dated this solution back to 1732.[108] But we have no means of verifying independently that the derivation of the brachistochrone in a vacuum which is printed in Fontaine's complete works of 1764 is exactly the same as Fontaine's derivation of 1732 of the brachistochrone in a vacuum. Assuming that the two derivations are essentially the same, we must still allow for the possibility that Fontaine revised his original solution to the problem of finding the brachistochrone in a vacuum before publishing it in 1764. For later in this chapter we shall see that Fontaine slightly modified some work that he did after he had entered the Paris Academy, which appears in the Academy's unpublished proceedings for the 1730s, but which he originally withheld from publication. Fontaine made changes using improvements and simplifications of this work which someone else had found. When Fontaine finally published this work in his complete works of 1764, he published the modified version of it, and he did so without acknowledging the other person. Nevertheless, the Paris Academy's unpublished proceedings, as well as the Academy's *Histoire*, do tell us that in 1732 Fontaine attacked isoperimetrical problems, such as the problem of finding the path of the motion of "least action" of a body along an arbitrarily given surface when the motion of the body is caused by central forces and where the possible paths of motion of the body are restricted to fulfill various side conditions (for example, the possible paths of motion of the body should all have the same given, fixed length and should all have the same endpoints, which are specified in advance[109]). Moreover, Fontaine's paper on finding the brachistochrone in a vacuum which was not published until 1764 has many features in common with Fontaine's paper on finding tautochrones published in the Paris Academy's *Mémoires* for 1734, as we shall see shortly. These two facts allow us to be fairly certain that Fontaine's paper on finding the brachistochrone in a vacuum published in 1764 gives a pretty good idea of the method that Fontaine did actually employ in 1732 to find the brachistochrone in a vacuum.

First I expand the formalism of Fontaine's "fluxio-differential method" as Fontaine utilized it to derive the brachistochrone in a vacuum. He based his solution of the problem upon two principles. First, he assumed that the path of quickest descent in a vacuum which lies in a vertical plane and which joins two points in the vertical plane which are neither situated on a horizontal line nor on a vertical line in the plane will be composed of local paths of quickest descent in a vacuum. In other words, if we consider two points located on a path of quickest descent in a vacuum which are very close to each other, Fontaine assumed that the portion of the path of quickest descent in a vacuum which has these two points as endpoints will also necessarily be the plane path of quickest descent in a vacuum which connects these two points (see Figure 22). Thus Fontaine presupposed that it sufficed to solve the problem for infinitesimal plane paths that a body traverses in a vacuum in infinitesimal amounts of time.

Fontaine wrote the minimal condition that determines the solution of the problem formulated this way as follows:

$$\frac{\dot{s}}{\sqrt{2gx}} + \frac{\dot{s}'}{\sqrt{2gx'}} = \text{a minimum}, \qquad (7.3.3.3)$$

where $AP = x$, $Ap = x'$, $A\pi = x''$, $Mm = \dot{s}$, and $mu = \dot{s}'$. He determined this condition in the following way. In a vacuum the speeds v and v', respectively, that a body initially at rest, whose motion is produced by a uniform field of gravity whose magnitude is g, has at the instants that it has traversed paths that lie in a vertical plane and that entail vertical descents whose distances are x and x', respectively, are

$$\frac{\dot{s}}{\dot{t}} = v = \sqrt{2gx}, \quad \frac{\dot{s}'}{\dot{t}'} = v' = \sqrt{2gx'}, \qquad (7.3.3.4)$$

respectively. It follows from the two equations in (7.3.3.4) that in a vacuum $\dot{t}$ is the infinitesimal amount of time that it takes a body to traverse the infinitesimal path Mm that lies in a vertical plane and that in a vacuum $\dot{t}'$ is the infinitesimal amount of time that it takes a body to traverse the infinitesimal path mu that lies in the same vertical plane as Mm. Thus condition (7.3.3.3) expresses the fact that in a vacuum the infinitesimal amount of time $\dot{t} + \dot{t}'$ that it takes a body to traverse the

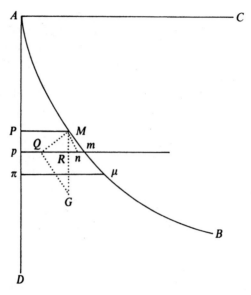

Figure 22. The brachistochrone in a vacuum (Fontaine 1764b: "Page 22, Plate 1" located at the end).

infinitesimal path $Mm + mu = Mmu$ that lies in the vertical plane and that joins the two fixed points M and u in the vertical plane which are very close to each other and which are neither situated on a horizontal line nor on a vertical line in the plane should be a minimum, meaning that it should take a body less time to traverse this infinitesimal path than to traverse any other infinitesimal path that lies in the vertical plane and that connects M and u, if Mmu is the infinitesimal plane path of quickest descent in a vacuum which joins M and u (that is, an infinitesimal brachistochrone in a vacuum). Here "·"-differentiation has been applied because all of the infinitesimal differences are taken *along* individual infinitesimal paths in the vertical plane.

Second, Fontaine interpreted the minimal condition (7.3.3.3) with respect to time to mean that

$$d\left(\frac{\dot{s}}{x^{1/2}} + \frac{\dot{s}'}{x'^{1/2}} \right) = 0, \qquad (7.3.3.5)$$

where it is evident that "d" stands for infinitesimal differences in the amounts of time that it takes a body to traverse in a vacuum *different*, neighboring infinitesimal paths that lie in the same vertical plane and that connect the same two points in the vertical plane which are very close to each other and which are neither located on a horizontal line nor on a vertical line in the plane. In other words, according to condition (7.3.3.5), in order for an infinitesimal path that lies in a vertical plane to be an infinitesimal path of quickest descent in a vacuum, the amounts of time that it takes a body to traverse in a vacuum infinitesimal paths that differ very slightly from the infinitesimal path in question, that lie in the same vertical plane as the infinitesimal path in question, and that join the same two points in the vertical plane as the infinitesimal path in question must be stationary.

Fontaine wrote as a sum of two terms the amount of time that it takes a body to traverse in a vacuum certain kinds of infinitesimal paths that lie in the same vertical plane and that connect the same two points in the vertical plane which are very close to each other and which are neither located on a horizontal line nor on a vertical line in the plane. He did this so that he could consider infinitesimal variations of an infinitesimal path of quickest descent in a vacuum which lie in the same vertical plane as that path and which connect the same two points as the path (that is, infinitesimal variations of an infinitesimal brachistochrone in a vacuum).

Specifically, Fontaine derived a "fluxio-differential" equation that expressed the condition that the amounts of time that it takes a body to traverse in a vacuum infinitesimal variations of an infinitesimal path of quickest descent in a vacuum which lie in the same vertical plane as that path and which join the same two points in the vertical plane as the path be stationary, when the infinitesimal variations of the infinitesimal path of quickest descent in a vacuum have the following form: Fontaine allowed the abscissa $pm = y'$ of m to vary in order to produce infinitesimal paths Mnu which differ very slightly from an infinitesimal path of quickest descent Mmu in a vacuum and which have the same endpoints M and u as the infinitesimal path of quickest descent Mmu in a vacuum. He held all other coordinates constant. In effect Fontaine determined an infinitesimal path of quickest descent in a vacuum among infinitesimal paths of this type and only this type. The operation "d" acts according to the rules specified in (7.3.3.1), hence

$$d\left(\frac{\dot{s}}{x^{1/2}} + \frac{\dot{s}'}{x'^{1/2}}\right) = \frac{d\dot{s}}{x^{1/2}} + \frac{d\dot{s}'}{x'^{1/2}},$$

since the ordinate x of M and the ordinate x' of m do not vary (that is, the endpoint M and the ordinate x' of m are fixed). Thus

$$\frac{d\dot{s}}{x^{1/2}} + \frac{d\dot{s}'}{x'^{1/2}} = 0 \tag{7.3.3.6}$$

is the relevant fluxio-differential equation that expresses condition (7.3.3.5) for variations of an infinitesimal path of quickest descent in a vacuum which have the form stipulated.

Since

$$\dot{s}^2 = \dot{x}^2 + \dot{y}^2 \quad \text{and} \quad \dot{s}'^2 = \dot{x}'^2 + \dot{y}'^2, \tag{7.3.3.7}$$

where $\dot{x} = x' - x$, $\dot{y} = y' - y$, $\dot{x}' = x'' - x'$, and $\dot{y}' = y'' - y'$, it follows from applying the operation "d" to the equations in (7.3.3.7) that

$$2\dot{s}d\dot{s} = 2\dot{x}d\dot{x} + 2\dot{y}d\dot{y} \qquad (7.3.3.8)$$

and

$$2\dot{s}'\,d\dot{s}' = 2\dot{x}'d\dot{x}' + 2\dot{y}'d\dot{y}' \qquad (7.3.3.9)$$

are true. Now, within equation (7.3.3.8)

$$d\dot{x} = d(x' - x) = dx' - dx = 0 - 0 = 0$$

is true, since neither the ordinate x of M nor the ordinate x' of m changes in forming the variation Mn of Mm. (Again the last statement is true because the endpoint M and the ordinate x' of m are fixed.) At the same time, within equation (7.3.3.8)

$$d\dot{y} = d(y' - y) = dy' - dy = dy'$$

is also true, because the abscissa $PM = y$ of M does not change in forming the variation Mn of Mm. (The last statement is true because the endpoint M is fixed.) Hence equation (7.3.3.8) reduces to

$$2\dot{s}d\dot{s} = 2\dot{y}dy',$$

which can be written as

$$d\dot{s} = \frac{\dot{y}}{\dot{s}}dy' \qquad (7.3.3.10)$$

instead. Similarly, within equation (7.3.3.9)

$$d\dot{x}' = d(x'' - x') = dx'' - dx' = 0 - 0 = 0$$

is true, since neither the ordinate x' of m nor the ordinate x'' of u changes in forming the variation nu of mu (that is, the ordinate x' of m and the endpoint u are fixed). At the same time, within equation (7.3.3.9)

$$d\dot{y}' = d(y'' - y') = dy'' - dy' = -dy'$$

is also true, because the abscissa $\pi u = y''$ of u does not change in forming the variation nu of mu. (The last statement is true because the endpoint u is fixed.) Therefore equation (7.3.3.9) reduces to

$$2\dot{s}'\,d\dot{s}' = -2\dot{y}'dy',$$

which can be written as

$$d\dot{s}' = -\frac{\dot{y}'}{\dot{s}'}dy' \qquad (7.3.3.11)$$

instead. If the expression on the right-hand side of equation (7.3.3.10) is substituted for $d\dot{s}$ in equation (7.3.3.6) and if the expression on the right-hand side of equation (7.3.3.11) is substituted for $d\dot{s}'$ in equation (7.3.3.6), the fluxio-differential equation

$$\left(\frac{\dot{y}}{x^{1/2}\dot{s}} - \frac{\dot{y}'}{x'^{1/2}\dot{s}'}\right)dy' = 0 \qquad (7.3.3.6')$$

results. If both sides of equation (7.3.3.6') are now divided by the arbitrarily small, nonzero factor dy', the equation

$$\frac{\dot{y}}{x^{1/2}\dot{s}} - \frac{\dot{y}'}{x'^{1/2}\dot{s}'} = 0 \qquad (7.3.3.6'')$$

follows. From equation (7.3.3.6'') Fontaine concluded that

$$\frac{\dot{y}}{x^{1/2}\dot{s}} = \text{a constant.} \qquad (7.3.3.12)$$

He recognized equation (7.3.3.12) to be "the equation of the cycloid."[110]

Let us note that equation (7.3.3.12) for the cycloid is an equation among differentials. The left-hand side of the equation is a quotient in whose numerator the differential $\dot{y}$ appears and in whose denominator the differential $\dot{s}$ appears. The equation is literally a "differential" equation. We must remember this fact in order to understand the way that Fontaine's fluxio-differential method subsequently evolved.

I quickly summarize the basic elements of Fontaine's derivation above as well as explain the essential ideas that underlie the formalism. Employing his calculus that involved two independent, differential operators, Fontaine found in 1732 the local (that is, the infinitesimal) brachistochrone in a vacuum. He used his calculus to reduce the problem to the first-order ordinary "differential" equation for the cycloid. In order to do so he introduced alternate infinitesimal paths of motion which differ very slightly from an infinitesimal path of quickest descent in a vacuum which is the solution of the problem. These variations of the infinitesimal path of quickest descent in a vacuum lie in the same vertical plane as the infinitesimal path of quickest descent in a vacuum, and they join the same two points in the vertical plane as the infinitesimal path of quickest descent in a vacuum. The two points that are the common endpoints of the infinitesimal path of quickest descent in a vacuum and its variations are very close to each other, and the points are neither situated on a horizontal line nor on a vertical line in the plane. The variations of the infinitesimal path of quickest descent, together with the infinitesimal path of quickest descent itself, make up a "family of infinitesimal segments of plane curves" which all lie in the same vertical plane (Figure 23). Motion of a body along each member of the family takes place in accordance with the same "differential" equation of motion $G(\dot{x}, v, \dot{v}) \equiv v\dot{v} - g\dot{x} = 0$, the "differential" equation of free fall in a vacuum, whose first integral is $v = \sqrt{2gx}$ (equation (7.3.3.4)), where x is the vertical distance that a body moving according to this

Figure 23. Infinitesimal variations of the local brachistochrone in a vacuum.

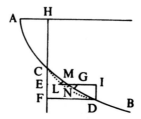

Figure 24. The brachistochrone in a vacuum (Jakob Bernoulli 1697: Table IV, Figure VI).

"differential" equation of motion has fallen when it reaches some particular point on its path of motion (the ordinate of the point in rectangular coordinates) and v is the speed of the body at that point. Fontaine applied the operator "d" in his fluxio-differential method, as well as the relevant minimal condition (7.3.3.3), where the s in this condition stands for arc length, to the members of a judiciously chosen subfamily of the family that consists of the variations of the infinitesimal path of quickest descent in a vacuum and the infinitesimal path of quickest descent in a vacuum itself. Taking (7.3.3.5) to express the minimal condition (7.3.3.3), Fontaine arrived at the "differential" equation (7.3.3.12) for the cycloid. Except for the particular notations that Fontaine chose to symbolize the two kinds of differentials, Fontaine's method of determining the brachistochrone in a vacuum closely resembles the way that Lagrange solved the same problem later. Lagrange's method is described by Woodhouse.[111]

Fontaine did not utilize any principles to determine the brachistochrone in a vacuum which had not already been used before. If we turn, for example, to Jakob Bernoulli's solution of the same problem, which is published in *Acta Eruditorum* for May 1697 (see Figure 24), we find that Bernoulli based his solution on the following statement: "since by hypothesis, the time through $CG + GD$, is to be a minimum, and since quantities at or near their state of minimum may be considered constant (for their increments or decrements are very small), we have

$$t.CG + t.GD = t.CL + t.LD$$

[$t.CG$ abrigedly representing the time through CG].[112] Although Bernoulli used infinitesimal geometry to find the brachistochrone in a vacuum, not a purely algebraic calculus of differential operators, he clearly deduced his solution from the same two principles that Fontaine made use of in applying his fluxio-differential method to solve the problem – namely, that the property that characterizes the plane curve that is the solution to the problem (that the curve be a path of quickest descent) is shared by each infinitesimal segment of the plane curve, and at the same time first-order differences in the amounts of time that it takes a body to traverse different variations of any infinitesimal segment of the plane curve that is the solution to the problem which lie in the same vertical plane as the plane curve must vanish in a neighborhood of the segment. In his *Methodus incrementorum directa et inversa* (1715), Brook Taylor used similar principles, together with notation that differed from Bernoulli's notation, in order to solve isoperimetrical problems. Fontaine maintained, however, years after he had determined the brachistochrone in a vacuum, that at the time he did his work he "had only read the *Livre des infiniments petits* by the Marquis de L'Hôpital," and that he "did not know the methods of Monsieur Jacques [that is, Jakob] Bernoulli, of Monsieur Taylor, etc., when it was proposed that I find the shortest line between two given points on an arbitrary surface."[113] The "proposal" in question was really a "challenge" to solve the problem at issue. It came at a time when Fontaine was a candidate for election to the Paris Academy.

In February of 1734, nine months after he had joined the Paris Academy, Fontaine read before the Academy a paper in which he solved problems of finding tautochrones. He also found his solutions to these problems by using his fluxio-differential method. Fontaine's paper was published in the Paris Academy's *Mémoires* for 1734.[114] Fontaine's desire to better Johann I Bernoulli's work on problems of finding tautochrones which was published in the Paris Academy's *Mémoires* for 1730 and which I mentioned above is what prompted Fontaine to take up such problems. Although the formalism of the fluxio-differential method remained the same, as we shall see in a moment, Fontaine utilized additional concepts that aided him in solving problems of finding tautochrones. Most important, he introduced differential coefficients. Unlike differentials, differential coefficients are finite quantities. First he defined a suitable, convenient set of differential coefficients, and then in effect he used his fluxio-differential method to determine other differential coefficients. In the end he arrived at equations that interrelated differential coefficients, which are precursors of our modern differential equations. (I shall call such equations "differential equations" here.)

How did Fontaine envisage problems of tautochrones? He imagined a tautochrone, a plane curve that lies in a vertical plane, to be composed of a family of segments MA, $M'A$, BA, etc., all of which had a common endpoint A, where X designated the finite arc length of a segment in the family. He also imagined each segment in the family to be made up of a family of subsegments all of which had a common endpoint, the same common endpoint A as the segments in the family

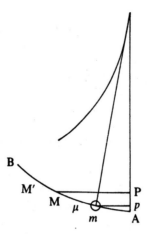

Figure 25. The tautochrone (Fontaine 1734b: figure on p. 371).

whose members made up the tautochrone, where x stood for the finite arc length of a member of this family. The finite arc length X of a segment MA in the family whose members made up the tautochrone was determined by an initial condition $y(X, X) = 0$, where $y(x, X)$ is a body's speed after the body has traversed a subsegment mA of this segment whose finite arc length is x, where $0 \leqslant x \leqslant X$. Each subsegment mA of a segment MA in the family whose members make up the tautochrone can itself obviously be thought of as a segment in the family whose members make up the tautochrone too. This view of the tautochrone enabled Fontaine to treat the tautochrone as a "family of segments of a plane curve," and he applied his "·" and "d" operations to the members of this family (see Figure 25).

The essential features of Fontaine's fluxio-differential method applied to problems of finding tautochrones can be seen in Fontaine's solution to the problem of finding the tautochrone in a medium that resists a body's motion directly as $(y^2 + my)/n$, where y designates the body's speed and m and $n \neq 0$ are nonnegative constants that can be arbitrarily chosen. Until Fontaine presented his paper before the Paris Academy in 1734, the problem had remained unsolved. (In his Paris Academy *mémoire* of 1730, Johann I Bernoulli solved the special case where $m = 0$.) Defining differential coefficients p, q, r, etc. by

$$\dot{z} = p\dot{x}, \qquad (7.3.3.13)$$

$$\dot{p} = q\dot{x}, \qquad (7.3.3.14)$$

$$\dot{q} = r\dot{x}, \qquad (7.3.3.15)$$

and so on, where z stands for the ordinate of a point in rectangular coordinates, Fontaine wrote the equation of motion as

$$\left(-gp \pm \frac{1}{n}y^2 \pm \frac{m}{n}y\right)\frac{\dot{x}}{y} = \dot{y},$$

which he then rewrote as

$$gp\dot{x} \mp \frac{1}{n}y^2\dot{x} \mp \frac{m}{n}y\dot{x} + y\dot{y} = 0, \qquad (7.3.3.16)$$

where g designates the magnitude of a uniform field of gravity. Then if the first-order differential operator "d" is applied to equation (7.3.3.16), the fluxio-differential equation

$$0 = d(0) = d(gp\dot{x} \mp \frac{1}{n}y^2\dot{x} \mp \frac{m}{n}y\dot{x} + y\dot{y}) = gpd\dot{x} + g\dot{x}dp \mp \frac{1}{n}y^2d\dot{x} \mp \frac{2}{n}y\dot{x}dy$$

$$\mp \frac{m}{n}yd\dot{x} \mp \frac{m}{n}\dot{x}dy + yd\dot{y} + \dot{y}dy \qquad (7.3.3.17)$$

follows by making use of the rules given in (7.3.3.1) for the first-order differential operator "d" which apply to sums and products of finite and/or infinitesimal quantities. Now, having defined q by means of (7.3.3.14), Fontaine assumed, without providing the reader with any explanation, that

$$dp = qdx \qquad (7.3.3.18)$$

is true as well. Around twenty-seven years later Lagrange explicitly stipulated, for the first time ever, that the supposition that (7.3.3.14) and (7.3.3.18) both hold at the same time is a fundamental rule of the calculus of variations.[115] Then if qdx is substituted for dp in equation (7.3.3.17), the fluxio-differential equation

$$gpd\dot{x} + gq\dot{x}dx \mp \frac{1}{n}y^2d\dot{x} \mp \frac{2}{n}y\dot{x}dy \mp \frac{m}{n}yd\dot{x} \mp \frac{m}{n}\dot{x}dy + yd\dot{y} + \dot{y}dy = 0 \quad (7.3.3.19)$$

results. The upper and lower signs for addition and subtraction in equation (7.3.3.19) apply according as bodies descend or ascend, respectively.

Now, the condition that expresses tautochronous motion is

$$FL\left(\frac{\dot{x}}{y}\right) = \text{the amount of time } T \text{ that it takes a body to traverse a segment of a tautochrone whose arc length is } X$$

$$= \text{a constant for all } X. \qquad (7.3.3.20)$$

Here $\dot{x}/y$ is just the infinitesimal amount of time that it takes a body to traverse at a speed y an infinitesimal segment of a plane curve whose infinitesimal arc length is $\dot{x}$. The notation FL stands for "fluent," which does not mean the fluent of Newton's infinitesimal calculus, but the operator that is the inverse of the Leibnizian differential "$\cdot$" instead. $FL(\dot{x}/y)$ on the left-hand side of (7.3.3.20) just expresses the amount of time that it takes a body whose motion is governed by the equation of motion (7.3.3.16) to traverse a segment of a plane curve whose finite arc length is X, where the segments of the plane curve all have a common endpoint. Expressed in modern notation,

$$FL\left(\frac{\dot{x}}{y}\right) = \int_0^X \frac{1}{y(x, X)}dx.$$

Then it follows from (7.3.3.20) that

$$d\,FL\left(\frac{\dot{x}}{y}\right) = 0,$$ (7.3.3.21)

from which Fontaine concluded that

$$d\left(\frac{\dot{x}}{y}\right) = 0.$$ (7.3.3.22)

Equation (7.3.3.22) can be deduced from equation (7.3.3.21) as follows: equation (7.3.3.21) requires that

$$d\dot{F}L\left(\frac{\dot{x}}{y}\right) = 0$$

holds as well–that is, that

$$\overline{d\,FL\left(\frac{\dot{x}}{y}\right)}^{\cdot} = 0.$$

Then if the difference operators "$\cdot$" and "d" are commuted, it follows that

$$d\dot{F}L\left(\frac{\dot{x}}{y}\right) = 0 \text{ — in other words, that } \overline{d\,FL\left(\frac{\dot{x}}{y}\right)}^{\cdot} = 0. \quad \text{But} \quad \dot{F}L\left(\frac{\dot{x}}{y}\right) = \frac{\dot{x}}{y},$$

because the fluent is the inverse of the differential operator "$\cdot$". Hence equation (7.3.3.22) results. Now, according to the rule given in (7.3.3.1) for the first-order differential operator "d" which applies to quotients of finite and/or infinitesimal quantities, equation (7.3.3.22) can be written as

$$\frac{yd\dot{x} - \dot{x}dy}{y^2} = 0,$$ (7.3.3.23)

and equation (7.3.3.23) reduces to

$$dy = \frac{y}{\dot{x}}d\dot{x}.$$ (7.3.3.24)

Fontaine then introduced the unspecified transformation ϕ defined by

$$dx = \phi dX,$$ (7.3.3.25)

where Fontaine stipulated that $\phi = \phi(x, X)$ fulfills the following two conditions: $\phi(0, X) = 0$ and $\phi(X, X) = 1$. Defining more differential coefficients γ, λ, μ, etc. by

$$\dot{\phi} = \gamma\dot{x},$$ (7.3.3.26)

$$\dot{\gamma} = \lambda\dot{x},$$ (7.3.3.27)

$$\dot{\lambda} = \mu\dot{x},$$ (7.3.3.28)

and so on, Fontaine found that

$$dx = \dot{dx} = \dot{\phi}dX = \gamma\dot{x}dX. \tag{7.3.3.29}$$

Two things are essential in order to arrive at equation (7.3.3.29): the assumption that the two differential operators "·" and "d" *commute*, and treating dX as a constant with respect to the "·" operator. I will explain shortly why the second of the these two steps is permitted. Now, if the expression on the right-hand side of equation (7.3.3.29) is substituted for $d\dot{x}$ in equation (7.3.3.24), the equation

$$dy = \gamma y dX \tag{7.3.3.30}$$

results. Then if the first-order differential operator "·" is applied to equation (7.3.3.30) in conjunction with (7.3.3.27), the equation

$$d\dot{y} = \dot{dy} = \{\dot{\gamma y}\}dX = \{\gamma\dot{y} + \dot{\gamma}y\}dX = \gamma\dot{y}dX + \lambda\dot{x}ydX \tag{7.3.3.31}$$

follows. Commuting the differential operators "·" and "d", treating dX as a constant with respect to the "·" operator, and the rule given in (7.3.3.1) for the first-order differential operator "·" which applies to products of finite and/or infinitesimal quantities have all been utilized to arrive at equation (7.3.3.31). Then if the expression for dx on the right-hand side of equation (7.3.3.25), the expression for $d\dot{x}$ on the right-hand side of equation (7.3.3.29), the expression for dy on the right-hand side of equation (7.3.3.30), and the expression for $d\dot{y}$ on the right-hand side of equation (7.3.3.31) are substituted for dx, for $d\dot{x}$, for dy, and for $d\dot{y}$, respectively, in the fluxio-differential equation (7.3.3.19), the equation

$$\left(gp\gamma\dot{x} + gq\phi\dot{x} \mp \frac{3}{n}\gamma y^2\dot{x} \mp \frac{2m}{n}\gamma y\dot{x} + 2\gamma y\dot{y} + \lambda y^2\dot{x} \right)dX = 0.$$

follows. If both sides of this last equation are now divided by the arbitrarily small, nonzero factor dX, an equation results which can be rewritten as

$$g\left(\frac{p\gamma + q\phi}{2\gamma} \right)\dot{x} + \left(\frac{\mp\frac{3}{n}\gamma + \lambda}{2\gamma} \right)y^2\dot{x} \mp \frac{m}{n}y\dot{x} + y\dot{y} = 0. \tag{7.3.3.32}$$

Finally, if equation (7.3.3.32) and the equation of motion (7.3.3.16) are assumed to be the same, and if these two equations are compared term by term, the two equations

$$\frac{p\gamma + q\phi}{2\gamma} = p \tag{7.3.3.33}$$

and

$$\frac{\mp\frac{3}{n}\gamma + \lambda}{2\gamma} = \mp\frac{1}{n} \tag{7.3.3.34}$$

follow. In other words, from the two equations (7.3.3.16) and (7.3.3.32) in the differentials $\dot{x}$ and $\dot{y}$ ("differential" equations), Fontaine derived the two equations (7.3.3.33) and (7.3.3.34) in differential coefficients ("differential equations").

Equation (7.3.3.34) can be treated like a second-order ordinary "differential equation" for ϕ. It can be solved for the unknown function $\phi = \phi(x, X)$ such that $\phi(0, X) = 0$ and $\phi(X, X) = 1$ and for γ by integrating the equation twice, which turns out can be done with no difficulty. Once $\phi(x, X)$ and γ have been found, they can be substituted into the "differential equation" (7.3.3.33). The equation that results when this is done is an equation that can be handled like a second-order ordinary "differential equation" for the ordinates $z = z(x)$ of points on a tautochrone as a function of arc length x, which can be solved for $z = z(x)$ by integrating the equation twice. Again it turns out that this can be effected with no difficulty. Then the abscissas $w = w(x)$ of points on the tautochrone as a function of arc length x can be found in terms of $z = z(x)$ by integrating the relation

$$\dot{w} = \sqrt{\dot{x}^2 - \dot{z}^2}, \qquad (7.3.3.35)$$

a first-order ordinary "differential" equation of first degree whose variables w and x are separated after the expression for $z = z(x)$ is substituted for z in it. In this way a parametric representation of a tautochrone, where x is the parameter, can be determined. However, it is enough to repeat Fontaine's own words. Having already applied his fluxio-differential method earlier in his Paris Academy *mémoire* of 1734 to find the tautochrone of a body that moves through a medium that resists the body's motion directly as y^2/n and having arrived at the same two equations (7.3.3.33) and (7.3.3.34) in differential coefficients, Fontaine concluded that this "proves that its tautochrone [where the medium resists a body's motion directly as y^2/n] is also the tautochrone of this one [where the medium resists a body's motion directly as $(y^2 + my)/n$].[116]

As I mentioned, in 1730 Johann I Bernoulli had solved the problem of finding the tautochrones in a medium that resists a body's motion directly as the square of the body's speed. But in order to do so he had to integrate the corresponding equation of motion once, because he needed a first integral of that equation, which is an expression for the body's speed, in order to make his method of solving the problem work. However, in the more general case that Fontaine considered, the corresponding equation of motion *cannot* be integrated. That is, a first integral of this equation cannot be found when $m \neq 0$ in $(y^2 + my)/n$. Bernoulli would not have been able to solve the problem of finding the tautochrones in this case using the method that he had applied in 1730. His approach to such a problem failed whenever the corresponding equation of motion could not be integrated to produce an expression for the speed. Like the problem of finding the plane curves orthogonal to a one-parameter family of transcendental plane curves that all lie in the same plane, finding tautochrones was an example of another problem where difficulties of the integral calculus could obstruct progress in solving the problem. The same troubles inevitably arise in problems of finding brachistochrones in resistant media for the same reasons. I have already mentioned that Fontaine's equations (7.3.3.4) are in fact just first integrals of the "differential" equation of motion $v\dot{v} = g\dot{x}$ in a vacuum, which is the "differential" equation of free fall in a vacuum, but first integrals of the corresponding equations of motion in cases where the motion is resisted cannot in general be found.[117]

By introducing two independent differential operations, Fontaine managed to surmount the obstacles to solving problems of finding tautochrones in resistant media which difficulties of the integral calculus created. In effect he reduced the problem of deriving ordinary "differential equations" for tautochrones in resistant media to the differential calculus. The fact that he was able to do this should be judged to have been one of the great achievements of the 1730s in the infinitesimal calculus.

When told by Maupertuis in mid-February of 1734 what Fontaine had apparently accomplished, a skeptical Johann I Bernoulli retorted early in March of 1734 that "either Monsieur Fontaine has the mind of an angel or else I am as dumb as an ox."[118] In 1742, Euler told Fontaine's Parisian colleague Clairaut:

with regard to Monsieur Fontaine, I greatly admired his fluxio-differential method for finding tautochrones. Although I worked hard on the problem of finding tautochrones myself, I must admit that he made much more progress and advanced the solving of the problem much further than I did. I beg your indulgence. Please tell him what I have said and assure him of the high esteem that I have for his profound meditations.[119]

Likewise, in the *Encyclopédie*, D'Alembert called Fontaine's Paris Academy *mémoire* on finding tautochrones "so excellent a work, that it can be regarded as one of the best to be found in the *Mémoires* of the Royal Academy of Sciences."[120]

I shall now try to elucidate the main ideas that underlie the formalism that Fontaine used to determine tautochrones. The way that he solved such problems appears rather unusual from a modern standpoint. Fontaine began with a "differential" equation of motion which has the form

$$G(\dot{z}, \dot{x}, y, \dot{y}) = 0, \qquad (7.3.3.36)$$

where z designates the ordinate of a point in rectangular coordinates, x symbolizes arc length, and y stands for speed, together with a condition

$$FL\left(\frac{\dot{x}}{y}\right) = T(X), \qquad (7.3.3.37)$$

where FL ("Fluent") stands for the inverse of the differential operator "·". Again, $\dot{x}/y$ here is just the infinitesimal amount of time that it takes a body to traverse at a speed y an infinitesimal segment of a plane curve whose infinitesimal arc length is $\dot{x}$. $FL(\dot{x}/y)$ on the left-hand side of (7.3.3.37) just expresses the amount of time that it takes a body whose motion is governed by an equation of motion which has the form (7.3.3.36) to traverse a segment of a plane curve whose finite arc length is X, where the segments of the plane curve all have a common endpoint. Expressed in modern notation,

$$FL\left(\frac{\dot{x}}{y}\right) = \int_0^x \frac{1}{y(x, X)} dx. \qquad (7.3.3.37')$$

$T(X)$ on the right-hand side of (7.3.3.37) designates a given function of the arc lengths X of segments of the plane curve which have a common endpoint. Thus $T(X)$ is a given expression in X which specifies the amount of time that it takes a body to traverse a segment whose arc length is X of the unknown plane curve that is the solution to a problem of the type expressed by means of equations that have the form (7.3.3.36) and (7.3.3.37). In particular, $T(X) \equiv$ a constant is the condition for tautochronous motion. That is, tautochronous motion occurs when $T(X) \equiv$ a constant for all such segments of the plane curve [equation (7.3.3.20)]. In his Paris Academy *mémoire* of 1734, Fontaine in fact did not try to solve problems of the kind in question in which $T(X)$ is not equal to a constant but is a nonconstant function of X instead. I will explain the reasons why he did not do so later in this section of this chapter. But it is enough to consider the problem of finding tautochrones in order to understand how and why he applied his fluxio-differential method to solve problems formulated in terms of equations that have the form (7.3.3.36) and (7.3.3.37).

In order to make use of the fluxio-differential method to solve a problem, the problem must involve a "family of plane curves" that all lie in the same plane, or a "family of segments of plane curves" which all lie in the same plane, or a "family of segments of a plane curve." In the case of a tautochrone, a plane curve that lies in a vertical plane, the members of the relevant family are simply segments MA, $M'A$, BA, etc., of the tautochrone itself, even though finding the tautochrone is the very problem to be solved. That is, the members of the family all lie along the unknown plane curve to be sought. All of the segments have a common endpoint A, and Fontaine in effect imagined that initial conditions on the speed of ascent and descent of a body along the tautochrone determined each segment. In other words, if v designates the speed at which a body ascends or descends, say, along the segment MA of the tautochrone whose arc length is X, then $v = v(x, X)$ is the body's speed at the point m on the segment whose distance from the segment's endpoint A measured along the segment is arc length x (thus x is the arc length of the subsegment mA of segment MA), where $0 \leqslant x \leqslant X$, and where $v(x, X) = 0$ provided $x = X$ (see Figure 25). In this way Fontaine imagined the curve in question (the tautochrone) to be decomposed into a family of segments, where the arc lengths X of the segments functioned as a parameter. By this means he generated a "one-parameter family of segments of a plane curve" (Figure 26). This made the operator "d" in Fontaine's dual-operator calculus applicable to the problem envisioned this way, since Fontaine introduced this operator in order to differentiate between or across neighboring members of a family of plane curves that all lie in the same plane, or a family of segments of plane curves which all lie in the same plane, or a family of segments of a plane curve. In a stroke of genius,

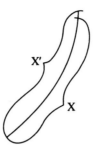

Figure 26. Plane curve (the tautochrone) decomposed into a family of segments.

Fontaine viewed a tautochrone, a plane curve along which the motion of a body takes place in accordance with an equation of motion which has the form (7.3.3.36), as *itself* made up of a "family of segments of a plane curve," each member of which is determined by an initial condition on the motion of a body that ascends or descends along it (specifically, an initial condition on the body's speed) and along which the motion of the body occurs according to an equation of motion which has the form (7.3.3.36). Thus, for example, the condition that expresses tautochronous motion is

$$FL\left(\frac{\dot{x}}{y}\right) = \text{the amount of time } T \text{ that it takes a body to traverse}$$
$$\text{a segment of a tautochrone whose arc length is } X$$
$$= \text{a constant for all } X. \tag{7.3.3.20}$$

This is a condition on the amount of time that it takes a body to traverse *different* segments that make up a tautochrone. Consequently the differential operator "*d*" is the correct differential operator in Fontaine's dual-operator calculus to apply to condition (7.3.3.20), since this operator is the one that is used to "differentiate from curve to curve." When this is done, the equation

$$dFL\left(\frac{\dot{x}}{y}\right) = 0 \tag{7.3.3.21}$$

follows.

In dealing with the problem of finding tautochrones, Fontaine introduced differential coefficients, which at that time were a rather novel device.[121] Unlike differentials, differential coefficients are finite quantities. Letting $z = z(x)$ designate the ordinate in rectangular coordinates of the endpoint of the segment of a tautochrone whose arc length is x, where the other endpoints of all such segments of the tautochrone are the same, Fontaine defined the differential coefficients p, q, and r, respectively, by means of (7.3.3.13), (7.3.3.14), and (7.3.3.15), respectively. He also introduced an unspecified transformation $\phi = \phi(x, X)$ defined by (7.3.3.25), where he stipulated that ϕ satisfy the conditions: $\phi(0, X) = 0$ and $\phi(X, X) = 1$. It is important to note that in a one-parameter family of plane curves that all lie in the same plane, or a one-parameter family of segments of plane curves which all lie in the same plane, or a one-parameter family of segments of a plane curve, the value of the parameter X which corresponds to a particular member of the family is a *constant* for that member. Hence this value of the parameter X is a constant quantity with respect to "·"-differentiation along that member. At the same time, differentiating between or across neighboring members of the family entails treating the parameter X as a *variable*. Indeed, the nonzero quantity dX comes into being as a result of this kind of differentiation. In other words, in determining tautochrones, Fontaine employed "differentiation with respect to a parameter" in his own, unique way. Using the transformation ϕ, he defined the differential coefficients γ, λ, and μ, respectively, by means of (7.3.3.26), (7.3.3.27), and (7.3.3.28), respectively. Applying his calculus of two differential operators, Fontaine reduced the problem of finding the tautochrone in a medium that resists a body's motion directly as $(y^2 + my)/n$, where y is the body's speed, to a pair of equations (7.3.3.33) and (7.3.3.34) in differential coefficients. These two equations can be combined to form a single, integrable, second-order ordinary differential equation for $z = z(x)$, expressed entirely in terms of differential coefficients (that is, an ordinary "differential equation" instead of an ordinary "differential" equation[122]). The abscissas $w = w(x)$ of points on the tautochrone as a function of arc length x could then be found in terms of $z = z(x)$ by integrating the relation

$$\dot{w} = \sqrt{\dot{x}^2 - \dot{z}^2},   (7.3.3.35)$$

a first-order ordinary "differential" equation of first degree whose variables w and x are separated after the expression for $z = z(x)$ is substituted for z in it.

In determining the equation for $z = z(x)$, Fontaine made use of

$$d\dot{X} = 0.   (7.3.3.38)$$

This result follows from the fact that "·"-differentiation at points situated on individual segments of a tautochrone is taken along the *individual* segments of the tautochrone. But X is a *constant* quantity for each of the segments in the family that makes up the tautochrone, so that $\dot{X} = 0$ along each of these segments for the reason that I just mentioned, namely, the value of the parameter that corresponds to a particular member of a one-parameter family of plane curves that all lie in the same plane, or a one-parameter family of segments of plane curves which all lie in the same plane, or a one-parameter family of segments of a plane curve is a constant with respect to "·"-differentiation along that member. But then $d(\dot{X}) = d(0) = 0$ is true as well, from which $d\dot{X} = d(\dot{X}) = 0$ follows by commuting the "·" and "d" operations. This explains why Fontaine treated dX as a constant with respect to "·"-differentiation in determining tautochrones.

Viewed from the standpoint of mathematicians who had already worked on the problems of finding brachistochrones and tautochrones, it may appear surprising and strange that Fontaine interpreted these problems as problems in a "family" of objects in a plane of one of the following kinds at all: plane curves that all lie in the same plane; segments of plane curves which all lie in the same plane; and segments of a plane curve. As I mentioned earlier, finding brachistochrones was not one of the problems that led Leibniz and Johann I Bernoulli to invent at the end of the seventeenth century the partial differential calculus to tackle problems that seemed to call for differentiation from curve to curve. They evidently did not regard the problem of finding brachistochrones to be a problem like the problem of finding the plane curves orthogonal to a one-parameter family of transcendental plane curves that all lie in the same plane (the problem of orthogonal trajectories), which does involve, as we saw earlier in this chapter, a "family of plane curves" that all lie in the same plane. Johann I Bernoulli's method of finding tautochrones, which appears in his Paris Academy *mémoire* of 1730, does not involve a family of plane curves that all lie in the same plane, so that it would appear that this problem was not customarily looked at in terms of such a family, either. Yet Fontaine took both the problems of finding brachistochrones and finding tautochrones as models or prototypes of problems in a "family" of objects in a plane of any one of the kinds just described (plane curves that all lie in the same plane, segments of plane curves which all lie in the same plane, segments of a plane curve). In that sense, Fontaine generalized problems in a "family of plane curves" that all lie in the same plane to problems that can be envisaged in terms of a more general class of families composed of objects in a plane of any one of the types just described.

One particular feature of both the problems of finding brachistochrones and tautochrones may help explain why Fontaine visualized these problems the way that he did. In neither problem is the relevant family of segments of plane curves which all lie in the same plane (in the case of the brachistochrone in a vacuum) nor the family of segments of a plane curve (in the case of a tautochrone) actually known at the beginning–no more so than the plane curves that are the solutions to each of the two problems. Therefore, in a sense, Fontaine interpreted these two kinds of problems as *converses* of the standard problems in a family of plane curves that all lie in the same plane. In any particular standard problem, the members of such a family are always *given either explicitly or implicitly at the start*, and then the problem is to find a plane curve or plane curves fulfilling some condition or having some property determined by the members of the family which is specified beforehand. Problems of finding orthogonal trajectories are examples of the standard problem. In this case the plane curves that are solutions to a problem of this type intersect at right angles all of the members of a given one-parameter family of plane curves that all lie in the same plane. The variations of the infinitesimal brachistochrone in a vacuum and the segments that make up a tautochrone are, of course, families of objects in a plane of one of the kinds just described (segments of plane curves which all lie in the same plane in the case of the infinitesimal brachistochrone in a vacuum, and segments of a plane curve in the case of a tautochrone) which do determine the plane curves in question in each of these two problems. However, the point to bear in mind is that, unlike a standard problem, where the members of a family of plane curves that all lie in the same plane are always *given either explicitly or implicitly at the start*, the members of the relevant family of objects in a plane are *not actually given either explicitly or implicitly at the outset*, either in the case of a problem of finding brachistochrones or in the case of a problem of finding tautochrones. With a great show of imagination, Fontaine envisioned *converses* of the standard problems, in which families of objects in a plane of any one of the types just described *do exist*, at least in cases where the problems in question actually have solutions, but *they are families whose members are no more known at the start than are the plane curves that are the solutions of the problems in each case.* By envisaging the problems of finding brachistochrones and tautochrones the way that he did, Fontaine turned these problems into problems that he could apply his fluxio-differential method to, because he specifically devised this method to solve problems in families of objects in a plane of any one of the kinds just described (plane curves that all lie in the same plane, segments of plane curves which all lie in the same plane, segments of a plane curve). Later in this chapter we will discover additional evidence of this fact.

I make some final observations concerning the formal aspects of Fontaine's calculus of two differential operations. The Leibnizian differential operator was not a function that uniquely assigned a specific numerical value to a quantity to which it was applied. Instead it just determined an "order of magnitude" of a quantity–namely, an infinitesimal quantity of first order with respect to a quantity to which it was applied. Thus the Leibnizian differential calculus was a calculus of "infinitesimal variables" that were only determined up to orders of magnitude. The infinitesimal variables were produced by successive applications of the Leibnizian differential operator to finite variables and other infinitesimal variables. But "differential" equations formally derived in this way are not entirely meaningful, because they include quantities that are not uniquely determined. Since the differential operator in Leibniz's differential calculus did not assign specific numerical values to any variables, some particular first-order differential ordinarily had to be chosen to be constant, in order to be able to interpret completely the "differential" equations formally arrived at using the calculus. How to interpret any particular "differential" equation depended critically on the particular choice of the first-order differential that was held constant in it. (Setting some first-order differential equal to a constant turns out to correspond to choosing a variable to be an independent variable in the modern differential calculus of functions of independent variables.)

Fontaine's calculus involved the use of two independent, first-order Leibnizian differential operations. Normally Fontaine would have had to stipulate that *both* a first-order "·"-differential *and* a first-order "*d*"-differential be constant when applying his fluxio-differential method to solve problems, otherwise fluxio-differential equations derived using his calculus of two differential operators would not be completely determined. But in applying his method to find tautochrones, Fontaine did not stipulate, for example, that $\dot{x}$ and that dX be constant. Moreover, although dX is constant with respect to "·"-differentiation [because $d\dot{X} = d(\dot{X}) = d(0) = 0$], neither $\dot{x}$ nor $\dot{y}$ is constant with respect to "d"-differentiation, otherwise $d(\dot{x})$ [(7.3.3.29)] or $d(\dot{y})$ [(7.3.3.31)] would equal zero in the calculations that Fontaine did to find tautochrones, which neither does. Fontaine did not state why the mixed differentials $d(\dot{x})$ and $d(\dot{y})$ cannot equal zero.

We can, however, imagine a possible way to explain why Fontaine did not stipulate that any first-order differentials be constant when applying his fluxio-differential method to determine tautochrones and why neither of the mixed differentials $d(\dot{x})$ and $d(\dot{y})$ can equal zero. Particular choices of constant first-order "·"-differentials and "d"-differentials, like $\dot{x}$ or $\dot{y}$ or dx or dy or dX, would only really have an effect on the way equations that include unmixed differentials of higher order which result from applying each of the operations "·" and "d" by itself more than one time to finite quantities are to be interpreted. For example, if first-order differentials $\dot{x}$ and dX were chosen to be constant, then any second-order differentials $\ddot{x}$ and ddX included in such equations would equal zero. But equations that only include differentials of first order have a unique property. They are "invariant," in the sense that even though the Leibnizian differential calculus leaves first-order differentials indeterminate, just as it leaves differentials of all orders indeterminate, nevertheless in an equation arrived at by formally applying the Leibnizian differential operator and that includes differentials of first order exclusively, the relations among the first-order differentials in such an equation turn out to be unambiguous. In other words, the relations are well determined. As a result, no first-order differential need be chosen to be constant in an equation that includes differentials of first order but that does not include any unmixed differentials of higher order, because choosing different first-order differentials to be constant in such an equation would not affect how the equation is to be interpreted.

But in the equation that Fontaine derived in order to find tautochrones, all unmixed differentials that result from applying each of the operations "·" and "d" alone to finite quantities are differentials of first order. Hence there is no need to stipulate that $\dot{x}$ or $\dot{y}$ or dx or dy or dX be constant. Why are all of the unmixed differentials that result from applying each of the operations "·" and "d" by itself to finite quantities differentials of first order in Fontaine's equations? By judiciously introducing first-order and *higher*-order *differential coefficients* (and, accordingly, higher-order "differential equations" like (7.3.3.33) and (7.3.3.34)), Fontaine in effect took a step toward reducing *higher*-order "differential" equations to *first*-order "differential" equations, where the higher-order "differential equations" act as a link between the higher-order "differential" equations and the first-order "differential" equations. The author of the definitive study of the Leibnizian differential calculus maintains that such a reduction became a very important activity in eighteenth-century mathematical analysis, once Euler began to carry out this reduction systematically in order to rid higher-order "differential" equations of their indeterminacy.[123] In finding tautochrones in the manner that he did, using his fluxio-differential method, Fontaine headed in the same direction that Euler did. It turns out that the way that Fontaine defined the differential coefficients p, q, and r, respectively, by means of (7.3.3.13), (7.3.3.14), and (7.3.3.15), respectively, and the differential coefficients γ, λ, and μ, respectively, by means of (7.3.3.26), (7.3.3.27), and (7.3.3.28), respectively, in determining tautochrones amounts to choosing x to be an independent variable in the modern differential calculus of functions of independent variables. Stipulating that $\dot{x}$ be constant with respect to "·"–differentiation would have the same effect, but, as I say, Fontaine had no need to do this, since none of the equations that he derived in order to determine tautochrones include unmixed differentials of higher order which result from applying the "·" operation alone to x more than one time.

In introducing the unspecified transformation ϕ defined by

$$dx = \phi \, dX, \qquad\qquad (7.3.3.25)$$

where he stipulated that $\phi = \phi\,(x, X)$ fulfill the two conditions: $\phi(0, X) = 0$ and $\phi(X, X) = 1$, Fontaine in effect also treated X as an independent variable, which means that he implicitly took dX to be constant with respect to "d"-differentiation. Now, we have seen that Fontaine used in his calculations the fact that dX is a constant with respect to "·"-differentiation, but this fact follows from the rules that govern the way that the two differential operators in the fluxio-differential method act [specifically, $d\dot{X} = d(\dot{X}) = d(0) = 0$]. But Fontaine did not have to stipulate that dX also be constant with respect to "d"-differentiation, because none of the equations that he derived in order to find tautochrones include unmixed differentials of higher order which result from applying the "d" operation by itself to X more than one time.)

My discussion here may help to explain why Fontaine did not stipulate that first-order differentials $\dot{x}$ or $\dot{y}$ or dx or dy or dX be constant when applying his fluxio-differential method to find tautochrones.

To understand why neither of the mixed differentials $d(\dot{x})$ nor $d(\dot{y})$ equals zero when Fontaine's fluxio-differential method is applied to determine tautochrones, it is probably enough to recall that $d(\dot{x}) = \dot{\overline{dx}}$ and $d(\dot{y}) = \dot{\overline{dy}}$ are true, because the differential operators "·" and "d" commute. Now, we remember that the fluxio-differential method can be used to solve problems that involve a family of objects in a plane of any one of the following kinds: plane curves that all lie in the same plane; segments of plane curves which all lie in the same plane; and segments of a plane curve. Furthermore, we recall that the dot "·" indicates differentiation of a quantity defined at points located on a member of such a family *along* that member of the family and that the "d" signifies differentiation of a quantity defined at points situated on a member of such family *from that* member of the family *to another* member of the family which is located nearby. In other words,

"·"-differentiation of a quantity defined at points located on a member of such a family means differentiation of the quantity *along* that member of the family, and "*d*"-differentiation of a quantity defined at points situated on a member of such a family means differentiation of the quantity *between* or *across* neighboring members of the family. In the case of a problem that involves any one of the three kinds of families just described (the problem of orthogonal trajectories involves a family of the first kind, the problem of finding the brachistochrone in a vacuum involves a family of the second kind, and the problem of finding a tautochrone involves a family of the third kind), if x is a coordinate for the members of the family (say finite arc length) and $\dot{x}$ is chosen to be constant, meaning that $\dot{x} = 0$, and X is the parameter of the family (which also designates the finite arc lengths of the members of the relevant family in the case of the tautochrone), then in general dx will *not* be constant at different points along a member of the family whose coordinates x are different. Consequently $\dot{\overline{dx}}$ will *not* equal zero at points along a member of the family, which explains why $d(\dot{x}) \neq 0$, since $d(\dot{x}) = \dot{\overline{dx}}$. In other words, the fact that $\dot{x}$ may be chosen to be constant with respect to "·"-differentiation does not entail that dx need be constant with respect to "·"-differentiation, in a calculus like Fontaine's which involves two independent Leibnizian differential operators.

One consequence of this result is the following one. Since neighboring members of the relevant family in the case of the tautochrone all but coincide with each other, one might think that the "·" and "d" operations "cannot tell each other apart," so to speak, where differences in arc length x along a member of the family or "between" neighboring members of the family are concerned. That is, in the case of such a family whose neighboring members overlap, there is nothing "between" neighboring members, so that one might presume that $\dot{x}$ and dx virtually cannot be distinguished from each other in the case of neighboring members of the family. In other words, one might be tempted to conclude that $\dot{x} = dx$ for neighboring members of the family (Figure 27). It would, however, be an error to draw such a conclusion. For $\dot{x}$ is constant with respect to "·"-differentiation, which means that $\dot{x} = 0$. Therefore if $\dot{x} = dx$ were true, it would follow that $0 = \dot{x} = \dot{\overline{dx}} = d(\dot{x})$, which is false.

Similarly, if y designates, say, the speed of a body that traverses a member of the relevant family in accordance with an equation of motion that has, say, the form (7.3.3.36), then in general dy will also not be constant at different points along a member of the family whose coordinates x are different. Therefore $\dot{\overline{dy}}$ will also not equal zero at points along a member of the family, which explains why $d(\dot{y}) \neq 0$, since $d(\dot{y}) = \dot{\overline{dy}}$. In other words, the fact that $\dot{y}$ may be chosen to be constant with respect to "·"-differentiation does not entail that dy need be constant with respect to "·"-differentiation, in a calculus like Fontaine's which involves two independent Leibnizian differential operators.[124]

I shall now demonstrate that Fontaine's solution to the problem of finding tautochrones can be arrived at by using the modern differential calculus of two

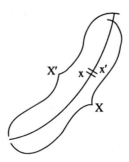

Figure 27. Plane curve decomposed into a family of segments.

variables. I do this because the exercise will help shed light upon Fontaine's subsequent work. I must make one thing very clear at the beginning. My solution is merely a translation of Fontaine's solution. I emphasize that I would never have found my solution if I had not studied Fontaine's solution. I shall employ here modern notation for derivatives, partial derivatives, and integrals.

I begin with Fontaine's equation (7.3.3.37), which generalizes tautochronous motion. I rewrite this equation as

$$T(X) = \int_0^X \frac{dx}{y(x, X)}.$$ (7.3.3.37′)

It follows from logarithmic differentiation that

$$\frac{d}{dX} \ln T(X) = \frac{1}{T(X)} \frac{d}{dX} T(X),$$

or

$$\frac{d}{dX} T(X) = T(X) \frac{d}{dX} \ln T(X),$$

which becomes after making substitutions

$$\frac{d}{dX} \int_0^X \frac{dx}{y(x, X)} = \left(\frac{d}{dX} \ln T(X) \right) \int_0^X \frac{dx}{y(x, X)} = \int_0^X \frac{\frac{d}{dX} \ln T(X)}{y(x, X)} \, dx,$$ (7.3.3.39)

for all X. Now, the left-hand side of (7.3.3.39) can be rewritten as

$$\frac{d}{dX} \int_0^X \frac{dx}{y(x, X)} = \int_0^X \frac{\partial}{\partial X} \left(\frac{1}{y(x, X)} \right) dx + \left[\frac{\partial}{\partial u} \int_0^u \frac{dx}{y(x, X)} \right]_{u = X}$$

(7.3.3.40)

$$= \int_0^X \frac{-1}{y(x, X)^2} \frac{\partial y}{\partial X} (x, X) \, dx + \left[\frac{\partial}{\partial u} \int_0^u \frac{dx}{y(x, X)} \right]_{u = X},$$

Figure 28. Segments of a plane curve with common endpoint (where segments are determined by initial conditions of motion along the curve).

where the first term on the right-hand side of the first equality in (7.3.3.40) follows from Leibniz's inversion of differentiation and integration.

Now, the fundamental theorem of the calculus cannot be used to evaluate the term

$$\left[\frac{\partial}{\partial u}\int_0^u \frac{dx}{y(x, X)}\right]_{u=x} \tag{7.3.3.41}$$

within (7.3.3.40), because

$$\left[\frac{1}{y(u, X)}\right]_{u=x} = \infty,$$

since $y(X, X) = 0$. In order to handle (7.3.3.41), we take Figure 28 and suppose that an arc of length x along a segment whose arc length is u corresponds to an arc of length x' along a segment whose arc length is X through

$$x = x' + \varphi(x', X)\{u - X\},$$

where $\varphi(x', X)$ is a function of x' along segment X which has yet to be determined and for which $\varphi(0, X) = 0$ (in order that $x = x' = 0$ at the start) and $\varphi(X, X) = 1$ (so that $x = u = X + \{u - X\} = x' + \{u - X\}$ correspond at the end). Then, using this as a change of variables,

$$x = X' + \varphi(x', X)\{u - X\},$$

$$dx = dx' + \frac{\partial\varphi}{\partial x'}(x', X)\, dx'\{u - X\} = \left(1 + \frac{\partial\varphi}{\partial x'}(x', X)\{u - X\}\right)dx', \tag{7.3.3.42}$$

for fixed u and X, we have $x' = X$, when $x = u$, and

$$\int_0^u \frac{dx}{y(x, X)} = \int_0^X \frac{(1 + (\partial\varphi/\partial x')(x', X)\{u - X\})}{y(x' + \varphi(x', X)\{u - X\}, X)}\, dx',$$

for fixed u and X. Hence

$$\frac{\partial}{\partial u}\int_0^u \frac{dx}{y(x, X)} = \frac{\partial}{\partial u}\int_0^X \frac{(1 + (\partial\varphi/\partial x')(x', X)\{u - X\})}{y(x' + \varphi(x', X)\{u - X\}, X)}\, dx'$$

holds for all u and X where $u \neq X$. Suppose we make the assumption that this equality holds as well when $u = X$, despite the singularity in the integrand of the integral on the left-hand side when $u = X$ and the singularity in the integrand of the integral on the right-hand side when $x' = X$ and $u = X$. Then in this case

$$\left[\frac{\partial}{\partial u}\int_0^u \frac{dx}{y(x, X)}\right]_{u=x} = \left[\frac{\partial}{\partial u}\int_0^X \frac{(1 + (\partial\varphi/\partial x'))(x', X)\{u - X\})}{y(x' + \varphi(x', X)\{u - X\}, X)}dx'\right]_{u=x}.$$

Since x' can be treated as a dummy variable in the integral on the right-hand side of this equation, we can replace it by x and thereby obtain

$$\left[\frac{\partial}{\partial u}\int_0^u \frac{dx}{y(x, X)}\right]_{u=x} = \left[\frac{\partial}{\partial u}\int_0^X \frac{(1 + (\partial\varphi/\partial x))(x, X)\{u - X\})}{y(x + \varphi(x, X)\{u - X\}, X)}dx\right]_{u=x}. \quad (7.3.3.43)$$

Applying Leibniz's inversion of differentiation and integration to the right hand side of (7.3.3.43), which we presume to hold up to and including the singularity in the integrand at $u = X$ when $x = X$, we obtain

$$\left[\frac{\partial}{\partial u}\int_0^u \frac{dx}{y(x, X)}\right]_{u=x} = \left[\int_0^X \frac{\partial}{\partial u}\left[\frac{(1 + (\partial\varphi/\partial x))(x, X)\{u - X\})}{y(x + \varphi(x, X)\{u - X\}, X)}\right]dx\right]_{u=x}. \quad (7.3.3.44)$$

If we assume the continuity of the integrand of the integral on the right-hand side of (7.3.3.44) as a function of (x, u) for each fixed X, despite the singularity in the integrand when $u = X$ and $x = X$, then (7.3.3.44) can be rewritten as

$$\left[\frac{\partial}{\partial u}\int_0^u \frac{dx}{y(x, X)}\right]_{u=x} = \int_0^X \frac{\partial}{\partial u}\left[\frac{(1 + (\partial\varphi/\partial x))(x, X)\{u - X\})}{y(x + \varphi(x, X)\{u - X\}, X)}\right]_{u=x}dx. \quad (7.3.3.45)$$

Now, the right-hand side of (7.3.3.45) equals

$$\int_0^X \left[\frac{1}{y(x + \varphi(x, X)\{u - X\}, X)^2}\left\{y(x + \varphi(x, X)\{u - X\}, X)\frac{\partial\varphi}{\partial x}(x, X)\right.\right.$$
$$\left.\left.-\varphi(x, X)\left(1 + \frac{\partial\varphi}{\partial x}(x, X)\{u - X\}\right)\frac{\partial y}{\partial g}(g, X)\Big|_{g=x+\varphi(x,X)\{u-X\}}\right\}\right]_{u=x}dx$$
$$= \int_0^X \left[\frac{(\partial\varphi/\partial x)(x, X)}{y(x, X)} - \frac{1}{y(x, X)^2}\frac{\partial y}{\partial x}(x, X)\varphi(x, X)\right]dx.$$

Hence the left-hand side of (7.3.3.39) equals

$$\frac{d}{dX}\int_0^X \frac{dx}{y(x, X)} = \int_0^X \left[\frac{-1}{y(x, X)^2}\frac{\partial y}{\partial X}(x, X)\right.$$
$$\left.+ \frac{(\partial\varphi/\partial x)(x, X)}{y(x, X)} - \frac{1}{y(x, X)^2}\frac{\partial y}{\partial x}(x, X)\varphi(x, X)\right]dx.$$

Equation (7.3.3.39) can thus be written as

$$\int_0^X \left[\frac{-1}{y(x, X)^2}\frac{\partial y}{\partial X}(x, X) + \frac{(\partial\varphi/\partial x)(x, X)}{y(x, X)} - \frac{1}{y(x, X)^2}\frac{\partial y}{\partial x}(x, X)\varphi(x, X)\right]dx$$
$$= \int_0^X \frac{(d/dX)\ln T(X)}{y(x, X)}dx, \quad (7.3.3.46)$$

for all X. If we equate the integrands on both sides of (7.3.3.46) for all x and X, a step that cannot in general be justified,[125] we obtain

$$-\frac{1}{y^2}\frac{\partial y}{\partial X} + \frac{(\partial\varphi/\partial x)}{y} - \frac{1}{y^2}\frac{\partial y}{\partial x}\varphi = \frac{(d/dX)\ln T(X)}{y},$$

or, equivalently,

$$\frac{\partial y}{\partial X} = y\frac{\partial\varphi}{\partial x} - \varphi\frac{\partial y}{\partial x} - y\frac{d}{dX}\mathrm{l.}\ T(X), \tag{7.3.3.47}$$

for all x and X. Then if we differentiate (7.3.3.47) with respect to x, we obtain

$$\frac{\partial}{\partial x}\left(\frac{\partial y}{\partial X}\right) = \frac{\partial}{\partial x}\left(y\frac{\partial\varphi}{\partial x} - \varphi\frac{\partial y}{\partial x} - y\frac{d}{dX}\ln T(X)\right)$$

$$= y\frac{\partial^2\varphi}{\partial x^2} - \varphi\frac{\partial^2 y}{\partial x^2} - \frac{\partial y}{\partial x}\frac{d}{dX}\ln T(X). \tag{7.3.3.48}$$

Now, the equation of motion through a medium that resists as $(y^2 + my)/n$, where $y = y(x, X)$ is speed, is (7.3.3.16), which I rewrite as

$$\frac{\partial y}{\partial x} = \frac{-g(dz/dx)}{y} \pm \frac{y}{n} \pm \frac{m}{n}, \tag{7.3.3.16'}$$

where $z = z(x)$ designates the ordinate in rectangular coordinates and g designates uniform gravity. Differentiating (7.3.3.16') with respect to x results in

$$\frac{\partial^2 y}{\partial x^2} = \frac{\partial}{\partial x}\left(\frac{-g(dz/dx)}{y} \pm \frac{y}{n} \pm \frac{m}{n}\right)$$

$$= \frac{-gy(d^2z/dx^2) + g(\partial y/\partial x)(dz/dx)}{y^2} \pm \frac{1}{n}\frac{\partial y}{\partial x}. \tag{7.3.3.49}$$

If we replace $\partial^2 y/\partial x^2$ in the right-hand side of (7.3.3.48) by the right-hand side of (7.3.3.49), we obtain

$$\frac{\partial}{\partial x}\left(\frac{\partial y}{\partial X}\right) = \frac{1}{y^2}\left\{y^3\frac{\partial^2\varphi}{\partial x^2} + gy\varphi\frac{d^2z}{dx^2} - g\varphi\frac{\partial y}{\partial x}\frac{dz}{dx}\right.$$

$$\left.\mp\frac{1}{n}y^2\varphi\frac{\partial y}{\partial x} - y^2\frac{\partial y}{\partial x}\frac{d}{dX}\ln T(X)\right\}. \tag{7.3.3.50}$$

Now, if we differentiate (7.3.3.16') again, but this time with respect to X, we obtain

$$\frac{\partial}{\partial X}\left(\frac{\partial y}{\partial x}\right) = \frac{\partial}{\partial X}\left(\frac{-g(dz/dx)}{y} \pm \frac{y}{n} \pm \frac{m}{n}\right)$$

$$= \frac{g(dz/dx)(\partial y/\partial X)}{y^2} \pm \frac{1}{n}\frac{\partial y}{\partial X}. \tag{7.3.3.51}$$

If we replace $\partial y/\partial X$ in the right-hand side of (7.3.3.51) by the right-hand side of

(7.3.3.47), we obtain

$$\frac{\partial}{\partial X}\left(\frac{\partial y}{\partial x}\right) = \frac{1}{y^2}\left\{gy\frac{dz}{dx}\frac{\partial\varphi}{\partial x} - g\varphi\frac{dz}{dx}\frac{\partial y}{\partial x} - gy\frac{dz}{dx}\frac{d}{dX}\ln T(X)\right.$$
$$\left.\pm\frac{1}{n}y^3\frac{\partial\varphi}{\partial x}\mp\frac{1}{n}y^2\varphi\frac{\partial y}{\partial x}\mp\frac{1}{n}y^3\frac{d}{dX}\ln T(X)\right\}. \qquad (7.3.3.52)$$

Like many of the other equations above, neither (7.3.3.50) nor (7.3.3.52) holds when $y = 0$–that is, when $x = X$ in $y(x, X)$, since $y(X, X) = 0$. However, since I have already assumed conditions that I did not show to be true at $y(X, X) = 0$ (Leibniz's inversion of differentiation and integration, the continuity of integrands), I will continue for the moment just to calculate formally without paying attention to the singularities. If we now assume that

$$\frac{\partial}{\partial x}\left(\frac{\partial y}{\partial X}\right) = \frac{\partial}{\partial X}\left(\frac{\partial y}{\partial x}\right), \qquad (7.3.3.53)$$

then we can equate the right-hand sides of (7.3.3.50) and (7.3.3.52). If we multiply both sides of the equation that results by y^2, then cancel the terms that the two sides of the new equation have in common, we arrive at the following equation

$$y^3\frac{\partial^2\varphi}{\partial x^2} + gy\varphi\frac{d^2z}{dx^2} - y^2\frac{\partial y}{\partial x}\frac{d}{dX}\ln T(X) = gy\frac{dz}{dx}\frac{\partial\varphi}{\partial x}$$
$$- gy\frac{dz}{dx}\frac{d}{dX}\ln T(X) \pm \frac{1}{n}y^3\frac{\partial\varphi}{\partial x}\mp\frac{1}{n}y^3\frac{d}{dX}\ln T(X). \qquad (7.3.3.54)$$

Now we consider two cases:

1. the tautochrone. In this case $T(X) \equiv a$ constant, and therefore $(d/dX)\ln T(X) \equiv 0$. If we eliminate the terms in (7.3.3.54) which are zero because of this, both sides of the equation that remains can be divided through by y, and the equation that results can be rewritten in the following way

$$\left(\frac{\partial^2\varphi}{\partial x^2}\mp\frac{1}{n}\frac{\partial\varphi}{\partial x}\right)y^2 + g\varphi\frac{d^2z}{dx^2} - g\frac{dz}{dx}\frac{\partial\varphi}{\partial x} = 0, \qquad (7.3.3.55)$$

which is a quadratic equation in the speed y whose coefficients do not depend upon time. If Fontaine had eliminated $\dot{y}$ between his equations (7.3.3.16) and (7.3.3.32), and then divided through the equation that results by the nonzero factor $\dot{x}$ that all of its terms have in common, he would have arrived at the following equation

$$\left(\frac{\mp(3/n)\gamma + \lambda}{2\gamma}\pm\frac{1}{n}\right)y^2 + g\left(\frac{p\gamma + g\varphi}{2\gamma} - p\right) = 0. \qquad (7.3.3.56)$$

Instead, however, he compared equations (7.3.3.16) and (7.3.3.32) term by term, and he equated the coefficients of the terms in each which correspond. This is the same condition as having the coefficients of the quadratic equation (7.3.3.56) in y vanish identically. Fontaine did not justify comparing his equations (7.3.3.16) and (7.3.3.32) term by term and equating the coefficients of corresponding terms – or, what amounts to the same thing, setting the coefficients of equation (7.3.3.56) identically equal to zero. In general, the coefficients of equation (7.3.3.56) need not vanish identically, because the coefficients of the quadratic polynomial on the left-hand side of equation (7.3.3.56) are functions of x and X, not constants. All of the speeds y with which motion takes place in accordance with (7.3.3.16) and (7.3.3.20) solve equation (7.3.3.56), which means that this quadratic equation has more than two solutions. However, because its coefficients are not constants but functions of x and X instead, the rule that the coefficients of a quadratic polynomial with more than two roots must vanish identically does not apply. Fontaine did note that comparing his two equations (7.3.3.16) and (7.3.3.32) term by term was *not* in general valid – namely, it was not valid when the forces causing the motion had a form more general than the ones that he considered in his paper.[126] If we set the coefficient in (7.3.3.55) equal to zero, we obtain two equations

$$\frac{\partial^2 \varphi}{\partial x^2} \mp \frac{1}{n}\frac{\partial \varphi}{\partial x} = 0 \qquad (7.3.3.57)$$

$$\varphi \frac{d^2 z}{dx^2} - \frac{dz}{dx}\frac{\partial \varphi}{\partial x} = 0. \qquad (7.3.3.58)$$

If we set the coefficients in (7.3.3.56) equal to zero, we just obtain Fontaine's two equations (7.3.3.33) and (7.3.3.34). The last two equations can be simplified:

$$\varphi q - p\gamma = 0 \qquad (7.3.3.33')$$

$$\lambda \mp \frac{1}{n}\gamma = 0. \qquad (7.3.3.34')$$

But (7.3.3.33') is (7.3.3.58), while (7.3.3.34') is (7.3.3.57). The derivatives and partial derivatives that appear in (7.3.3.57) and (7.3.3.58) are the same as the differential coefficients that appear in (7.3.3.33') and (7.3.3.34').

2. Curves other than the tautochrone. In this case $T(X) \not\equiv$ a constant hence $(d/dX) \ln T(X) \not\equiv 0$. The left-hand side of equation (7.3.3.54) includes a term that has $\partial y/\partial x$ as a factor, and we can solve (7.3.3.54) for $\partial y/\partial x$:

$$\frac{\partial y}{\partial x} = \frac{1}{y^2 (d/dX)\ln T(X)}\left\{ y^3 \frac{\partial^2 \varphi}{\partial x^2} + gy\varphi \frac{d^2 z}{dx^2} - gy\frac{dz}{dx}\frac{\partial \varphi}{\partial x} + gy\frac{dz}{dx}\frac{d}{dX}\ln T(X) \right.$$

$$\left. \mp \frac{1}{n}y^3 \frac{\partial \varphi}{\partial x} \pm \frac{1}{n}y^3 \frac{d}{dX}\ln T(X) \right\}. \qquad (7.3.3.59)$$

If we equate the right-hand sides of (7.3.3.59) and (7.3.3.16′), the equation that results can be reduced to the following equation:

$$\left(\frac{\partial^2 \varphi}{\partial x^2} \mp \frac{1}{n}\frac{\partial \varphi}{\partial x}\right)y^2 \mp \frac{m}{n}y\frac{d}{dX}\ln T(X) + g\varphi\frac{d^2 z}{dx^2}$$

$$- g\frac{dz}{dx}\frac{\partial \varphi}{\partial x} + 2g\frac{dz}{dx}\frac{d}{dX}\ln T(X) = 0. \qquad (7.3.3.60)$$

As I indicated earlier, Fontaine began by considering paths more general than the tautochrone. Specifically, he assumed that segments of his curves whose arc lengths are X are traversed in time

$$T(X) = \frac{X^k}{A^k}.$$

The constant of proportionality A^k took the form that it did in order to preserve the dimensional homogeneity that is a vestige of seventeenth-century geometrical algebra. When $k = 0$ the curve is a tautochrone. When Fontaine treated motion of bodies through media that resist as $(y^2 + my)/n$, he only investigated the case where $k = 0$, and it is not difficult to figure out why. Any calculations that Fontaine did which involved curves that generalize the tautochrone would have undoubtedly led him to results that amount to having the coefficients of the quadratic polynomial in y on the left-hand side of (7.3.3.60) become equal to zero. This means that the coefficient $\mp (m/n)(d/dX)\ln T(X)$ of the y-term in the quadratic polynomial must vanish. But m does not equal zero. Hence the only possibility is that $(d/dX)\ln T(X) = 0$, in which case $T(X) = $ a constant, or the condition for tautochronous motion in other words. And of course (7.3.3.60) reduces to (7.3.3.55), or case 2 reduces to case 1.

What does my derivation of the tautochrone in a medium that resists the motion of a body directly as $(y^2 + my)/n$, where y is the speed of the body, show? It demonstrates that Fontaine's solution to the problem of finding tautochrones, and his fluxio-differential method more generally, *must be closely related to the partial differential calculus.* I have presented my derivation of the tautochrone in a medium that resists the motion of a body directly as $(y^2 + my)/n$, where y is the speed of the body, as the first piece of evidence that connections between the two calculuses exist. Recognizing the fact that the two calculuses are closely related is important for understanding the work that Fontaine did using the partial differential calculus after 1734, which I shall begin to discuss shortly.

Before I do that I shall make a few additional observations regarding Fontaine's fluxio-differential method which will turn out to be useful for understanding the relations between this calculus and the partial differential calculus. The following rules apply to the first-order differential operators "·" and "d" of the fluxio-differential method when these two operators act on *finite* quantities $F = F(x, a)$ expressed exclusively in terms of two *finite* variables x and a. If

$F = F(x, a)$ is a finite quantity defined in terms of finite variables x and a, then

$$dF = Adx + Bda \quad \text{and} \quad \dot{F} = A\dot{x} + B\dot{a}, \qquad (7.3.3.61)$$

where A and B are differential coefficients. We recall that, unlike differentials, differential coefficients are finite quantities. A can be found by applying the differential operators "d" to $F = F(x, a)$, using the rules given in (7.3.3.1) which govern the way that "d" acts, while holding a constant in the expression $F = F(x, a)$. Similarly, B can be found by applying the differential operator "d" to $F = F(x, a)$, again using the rules given in (7.3.3.1) which govern the way that "d" acts, while holding x constant in the expression $F = F(x, a)$. When the two differentials arrived at this way are added together, the result is the differential dF. A and B can also be determined by applying the same technique to the differential operator "$\cdot$" instead. I shall explain in a moment why the differential coefficients of dx and $\dot{x}$, respectively, in the expressions for dF and $\dot{F}$, respectively, in (7.3.3.61) must be the same and, similarly, why the differential coefficients of da and $\dot{a}$, respectively, in the expressions for dF and $\dot{F}$, respectively, in (7.3.3.61) must also be the same.

The rules in (7.3.3.61) can be derived rather easily. (The method for deriving them is essentially the same as the one now used to determine the total differential of a function of two independent variables which has continuous partial derivatives in the two variables.) But it is enough to check that these rules accord with the rules given in (7.3.3.1) which govern the ways that the differential operators "d" and "$\cdot$" act.

For example, let $F(x, a) \equiv x + a$. Then according to the rule given in (7.3.3.1) for "d" which applies to sums, dx is the "d"-differential of F when a is held constant in $F = F(x, a)$. Hence $Adx = dx$ implies that $A = 1$. Moreover, according to the same rule given in (7.3.3.1), da is the "d"-differential of F when x is held constant in $F = F(x, a)$, in which case $Bda = da$ means that $B = 1$. Then if these two "d"-differentials are added together, the result is the expression $Adx + Bda = dx + da$, which equals dF when $F(x, a) \equiv x + a$, according to the rule given in (7.3.3.1) which governs the way "d" operates when it is applied to sums.

If $F(x, a) \equiv xa$ instead, then according to the rule given in (7.3.3.1) for "d" which applies to products, adx is the "d"-differential of F when a is held constant in $F = F(x, a)$. Therefore $Adx = adx$ implies that $A = a$. Furthermore, according to the same rule given in (7.3.3.1), xda is the "d"-differential of F when x is held constant in $F = F(x, a)$. But $Bda = xda$ means that $B = x$. Then if these two "d"-differentials are added together, the result is the expression $Adx + Bda = adx + xda$, which equals dF when $F(x, a) \equiv xa$, according to the rule given in (7.3.3.1) which governs the way that "d" operates when it is applied to products.

Finally, if $F(x, a) \equiv x/a$, then according to the rule given in (7.3.3.1) for "d" which applies to quotients, $(adx)/a^2 = (1/a)dx$ is the "d"-differential of F when a is held constant in $F = F(x, a)$. Hence $Adx = (1/a)dx$ implies that $A = 1/a$. In addition, according to the same rule given in (7.3.3.1), $(-xda)/a^2 = -(x/a^2)da$ is the "d"-differential of F when x is held constant in $F = F(x, a)$, in which case $Bda = -(x/a^2)da$ means that $B = -(x/a^2)$. Then if these two "d"-differentials are added together, the result is the expression

$$Adx + Bda = \frac{1}{a}dx - \frac{x}{a^2}da = \frac{adx - xda}{a^2},$$

which equals dF when $F(x, a) \equiv x/a$, according to the rule given in (7.3.3.1) which governs the way that "d" operates when it is applied to quotients.

Now, if $F = F(x, a)$, we find using analogous arguments that $\dot{F} = C\dot{x} + D\dot{a}$, where C can be found by applying the differential operator "$\cdot$" to $F = F(x, a)$, using the rules given in (7.3.3.1) which govern the way that "$\cdot$" acts, while holding a constant in the expression $F = F(x, a)$, and D can be found by applying the differential operator "$\cdot$" to $F = F(x, a)$, again using the rules given in (7.3.3.1) which govern the way that "$\cdot$" acts, while holding x constant in the expression $F = F(x, a)$. When the two differentials arrived at this way are added together, the result is the differential $\dot{F}$. But we can actually conclude even more. The rules given in (7.3.3.1) which govern the way that "d" operates when applied to sums, products, and quotients, respectively, are exactly the same as the rules given in (7.3.3.1) which govern the way that "$\cdot$" operates when applied to sums, products, and quotients, respectively. Consequently it must follow that $A = C$ and $B = D$. In other words, $\dot{F} = A\dot{x} + B\dot{a}$ must also be true if $dF = Adx + Bda$ is true. This explains why the differential coefficients of dx and $\dot{x}$, respectively, in the expressions for dF and $\dot{F}$, respectively, in (7.3.3.61) must be the same and why the differential coefficients of da and $\dot{a}$, respectively, in the expressions for dF and $\dot{F}$, respectively, in (7.3.3.61) must also be the same.

In case $F(x, a) \equiv F(x)$, then

$$dF = Adx \quad \text{and} \quad \dot{F} = A\dot{x}, \tag{7.3.3.61'}$$

which accords with Fontaine's assumption that both (7.3.3.14) and (7.3.3.18) hold simultaneously.

Now, let us suppose in particular that x designates some coordinate defined for points situated on plane curves in a family of plane curves that all lie in the same plane, or defined for points located on segments of plane curves in a family of segments of plane curves which all lie in the same plane, or defined for points situated on segments of a plane curve in a family of segments of a plane curve. Let us also assume that a stands for the parameter of the particular family among those just described which is relevant to a particular problem, in the case that the members of that family are distinguished from one another by means of a parameter (which need not always be true). We recall that "$\cdot$" signifies differentiation of a quantity defined at points located on a member of such a one-parameter family *along* that member of the family. We also remember that the value of the parameter a that corresponds to a particular member of such a one-parameter family is *constant* for that member of the family. Hence the parameter a is constant with respect to "$\cdot$"-differentiation along that member of the one-parameter family. Consequently

$$\dot{a} = 0 \qquad (7.3.3.62)$$

is always true [(7.3.3.38) is a special case of this result.] Thus in a case of this kind the rules given in (7.3.3.61) reduce to

$$dF = Adx + Bda \quad \text{and} \quad \dot{F} = A\dot{x}. \qquad (7.3.3.61'')$$

In other words, in an example of this type the right-hand side of the second relation $\dot{F} = A\dot{x} + B\dot{a}$ in (7.3.3.61'') includes no second term $B\dot{a}$, because a non-zero differential $\dot{a}$ does not exist. The first relation $dF = Adx + Bda$ in (7.3.3.61'') behaves like a total differential of the finite quantity F with respect to the coordinate x and the parameter a. In fact, like other relations that already exist in Fontaine's calculus, the relation $dF = Adx + Bda$ in (7.3.3.61'') is just another relation that connects differentials dF, dx, and da which already exist in Fontaine's calculus by means of other differential coefficients like those that also are already present in Fontaine's calculus. Thus the relation $dF = Adx + Bda$ in (7.3.3.61'') adds nothing conceptually new to what we have seen thus far.

In a little while, however, we will discover evidence that at some point during the years 1734–37, Fontaine apparently decided that although the formalism in (7.3.3.61'') is a coherent one, it also included superfluous elements. That is, he evidently decided that a *single* differential operator, which I shall designate by the Greek letter "δ" for the time being, defined by

$$\delta F = A\delta x + B\delta a, \quad \text{if} \quad F = F(x, a), \qquad (7.3.3.63)$$

sufficed to express differentiation of finite quantities $F = F(x, a)$ defined in terms of a coordinate x and a parameter a *both along* a member of any one of the kinds of one-parameter families described above *as well as between* or *across* neighboring members of any one of the kinds of one-parameter families described above. I say that Fontaine evidently came to this conclusion, because we shall soon see examples of the way that Fontaine used such a single differential operator "δ" after 1734 to solve problems expressed in terms of such one-parameter families. In proceeding this way, what Fontaine did in effect was utilize the partial differential calculus instead of his fluxio-differential method to solve such problems.

One way to come to the conclusion that the formalism in (7.3.3.61'') includes redundant elements is to observe that in limiting cases the two relations given in (7.3.3.61'') are transformed into each other. Or to express the matter another way, the two relations reduce to a single relation. However, in the argument that follows in which I demonstrate that this happens, we must remember that the Leibnizian differential operator was not a function that uniquely assigned a specific numerical value to a quantity to which it was applied. Instead it just determined an "order of magnitude" of a quantity – namely, an infinitesimal quantity of first order with respect to the quantity to which it was applied. Thus

$dF, \dot{F}, dx, \dot{x}$, and da do not have precise numerical values. Consequently my use of limiting processes now to determine what happens when we examine members of any one of the kinds of one-parameter families described which are nearer and nearer to each other must be regarded as more heuristic than anything else. With that in mind, if we look at members of any one of the kinds of one-parameter families described which are closer and closer to each other, it follows that $da \to 0$, in which case $dF = Adx + Bda \to Adx$ in (7.3.3.61″).The question is: Does it follow that $dF \to \dot{F}$, too, as it seems that dF should do? The answer is yes, because if we look at members of any one of the kinds of one-parameter families described which are nearer and nearer to each other, it also follows that $dx \to \dot{x}$ as well, in which case $dF = Adx + Bda \to Adx \to A\dot{x} = \dot{F}$ in (7.3.3.61″), too. Thus "in the limit" the two differential operations "·" and "d" reduce to a single differential operation. In other words, where differentiation of finite quantities $F = F(x,a)$ expressed in terms of the particular kinds of finite variables x and a specified is concerned (namely, a coordinate x and a parameter a of any one of the kinds of one-parameter families described), it should not be necessary to have to use two different differential operators to distinguish differentiation along individual members of any one of the kinds of one-parameter families described from differentiation between or across neighboring members of any one of the kinds of one-parameter families described. A single relation like (7.3.3.63) should in principle suffice to express both kinds of differentiation; it is unnecessary to use two different differential operators to distinguish differentiation along individual members of any one of the kinds of one-parameter families described from differentiation between or across neighboring members of any one of the kinds of one-parameter families described.

This argument is equally valid when F is independent of the parameter a – in other words, when $F(x,a) \equiv F(x)$. In this case

$$dF = Adx \text{ and } \dot{F} = A\dot{x}. \tag{7.3.3.61″}$$

Then if we examine members of any one of the kinds of one-parameter families described which are closer and closer to each other, it follows that $da \to 0$ and that $dx \to \dot{x}$. But then it follows from (7.3.3.61′) that $dF = Ad\dot{x} \to A\dot{x} = \dot{F}$, too, as dF should do.

This property of the two relations given in (7.3.3.61′) to reduce "in the limit" to a single relation in the particular case of an expression $F(x,a) \equiv F(x)$ may appear at first sight to contradict what I said previously concerning the two relations given in (7.3.3.61′). The point is that in general when the fluxio-differential method is used to solve a problem, dx never actually becomes equal to $\dot{x}$ and dF never actually becomes equal to $\dot{F}$. In other words, in general when the two relations in (7.3.3.61′) arise specifically from the application of the fluxio-differential method to a problem, these two relations do not transform into each other. Or to express the matter another way, the two relations do not reduce to a single relation. For example, we have seen that when the fluxio-differential method is applied to the problem of finding tautochrones, it cannot be true that $dx = \dot{x}$. Accordingly, we remember that the finite arc length X of a member of the particular one-parameter family relevant to the problem of finding a tautochrone is the parameter of that family and that in order to arrive at equation (7.3.3.32) dX can be chosen to be arbitrarily small in the equation above which immediately precedes equation (7.3.3.32) but that dX cannot be allowed to equal zero in that equation.

Fontaine in fact went much further in developing the notion of a single differential operator that can be used to differentiate in various ways. Eventually choosing the letter "d" to stand for the single differential operator, he defined partial differential coefficients, which he wrote as

$$\frac{dF}{dx}, \frac{dF}{dy}, \cdots$$

in whole-symbol notations, to be the coefficients of the differentials dx, dy, ... in the total differential

$$dF = \frac{dF}{dx} dx + \frac{dF}{dy} dy + \cdots \tag{7.3.3.64}$$

of a finite quantity $F = F(x, y, \ldots)$ expressed in terms of several finite variables $x, y, \ldots$ Fontaine substituted each of these particular coefficients for F in (7.3.3.64), and by means of such iteration he defined the full range of partial differential coefficients, both mixed and unmixed, of various orders. He introduced whole-symbol notations

$$\frac{dF}{dx}, \frac{d^2F}{dx^2}, \frac{d^2F}{dxdy}, \frac{d^3F}{dx^3}, \frac{dF}{dy}, \frac{d^2F}{dy^2}, \frac{d^3F}{dy^3}$$

and so forth to symbolize these partial differential coefficients of different orders.[127] He carefully distinguished these notations for partial differential coefficients from quotients of differentials, and his notations for partial differential coefficients became the standard notations for these coefficients and remained the standard notations for them for quite some time. In 1752 Euler stated that "these notations, first introduced by Fontaine, facilitate calculation."[128]

I mentioned above that we would soon see examples of the way that Fontaine used a single differential operator after 1734 to solve problems expressed in terms of the kinds of one-parameter families described. I also observed that in doing this he in effect employed the partial differential calculus instead of his fluxio-differential method to solve such problems. But we are not quite ready to put the fluxio-differential method out of mind and move on. For I contend that Fontaine's subsequent work involving the partial differential calculus, which I shall begin to examine shortly, *grew specifically out of the fluxio-differential method.* I meant to suggest as much in deriving the tautochrone in a medium that resists the motion of a body directly as $(y^2 + my)/n$, where y is the speed of the body, by using the partial differential calculus. I also intended to create this impression in arguing that the formalism in (7.3.3.61″) includes superflous elements. My demonstration of this statement also suggests how Fontaine might have derived the partial differential calculus from the fluxio-differential method. The demonstration enables us to imagine how this could have happened. But the demonstration *only* suggests *how* this *might* have occurred, and my derivation of the tautochrone *merely* suggests *that* such a thing *might* have happened. We need further evidence that will help support my claim that *this is indeed what actually occurred.*

Thus before I leave Fontaine's solution to the problem of finding tautochrones and turn to the work that he did afterward using the partial differential calculus, I want to examine his solution from still another standpoint. It will help us to understand better the work that he subsequently did using the partial differential calculus, as well as help make plausible certain claims that Fontaine made years later about connections between his fluxio-differential method and the work that he did afterward using the partial differential calculus which at first sight might appear improbable.

First, however, I shall replace Fontaine's notations "·" and "d", respectively, for his two differential operations by symbols "d" and "δ," respectively. I do this because Fontaine's notation "·" is not suitable for constructing whole-symbol notations for differential coefficients, which probably explains why Fontaine ultimately discarded the notation "·".

By applying his fluxio-differential method to equation (7.3.3.16) and to condition (7.3.3.20), Fontaine arrived at the following set of equations. Here I only indicate the forms that the individual equations in the set have, and I write these forms using the new notations "d" and "δ":

$$H(dz, y, dx, dy) = 0 \qquad\qquad (7.3.3.36')$$
$$dz = p\,dx \qquad\qquad (7.3.3.13')$$
$$dp = q\,dx \qquad\qquad (7.3.3.14')$$
$$G(p, q, y, dx, dy, \delta x, \delta y, \delta dx, \delta dy) = 0 \qquad\qquad (7.3.3.19')$$
$$\delta y = J(y, dx)\delta dx \qquad\qquad (7.3.3.24')$$
$$\delta x = \phi(x, X)\delta X \qquad\qquad (7.3.3.25')$$
$$d\phi = \gamma\,dx \qquad\qquad (7.3.3.26')$$
$$d\gamma = \lambda\,dx \qquad\qquad (7.3.3.27')$$
$$\delta dx = \psi(\gamma, dx)\delta X \qquad\qquad (7.3.3.29')$$
$$\delta dy = \Omega(\gamma, \lambda, y, dx, dy)\delta X \qquad\qquad (7.3.3.31')$$

The equations in this set do not all have the same functions or serve the same purposes. Equation (7.3.3.36′) is the original equation of motion. When it is combined with (7.3.3.13′), the equation that results is an equation of motion which has the form

$$K(p, y, dx, dy) = 0. \qquad\qquad (7.3.3.16')$$

Equations (7.3.3.13′), (7.3.3.14′), (7.3.3.25′), (7.3.3.26′), and (7.3.3.27′) *define differential coefficients* p, q, ϕ, γ *and* λ, *which are finite quantities, by means of differentials*, which are infinitesimal variables. Other equations are derived using the rules in (7.3.3.1) which govern the ways that the first-order differential operators "d" and "δ" act when these operators are applied to sums and products. Arriving at (7.3.3.29′) and (7.3.3.31′) involves using the fact that

$$d\delta X = 0. \tag{7.3.3.38′}$$

Deriving (7.3.3.29′) and (7.3.3.31′), respectively, also requires employing the equalities $\delta dx = d\delta x$ and $\delta dy = d\delta y$, respectively, which follow from the assumption that the two differential operators "d" and "δ" commute.

This set of equations can be treated as a system of equations in the differentials $dz, dx, dy, \delta x, \delta y, \delta X, \delta dx$, and δdy, which are all infinitesimal variables. (The differential δz does not appear in any of the equations in the system.) The differentials $dz, dy, \delta x, \delta y, \delta dx$, and δdy can then be eliminated from this system of equations, and a single equation

$$\left(g \left[\frac{p\gamma + q\phi}{2\gamma} - p \right] + \left[\frac{\mp (3/n)\gamma + \lambda}{2\gamma} \pm \frac{1}{n} \right] y^2 \right) dx \delta X = 0 \tag{7.3.3.65}$$

in two remaining differentials dx and δX results. Here I have written equation (7.3.3.65) explicitly, instead of simply indicating its form alone. We now consider the expression

$$T(y, p, q, \phi, \gamma, \lambda) \equiv g \left(\frac{p\gamma + q\phi}{2\gamma} - p \right) + \left(\frac{\mp (3/n)\gamma + \lambda}{2\gamma} \pm \frac{1}{n} \right) y^2. \tag{7.3.3.66}$$

Setting $dx \neq 0$ and $\delta X \neq 0$ in equation (7.3.3.65), we can think of

$$T(y, p, q, \phi, \gamma, \lambda) = 0 \tag{7.3.3.67}$$

as a "conditional equation" in y and the differential coefficients p, q, ϕ, γ, and λ which must hold in order for equation (7.3.3.65) to hold, hence in order for the system of equations in the differentials $dz, dx, dy, \delta x, \delta y, \delta X, \delta dx$, and δdy, which reduces to equation (7.3.3.65), to have a nontrivial solution, hence in order for the equation (7.3.3.36′) and the condition (7.3.3.20) from which this system of equations was derived and which express the original problem (finding a tautochrone) to have a solution, too. I call the reader's attention to this fact here because we will discover later in this chapter that conditional equations played an important part in some work in the integral calculus which Fontaine did at the end of the 1730s and which involved the use of the partial differential calculus.

In a brief paragraph that ended his Paris Academy *mémoire* of 1734 on finding tautochrones, Fontaine made some obscure remarks that suggest that he intended the fluxio-differential method to be used to solve a range of problems much broader than the one that he had covered in detail in the *mémoire*. For the moment I return to Fontaine's original notations "$\cdot$" and "d" to explain these problems. Fontaine vaguely illustrated what he meant as follows: if x and y designate the abscissa of a point and the ordinate of that point, respectively, in rectangular coordinates, and if any equation that has the form $G(x,y,X,A) = 0$ (that is, any finite equation) or the form $G(x,y,\dot{x},\dot{y},X,A) = 0$ (in other words, any fluxional equation in Fontaine's sense, which means any ordinary "differential" equation, which Fontaine wrote using the notation "$\cdot$" because the differentials at points situated on plane curves that all lie in the same plane and that are the solutions of the equation are of course all taken *along* the individual plane curves) is given, where X stands for the abscissa of one of the endpoints in rectangular coordinates of some particular segment of a plane curve and A is the value of the parameter for that plane curve, then according to Fontaine the fluxio-differential method would be equally useful for determining what expressions like $FL(\dot{x}^2 + \dot{y}^2)^{1/2}, FL(y\dot{x})$, etc., considered as functions of x, y, and A, become when x has the value X, or, conversely, for determining what has to be done in order to make expressions like $FL(\dot{x}^2 + \dot{y}^2)^{1/2}, FL(y\dot{x})$, etc., respectively, equal to a given quantity or to some other, given expression, when x has the value X. As Fontaine expressed the matter, an "infinity" of such problems "can depend upon the fluxio-differential method."[129]

In his Paris Academy *mémoire* of 1734 Fontaine did not provide the reader with any further explanation of what he was talking about here. In fact, the examples that he vaguely described in one short paragraph at the end of his *mémoire* furnish us with enough clues, however, for us to determine what he meant. Given the meanings assigned to the letters above, $FL(\dot{x}^2 + \dot{y}^2)^{1/2}$ just stands for the arc length of a segment of a plane curve in the notation of fluxions and fluents [which is written as

$$\int_0^X \sqrt{dx^2 + dy^2} = \int_0^X \sqrt{1 + \left(\frac{dy}{dx}\right)^2}\, dx$$

using modern notation], and $FL(y\dot{x})$ just symbolizes the area under a segment of a plane curve in the notation of fluxions and fluents [which is written as

$$\int_0^X y(x)dx$$

using modern notation]. Fontaine was evidently interested in problems in which a one-parameter family of plane curves that all lie in the same plane was given at the beginning either explicitly or implicitly by means of a finite equation or else through a first-order ordinary "differential" equation, where it was required to determine how the arc length of a particular segment of a particular plane curve in the one-parameter family or the area under such a segment (say, a segment along a plane curve in the one-parameter family whose value of the parameter is A, where X stands for the abscissa of one of the endpoints of the segment in rectangular coordinates) were expressed as functions of A and X (that is, as functions of the particular plane curves in the one-parameter family as well as the upper limits of integration of the integrals that express the arc lengths of segments of the plane curves in the one-parameter family or the areas under segments). Moreover, Fontaine was also interested in the converses of these kinds of problems: What must be done to make the arc lengths of segments of each plane curve in the one-parameter family or the areas under such segments, when these arc lengths or areas are considered as functions of the value A of the parameter for each plane curve in the one-parameter family as well as the upper limits of integration X of the integrals that express the arc lengths or areas, equal to a given quantity or to some other, given expression? [I note that the X appearing in $G(x, y, X, A) = 0$ is meaningless and that the X and the A appearing in $G(x, y, \dot{x}, \dot{y}, X, A) = 0$ are meaningless. The X does not belong in $G(x, y, X, A) = 0$, and the X and the A do not belong in $G(x, y, \dot{x}, \dot{y}, X, A) = 0$.]

It is useful to compare these kinds of problems with the problem of finding tautochrones. In his Paris Academy *mémoire* of 1734, Fontaine actually applied his fluxio-differential method to solve problems expressed by means of equations that have the form

$$G(\dot{z}, \dot{x}, y, \dot{y}) = 0 \text{ (an equation of motion)} \tag{7.3.3.36}$$

$$FL\left(\frac{\dot{x}}{y}\right) = T(X), \tag{7.3.3.37}$$

where z designates the ordinate of a point in rectangular coordinates, x symbolizes arc length, y stands for speed, $\dot{x}/y$ is the infinitesimal amount of time that it takes a body to traverse at a speed y an infinitesimal segment of a plane curve whose infinitesimal arc length is $\dot{x}$, $FL(\dot{x}/y)$ on the left-hand side of (7.3.3.37) expresses the amount of time that it takes a body whose motion is governed by an equation of motion which has the form (7.3.3.36) to traverse a segment of a plane curve whose finite arc length is X, where the segments of the plane curve all have a common endpoint [hence when expressed in terms of modern notation,

$$FL\left(\frac{\dot{x}}{y}\right) = \int_0^X \frac{1}{y(x, X)} dx \bigg], \tag{7.3.3.37'}$$

and $T(X)$ on the right-hand side of (7.3.3.37) designates a given function of the arc lengths X of segments of the plane curve which have a common endpoint. Thus $T(X)$ is a given expression in X which specifies the amount of time that it takes a body to traverse a segment whose arc length is X of the unknown plane curve that is the solution to a problem of the type expressed by means of equations that have the form (7.3.3.36) and (7.3.3.37). In particular, $T(X) \equiv$ a constant is the condition for tautochronous motion. That is, tautochronous motion occurs when $T(X) \equiv$ a constant for all such segments of the plane curve [equation (7.3.3.20)]. [As I mentioned earlier in this section of this chapter, the case that $T(X) \equiv$ a constant is the only case that Fontaine actually solved in his Paris Academy *mémoire* of 1734, and I also explained the reasons why.] Fontaine decomposed the unknown plane curve that is the solution to such a problem into a one-parameter family of overlapping segments of this unknown plane curve, where the parameter of the family is the arc lengths X of the segments in the family. As a result he could apply his fluxio-differential method to neighboring members of the family in order to try to solve a problem of this kind. Assuming that a particular problem of this type actually does have a solution, the family of segments of the plane curve that is the solution to the problem *does truly exist, but the segments in this family are unknown at the start, because these segments make up the very plane curve that is the solution to the problem, which is of course unknown at the outset.*

Fontaine's application of his fluxio-differential method in his Paris Academy *mémoire* of 1734 to solve problems envisaged this way should be contrasted with the remarks that Fontaine made at the end of his *mémoire*, where he contemplated applying his fluxio-differential method to one-parameter families of plane curves, where the plane curves in such a family all lie in the same plane and are *given at the beginning either explicitly or implicitly* through equations that have the forms

$$G(x, y, X, A) = 0 \quad \text{or} \quad G(x, y, \dot{x}, \dot{y}, X, A) = 0, \qquad (7.3.3.68)$$

where x and y designate the abscissa of a point and the ordinate of that point, respectively, in rectangular coordinates, where X stands for the abscissa of one of the endpoints in rectangular coordinates of some particular segment of one of the plane curves in such a one-parameter family, and where A denotes the parameter [in other words, finite equations or first-order ordinary "differential" equations for one-parameter families of plane curves, where the plane curves all lie in the same plane and where again I mention that the X appearing in $G(x, y, X, A) = 0$ is meaningless and that the X and the A appearing in $G(x, y, \dot{x}, \dot{y}, X, A) = 0$ are meaningless; the X does not belong in $G(x, y, X, A) = 0$, and the X and the A do not belong in $G(x, y, \dot{x}, \dot{y}, X, A) = 0$], in conjunction with expressions like

$$FL(\dot{x}^2 + \dot{y}^2)^{1/2} \quad \text{or} \quad FL(y\dot{x}), \qquad (7.3.3.69)$$

where the problem is to determine what is required to make expressions like those in (7.3.3.69) equal to given quantities or to some other, given expressions, in the particular case that the expressions in (7.3.3.69) are applied to one-parameter families of plane curves, where the plane curves in such a family all lie in the same plane and are given through equations that have either of the forms in (7.3.3.68). In such a case the abscissa X of one of the endpoints in rectangular coordinates of some particular segment of one of the plane curves in such a one-parameter family is an upper limit of integration for expressions like those in (7.3.3.69), just as the X that designates the arc length of a segment of the unknown plane curve that is the solution to a problem of the kind expressed by means of equations that have the form (7.3.3.36) and (7.3.3.37), where all such segments of the unknown plane curve have a common endpoint, is an upper limit of integration in the expression $FL(\dot{x}/y)$ on the left-hand side of (7.3.3.37).

Expressed in modern notation,

$$FL\left(\frac{\dot{x}}{y}\right) = \int_0^x \frac{1}{y(x, X)} dx.$$ (7.3.3.37′)

Here the X stands as well for the parameter of the one-parameter family of segments that make up the unknown plane curve that is the solution to a problem of this type, and $T(X)$ on the right-hand side of (7.3.3.7) designates a given function of the arc lengths X of segments of the plane curve which have a common endpoint. Hence $T(X)$ is a given expression in X which specifies the amount of time that it takes a body to traverse a segment whose arc length is X of the unknown plane curve that is the solution to a problem of the kind expressed by means of equations that have the form (7.3.3.36) and (7.3.3.37). (As I have already emphasized, assuming that a particular problem of this type actually does have a solution, the family of segments of the plane curve that is the solution to the problem does truly exist, but the segments in this family are unknown at the start, because these segments make up the very plane curve that is the solution to the problem, which is of course unknown at the beginning.) In cases where the problem is to determine what has to be done in order to make the expressions like (7.3.3.69) equal to given quantities or to some other, given expressions, it is natural that the given expressions have the form $T(X, A)$.

In his Paris Academy *mémoire* of 1734, Fontaine solved no problems expressed by means of equations that have either of the forms in (7.3.3.68). Nor did he even describe very clearly in his *mémoire* the problems of this kind which he had in mind. Nevertheless, he did give enough hints that allow us to infer that he also meant his fluxio-differential method to be utilized to solve problems expressed in terms of given one-parameter families of plane curves, where the plane curves in such a family all lie in the same plane. In other words, he also intended that his fluxio-differential method be used to solve what I have called standard problems in a family of plane curves that all lie in the same plane. In any particular standard problem, the members of such a one-parameter family are always given at the start either explicitly or implicitly, and then the problem is to find a plane curve or plane curves fulfilling some condition or having some property determined by the members of the given family which is specified beforehand. Finding orthogonal trajectories are examples of a standard problem. In this case the plane curves that are solutions to a problem of this kind intersect at right angles all of the members of a given one-parameter family of plane curves that all lie in the same plane. Evidently, the differential operator "·" would be used to differentiate quantities defined at points situated on a plane curve in the one-parameter family that is associated with such a standard problem *along* that plane curve, and the differential operator "d" would be used to differentiate quantities defined at points located on plane curves in the one-parameter family that is associated with such a standard problem *between* or *across* neighboring plane curves in the family. In other words, Fontaine also meant that his fluxio-differential method be employed to solve the kinds of problems expressed by means of given one-parameter families of plane curves, where the plane curves in such a family all lie in the same plane, which had induced Leibniz, Johann I Bernoulli, and Nikolaus I Bernoulli to create the partial differential calculus. Thus we should not be surprised to find that certain affinities exist between Fontaine's fluxio-differential method of 1732–34 and the work that Fontaine did afterward using the partial differential calculus. Indeed, as I suggested, we might expect to find that the connections exist between the fluxio-differential method and the partial differential calculus *themselves*.

From mid-March till the end of June 1737, Nicole read a paper before the Paris Academy in which he solved problems that, like the problem of finding orthogonal trajectories, required that plane curves be found which fulfilled conditions determined by means of a one-parameter family of curves that all lie in the same plane which was given at the start. The paper was published in the Paris Academy's *Mémoires* for 1737.[130] In each of the problems that Nicole considered, the members of the given one-parameter family of plane curves all meet at one point. In one problem the object was to find the plane curve that intersects the members of the given one-parameter family in such a way that the segments along all plane curves in the family which join the points of intersection to the point that all of the plane curves in the one-parameter family have in common are traversed in a vacuum by a body in equal amounts of time (the problem of finding synchrones). In another problem the plane curve to be found had to intersect the members of the given one-parameter family in such a way that the segments along all of the plane curves in the family which connect the points of intersection to the point where the plane curves in the family meet bound equal areas. In yet another problem, the segments of the members of the given one-parameter family intercepted by the plane curve to be found which join the points of intersection to the point where the plane curves in the family meet all had to have the same arc length.

These kinds of problems also belong to the general class of problems called problems of trajectories. In the *Journal des Savants* for 1697, Johann I Bernoulli had proposed several different problems of trajectories of these kinds. In *Acta Eruditorum* for January 1698, L'Hôpital published the plane curves that were the solutions to two of the problems that Johann I Bernoulli had formulated. In the *Acta* for May 1698, Jakob Bernoulli published his solutions to the same two problems. Like all mathematicians who participated in the late-seventeenth-century challenges, L'Hôpital and Jakob Bernoulli only published their solutions. They withheld the methods that they had used to find their solutions. In the *Acta* for May 1697, Johann I Bernoulli published a solution of sorts to one of the other problems that he had proposed. Johann only stated a property that the plane curve that solved that problem must have, namely, that it must intersect at right angles all of the members of the given one-parameter family of plane curves that all lie in the same plane, a certain family of cycloids in this particular case, at right angles. Nicole conjectured, probably correctly, that the problem of determining the plane curves orthogonal to the generalized brachistochrone cycloids had its basic origins in this particular property that Johann enunciated in 1697.

To solve these other kinds of problems of trajectories Nicole made use of exactly the same kinds of techniques that he had employed in order to find the plane curves orthogonal to the generalized brachistochrone cycloids in his Paris Academy *mémoire* of 1725. Namely, in each of the particular problems of trajectories which he tackled, Nicole started with an equation whose right-hand side was a nonalgebraic integral that made it impossible to solve the equation explicitly for the parameter of the given one-parameter family of plane curves that all lie in the same plane. He held the parameter constant; he expanded in an infinite series the integrand of the nonalgebraic integral; and then he integrated term by term this infinite series. Each problem of trajectories which Nicole attacked also involved a second equation that was algebraic and which could be solved explicitly for the parameter. Nicole solved this equation for the parameter, and he substituted the resulting expression for the parameter into the infinite series that followed from the term-by-term integration just mentioned. In this way he eliminated the parameter from the problem. He then manipulated the first equation transformed in the way that I just described. Specifically, he multiplied both sides of this equation, whose right-hand side was now an infinite series, by a judiciously chosen factor. He then differentiated once the new equation that followed from this operation. This involved differentiating term by term the infinite series that is the right-hand side of this new equation. In this way Nicole arrived at another equation whose right-hand side was also an infinite series but one whose sum he recognized. The equation accordingly reduced to an equation that had a finite number of terms. The equation simplified this way was in fact a first-order ordinary "differential" equation of first degree for the plane curve that was a solution to the particular problem of trajectories in question. By making a suitable change of variables Nicole was able to convert this equation into a first-order ordinary "differential" equation of first degree whose variables were separated. This second first-order ordinary "differential" equation of first degree he could consequently integrate, or, if not, he could at least reduce the equation to quadratures. Just as in Nicole's Paris Academy *mémoire* of 1725, the operations and procedures that Nicole utilized to derive the first-order ordinary "differential" equation of first degree for the plane curve that was the solution to each particular problem of trajectories depended entirely upon the particular one-parameter family of plane curves specified at the beginning, as well as upon the particular condition to be fulfilled which was determined by means of the given one-parameter family of plane curves.

Early in July of 1737, or less than a week after Nicole had finished reading his paper, Fontaine read a paper of his own before the Paris Academy in which he unified the individual problems that Nicole had solved and slightly generalized them.[131] The paper was never published in the Paris Academy's *Mémoires*. In the paper Fontaine observed that all of Nicole's problems could be stated in the following general way: solve a pair of equations that have the forms

$$F(x, y, p) = \phi(x, p), \qquad (7.3.3.70)$$

$$H(x, y, p) = 0, \qquad (7.3.3.71)$$

where

$$\phi(x,p) \equiv \int_0^x f(x^*, p)\, dx^*. \qquad (7.3.3.72)$$

Here $F(x, y, p)$, $H(x, y, p)$, and $f(x, p)$ designate given, finite algebraic expressions in rectangular coordinates x and y and a parameter p, where x is the abscissa and y is the ordinate of a point in rectangular coordinates; the integral (7.3.3.72) cannot be expressed algebraically; and one of the equations (7.3.3.70) or (7.3.3.71) represents a one-parameter family of plane curves that all lie in the same plane, which all meet at one point, and whose parameter is p, and the other equation acts as a condition that specificies the way that the plane curve that is the solution to the problem must intersect the plane curves in the given one-parameter family.

Nicole had treated the following kinds of specific cases of these problems: (i) $F(x, y, p) \equiv$ a constant in equation (7.3.3.70), in which case equation (7.3.3.70) was a nonalgebraic equation that functioned as a condition that specified how the plane curve that is the solution to the problem intersects the members of a given one-parameter family of algebraic plane curves expressed by means of equation (7.3.3.71) which all lie in the same plane, which all meet at one point, and whose parameter is p; (ii) $F(x, y, p) \equiv y$ in equation (7.3.3.70), in which case equation (7.3.3.71) was an algebraic equation that served as a condition that determined how the plane curve that is the solution to the problem intersects the members of a given one-parameter family of nonalgebraic plane curves expressed through equation (7.3.3.70) which all lie in the same plane, which all meet at one point, and whose parameter is p.

Again, like some of the more difficult problems of finding orthogonal trajectories, we have two equations, one of which is not algebraic [equation (7.3.3.70)]. The equation that is not algebraic cannot in general be solved explicitly for the parameter p, in which case the parameter p cannot easily be eliminated from the problem. Once again difficulties of the integral calculus created obstacles to solving these other kinds of problems of trajectories. To solve such problems what Nicole did in effect was to hold the parameter p constant in equation (7.3.3.70) and in (7.3.3.72), to expand in an infinite series the integrand of (7.3.3.72), to integrate term by term the resulting infinite series, to solve the algebraic equation (7.3.3.1) explicitly for p in terms of x and y, and then to substitute the resulting expression for p into the infinite series arrived at by term-by-term integration. In this way Nicole eliminated the parameter from the problem. Fontaine aimed to show how to deal with such problems of trajectories without having to resort to the use of infinite series the way that Nicole had done in his Paris Academy *mémoire* of 1725 and had done again in the paper that he had just finished reading.

In order to understand why Fontaine approached these other kinds of problems of trajectories in the manner that he did, we must recall that at the end of his Paris Academy *mémoire* of 1734 on finding tautochrones, Fontaine described very obscurely problems of determining how expressions like $FL(\dot{x}^{*2} + \dot{y}^{*2})^{1/2}$ (the arc length

$$\int_0^x \sqrt{dx^{*2} + dy^{*2}} = \int_0^x \sqrt{1 + \left(\frac{dy}{dx^*}\right)^2}\, dx^*$$

of a segment of a plane curve in modern notation, where x stands for the abscissa of one of the endpoints of the segment) and $FL(y\dot{x}^*)$ (the area

$$\int_0^x y(x^*)\, dx^*$$

under a segment of a plane curve in modern notation, where x again stands for the abscissa of one of the endpoints of the segment) behave as functions of the upper limit of integration x and the parameter p, when an equation that has the form

$$H(x^*, y, x, p) = 0 \qquad (7.3.3.73)$$

(that is, any finite equation) or the form

$$H(x^*, y, \dot{x}^*, \dot{y}, x, p) = 0 \qquad (7.3.3.74)$$

(in other words, any first-order ordinary "differential" equation) in rectangular coordinates x^* and y is given at the beginning, where x^* is the abscissa and y is the ordinate of a point in rectangular coordinates and where the given equation is an equation for a one-parameter family of plane curves that all lie in the same plane and whose parameter is p. [Note that in accord with what I said earlier about equations that have the forms in (7.3.3.68), the x appearing in equation (7.3.3.73) is meaningless, and the x and the p appearing in equation (7.3.3.74) are meaningless. The x does not belong in equation (7.3.3.73), and the x and the p do not belong in equation (7.3.3.74).]

If the one-parameter family of plane curves given at the start by means of an equation that has the form (7.3.3.73) or an equation that has the form (7.3.3.74) all meet at one point and can be expressed as $y = r(x^*, p)$, where $r(x^*, p)$ is a finite algebraic expression in x^* and p, then the arc lengths of segments of plane curves in the family which all have the point in common as one endpoint, the areas under segments of plane curves in the family which all have the point in common as one endpoint, etc., have the form

$$\int_0^x f(x^*, p)\, dx^*,$$

or precisely the form that appears in (7.3.3.72); for example,

$$\int_0^x \sqrt{dx^{*2} + dy^2} = \int_0^x \sqrt{1 + \left(\frac{dr}{dx^*}(x^*, p)\right)^2}\, dx^*, \qquad (7.3.3.75)$$

and

$$\int_0^x y(x^*)\, dx^* = \int_0^x r(x^*, p)\, dx^*. \qquad (7.3.3.76)$$

At the end of his Paris Academy *mémoire* of 1734 on finding tautochrones Fontaine had also vaguely described problems that are converses of these problems: if the arc lengths (7.3.3.75) and areas (7.3.3.76) are set equal at the beginning to some constant quantity or given expression in x and p, what does this entail? One could guess that doing this acts as a condition that determines a plane curve that intersects in a certain way the members of the one-parameter family of plane curves that all lie in the same plane and that are given through an equation that has the form (7.3.3.73) or the form (7.3.3.74).

In fact this is precisely what happens when an equation that has the form (7.3.3.73) expresses a one-parameter family of algebraic plane curves that all lie in the same plane, which all meet at one point, whose parameter is p, and which can be reexpressed in the form $y = r(x^*, p)$, but the corresponding expressions (7.3.3.75) and (7.3.3.76) turn out not to be algebraic. If in this case we set (7.3.3.75) or (7.3.3.76) equal to a constant, we arrive at an example of a problem of type (i) above where equation (7.3.3.71) expresses a given one-parameter family of algebraic plane curves that all lie in the same plane, which all meet at one point, and whose parameter is p, and where equation (7.3.3.70) is a nonalgebraic equation that functions as a condition that determines a plane curve that intersects the members of the given one-parameter family of algebraic plane curves a certain way – namely, such that the segments of the plane curves in the given one-parameter family which are intercepted by the plane curve to be found and which all have the point in common as one endpoint all bound equal areas or all have the same arc lengths. Here equation (7.3.3.73) plays the role of equation (7.3.3.71), and (7.3.3.75) or (7.3.3.76) set equal to a constant functions as equation (7.3.3.70). [More generally, if in this case we set (7.3.3.75) or (7.3.3.76) equal to a given expression in x and p, then the plane curve to be found intercepts segments of plane curves in the given one-parameter family which all have the point in common as one endpoint and bound areas equal to the given expression in x and p or which all have the point in common as one endpoint and have arc lengths equal to the given expression in x and p.] In short, the kinds of problems that Fontaine vaguely described at the end of his Paris Academy *mémoire* of 1734 on finding tautochrones and claimed could be solved using his fluxio-differential method were included among the ones that Nicole tackled in his Paris Academy *mémoire* of 1737.

The particular problem of finding synchrones which Nicole formulated and solved in his Paris Academy *mémoire* of 1737 is also a problem of type (i). It too can be expressed the same way as the preceding problems by deriving the relevant nonalgebraic integral ϕ that appears in (7.3.3.72) from equation (7.3.3.73) in conjunction with the equation of motion of free fall in a vacuum, when equation (7.3.3.73) expresses a given one-parameter family of algebraic plane curves which all lie in the same plane, which all meet at one point, and whose parameter is p, and then setting the nonalgebraic expression equal to a constant. In this case equation (7.3.3.73) again plays the role of equation (7.3.3.71), and the nonalgebraic expression set equal to a constant again functions as equation (7.3.3.70).

In the problem of type (ii) which Nicole formulated and solved, the relevant nonalgebraic integral ϕ that appears in (7.3.3.72) was given at the beginning, and the condition that the segments of the members of the particular corresponding one-parameter family of nonalgebraic plane curves that all lie in the same plane which are given through equation (7.3.3.70), which all meet at one point, whose parameter is p, and which are intercepted by the plane curve to be found all bound equal areas happens in this particular case to be expressed by an algebraic equation that has the form (7.3.3.71).

To handle such expressions as (7.3.3.75) and (7.3.3.76), and, more generally, the kinds of nonalgebraic integrals ϕ that appear in (7.3.3.72), Fontaine utilized a theorem that he had probably discovered while applying the fluxio-differential method to determine how these expressions behave as functions of the upper limit of integration x and the parameter p. At least, Fontaine maintained years later that he had found the theorem "immediately after solving the problem of the tautochrones," and, moreover, specifically "as a result of using the [fluxio-differential] method to solve that problem."[132] I shall now demonstrate this theorem the way that Fontaine very likely first did.

To do so I shall once again replace the notations "·" and "d", respectively, for the two first-order differential operators in Fontaine's fluxio-differential method by the symbols "d" and "δ," respectively, in order to be able to use whole symbol notation for differential coefficients. Thus "d" designates an infinitesimal difference in values of a quantity defined at two neighboring points situated on a single plane curve, and "δ" stands for an infinitesimal difference in values of a quantity defined at two neighboring points located on two different, neighboring plane curves that lie in the same plane. In the case of a continuum of plane curves that all lie in the same plane and that are distinguished from each other by means of different values of a parameter p, let p' and p designate the values of the parameter which correspond to two different plane curves in this one-parameter family of plane curves that all lie in the same plane and which come infinitesimally close to each other, so that $p' - p \equiv \delta p$ is a very small number but is not equal to zero (see

Figure 29. *Infinitesimally close members of a family of plane curves in the same plane whose parameter is p.*

Figure 29). Let x, x', and x'' be upper limits of integration of

$$\phi(x, p) \equiv \int_0^x f(x^*, p) \, dx^* \qquad (7.3.3.72)$$

which are very near each other.

Using the two new symbols "d" and "δ" for the two first-order differential operations in Fontaine's fluxio-differential method, and replacing F by G and a by p in (7.3.3.61''), we find that

$$\delta G = G(x', p') - G(x, p) = \frac{\delta G}{\delta x} \delta x + \frac{\delta G}{\delta p} \delta p$$

and

$$dG = G(x^*, p) - G(x, p) = \frac{dG}{dx} dx,$$

where

$$\frac{\delta G}{\delta x} = \frac{dG}{dx}$$

is the way that (7.3.3.61'') appears using whole-symbol notation for differential coefficients. There is only one term on the right-hand side of the second of the three preceding equations, because "d"-differentials are taken *along* one plane curve in the one-parameter family, and the value of the parameter p is *constant* for any one plane curve in the one-parameter family.

If we take the ϕ that appears in (7.3.3.72) to be G in the three preceding equations, then

$$\delta\phi = f(x, p)\delta x + e(x, p)\delta p \qquad (7.3.3.77)$$

and

$$d\phi = f(x, p) \, dx \qquad (7.3.3.78)$$

in this case, where $e(x, p)$ symbolizes the differential coefficient of δp with respect to $\delta\phi$. We now limit the values of x, x', p, and p' to those that fulfill the relation $x'/x = p'/p$. Then

$$\frac{x' - x}{x} = \frac{p' - p}{p}. \qquad (7.3.3.79)$$

In this case all nonzero first-order differentials will be "δ"-differentials, because if x and x' were to be situated on the same plane curve, in which case $p' = p$ and consequently

$$\delta p \equiv p' - p = p - p = 0,$$

then x would have to coincide with x' in order for (7.3.3.79) to hold. In other words $dx \equiv x' - x = x - x = 0$ would have to be true. Hence (7.3.3.79) can be written as

$$\frac{\delta x}{x} = \frac{\delta p}{p}, \tag{7.3.3.79'}$$

or

$$\delta x = \frac{x}{p} \delta p. \tag{7.3.3.79''}$$

If we substitute the right-hand side of (7.3.3.79'') for δx in the expression for $\delta \phi$ on the right-hand side of (7.3.3.77), we find that

$$\delta \phi = f \delta x + e \delta p = f \left(\frac{x}{p} \delta p \right) + e \delta p = \left(f \frac{x}{p} + e \right) \delta p \tag{7.3.3.80}$$

with x, x', p, and p' restricted in the way described.

We now consider expressions $\phi(x, p)$ which are homogeneous expressions of degree n in x and p. Then independently of the preceding calculations, we conclude that

$$\frac{\phi(x', p')}{\phi(x, p)} = \frac{p'^n}{p^n} \tag{7.3.3.81}$$

is true for such expressions when x', x, p', and p satisfy the relation $x'/x = p'/p$. This follows from the property

$$\phi(\lambda x, \lambda p) = \lambda^n \phi(x, p) \tag{7.3.3.82}$$

which characterizes homogeneous functions of degree n in two finite variables x and p. We can also write (7.3.3.81) as

$$\frac{\phi(x', p') - \phi(x, p)}{\phi(x, p)} = \frac{p'^n - p^n}{p^n}. \tag{7.3.3.81'}$$

If $x' - x$ and $p' - p$ now become very small in ways that accord with the relation $x'/x = p'/p$, then (7.3.3.81') becomes

$$\frac{\delta \phi}{\phi} = \frac{\delta p^n}{p^n} = \frac{n p^{n-1} \delta p}{p^n} = \frac{n}{p} \delta p, \tag{7.3.3.83}$$

or

$$\delta \phi = \frac{n}{p} \phi \delta p. \tag{7.3.3.83'}$$

[Again, all nonzero first-order differentials are "δ"-differentials for the same reason that I mentioned. The relation $\delta p^n = np^{n-1} \, \delta p$ can be derived by using induction and by applying the rule given in (7.3.3.1) for taking differentials of products to $p^n = p^{n-1} \times p$.] If we then combine the two expressions (7.3.3.80) and (7.3.3.83') for $\delta \phi$, which we determined independently of each other and both of which were derived having assumed that x, x', p, and p' fulfill the same condition $x'/x = p'/p$, then the equation

$$\frac{n\phi}{p} \delta p = \delta \phi = \left(f\frac{x}{p} + e \right) \delta p \qquad (7.3.3.84)$$

follows. Finally, if both sides of the equation (7.3.3.84) are divided by the very small, nonzero factor δp, an equation

$$\frac{n\phi}{p} = \frac{fx}{p} + e \qquad (7.3.3.85)$$

results, which can be rewritten as

$$n\phi = fx + ep. \qquad (7.3.3.85')$$

It seems that by this time, however, Fontaine had adopted the formalism (7.3.3.63), for in the paper that he read before the Paris Academy early in July of 1737, he derived (7.3.3.85') exactly the same way that I just did, except that he only introduced a *single* first-order differential operator to do so. Let us note that (7.3.3.78) has no function in the demonstration that I just gave; only the differential operator "δ" does any work. All differentials in the preceding demonstration are "δ"-differentials. As a result the preceding derivation of (7.3.3.85') appears like an application of the partial differential calculus of two variables to carry out differentiation with respect to the parameter p. If we let the letter "d" stand for this single differential operator, as Fontaine himself ultimately did, then the theorem just derived can be stated as follows using whole-symbol notation for partial differential coefficients: If $\phi(x, p)$ is a homogeneous function of degree n in two finite variables x and p, where

$$d\phi = \frac{d\phi}{dx} dx + \frac{d\phi}{dp} dp, \qquad (7.3.3.86)$$

then

$$n\phi = \frac{d\phi}{dx} x + \frac{d\phi}{dp} p. \qquad (7.3.3.87)$$

This result is called the "homogeneous function theorem" in two variables. It is one of the elementary theorems of the partial differential calculus of two variables.

Why did Fontaine introduce this theorem? In all of the examples of problems of types (i) and (ii) which Nicole solved, the troublesome nonalgebraic integrals ϕ

in (7.3.3.72) happen to be homogeneous expressions in x and p. Replacing "δ" by "d", equation (7.3.3.77) can be rewritten as

$$d\phi = f(x, p)\, dx + e(x, p)\, dp. \qquad (7.3.3.77')$$

Ordinarily this equation would not be very useful, because in general the nonalgebraic integral ϕ in (7.3.3.72) is unwieldy enough to make it impossible to determine $e(x, p)$. [For example, Leibniz's inversion (7.1.1.1) of differentiation and integration usually does not help.] However, in case the nonalgebraic integral ϕ in (7.3.3.72) is homogeneous in x and p, which ϕ is in all of the problems of trajectories which Nicole solved, equation (7.3.3.85') follows from equation (7.3.3.77'). Then equation (7.3.3.85') can be solved for $e(x, p)$:

$$e(x, p) = \frac{n\phi - fx}{p}. \qquad (7.3.3.85'')$$

Hence equation (7.3.3.77') becomes

$$d\phi = f\,dx + \left(\frac{n\phi - fx}{p}\right) dp. \qquad (7.3.3.88)$$

Next Fontaine used (7.3.3.70) to eliminate the unmanageable, nonalgebraic integral ϕ in (7.3.3.72) altogether from such problems. Replacing ϕ by F in equation (7.3.3.88), equation (7.3.3.88) becomes

$$dF = f\,dx + \left(\frac{nF - fx}{p}\right) dp. \qquad (7.3.3.89)$$

Now, the finite algebraic equation (7.3.3.71) was not treated as a problem in its own right here, in the sense that in practice it was of a degree low enough in p (meaning of degree less than or equal to four in p) to be solved for p. Thus in practice Fontaine could reexpress the equation (7.3.3.71) in the form

$$p = g(x, y). \qquad (7.3.3.90)$$

By this means he transformed the original pair of equations (7.3.3.70) and (7.3.3.71) into (7.3.3.89), (7.3.3.90) and

$$dp = \frac{dg}{dx}\,dx + \frac{dg}{dy}\,dy, \qquad (7.3.3.91)$$

where

$$dF = \frac{dF}{dx}\,dx + \frac{dF}{dy}\,dy + \frac{dF}{dp}\,dp. \qquad (7.3.3.92)$$

The four equations (7.3.3.89)–(7.3.3.92) only included expressions that were either given or else could be found. Replacing p and dp, respectively, in (7.3.3.89) and (7.3.3.92) by the expressions for them in x, y, dx, and dy on the right-hand sides of (7.3.3.90) and (7.3.3.91), respectively, and then equating the right-hand side of (7.3.3.89) and the right-hand side of (7.3.3.92), Fontaine arrived at a first-order ordinary "differential" equation in x and y of first degree for the plane curve that was the solution to the problem that the two equations (7.3.3.70) and (7.3.3.71) determined. We note that in the general case of Fontaine's first-order ordinary "differential" equation in x and y of first degree for a trajectory the variables need not be separable. As we shall soon see this became an important consideration in Fontaine's later research.

If the homogeneous function theorem in two variables limited the range of problems to which Fontaine could apply his method, Fontaine nevertheless tried to approach the problems using procedures that were general in a way that Nicole's techniques were not. Moreover, by letting $F = F(x, y, p)$ in (7.3.3.70), Fontaine had evidently also decided by this time that expressions in *more* than two finite variables could be differentiated with respect to each of its variables, as in (7.3.3.92).

The homogeneous function theorem is normally attributed to Euler, who, in fact, introduced the theorem in two variables early in the 1730s to attack the same kinds of problems of trajectories which Fontaine did in 1737 and in exactly the same way.[133] Euler published his solutions for the first time in 1736. I shall have occasion to recall that fact again shortly. Because Fontaine employed the partial differential calculus in 1737 to differentiate with respect to the parameter of a one-parameter family of plane curves that all lie in the same plane, one can ask what he borrowed from his predecessors. He could not have imitated very much of what mathematicians had done before him. As I already noted earlier in this chapter, the partial differential calculus developed by 1720 was not clearly explained in *Acta Eruditorum* in the years after Leibniz challenged the British mathematicians to solve problems of orthogonal trajectories late in November of 1715 and early in April of 1716.

In his article published in the *Acta* for 1719, Nikolaus I Bernoulli proposed the generalized "completion problem," but he did not solve it there.[134] As I mentioned previously in this chapter, in the same article in the *Acta* Nikolaus stated rules for constructing the plane curves orthogonal to the generalized brachistochrone cycloids, but he withheld much of the analysis that he had used to find his solution to the problem.[135] In particular, the corresponding "variable-parameter equation" appears in no explicit, clear-cut way in his article. Indeed, Nikolaus's uncle Johann I Bernoulli did not realize while reading his nephew's article in the *Acta* for 1719 that his nephew had solved the problem of determining the plane curves orthogonal to the generalized brachistochrone cycloids by integrating the corresponding variable-parameter equation. Johann assumed that Nikolaus had utilized a method similar to his own, which did not involve the use of the partial differential calculus (that is, differentiation from curve to curve and the variable-parameter equation), to solve the problem.[136] Nikolaus phrased obscurely and abstrusely the matters that he treated in his article in the *Acta* for

1719, whether he meant to do so or not.[137] Only in unpublished manuscripts did Nikolaus make clear how he solved the problem of finding the plane curves orthogonal to the generalized brachistochrone cycloids specifically by integrating the corresponding variable-parameter equation.[138] More generally, only in this unpublished work did he reveal just how far the partial differential calculus had progressed in his hands beyond what Leibniz and Johann I Bernoulli had done.[139] Nor did Nikolaus communicate any of the partial differential calculus appearing in his unpublished work to his correspondents.[140] Not even his trustworthy correspondent Montmort, who acted as go-between during the challenges and who kept strictly confidential everything that he learned from the participants, was an exception. Other Paris Academicians knew even less. For example, all that Mairan knew about Nikolaus I Bernoulli was that he was a young mathematician whom Montmort, while alive, had been very excited about. Mairan had absolutely no substantive knowledge whatever of Nikolaus's work.[141]

The article by Johann I Bernoulli's eldest son Nikolaus II Bernoulli, which appeared in the *Acta* for 1720 and which Johann supplied the mathematics for, is the work concerning the problem of determining orthogonal trajectories which was cited most often in the years that followed its publication. The article surveyed the various methods for constructing orthogonal trajectories. In it Leibniz's inversion (7.1.1.1) of differentiation and integration and the variable-parameter equation were published for the first time.[142] Johann I Bernoulli both demonstrated the inversion of differentiation and integration and derived the variable-parameter equation in the article, only to abandon the equation at the end as hopeless for solving problems of finding orthogonal trajectories in general. In effect, the older Bernoulli renounced the partial differential calculus in this article.[143] This illustrates his failure to see through Nikolaus I Bernoulli's solution to the problem of finding the plane curves orthogonal to the generalized brachistochrone cycloids published in the *Acta* for 1719, which in fact Johann's nephew had found using the partial differential calculus. Nikolaus I Bernoulli also communicated a solution of the generalized completion problem to his uncle Johann, who then added it to the article published under his eldest son's name in the *Acta* for 1720. But Johann's nephew's solution of this problem was obscure and esoteric, too.[144] It did not clearly reveal the inner workings of the partial differential calculus upon which it was based.[145] In short, Nikolaus II Bernoulli's article in the *Acta* for 1720 shed no more real light on the partial differential calculus than did any of the articles that had appeared earlier in that journal.[146] All of this goes to show that where the partial differential calculus is concerned, Fontaine's predecessors did not leave too many clues in the published literature for him to find, if they left any clues at all.

There is evidence, both circumstantial and internal, that Fontaine had looked at Nikolaus II Bernouli's article in the *Acta* for 1720 and the supplements to it. In his Paris Academy *mémoire* of 1730 on finding tautochrones, Johann I Bernoulli mentioned some of his earlier work that had appeared in one of the supplements to his son Nikolaus's article in the *Acta* for 1720 (volume 7 of the supplements to the *Acta*, published in 1721).[147] Fontaine would have had every reason to check this reference, because in 1734 he regarded Johann's *mémoire* of 1730 on finding tautochrones as an achievement to be improved upon. In his *mémoire* on finding tautochrones, Johann cited the supplement to his eldest son's article in the *Acta* for 1720 in connection with what he called "similar functions," which in fact are the same as the homogeneous functions that I talked about. Johann exploited the properties of such functions to find tautochrones in media that resist a body's motion directly as the square of the body's speed. Fontaine, however, thought the use of such functions to be too restricting where solving problems of finding tautochrones were concerned. He maintained that the application of these functions to solve such problems required that conditions be satisfied which in fact need not hold in general. In addition, as I have already observed, Johann also needed an explicit expression for speed in order to make his method of finding tautochrones work, which meant that he had to integrate the equation of motion once. But Fontaine easily constructed examples where a first integral of the equation of motion could not be determined. A medium that resists a body's motion directly as $(y^2 + my)/n$, where the constant $m \neq 0$ and where y is the body's speed, is one such case. Fontaine's fluxio-differential method, which involved two independent first-order differential operations, made such limiting assumptions as Johann's unnecessary.

But in his paper of 1737 on problems of trajectories, Fontaine utilized homogeneous functions himself, as we have seen. He began that paper by defining similar figures ("figures semblables"), similar curves ("courbes semblables"), similar functions ("fonctions semblables"), etc., and he stated a "theorem" that "homologous quantities ("quantitiés homologues") in similar figures, similar curves, similar functions, etc. are to each other as homogeneous powers of their parameters."[148] This "theorem," in effect, is just the property that defines what we mean by a homogeneous function in several finite variables. Equation (7.3.3.82) defines a homogeneous function of degree n in two finite variables. More generally,

$$f(\lambda x, \lambda y, \lambda z, \ldots) = \lambda^n f(x, y, z, \ldots) \qquad (7.3.3.93)$$

defines a homogeneous function of degree n in the finite variables $x, y, z, \ldots$. Now, the particular page of the supplement to Nikolaus II Bernoulli's article which Johann I Bernoulli called the reader's attention to in his Paris Academy *mémoire* of 1730 on finding tautochrones contains practically the same discussion of "similarity," including the vocabulary, as the discussion of similarity at the beginning of Fontaine's paper of 1737 on problems of trajectories, which I just quoted an extract from. Fontaine must have translated the technical terms into French from the Latin that the supplement to Nikolaus II Bernoulli's article is written in. The term "homologa" which is located on the page in question of the supplement is particularly revealing.[149] Unlike Fontaine, Johann used no equivalent French term in his *mémoire* of 1730 on finding tautochrones to characterize "similar functions." Moreover, like Fontaine, the author of the relevant page of the supplement to Nikolaus II Bernoulli's article defined the concept of similarity in a general way,[150] whereas Johann confined his discussion to similar functions alone in his *mémoire* of 1730 on finding tautochrones. In short, Fontaine evidently looked at more than just Johann's Paris Academy *mémoire* of 1730 on finding tautochrones. He undoubtedly at least glanced at the supplement to the article in the *Acta* for 1720 which had been published under Johann's eldest son Nikolaus's name.

Euler, like Fontaine, had doubtlessly read some of the literature published in the *Acta* concerning problems of finding orthogonal trajectories too.[151] But as I have already stated, that literature did not clarify the partial differential calculus developed by 1720. Hence Euler, for example, only learned years later from Nikolaus I Bernoulli, in a letter that Nikolaus wrote to Euler in 1743, how Nikolaus had found the plane curves orthogonal to the generalized brachistochrone cycloids, whose rules for construction he had published in the *Acta* for 1719 without revealing the method that he had used to solve the problem, by actually integrating the corresponding variable parameter equation.[152] As it turns out, in the 1730s Euler would have independently reproduced Nikolaus's solution, had he not made a careless error.[153] We can conclude that Nikolaus evidently had not put Euler on the right track to that solution. The literature published in *Acta Eruditorum* concerning problems of finding orthogonal trajectories hardly pointed the attentive reader in any one particular direction. Thus we should not be surprised to find that Euler and Fontaine, for example, did not react the same way after reading this literature. They went off in different directions instead. Even though their work does overlap in places, as I have already noted, the two mathematicians did not follow the same routes on the whole to the partial differential calculus. That calculus accordingly advanced further in their hands in ways that differed from each other by and large, as I shall now illustrate.

Euler introduced what he called "modular equations" to try to make the completion problem more useful. Solutions of the *generalized* completion problem had proven thus far to be pretty worthless for determining orthogonal trajectories. That is to say, finding an expression $q(x, y, a)$ that satisfies the total "differential" equation

$$dx = p(x, y, a)\, dy + q(x, y, a)\, da \qquad (7.3.3.94)$$

(which is the same as equation (7.1.2.5)) when

$$d_y x = p(x, y, a)\, dy \qquad (7.3.3.95)$$

[which is the same as equation (7.1.2.4)] did not help carry out the task of integrating the variable-parameter equation for a given one-parameter family of plane curves that all lie in the same plane expressed through the equation (7.3.3.95), where a is the parameter of the family. (We recall that Nikolaus I Bernoulli used his solution of the *restricted* completion problem to find the plane curves orthogonal to the generalized brachistochrone cycloids by integrating the corresponding variable-parameter equation.) In Nikolaus II Bernoulli's article published in *Acta Eruditorum* for 1720, Nikolaus I Bernoulli solved the generalized completion problem, which he had formulated in his article in the *Acta* for 1719, for $q(x, y, a)$.[154] We remember that he had transmitted his solution to his uncle Johann, who inserted it in the article published under his eldest son Nikolaus's name in the *Acta* for 1720. In fact, in the same article Johann accomplished what amounted to the same thing. Namely, he derived the variable-parameter equation for a given one-parameter family of plane curves that all lie in the same plane expressed by means of equation (7.3.3.95), in which appears the same expression for $q(x, y, a)$ as his nephew determined in solving the generalized completion problem.[155] But, as I have already said, the older Bernoulli ultimately abandoned the partial differential calculus in this article. The variable-parameter equation that he found for the one-parameter family of plane curves that all lie in the same plane given by equation (7.3.3.95) appeared to him to be so hideous that he dismissed at one fell swoop all of the results concerning differentiation from curve to curve and variable parameters which he and Leibniz had discovered in the 1690s. These results he now contemptuously called "general abstract nonsense."[156]

Euler tried out a new strategy; he worked out the completion problem "backwards," so to speak. Let $dy = Q(x, a)\, dx + R(x, a)\, da$ be the complete "differential" equation for a given one-parameter family $y = P(x, a)$ of transcendental plane curves that all lie in the same plane and whose parameter is a. [Note: again following Engelsman (1984: 143), I have interchanged x and y.] Thus, for example, if the one-parameter family is given by

$$y = P(x, a) \equiv \int_{x_0}^{x} Q(x, a)\, dx,$$

then, according to Leibniz's inversion (7.1.1.1) of differentiation and integration,

$$R(x, a) = \int_{x_0}^{x} Q_a(x, a)\, dx.$$

We recall that in trying to find the plane curves orthogonal to all of the members of a one-parameter family $y = P(x, a)$ of plane curves that all lie in the same plane when the equation $y = P(x, a)$ could not be solved explicitly for the parameter a as a function of x and y, which was usually true when the one-parameter family was composed of transcendental plane curves, Johann I Bernoulli had introduced the variable-parameter equation for the one-parameter family. However, in the past the variable-parameter equation could not be integrated in this case in the event that $R(x, a)$ could not be expressed algebraically in terms of x and a. [Nikolaus I Bernoulli's great achievement was, of course, to have carried out this very integration for the family of generalized brachistochrone cycloids, even though $R(x, a)$ is a transcendental expression in this case As I mentioned, however, Euler only learned of this from Nikolaus in 1743.] But suppose that such an $R(x, a)$ could at least be expressed algebraically in terms of x, a, and $y = P(x, a)$. Then in that case the complete "differential" equation for $y = P(x, a)$ has the form

$$dy = Q(x, a)dx + R(x, y, a)da.$$

$$\uparrow \qquad\qquad \uparrow$$

algebraic   algebraic

Euler called such a complete "differential" equation a first-order "modular equation." He endeavored to solve problems of trajectories which involved a given one-parameter family

$$y = P(x, a) \equiv \int_{x_0}^{x} Q(x, a)\, dx$$

of transcendental plane curves that all lie in the same plane and whose parameter is a by means of the first-order modular equation for $y = P(x, a)$ and procedures for eliminating the parameter a from this equation. This usually required eliminating the integrals

$$\int_{x_0}^{x} Q(x, a)dx, \qquad\qquad (7.3.3.95^*)$$

which include the parameter a, from such problems. These integrals caused difficulty because they often prevented the equation

$$y = P(x, a) \equiv \int_{x_0}^{x} Q(x, a)dx$$

from being solved explicitly for the parameter a as a function of x and y. As I have said, these integrals usually made it impossible to do this when the given one-parameter family consisted of transcendental plane curves. Consequently the integrals

$$\int_{x_0}^{x} Q(x, a)dx$$

acted as obstacles to eliminating the parameter a from such problems. This stumbling block is what originally induced Euler to introduce such first-order modular equations. That is, with all coefficients of the first-order modular equation for $y = P(x, a)$ algebraic, Euler hoped that he could eliminate from problems of trajectories both the parameter a and the troublesome integrals

$$\int_{x_0}^{x} Q(x, a)dx$$

that acted as obstacles to eliminating the parameter a, by using the auxiliary equation determined by each problem of trajectories, which depended upon the particular problem. By using the first-order modular equation for $y = P(x, a)$ and the auxiliary equation determined by a particular problem of trajectories, Euler hoped that he could eliminate from that problem the parameter a as well as the unmanageable integral

$$\int_{x_0}^{x} Q(x, a)dx$$

that acted as an obstacle to eliminating the parameter a and that an ordinary "differential" equation in x and y for the plane curve that is a solution to that problem of trajectories would result. In fact, eliminating both the parameter a and the troublesome integrals

$$\int_{x_0}^{x} Q(x, a)dx$$

that acted as obstacles to eliminating the parameter a is basically the same idea that Fontaine put to use in his paper of 1737 on problems of trajectories, although Fontaine did not envisage such problems in terms of Euler's general concept of a first-order modular equation.

In working backward, Euler assumed that $R(x, y, a)$ could be expressed in some convenient form $R = R(x, y, Q, a)$ that was algebraic in x, y, Q, and a. [When $Q(x, a)$ is algebraic in x and a, R is then algebraic in x, y, and a.] He then determined what conditions such a supposition required that $Q(x, a)$ fulfill.

The most useful expression $R(x, y, Q, a)$ has the form $R \equiv (ny - Qx)/a$, in which case it turns out that Q must be a homogeneous function of degree $n - 1$ in x and a. This result is just the converse of the homogeneous function theorem in two variables. Euler first introduced this theorem and its converse in connection with solving problems of trajectories – in other words, for exactly the same reason that Fontaine first employed the homogeneous function theorem in two variables. Both mathematicians used the partial differential calculus, including the homogeneous function theorem, to reduce the kinds of problems of trajectories which Nicole treated in his Paris Academy *mémoire* of 1737 to first-order ordinary "differential" equations of first degree. They attacked these problems in a way that was more general than Nicole's approach to them. Using essentially the same idea Euler also found a second-order ordinary "differential" equation for the plane curves orthogonal to a one-parameter family of transcendental plane curves that all lie in the same plane given by

$$y = P(x, a) \equiv \int_{0}^{x} Q(x, a)dx,$$

when the integral

$$\int_{0}^{x} Q(x, a)dx$$

is a homogeneous expression in x and a. Unlike Fontaine, however, Euler made the homogeneous function theorem only a small part of a larger scheme for using first-order and higher-order modular equations to tackle problems of trajectories and, ultimately, to solve ordinary "differential" equations.[157] Fontaine did eventually become interested in the problem of solving first-order ordinary "differential" equations of first degree, as we shall see shortly, but he dealt with that problem in a way that totally differed from Euler's methods of treating ordinary differential equations which involved modular equations.

It is unlikely that the work that Euler did during the 1730s using the partial differential calculus influenced the work described which Fontaine did using the partial differential calculus. The volumes of the St. Petersburg Academy's *Commentarii* for the 1730s, in which Euler's work appears, were not published until 1740 and 1741. Euler's *Mechanica*, published in two volumes by the St. Petersburg Academy in 1736, also included some partial differential calculus, as I have already mentioned. In volume II, for example, Euler solved the problem of finding synchrones exactly the same way that Fontaine did, by utilizing the homogeneous function theorem in two variables.[158] However, there is ample evidence to show that if the *Mechanica* was available in Paris in 1737, it did not begin to influence able Paris Academy mathematicians like Fontaine and Clairaut much before 1740. Clairaut and Maupertuis were in Lapland when the *Mechanica* appeared. Clairaut remained unfamiliar with the contents of the treatise until he began to correspond with Euler in September of 1740. (I shall have occasion to call attention to that fact again later.) Maupertuis knew about the *Mechanica* when he began to correspond with Euler in May of 1738, but if Maupertuis had actually seen that work by then, he certainly had not read it. He was simply too mediocre a mathematician to have been able to understand it.[159] Detailed reviews of the *Mechanica* were only published in French journals for the first time in 1740.[160]

Fontaine's route to the partial differential calculus differed from Euler's. In Fontaine's case the fluxio-differential method of 1732–34 is the key. Whatever knowledge Fontaine had of Nikolaus II Bernoulli's article in the *Acta* for 1720 and its supplements, and it certainly appears that Fontaine had some, had at most an incidental part in Fontaine's own work.[161] Formal, superficial differences at most distinguish Fontaine's fluxio-differential method of 1732–34 from Fontaine's partial differential calculus of 1737. In a nutshell, if the variations of a plane curve that is a solution to a problem in the calculus of variations all lie in the same plane and can be distinguished from one another by means of a parameter, then the variations make up a one-parameter family of plane curves that all lie in the same plane, and the problem can be dealt with solely by using the partial differential calculus, exactly as in the case of problems of trajectories.[162] Indeed, the famous Euler–Lagrange equation that expresses a necessary condition for certain problems in the calculus of variations to have solutions is commonly derived today by viewing variations precisely this way and then by applying the partial differential calculus to them. In other words, Fontaine's fluxio-differential method, which involves two independent first-order differential operators "·" and "d," is neither exclusively nor intrinsically associated with problems of determining extrema or isoperimetrical problems or problems of finding tauto-chrones. It is not exclusively related to them, because the method not only can be applied to problems like finding brachistochrones and tautochrones, but it can be applied in addition to problems determined by given families of plane curves, where the plane curves in such a family all lie in the same plane, which include problems of trajectories, where the members of a given family may or may not be distinguished from one another by a parameter. Nor is Fontaine's fluxio-differential method intrinsically connected with problems of determining extrema or isoperimetrical problems, because in case variations in a problem of finding extrema or an isoperimetrical problem can be distinguished from each other by means of a parameter, the method can be done away with altogether, and the partial differential calculus can be used instead to solve the problem. The method is not intrinsically connected with problems of finding tautochrones, either, because as I demonstrated, such problems can be treated like problems in one-parameter families of plane curves, where the plane curves in such a family all lie in the same plane, and can consequently be solved using the partial differential calculus instead. In other words, the differential operators "·" and "d" in Fontaine's fluxio-differential method do nothing that cannot also be achieved by using the partial differential calculus. These connections between the two calculuses, the obscure statements that Fontaine made at the end of his Paris Academy *mémoire* of 1734 on finding tautochrones about using his fluxio-differential method to solve what we recognize to be the kinds of problems of trajectories that he and Nicole attacked in 1737 and to which Fontaine applied the partial differential calculus in 1737, and the chronology of Fontaine's work during the years 1732–37 make all the more plausible the thesis that Fontaine's applications of the partial differential calculus in 1737 to tackle problems of trajectories *grew out of* his fluxio-differential method of 1732–34. In the next part of this section of this chapter I shall provide other evidence that Fontaine's additional uses of the partial differential calculus after 1737 had their origins in his fluxio-differential method of 1732–34.

As I have recently discovered, relations between the two calculuses have been described clearly and succinctly by the authors of a relatively modern textbook in mechanics. The authors of the book do this in connection with the famous "δ method" that Lagrange introduced to solve problems in the calculus of variations and whose rules to all intents and purposes cannot be distinguished from those of Fontaine's fluxio-differential method. The authors emphasize that in solving problems of determining extrema or isoperimetrical problems, the Lagrangean "δ" and "d" operations, provided they are "used intelligently," can greatly simplify calculations that are clumsy to write using the partial differential calculus.[163] My long solution to the problem of finding the tautochrone in media that resist a body's motion directly as $(y^2 + my)/n$, where $m \neq 0$ and where y is the body's speed, in which I employed the partial differential calculus, probably illustrates their point very well. Fontaine's solution to the same problem, based on the fluxio-differential method and which I outlined above, involves calculations scattered in his Paris Academy *mémoire* of 1734 which, if collected together, would fill about one page of the Paris Academy's *Mémoires*!

One apparent difference between the fluxio-differential method and the partial differential calculus is, of course, the following one. The two independent, first-order Leibnizian differential operators involved in the fluxio-differential method can act on *both finite and infinitesimal quantities*, whereas the two differential operators in the formalism (7.3.3.61″) are the two differential operators involved in the fluxio-differential method restricted to acting on *finite quantities F, x, and a only.*

The real difference between the fluxio-differential method and the partial differential calculus – at least, when considered from the standpoint of continental mathematicians of the 1730s – lies elsewhere, however. The geometric entity, the individual plane curve, served as the foundation of the fluxio-differential method, for the same reasons that it had played a fundamental part in Leibniz's creation of the original differential calculus. In the partial differential calculus, on the other hand, the individual plane curve became an object of secondary importance. In this calculus variables replaced the plane curve as the things of primary importance instead. In Fontaine's shift from two independent first-order differential operations to a single first-order differential operation, coordinates or variables took over the position of primary importance which the individual plane curve had enjoyed in the fluxio-differential method. The distinction between taking differentials *along* a plane curve and taking differentials *between* or *across* neighboring plane curves that lie in the same plane disappeared. This fact evidences that the foundations of the infinitesimal calculus underwent significant changes during the eighteenth century. It illustrates one of the ways in which the infinitesimal calculus lost during the eighteenth century the geometric foundations that it had in the beginning.

At the end of November 1738 Fontaine read a paper before the Paris Academy in which he proposed a method for integrating in finite terms or reducing to quadratures the general first-order ordinary "differential" equation of first degree[164]

$$dx + \alpha(x, y)dy = 0. \tag{7.3.3.96}$$

Why did Fontaine decide to attack this problem in integral calculus in all its generality? There are at least three factors to consider. First, we recall that in volume I of the St. Petersburg *Commentarii* Johann I Bernoulli had shown how to integrate or reduce to quadratures equations of form (7.3.3.96) when $\alpha(x,y)$ is a homogeneous expression of degree zero in x and y (meaning that the numerator and denominator of $\alpha(x,y)$ are homogeneous expressions of the same degree in x

and y), hence equations whose variables could easily be separated. As we saw earlier in this chapter, Maupertuis had called the attention of the members of the Paris Academy to what Bernoulli had done at least twice during the 1730s (first in the spring of 1731 and then again in March of 1736), and at the same time Maupertuis had stressed that the condition on $\alpha(x, y)$ that had to hold in order to be able to use Bernoulli's method severely restricted the range of the equations of form (7.3.3.96) which the method could be applied to. The equations of form (7.3.3.96) which interested Maupertuis the most were ones in which $\alpha(x, y)$ was not homogeneous of degree zero in x and y, hence equations whose variables could not readily be separated, if they could be separated at all. On the same occasions that Maupertuis emphasized the limitations of Bernoulli's method before the Paris Academy, he appealed to the members of the Academy to take seriously the problem of integrating in finite terms or reducing to quadratures such equations, and he urged them to try to make some progress in solving the problem. At long last late in 1738 a mathematician in the Paris Academy, Fontaine, who did not yet belong to the Academy in 1731 when Maupertuis had made his first plea, resolved to take up Maupertuis's challenge that had gone unanswered until then.

Second, Fontaine had grown up in the southern part of France and had gone to a Jesuit school there (the Collège de Tournon). When Fontaine came to Paris for the first time at the age of twenty, thus in 1724 or 1725, he sought the aid and counsel of Louis-Bertrand Castel,[165] who, we remember, was a Jesuit who taught mathematics at the Jesuit Collège de Louis-le-Grand. Castel succeeded in getting Fontaine interested in mathematics. While he was in Paris, Fontaine also read Fontenelle's *Eléments de la géométrie de l'infini*, published in 1727, which inspired him with enthusiasm and excited his interest in mathematics even more.[166] Castel very likely gave Fontaine the book to read. Although the Paris Academy mathematicians despised Castel, Castel and Fontenelle admired each other. The two men had the same basic objectives and the same fundamental concerns: to make science accessible to the layman. The correspondence between Castel and Fontenelle evidences their rapport and their mutual interests.[167] Castel praised Fontenelle's book. Fontenelle talked more about what we would now call the foundations of the infinitesimal calculus than he did about the infinitesimal calculus itself, which is why the book did not appear to mathematicians to be a work written by a mathematician. In Castel's opinion, Fontenelle took what had become in the hands of mathematicians, including their deceased colleague L'Hôpital, a "routine" of doing the same things over and over again with "a and x" by "trial and error" and reinvigorated it by returning to what Castel called the philosophical ideas that underlie infinity and the infinite. Castel thought that Fontenelle approached the subject of the infinite in what he considered to be a more "systematic" way, which included highlighting what Castel called the "paradoxes" of the infinite, by which Castel did not mean aspects of infinity which contradicted each other. More than one hundred years later these sorts of paradoxes would give rise, for example, to the idea of infinitely large cardinal numbers whose magnitudes are different.[168] Fontenelle reciprocated by applauding the subsequent publication of Castel's *Mathématique universelle abrégée*, which I talked about earlier in this chapter.[169]

In 1728 Fontaine had to leave Paris and return home, where he stayed until his older brother died. He then sold the land that he inherited and bought an estate near Compiègne, northeast of Paris. Upon his return to Paris he met Maupertuis and Clairaut.[170]

Although mathematicians did not consider Fontenelle's book of 1727 to be a mathematical work, Fontaine nevertheless took it as a starting point for his first piece of mathematical research. In 1731 Fontaine presented a new formula for the curvature of plane curves. That is, his formula differed from the usual formula that involves the radius of the evolute of a plane curve at a point on the plane curve, which we now call the radius of curvature of a plane curve at a point on the plane curve. In his book Fontenelle had found curvatures of plane curves using the sines of angles of contingence. Fontaine did not repeat what Fontenelle had done, but he got the idea for his formula for the curvature of plane curves from Fontenelle. The members of the Paris Academy commended Fontaine for his new formula and his applications of it to several plane curves. They judged what Fontaine did to be the work of an individual highly skilled in the infinitesimal calculus.[171]

Fontaine's first biographer the Marquis de Condorcet wrote that Castel

was the most famous mathematician in the Society [of Jesuits]. Nature had given him a fervent imagination and a bold and lofty intelligence. Perhaps he would have had a reputation that endured, had the certainty of being able to count as many admirers as there were Jesuits not killed in him the anxiety that makes us demanding with regard to what we produce and without which even genius rarely rises to the occasion. This is so true in more ways than one that there are no enemies of talent more dangerous than individuals who extol and glorify it.[172]

Castel was not quite the ignoramus that some Paris Academy mathematicians and Johann I Bernoulli made him appear to be. It seems unlikely that Castel would have understood the problems of finding tautochrones in resistant media and Fontaine's solutions to these any better than he had grasped isoperimetrical problems and Bernoulli's solutions to those, which we remember Castel had completely misunderstood. If Castel could not possibly have followed Fontaine's mathematics, one thing would have given him much pleasure nevertheless. Fontaine had not imitated *anyone's* calculations. Fontaine did a highly imaginative piece of mathematical research in determining tautochrones in resistant media. He did not mindlessly manipulate the formalisms of the infinitesimal calculus developed until then. Instead Fontaine did some mathematics that fits the description that we read earlier in this chapter of what Castel meant by originality, which Castel had stated in his *Mathématique universelle abrégée*.

Castel and Maupertuis viewed the integral calculus, and how to treat ordinary "differential" equations in particular, in ways that totally conflicted. At one time or another, both individuals directly influenced Fontaine. Having been exposed to opinions as far apart as Castel's and Maupertuis's, it seems likely that Fontaine might be tempted to tackle the problem of integrating in finite terms or reducing to quadratures first-order ordinary "differential" equations of first degree in generality. Someone conscious of the clash of views would have been inclined to consider the methods of integrating in finite terms or reducing to quadratures first-order ordinary "differential" equations of first degree to be a major mathematical issue.

Finally, in the special cases of problems of trajectories which Nicole solved in his Paris Academy *mémoire* of 1737, Nicole always managed to separate the variables in the first-order ordinary "differential" equations of first degree which he found for the plane curves that were the solutions to these problems of trajectories by making suitable changes of variables. This enabled him to integrate in finite terms these first-order ordinary "differential" equations of first degree or, if not, at least reduce these equations to quadratures. But the variables in the first-order ordinary "differential" equations of first degree which Fontaine found for the plane curves that are solutions of the problem of trajectories which he formulated by unifying and generalizing the particular problems of trajectories which Nicole solved, which are formed by equating the right-hand side of (7.3.3.89) and the right-hand side of (7.3.3.92), need not be separable. Thus doubtlessly would also have led Fontaine to consider the part of his more general problem of trajectories which remained to be solved to be an important one, namely, to integrate in finite terms or reduce to quadratures the first-order ordinary "differential" equations of first degree for the trajectories. Moreover, such an attitude would have accorded perfectly with the point that Maupertuis had tried to make at least twice before the Paris Academy in the 1730s but seemingly in vain each time.

The paper that Fontaine read before the Paris Academy at the end of November 1738 was not copied into the Academy's unpublished proceedings (the *Registres des Procès-Verbaux*). The paper was never published in the Academy's *Mémoires* either. Fontaine's manuscript does not survive, as far as we know. Nor is the version of the paper which Fontaine published in his complete works of 1764 of any use. We shall see later that this version is not a faithful reproduction, down to the last detail, of the paper that he wrote in 1738. We can determine that this is true because Nicole and Clairaut were appointed to examine Fontaine's paper and to review it. Fortunately, the report that Nicole and Clairaut wrote, which the two mathematicians read before the Academy in February of 1739, was recorded in the Academy's unpublished proceedings. This report includes vital information for reconstructing the contents of Fontaine's paper of late 1738.

Clairaut and Nicole stated that Fontaine proposed to integrate all first-order "differential" equations that can be expressed in the form $dx + \alpha dy = 0$, where α

stands for any quotient whose numerator and denominator are expressions of the same degree in x, y, and a parameter p. The problem was to find the third term that would have existed in a given "differential" equation that has this form, had the parameter p in the integral of the equation not been held constant but been treated as a variable instead. Differentiating the expression that designates the integral would have led to a "differential" equation that has three terms in that case. According to the two reviewers, Fontaine wrote πdp to express the third term, where π symbolizes an unknown expression in x, y, and p whose numerator and denominator, like those of α, are expressions of the same degree in x, y, and p. By means of a certain theorem, as well as several complicated equations that Fontaine derived using the theorem, Fontaine arrived at three general equations – equations that Nicole and Clairaut did not specify in their report. The two reviewers did remark that the integral in question could be found by utilizing the first two equations, once the unknown function π had been determined. They mentioned in addition that the third equation related α and π in such a way that once the general form of π was known, π could then be found by means of the "method of undetermined quantities" ("méthode des indéterminées") applied to the third equation. Although Clairaut and Nicole discovered that Fontaine's method of integrating "differential" equations that have the form specified above had limitations and could not always be made to work, they also acknowledged that his approach to the problem of integrating in finite terms or reducing to quadratures first-order ordinary "differential" equations of first degree was "exceedingly original."[173] Using their report together with a few remarks that Clairaut made in some work that he published in the Paris Academy's *Mémoires*, I shall attempt to reconstruct Fontaine's method.

Since it was already known that equation (7.3.3.96) could always be integrated in finite terms or reduced to quadratures in case $\alpha(x, y)$ was a homogeneous expression of degree zero in x and y, Fontaine considered expressions $\alpha(x, y)$ in equation (7.3.3.96) which were not homogeneous of degree zero in x and y. By suitably choosing a constant p that appears in α, Fontaine converted α into a homogeneous expression of degree zero in x, y, and p *now treated as a variable*, meaning that the numerator and denominator of α are homogeneous expressions of the same degree in x, y, and p. The problem according to Fontaine was then to determine a homogeneous expression π of degree zero in three variables x, y, and p such that the equation

$$dx + \alpha(x, y, p)dy + \pi(x, y, p)dp = 0 \qquad (7.3.3.97)$$

is integrable. By this he meant that an expression $\phi(x, y, p)$ in three variables x, y, and p could be found for this choice of π such that

$$\frac{d\phi}{dx} = u, \quad \frac{d\phi}{dy} = u\alpha, \quad \text{and} \quad \frac{d\phi}{dp} = u\pi \qquad (7.3.3.98)$$

for some integrating factor $u(x, y, p)$, because in this event if u, $u\alpha$, and $u\pi$, respectively, are substituted for $d\phi/dx$, $d\phi/dy$, and $d\phi/dp$, respectively, in

$$d\phi = \frac{d\phi}{dx}dx + \frac{d\phi}{dy}dy + \frac{d\phi}{dp}dp, \qquad (7.3.3.99)$$

then it follows that (7.3.3.99) is just

$$d\phi(x, y, p) = u(x, y, p)(dx + \alpha(x, y, p)dy + \pi(x, y, p)\,dp) \qquad (7.3.3.100)$$

in this case. If ϕ is then set equal to k where k is a constant:

$$\phi(x, y, p) = k, \qquad (7.3.3.101)$$

and if equation (7.3.3.101) is then differentiated [that is, if the differential of equation (7.3.3.101) is taken], the equation

$$0 = dk = d\phi = u(dx + \alpha dy + \pi dp) \qquad (7.3.3.102)$$

results, from which equation (7.3.3.97) follows after dividing both sides of equation (7.3.3.102) by $u(x, y, p)$. In other words, equation (7.3.3.101) in x, y, and p will be an integral of equation (7.3.3.97). If p is now made constant again, then equation (7.3.3.101) will clearly be an integral of equation (7.3.3.96). [When p is constant in the expression ϕ in equation (7.3.3.101), ϕ may either be integrable in finite terms, or else it may at most be reducible to quadratures.] Varying the constant k in equation (7.3.3.101) produces a one-parameter family of solutions of equation (7.3.3.96). (We will see why α and π were chosen to be homogeneous expressions of degree *zero* in x, y, and p shortly.)

This procedure entails the following interpretation of an integral of equation (7.3.3.96). The objective is to find an expression $\phi(x, y)$ in x and y such that

$$\frac{d\phi}{dx} = u(x, y) \quad \text{and} \quad \frac{d\phi}{dy} = u(x, y)\alpha(x, y) \qquad (7.3.3.103)$$

for some integrating factor $u(x, y)$, because in this event if $u(x, y)$ and $u(x, y)\alpha(x, y)$, respectively, are substituted for $d\phi/dx$ and $d\phi/dy$, respectively, in

$$d\phi = \frac{d\phi}{dx}dx + \frac{d\phi}{dy}dy, \qquad (7.3.3.104)$$

then it follows that (7.3.3.104) is just

$$d\phi(x, y) = u(x, y)dx + u(x, y)\alpha(x, y)dy \qquad (7.3.3.105)$$

in this case. If the expression $\phi(x, y)$ is then set equal to a constant k:

$$\phi(x, y) = k, \qquad (7.3.3.106)$$

then differentiating (that is, taking the differential of) equation (7.3.3.106) produces the equation:

$$0 = dk = d\phi(x, y) = u(x, y)dx + u(x, y)\alpha(x, y)dy. \qquad (7.3.3.107)$$

If both sides of equation (7.3.3.107) are then divided by $u(x, y)$, equation (7.3.3.96) follows. In other words, the equation (7.3.3.106) in x and y will be an integral of equation (7.3.3.96). [The expression ϕ in equation (7.3.3.106) may either be integrable in finite terms, or else it may at most be reducible to quadratures.] Varying the constant k in equation (7.3.3.106) produces a one-parameter family of solutions of equation (7.3.3.96). Fontaine in effect tacitly introduced a new definition of integrals of equation (7.3.3.96). It was one based on the idea of a total differential

$$dF(x, y) = \frac{dF}{dx}dx + \frac{dF}{dy}dy \qquad (7.3.3.108)$$

of an expression $F(x, y)$ in two finite variables x and y. Such a total differential appears in (7.3.3.104). In fact, as we saw, Fontaine had already employed total differentials in two finite variables in his paper of 1737 on problems of trajectories. As I mentioned, he pioneered the use of whole-symbol notations to express the differential coefficients in such differentials. One of his new ideas in 1738 was to make use of total differentials to integrate first-order ordinary "differential" equations of first degree. In addition, we note that in (7.3.3.99) Fontaine utilized the idea of a total differential

$$dF = \frac{dF}{dx}dx + \frac{dF}{dy}dy + \frac{dF}{dp}dp \qquad (7.3.3.109)$$

of an expression $F(x, y, p)$ in three finite variables x, y, and p. We saw signs above that Fontaine had already had this idea by the time that he wrote his paper of 1737 on problems of trajectories too.

In introducing the expression $\pi(x, y, p)$, Fontaine in effect attacked the problem of integrating in finite terms or reducing to quadratures equation (7.3.3.96) by means of a completion problem that was even more general than any of the earlier ones that we examined – namely, given an equation

$$dx + \alpha(x, y, p) = 0 \qquad (7.3.3.110)$$

where the value of p is constant, determine $\pi(x, y, p)$ such that equation (7.3.3.97) is an integrable equation in three variables x, y, and p, meaning that an expression $\phi(x, y, p)$ in the three variables x, y, and p can be found whose total differential $d\phi(x, y, p)$ satisfies

$$d\phi(x, y, p) = u(x, y, p)(dx + \alpha(x, y, p)dy + \pi(x, y, p)dp) \qquad (7.3.3.100)$$

for some integrating factor $u(x, y, p)$. As we shall see in a moment, finding a suitable expression $\pi(x, y, p)$ could not always be done easily, if it could be done at all, which is what Clairaut and Nicole meant when they said that Fontaine's method could not always be made to work. Nevertheless, for the first time, a direct approach to a generalized completion problem, which is to be contrasted with Euler's maneuvers backward in treating such a problem, was shown to have some use. This is one example that illustrates the statement that I made earlier that the partial differential calculus progressed in the hands of Euler and Fontaine in ways that differed.

In connection with Fontaine's paper, Clairaut later mentioned in his own published work what must have been the third of the three equations that he and Nicole had alluded to in their report on Fontaine's paper. This is the equation[174]

$$\alpha \frac{d\pi}{dx} - \pi \frac{d\alpha}{dx} + \frac{d\alpha}{dp} - \frac{d\pi}{dy} = 0. \qquad (7.3.3.111)$$

It expresses a condition for the integrability, in the sense described, of equation (7.3.3.97). For a given $\alpha(x, y, p)$, equation (7.3.3.111) is a first-order linear partial "differential equation" in three variables x, y, and p for $\pi(x, y, p)$, meaning that it is an equation that includes differential coefficients but no differentials. Fontaine's method of integrating equation (7.3.3.96) evidently involved solving equation (7.3.3.111) for a particular solution. Clairaut and Nicole concluded in their report that Fontaine had not found a sufficiently general way of accomplishing this.

Now, ordinarily equation (7.3.3.97) cannot possibly be easier to integrate in finite terms or reduce to quadratures than equation (7.3.3.96). Fontaine showed, however, that if $\alpha(x, y, p)$ and $\pi(x, y, p)$ are *homogeneous* expressions of degree zero in x, y, and p, then in that case equation (7.3.3.97) can be integrated in finite terms or reduced to quadratures rather easily. To demonstrate this Fontaine must have used the theorem that Clairaut and Nicole alluded to in their report. And this theorem must be the "homogeneous function theorem"

$$nF = \frac{dF}{dx}x + \frac{dF}{dy}y + \frac{dF}{dp}p \qquad (7.3.3.112)$$

which holds when the expression $F(x, y, p)$ in (7.3.3.109) is a homogeneous expression of degree n in the three finite variables x, y, and p, which means that

$$F(\lambda x, \lambda y, \lambda p) = \lambda^n F(x, y, p) \qquad (7.3.3.112^*)$$

is true. Clairaut also mentioned this theorem later in his published work in connection with Fontaine's paper.[175] Fontaine evidently saw that the homogeneous function theorem (7.3.3.87) in two finite variables, which is the theorem that he applied in his paper of 1737 on problems of trajectories, could easily be extended to include expressions in three or more finite variables.

If we ignore the origins of equation

$$e\phi = \frac{d\phi}{dx}x + \frac{d\phi}{dy}y + \frac{d\phi}{dp}p \qquad (7.3.3.113)$$

for the moment and simply consider the form that equation (7.3.3.113) has, then it clearly can be seen that if $d\phi/dx$, $d\phi/dy$, and $d\phi/dp$ on the right-hand side of equation (7.3.3.113) all happen to be homogeneous expressions of degree $e - 1$ in x, y, and p, then ϕ on the left-hand side of equation (7.3.3.113) must be a homogeneous expression of degree e in x, y, and p. On the other hand, if we start with a homogeneous expression ϕ of degree e in x, y, and p, then equation (7.3.3.113) expresses the homogeneous function theorem in this case. But we can also conclude even more. Because of the form that the right-hand side of equation (7.3.3.113) has, it seems rather obvious that $d\phi/dx$, $d\phi/dy$, and $d\phi/dp$ must all be homogeneous expressions of degree $e - 1$ in x, y, and p, if ϕ is a homogeneous expression of degree e in x, y, and p. In fact this can formally be demonstrated to be true.[176] Then from the three equations in (7.3.3.98) we can see why α and π in this case must have the same degree of homogeneity in x, y, and p. Indeed, we can deduce even more. From the first of the three equations in (7.3.3.98):

$$\frac{d\phi}{dx} = u,$$

we can infer that the integrating factor $u(x, y, p)$ must itself be a homogeneous expression of degree $e - 1$ in x, y, and p. Then from the other two equations in (7.3.3.98):

$$\frac{d\phi}{dy} = u\alpha \quad \text{and} \quad \frac{d\phi}{dp} = u\pi,$$

we can see clearly why α and π both must be homogeneous expressions of degree *zero* in x, y, and p.

Fontaine apparently treated equation (7.3.3.111) as a sufficient condition for the existence of an integrating factor $u(x, y, p)$ that is homogeneous in x, y, and p for the differential $dx + \alpha\,dy + \pi\,dp$ whose coefficients α and π are homogeneous expressions of degree zero in x, y, and p, meaning that an expression $\phi(x, y, p)$ exists such that

$$d\phi = udx + u\alpha dy + u\pi dp. \qquad (7.3.3.100)$$

Now, such an expression $\phi(x, y, p)$ is not unique; it is determined up to an arbitrary additive constant. Fontaine must have then applied the converse of the homogeneous function theorem (7.3.3.112) in order to conclude that if such an integrating factor $u(x, y, p)$ is homogeneous of degree $e - 1$ in x, y, and p, then an expression $\phi(x, y, p)$ that satisfies (7.3.3.100) can be chosen which is itself homogeneous of degree e in x, y, and p such that the equation

$$e\phi = ux + u\alpha y + u\pi p \qquad (7.3.3.114)$$

holds. Consequently if an integrating factor $u(x, y, p)$ that is homogeneous of degree $e - 1$ in x, y, and p for the differential $dx + \alpha dy + \pi dp$ could actually be found in some way, then the particular homogeneous expression $\phi(x, y, p)$

in question could also be found. Namely, if the integrating factor $u(x, y, p)$ that is homogeneous of degree $e - 1$ in x, y and p is substituted for u in equation (7.3.3.114), then equation (7.3.3.114) determines a homogeneous function $\phi(x, y, p)$ of degree e in x, y, and p, expressed in terms of α, π, and this integrating factor u that is homogeneous of degree $e - 1$ in x, y, and p, whose total differential $d\phi$ equals $udx + u\alpha dy + u\pi dp$ [equation (7.3.3.100)]. Evidently equation (7.3.3.114) was the first of the three equations that Clairaut and Nicole alluded to in their report on Fontaine's paper.

Now, if $x + \alpha y + \pi p$ happens to be zero, then e equals zero on the left-hand side of equation (7.3.3.114), in which case (7.3.3.114) cannot be used to determine the homogeneous expression ϕ of degree $e = 0$ in x, y, and p which corresponds to this particular case. In fact, it is easy to verify directly that

$$\pi \equiv \frac{-x - \alpha y}{p} \tag{7.3.3.115}$$

is a homogeneous expression of degree zero in x, y, and p which is indeed a solution of equation (7.3.3.111). [That expression (7.3.3.115) is a solution of equation (7.3.3.111) follows from the homogeneous function theorem (7.3.3.112) applied to the homogeneous expression α of degree $n = 0$ in x, y, and p.] Consequently this choice of π must be rejected and another found instead, in order for Fontaine's method to work. By the same token, this would also appear to limit Fontaine's technique of integrating equation (7.3.3.97) with homogeneous coefficients α and π of degree zero in x, y, and p to ones whose corresponding homogeneous expressions ϕ are homogeneous of degree $e \neq 0$ in x, y, and p. [In other words, the π that appears in equation (7.3.3.97) cannot be chosen to be expression (7.3.3.115).] In fact, there is no such restriction, as Fontaine demonstrated shortly afterward. Namely, if equation (7.3.3.111) is first solved for π, where

$$\pi \not\equiv \frac{-x - \alpha y}{p},$$

and if ϕ is the corresponding homogeneous expression of degree $e \neq 0$ in x, y, and p determined by equation

$$e\phi = ux + u\alpha y + u\pi p \tag{7.3.3.114}$$

for this choice of π, then

$$d\phi = udx + u\alpha dy + u\pi dp \tag{7.3.3.100}$$

is true for this choice of π. And if this expression ϕ is set equal to a constant k, then the equation

$$\phi = k \tag{7.3.3.101}$$

is just the integral of the equation (7.3.3.97) for this choice of π. Now, since ϕ is a homogeneous expression of degree $e \neq 0$ in x, y, and p, it follows that ϕ/p^e is a homogeneous expression of degree zero in x, y, and p. Then if the expression ϕ/p^e is now set equal to the constant k:

$$\frac{\phi}{p^e} = k, \tag{7.3.3.116}$$

and if equation (7.3.3.116) is differentiated (that is, if the differential of equation (7.3.3.116 is taken), it follows that

$$0 = dk = d\left(\frac{\phi}{p^e}\right) = \frac{d}{dx}\left(\frac{\phi}{p^e}\right)dx + \frac{d}{dy}\left(\frac{\phi}{p^e}\right)dy + \frac{d}{dp}\left(\frac{\phi}{p^e}\right)dp$$

$$= \frac{d\phi}{dx}\frac{1}{p^e}dx + \frac{d\phi}{dy}\frac{1}{p^e}dy + \left(\frac{p^e(d\phi/dp) - e\phi p^{e-1}}{p^{2e}}\right)dp. \tag{7.3.3.117}$$

But

$$d\phi = u\,dx + u\alpha\,dy + u\pi\,dp, \tag{7.3.3.100}$$

hence

$$\frac{d\phi}{dx} = u, \quad \frac{d\phi}{dy} = u\alpha, \quad \text{and} \quad \frac{d\phi}{dp} = u\pi. \tag{7.3.3.98}$$

Then if u, $u\alpha$, and $u\pi$, respectively, are substituted for $d\phi/dx$, $d\phi/dy$, and $d\phi/dp$, respectively, in equation (7.3.3.117), that equation can be written as

$$0 = u\frac{1}{p^e}dx + u\alpha\frac{1}{p^e}dy + \left(\frac{p^e u\pi - e\phi p^{e-1}}{p^{2e}}\right)dp. \tag{7.3.3.118}$$

If both sides of equation (7.3.3.118) are now divided by u/p^e, then equation

$$0 = dx + \alpha\,dy + \left(\pi - \frac{e\phi}{up}\right)dp \tag{7.3.3.119}$$

results. But equation (7.3.3.114) can be rewritten as

$$\pi - \frac{e\phi}{up} = \frac{-x - \alpha y}{p}. \tag{7.3.3.120}$$

If the coefficient of dp in equation (7.3.3.119) is replaced by the expression on the right-hand side of equation (7.3.3.120), the equation

$$0 = dx + \alpha\,dy + \left(\frac{-x - \alpha y}{p}\right)dp \tag{7.3.3.121}$$

follows. In short, equation (7.3.3.116) is an integral of equation (7.3.3.121).[177]

Finally, the second equation that Clairaut and Nicole alluded to in their report on Fontaine's paper must have been an equation that actually enabled an integrating factor $u(x, y, p)$ that is homogeneous of degree $e - 1 \neq -1$ in x, y, and p for the differential $dx + \alpha dy + \pi dp$ to be determined, for if such an integrating factor could be found, then the problem would be completely solved. The homogeneous expression ϕ of degree $e \neq 0$ could be constructed in terms of α, π, and u by using equation (7.3.3.114). Later in his own published work, Clairaut made a remark that is a useful hint. He mentioned Fontaine's "integration of his equation[178]

$$\frac{du}{u} = \frac{(e-1)[dx]}{x + \alpha y + \pi p} \text{ etc.}" \tag{7.3.3.122}$$

Now, if $d\phi$ could be eliminated from equation

$$d\phi = udx + u\alpha dy + u\pi dp, \tag{7.3.3.100}$$

an equation for $u(x, y, p)$ could be determined. But it follows from differentiating (that is, taking the differential of) equation

$$e\phi = ux + u\alpha y + u\pi p \tag{7.3.3.114}$$

that

$$ed\phi = ud(x + \alpha y + \pi p) + (x + \alpha y + \pi p)du. \tag{7.3.3.123}$$

If both sides of equation (7.3.3.123) are divided by e, which can be done when $e \neq 0$, and if the right-hand side of the resulting equation and the right-hand side of equation (7.3.3.100) are then equated, the equation

$$\frac{du}{u} = \frac{(e-1)\,dx}{x + \alpha y + \pi p} + \frac{e\alpha dy}{x + \alpha y + \pi p} + \frac{e\pi dp}{x + \alpha y + \pi p}$$
$$- \frac{d(\alpha y + \pi p)}{x + \alpha y + \pi p} \tag{7.3.3.124}$$

for $u(x, y, p)$ follows. If we compare equation (7.3.3.124) with equation (7.3.3.122), we conclude that equation (7.3.3.124) is doubtlessly the second of the three equations that Clairaut and Nicole alluded to in their report on Fontaine's paper.

[Note that neither a solution π of equation (7.3.3.111) nor a solution u of equation (7.3.3.124) need be an integral that can be expressed in finite terms; one or both of π and u may at most be reducible to quadratures, which in this case, unlike the notion of quadratures discussed earlier in this chapter, can mean reducible to integrals in three finite variables, x, y, and p which cannot be expressed in finite terms, instead of to integrals in one finite variable which cannot be expressed in finite terms.]

One question remains to be answered: how did Fontaine arrive at the "conditional equation" (7.3.3.111) for the integrability of equation (7.3.3.97)? Once again, Clairaut provides a useful clue. In his published work he mentioned Fontaine's use of "nine very complicated equations" to derive equation

(7.3.3.111).[179] Clairaut did not exhibit these nine equations in the published work in question. However equation (7.3.3.111) can only be determined by using what appears to be a very complicated approach in one way, namely, by employing the fluxio-differential method. Indeed, we shall see that equation (7.3.3.111) can be derived using the fluxio-differential method in a manner that appears rather involved but at the same time brings to mind Fontaine's solution to the problem of finding tautochrones.

We have seen that integrals of integrable equations (7.3.3.97) generally have the form of equation (7.3.3.101). We also know that Fontaine had been interested in problems that involve one-parameter families of plane curves

$$x = x(y, p), \qquad (7.3.3.125)$$

where the plane curves in such a family all lie in the same plane and where p is the parameter, at least since 1734. We also know that he was necessarily interested in the first-order ordinary "differential" equations of first degree which certain kinds of problems involving such families give rise to, because these equations must be solved in order to solve the problems that give rise to them. It is consequently plausible that Fontaine first hit upon equation (7.3.3.111) while solving equations (7.3.3.97) for one-parameter families of plane curves, where the plane curves in such a family all lie in the same plane and where p is the parameter. For example, if $\phi(x, y, p)$ in equation (7.3.3.101) is the homogeneous expression $x - \sqrt{p^2 - y^2}$ of degree 1 in x, y, and p, then for each constant value of k, equation

$$x - \sqrt{p^2 - y^2} = k \qquad (7.3.3.101')$$

is an integral of the equation

$$dx + \frac{y}{\sqrt{p^2 - y^2}} dy - \frac{p}{\sqrt{p^2 - y^2}} dp = 0 \qquad (7.3.3.126)$$

whose coefficients are homogeneous expressions of degree 0 in x, y, and p. But equation (7.3.3.101') can be written as

$$x = \sqrt{p^2 - y^2} + k. \qquad (7.3.3.101'')$$

In other words, equation (7.3.3.101') can be written so as to express a one-parameter family of curves (7.3.3.125) that all lie in the same plane and whose parameter is p. If $k = 0$, then (7.3.3.101'') is an integral of the equation

$$dx + \frac{y}{x} dy - \frac{p}{x} dp = 0. \qquad (7.3.3.126')$$

It is easy to see that integrals (7.3.3.101) of integrable equations (7.3.3.97) generally need *not* have the form (7.3.3.125).[180] Assuming that they do have this form, however, the fluxio-differential method can be used to deal with the problem of integrating integrable equations (7.3.3.97), since this calculus that involves the use of two independent, first-order Leibnizian differential operators can be applied in general to *any* problems that involve one-parameter families of plane curves, where the plane curves in such a family all lie in the same plane. If we hypothesize that equation (7.3.3.97) has integrals that have form (7.3.3.125), then equation (7.3.3.97) can be written as

$$dx(y, p) = - \alpha(x(y, p), y, p)dy - \pi(x(y, p), y, p)dp, \qquad (7.3.3.127)$$

and the question becomes: what, if any, conditions of integrability does equation (7.3.3.127) involve?

Fontaine at first very likely thought of equation (7.3.3.127) as having the following alternate form expressed by means of the formalism (7.3.3.61″), namely, as

$$dx(y, p) = - \alpha(x(y, p), y, p)dy - \pi(x(y, p), y, p)dp, \qquad (7.3.3.128)$$

$$\dot{x}(y, p) = - \alpha(x(y, p), y, p)\dot{y}, \qquad (7.3.3.129)$$

using the notations "·" and "*d*" of Fontaine's fluxio-differential method, where F is replaced by x, x is replaced by y, a is replaced by p, A is replaced by $-\alpha$, and B is replaced by $-\pi$ in (7.3.3.61″).

Using the fluxio-differential method I deduce the following general result: Let

$$dA = Cdx + Dda, \ \dot{A} = C\dot{x} \quad \text{and} \quad dB = Edx + Gda, \ \dot{B} = E\dot{x} \qquad (7.3.3.130)$$

define the differential coefficients $C, D, E,$ and G in terms of the differential coefficients A and B that appear in (7.3.3.61″). Assuming as usual that the differential operators "·" and "*d*" commute, in which case $dx = d(\dot{x})$, $da = d(\dot{a})$, and $dF = d(\dot{F})$ in particular, then

$$D = E. \qquad (7.3.3.131)$$

This can easily be demonstrated as follows: According to the rules given in (7.3.3.1) which govern the way that "·" operates when it is applied to sums and products, it follows from $dF = Adx + Bda$ that $d\dot{F} = \dot{A}dx + A\dot{\overline{dx}} + \dot{B}da + B\dot{\overline{da}}$. Then by making substitutions using (7.3.3.130), it follows that $\dot{A}dx + A\dot{\overline{dx}} + \dot{B}da + B\dot{\overline{da}} = C\dot{x}dx + A\dot{\overline{dx}} + E\dot{x}da + B\dot{\overline{da}}$. Likewise, according to the rule given in (7.3.3.1) which governs the way that "*d*" operates when it is applied to products, it follows from $\dot{F} = A\dot{x}$ that $d(\dot{F}) = \dot{x}dA + Ad(\dot{x})$. Again by making substitutions using (7.3.3.130), it follows that $\dot{x}dA + Ad(\dot{x}) = (Cdx + Dda)\dot{x} + Ad(\dot{x}) = C\dot{x}dx + D\dot{x}da + Ad(\dot{x})$. Then if the left-hand sides and the right-hand sides, respectively, of the two equations

$$d\dot{F} = C\dot{x}dx + A\dot{\overline{dx}} + E\dot{x}da + B\dot{\overline{da}}$$

and

$$d(\dot{F}) = C\dot{x}dx + D\dot{x}da + Ad(\dot{x})$$

are subtracted from each other, the equation

$$d\dot{F} - d(\dot{F}) = A(d\dot{x} - d(\dot{x})) + (E - D)\dot{x}da + Bd\dot{a}$$

results. But $d\dot{F} = d(\dot{F})$ and $d\dot{x} = d(\dot{x})$, in which case the preceding equation reduces to the equation

$$(E - D)\dot{x}da + Bd\dot{a} = 0.$$

Finally, $\dot{a} = 0$ ((7.3.3.62)) and $d\dot{a} = d(\dot{a})$. Hence $d\dot{a} = d(\dot{a}) = d(0) = 0$. Thus the preceding equation reduces to the equation

$$(E - D)\dot{x}da = 0.$$

Now, if we choose $\dot{x} \neq 0$ and $da \neq 0$ in this last equation, it follows that $E - D$ must equal zero. In other words, the equality $D = E$, equation (7.3.3.131), is true.

If F is replaced by x, x is replaced by y, a is replaced by p, A is replaced by $-\alpha$, and B is replaced by $-\pi$ in (7.3.3.61″) and in (7.3.3.130), then equality (7.3.3.131) ultimately expresses a condition for the integrability of equations (7.3.3.128) and (7.3.3.129), but it is impossible to recognize this condition when the symbols "·" and "d" are used. Therefore I once again replace the symbols "·" and "d", respectively, by the symbols "d" and "δ", respectively, in which case (7.3.3.61″) becomes

$$\delta F = A\delta x + B\delta a \quad \text{and} \quad dF = Adx, \qquad (7.3.3.61''')$$

and (7.3.3.130) becomes

$$\delta A = C\delta x + D\delta a, \, dA = Cdx \quad \text{and} \quad \delta B = E\delta x + G\delta a, \, dB = Edx. \qquad (7.3.3.130')$$

This change of symbols enables us to exploit to good advantage whole-symbol notations for differential coefficients. When expressed through (7.3.3.61‴) and (7.3.3.130′), (7.3.3.131) appears as

$$\frac{\delta}{\delta a}\left(\frac{dF}{dx}\right) = \frac{d}{dx}\left(\frac{\delta F}{\delta a}\right) \quad \text{or} \quad \frac{\delta}{\delta a}\left(\frac{\delta F}{\delta x}\right) = \frac{\delta}{\delta x}\left(\frac{\delta F}{\delta a}\right) \qquad (7.3.3.131')$$

using whole-symbol notations for differential coefficients. Using the new notations "d" and "δ", equations (7.3.3.128) and (7.3.3.129) appear as

$$\delta x(y, p) = -\alpha(x(y, p), y, p)\delta y - \pi(x(y, p), y, p)\delta p, \qquad (7.3.3.128')$$

$$dx(y, p) = -\alpha(x(y, p), y, p)dy, \qquad (7.3.3.129')$$

and when expressed through (7.3.3.61‴) and (7.3.3.130′), (7.3.3.131) appears as

$$\frac{\delta}{\delta p}\left(\frac{dx}{dy}\right) = \frac{d}{dy}\left(\frac{\delta x}{\delta p}\right) \quad \text{or} \quad \frac{\delta}{\delta p}\left(\frac{\delta x}{\delta y}\right) = \frac{\delta}{\delta y}\left(\frac{\delta x}{\delta p}\right) \qquad (7.3.3.131'')$$

using whole-symbol notations for differential coefficients, where F is replaced by x, x is replaced by y, a is replaced by p, A is replaced by $-\alpha$, and B is replaced by $-\pi$ in (7.3.3.61‴), in (7.3.3.130′), and in (7.3.3.131′).

I now show that the condition (7.3.3.131″) for the integrability of equations (7.3.3.128′) and (7.3.3.129′) can be expanded using the fluxio-differential method. To do this I replace F by x, x by y, and a by p in (7.3.3.61‴) and then rewrite (7.3.3.61‴) using whole-symbol notations for differential coefficients as

$$\delta x = \frac{\delta x}{\delta y}\delta y + \frac{\delta x}{\delta p}\delta p \qquad (7.3.3.132)$$

and

$$dx = \frac{dx}{dy}dy, \qquad (7.3.3.133)$$

where

$$\frac{\delta x}{\delta y} = \frac{dx}{dy}. \qquad (7.3.3.134)$$

Using whole-symbol notations for differential coefficients, equations (7.3.3.128′) and (7.3.3.129′) appear as

$$\frac{\delta x(y, p)}{\delta y} = -\alpha(x(y, p), y, p), \qquad (7.3.3.135)$$

$$\frac{\delta x(y, p)}{\delta p} = -\pi(x(y, p), y, p), \qquad (7.3.3.136)$$

$$\frac{dx(y, p)}{dy} = -\alpha(x(y, p), y, p). \qquad (7.3.3.137)$$

I shall now apply the algebra of differential operators "d" and "δ" to equations (7.3.3.128′) and (7.3.3.129′) in much the same way that Fontaine applied his fluxio-differential method to solve (7.3.3.36) and (7.3.3.37) in his Paris Academy *mémoire* of 1734 on finding tautochrones.

The differentials $\delta\alpha, \delta x, \delta y$, and δp are related through differential coefficients as follows:

$$\delta\alpha(x, y, p) = \frac{\delta\alpha}{\delta x}\delta x + \frac{\delta\alpha}{\delta y}\delta y + \frac{\delta\alpha}{\delta p}\delta p. \qquad (7.3.3.138)$$

Substituting the right-hand side of equation (7.3.3.128′), which is satisfied by $x = x(y, p)$, for δx in equation (7.3.3.138), we find that the equation

$$\delta\alpha = \left(\frac{\delta\alpha}{\delta y} - \alpha\frac{\delta\alpha}{\delta x}\right)\delta y + \left(\frac{\delta\alpha}{\delta p} - \pi\frac{\delta\alpha}{\delta x}\right)\delta p, \qquad (7.3.3.139)$$

which is satisfied by $\alpha(x(y, p), y, p)$ and $\pi(x(y, p), y, p)$, results. If we apply the "δ" differential operator to equation (7.3.3.129′), which is satisfied by $x = x(y, p)$, the equation

$$\delta dx = -\delta\alpha dy - \alpha\delta dy, \qquad (7.3.3.140)$$

which is satisfied by $x = x(y, p)$ and $\alpha(x(y, p), y, p)$, follows. If we substitute the right-hand side of equation (7.3.3.139) for $\delta\alpha$ in (7.3.3.140), the equation

$$\delta dx = \left(\alpha\frac{\delta\alpha}{\delta x} - \frac{\delta\alpha}{\delta y}\right)\delta y dy + \left(\pi\frac{\delta\alpha}{\delta x} - \frac{\delta\alpha}{\delta p}\right)\delta p dy - \alpha\delta dy, \qquad (7.3.3.141)$$

which is satisfied by $x = x(y, p)$, $\alpha(x(y, p), y, p)$, and $\pi(x(y, p), y, p)$, results. Likewise, the differentials $d\alpha, dx$, and dy are related through differential coefficients as follows:

$$d\alpha(x, y, p) = \frac{d\alpha}{dx}dx + \frac{d\alpha}{dy}dy. \qquad (7.3.3.142)$$

Substituting the right-hand side of equation (7.3.3.129′), which is satisfied by $x = x(y, p)$, for dx in equation (7.3.3.142), we find that the equation

$$d\alpha = \left(\frac{d\alpha}{dy} - \alpha\frac{d\alpha}{dx}\right)dy, \qquad (7.3.3.143)$$

which is satisfied by $\alpha(x(y, p), y, p)$, results. Similarly, the differentials $d\pi, dx$ and dy are related through differential coefficients as follows:

$$d\pi(x, y, p) = \frac{d\pi}{dx}dx + \frac{d\pi}{dy}dy. \qquad (7.3.3.144)$$

If we substitute the right-hand side of equation (7.3.3.129′), which is satisfied by $x = x(y, p)$, for dx in equation (7.3.3.144), the equation

$$d\pi = \left(\frac{d\pi}{dy} - \alpha\frac{d\pi}{dx}\right)dy, \qquad (7.3.3.145)$$

which is satisfied by $\alpha(x(y, p), y, p)$ and $\pi(x(y, p), y, p)$, results. The fact that the right-hand sides of equations (7.3.3.142) and (7.3.3.144) only have two terms instead of three, unlike the right-hand side of equation (7.3.3.138), which has three terms, follows from the fact that a nonzero differential dp does not exist ("d"-differentials of quantities defined at points situated on plane curves in a one-parameter family of plane curves that all lie in the same plane, where p is the parameter, are always taken *along individual* plane curves in the family, and the parameter p is a *constant* quantity for points located on any *one* plane curve in the family). If we apply the "d" differential operator to equation (7.3.3.128′), which is satisfied by $x = x(y, p)$, the equation

$$d\delta x = -d\alpha\delta y - \alpha d\delta y - d\pi\delta p - \pi d\delta p, \qquad (7.3.3.146)$$

which is satisfied by $x = x(y, p)$, $\alpha(x(y, p), y, p)$, and $\pi(x(y, p), y, p)$, results. It also follows from $dp = 0$ that

$$d\delta p = \delta dp = \delta(0) = 0, \qquad (7.3.3.147)$$

because the differential operators "d" and "δ" commute. Substituting the right-hand side of equation (7.3.3.143), the right-hand side of equation (7.3.3.145), and

$$d\delta p = 0, \qquad (7.3.3.147)$$

respectively, for $d\alpha$, $d\pi$, and $d\delta p$, respectively, in equation (7.3.3.146), we find that the equation

$$d\delta x = \left(\alpha\frac{d\alpha}{dx} - \frac{d\alpha}{dy}\right)dy\delta y - \alpha d\delta y + \left(\alpha\frac{d\pi}{dx} - \frac{d\pi}{dy}\right)dy\delta p, \qquad (7.3.3.148)$$

which is satisfied by $x = x(y,p)$, $\alpha(x(y,p), y, p)$, and $\pi(x(y,p), y, p)$, results. Next, if we subtract the left-hand side and right-hand side, respectively, of equation (7.3.3.148) from the left-hand side and the right-hand side, respectively, of equation (7.3.3.141), the equation

$$\delta dx - d\delta x = \left[\left(\alpha\frac{\delta\alpha}{\delta x} - \frac{\delta\alpha}{\delta y}\right) - \left(\alpha\frac{d\alpha}{dx} - \frac{d\alpha}{dy}\right)\right]\delta y dy + \alpha(d\delta y - \delta dy)$$

$$+ \left[\left(\pi\frac{\delta\alpha}{\delta x} - \frac{\delta\alpha}{\delta p}\right) - \left(\alpha\frac{d\pi}{dx} - \frac{d\pi}{dy}\right)\right]\delta p dy, \qquad (7.3.3.149)$$

which is satisfied by $x = x(y,p)$, $\alpha(x(y,p), y, p)$, and $\pi(x(y,p), y, p)$, follows. Now, if we apply

$$\delta F = \frac{\delta F}{\delta y}\delta y + \frac{\delta F}{\delta p}\delta p, \quad dF = \frac{dF}{dy}dy, \quad \text{where} \quad \frac{\delta F}{\delta y} = \frac{dF}{dy} \qquad (7.3.3.150)$$

(that is, (7.3.3.61''') expressed using whole-symbol notations for differential coefficients, where y replaces p and x replaces a in (7.3.3.61''')) to $F(y,p) \equiv \alpha(x(y,p), y, p)$, remembering at the same time that $\pi(x(y,p), y, p)$ has the form $\pi^*(y, p)$ too, then it follows from equations (7.3.3.139) and (7.3.3.143) that

$$\frac{\delta\alpha}{\delta y} - \alpha\frac{\delta\alpha}{\delta x} = \frac{\delta F}{\delta y} = \frac{dF}{dy} = \frac{d\alpha}{dy} - \alpha\frac{d\alpha}{dx}. \qquad (7.3.3.151)$$

Consequently, equation (7.3.3.149) reduces to equation

$$\delta dx - d\delta x = \alpha(d\delta y - \delta dy) + \left[\left(\pi\frac{\delta\alpha}{\delta x} - \frac{\delta\alpha}{\delta p}\right) - \left(\alpha\frac{d\pi}{dx} - \frac{d\pi}{dy}\right)\right]\delta p dy. \qquad (7.3.3.152)$$

By this point the similarity between this demonstration and the demonstration used above to arrive at equality (7.3.3.131) should be apparent. Since the differential operators "d" and "δ" commute, it follows that

$$\delta dx = d\delta x, \quad \text{and} \quad \delta dy = d\delta y.$$

Hence (7.3.3.152) simplifies further to equation

$$0 = \left[\left(\pi\frac{\delta\alpha}{\delta x} - \frac{\delta\alpha}{\delta p}\right) - \left(\alpha\frac{d\pi}{dx} - \frac{d\pi}{dy}\right)\right]\delta p dy. \qquad (7.3.3.153)$$

Now, let us consider the expression

$$T\left(\pi, \alpha, \frac{\delta\alpha}{\delta x}, \frac{\delta\alpha}{\delta p}, \frac{d\pi}{dx}, \frac{d\pi}{dy}\right) \equiv \left(\pi\frac{\delta\alpha}{\delta x} - \frac{\delta\alpha}{\delta p}\right) - \left(\alpha\frac{d\pi}{dx} - \frac{d\pi}{dy}\right). \quad (7.3.3.154)$$

Setting $\delta p \neq 0$ and $dy \neq 0$ in equation (7.3.3.153), we arrive at the equation

$$T\left(\pi, \alpha, \frac{\delta\alpha}{\delta x}, \frac{\delta\alpha}{\delta p}, \frac{d\pi}{dx}, \frac{d\pi}{dy}\right) = 0. \quad (7.3.3.155)$$

In fact, equation (7.3.3.155) is just condition (7.3.3.131″) expanded. It is a condition for the integrability of equations (7.3.3.128′) and (7.3.3.129′). We can think of equation (7.3.3.155) as a conditional equation in α, π, and the differential coefficients $\delta\alpha/\delta x, \delta\alpha/\delta p, d\pi/dx$, and $d\pi/dy$ which must hold in order for equation (7.3.3.153) to hold, hence in order for the system of equations (7.3.3.128′), (7.3.3.129′), (7.3.3.138), (7.3.3.140), (7.3.3.142), (7.3.3.144), (7.3.3.146), (7.3.3.150), and (7.3.3.156) in the differentials $dx, dy, \delta x, \delta y, \delta p, d\delta x$, and $d\delta y$, which are all infinitesimal variables, which reduces to equation (7.3.3.153), to have a nontrivial solution $x = x(y, p)$, hence in order for the original pair of equations (7.3.3.128′) and (7.3.3.129′) from which this system of equations was derived to have a solution as well. Moreover, equation (7.3.3.155) is a conditional equation *in exactly the same sense* that the quadratic equation (7.3.3.67) in y whose coefficients are composed of the expressions used to form the equations for Fontaine's solution to the problem of finding the tautochrone in a medium that resists a body's motion directly as $(y^2 + my)/n$, where $m \neq 0$ and y is the body's speed, is a conditional equation in y and the differential coefficients p, q, ϕ, γ, and λ which must hold in order for equation (7.3.3.65) to hold, hence in order for the system of equations (7.3.3.36′), (7.3.3.13′), (7.3.3.14′), (7.3.3.19′), (7.3.3.24′)–(7.3.3.27′), (7.3.3.29′), and (7.3.3.31′) in the differentials $dz, dx, dy, \delta x, \delta y, \delta X, \delta dx$, and δdy, which reduces to equation (7.3.3.65), to have a nontrivial solution, hence in order for the equation (7.3.3.36′) and the condition (7.3.3.20) from which this system of equations was derived and which express the original problem (finding the tautochrone in question) to have a solution too!

Now, we can reduce equation (7.3.3.155) to a form that is still simpler. The differentials $\delta\pi, \delta x, \delta y$, and δp are related through differential coefficients as follows:

$$\delta\pi(x, y, p) = \frac{\delta\pi}{\delta x}\delta x + \frac{\delta\pi}{\delta y}\delta y + \frac{\delta\pi}{\delta p}\delta p. \quad (7.3.3.156)$$

If we substitute the right-hand side of equation (7.3.3.128′), which is satisfied by $x = x(y, p)$, for δx in equation (7.3.3.156), the equation

$$\delta\pi = \left(\frac{\delta\pi}{\delta y} - \alpha\frac{\delta\pi}{\delta x}\right)\delta y + \left(\frac{\delta\pi}{\delta p} - \pi\frac{\delta\pi}{\delta x}\right)\delta p \quad (7.3.3.157)$$

which is satisfied by $\alpha(x(y,p),y,p)$ and $\pi(x(y,p),y,p)$ results. Then, if we apply

$$\delta F = \frac{\delta F}{\delta y}\delta y + \frac{\delta F}{\delta p}\delta p, \quad dF = \frac{dF}{dy}dy, \quad \text{where} \quad \frac{\delta F}{\delta y} = \frac{dF}{dy} \qquad (7.3.3.150)$$

[that is, (7.3.3.61''') expressed using whole-symbol notations for differential coefficients, where y replaces p and x replaces a in (7.3.3.61''')] to $F(y,p) \equiv \pi(x(y,p),y,p)$, remembering at the same time that $\alpha(x(y,p),y,p)$ has the form $\alpha^*(y,p)$ too, then it follows from equations (7.3.3.145) and (7.3.3.157) that

$$\frac{d\pi}{dy} - \alpha\frac{d\pi}{dx} = \frac{dF}{dy} = \frac{\delta F}{\delta y} = \frac{\delta\pi}{\delta y} - \alpha\frac{\delta\pi}{\delta x}. \qquad (7.3.3.158)$$

Now, equation (7.3.3.155) can be rewritten as

$$\pi\frac{\delta\alpha}{\delta x} - \frac{\delta\alpha}{\delta p} = \alpha\frac{d\pi}{dx} - \frac{d\pi}{dy}, \qquad (7.3.3.155')$$

and equation (7.3.3.158) can be rewritten as

$$\alpha\frac{\delta\pi}{\delta x} - \frac{\delta\pi}{\delta y} = \alpha\frac{d\pi}{dx} - \frac{d\pi}{dy}. \qquad (7.3.3.158')$$

Then if we subtract the left-hand side and the right-hand side, respectively, of equation (7.3.3.158') from the left-hand side and the right-hand side, respectively, of equation (7.3.3.155'), the equation

$$\left(\pi\frac{\delta\alpha}{\delta x} - \frac{\delta\alpha}{\delta p}\right) - \left(\alpha\frac{\delta\pi}{\delta x} - \frac{\delta\pi}{\delta y}\right) = 0 \qquad (7.3.3.159)$$

results. Finally, we remember that I chose the symbol δ arbitrarily. The letter d can be substituted for the symbol δ. If we make this substitution, we see that the condition (7.3.3.159) for the integrability of equations (7.3.3.128') and (7.3.3.129') cannot be distinguished from the condition (7.3.3.111) for the integrability of equation (7.3.3.127). Furthermore, the condition for the integrability of equation (7.3.3.97) and the condition for the integrability of equation (7.3.3.127) turn out to be identical.

Thus it seems likely that Fontaine derived the conditional equation (7.3.3.111) for the integrability of equation (7.3.3.97) using his fluxio-differential method. As we have just seen, this in fact can be done in a way that closely parallels the manner in which Fontaine solved the problem of finding tautochrones. But in the case of equation (7.3.3.111), the derivation appears "very complicated," to use Clairaut's words. The nine equations that Clairaut alluded to could have been a system of equations equivalent to the system of equations (7.3.3.128'), (7.3.3.129'), (7.3.3.138), (7.3.3.140), (7.3.3.142), (7.3.3.144), (7.3.3.146), (7.3.3.150), and (7.3.3.156) in the differentials dx, dy, δx, δy, δp, $d\delta x$, and $d\delta y$. Equation (7.3.3.159) is then a conditional equation, meaning that it expresses a condition that the coefficients in the system of equations must satisfy, much like the conditions on determinants of $n \times n$ systems of linear equations, in order for the systems of equations to be solvable. The preceding derivation of equation (7.3.3.111) helps make intelligible Fontaine's seemingly peculiar assertion, which Fontaine made at the beginning of his complete works published in 1764, that "every theorem to be used in the integral calculus [that is, in Fontaine's methods of integrating "differential" equations which appear in his complete works] is hidden" in his Paris Academy *mémoire* of 1734 on finding tautochrones.[181] The parallel that I drew between conditional equation (7.3.3.159) and conditional equation (7.3.3.67) for the existence of the tautochrone in a medium that resists a body's motion directly as $(y^2 + my)/n$, where $m \neq 0$ and y is the body's speed, helps make such a statement comprehensible. Moreover, we saw earlier in this section of this chapter that the fluxio-differential method very likely gave rise to the homogeneous function theorem in Fontaine's work, just as Fontaine maintained at the beginning of his complete works published in 1764 that the fluxio-differential method did. And Fontaine used this theorem from the partial differential calculus in developing his method of 1738 of integrating or reducing to quadratures first-order ordinary "differential" equations of first degree. In Fontaine's mathematical work the partial differential calculus and the fluxio-differential method are intrinsically related. In following closely the chronology of the mathematical work that Fontaine did during the years 1732–38 and in taking into account the obscure statements that he made at the end of his Paris Academy *mémoire* of 1734 on finding tautochrones, which relate his fluxio-differential method to the problems of trajectories which he subsequently treated in July of 1737 using the partial differential calculus, as well as the hints that appear at the beginning of his complete works of 1764 and the clues that can be found in some of Clairaut's own published work, we conclude that the partial differential calculus that Fontaine used during the period 1732–38 *grew out of* the fluxio-differential method.

Accordingly Fontaine's route to the partial differential calculus differed significantly from the one that Leibniz and the Bernoullis followed. This fact also helps shed light upon another incident that appears incredible at first sight. When Maupertuis sent Johann I Bernoulli a paper in which Fontaine explained his method of integrating first-order ordinary "differential" equations of first degree, Bernoulli replied that he could not understand the author's point of view. Moreover, Bernoulli showed the paper to his son Daniel and to his nephew Nikolaus, in whose hands, we recall, the partial differential calculus advanced further than it had in the hands of his uncle, and they, too, could not understand Fontaine's paper and the calculations upon which it was based either![182]

7.3.4. First French advances in the partial differential calculus II: Clairaut (1739–1741). Clairaut had not yet returned to Paris from Lapland when Fontaine first introduced the partial differential calculus to the Paris Academy early in July of 1737, in the paper in which Fontaine treated problems of trajectories by utilizing this calculus. When Clairaut did return to Paris in August of 1737, Fontaine left Paris almost immediately, in September of 1737, in order to work in solitude at his estate near Compiègne. He did not return to Paris to communicate new research until late November of 1738, when he presented his work on integral calculus (integrating or reducing to quadratures first-order ordinary "differential" equations of first degree) before the Paris Academy. Shortly afterward, Clairaut and Nicole read their referees' report on this work. At the same time, Clairaut did not delay getting busy in the "new world"[183] that Fontaine had founded in Paris. Where the story told in these pages is concerned, it is important to consider the fact that Clairaut sent his second paper on the theory of the earth's shape to the Royal Society in October of 1738,[184] nearly two months before he learned of Fontaine's work on the partial differential calculus and Fontaine's use of it to integrate or reduce to quadratures first-order ordinary "differential" equations of first degree. In other words, Clairaut had no knowledge whatever of Fontaine's latest work in mathematics when he wrote his paper of 1738 on the theory of the earth's shape. In Chapter 9 we shall see the extent to which the partial differential calculus subsequently influenced Clairaut's further work on the theory of the earth's shape. Furthermore, we will realize just how much the lack of suitable mathematics had obstructed Clairaut's efforts to make progress in dealing with the mechanics problem in his earlier work.

Early in March of 1739 Clairaut read a paper before the Paris Academy in which he described a method for integrating in finite terms or reducing to quadratures first-order ordinary "differential" equations of first degree.[185] The paper was published in the Paris Academy's *Mémoires* for 1739.[186] In developing his method, Clairaut made use of the partial differential calculus, but the method differed from the one that appears in the paper that Fontaine read before the Paris Academy in November of 1738.

Clairaut used *partial differential coefficients* to the best advantage in putting together his method. Fontaine, of course, employed partial differential coefficients too. Indeed, Fontaine first introduced differential coefficients in Paris and demonstrated their utility in his Paris Academy *mémoire* of 1734 on finding tautochrones, in the paper that he read before the Paris Academy in July of 1737 in which he attacked problems of trajectories by using the homogeneous function theorem, and in the paper that he read before the Paris Academy in November of 1738 on integrating or reducing to quadratures first-order ordinary "differential" equations of first degree in which he stated the homogeneous function theorem again, as well as presented and derived the conditional equation (7.3.3.111) for the integrability of equation (7.3.3.97). Moreover, Fontaine introduced whole-symbol notations to designate partial differential coefficients as well. But he could not have used such differential coefficients *alone* to derive the conditional equation (7.3.3.111) in a manner that Clairaut found to be long and involved.

We saw in the preceding section of this chapter how Fontaine very likely arrived at equation (7.3.3.111) by using the fluxio-differential method. In effect,

Fontaine showed that the equations

$$\frac{\delta}{\delta p}\left(\frac{\delta x}{\delta y}\right) = \pi \frac{\delta\alpha}{\delta x} - \frac{\delta\alpha}{\delta p} \tag{7.3.4.1}$$

and

$$\frac{\delta}{\delta y}\left(\frac{\delta x}{\delta p}\right) = \alpha \frac{\delta\pi}{\delta x} - \frac{\delta\pi}{\delta y} \tag{7.3.4.2}$$

are both true, where equality (7.3.3.131″) holds and where $x = x(y, p)$ satisfies equations (7.3.3.128′) and (7.3.3.129′).

But we also know that equation (7.3.3.129′) is superflous. The coefficients $\delta F/\delta y$ and dF/dy in (7.3.3.150) are the same, and the first equation in (7.3.3.150) alone consequently contains all of the information necessary to fix the problem in question in this case. In July of 1737 Fontaine evidently realized that the same was true of equations (7.3.3.77) and (7.3.3.78), because in the paper that he read before the Paris Academy at that time he treated problems of trajectories by using a single differential operator "d" to determine total differentials of expressions in more than one finite variable. We also saw that he utilized the idea of total differentials in the paper that he read before the Paris Academy in November of 1738 on integrating or reducing to quadratures first-order ordinary "differential" equations of first degree, and, as I say, he first introduced whole-symbol notations to stand for the partial differential coefficients that appear in such differentials.

But Fontaine could not *only* have used partial differential coefficients to arrive at equation (7.3.3.111) in his paper of November 1738, because any demonstration of equation (7.3.3.111) which involves partial differential coefficients *alone* cannot be long and complicated. For example, using the modern theory of functions, the chain rule for differentiating functions of functions, and the partial derivatives that have their origins in partial differential coefficients and which eventually replaced these coefficients, equation (7.3.4.1) can easily be derived from equations (7.3.3.128′) and (7.3.3.135) as follows:

$$\frac{\delta}{\delta p}\left(\frac{\delta x(y, p)}{\delta y}\right) = \frac{\delta(-\alpha(x(y, p), y, p))}{\delta p} = -\left(\frac{\delta\alpha}{\delta x}\frac{\delta x(y, p)}{\delta p} + \frac{\delta\alpha}{\delta p}\right)$$

$$= -\left(\frac{\delta\alpha}{\delta x}(-\pi) + \frac{\delta\alpha}{\delta p}\right) = \pi\frac{\delta\alpha}{\delta x} - \frac{\delta\alpha}{\delta p}.$$

Here I have treated the symbol "δ" like the standard symbol "∂" used in writing partial derivatives. Similarly, equation (7.3.4.2) can be derived from equations (7.3.3.128′) and (7.3.3.136) as follows:

$$\frac{\delta}{\delta y}\left(\frac{\delta x(y, p)}{\delta p}\right) = \frac{\delta(-\pi(x(y, p), y, p))}{\delta y} = -\left(\frac{\delta\pi}{\delta x}\frac{\delta x(y, p)}{\delta y} + \frac{\delta\pi}{\delta y}\right)$$

$$= -\left(\frac{\delta\pi}{\delta x}(-\alpha) + \frac{\delta\pi}{\delta y}\right) = \alpha\frac{\delta\pi}{\delta x} - \frac{\delta\pi}{\delta y}.$$

Hence we see that in one case at least, a result that Fontaine laboriously determined using the fluxio-differential method can ultimately be arrived at in a couple of lines.

Let us note that the partial differential coefficients in total differentials obey the same rules that govern the ways that the differential operator acts when it is applied to sums, products, and quotients of finite quantities. For example, if $F = F(y, p)$ and $G = G(y, p)$ are finite quantities expressed in terms of the finite variables y and p, and if "d" stands for the differential operator, then it follows from $d(F + G) = dF + dG$ that

$$\frac{d(F + G)}{dy} = \frac{dF}{dy} + \frac{dG}{dy}$$

and that

$$\frac{d(F + G)}{dp} = \frac{dF}{dp} + \frac{dG}{dp}.$$

Likewise,

$$\frac{d(FG)}{dy} = F\frac{dG}{dy} + G\frac{dF}{dy} \quad \text{and} \quad \frac{d(FG)}{dp} = F\frac{dG}{dp} + G\frac{dF}{dp}$$

follow from $d(FG) = FdG + GdF$. Similarly, it follows from

$$d\left(\frac{F}{G}\right) = \frac{GdF - FdG}{G^2}$$

that

$$\frac{d(F/G)}{dy} = \frac{G(dF/dy) - F(dG/dy)}{G^2} \quad \text{and} \quad \frac{d(F/G)}{dp} = \frac{G(dF/dp) - F(dG/dp)}{G^2}.$$

This means that the basic rules that govern the ways that the differential operator acts translate into analogous rules that the partial differential coefficients in total differentials obey. Clairaut fully exploited this fact to the best advantage. He saw how to apply the partial differential calculus to the problem of integrating in finite terms or reducing to quadratures first-order ordinary "differential" equations of first degree by *discarding differentials* and *calculating with partial differential coefficients exclusively*. In this way Clairaut simplified the partial differential calculus.

Clairaut let

$$M(x, y)\, dx + N(x, y)\, dy = 0, \tag{7.3.4.3}$$

where $M(x, y)$ and $N(x, y)$ are expressions in x, y, and constants which have no common factors and no denominators represent the general first-order ordinary "differential" equation of first degree. According to Clairaut the problem was to find an expression $u(x, y)$ in x and y and a corresponding expression $\phi(x, y)$ in x and y whose total differential equals $uMdx + uNdy$. In other words

$$d\phi = uMdx + uNdy.$$

For if $\phi(x, y)$ is then set equal to a constant k:

$$\phi(x, y) = k, \tag{7.3.4.4}$$

and if equation (7.3.4.4) is then differentiated (that is, if the differential of equation (7.3.4.4) is taken), the equation

$$0 = dk = d\phi = uMdx + uNdy \tag{7.3.4.5}$$

results, from which equation (7.3.4.3) follows after dividing both sides of equation (7.3.4.5) by the integrating factor $u(x, y)$. Thus equation (7.3.4.4) is an integral of equation (7.3.4.3). [The expression ϕ in equation (7.3.4.4) may either be integrable in finite terms, or else it may at most be reducible to quadratures.] Varying the constant k in equation (7.3.4.4) produces a one-parameter family of plane curves that are solutions to equation (7.3.4.3). Up to this point Clairaut followed the interpretation of integrals of first-order ordinary "differential" equations of first degree which Fontaine had introduced. Clairaut differed from Fontaine from this point on, however, because Clairaut considered it to be "quite unnecessary" to introduce a third variable in order to integrate these equations in finite terms or to reduce them to quadratures.[187]

First Clairaut determined a condition that is necessary for a differential $A(x, y)\,dx + B(x, y)\,dy$ in two finite variables x and y to be what he called a "complete differential" ("différentielle complète"). He meant by this that an expression $\phi(x, y)$ in x and y could be found whose total differential is the differential $Adx + Bdy$. In other words

$$d\phi = Adx + Bdy.$$

The condition necessary for such an expression $\phi(x, y)$ to exist is that the equality

$$\frac{dA}{dy} = \frac{dB}{dx}, \tag{7.3.4.6}$$

where the symbols in the equality (7.3.4.6) express partial differential coefficients in Fontaine's whole-symbol notations for such coefficients hold. Clairaut deduced condition (7.3.4.6) from the equality

$$\frac{d^2\phi}{dydx} = \frac{d^2\phi}{dxdy} \tag{7.3.4.7}$$

of mixed, second-order partial differential coefficients. Equality (7.3.4.6) follows from equality (7.3.4.7) if an expression $\phi(x, y)$ exists such that $d\phi = Adx + Bdy$, because in that case $d\phi/dx = A$ and $d\phi/dy = B$.

[In the preceding section of this chapter I contended that in deriving the "conditional equation" (7.3.3.111) in November of 1738, Fontaine in effect demonstrated what amounts to equality (7.3.4.7) himself using his fluxio-differential method. Namely, he did this in deriving equalities (7.3.3.131') and (7.3.3.131"). However, whether anyone including Fontaine himself realized this at the time can be questioned. A crucial assumption in Fontaine's derivation of equalities (7.3.3.131') and (7.3.3.131") is that the two independent, first-order Leibnizian differential operators involved in his fluxio-differential method *commute*.]

Clairaut demonstrated equality (7.3.4.7) in the following way. First he showed that the equality held for monomials $\phi(x, y) \equiv rm^m y^n$, where r is a constant. He concluded from this result that the equality held for expressions $\phi(x, y)$ in general, since "every function [of x and y] being reduced to series will be made up of terms that have this form [that is, monomials], and if the theorem is true for one term, it will be true for an infinity of terms"[188] by differentiating term by term the terms in the infinite series.

Moreover, Clairaut showed that equality (7.3.4.6) was also a sufficient condition for an expression $\phi(x, y)$ to exist such that $d\phi = Adx + Bdy$. Indeed, he did this by demonstrating that equality (7.3.4.6) together with the inversion (7.1.1.1) of differentiation and integration could actually be used to construct $\phi(x, y)$ in the following manner. Assume that $\phi(x, y)$ exists such that

$$d\phi(x, y) = A(x, y) \, dx + B(x, y) \, dy.$$

Suppose tht ϕ can be expressed as

$$\phi = \int Adx + Y. \tag{7.3.4.8}$$

What can be said about Y? Well, by differentiating (that is, by taking differentials) and by applying the rules that differential coefficients obey, we can see that

$$\frac{dY}{dx} = \frac{d}{dx}\left(\phi - \int Adx\right) = \frac{d\phi}{dx} - \frac{d}{dx}\left(\int Adx\right) = A - A = 0.$$

Consequently if Y actually exists (in other words, if hypothesis (7.3.4.8) is true), then Y must be an expression that is independent of x. In other words, $Y = Y(y)$. Clairaut wrote $Y(y)$ as $[y \cdot p]$, where p stands for constants. Next he determined $[y \cdot p]$. This he accomplished as follows: again by differentiating (that is, by taking differentials) and by applying the rules that differential coefficients obey, Clairaut found that

$$\frac{d}{dy}[y \cdot p] = \frac{d}{dy}\left(\phi - \int Adx\right) = \frac{d\phi}{dy} - \frac{d}{dy}\int Adx = B - \frac{d}{dy}\int Adx.$$

Now,

$$B - \frac{d}{dy}\int Adx$$

is expressed in terms of quantities that are given (namely, A and B). Since $[y \cdot p]$ is a function of y alone, the differential coefficient $d[y \cdot p]/dy$ must also be a function of y alone. Hence the question reduces to the following one: what conditions are necessary and sufficient for

$$B - \frac{d}{dy} \int A dx$$

to be a function of y alone? Clairaut assumed that differentiation and integration can be inverted. Thus

$$B - \frac{d}{dy} \int A dx = B - \int \frac{dA}{dy} dx.$$

In order for this expression to be a function of y alone, hence independent of x, the differential coefficient

$$\frac{d}{dx}\left(B - \int \frac{dA}{dy} dx \right)$$

must equal zero. Then by differentiating (that is, by taking differentials) and by applying the rules that differential coefficients obey, Clairaut discovered that

$$\frac{d}{dx}\left(B - \int \frac{dA}{dy} dx \right)$$

equals zero if and only if

$$0 = \frac{d}{dx}\left(B - \int \frac{dA}{dy} dx \right) = \frac{dB}{dx} - \frac{d}{dx} \int \frac{dA}{dy} dx = \frac{dB}{dx} - \frac{dA}{dy}.$$

Hence equality (7.3.4.6) is both a necessary and sufficient condition for

$$B - \frac{d}{dy} \int A dx$$

to be a function of y alone (assuming, of course, that differentiation and integration can be inverted). Finally, since

$$\frac{d}{dy}[y \cdot p] = B - \frac{d}{dy} \int A dx$$

it follows that

$$[y \cdot p] = \int \left(B - \frac{d}{dy} \int A dx \right) dy.$$

Therefore

$$\phi(x, y) = \int A dx + \int \left(B - \frac{d}{dy} \int A dx \right) dy \qquad (7.3.4.9)$$

is true. The two indefinite integrals

$$\int A\,dx \quad \text{and} \quad \int \left(B - \frac{d}{dy}\int A\,dx \right) dy$$

on the right-hand side of (7.3.4.9) are not unique; they are determined up to arbitrary additive constants. If one or the other of these two integrals cannot be integrated in finite terms, then in that event $\phi(x, y)$ has at least been reduced to quadratures, which in this case, unlike the notion of quadratures discussed earlier in this chapter, means reduced to integrals in two finite variables x and y which cannot be expressed in finite terms, instead of to integrals in one finite variable which cannot be expressed in finite terms.

If the left-hand side of equation (7.3.4.3) happens to be a complete differential, then $\phi(x, y)$ can be found in the manner just described, and equation (7.3.4.4) is then an integral of equation (7.3.4.3).

In case the left-hand side of equation (7.3.4.3) turns out not to be a complete differential, which means that the equality

$$\frac{dM}{dy} = \frac{dN}{dx} \qquad (7.3.4.10)$$

does not hold, Clairaut stated that the objective in that event was to find an expression $u(x, y)$ that makes the differential $uM\,dx + uN\,dy$ be a complete differential. For if such an integrating factor $u(x, y)$ could be determined, then an expression $\phi(x, y)$ such that

$$d\phi = uM\,dx + uN\,dy$$

could be constructed using the method described. And for each constant value of k, equation

$$\phi(x, y) = k \qquad (7.3.4.4)$$

would be an integral of equation (7.3.4.3). When the condition (7.3.4.6) is applied to the differential $uM\,dx + uN\,dy$, where $u = u(x, y)$ is unknown, it follows that

$$\frac{d}{dy}(uM) \quad \text{must equal} \quad \frac{d}{dx}(uN). \qquad (7.3.4.11)$$

According to the rules that partial differential coefficients obey, equality (7.3.4.11) can be written as

$$u\frac{dM}{dy} + M\frac{du}{dy} = u\frac{dN}{dx} + N\frac{du}{dx}, \qquad (7.3.4.11')$$

and equality (7.3.4.11') can be rewritten as

$$u\frac{dM}{dy} + M\frac{du}{dy} - u\frac{dN}{dx} - N\frac{du}{dx} = 0. \qquad (7.3.4.11'')$$

Here we see Clairaut taking full advantage of the rules that partial differential coefficients obey. Equality (7.3.4.11″) is a linear, first-order partial "differential equation" in two variables x and y for $u(x, y)$. Hence the problem of integrating equation (7.3.4.3) in finite terms or reducing it to quadratures reduces to solving equation (7.3.4.11″) for a particular solution by means of "the method of undetermined quantities."[189]

[Again we note that a solution $u(x, y)$ of equation (7.3.4.11″) need not be an integral that can be expressed in finite terms. The solution may at most be reducible to quadratures, which in this case, unlike the notion of quadratures discussed earlier in this chapter, can mean reducible to integrals in two finite variables x and y which cannot be expressed in finite terms, instead of to integrals in one finite variable which cannot be expressed in finite terms.]

For this reason Clairaut found it "quite unnecessary" for Fontaine to have introduced a third variable in order to integrate in finite terms or reduce to quadratures first-order ordinary "differential" equations of first degree. Clairaut did, however, regard Fontaine's method to be useful for integrating in finite terms or reducing to quadratures integrable first-order "differential" equations of first degree

$$dx + \alpha dy + \pi dp + \lambda dq + \text{etc.} = 0 \qquad (7.3.4.12)$$

in three or more variables x, y, p, q, etc., where α, π, λ, etc. are homogeneous expressions of degree zero in the variables x, y, p, q, etc. But in order to demonstrate the utility of Fontaine's method for this purpose, Clairaut first had to simplify the method. It turns out that Fontaine presented his method in a way that was unnecessarily complicated.

Let us consider Fontaine's equation

$$dx + \alpha dy + \pi dp = 0, \qquad (7.3.3.97)$$

where α and π are homogeneous expressions of degree zero in x, y, and p which satisfy the conditional equation (7.3.3.111). Clairaut realized that Fontaine's equation (7.3.3.122) for an integrating factor $u(x, y, p)$ that is homogeneous in x, y, and p for the differential $dx + \alpha dy + \pi dp$ on the left-hand side of equation (7.3.3.97) was totally unnecessary. Equation (7.3.3.122) can be eliminated altogether, which probably explains why Clairaut gave us just barely enough information to determine the equation. If $\phi(x, y, p)$ is Fontaine's homogeneous expression of degree e in x, y, and p which satisfies the equation

$$d\phi = udx + u\alpha \, dy + u\pi \, dp, \qquad (7.3.3.100)$$

where $u(x, y, p)$ is a homogeneous expression of degree $e - 1$ in x, y, and p, Clairaut simply divided the left-hand of equation (7.3.3.100) by the left-hand side of equation

$$e\phi = ux + u\alpha y + u\pi p, \qquad (7.3.3.114)$$

and he divided the right-hand side of equation (7.3.3.100) by the right-hand side of equation (7.3.3.114), when $x + \alpha y + \pi p \neq 0$ (in other words, when $e \neq 0$). The factor $u(x, y, p)$ in the numerator and the denominator of the right-hand side of the resulting equation cancel out, and the equation

$$d\left(\frac{l\phi}{e}\right) = \frac{d\phi}{e\phi} = \frac{dx + \alpha dy + \pi dp}{x + \alpha y + \pi p} \qquad (7.3.4.13)$$

follows, where $l\phi$ is the notation for the logarithm of ϕ which Leonhard Euler introduced.[190] Now, the left-hand side of equation (7.3.4.13) is the differential of the expression $l\phi/e$, which means that the right-hand side of equation (7.3.4.13) must just be the total differential of the expression $l\phi/e$. In other words, the right-hand side of equation (7.3.4.13) must be a complete differential in the three variables x, y, and p. In short, dividing the two sides of one of Fontaine's equations by the two sides of one of Fontaine's other equations immediately produces an integrating factor

$$\frac{1}{x + \alpha y + \pi p}$$

for the differential $dx + \alpha\,dy + \pi\,dp$ on the left-hand side of equation (7.3.3.97). There is no need to solve any equation for an integrating factor. Let $\psi(x, y, p)$ be an expression in x, y, and p whose differential $d\psi$ equals the complete differential

$$\frac{dx + \alpha dy + \pi dp}{x + \alpha y + \pi p}.$$

Then if ψ is set equal to a constant k, and if the equation $\psi = k$ is then differentiated (that is, if the differential of the equation $\psi = k$ is taken), the equation

$$0 = d\psi = \frac{dx + \alpha dy + \pi dp}{x + \alpha y + \pi p}$$

results. If both sides of the preceding equation are multiplied by the expression $x + \alpha y + \pi p$, the equation

$$dx + \alpha dy + \pi dp = 0 \qquad (7.3.3.97)$$

follows. In short, the equation $\psi = k$ must be an integral of equation (7.3.3.97).

Likewise, when $x + \alpha y + \pi p + \lambda q \neq 0$, the expression

$$\frac{1}{x + \alpha y + \pi p + \lambda q}$$

must be an integrating factor for an integrable first-order "differential" equation of first degree

$$dx + \alpha\,dy + \pi dp + \lambda dq = 0,$$

where α, π, and λ are homogeneous expressions of degree zero in x, y, p, and q and where the word "integrable" here again means either integrable in finite terms or reducible to quadratures. Clairaut sketched without going into details how an expression $\Omega(x, y, p, q)$ in x, y, p, and q whose differential $d\Omega$ is the complete differential

$$\frac{dx + \alpha dy + \pi dp + \lambda dq}{x + \alpha y + \pi p + \lambda q} \tag{7.3.4.14}$$

could be found. Clairaut assumed that Ω could be expressed as

$$\Omega = \int \frac{1}{x + \alpha y + \pi p + \lambda q} dx + Y. \tag{7.3.4.15}$$

He then asked: what can be said about Y? By differentiating (that is, by taking differentials) and by applying the rules that differential coefficients obey, Clairaut found that

$$\frac{dY}{dx} = \frac{d}{dx}\left(\Omega - \int \frac{1}{x + \alpha y + \pi p + \lambda q} dx\right)$$

$$= \frac{d\Omega}{dx} - \frac{d}{dx}\int \frac{1}{x + \alpha y + \pi p + \lambda q} dx$$

$$= \frac{1}{x + \alpha y + \pi p + \lambda q} - \frac{1}{x + \alpha y + \pi p + \lambda q} = 0,$$

from which it follows that if Y actually exists (in other words, if hypothesis (7.3.4.15) is true), then Y must be an expression that is independent of x. Hence in this case $\Omega(x, y, p, q)$ must have the form

$$\Omega = \int \frac{1}{x + \alpha y + \pi p + \lambda q} dx + [y \cdot p \cdot q], \tag{7.3.4.15'}$$

where $[y \cdot p \cdot q]$ designates a function of y, p, and q which is independent of x. Clairaut now simply stated that the total differential $d[y \cdot p \cdot q]$ of the function $[y \cdot p \cdot q]$ in (7.3.4.15') which remained to be determined must be

$$\left(\frac{\alpha}{x + \alpha y + \pi p + \lambda q} - \frac{d}{dy}\int \frac{1}{x + \alpha y + \pi p + \lambda q} dx\right) dy$$

$$+ \left(\frac{\pi}{x + \alpha y + \pi p + \lambda q} - \frac{d}{dp}\int \frac{1}{x + \alpha y + \pi p + \lambda q} dx\right) dp$$

$$+ \left(\frac{\lambda}{x + \alpha y + \pi p + \lambda q} - \frac{d}{dq}\int \frac{1}{x + \alpha y + \pi p + \lambda q} dx\right) dq. \tag{7.3.4.16}$$

Clairaut did not explain at this time why this was true. He did not show that the coefficients of dy, dp, and dq in the differential (7.3.4.16) are independent of x; he did not demonstrate that the differential (7.3.4.16) is a complete differential in the three variables y, p, and q, which the differential (7.3.4.16) must be in order for it to be the total differential of a function $[y \cdot p \cdot q]$ of y, p, and q; nor did he explain how to construct the function $[y \cdot p \cdot q]$ in (7.3.4.15') from the differential (7.3.4.16), assuming that the differential (7.3.4.16) is a complete differential in the three variables y, p, and q. Let us note, however, the likeness between the right-hand side of (7.3.4.8) and the right-hand side of (7.3.4.15) and also the parallel between the differential

$$\left(B - \frac{d}{dy} \int A\,dx \right) dy,$$

used to construct the second term on the right-hand side of (7.3.4.9), and the differential (7.3.4.16). The differential

$$\left(B - \frac{d}{dy} \int A\,dx \right) dy$$

has one variable less and one term less than the complete differential

$$A(x, y)\,dx + B(x, y)\,dy.$$

Similarly, the differential (7.3.4.16) has one variable less and one term less than the complete differential (7.3.4.14). Clairaut's brief discussion here suggests that there is a way of integrating or reducing to quadratures complete differentials in three or more variables which generalizes his method of integrating or reducing to quadratures complete differentials in two variables – namely, by reducing the original problem to a problem of the same kind, but one that involves one less variable and one less term. By repeating this process of reduction over and over again, the expression whose differential is a given, complete differential in any number of variables could be found by systematic reduction step by step to expressions that could be integrated in finite terms or, if not, at least reduced to quadratures. But Clairaut did not clarify the details of this procedure in his paper.

Clairaut employed Fontaine's method as he had simplified it to integrate or reduce to quadratures all first-order ordinary "differential" equations of first degree

$$dx + \alpha\,dy = 0$$

in two variables x and y, where α is any homogeneous expression of degree zero in x and y except $\alpha \equiv -x/y$ (that is, except the case when $x + \alpha y \equiv 0$). We recall that Johann I Bernoulli had already solved this problem in the first volume of the St. Petersburg *Commentarii*, but Clairaut needed an easy example to illustrate Fontaine's method simplified. In this case it follows from equation

$$d\left(\frac{l\phi}{e} \right) = \frac{d\phi}{e\phi} = \frac{dx + \alpha\,dy + \pi\,dp}{x + \alpha y + \pi p}, \qquad (7.3.4.13)$$

with $p = 0$, that the differential

$$\frac{dx + \alpha dy}{x + \alpha y}$$ (7.3.4.17)

is a complete differential. Clairaut utilized his method of integrating or reducing to quadratures complete differentials in two variables to determine $\gamma(x, y)$ such that

$$d\gamma = \frac{dx + \alpha dy}{x + \alpha y}.$$ (7.3.4.18)

If γ is then set equal to a constant k, the equation

$$\gamma = k$$

is an integral of equation

$$dx + \alpha dy = 0.$$

It is worth noting that $\gamma(x, y)$ *cannot* be a *homogeneous* expression in x and y.[191]

Let us observe that independently of equation (7.3.4.13), the differential (7.3.4.17) can be shown to be a complete differential by using the condition (7.3.4.6). That is, by applying the rules that partial differential coefficients obey, we find that

$$\frac{d}{dy}\left(\frac{1}{x + \alpha y}\right) = \frac{d}{dx}\left(\frac{\alpha}{x + \alpha y}\right)$$ (7.3.4.6*)

is true, which means that equality

$$\frac{dA}{dy} = \frac{dB}{dx}$$ (7.3.4.6)

holds, when

$$\frac{dx + \alpha dy}{x + \alpha y} \equiv A dx + B dy.$$

Equality (7.3.4.6*) follows from the homogeneous function theorem applied to the homogeneous expression α of degree zero in x and y, which means that the equation

$$0 = (0)\alpha = \frac{d\alpha}{dx}x + \frac{d\alpha}{dy}y$$

is true.

[Fontaine showed later that if $x + \alpha y \equiv 0$ (in other words, if $\alpha \equiv -x/y$), then *any* homogeneous expression $\phi(x, y)$ of degree zero in x and y can be used to find an integral of the equation $dx + \alpha dy = 0$ in this particular case. This is easy to see as follows. If $udx + vdy$ is the total differential of such an expression $\phi(x,y)$, then if ϕ is set equal to a constant k and if the resulting equation $\phi = k$ is then differentiated (that is, if the differential of the equation is taken), the equation

$$0 = dk = d\phi = udx + vdy = dx + \frac{v}{u}dy$$

results. Now, if the homogeneous function theorem is applied to the homogeneous expression ϕ of degree zero in x and y, the equation

$$0 = (0)\phi = ux + vy$$

results. But it follows from the preceding equation that

$$\frac{v}{u} = \frac{-x}{y} \equiv \alpha,$$

hence

$$0 = dx + \frac{v}{u}dy = dx + \alpha dy$$

is true. That is, the equation $\phi = k$ is an integral of the equation $0 = dx + \alpha dy$.]

As I just mentioned, the expression $\gamma(x, y)$ such that

$$d\gamma = \frac{dx + \alpha dy}{x + \alpha y}, \tag{7.3.4.18}$$

when $x + \alpha y \neq 0$, cannot be a homogeneous expression in x and y. In effect, in modifying Fontaine's method, Clairaut appears to have done more than just simplify it. Fontaine's method and Clairaut's reduction of it involve different complete differentials and, accordingly, different expressions whose total differentials are these different complete differentials. We recall that in order to integrate an integrable equation

$$dx + \alpha dy + \pi dp = 0, \tag{7.3.3.97}$$

where α and π are homogeneous expression of degree zero in x, y, and p, Fontaine determined an expression $\phi(x, y, p)$ which was homogeneous of degree $e \neq 0$ in x, y, and p and which satisfied an equation

$$d\phi = udx + u\alpha dy + u\pi dp, \tag{7.3.3.100}$$

where u is a homogeneous integrating factor of degree $e - 1$ in x, y, and p for the differential

$$dx + \alpha dy + \pi dp.$$

Now, in general, if α, π, λ, etc. are homogeneous expressions of degree zero in the variables x, y, p, q, etc. in the differential

$$\frac{dx + \alpha dy + \pi dp + \lambda dq + \text{etc.}}{x + \alpha y + \pi p + \lambda q + \text{etc.}}, \tag{7.3.4.19}$$

when $x + \alpha y + \pi p + \lambda q + $ etc. $\neq 0$, then it is easy to see that the coefficients of dx, dy, dp, dq, etc. in the differential (7.3.4.19) are all homogeneous expressions of degree -1 in x, y, p, q, etc. With (7.3.4.18) in mind, we can ask more generally: if the differential (7.3.4.19) is a complete differential, does an expression Γ that is homogeneous in x, y, p, q, etc. exist whose total differential is (7.3.4.19)? The answer is no. For if such a homogeneous expression Γ did exist, whose degree of homogeneity in x, y, p, q, etc. is, say, e, then according to the homogeneous function theorem applied to Γ, the equation

$$e\Gamma = \frac{x + \alpha y + \pi p + \lambda q + \text{etc.}}{x + \alpha y + \pi p + \lambda q + \text{etc.}} = 1$$

must be true, which is impossible. (If $e = 0$, the preceding equation reduces to $0 = 1$, which is false; if $e \neq 0$, the preceding equation requires that Γ be constant, which is also false.) This is consequently one case where the converse of the homogeneous function theorem does not hold. Indeed, no homogeneous expression Γ of any degree in x, y, p, q, etc. exists whose total differential is (7.3.4.19), much less an expression Γ that is homogeneous of degree zero in x, y, p, q, etc., whereas such a homogeneous expression Γ of degree zero in x, y, p, q, etc. would exist if the converse of the homogeneous function theorem held in this case.

In fact, in discussing Fontaine's method, I already ruled out this case on other grounds. The trouble here is that in (7.3.4.19) the expression

$$\frac{1}{x + \alpha y + \pi p + \lambda q + \text{etc.}}$$

is an integrating factor for the differential

$$dx + \alpha dy + \pi dp + \lambda dq + \text{etc.}$$

which is homogeneous of degree $e - 1 = -1$ in x, y, p, q, etc. But as we already saw in the preceding section of this chapter, in applying the converse of the homogeneous function theorem in the case where there are three variables x, y, and p, which is equation

$$e\phi = ux + u\alpha y + u\pi p, \tag{7.3.3.114}$$

in order to determine a homogeneous expression $\phi(x, y, p)$ of degree e in x, y, and p such that

$$d\phi = udx + u\alpha dy + u\pi dp, \tag{7.3.3.100}$$

the homogeneous integrating factor u of degree $e - 1$ in x, y, and p must not be homogeneous of degree $e - 1 = -1$ in x, y, and p. For if it were, the homogeneous expression ϕ in equation (7.3.3.114) would be homogeneous of degree $e = 0$ in x, y, and p, in which case equation (7.3.3.114) could not be used to determine ϕ. Thus u cannot be taken to be

$$\frac{1}{x + \alpha y + \pi p}$$

in equations (7.3.3.100) and (7.3.3.114). In other words, Fontaine's method does not work when complete differentials have the form

$$\frac{dx + \alpha dy + \pi dp + \lambda dq + \text{etc.}}{x + \alpha y + \pi p + \lambda q + \text{etc.}}, \tag{7.3.4.19}$$

where α, π, λ, etc. are homogeneous expression of degree zero in the variables x, y, p, q, etc. and where $x + \alpha y + \pi p + \lambda q + \text{etc.} \neq 0$. That is to say, his method is not designed to produce integrals of complete differentials of this sort. Another way to see this is as follows. Again let us consider the case where there are three variables x, y, and p. If the expression $l\phi/e$ on the left-hand side of equation

$$d(l\phi/e) = \frac{d\phi}{e\phi} = \frac{dx + \alpha dy + \pi dp}{x + \alpha y + \pi p} \tag{7.3.4.13}$$

were a homogeneous expression Γ of degree zero in x, y, and p, then $l\phi$, like Γ, would also have to be a homogeneous expression of degree zero in x, y, and p. But this can only occur if ϕ itself is a homogeneous expression of degree zero in x, y, and p. (In fact, $l\phi$ can *never* be homogeneous in its variables *unless* ϕ is homogeneous of degree zero in its variables, in which case $l\phi$ is homogeneous of degree zero in its variables.) However, as we saw previously, Fontaine found expressions $\phi(x, y, p)$ that were homogeneous of degree $e \neq 0$. Thus in simplifying Fontaine's method, Clairaut also changed it somewhat.

As I have already mentioned, Fontaine derived the "conditional equation" (7.3.3.111) in a way that Clairaut found very long and very difficult. Clairaut discovered a much shorter and simpler way to arrive at this equation, which he presented at the end of his Paris Academy *mémoire* of 1739 on integrating in finite terms or reducing to quadratures first-order ordinary "differential" equations of first degree. He did not explain the foundations of his derivation of the equation very clearly; he simply noted that the derivation involved condition (7.3.4.6). Clairaut's calculations, however, enable us to reconstruct his argument. Suppose an integrating factor $u(x, y, p)$ exists which makes the right-hand side of equation

$$d\phi = udx + u\alpha dy + u\pi dp \tag{7.3.3.100}$$

be a complete differential. Clairaut tacitly assumed that in order for such a differential in three variables x, y, and p to be a complete differential in the three variables, each of the three differentials in two variables formed by individually setting x, y, and p equal to constants and, accordingly, setting $dx = 0$, $dy = 0$, and $dp = 0$ must be a complete differential in two variables. Clairaut applied condition (7.3.4.6) to each of these three complete differentials in two variables, and he arrived at the following three equations as a result:

$$\frac{du}{dy} = \frac{d}{dx}(u\alpha) = u\frac{d\alpha}{dx} + \alpha\frac{du}{dx} \tag{7.3.4.20}$$

$$u\frac{d\alpha}{dp} + \alpha\frac{du}{dp} = \frac{d}{dp}(u\alpha) = \frac{d}{dy}(u\pi) = u\frac{d\pi}{dy} + \pi\frac{du}{dy} \tag{7.3.4.21}$$

and

$$\frac{du}{dp} = \frac{d}{dx}(u\pi) = u\frac{d\pi}{dx} + \pi\frac{du}{dx}. \tag{7.3.4.22}$$

Here Clairaut once again used to the best advantage the rules that partial differential coefficients obey. When Clairaut eliminated $u(x, y, p)$ and the partial differential coefficients du/dx, du/dy, and du/dp from equations (7.3.4.20), (7.3.4.21), and (7.3.4.22), equation (7.3.3.111) resulted. In this way Clairaut arrived at this conditional equation much more easily than Fontaine had.

Nevertheless, as I indicated earlier, the fluxio-differential method can just as easily simplify calculations as make them more complex. After all, as we saw earlier in this chapter, Fontaine found the tautochrone in a medium that resists a body's motion directly as $(y^2 + my)/n$, where $m \neq 0$ and y is the body's speed, using the fluxio-differential method much more readily than I did using the partial differential calculus. Which of the two calculuses leads to the shorter demonstration depends a great deal on the particular problem.

Had Fontaine introduced the equation

$$d\left(\frac{1}{e}l\phi\right) = \frac{d\phi}{e\phi} = \frac{dx + \alpha dy + \pi dp}{x + \alpha y + \pi p} \tag{7.3.4.13}$$

in his paper of November 1738, as well as the condition (7.3.4.6), he could have found the conditional equation (7.3.3.111) for the integrability of equations

$$dx + \alpha(x, y, p)dy + \pi(x, y, p)dp, \tag{7.3.3.97}$$

where $\alpha(x, y, p)$ and $\pi(x, y, p)$ are homogeneous expressions of degree zero in x, y, and p, in a way that is simpler than the method that I suggested that he utilized. This he could have accomplished as follows. Because of the form that equation (7.3.4.13) has, we suppose that its right-hand side is a complete differential in three variables, which I rewrite as

$$\frac{1}{x + \alpha y + \pi p}dx + \frac{\alpha}{x + \alpha y + \pi p}dy + \frac{\pi}{x + \alpha y + \pi p}dp. \tag{7.3.4.23}$$

Imitating what Clairaut did (as shown above), I now apply condition (7.3.4.6) to the three terms in the differential (7.3.4.23) taken two at a time. First, according to the rules that partial differential coefficients obey, it follows that

$$\frac{d}{dy}\left(\frac{1}{x + \alpha y + \pi p}\right) = \frac{d}{dx}\left(\frac{\alpha}{x + \alpha y + \pi p}\right), \tag{7.3.4.24}$$

which is condition (7.3.4.6) applied to the first and second terms of the differential (7.3.4.23), can be written as

$$x\frac{d\alpha}{dx} + y\frac{d\alpha}{dy} + p\left(\frac{d\pi}{dy} + \pi\frac{d\alpha}{dx} - \alpha\frac{d\pi}{dx}\right) = 0. \tag{7.3.4.24'}$$

At the same time, the equation

$$x\frac{d\alpha}{dx} + y\frac{d\alpha}{dy} = -p\frac{d\alpha}{dp} \qquad (7.3.4.25)$$

follows from the homogeneous function theorem (7.3.3.112) applied to the expression $\alpha(x, y, p)$ which is homogeneous of degree zero in x, y, and p. Finally, if the expression

$$x\frac{d\alpha}{dx} + y\frac{d\alpha}{dy}$$

on the left-hand side of equation (7.3.4.24') is replaced by the right-hand side of equation (7.3.4.25) and then if both sides of the resulting equation are divided by the factor p, equation (7.3.3.111) results. Alternatively, according to the rules that partial differential coefficients obey, it follows that

$$\frac{d}{dp}\left(\frac{1}{x + \alpha y + \pi p}\right) = \frac{d}{dx}\left(\frac{\pi}{x + \alpha y + \pi p}\right), \qquad (7.3.4.26)$$

which is condition (7.3.4.6) applied to the first and third terms of the differential (7.3.4.23), can be written as

$$x\frac{d\pi}{dx} + p\frac{d\pi}{dp} + y\left(\frac{d\alpha}{dp} + \alpha\frac{d\pi}{dx} - \pi\frac{d\alpha}{dx}\right) = 0. \qquad (7.3.4.26')$$

At the same time, the equation

$$x\frac{d\pi}{dx} + p\frac{d\pi}{dp} = -y\frac{d\pi}{dy} \qquad (7.3.4.27)$$

follows from the homogeneous function theorem (7.3.3.112) applied to the expression $\pi(x, y, p)$ which is homogeneous of degree zero in x, y, and p. Then if the expression

$$x\frac{d\pi}{dx} + p\frac{d\pi}{dp}$$

on the left-hand side of equation (7.3.4.26') is replaced by the right-hand side of equation (7.3.4.27) and if both sides of the resulting equation are divided by the factor y, equation (7.3.3.111) again results. If we applied condition (7.3.4.6) to the second and third terms of the differential (7.3.4.23), together with the homogeneous function theorem (7.3.3.112) applied to the expressions $\alpha(x, y, p)$ and $\pi(x, y, p)$ which are homogeneous of degree zero in x, y, and p, equation (7.3.3.111) would again follow as the result.

But Fontaine did *not* write equation (7.3.4.13) in his paper of November 1738, nor did he introduce condition (7.3.4.6) in this paper – at least, not in a way that anyone recognized. Moreover, the derivation of equation (7.3.3.111) that I just gave, which makes use of equation (7.3.4.13), condition (7.3.4.6), and, like Clairaut's derivation, partial differential coefficients and no differentials, hardly involves "nine very complicated equations," while Clairaut stated that Fontaine's derivation did include nine such equations. Thus it seems unlikely that Fontaine arrived at equation (7.3.3.111) by using equation (7.3.4.13), condition (7.3.4.6), and partial differential coefficients but no differentials. In short, there seem to be no alternatives to the conclusion that Fontaine employed his fluxio-differential method to find equation (7.3.3.111).

By September of 1740 Clairaut had decided to have the paper on integral calculus which he had read before the Paris Academy in March of 1739 published in the Academy's *Mémoires*. In the meantime he became concerned about priority. Had he merely repeated what mathematicians elsewhere had already done? Since he planned to publish, he needed this question answered. The question came up in the following way.

In August of 1739, Clairaut sent a copy of his manuscript on integral calculus to Daniel Bernoulli in Basel, Johann I Bernoulli's second son, whom I shall have much to say about in Chapter 8. Bernoulli told Clairaut that Euler, who still belonged to the St. Petersburg Academy since having joined it in 1727, knew condition (7.3.4.6) for a differential in two variables to be complete. Clairaut, we remember, had also inverted differentiation and integration to determine the expression (7.3.4.9) whose total differential is a given, complete differential in two variables. I mentioned in Chapter 4 that Bouguer had been the first mathematician in Paris to reverse differentiation and integration, which he did in a Paris Academy *mémoire* of 1733 in order to solve a problem in mechanics. Fontaine subsequently stated the result in generality and demonstrated it. But Bernoulli informed Clairaut that Bouger had not been the first person to apply the technique, as we already know from the discussion that began this chapter. (Bernoulli had told Euler that Clairaut attributed the procedure to Bouguer, and Euler replied that Clairaut was mistaken.)

Early in January of 1740, Daniel Bernoulli sent Euler a copy of Clairaut's manuscript on integral calculus. Bernoulli thereby paved the way for Clairaut, who was now filled with concern since he intended to publish, to ask Euler the questions about priority which the information that Bernoulli had given Clairaut raised.

In September of 1740, Clairaut wrote Euler for the first time in order to ask him three questions:

1. Where did condition (7.3.4.6) appear in his work?
2. Since Bouguer was not the first person to invert differentiation and integration, who did utilize the technique originally?
3. Did he know the homogeneous function theorem (7.3.3.112)?[192]

Euler replied in October of 1740.[193] He told Clairaut that he had known "for a long time" that differentiation and integration could be inverted, but he also admitted that he had "not ... perceived with all the requisite clarity its great utility for determining integrals" of complete differentials in two variables which Clairaut had demonstrated in his manuscript on integral calculus. Euler explained that the inversion of differentiation and integration (7.1.1.1) was first published in the articles in *Acta Eruditorum* on problems of finding orthogonal trajectories. Thus in 1740 Clairaut still remained unfamiliar with this literature, and I shall have occasion in the last section of this chapter to bring up again his ignorance of it at this late date. Euler mentioned that he discussed the technique in his own work published in Volume VII of the St. Petersburg Academy's *Commentarii*, which he told Clairaut had just appeared. This volume contained articles that had been read before the St. Petersburg Academy during the years 1734–35.

Euler also stated that the equality (7.3.4.7) of mixed, second-order partial differential coefficients could also be found in his work published in Volume VII of the St. Petersburg Academy's *Commentarii*. He told Clairaut that the inversion of differentiation and integration "depended upon" equality (7.3.4.7). [Here Euler erred; as we shall see later, it is the converse that is actually true. Euler might have meant to say that condition (7.3.4.6) depends upon the equality (7.3.4.7), for Euler strangely neglected to answer Clairaut's first question above, although in fact condition (7.3.4.6) also appears in Euler's work published in Volume VII of the St. Petersburg Academy's *Commentarii*. Naturally, Euler did not use Fontaine's compact whole-symbol notations for partial differential coefficients, which he did not know about at the time, to express condition (7.3.4.6) or equality (7.3.4.7) in his work published in that volume.] The foundations of Euler's demonstration of equality (7.3.4.7) differed form the foundations of Clairaut's demonstration of the same equality. Clairaut had assumed that all expressions in two variables could be represented as infinite series of monomials in two variables. Euler supposed that the same second-order differentials of expressions in two variables result, no matter what order the two variables are held constant while taking successive differentials of the expressions with respect to the variable not held constant. As I just mentioned, Euler did not use Fontaine's compact whole-symbol notations for partial differential coefficients, which of course he did not know about at the time, to establish equality (7.3.4.7). Instead he needed to introduce three total differentials in two variables in order to demonstrate and to express equality (7.3.4.7). Just to state equality (7.3.4.7) this way takes up more space than is necessary, and this makes obvious the advantages of Fontaine's whole-symbol notations for partial differential coefficients. Euler did not joke when, as I mentioned earlier in this chapter, he stated in 1752 that Fontaine's notations "facilitate calculation."

Finally, Euler mentioned that he had applied the homogeneous function theorem in two variables in Proposition 14 of Volume II of his *Mechanica*. (I remarked earlier in this chapter that he used it there to attack problems of finding synchrones exactly the same way that Fontaine did in his paper of July 1737.) Euler noted in addition that he discussed the general theorem for any number of variables in his work appearing in Volume VII of the St. Petersburg Academy's *Commentarii*. His proof of the theorem, in which he made use of a change of variables, differs from Fontaine's proof of it.

All of the results that Euler told Clairaut were published in Volume VII of the St. Petersburg Academy's *Commentarii* can actually be found in the same two articles included in that volume. [194] But the publication of the first volumes of the *Commentarii* was often slow. Some of these volumes appeared as much as eleven years late. [195] The delays in publication of the volumes for the 1730s account in large part for the ignorance of Clairaut, Fontaine, and Bouguer in the 1730s of the work in mathematics and mechanics which Euler did during that decade. As I explained earlier in this chapter, mathematicians in Paris in the 1730s who were sufficiently capable of understanding Euler's *Mechanica*, published in 1736, only began to pay attention to that work around 1740. When Euler left St. Petersburg in 1741 and entered the Berlin Academy, whose annual volume was published in French instead of in Latin, his work in mathematics and mechanics became much better known to mathematicians in Paris at that time.

Late in May and early in June of 1741, Clairaut read a second paper on integral calculus before the Paris Academy in which he continued and generalized what he had begun in 1739 in his first paper on the subject. [196] The second paper was published in the Paris Academy's *Mémoires* for 1740. [197]

Soon after Clairaut had finished reading his first paper on integral calculus before the Paris Academy in March of 1739, Fontaine left Paris and stayed away from the city much of the time during the years that followed. (I shall discuss the reasons for this later in this chapter.) Fontaine was at his estate near Compiègne when Clairaut read his second paper on integral calculus before the Paris Academy. Somehow, however, Fontaine learned of Clairaut's new paper almost immediately. Eight days after Clairaut finished reading his paper, Fontaine sent a letter to Mairan from his estate in which he announced that he was having a paper on integral calculus delivered to the Paris Academy in a sealed envelope (*pli cacheté*). Mairan, who replaced Fontenelle as the Academy's *secrétaire perpétuel* in 1741, received the envelope two days later from the Academy's courier Louis-Léon Pajot, comte D'Onsenbray. The contents of the envelope were only supposed to be shown to the members of the Academy when the sender requested that the envelope be opened. Jean LeRond D'Alembert and the Abbé Jean-Paul de Gua de Malves, two other members of the new generation of Paris Academy mathematicians, were appointed to examine the contents of the envelope and to review the contents, if and when Fontaine asked that the envelope be opened. (In fact, Fontaine specified in his letter to Mairan that the envelope not be opened before Clairaut's paper was printed in the Academy's *Mémoires*. [198])

In beginning his second paper on integral calculus, Clairaut primarily restated what he had said concerning the integration of first-order ordinary "differential" equations (7.3.4.3) of first degree in two variables x and y in his first paper on integral calculus, although he explained certain aspects of what he had done in that first paper more clearly this time. In a footnote that should have been added to his first Paris Academy *mémoire* on integral calculus, published in 1741, but for some unexplained reason was joined to his second Paris Academy *mémoire* on the same subject, published in 1742, instead, Clairaut attributed the independent discovery of condition (7.3.4.6) both to Fontaine and to Euler. According to the footnote, Fontaine had presented a written demonstration of condition (7.3.4.6) to the Paris Academy on the same day that Clairaut read his first paper on integral calculus before the Academy in March of 1739. [Fontaine must have realized by that day in March of 1739 that equality (7.3.4.7), from which condition (7.3.4.6) follows, existed in the form of equalities (7.3.3.131′) and (7.3.3.131″) in the paper that he had read before the Academy in November of 1738.] Although in his reply to the letter that Clairaut had sent him in September of 1740 Euler did not answer the first of the three questions that Clairaut had asked him in the letter, Clairaut evidently guessed correctly that condition (7.3.4.6), which Daniel Bernoulli had told Clairaut that Euler knew, also could be found in Euler's work that was published in Volume VII of the St. Petersburg Academy's *Commentarii*, for Clairaut alluded in his footnote to this volume and to Euler's work appearing in the volume in connection with condition (7.3.4.6).

Clairaut really only added one new thing at the beginning of his second Paris Academy *mémoire* on integral calculus to supplement the discussion of integrating equation (7.3.4.3) which appears in his first Paris Academy *mémoire* on integral calculus. In addition to his demonstration of condition (7.3.4.6) by means of what he now called "induction," which he had utilized in his first Paris Academy *mémoire* on integral calculus and repeated in his second Paris Academy *mémoire* on the same subject and in which he made use of infinite series of monomials in x and y, he included as well in the second *mémoire* what he called a demonstration of condition (7.3.4.6) "*à priori*."[199] This he did presumably to avoid the question whether all expressions in x and y can actually be represented as infinite series of such monomials. In fact he simply applied his method for deriving (7.3.4.9), which he employed in his first Paris Academy *mémoire* on integral calculus, backward. That is, assume that the differential

$$A(x, y)dx + B(x, y)dy$$

is a complete differential in the two variables x and y. Then an expression $\phi(x, y)$ exists such that

$$d\phi(x, y) = A(x, y)dx + B(x, y)dy.$$

Suppose that ϕ can be expressed as

$$\phi = \int A\,dx + Y. \tag{7.3.4.8}$$

Then we know that according to the rules that differential coefficients obey, it follows that

$$\frac{dY}{dx} = \frac{d}{dx}\left(\phi - \int A\,dx\right) = \frac{d\phi}{dx} - \frac{d}{dx}\int A\,dx = A - A = 0,$$

from which we conclude that Y is independent of x. That is $Y = Y(y)$. In fact, $Y(y)$ has the form

$$\int\left(B - \frac{d}{dy}\int A\,dx\right)dy,$$

because, by applying the rules that differential coefficients obey, we have already seen that

$$\frac{dY}{dy} = \frac{d}{dy}\left(\phi - \int A\,dx\right) = \frac{d\phi}{dy} - \frac{d}{dy}\int A\,dx = B - \frac{d}{dy}\int A\,dx.$$

Clairaut again assumed that differentiation and integration can be inverted, in which case

$$\frac{dY}{dy} = B - \frac{d}{dy}\int A\,dx = B - \int\frac{dA}{dy}\,dx.$$

Finally, since Y is a function of y alone, dY/dy is independent of x too. Hence by applying again the rules that differential coefficients obey, we find that

$$0 = \frac{d}{dx}\left(\frac{dY}{dy}\right) = \frac{d}{dx}\left(B - \int\frac{dA}{dy}\,dx\right) = \frac{dB}{dx} - \frac{d}{dx}\int\frac{dA}{dy}\,dx = \frac{dB}{dx} - \frac{dA}{dy}.$$

In other words, equality (7.3.4.6) is a necessary condition for

$$A(x, y)\,dx + B(x, y)\,dy$$

to be a complete differential in the two variables x and y, assuming that differentiation and integration can be reversed.

Clairaut evidently searched for and located Euler's references to the use of the inversion of differentiation and integration which appear in the articles published in *Acta Eruditorum* on problems of finding orthogonal trajectories, because Clairaut now cited the supplement to Nikolaus II Bernoulli's article in the *Acta* for 1720, which was published in volume 7 of the supplements to the *Acta*, as the source of a much earlier application of the technique.

Let us also note that if A and B in the total differential

$$d\phi(x, y) = A(x, y)\,dx + B(x, y)\,dy$$

of $\phi(x, y)$ are expressed using Fontaine's whole-symbol notations for partial differential coefficients, then Clairaut's demonstration above shows in effect that Leibniz's inversion (7.1.1.1) of differentiation and integration is a sufficient condition for the equality (7.3.4.7) of mixed, second-order partial differential coefficients to hold. [Euler, we recall, had mistakenly told Clairaut in his letter to Clairaut of October of 1740 that (7.1.1.1) was a necessary condition for equality (7.3.4.7) to hold.[200]]

Clairaut simply remarked that if the preceding demonstration is reversed, it proves that equality (7.3.4.6) is also a sufficient condition for the differential

$$A(x, y)dx + B(x, y)dy$$

to be "a complete differential – that is to say, some function of x and y exists, which is algebraic or depends upon quadratures, which will be its integral."[201] It was enough for Clairaut just to mention here that this was true, because this in fact is what Clairaut had already shown in detail to be the case in deriving (7.3.4.9) in his first Paris Academy *mémoire* on integral calculus.

In the next part of his second Paris Academy *mémoire* on integral calculus, Clairaut examined first-order "differential" equations of first degree

$$M(x, y, z)dx + N(x, y, z)dy + P(x, y, z)dz = 0 \qquad (7.3.4.28)$$

in three variables x, y, and z. As we saw before, Clairaut had already made use of some results concerning such equations in his first Paris Academy *mémoire* on integral calculus, but he merely sketched out his ideas involving these results in that *mémoire*. He did not really clarify what he was doing there, nor did he prove the results that he utilized. Now he explained himself very clearly as well as demonstrated the results in question.

Clairaut now called a differential

$$M(x, y, z)dx + N(x, y, z)dy + P(x, y, z)dz$$

in three finite variables x, y, and z an "exact differential" ("différentielle exacte") if an expression $\phi(x, y, z)$ in x, y, and z could be found whose total differential is $Mdx + Ndy + Pdz$.[202] That is,

$$d\phi = Mdx + Ndy + Pdz.$$

Thus Clairaut meant by an exact differential what he had been calling until now a complete differential. He stated that the differential on the left-hand side of equation (7.3.4.28) cannot be such a differential unless the three equalities

$$\frac{dM}{dy} = \frac{dN}{dx}, \quad \frac{dM}{dz} = \frac{dP}{dx}, \quad \text{and} \quad \frac{dN}{dz} = \frac{dP}{dy}, \qquad (7.3.4.29)$$

where the symbols in the three equalities (7.3.4.29) express partial differential coefficients in Fontaine's whole-symbol notations for such coefficients, hold. The truth of this statement follows easily from equality (7.3.4.6), because if the differential on the left-hand side of equation (7.3.4.28) is the total differential of some expression $\phi(x, y, z)$ in the three variables x, y, and z, then holding x, y, and z constant one at a time in $\phi(x, y, z)$ produces three expressions whose total differentials are the three differentials in two variables which result from individually setting $x = $ a constant and $dx = 0$, $y = $ a constant and $dy = 0$, and $z = $ a constant and $dz = 0$ in the differential on the left-hand side of equation (7.3.4.28). Clairaut then simply applied equality (7.3.4.6) to each of these three exact differentials in two variables and thereby arrived at the three equalities (7.3.4.29).

Conversely, Clairaut showed that if the three conditions (7.3.4.29) hold, then the differential on the left-hand side of equation (7.3.4.28) must be an exact differential in the three variables x, y, and z. Indeed, he showed how conditions (7.3.4.29) could be used to construct an expression $\phi(x, y, z)$ whose total differential is the differential on the left-hand side of equation (7.3.4.28). His method generalizes his procedure for deriving (7.3.4.9) – that is, his procedure for determining a ϕ in two variables x and y, expressed through indefinite integrals that can be integrated in finite terms and/or reduced at most to quadratures, whose total differential is a given, complete differential in two variables. Clairaut described this technique in his first Paris Academy *mémoire* on integral calculus, where he hinted that it could be expanded to cover complete differentials in more than two variables. But he did not prove in that *mémoire* that this is true. Assume that an expression $\phi(x, y, z)$ exists whose total differential is the differential on the left-hand side of equation (7.3.4.28). Suppose that ϕ has the form

$$\phi = \int M dx + Y. \tag{7.3.4.30}$$

Then Y must be a function of y and z alone, because if we apply the rules that differential coefficients obey, we find that

$$\frac{dY}{dx} = \frac{d}{dx}\left(\phi - \int M dx\right) = \frac{d\phi}{dx} - \frac{d}{dx}\int M dx = M - M = 0,$$

which means that Y does not depend upon x. In order to determine the form that Y has, we again note that according to the rules that differential coefficients obey, it follows that

$$\frac{dY}{dy} = \frac{d}{dy}\left(\phi - \int M dx\right) = \frac{d\phi}{dy} - \frac{d}{dy}\int M dx = N - \frac{d}{dy}\int M dx \tag{7.3.4.31}$$

and that

$$\frac{dY}{dz} = \frac{d}{dz}\left(\phi - \int M dx\right) = \frac{d\phi}{dz} - \frac{d}{dz}\int M dx = P - \frac{d}{dz}\int M dx. \tag{7.3.4.32}$$

That is to say, $Y = Y(y, z)$ should satisfy the equation

$$dY = \left(N - \frac{d}{dy}\int M dx\right) dy + \left(P - \frac{d}{dz}\int M dx\right) dz. \tag{7.3.4.33}$$

Hence in order that Y exist, the differential on the right-hand side of equation (7.3.4.33) must be an exact differential in the two variables y and z, in which case this exact differential is the total differential of Y, and Y is just an integral of this exact differential. (Again, such an integral Y is not unique; it is determined up to an arbitrary additive constant.) Now, in order for the differential on the right-hand side of (7.3.4.33) even to be a differential in the two variables y and z alone, the expressions on the right-hand sides of (7.3.4.31) and (7.3.4.32) must not depend upon x. Clairaut again assumed that differentiation and integration can be inverted. Therefore according to the rules that differential coefficients obey, it follows that

$$\frac{d}{dx}\left(N - \frac{d}{dy}\int M dx\right) = \frac{d}{dx}\left(N - \int \frac{dM}{dy} dx\right)$$

$$= \frac{dN}{dx} - \frac{d}{dx}\int \frac{dM}{dy} dx = \frac{dN}{dx} - \frac{dM}{dy} = 0$$

and that

$$\frac{d}{dx}\left(P - \frac{d}{dz}\int M dx\right) = \frac{d}{dx}\left(P - \int \frac{dM}{dz} dx\right)$$

$$= \frac{dP}{dx} - \frac{d}{dx}\int \frac{dM}{dz} dx = \frac{dP}{dx} - \frac{dM}{dz} = 0,$$

where

$$\frac{dN}{dx} - \frac{dM}{dy} = 0$$

and

$$\frac{dP}{dx} - \frac{dM}{dz} = 0$$

follow from the first two of the three equalities in (7.3.4.29). Thus the expressions on the right-hand sides of (7.3.4.31) and (7.3.4.32) are indeed independent of x, and the differential on the right-hand side of equation (7.3.4.33) is consequently a differential in the two variables y and z. Finally, using condition (7.3.4.6) for exact differentials in two variables, we see that the equality

$$\frac{d}{dz}\left(N - \frac{d}{dy}\int M dx\right) = \frac{d}{dy}\left(P - \frac{d}{dz}\int M dx\right) \qquad (7.3.4.6')$$

of the partial differential coefficients

$$\frac{d}{dz}\left(N - \frac{d}{dy}\int M dx\right) \quad \text{and} \quad \frac{d}{dy}\left(P - \frac{d}{dz}\int M dx\right)$$

is both a necessary and sufficient condition for the differential on the right-hand side of equation (7.3.4.33) to be an exact differential in the two variables y and z.

Now, the left-hand side of equality (7.3.4.6′) can be written as

$$\frac{dN}{dz} - \frac{d}{dz}\left(\frac{d}{dy}\int M dx\right)$$

and the right-hand side of equality (7.3.4.6′) can be written as

$$\frac{dP}{dy} - \frac{d}{dy}\left(\frac{d}{dz}\int M dx\right).$$

Consequently, since the equality

$$\frac{dN}{dz} = \frac{dP}{dy}$$

holds [this is the third equality in (7.3.4.29)], equality (7.3.4.6′) reduces to the condition

$$\frac{d}{dz}\left(\frac{d}{dy}\int M dx\right) = \frac{d}{dy}\left(\frac{d}{dz}\int M dx\right). \tag{7.3.4.6″}$$

But condition (7.3.4.6″) indeed holds by the equality (7.3.4.7) of mixed, second-order partial differential coefficients of the expression $\int M dx$ with respect to y and z. Thus Clairaut reduced the problem of finding $\phi(x, y, z)$ on the left-hand side of (7.3.4.30) to determining an expression $Y(y, z)$ whose total differential is the exact differential in the two variables y and z on the right-hand side of equation (7.3.4.33). But Clairaut had already solved this particular problem in his first Paris Academy *mémoire* on integral calculus, in deriving (7.3.4.9), as well as again in the first part of his second Paris Academy *mémoire* on integral calculus. The expression $\phi(x, y, z)$ on the left-hand side of (7.3.4.30) could therefore be constructed. When the differential on the left-hand side of equation (7.3.4.28) is an exact differential, the total differential of the expression $\phi(x, y, z)$ constructed this way is the differential on the left-hand side of equation (7.3.4.28). If $\phi(x, y, z)$ is then set equal to a constant k, the equation $\phi = k$ is an integral of the equation (7.3.4.28).

Independently of the form of equation (7.3.4.13), the differential (7.3.4.23) on the right-hand side of equation (7.3.4.13) can be shown to be an exact differential in the three variables x, y, and p using Clairaut's results here, when $\alpha(x, y, p)$ and $\pi(x, y, p)$ are homogeneous expressions of degree zero in x, y, and p such that Fontaine's "conditional equation" (7.3.3.111) is satisfied. Namely, (7.3.4.24′) follows from (7.3.3.111) and (7.3.4.25), and (7.3.4.24) is easily seen to be the same as (7.3.4.24′) after the rules that partial differential coefficients obey have been applied to (7.3.4.24). Likewise, (7.3.4.26′) follows from (7.3.3.111) and (7.3.4.27), and (7.3.4.26) is also easily seen to be the same as (7.3.4.26′) after the rules that partial differential coefficients obey have been applied to (7.3.4.26). But (7.3.4.24) and (7.3.4.26) are two of the three equalities in (7.3.4.29) applied to the differential (7.3.4.23). That the third equality in (7.3.4.29) applied to the differential (7.3.4.23) holds follows from a similar argument. This demonstration actually proves that if α and π are homogeneous expressions of degree zero in x, y, and p, then (7.3.3.111) is a sufficient condition for the differential (7.3.4.23) on the right-hand side of equation (7.3.4.13) to be an exact differential in the three variables x, y, and p. To express the matter another way, the demonstration shows that

$$\frac{1}{x + \alpha y + \pi p}$$

must be an integrating factor for equation (7.3.3.97), when $x + \alpha y + \pi p \neq 0$. In other words, equation (7.3.3.97) must be integrable when α and π are homogeneous expressions of degree zero in x, y, and p which satisfy the conditional equation (7.3.3.111).

In case the differential on the left-hand side of equation (7.3.4.28) does not satisfy the three conditions (7.3.4.29), hence is not an exact differential, Clairaut multiplied both sides of equation (7.3.4.28) by an integrating factor $u(x, y, p)$ and then applied the three conditions (7.3.4.29) to the exact differential

$$uMdx + uNdy + uPdz.$$

The three equalities

$$\frac{d(uM)}{dy} = \frac{d(uN)}{dx}, \tag{7.3.4.34}$$

$$\frac{d(uM)}{dz} = \frac{d(uP)}{dx}, \tag{7.3.4.35}$$

and

$$\frac{d(uN)}{dz} = \frac{d(uP)}{dy} \tag{7.3.4.36}$$

in partial differential coefficients resulted. Clairaut then once again exploited to the best advantage the rules that partial differential coefficients obey. In so doing, equalities (7.3.4.34)–(7.3.4.36) became equations

$$M\frac{du}{dy} + u\frac{dM}{dy} = N\frac{du}{dx} + u\frac{dN}{dx}, \tag{7.3.4.34'}$$

$$M\frac{du}{dz} + u\frac{dM}{dz} = P\frac{du}{dx} + u\frac{dP}{dx}, \tag{7.3.4.35'}$$

and

$$N\frac{du}{dz} + u\frac{dN}{dz} = P\frac{du}{dy} + u\frac{dP}{dy}. \tag{7.3.4.36'}$$

Clairaut observed that equations (7.3.4.34')–(7.3.4.36'), which are three linear, first-order partial "differential equations" for $u(x, y, z)$, need not always be consistent. That is, they may have no solution $u(x, y, z) \neq 0$. To show this Clairaut eliminated $u(x, y, z)$ and the partial differential coefficients du/dx, du/dy, and du/dz from equations (7.3.4.34′)–(7.3.4.36′). When he did this a single equation

$$N\frac{dP}{dx} - P\frac{dN}{dx} + M\frac{dN}{dz} - N\frac{dM}{dz} - M\frac{dP}{dy} + P\frac{dM}{dy} = 0, \qquad (7.3.4.37)$$

which involves M, N, and P alone, resulted. Whether equation (7.3.4.37) holds or not depends upon M, N, and P. Equation (7.3.4.37) is consequently a conditional equation for the integrability of equation (7.3.4.28). It is a necessary condition that must be fulfilled in order that equation (7.3.4.28) be integrable.

Clairaut had in fact already applied the same kind of argument in his first Paris Academy *mémoire* on integral calculus, when he derived equations (7.3.4.20)–(7.3.4.22) and then used these three equations to arrive at Fontaine's conditional equation (7.3.3.111) for the integrability of equation (7.3.3.97). [In Chapter 9 we will see why equation (7.3.3.111) is a conditional equation for the integrability of equation (7.3.3.97) if and only if equation (7.3.4.37) is a conditional equation for the integrability of equation (7.3.4.28).]

Clairaut also noted that if the conditional equation (7.3.4.37) holds, then equations (7.3.4.34′)–(7.3.4.36′) are not independent. They are related through equation (7.3.4.37), which I advantageously rewrite as

$$M\left(\frac{dN}{dz} - \frac{dP}{dy}\right) + N\left(\frac{dP}{dx} - \frac{dM}{dz}\right) + P\left(\frac{dM}{dy} - \frac{dN}{dx}\right) = 0. \qquad (7.3.4.37′)$$

Consequently, Clairaut remarked, when equation (7.3.4.37) holds, it sufficed to solve any two of the three equations (7.3.4.34′)–(7.3.4.36′) for a particular solution $u(x, y, z)$ using the "method of undetermined quantities." An expression $u(x, y, z)$ found this way would automatically satisfy the third equation as well. The easiest way to see this is as follows. If

$$uM dx + uN dy + uP dz \qquad (7.3.4.38)$$

is an exact differential, then the conditional equation (7.3.4.37′) is certainly satisfied, which in this particular case is the equation

$$uM\left(\frac{d}{dz}(uN) - \frac{d}{dy}(uP)\right) + uN\left(\frac{d}{dx}(uP) - \frac{d}{dz}(uM)\right)$$
$$+ uP\left(\frac{d}{dy}(uM) - \frac{d}{dx}(uN)\right) = 0. \qquad (7.3.4.37″)$$

Now, since the differential (7.3.4.38) is an exact differential, equalities (7.3.4.34)–(7.3.4.36) all hold. This means that each of the three expressions

$$\frac{d}{dz}(uN) - \frac{d}{dy}(uP)$$

$$\frac{d}{dx}(uP) - \frac{d}{dz}(uM)$$

and

$$\frac{d}{dy}(uM) - \frac{d}{dx}(uN)$$

in parentheses in equation (7.3.4.37″) is individually equal to zero. However, we can easily see that if any two of these expressions in parentheses in equation (7.3.4.37″) are equal to zero, then the third expression in parentheses in equation (7.3.4.37″) must equal zero as well.[203] This shows how the three equations (7.3.4.34′)–(7.3.4.36′) are interrelated through the conditional equation (7.3.4.37).

When $u(x, y, z)$ has been found, an expression $\phi(x, y, p)$ whose total differential is the complete differential (7.3.4.38) can be determined by using the procedure described. If this expression $\phi(x, y, p)$ is then set equal to a constant k, the equation $\phi = k$ is an integral of equation (7.3.4.28).

[Again we note that a solution $u(x, y, z)$ of equations (7.3.4.34′)–(7.3.4.36′) need not be an integral that can be expressed in finite terms. The solution may at most be reducible to quadratures, which in this case, unlike the notion of quadratures discussed earlier in this chapter, can mean reducible to integrals in three finite variables x, y, and z which cannot be expressed in finite terms, instead of to integrals in one finite variable which cannot be expressed in finite terms.]

We have seen that if α and π are homogeneous expressions of degree $e = 0$ in x, y, and p, then equation (7.3.3.111) is a sufficient condition as well as a necessary condition for the integrability of equation (7.3.3.97). Historians in the past have made conflicting statements about the question whether Clairaut showed that equation (7.3.4.37) is not only a necessary condition for equation (7.3.4.28) to be integrable in finite terms or reducible to quadratures but also demonstrated that equation (7.3.4.37) is a sufficient condition for equation (7.3.4.28) to be integrable in finite terms or reducible to quadratures too.[204] In fact Lagrange seems to have been the first mathematician to demonstrate adequately the sufficiency of condition (7.3.4.37) for the integrability of equation (7.3.4.28). This he did in his well-known Berlin Academy *mémoire* of 1772 on reducing the problem of solving the general first-order partial "differential equation" in two variables to the problem of solving a linear first-order partial "differential equation" in three variables.[205]

In his second Paris Academy *mémoire* on integral calculus, Clairaut also showed how to extend his method described previously of treating first-order ordinary "differential" equations of first degree and integrable first-order "differential" equations of first degree in three variables to cover integrable first-order "differential" equations of first degree

$$M dx + N dy + P dz + Q du + R ds + \text{etc.} = 0 \qquad (7.3.4.39)$$

in four or more variables x, y, z, u, s, etc., where the word "integrable" here again means either integrable in finite terms or reducible to quadratures. I shall not discuss this part of Clairaut's second Paris Academy *mémoire* on integral calculus, because it is not needed to understand how Clairaut applied the partial differential calculus to tackle the problem in fluid mechanics which he first raised at the end of his second paper on the theory of the earth's shape and which I discussed at the end of Chapter 6. His use of the partial differential calculus to treat this problem, which I shall examine in Chapter 9, only involves first-order ordinary "differential" equations of first degree and integrable first-order "differential" equations of first degree in three variables.

Moreover, Clairaut's method of handling integrable first-order "differential" equations of first degree (7.3.4.39) in four or more variables can easily be understood without having to go into the method in detail. We recall that in his first Paris Academy *mémoire* on integral calculus, Clairaut hinted that there is a way of integrating in finite terms or reducing to quadratures complete differentials in three or more variables which generalizes his method of integrating in finite terms or reducing to quadratures complete differentials in two variables – namely, by reducing the original problem to a problem of the same kind, but one which involves one less variable and one less term. By repeating this process of reduction over and over again, the expression whose differential is a given, complete differential in any number of variables could be found by systematic reduction step by step to expressions that could be integrated in finite terms or, if not, could at least be reduced to quadratures. We saw earlier how Clairaut used this process in his second Paris Academy *mémoire* on integral calculus to reduce the problem of integrating in finite terms or reducing to quadratures integrable first-order "differential" equations of first degree

$$M(x, y, z)dx + N(x, y, z)dy + P(x, y, z)dz = 0 \qquad (7.3.4.28)$$

in three variables x, y, and z to the problem of integrating in finite terms or reducing to quadratures first-order ordinary "differential" equations

$$M(x, y)dx + N(x, y)dy = 0 \qquad (7.3.4.3)$$

of first degree. Clairaut showed as well in his second Paris Academy *mémoire* on integral calculus that the method can easily be generalized so as to reduce the problem of integrating in finite terms or reducing to quadratures integrable first-order "differential" equations (7.3.4.39) in n variables to the problem of integrating in finite terms or reducing to quadratures integrable first-order "differential" equations (7.3.4.39) in $n - 1$ variables, where $4 \leqslant n$. He thereby clarified what he had already hinted at in his first Paris Academy *mémoire* on integral calculus.

There is one additional consideration needed to understand the integration of integrable equations (7.3.4.39), which I shall just mention here. The number of conditional equations of type (7.3.4.37) increases as the number of variables in integrable equations (7.3.4.39) increases. It is easy to see how to determine the number of independent conditional equations of type (7.3.4.37) for the integrability of an equation (7.3.4.39) by setting M equal to one in equation (7.3.4.39), in which case equation (7.3.4.39) becomes equation

$$dx + Ndy + Pdz + Qdu + Rds + \text{etc.} = 0. \qquad (7.3.4.39')$$

Then the conditional equation of type (7.3.4.37) associated with, say, the equation

$$Ndy + Pdz + Qdu = 0$$

in the three variables y, z, and u gotten by holding constant all of the other variables in equation (7.3.4.39') can be interpreted as one of the $_4C_3 = \text{four}$ conditional equations of type (7.3.4.37) for the equation

$$dx + Ndy + Pdz + Qdu = 0$$

in four variables. The reason that it is convenient to look at things in this way is that, of these four conditional equations, only three turn out to be independent. Hence any one of the four equations can be eliminated. This equation can be chosen to be the one associated with the equation

$$Ndy + Pdz + Qdu = 0$$

in three variables. In that case the three conditional equations of type (7.3.4.37) individually associated with each of the three equations

$$dx + Ndy + Pdz = 0, \quad dx + Ndy + Qdu = 0, \quad \text{and} \quad dx + Pdz + Qdu = 0$$

in three variables, respectively, are kept as the three independent conditional equations of type (7.3.4.37) for the equation

$$dx + Ndy + Pdz + Qdu = 0$$

in four variables. Hence it is easy to see from looking at things this way that it is enough to count the number of conditional equations of type (7.3.4.37) which are individually associated with equations in three variables which have the form

$$dx + Ndy + Pdz = 0, \quad dx + Ndy + Qdu = 0, \quad dx + Pdz + Qdu = 0,$$

and so forth gotten by holding constant all of the other variables in equation (7.3.4.39') in order to determine the number of independent conditional equations of type (7.3.4.37) for equation (7.3.4.39'), hence for equation (7.3.4.39). But this number is just equal to $_{n-1}C_2 = ((n-1)(n-2))/2$ for an equation (7.3.4.39) in n variables, which is the same number that Clairaut found by a similar use of combinations.

In his first Paris Academy *mémoire* on integral calculus, Clairaut showed that in the case of an integrable equation (7.3.3.97) in three variables x, y, and p, no equations need to be solved for an integrating factor $u(x, y, p)$, when $\alpha(x, y, p)$ and $\pi(x, y, p)$ are homogeneous expressions of degree zero in x, y, and p. Namely, he showed that the differential on the right-hand side of equation (7.3.4.13) is a complete differential in x, y, and p. In other words, the expression

$$\frac{1}{x + \alpha y + \pi p}$$

in x, y, and p is an integrating factor for equation (7.3.3.97) in this case. Clairaut ended his second Paris Academy *mémoire* on integral calculus by stating that if equation (7.3.4.39) is an integrable first-order "differential" equation in any number of variables x, y, z, u, s, etc., where M, N, P, Q, R, etc. are homogeneous expressions of the same degree in x, y, z, u, s, etc., then no equations need to be solved for an integrating factor $U(x, y, z, u, s$, etc.) either. He demonstrated this specifically for the case of an integrable equation (7.3.4.28) in three variables x, y, and z. In other words, he showed that if M, N, and P are homogeneous expressions of the same degree m in the three variables x, y, and z, then an integrable equation (7.3.4.28) could be integrated without having to utilize two of the three equations (7.3.4.34')–(7.3.4.36') to find an integrating factor $u(x, y, z)$. However, instead of beginning with the elements of Fontaine's method of integration, which is what he did in his first Paris Academy *mémoire* on integral calculus, Clairaut used what he called a method of integrating which "resembles the one ordinarily employed for [integrating or reducing to quadratures] homogeneous [first-order "differential"] equations [of first degree] in two variables."[206]

Clairaut doubtlessly had the following idea in mind when he made this statement. Suppose that $M(x, y)$ and $N(x, y)$ in equation (7.3.4.3) are homogeneous expressions of the same degree n in the variables x and y. If we make a change of variables

$$y = xu, \quad dy = d(xu) = xdu + udx$$

and use (7.3.3.82) which defines a homogeneous expression of degree n in two variables, then equation (7.3.4.3) becomes

$$0 = M(x, y)dx + N(x, y)dy = M(x, xu)dx + N(x, xu)\{xdu + udx\}$$

$$= x^n M(1, u)dx + x^n N(1, u)\{xdu + udx\}$$

$$= \{x^n M(1, u) + x^n u N(1, u)\}dx + x^{n+1} N(1, u)du \qquad (7.3.4.40)$$

$$= \frac{1}{x}dx + \frac{N(1, u)}{M(1, u) + uN(1, u)}du.$$

But the variables x and u in the differential on the right-hand side of equation (7.3.4.40) are separated in the sense that I talked about earlier in this chapter. In other words, after making a suitable change of variables in equation (7.3.4.3), when $M(x, y)$ and $N(x, y)$ in equation (7.3.4.3) are homogeneous expressions of the same degree n in the variables x and y, the new variables in the corresponding first-order ordinary "differential" equation of first degree are separated, hence this first-order ordinary "differential" equation of first degree can be integrated in finite terms or, if not, can at least be reduced to quadratures.

In his letter to Clairaut of October 1740, Euler had used the same change of variables to prove the homogeneous function theorem for homogeneous express-ions $\phi(x, y)$ in two variables x and y. That is, in this letter Euler demonstrated the homogeneous function theorem in two variables by separating the variables. Clairaut now put this special feature of homogeneous functions, namely, that they separate variables when suitable changes of variables are made, further to work.

Namely, assuming that N, M, and P are all homogeneous expressions of degree m in x, y, and z in an integrable equation (7.3.4.28), Clairaut made the changes of variables

$$y = xu, \quad dy = d(xu) = xdu + udx$$

and

$$z = xt, \quad dz = d(xt) = xdt + tdx$$

in equation (7.3.4.28) and thereby arrived at the equation

$$0 = M(x, xu, xt)\,dx + N(x, xu, xt)\{xdu + udx\} + P(x, xu, xt)\{xdt + tdx\}$$
$$(7.3.4.41)$$
$$= x^m M(1, u, t)dx + x^m N(1, u, t)\{xdu + udx\} + x^m P(1, u, t)\{xdt + tdx\}.$$

Letting $M(1, u, t) \equiv F(u, t)$, $N(1, u, t) \equiv G(u, t)$, and $P(1, u, t) \equiv H(u, t)$, and then rear-ranging terms, Clairaut rewrote equation (7.3.4.41) as

$$x^m(F + Gu + Ht)\,dx + x^{m+1}\,Gdu + x^{m+1}\,Hdt = 0. \qquad (7.3.4.42)$$

Dividing both sides of equation (7.3.4.42) by $x^{m+1}(F + Gu + Ht)$, Clairaut arri-ved at the equation

$$\frac{dx}{x} + \frac{Gdu + Hdt}{F + Gu + Ht} = 0. \qquad (7.3.4.43)$$

We clearly see that in equation (7.3.4.43), a new form of equation (7.3.4.28) which results from making the changes of variables

$$y = xu, \quad dy = d(xu) = xdu + udx$$

and

$$z = xt, \quad dz = d(xt) = xdt + tdx$$

in equation (7.3.4.28), the variable x and its differential dx have been separated from the other terms on the left-hand side of the equation. Now, equation (7.3.4.28) is integrable in finite terms or reducible to quadratures if and only if equation (7.3.4.43) is integrable in finite terms or reducible to quadratures. But equation (7.3.4.43) has the form

$$L(x)dx + A(u, t)du + B(u, t)dt = 0. \qquad (7.3.4.43')$$

If we write the conditional equation (7.3.4.37') for the integrability of equation (7.3.4.43'), we easily see that it reduces to the equality

$$\frac{dA}{dt} = \frac{dB}{du}, \qquad (7.3.4.44)$$

because the partial differential coefficients

$$\frac{dB}{dx}, \quad \frac{dL}{dt}, \quad \frac{dL}{du} \quad \text{and} \quad \frac{dA}{dx}$$

are all equal to zero.[207] But equality (7.3.4.44) means that the differential

$$A(u, t)du + B(u, t)dt$$

must be a complete differential in the two variables u and t. In other words equation (7.3.4.43) is an integrable equation if and only if the differential

$$\frac{Gdu + Hdt}{F + Gu + Ht} \qquad (7.3.4.45)$$

is a complete differential in the two variables u and t. Thus if equation (7.3.4.43) is an integrable equation, it requires no integrating factor to be integrated. The differential on the left-hand side of equation (7.3.4.43) will be a complete differential in the three variables $x, u,$ and t. That is, an expression $\phi(x, u, t)$ in $x, u,$ and t will exist whose total differential is the differential on the left-hand side of equation (7.3.4.43). This expression ϕ is

$$\phi = lx + \Pi(u, t), \qquad (7.3.4.45')$$

where $\Pi(u, t)$ is an expression in the two variables u and t whose total differential is the complete differential (7.3.4.45). In fact this is just a particular case of Clairaut's technique (7.3.4.30) and (7.3.4.33) for integrating in finite terms or reducing to quadratures complete differentials in three variables by reducing the problem to one of finding the expression in two variables whose total differential is a certain complete differential in those two variables, where the first of the two terms on the right-hand side of (7.3.4.45') has a particularly simple form because the variable x and its differential dx have been separated from the other terms on the left-hand side of equation (7.3.4.43) as a result of making the changes of variables

$$y = xu, \quad dy = d(xu) = xdu + udx$$

and

$$z = xt, \quad dz = d(xt) = xdt + tdx$$

in equation (7.3.4.28). Clairaut observed that in order to find out whether equation (7.3.4.28) in three variables is integrable in finite terms or reducible to quadratures or not, one can either examine the conditional equation (7.3.4.37) to determine if that equation holds or not, or else one can make the changes of variables

$$y = xu, \quad dy = d(xu) = xdu + udx$$

and

$$z = xt, \quad dz = d(xt) = xdt + tdx$$

in equation (7.3.4.28) to check whether the differential (7.3.4.45) is a complete differential in the two variables u and t or not. He noted that if the differential (7.3.4.45) is a complete differential, then $\Pi(u, t)$ can always be found using the technique for integrating in finite terms or reducing to quadratures complete differentials in two variables which he had discussed at the beginning of his second Paris Academy *mémoire* on integral calculus and had first introduced as (7.3.4.9) in his first Paris Academy *mémoire* on integral calculus.

Now, as I have said, equation (7.3.4.28) is integrable in finite terms or reducible to quadratures if and only if the differential on the left-hand side of equation (7.3.4.43) is a complete differential. But this differential can be rewritten as

$$\frac{x^m dx(F + Gu + Ht) + x^{m+1} Gdu + x^{m+1} Hdt}{x^{m+1}(F + Gu + Ht)}. \tag{7.3.4.46}$$

If we now change back to the original variables x, $y = xu$, and $z = xt$, we find that the differential (7.3.4.46) becomes

$$\frac{Mdx + Ndy + Pdz}{Mx + Ny + Pz}. \tag{7.3.4.47}$$

But the differential (7.3.4.47), like the differential (7.3.4.46), must be a complete differential. In short, if equation (7.3.4.28) is integrable in finite terms or reducible to quadratures, no equations need to be solved for an integrating factor $u(x, y, p)$. It is assured that the expression

$$\frac{1}{Mx + Ny + Pz}$$

in x, y, and z must be an integrating factor for equation (7.3.4.28). In fact this result is equivalent to the differential on the right-hand side of equation (7.3.4.13) being a complete differential. In his first Paris Academy *mémoire* on integral calculus Clairaut had shown by simplying Fontaine's method of integrating in finite terms or reducing to quadratures first-order ordinary "differential" equations of first degree that the differential on the right-hand side of equation (7.3.4.13) is a complete differential. In his second Paris Academy *mémoire* on integral calculus Clairaut bypassed and avoided Fontaine's equations altogether in determining that the differential (7.3.4.47) must be a complete differential. Instead Clairaut exploited to the best advantage the idea of complete differentials together with the special feature that homogeneous functions have, namely, that they separate variables when suitable changes of variables are made.

Finally, from equations (7.3.4.28), (7.3.4.41), and (7.3.4.42), we see that the differential

$$Mdx + Ndy + Pdz \tag{7.3.4.48}$$

on the left-hand side of equation (7.3.4.28) becomes the differential

$$x^m(F + Gu + Ht)dx + x^{m+1}Gdu + x^{m+1}Hdt \tag{7.3.4.49}$$

on the left hand side of equation (7.3.4.42) after making the changes of variables

$$y = xu, \quad dy = d(xu) = xdu + udx$$

and

$$z = xt, \quad dz = d(xt) = xdt + tdx$$

in equation (7.3.4.28). If the differential (7.3.4.48) is a complete differential, then the differential (7.3.4.49) must be a complete differential too. In this case Clairaut stated that the expression $\phi(x, u, t)$ whose total differential is the complete differential (7.3.4.49) must have the form

$$\phi(x, u, t) = \int x^m(F + Gu + Ht)dx + T(u, t)$$

$$= \frac{x^{m+1}}{m+1}(F + Gu + Ht) + T(u, t). \tag{7.3.4.50}$$

This follows from Clairaut's technique (7.3.4.30) and (7.3.4.33) for integrating in finite terms or reducing to quadratures complete differentials in three variables by reducing the problem to one of finding the expression in two variables whose total differential is a certain complete differential in those two variables. Clairaut determined by inspection that $T(u, t)$ in (7.3.4.50) must equal zero. This can be understood as follows. $T(u, t)$ must be such that the partial differential coefficients dT/du and dT/dt have the forms

$$\frac{dT}{du}(u, t) = \frac{d}{du}\left(\phi - \frac{x^{m+1}}{m+1}(F + Gu + Ht)\right) = \frac{d\phi}{du} - \frac{d}{du}\left(\frac{x^{m+1}}{m+1}(F + Gu + Ht)\right)$$

$$= x^{m+1}G - \frac{x^{m+1}}{m+1}J(u, t) = x^{m+1}R(u, t) \tag{7.3.4.51}$$

and

$$\frac{dT}{dt}(u, t) = \frac{d}{dt}\left(\phi - \frac{x^{m+1}}{m+1}(F + Gu + Ht)\right) = \frac{d\phi}{dt} - \frac{d}{dt}\left(\frac{x^{m+1}}{m+1}(F + Gu + Ht)\right)$$

$$= x^{m+1}H - \frac{x^{m+1}}{m+1}K(u, t) = x^{m+1}L(u, t). \tag{7.3.4.52}$$

But the left-hand sides of both (7.3.4.51) and (7.3.4.52) do not depend upon x, hence the right-hand sides of both (7.3.4.51) and (7.3.4.52) must also be independent of x. But this can only occur if the expression $R(u, t)$ on the right-hand side of ·(7.3.4.51) and the expression $L(u, t)$ on the right-hand side of (7.3.4.52) are both identically equal to zero. Consequently $dT/du \equiv 0$ and $dT/dt \equiv 0$ must both be true as well, which means that $T(u, t) = $ a constant k. Therefore the expression ϕ in (7.3.4.50) must have the form

$$\phi(x, u, t) = \int x^m(F + Gu + Ht)dx + k$$

$$= \frac{x^{m+1}}{m+1}(F + Gu + Ht) + k. \qquad (7.3.4.50')$$

Changing back to the original variables x, $y = xu$, and $z = xt$, Clairaut found that (7.3.4.50') becomes

$$\phi(x, y, z) = \frac{Mx + Ny + Pz}{m+1} + k. \qquad (7.3.4.53)$$

Now, the expression on the right-hand side of (7.3.4.53) is a homogeneous expression of degree $m + 1$ in x, y, and z, where m cannot equal -1, if and only if $k = 0$. Thus Clairaut took k to be zero, in which case (7.3.4.53) becomes

$$\phi(x, y, z) = \frac{Mx + Ny + Pz}{m+1}. \qquad (7.3.4.53')$$

In other words, (7.3.4.53') is, Clairaut concluded, "the integral of [the complete differential] $Mdx + Ndy + Pdz$, as Monsieur Fontaine's theorem also shows."[208] Having arrived at this result, Clairaut ended his second Paris Academy *mémoire* on integral calculus. In a footnote to the *mémoire* Clairaut repeated what Euler had told him by letter in October of 1740 concerning the location of the homogeneous function theorem in Euler's own work. Clairaut mentioned where the homogeneous function theorem could be found in Euler's work.

In effect what Clairaut proved is that if the differential $Mdx + Ndy + Pdz$ is a complete differential in the three variables x, y, and z, where M, N, and P are all homogeneous expressions of degree m in x, y, and z, then (7.3.4.53'), which is a homogeneous expression of degree $m + 1$ in x, y, and z, is an integral of the complete differential $Mdx + Ndy + Pdz$. In other words, (7.3.4.53') is an expression in x, y, and z whose total differential is the complete differential $Mdx + Ndy + Pdz$. Again we note, however, that m must not equal -1. This result is actually the converse of the homogeneous function theorem in three variables.[209] That is, according to the homogeneous function theorem in three variables, if $\phi(x, y, z)$ is a homogeneous expression of degree $m + 1$ in the three variables x, y, and z whose total differential is the differential $Mdx + Ndy + Pdz$, then

$$(m + 1)\phi = Mx + Ny + Pz,$$

where M, N, and P are all homogeneous expressions of degree m in x, y, and z. Clairaut showed that, conversely, if M, N, and P are homogeneous expressions of degree m in the three variables x, y, and z such that the differential $M dx + N dy + P dz$ is a complete differential, then provided m does not equal -1, the homogeneous expression

$$\phi(x, y, z) = \frac{Mx + Ny + Pz}{m + 1} \qquad (7.3.4.53')$$

of degree $m + 1$ in x, y, and z is an expression $\phi(x, y, z)$ whose total differential is the differential $M dx + N dy + P dz$. In fact, Fontaine himself actually made use of the converse of the homogeneous function theorem in three variables x, y, and p in his own method of integrating in finite terms or reducing to quadratures first-order ordinary differential equations of first degree, namely, in constructing $\phi(x, y, p)$ from $u(x, y, p)$, $\alpha(x, y, p)$, and $\pi(x, y, p)$ by using equation (7.3.3.114).

We saw earlier in this section of this chapter that the converse of the homogeneous function theorem in three variables x, y, and p cannot hold for complete differentials that have the form

$$\frac{dx + \alpha dy + \pi dp}{x + \alpha y + \pi p},$$

where α and π are homogeneous expressions of degree zero in x, y, and p. Clairaut's derivation of $(7.3.4.53')$ provides us with another explanation why this is true. Namely, if

$$M(x, y, p)dx + N(x, y, p)dy + P(x, y, p)dp \equiv \frac{dx + \alpha dy + \pi dp}{x + \alpha y + \pi p},$$

then M, N, and P are homogeneous expressions of degree $m = -1$ in x, y, and p, in which case Clairaut's derivation of $(7.3.4.53')$ does not hold. In other words, the converse of the homogeneous function theorem in three variables x, y, and p cannot hold in this case.

Thus Clairaut ended his second Paris Academy *mémoire* on integral calculus by demonstrating that Fontaine's method of integrating in finite terms or reducing to quadratures first-order ordinary "differential" equations of first degree included no theorems or results that his own method of integrating in finite terms or reducing to quadratures integrable first-order "differential" equations of first degree in three variables could not account for. Clairaut appeared to reduce the foundations of Fontaine's method of integrating in finite terms or reducing to quadratures first-order ordinary "differential" equations of first degree to special cases of his own method of integrating in finite terms or reducing to quadratures integrable first-order "differential" equations of first degree in three variables, by making use of the special property that homogeneous functions have, namely, that they separate variables when suitable changes of variables are made.

7.3.5. First French advances in the partial differential calculus III: Fontaine versus Clairaut. Clairaut considered his method of integrating in finite terms or reducing to quadratures first-order ordinary "differential" equations of first degree

$$M(x, y)dx + N(x, y)dy = 0 \qquad (7.3.4.3)$$

to be "much simpler" than Fontaine's method for doing the same thing to first-order ordinary "differential" equations of first degree[210]

$$dx + \alpha(x, y) = 0, \qquad (7.3.3.96)$$

and Clairaut was probably right. [It is easy to see that equations (7.3.4.3) and (7.3.3.96) are equivalent by simply dividing both sides of equation (7.3.4.3) by $M(x, y) \neq 0$ so that $\alpha(x, y) \equiv N(x, y)/M(x, y)$.] After all, after Clairaut simplified Fontaine's method, the method still required that the conditional equation

$$\alpha \frac{d\pi}{dx} - \pi \frac{d\alpha}{dx} + \frac{d\alpha}{dp} - \frac{d\pi}{dy} = 0 \qquad (7.3.3.111)$$

for the integrability of equation

$$dx + \alpha(x, y, p)dy + \pi(x, y, p)dp = 0, \qquad (7.3.3.97)$$

which is a linear, first-order partial "differential equation" in three variables x, y, and p, be solved for a suitable expression $\pi(x, y, p)$ that corresponded to a given expression $\alpha(x, y, p)$. Then a complete differential

$$\frac{1}{x + \alpha y + \pi p}dx + \frac{\alpha}{x + \alpha y + \pi p}dy + \frac{\pi}{x + \alpha y + \pi p}dp \qquad (7.3.4.23)$$

in three variables x, y, and p had to be integrated in finite terms or reduced to quadratures. Clairaut's method involved instead solving a linear, first-order partial "differential equation"

$$u\frac{dM}{dy} + M\frac{du}{dy} - u\frac{dN}{dx} - N\frac{du}{dx} = 0 \qquad (7.3.4.11'')$$

in two variables x and y for an integrating factor $u(x, y)$ for the first-order ordinary "differential" equation (7.3.4.3) of first degree, when the differential $Mdx + Ndy$ on the left-hand side of equation (7.3.4.3) was not a complete differential in the two variables x and y. Next a complete differential $uMdx + uNdy$ in two variables x and y had to be integrated in finite terms or reduced to quadratures using an uncomplicated technique that never failed. [In case the differential $Mdx + Ndy$ on the left-hand side of equation (7.3.4.3) is a complete differential in the two variables x and y, then $u(x, y) \equiv 1$.] Clairaut's method by contrast with Fontaine's seemed easy. As Clairaut maintained, there was indeed no need to introduce a third variable p to treat the problem.

Yet both methods did require that linear, first-order partial "differential equations" of one sort or another be solved. Neither method required solving the linear, first-order partial "differential equation" relevant in each case for a general solution. A particular solution was all that was needed. But no alternative to proceeding by trial and error to determine particular solutions of linear, first-order partial "differential equation" existed at the time. Neither Fontaine nor Clairaut could discover ways of finding particular solutions which always worked. Consequently, neither of their two methods of integrating in finite terms or reducing to quadratures first-order ordinary "differential" equations of first degree could be made to work all the time. Hence both Fontaine and Clairaut failed in their endeavor to integrate in finite terms or reduce to quadratures all first-order ordinary "differential" equations of first degree.

But we now know that it was inevitable that they would fail. We recall that I mentioned earlier in this chapter while discussing Maupertuis's appeals to the members of the Paris Academy to go beyond the search for methods of separating variables in first-order ordinary "differential" equations of first degree and to look for other ways to treat such equations that it was not conclusively demonstrated until the nineteenth century that not all first-order ordinary "differential" equations of first degree can be reduced to quadratures, much less be integrated in finite terms. Thus, for example, at the time that he wrote his second Paris Academy *mémoire* on integral calculus, Clairaut believed that "nothing that I know of stops us from thinking that an arbitrary "differential" equation in two variables can be gotten by differentiating some equation expressed in finite terms." Clairaut made this statement thinking that there was a basic difference of an analytic nature between first-order ordinary "differential" equations of first degree (7.3.4.3) and first-order "differential" equations of first degree

$$M(x, y, z)dx + N(x, y, z)dy + P(x, y, z)dz = 0 \qquad (7.3.4.28)$$

in three variables $x, y,$ and z. That is, he believed that all first-order ordinary "differential" equations of first degree (7.3.4.3) could be gotten in the way that he just described, whereas there were infinitely many first-order "differential" equations (7.3.4.28) of first degree in three variables which could not be gotten by differentiating any equation expressed in finite terms – namely, those equations (7.3.4.28) that failed to satisfy the conditional equation[211]

$$N\frac{dP}{dx} - P\frac{dN}{dx} + M\frac{dN}{dz} - N\frac{dM}{dz} - M\frac{dP}{dy} + P\frac{dM}{dy} = 0. \qquad (7.3.4.37)$$

But it was only shown once and for all during the nineteenth century that not even all first-order ordinary "differential" equations of first degree (7.3.4.3) can be gotten in the way that Clairaut just described.

Clairaut also regarded his method of integrating in finite terms or reducing to quadratures first-order ordinary "differential" equations of first degree (7.3.4.3) to be "more natural" than Fontaine's method for doing the same thing to first-order ordinary "differential" equations of first degree (7.3.3.96).[212] Clairaut found Fontaine's varying a parameter in the integral of a first-order ordinary "differential" equation of first degree to be an "uncommon consideration" ("considération singuliére"). Clairaut stated that "since the integral calculus is only the inverse of the differential calculus, and constants are treated as constants [not as variables] in the latter, the same should be true in the former as well."[213] In other words, since varying constants that appear in the primitive of an ordinary "differential" equation for a family of plane curves that all lie in the same plane does not enter in any way in determining the tangents to the members of the family by differentiating (that is, by taking differentials), Clairaut believed that the same should be true of constants when determining primitives that are integrals of this ordinary "differential" equation.

Indeed, if we differentiate (that is, we take differentials of) a primitive

$$\phi(x, y, p) = 0, \tag{7.3.5.1}$$

for each fixed value of a parameter p, we arrive at a first-order ordinary "differential" equation of form

$$G(x, y, dx, dy, p) = 0. \tag{7.3.5.2}$$

Then, in principle, p can be simultaneously eliminated from equations (7.3.5.1) and (7.3.5.2), and when this is done a first-order ordinary "differential" equation

$$H(x, y, dx, dy) = 0, \tag{7.3.5.3}$$

whose integral is equation (7.3.5.1), results *without varying the parameter p.*

Clairaut was at heart a geometer, in the contemporary sense of the word. "Varying parameters" seemed to him to be a mere contrivance, artifice, or subterfuge that was liable to cause people to confuse problems in two geometric dimensions with problems in three geometric dimensions. Clairaut's view of the part that parameters play in equations for plane curves that all lie in the same plane and equations for surfaces, namely, as *fixed* quantities, can be traced back at least as far as his first major publication, a treatise on skew curves, the first draft of which he composed when he was about fifteen years old.[214] Indeed, we recall that I mentioned at the beginning of this chapter that until Leibniz converted the parameter of a one-parameter family of plane curves that all lie in the same plane into a variable in order to find the "envelope" of such a family of plane curves by varying the parameter and differentiating with respect to the parameter (differentiating from curve to curve), the *only* way that the parameter of a one-parameter family of plane curves that all lie in the same plane had been meaningfully interpreted was through the *constant* values of the parameter which determine the individual plane curves in such a family.

However, when an ordinary "differential" equation does not have unique solutions to initial value problems, fixed values of constants of integration in an integral of the ordinary "differential" equation need *not* furnish *all* of the plane curves that lie in the same plane which satisfy the ordinary "differential" equation. Varying the constant of integration in the one-parameter family of solutions to such an ordinary "differential" equation can determine the "envelope" of the family, which is itself a "singular" solution to the ordinary "differential" equation which is not associated with any one particular value of the constant of integration. Thus it should come as no surprise that Clairaut, although the "geometer's geometer" of his day in Paris, nevertheless failed to realize that the singular solution that he discovered in 1734 to the famous first-order ordinary "differential" equation that bears his name[215] is the envelope of its one-parameter family of solutions. Clairaut rejected "varying parameters" in ordinary "differential" equations as ungeometric, when in fact he had simply failed to discover the way that geometry is connected with varying parameters in ordinary "differential" equations. Clairaut evidently knew nothing about Leibniz's first application of the partial differential calculus to find envelopes of one-parameter families of plane curves, where all of the plane curves in such a family lie in the same plane, by varying the parameters and differentiating with respect to the parameters (differentiating from curve to curve). (Nor for that matter does Leibniz's determining envelopes this way appear to have influenced developments in mathematics anywhere.[216])

Fontaine, on the other hand, encountered the problem of integrating in finite terms or reducing to quadratures first-order ordinary "differential" equations of first degree while treating problems in families of curves, like the various problems of trajectories, which required varying the parameters in one-parameter families of curves, where all of the plane curves in such a family lie in the same plane, in order to handle such problems. His "strange" way of attacking the problem of integrating in finite terms or reducing to quadratures first-order ordinary "differential" equations of first degree, by varying parameters in the integrals of such equations, certainly accords with the earlier "completion problems" that form part of the variable-parameter tradition.

The fact that Clairaut disliked "varying parameters" also suggests that he was unfamiliar with research done earlier by continental mathematicians. Part of his ignorance is to be expected. We remember that Nikolaus I Bernoulli found the plane curves orthogonal to the generalized brachistochrone cycloids by varying parameters. But Bernoulli only published his solution of the problem, in *Acta Eruditorum* for 1719. He never published the method that he used to solve the problem. Another part of Clairaut's ignorance has to do with the fact that mathematics went through a revival in Paris in the 1730s, as we saw earlier in this chapter. And the revival took place gradually. During the 1730s, mathematicians in Paris only began to become aware of much of the relevant older literature published in Germany in *Acta Eruditorum* concerning problems of finding orthogonal trajectories. Clairaut's own ignorance, which endured well into the 1730s, of some of this earlier literature, not to mention his ignorance of some mathematical work done more recently, appearing in the first two volumes of the St. Petersburg Academy's *Commentarii*, published in 1728 and 1729, respectively, is indicative.[217] Anyone who thought it an "uncommon consideration" to vary a parameter in an integral of a first-order ordinary "differential" equation of first degree was completely unaware of the "completion problems" that came up earlier in connection with trying to solve problems of finding orthogonal trajectories. Such a person could not possibly have read carefully, for example, Nikolaus I Bernoulli's article published in *Acta Eruditorum* for 1719 or Nikolaus II Bernoulli's article published in *Acta Eruditorum* for 1720 and its supplements. Clairaut continued to neglect the earlier literature until Euler first brought it to his attention toward the end of 1740. It is paradoxical that Fontaine is ordinarily portrayed as having paid little attention to work done by other mathematicians,[218] when in fact if any member of the new generation of Paris Academy mathematicians of the 1730s did not overlook the work done earlier by mathematicians outside France, it was he, at least during the early years of his career.

The partial differential calculus advanced differently in the hands of Clairaut than it did in the hands of the members of the variable-parameter tradition like Fontaine. In his second Paris Academy *mémoire* on integral calculus, Clairaut demonstrated a geometrical result which I have not discussed yet. Clairaut would soon find it useful for making progress in his research on the theory of the earth's shape. I shall just state Clairaut's discovery here and postpone explaining Clairaut's proof of it until Chapter 9, where I shall discuss the result in connection with the mechanics problem that Clairaut first raised at the end of his second paper on the theory of the earth's shape published in the *Philosophical Transactions* and that I talked about at the end of Chapter 6.

Clairaut ignored the two-parameter families of plane curves that all lie in the same plane which can be thought of as integrals of equation (7.3.3.97).[219] The geometry of such families is indeterminate, unless one parameter is chosen to be some arbitrary function of the other.[220] Clairaut showed that the following is true. Suppose that the first-order "differential" equation of first degree

$$dz - \omega(z, x, y)dx - \theta(z, x, y)dy = 0 \qquad (7.3.5.4)$$

in three variables $z, x,$ and y represents a surface in three-dimensional space. Clairaut demonstrated that the equation

$$\frac{d\theta}{dx} + \omega\frac{d\theta}{dz} = \frac{d\omega}{dy} + \theta\frac{d\omega}{dz} \qquad (7.3.5.5)$$

must hold in this case. Now, if we replace x by z, y by x, p by y, α by $-\omega$, and π by $-\theta$ in equation (7.3.3.97) and in Fontaine's conditional equation (7.3.3.111) for the integrability of equation (7.3.3.97), then equation (7.3.3.97) becomes equation (7.3.5.4) and equation (7.3.3.111) becomes the equation

$$\omega\frac{d\theta}{dz} - \theta\frac{d\omega}{dz} - \frac{d\omega}{dy} + \frac{d\theta}{dx} = 0. \qquad (7.3.5.6)$$

But equation (7.3.5.6) is the same as equation (7.3.5.5). In other words, as Clairaut observed himself, equation (7.3.5.5) is Fontaine's conditional equation for the integrability of equation (7.3.5.4). In other words, if equation (7.3.5.4) expresses a surface in three-dimensional space, then the conditional equation (7.3.5.5) for the integrability of equation (7.3.5.4) must be satisfied.[221]

Clairaut thought that there was an essential geometric difference between first-order ordinary "differential" equations of first degree and first-order "differential" equations of first degree in three variables. At the time that he wrote his second Paris Academy *mémoire* on integral calculus, he expressed what he believed to be the difference this way: "The difference between equations [(7.3.4.3)] in two variables and equations [(7.3.4.28)] in three variables is even greater than it appears to be as a result of the preceding demonstration [involving complete differentials]. For without knowing if all equations [(7.3.4.3)] in two variables can be gotten by some differentiation [which is to consider the matter analytically], it is easy to show that they always express some [plane] curve whose construction is possible. But all equations [(7.3.4.28)] in three variables which do not satisfy the [conditional] equation [(7.3.4.37)] cannot be constructed in any way."[222] Clairaut exaggerated when he made the statement that begins, "it is easy to show that." Clairaut's belief here concerning first-order ordinary "differential" equations of first degree was a matter of faith. He certainly did not prove that all such equations are geometrically meaningful – that is, that they always represent geometric entities in two dimensions, namely, plane curves. Euler was apparently the first mathematician to take up the question of the *existence* of the integrating factor $u(x, y)$ of a first-order ordinary "differential" equation (7.3.4.3) of first degree.[223] Eventually Euler produced a primitive version of what is usually called the "Cauchy–Lipschitz" existence theorem for a solution with a prescribed initial value of a first-order ordinary "differential" equation of first degree, in which Euler utilized a suitable sequence of polygonal approximations of the solution of the equation which converge to the solution.[224]

As I mentioned in the preceding section of this chapter, we will see in Chapter 9 why equation (7.3.3.111) is a conditional equation for the integrability of equation (7.3.3.97) if and only if equation (7.3.4.37) is a conditional equation for the integrability of equation (7.3.4.28). Both Clairaut and Euler believed that equations (7.3.4.28) that failed to satisfy the conditional equation (7.3.4.37) had no real significance. That is, they believed that such equations had no solutions that could be constructed in any way. That such equations were therefore meaningless was first disproved by Gaspard Monge, who first revealed the geometric import of equations (7.3.4.28) that do not satisfy the conditional equation (7.3.4.37).[225]

Using conditions for integrability, Clairaut shortly afterward analytically expressed by means of *level surfaces* of effective gravity necessary conditions for the relative equilibrium of homogeneous and stratified figures of revolution which revolve around their axes of symmetry and of homogeneous and stratified figures in three dimensions more generally – for example, a figure that revolves around an axis that is not an axis of symmetry of the figure. Specifically, to do this he utilized the three conditions (7.3.4.29) for the differential in three variables on the left-hand side of equation (7.3.4.28) to be a complete differential. This we shall examine in detail in Chapter 9.

In his second Paris Academy *mémoire* on integral calculus, Clairaut used his geometric interpretation of equation (7.3.5.4) to derive condition (7.3.4.6), thereby making condition (7.3.4.6) appear to be a special case of this geometric interpretation of equation (7.3.5.4). This he accomplished as follows. He said that if the differential $A(x, y)dx + B(x, y)dy$ in two variables x and y is a complete differential, then the equation

$$dz - A(x, y)dx - B(x, y)dy = 0 \qquad (7.3.5.7)$$

in three variables z, x, and y will "evidently" be an equation of some surface,[226] in which case the conditional equation for the integrability of equation (7.3.5.7) must hold. He also stated that, conversely, if equation (7.3.5.7) is an integrable equation, then the differential $A(x, y)dx + B(x, y)dy$ must be a complete differential, since A and B do not depend upon z.[227] Clairaut then determined the conditional equation for the integrability of equation (7.3.5.7). If we replace ω by A and θ by B in equation (7.3.5.5), the equation

$$\frac{dB}{dx} + A\frac{dB}{dz} = \frac{dA}{dy} + B\frac{dA}{dz} \qquad (7.3.5.8)$$

results, which is the conditional equation for the integrability of equation (7.3.5.7). But dA/dz and dB/dz are both equal to zero, since A and B are independent of z. Therefore equation (7.3.5.8) reduces to the equality

$$\frac{dA}{dy} = \frac{dB}{dx}, \qquad (7.3.5.8')$$

which is the same as condition (7.3.4.6). Clairaut had first found and demonstrated condition (7.3.4.6) in his first Paris Academy *mémoire* on integral calculus by using the equality (7.3.4.7) of mixed, second-order partial differential coefficients. Later, in his second Paris Academy *mémoire* on integral calculus, he established condition (7.3.4.6) a second time by inverting differentiation and integration ((7.3.1.1)).

During Gaspard Monge's late-eighteenth-century revival of analytic and differential geometry in Paris, following a long period during which these branches of mathematics declined in importance there, Clairaut's geometric demonstrations which appear in his second Paris Academy *mémoire* on integral calculus were seen in retrospect with admiration. For example, Sylvestre-François Lacroix, who had an important part in the revival where pedagogy is concerned, found the second of Clairaut's two published Paris Academy *mémoires* on integral calculus to be "especially interesting, because the equation

$$\frac{dA}{dy} = \frac{dB}{dx}$$

is derived in it by considering surfaces."[228]

If Clairaut appears to have outdone and bested Fontaine in developing a method of integrating in finite terms or reducing to quadratures first-order ordinary "differential" equations of first degree, the fact remains that Fontaine introduced the partial differential calculus in Paris, which is the mathematics that Clairaut utilized to put together his particular method. In other words, Fontaine handed Clairaut the mathematical tools that Clairaut used to invent his method. Clairaut more or less even tacitly acknowledged that this was true, when he wrote in his first Paris Academy *mémoire* on integral calculus that, were it not for the fact that his own method of integrating in finite terms or reducing to quadratures first-order ordinary "differential" equations of first degree was "much simpler" and "more natural" than Fontaine's method for doing the same thing, he "could be accused of having disguised his [Fontaine's] ideas."[229] Here Clairaut implicitly admitted that Fontaine had made innovations in mathematics. Moreover, we shall discover in Chapter 9 that Fontaine did this precisely at a time when Clairaut needed it done most.

Clairaut's father Jean-Baptiste Clairaut taught mathematics in Paris for a living. In the Paris Academy's *Histoire* for 1725, Fontenelle commended him for practicing his profession "with success."[230] A corresponding member of the Berlin Academy, Clairaut's father published in that academy's journal, entitled the *Miscellanea Berolinensia* at the time, what one scholar considers to be an elegant and concise treatment of the general theory of plane flexible lines.[231] Jean-Baptiste began to introduce his two gifted sons to mathematics when they were still very young. Unfortunately, Clairaut's younger brother died at the age of sixteen. Much later, in correcting Euler's mistaken impression that he had written his father's article on a new kind of tractrix, which was published in the *Miscellanea Berolinensia*, Clairaut made it clear that he owed a great deal to his father, who had taught him much.[232]

Clairaut made his debut at the Paris Academy one month before his thirteenth birthday. When he finished his book on skew curves in 1729 at the age of sixteen, which was published in 1731, the members of the Academy proposed that he be admitted to their society, despite the rules that required a minimum age of twenty for members. It is true that Clairaut had completed a man's work that no mathematician in the Academy at the time could have done. The king finally authorized Clairaut's election in 1731, a few months after Clairaut had reached the age of eighteen. No one had joined the Academy at the age of eighteen before. Since then the Academy has not broken its rules concerning the minimum age of members for anyone. Clairaut was the Academy's pet of his day.

As we clearly saw in Chapter 5, during the first three and a half decades of the eighteenth century the Paris Academy did not permit its members to criticize in their publications the work of other members of the Academy. And when such criticisms did get into print, the Academy did not condone them. Ultimately, however, the Academy allowed Clairaut, the Academy's darling, to criticize in his publications the work of its other members, beginning with Fontaine. Thus during the disagreements between Clairaut and Fontaine which occurred in the mid-1730s and which I first brought up in Chapter 5, Clairaut succeeded in having printed in his published Paris Academy *mémoires* of that period his criticisms of Fontaine's work. We recall that while Clairaut and Fontaine argued, Fontaine read a paper before the Paris Academy in which he called the attention of the members of the Academy to "oversights" in a paper that Clairaut had read before the Academy earlier. But all of Fontaine's references to "oversights" were left out of the version of Fontaine's reply to Clairaut's paper which was published in the Paris Academy's *Mémoires*, while the version of Clairaut's paper published in the Paris Academy's *Mémoires* includes no acknowledgment of Fontaine's corrections, either. Neither published *mémoire* includes any reference whatever to the other party in the controversy. The Academy, as usual, hushed up the affair.

Now, this particular dispute did not end there; it dragged on. In the next round Clairaut pointed out specifically to the members of the Academy what he considered to be defects in Fontaine's analysis. Only this time the Academy allowed Clairaut's remarks to be printed in the relevant Paris Academy *mémoire*.[233] Afterward Clairaut left Paris for Lapland. While he was away, Fontaine answered Clairaut's objections. He defended himself without mentioning Clairaut.[234] After Clairaut returned to Paris, Fontaine's latest response to Clairaut's complaints did not immediately come to Clairaut's knowledge. The news was delayed several months. (When Clairaut returned to Paris from Lapland, we remember that Fontaine left the city for his estate in the country-side.) When his colleague's reply finally did come to his attention, Clairaut struck again. He wrote another paper that criticized Fontaine's approach to the problem in question, and the paper was published in the Paris Academy's *Mémoires* with the criticisms included.[235] In other words, except for the first two published Paris Academy *mémoires* written by the two mathematicians concerning the particular problem in question at the time, all of Clairaut's subsequent, published replies to Fontaine which dealt with the problem at issue had an argumentative form, unlike Fontaine's published answers to Clairaut's faultfinding, which do not mention Clairaut. Clairaut and Fontaine conducted themselves in opposite ways during this affair.

On still another occasion during the mid-1730s, Clairaut referred in another of his published Paris Academy *mémoires* of the period to a dispute between himself and Fontaine about tractrices and to his having refuted Fontaine's claim that the path of a certain constrained motion was a tractrix.[236]

It is perhaps not wholly fortuitous that Clairaut began to harass Fontaine in his published Paris Academy *mémoires after* Fontaine had read his great paper on the problem of finding tautochrones in resistant media before the Paris Academy in February of 1734. We saw earlier in this chapter that both Euler and D'Alembert admired Fontaine's solution to that problem and praised him for having solved the problem. But Clairaut did not publicly congratulate Fontaine for having solved the problem.

We need to recall how Fontaine and Clairaut behaved during their arguments in the mid-1730s in order to understand what happened when Clairaut took up the partial differential calculus that Fontaine had introduced in Paris. Clairaut presented before the Paris Academy his first paper on integrating in finite terms or reducing to quadratures first-order ordinary "differential" equations of first degree in March of 1739 as more than just an alternative approach to the problem. He challenged Fontaine. He rejected the way that Fontaine had treated the problem. In the preface to his first published Paris Academy *mémoire* on integral calculus, Clairaut stated that his own method of integrating in finite terms or reducing to quadratures first-order ordinary "differential" equations of first degree was "as general as Monsieur Fontaine's" method for doing the same thing, as well as "much simpler in theory and easier in practice" than Fontaine's method. In the preface he also disapproved of what he called "the length and difficulty of Monsieur Fontaine's calculations."[237] I have already mentioned that Clairaut stated in the text of this *mémoire* that he judged his own method to be "much simpler" and "more natural" than Fontaine's method and that Fontaine's homogeneous function theorem in three variables was "quite unnecessary" for integrating in finite terms or reducing to quadratures first-order ordinary "differential" equations of first degree. Moreover, Clairaut also declared in the text of this same *mémoire* that "the route that Monsieur Fontaine took" in applying the homogeneous function theorem in three variables x, y, and p to integrate in finite terms or reduce to quadratures integrable first-order "differential" equations

$$dx + \alpha(x, y, p)dy + \pi(x, y, p)dp = 0 \qquad (7.3.3.97)$$

of first degree in three variables x, y, and p, when $\alpha(x, y, p)$ and $\pi(x, y, p)$ in equation (7.3.3.97) are homogeneous expressions of degree zero in x, y, and p, was "so long," whereas he discovered a way to use the homogeneous function theorem in three or more variables x, y, p, q, etc. to integrate in finite terms or reduce to quadratures integrable first-order "differential" equations of first degree

$$dx + \alpha dy + \pi dp + \lambda dq + \text{etc.} = 0 \qquad (7.3.4.12)$$

in the variables x, y, p, q, etc., where α, π, λ, etc. are homogeneous expressions of degree zero in x, y, p, q, etc., which was "extremely simple."[238]

There is good evidence that the argumentative tone of Clairaut's first Paris Academy *mémoire* on integral calculus which Clairaut had designed to demonstrate what he regarded as the superiority of his own use of the partial differential calculus to integrate in finite terms or reduce to quadratures first-order ordinary "differential" equations of first degree to Fontaine's application of the partial differential calculus to do the same thing annoyed Fontaine. As usual, however, Fontaine did not defend himself. Toward the end of June 1741 Fontaine wrote Mairan from his estate outside Paris to request that the Clairaut's first paper on the integral calculus, which Clairaut had decided by September of 1740 to have published, "be printed in its entirety and word for word as it is without removing from or changing anything in the preface or any other thing, conforming to the agreement that the Academy just made." In particular then, Fontaine *asked* that Clairaut's whole contentious, argumentative preface to the paper be published without change. Fontaine also asked for a copy of Clairaut's second paper on the integral calculus, which Clairaut had recently finished reading before the Paris Academy. Fontaine told Mairan that he needed to see the paper in order to help him decide what to do ultimately with the *pli cacheté* that he had just sent to the Academy.[239] After Mairan received Fontaine's letter, Clairaut read the preface to his first paper on integral calculus before the Academy, at the beginning of July 1741, in order to allow the members of the Academy to decide whether it was appropriate to publish the preface in its current form as Fontaine had asked. The members of the Academy concluded that the preface should be published unrevised.[240] In a footnote that he added to his first Paris Academy *mémoire* on integral calculus before it was published in 1741, Clairaut stated that Fontaine's paper on integral calculus should have appeared before his. But Fontaine had not submitted his paper for publication. Thus the Academy decided to publish Clairaut's paper first.[241] Clairaut's second Paris Academy *mémoire* on integral calculus did not have contentious parts.

Fontaine, we remember, sent his *pli cacheté* to the Paris Academy in mid-June of 1741, shortly after Clairaut had read his second paper on integral calculus before the Paris Academy. Fontaine returned to the Paris Academy at the end of November 1741, following one of his typically long absences, at which time he asked that his *pli cacheté* be opened. It was given to the reviewers D'Alembert and De Gua who examined it.[242] The two reviewers read their report before the Academy in mid January of 1742, on the very same day that the Academy's *Mémoires* for 1739, which included Clairaut's first *mémoire* on integral calculus, were distributed to the members of the Academy.[243] The timing here cannot be accidental. The Academy deliberately synchronized the two events. The two reviewers found that the *pli cacheté* contained a treatise on integral calculus and that Fontaine had devoted a substantial portion of the treatise to the problem of integrating "differential" equations of *any* order, homogeneous or inhomogeneous, in any number of variables. The treatise included the version of Fontaine's method of November 1738 simplified by Clairaut in March of 1739 for integrating in finite terms or reducing to quadratures first-order ordinary "differential" equations of first degree. It also included Clairaut's generalization of this same method, which Clairaut had used in March of 1739 in order to integrate in finite terms or reduce to quadratures integrable first-order "differential" equations of first degree

$$dx + \alpha dy + \pi dp + \lambda dq + \text{etc.} = 0 \qquad (7.3.4.12)$$

in three or more variables x, y, p, q, etc., where α, π, λ, etc. are homogeneous expressions of degree zero in the variables x, y, p, q, etc. Clairaut had already published both the simplified method and its generalization in his first Paris Academy *mémoire* on integral calculus. Most important, however, the two reviewers of Fontaine's treatise described methods for integrating "differential" equations of second, third, and fourth orders, in which they said that Fontaine made ingenious use of a generalized homogeneous function theorem that could be applied to expressions that are homogeneous in finite *and infinitesimal* variables. This led Fontaine to the "indeterminacy problem" for higher-order ordinary "differential" equations. I discussed this problem in Section 3.3 of this chapter. I will not discuss the two reviewers' account of Fontaine's treatise in any detail, because first-order ordinary "differential" equations of first degree and integrable first-order "differential" equations of first degree in three variables are the only kinds of "differential" equations that we need to know in order to understand the part of the story that remains to be told in these pages. The two reviewers did recommend that Fontaine's treatise be published with the Academy's approval.

But Clairaut's printed criticisms of Fontaine's first method of integrating in finite terms or reducing to quadratures first-order ordinary "differential" equations of first degree not only failed as usual to rouse the reticent mathematician to defend himself at Clairaut's expense, but in addition they totally silenced him this time – for the next twenty-two years! Fontaine published *nothing* that concerned the integral calculus until his complete works, which appeared in 1764. When Fontaine did publish for the first time, in his complete works of 1764, his method of November 1738 for integrating in finite terms or reducing to quadratures first-order order ordinary "differential" equations of first degree, he omitted equation (7.3.3.122) and substituted equation (7.3.4.13) instead.[244] [In fact, as I just implied, the report written by D'Alembert and De Gua on Fontaine's treatise that the *pli cacheté* contained indicates that Fontaine had already replaced equation (7.3.3.122) by equation (7.3.4.13) in that treatise.] But this is Fontaine's method simplified by Clairaut in March of 1739. However, in his complete works of 1764 Fontaine neglected to mention that Clairaut had modified Fontaine's method of November 1738 for integrating in finite terms or reducing to quadratures first-order order ordinary "differential" equations of first degree, which Fontaine now published for the first time, and that Clairaut had found equation (7.3.4.13) while simplifying Fontaine's method, not he. Fontaine also published in his complete works of 1764 Clairaut's generalization of this method, which Fontaine used to integrate in finite terms or reduce to quadratures integrable first-order "differential" equations of first degree

$$dx + \alpha dy + \pi dp + \lambda dq + \text{etc.} = 0 \qquad (7.3.4.12)$$

in three or more variables x, y, p, q, etc., where α, π, λ, etc. are homogeneous expressions of degree zero in the variables x, y, p, q, etc. Again, however, Fontaine failed to mention in his complete works of 1764 that it was Clairaut who had generalized the method, not he. These "oversights" make Fontaine's complete works of 1764 virtually worthless for determining how Fontaine first integrated in finite terms or reduced to quadratures first-order ordinary "differential" equations of first degree in November of 1738 or, for that matter, for reconstructing what Fontaine's treatise contained in the *pli cacheté* included.

While Fontaine's method of November 1738 for integrating in finite terms or reducing to quadratures first-order ordinary "differential" equations of first degree which appears in the Paris Academy's unpublished proceedings was well enough known at the time by those Parisian colleagues to whom it mattered, like Clairaut and D'Alembert, Fontaine has paid dearly for his silence where history is concerned. I shall say more about that in the epilogue in Chapter 10 which ends this story.

Why did Fontaine leave Paris in September of 1737 for his estate near Compiègne and then only return to the city more than a year later in order to communicate new research? Did he fear Clairaut, who had just returned to Paris from Lapland? Did he leave Paris in order to avoid further controversies with Clairaut, who always seemed to have the last word? This was neither the first nor the last time that Fontaine sought the privacy of his estate. In the 1740s and 1750s, Fontaine missed most of the Paris Academy's sessions during a period that lasted thirteen years! He only returned to Paris during this period when business at the Paris Academy concerned him personally. He thereby violated the Academy's residence requirement for pensioners, the grade to which Fontaine was promoted in 1742. What else should one expect of an individual quoted as having said: "one discovery was better than ten years of assiduity at the Academy."[245] On one occasion, while the Abbé Nollet jabbered endlessly about the prices of different commodities during a session of the Paris Academy which Fontaine did attend, the mathematician sniggered caustically and derisively: "this man knows the price of everything – except time."[246] Where the partial differential calculus was concerned, Fontaine very likely made up his mind that there wasn't time to waste! Fontaine's most reliable biographer tells us that Fontaine "would have nothing to do with any intrigues" ("*Etranger à toute brigue,...*").[247] In Fontaine's day plots, schemes, and conspiracies were rampant among the members of the Paris Academy. Hence it seems reasonable that someone like Fontaine would try to avoid the Paris Academy as much as possible.

Fontaine grew up in the provinces Dauphiné and Vivarais in the South of France (in what are today the *départements* Drôme and Ardèche). He came from a wealthy family, and this enabled him to buy the estate near Compiègne, northeast of Paris. His first sojourn in Paris, his meeting Castel during that first stay in Paris, and his reading Fontenelle's book during that first visit to Paris had all whetted his appetite for mathematics. After he returned home his older brother died. He inherited the family estate as a result, which he then sold in order to buy the estate outside Paris. There he could farm and work on mathematics in seclusion. But he originally bought an estate in the environs of Paris so that he could get to know the mathematicians in Paris. Fontaine entered the Paris Academy as assistant member in May of 1733, nearly two years after Clairaut had joined the Academy as assistant member. Fontaine was then twenty-eight years old; Clairaut had just turned twenty. Although he was eight years younger than Fontaine, Clairaut was far more worldly. A native Parisian, he enjoyed city life. As the darling of the Paris Academy, he took delight in his celebrity at that institution and thoroughly basked in the favour of its members. But the Academy's pet, who could only have been completely spoiled after having been pampered by the members of the Academy ever since he was twelve years old, may have also found Fontaine to be a far more able and imaginative mathematician and original thinker than he had anticipated and more than he had bargained for, which may partly explain why Clairaut began to harass Fontaine in his published Paris Academy *mémoires*. Fontaine, moreover, thought algebraically for the most part, while Clairaut mainly thought geometrically, which partly explains why the two mathematicians never seemed to see eye to eye regarding mathematics. They probably differed from each other in as many ways as, say, Monge and Cauchy or Monge and Lagrange did. As we shall see in Chapter 9, one thing is certain as far as the rest of the story to come goes: Clairaut made further progress thanks in large part to Fontaine.

8

Interlude II: The Paris Academy's contest on the tides (1740)

8.1. Daniel Bernoulli's essay

In the spring of 1738, the Paris Academy of Sciences announced the subject of its contest for the year 1740. Competitors had until the fall of 1739 to submit papers. The contestants were asked to explain the causes of the tides. These competitions, which the Academy instituted in 1720, were held annually. They helped the Paris Academy to raise its standards. Only individuals who did not belong to the Academy could compete for prizes. As a result, foreigners in France and abroad introduced many ideas into France which were new to the French. The Treaty of Utrecht (1713) and Louis XIV's death (1715) fostered a climate that favored this kind of international exchange. The Paris Academy's contests established in 1720 helped to sustain the revival of mathematics and mathematical science in Paris which began in 1730.

In the spring of 1740, the Academy announced four winners of the competition on the causes of the tides. One of them, the Jesuit P. Antoine Cavalleri, professor of mathematics at the University of Cahors, had recently won the Bordeaux Academy's prize of 1738 for his essay *Sur la cause de la diaphaneité, et de l'opacité des corps*, as well as the same Academy's prize of 1739 for his essay *Sur la cause de la chaleur et de la froideur des eaux minerales.* Cavalleri proposed a Cartesian vortex theory of the tides in his essay for the Paris Academy's prize of 1740. Late in 1743, Maupertuis recounted the following anecdote to his friend Johann II Bernoulli, Johann I Bernoulli's third son. Maupertuis told the younger Bernoulli that in order to pacify Réaumur, the diehard in favor of Cartesianism among the five members of the Academy who judged the essays submitted to the Academy for the contest of 1740, a Cartesian essay was randomly selected.[1] Given the fact that Cavalleri was momentarily conspicuous as a result of his recently having won two other contests, it was natural for the judges to choose his essay. (Another of the five judges, the Malebranchist Mairan, himself a former member of the Bordeaux Academy and a winner of three of that academy's prizes, probably exerted some influence here.) The other three winners of the competition of 1740 were distinguished foreign scholars: Leonhard Euler, then in St. Petersburg; Daniel Bernoulli in Basel; and Colin MacLaurin in Edinburgh. In one way or another, these three all took Newton's theories of the tides which appeared in the *Principia* as their starting points.[2]

Although Euler was designated *accessit* (runner-up) in the Paris Academy's contest of 1727, which Bouguer won, Euler did not win the first of his twelve Paris Academy prizes until 1738, the year that he first corresponded with Maupertuis. Again we see evidence of the delayed influence that Euler's scientific and mathematical work had in Paris.

Clairaut also served as one of the five judges of the contest of 1740. The essays by two of the other winners, Bernoulli and MacLaurin, influenced Clairaut's subsequent research on the theory of the earth's shape.

Daniel Bernoulli won the first of his ten Paris Academy prizes in 1725. He first met Clairaut during a short visit to Paris, in the fall of 1733, while returning home to Basel after having spent seven years in the St. Petersburg Academy. The French prodigy, who was thirteen years younger than Bernoulli, evidently impressed the Swiss physicist from the very beginning. During his visit to Paris, Bernoulli wrote to Joseph-Nicolas Delisle, the Paris Academy astronomer who had left Paris for the St. Petersburg Academy in 1725. In the letter Bernoulli announced that Fontenelle would soon retire as the Paris Academy's *secrétaire perpétuel* and that Mairan would replace him. Bernoulli thought it likely that Clairaut, then twenty years old, would fill the position of pensioner vacated by Mairan.[3]

(In fact, Mairan did not succeed Fontenelle until 1741, and Clairaut was promoted to pensioner in 1738 and not before spending a period as an associate member of the Academy. Although Bouguer did join the Academy in 1731 directly as associate member, slightly less than two months after Clairaut entered as assistant member, Bouguer was also fifteen years older than Clairaut. It seems unlikely that in 1733 the Academy would have advanced to pensioner an underage member who had only joined the society two years earlier. By the middle of the eighteenth century the Academy had set up a system of promotions based on seniority. The way that the system worked was tacitly understood; the laws that governed it were never written into the Academy's statutes. But seniority clearly did become the de facto basis for promotions within the Academy from about midcentury.[4])

When Clairaut accompanied Maupertuis to Basel to see Johann I Bernoulli in the fall of 1734, he and Daniel Bernoulli had a better opportunity to get to know each other. The two became close friends. Shortly after Clairaut left Basel to return to Paris, Daniel Bernoulli wrote Euler to recommend that he engage Clairaut as a member of the St. Petersburg Academy.[5] The subsequent correspondence between Clairaut and Daniel Bernoulli reveals an intimacy that Clairaut enjoyed with no other colleague.[6] Unfortunately, the correspondence for the 1730s is lost; only that from later years survives.[7] Nevertheless, from

Euler's correspondence with Daniel Bernoulli and with others, including Clairaut, we know that Bernoulli and Clairaut corresponded during the 1730s. For example, early in March of 1739, Bernoulli wrote Euler to inform him of Fontaine's method of November 1738 for integrating in finite terms or reducing to quadratures first-order ordinary "differential" equations of first degree, which Bernoulli said that Clairaut found to be very important but not general.[8] Clairaut obviously had told Bernoulli about Fontaine's work. We also recall that Clairaut sent Bernoulli a copy of his first Paris Academy *mémoire* on integral calculus while it was still in manuscript and that afterward Bernoulli furnished Euler with a copy of the manuscript, which paved the way for Clairaut's sending his first letter to Euler, in September of 1740.

Bernoulli influenced the work that Clairaut did in a number of areas of mechanics. The way that Bernoulli's essay of 1740 on the tides subsequently affected Clairaut's theory of the earth's shape is one example, as we shall see in the next chapter. In his colossal study of theories of attraction and the earth's shape, which is really a chronicle with textual glosses, not a history, Isaac Todhunter paid no attention to Bernoulli's essay. Although Todhunter acknowledged that the essay had a very important part in the history of tidal theories, he claimed that Bernoulli's essay had nothing whatever to do with the subject of his book. (He even judged Poleni's letter that I discussed in Chapter 3 to be "of little importance.") Todhunter maintained that Clairaut had already demonstrated in his first paper on the theory of the earth's shape published in the *Philosophical Transactions* all of the results that Bernoulli derived concerning attractions of homogeneous bodies shaped like ellipsoids of revolution which attract according to the universal inverse-square law and which appear in Bernoulli's essay of 1740 on the tides.[9] As we shall see in a moment, this is not quite true. More important however, in dismissing Bernoulli's essay for being irrelevant to the subject of his book, which Todhunter did as a result of having defined the scope of his study too narrowly, Todhunter overlooked critical, indirect connections.

Bernoulli developed a static-equilibrium theory of the tides in his essay, using Newton's universal inverse-square law of attraction. A solid, spherical earth composed of concentric, spherical, individually homogeneous strata which attract according to the universal inverse-square law is surrounded by an outer layer of homogeneous fluid which also attracts in accordance with the universal inverse-square law and which is concentric with the earth. The fluid layer is distorted into a stationary (or quasistatic) mass whose outer surface is an ellipsoid of revolution which is elongated at its poles and concentric with the earth. In his theory Bernoulli took into account the three influences that would cause the earth, were it in a state that would allow it to be deformed, to elongate along the two axes that join its center to the centers of the sun and moon. These three influences are as follows. First, the sun and moon attract the earth according to the universal inverse-square law. Secondly, the earth revolves around the sun and around the center of gravity of the system consisting of the earth and moon, which generate centrifugal forces of revolution. Finally, the earth and its seas and oceans attract themselves in accordance with the universal inverse-square law. (Bernoulli neglected the influence that the earth's rotation around its polar axis would have upon the earth if the earth were pliant – namely, to cause its polar axis to shorten as a result of the centrifugal force of rotation.) In effect, the three influences just specified cause the earth's oceans and seas to lengthen along the two axes just mentioned. As a consequence of the earth's diurnal motion and the proper motions of the sun and moon, these two axes continually change their positions. According to Bernoulli, the ebb and flow of the tides result from these changes.

(In fact, we shall have to wait until the next chapter to see that Bernoulli's theory really *can be a static-equilibrium* theory of the tides. That is, we shall have to wait until Chapter 9 to learn why the outer fluid layer truly can be a *homogeneous figure of equilibrium* in Bernoulli's theory. The reason is that, as we shall observe below, Bernoulli essentially applied the principle of balanced columns to determine the heights of tides. But we recall that Bouguer had demonstrated in his Paris Academy *mémoire* of 1734 that it was not enough for a homogeneous figure simply to satisfy the principle of balanced columns for it to be a figure of equilibrium. I say that the outer, fluid layer in Bernoulli's theory "can be" a homogeneous figure of equilibrium instead of "is" a homogeneous figure of equilibrium because, as I mentioned in Chapters 1 and 6, the part of the story told in these pages which has to do with figures of equilibrium mainly involves the discovery and use of various *necessary* conditions that must be fulfilled in order that a fluid mass be a figure of equilibrium. The story does not deal at all with the problem of determining conditions that *suffice* for figures of equilibrium to exist. We shall also discover when I turn to Clairaut's theories of homogeneous and stratified figures of equilibrium in Chapter 9 that Bernoulli made errors. Although Clairaut did not actually prove that Bernoulli's calculations turn out not to be valid when the densities of the individually homogeneous strata are not all the same – namely, not all equal to the density of the homogeneous outer, fluid layer – we shall see in Chapter 9 that what Clairaut did succeed in demonstrating suggests that this is true.)

Bernoulli began by considering the earth and a single heavenly body that attracts it. He took the earth's solid part to be a spherical figure made up of an infinite number of concentric, spherical, solid, individually homogeneous strata that are all spheres, which means that the earth's strata are surfaces, hence are "infinitesimally" thin. He assumed the surface of the earth's solid part to be covered with a thin layer of homogeneous fluid, whose depth was much smaller than the earth's radius. If no forces acted, he supposed that the shape of the outer surface of this outer, fluid layer would be spherical and that the layer would be uniformly deep. He then hypothesized that the attraction of the earth and the waters at the earth's surface for themselves, the heavenly body's attraction for the earth, and the centrifugal force of the earth's revolution around the heavenly

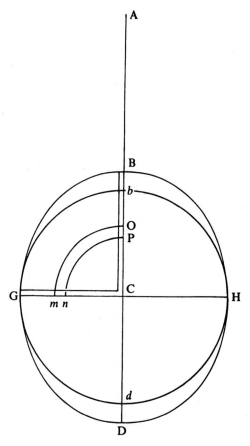

Figure 30. Daniel Bernoulli's figure used for calculating the height Bb of the earth's neap tide in the presence of a single, attracting body at A (Daniel Bernoulli 1740: figure on p. 157).

body would distort the shape of the outer surface of the earth's outer, fluid layer, causing it to change from a figure shaped like a sphere into a nearly spherical figure shaped like an ellipsoid of revolution which is elongated along the axis joining the earth's center to the center of the heavenly body and whose center is situated at the common center of the strata (see Figure 30). To find the height of the tides Bernoulli balanced columns from the center of this elongated figure to the surface of the figure along the axes of the figure. The distance *Bb* would be the height of the neap tide in the presence of a single, attracting heavenly body. Bernoulli noted that if the earth's solid part were taken to be homogeneous with the same density as the earth's seas and oceans, as Newton had assumed in the equilibrium theory of the tides which he presented in the *Principia*, then his formula for *Bb* gave "exactly the height in feet, inches, and lines which Monsieur

Newton indicated without showing the calculation of this height, or at least without making it [the calculation, meaning the calculation that leads to the formula for *Bb*] understandable, I won't say to everyone, but only to those who would like to make the effort necessary to go deeper into it" in the corollary to Proposition 36 of Book III of the *Principia*.[10]

In order to balance columns from the center of the elongated figure to the surface of the figure along the axes of the figure so as to determine *Bb*, Bernoulli had to calculate the magnitude of the attraction at a pole and at the equator of a homogeneous figure shaped like an ellipsoid of revolution which is elongated at its poles, whose ellipticity is infinitesimal, and which attracts in accordance with the universal inverse-square law. Although he and Clairaut worked on different problems at this time, the ways that they approached their problems closely resemble each other. Bernoulli's theory of the tides only held at the surfaces of figures whose outer, fluid layers had outer surfaces that were ellipsoids of revolution elongated at their poles and whose ellipticities were infinitesimal. Bernoulli consequently found that calculations only valid to first order sufficed.

Clairaut's method of calculating to terms of first order the magnitude of the attraction at a pole of a homogeneous figure shaped like an ellipsoid of revolution which is flattened at its poles, whose ellipticity is infinitesimal, whose density is 1, and which attracts according to the universal inverse-square law and Bernoulli's way of calculating to terms of first order the magnitude of the attraction at the equator of a homogeneous figure shaped like an ellipsoid of revolution which is elongated at its poles, whose ellipticity is infinitesimal, whose density is *u*, and which attracts according to the universal inverse-square law are very similar. In order to do their calculations, Clairaut and Bernoulli both partitioned their figures into figures shaped like spheres covered by caps shaped liked meniscuses situated directly opposite each other (see Figures 16 and 31).[11] The calculations are not identical, because in Clairaut's figure shaped like an ellipsoid of revolution which is flattened at its poles the polar axis *Aa* of the figure is an axis of symmetry of the figure. Hence the boundary of the cross section of the figure through *P* perpendicular to *Aa* is a *circle* through *P* with radius *PN*. In Bernoulli's figure shaped like an ellipsoid of revolution which is elongated at its poles the polar axis *BD* of the figure is an axis of symmetry of the figure. But the equatorial axis *GH* of Bernoulli's figure is *not* an axis of symmetry of the figure. The boundary of the cross section of his figure through *F* perpendicular to *GH* is a nearly circular *ellipse*, not a circle.

In his first paper on the theory of the earth's shape published in the *Philosophical Transactions*, Clairaut in fact did *not* determine the approximate value of the magnitude of the attraction at the *equator* of a homogeneous figure shaped like an ellipsoid of revolution which is elongated at its poles, whose ellipticity is infinitesimal, whose density is 1, and which attracts according to the universal inverse-square law. Thus Todhunter overstated what Clairaut had demonstrated in that paper. Contrary to what Todhunter declared in his book,

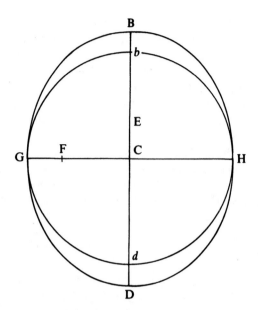

Figure 31. Daniel Bernoulli's figure used for determining the magnitude of the attraction at the equator of a homogeneous elongated ellipsoid of revolution which attracts according to the universal inverse-square law and whose ellipticity is infinitesimal (Daniel Bernoulli 1740: figure on p. 145).

Daniel Bernoulli was in fact the first person to demonstrate clearly how to do this calculation.

We also recall that in Chapter 6 I showed how Clairaut probably found the approximate value of the magnitude of the attraction at a pole of a homogeneous figure shaped like an ellipsoid of revolution which is elongated at its poles, whose ellipticity is infinitesimal, whose density is 1, and which attracts in accordance with the universal inverse-square law as a "corollary" of his method for calculating the approximate value of the magnitude of the attraction at a pole of a homogeneous figure shaped like an ellipsoid of revolution which is flattened at its poles, whose ellipticity is infinitesimal, whose density is 1, and which attracts according to the universal inverse-square law. We remember that Clairaut did not explain how the first calculation followed as a "corollary" of the second calculation. Bernoulli determined the approximate value of the magnitude of the attraction at a pole of a homogeneous figure shaped like an ellipsoid of revolution which is elongated at its poles, whose ellipticity is infinitesimal, whose density is u, and which attracts according to the universal inverse-square law (see Figure 32) more directly, by using Corollary 1 to Proposition 90 of Book I of the *Principia*. Bernoulli let $GC = b$ and $BC = b + \beta$. He tacitly defined the ellipticity of his homogeneous figure shaped like an elongated ellipsoid of revolution to be

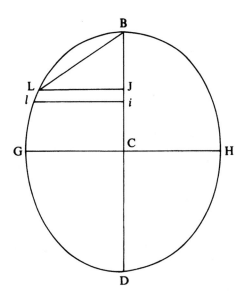

Figure 32. Daniel Bernoulli's figure used for determining the magnitude of the attraction at a pole of a homogeneous elongated ellipsoid of revolution which attracts according to the universal inverse-square law and whose ellipticity is infinitesimal (Daniel Bernoulli 1740: figure on p. 143).

$(GC - BC)/GC = -\beta/b$.[12] We note, however, that according to Clairaut's definition of ellipticity,

$$\frac{GC - BC}{BC} = -\frac{\beta}{b + \beta} = -\frac{\beta}{b}\left(\frac{1}{1 + (\beta/b)}\right)$$

is the ellipticity of the elongated figure. But $(\beta/b)^2 \ll \beta/b$, because the figure is nearly spherical. Hence

$$-\frac{\beta}{b}\left(\frac{1}{1 + (\beta/b)}\right) \simeq -\frac{\beta}{b}\left(1 - \frac{\beta}{b}\right) \simeq -\frac{\beta}{b}.$$

In other words, for nearly spherical figures shaped like ellipsoids of revolution, Clairaut's and Bernoulli's definitions of ellipticity cannot be distinguished up to terms of first order. Using Corollary 1 to Proposition 90 of Book I of the *Principia*, Bernoulli determined

$$\frac{2}{3}nub + \frac{2}{15}\left(\frac{\beta}{b}\right)nub, \tag{8.1.1}$$

where $n = 2\pi$, to be the approximate value of the magnitude of the attraction at a pole of a homogeneous figure shaped like an ellipsoid of revolution which is elongated at its poles, whose ellipticity is infinitesimal, whose density is u, and which attracts according to the universal inverse-square law. In Chapter 6 I stated that expression (8.1.1) agrees with the one found by applying Clairaut's "corollary." [This is easily seen by simply letting $u = 1$ in expression (8.1.1) and by replacing c by n, r by b, and a by β/b in expression (6.1.25''').]

In Chapter 1 we saw that Newton published in the *Principia* the value 501/500 of the ratio of the approximate values of the magnitudes of the attraction at the equator and at a pole of a homogeneous figure shaped like an ellipsoid of revolution flattened at its poles whose infinitesimal ellipticity is $\frac{1}{100}$ and which attracts according to the universal inverse-square law. We also remember that Newton did not actually demonstrate in the *Principia* that $\frac{501}{500}$ was the value of the ratio in this particular case. Newton used this numerical value in outlining his theory of the earth's shape in the *Principia*, which I discussed in Chapter 1. Bernoulli formed the ratio of the approximate values of the magnitudes of the attraction which he had calculated at the equator and at a pole of his homogeneous figure shaped like an ellipsoid of revolution which is elongated at its poles, whose ellipticity is infinitesimal, whose density is u, and which attracts according to the universal inverse-square law, and he found

$$1 + \frac{1}{5}\left(\frac{\beta}{b}\right) \tag{8.1.2}$$

to be the expression for this ratio valid to terms of first order. [The expression (8.1.2) for this ratio is obviously independent of the density u, because the density appears as a factor in both the numerator and the denominator of the ratio, and so the two factors cancel each other.] Then Bernoulli evidently substituted $\frac{1}{100}$ for β/b in expression (8.1.2), which gave

$$1 + \frac{1}{5}\left(\frac{1}{100}\right) = \frac{501}{500}$$

as the value of the ratio in this case. And Bernoulli noted that his value of the ratio agreed with the value that Newton gave without proof (that is, the value in Newton's numerical example) on page 380 of the second edition of the *Principia*, which Newton made use of to determine the ratio of the earth's axes (in other words, the earth's ellipticity). (Bernoulli did not mention that his figure and Newton's figure differed from each other. Newton's figure is *flattened* at its poles; Bernoulli's figure is *elongated* at its poles.) As for the absence in the *Principia* of any demonstration of the expression (8.1.2) for this ratio and what reasoning Newton might have used to arrive at the value $\frac{501}{500}$ of this ratio, Bernoulli wrote eloquently: "only he could have answered this question, for only he saw his reasoning clearly. The great man saw through a veil what someone else could only see, and just barely at that, with a microscope."[13] These words written by one of the greatest, if not the greatest, physicist of the eighteenth century confirms the thesis that I argued in Chapter 1. For if Daniel Bernoulli glorified Newton, he testified at the same time to the enormous gaps in the *Principia*.

It seems unlikely that Bernoulli saw either of Clairaut's two papers on the theory of the earth's shape published in the *Philosophical Transactions* before sending his essay on the tides to the Paris Academy, even though the first of Clairaut's two papers appears to have been printed by October of 1738.[14] In his essay on the tides, Bernoulli believed that by taking an ellipsoid of revolution elongated at its poles to be the shape of the outer surface of the homogeneous fluid layer which surrounded the solid, spherical earth, where the columns from the earth's center to the outer surface of the fluid layer are all assumed to balance or weigh the same, he had introduced an hypothesis that could not be demonstrated using Newton's universal inverse-square law of attraction.[15] But Clairaut had proven in his first paper on the theory of the earth's shape published in the *Philosophical Transactions* that a homogeneous figure shaped like a flattened ellipsoid of revolution which attracts according to the universal inverse-square law and whose ellipticity is infinitesimal is a figure whose columns from center to surface can all balance or weigh the same. It would appear that when Bernoulli wrote his essay he did not know that Clairaut had shown this in that paper.

(Bernoulli's figure is shaped like an ellipsoid of revolution elongated at its poles which attracts according to the universal inverse-square law and whose ellipticity is infinitesimal. If the strata that make up the solid part of the figure are all assumed to have the same density as the surrounding fluid layer, then the figure is a homogeneous figure shaped like an elongated ellipsoid of revolution which attracts according to the universal inverse-square law, whose ellipticity is infinitesimal, and whose columns from center to surface all balance or weigh the same. At first sight this appears paradoxical, for, as we shall discover in Chapter 9, if such a figure whose columns from center to surface all balance or weigh the same existed, the figure could be a figure of equilibrium, whereas I mentioned in Chapter 2 that elongated figures of equilibrium which attract according to the universal inverse-square law cannot possibly exist. However, we must remember that Bernoulli's figure not only attracts *itself* but that it is attracted by the sun and moon *as well*. What cannot exist are homogeneous figures elongated at their poles whose ellipticities are infinitesimal and which attract themselves *alone* according to the universal inverse-square law. As I mentioned, it turns out that Bernoulli's calculations are not valid when the strata that make up the solid part of the figure do not have the same density as the fluid layer that surrounds the strata, but I shall not discuss Bernoulli's mistake here. The error can only be understood with the help of Clairaut's theory of stratified figures of equilibrium which attract according to the universal inverse-square law and which I shall treat in Chapter 9. Thus I shall wait until the end of Chapter 9 to talk about Bernoulli's mistake. As I have already said, although Clairaut did not actually prove that Bernoulli's calculations turn out not to be valid when the densitites of the individually homogeneous strata are not all the same – namely, not all equal to the density of the homogeneous outer, fluid layer – we shall see in Chapter 9 that what Clairaut did succeed in demonstrating suggests that this is true.)

In addition, Bernoulli emphasized that his first-order approximations simplified what would otherwise be "extremely tedious calculations" in "pure analysis."[16] Bernoulli spoke as if his approximations had no precedents. Newton did not exhibit his calculations; consequently Bernoulli, like everyone else, had no idea how Newton arrived at his results. But Clairaut had clearly made use of such first-order approximations in his first paper on the theory of the earth's shape published in the *Philosophical Transactions*, and Bernoulli did not call the reader's attention to the approximations that appear in this paper written by his friend Clairaut. This too evidences that Bernoulli did not see either of Clairaut's two papers on the theory of the earth's shape published in the *Philosophical Transactions* before sending his essay on the tides to the Paris Academy.

Conversely, it is difficult to imagine how Bernoulli's calculations, which Bernoulli did independently of Clairaut and yet resembled Clairaut's calculations, could have failed to interest his friend Clairaut. Moreover, unlike the other winners of the Paris Academy's competition of 1740 on the tides, Bernoulli had carried out his theoretical investigation far enough and did enough calculations to allow him to construct tables of tides. As a result, his theory is often regarded as *the equilibrium* theory of the tides.[17] It has lately been judged a failure for not having taken variations from place to place into account.[18] But such a modern criticism tells us nothing whatever about the way that the theory functioned in its own time and the influence that it had at that time. Bernoulli's tables would have interested Clairaut, whose own work on the theory of the earth's shape, after all, involved numerical values – in particular, the value of the earth's infinitesimally small ellipticity.

The fact that Clairaut's and Bernoulli's approaches to two different problems resemble each other may not be entirely accidental. Bernoulli was by this time the Continent's physicist *par excellence*. So far as I know, Clairaut's approximations that appear in his two papers on the theory of the earth's shape published in the *Philosophical Transactions* have no antecedents or precursors in his earlier work. Such approximations certainly do not appear in any of Clairaut's publications before 1734, the year that Clairaut had the opportunity to get to know Bernoulli well. Until 1734, Clairaut had literally been the "geometer's geometer," who worked on the theory of curves, especially skew curves,[19] on isoperimetrical problems,[20] and the like. Only after 1734 do we find Clairaut utilizing approximations to solve problems in mechanics. Nor does it seem likely that he did this under the influence of, say, Johann I Bernoulli or Maupertuis. (We saw in Chapter 5 that Maupertuis seemed unable to make correctly all but the simplest first-order approximations.[21]

I give one example of the subtleties and connections that historians easily overlook when they simply partition the history of scientific ideas into histories of "unit ideas," like "the earth's shape" or "the tides." Todhunter, who did precisely this in writing his numerous histories, was led as a result of this approach to the problem of writing history to make assertions which are either false or altogether misleading, like the two concerning Daniel Bernoulli's essay on the tides which I mentioned previously. When Euler finally saw Clairaut's second paper on the theory of the earth's shape published in the *Philosophical Transactions*, he informed the author that from the theory developed in the paper together with the value $\frac{1}{177}$ for the earth's ellipticity determined by the members of the Lapland expedition, the earth's density could now be inferred to increase from the center of the earth to the surface of the earth "without risk," contrary to Newton's conjecture that the earth's density decreased from the center of the earth to the surface of the earth.[22] But observations of tides had led Daniel Bernoulli to conclude in his essay of 1740 that the earth's density did decrease from the center of the earth to the surface of the earth.[23] Whether Euler had seen Bernoulli's essay by this time, which seems unlikely, or not makes little difference.[24] The point is that Bernoulli's theory of the tides bore, at least indirectly, upon the theory of the earth's shape which Clairaut had been preoccupied with. Both Bernoulli's and Clairaut's theories related observable phenomena (heights of tides in the case of Bernoulli's theory and variations in the magnitude of effective gravity and degrees of latitude with latitude at the earth's surface in the case of Clairaut's theory) to the earth's internal structure. Another scholar has also taken the Paris Academy's contest of 1740 on the tides as evidence of the Academy's interest in equilibrium theories of the earth's shape at this time.[25]

[The tides had already been the subject of the Bordeaux Academy's contest of 1726, which Jacques Alexandre won. Alexandre's essay was entitled *Dissertation sur les causes du flux et reflux de la mer* (Bordeaux: R. Brun, 1726). It was republished with the title *Traité du flux et reflux de la mer* (Paris: Babuty, 1726). However, the Bordeaux Academy's contest comes much too early to explain the Paris Academy's interest in the tides in 1738, the year that the Academy announced the causes of the tides as the subject of its contest of 1740.]

Years later, in the commentaries appended to the French edition of the *Principia*, which Clairaut wrote for his friend the Marquise du Châtelet, who edited the edition, Clairaut praised his old friend, like-minded colleague, and most intimate correspondent Daniel Bernoulli. Clairaut stated how much he admired Bernoulli's essay of 1740 on the tides. He singled it out from the other essays whose authors shared the Paris Academy's prize of 1740 for special commendation. In the editor's introduction to this edition of the *Principia*, the writer noted that "Monsieur Daniel Bernoulli won the [Paris Academy's] contest concerning the tides. His dissertation that was awarded a prize forms the basis of the last section" of Clairaut's appendix. The writer observed as well that Bernoulli's essay was "augmented by various notes and clarifications that the author communicated."[26] Why did Clairaut select Bernoulli's essay on the tides to serve as the foundation for his commentary in the last section of his appendix? Clairaut asserted that he found "more order, clarity, and precision" in it than he did in the other essays whose authors shared the Paris Academy's prize of 1740,[27] not to mention that, by this time, Clairaut's particular taste and style were more like Bernoulli's than like Euler's or Maclaurin's.

Clairaut must have made this retrospective appraisal of Bernoulli's essay, which expresses the deep affection that he had for his close friend in Basel, with wisdom that he acquired as a result of another decade's worth of experiences. For in actuality, MacLaurin's essay of 1740 on the tides had, generally speaking, a greater immediate effect on Clairaut's subsequent research on the theory of the earth's shape than Bernoulli's essay did, as we shall now see.

8.2. Colin MacLaurin's essay and his subsequent correspondence with Clairaut

In the early 1720s MacLaurin visited France, where he spent at least two years (August 1722 to August 1724). During the 1720s he became Mairan's ally in the controversy over the conservation of *vis-viva*. He won the Paris Academy's prize of 1724 for his essay on the motions of "hard" bodies after colliding. In this competition he defeated Johann I Bernoulli among others. As I noted in Chapter 7, Bernoulli's failure to win either this Paris Academy contest or the Paris Academy contest of 1726 concerning the laws of collision of "elastic" bodies doubtlessly marked the decline of Bernoulli's influence in Paris. These losses resulted in large part from the deaths of Bernoulli's contacts among the mathematicians in the Paris Academy. The downward trend finally changed and Bernoulli's situation in Paris improved when Bernoulli befriended Maupertuis in 1729, and the two mathematicians together started the revival of mathematics in Paris which took place during the 1730s. After the Treaty of Utrecht was signed in 1713, which ended the wars between France and her neighbors, and after Louis XIV, whose desire to expand French territory had caused those wars to start, died in 1715, French and British scientists reopened the communication that had all but ceased during the wars. MacLaurin's stay in France was not the first sign of this resumption of contacts, but it does illustrate the renewal of communication. (We recall that Brook Taylor had already visited France in June of 1715. In addition, Montmort, Claude-Joseph Geoffroy, and the Chevalier de Louville went to London to observe a total eclipse of the sun in May of 1715 and to visit the Royal Society.) It is no accident that the reception in Paris of the world system that Newton presented in the *Principia*, not to mention the greater part of the diffusion of Newton's *Opticks* in Paris, did not occur before 1715.[28]

During his stay in France MacLaurin did not meet Clairaut, who was only a youngster at the time. The Scottish mathematician James Stirling first called MacLaurin's attention to Clairaut in 1738. As I mentioned in Chapter 6, Stirling had published a paper in the *Philosophical Transactions*, No. 438, for July–September of 1735 which was similar to Clairaut's first paper on the theory of the earth's shape published in the *Philosophical Transactions*. [Unlike Clairaut, who demonstrated in his paper that the columns from the center to the surface of a homogeneous fluid figure shaped like an ellipsoid of revolution which is flattened at its poles, which attracts according to the universal inverse-square law, and which revolves around its axis of symmetry (its polar axis) "slowly," meaning that the ratio ϕ of the magnitude of the centrifugal force per unit of mass to the magnitude of the attraction at the equator of the figure is infinitesimal ($\phi^2 \ll \phi$), hence whose ellipticity δ is infinitesimal as well ($\delta^2 \ll \delta$), can truly all balance or weigh the same, Stirling showed instead that the principle of the plumb line can hold at all points on the surface of such a figure too. Furthermore, Stirling did not really demonstrate his results; he described them in words.]

Clairaut had not yet seen Stirling's paper when he sent his own manuscript to the Royal Society from Lapland in 1737. Clairaut began his second paper on the theory of the earth's shape published in the *Philosophical Transactions* by apologizing to Stirling for having failed to acknowledge him in his first paper on the theory of the earth's shape published in the *Philosophical Transactions* and by explaining the circumstances that had caused him inadvertently to overlook Stirling's paper.[29]

Late in October of 1738, after Clairaut had sent his second manuscript to the Royal Society, Stirling wrote MacLaurin to say that he had heard recently from Clairaut, who had asked forgiveness for his oversight and for having failed to acknowledge him in his first paper on the theory of the earth's shape published in the *Philosophical Transactions* and who had mentioned as well in his letter having sent a second manuscript to the Royal Society. Stirling told MacLaurin that Clairaut had also asked him for his opinion of his two papers published or to be published in the *Philosophical Transactions*. Stirling said that he had found the first one, which he "barely saw before it was printed," to be "not of low rank," but that he had not yet seen the second one. Stirling asked MacLaurin if he would examine both papers, so that Stirling could comply with Clairaut's request for a judgment.[30]

After having seen a letter now lost from MacLaurin to MacLaurin's old ally Mairan, in which the Edinburgh mathematician talked about Clairaut, Clairaut promptly wrote MacLaurin in February of 1741, hoping to strike up a correspondence with him. In his letter Clairaut praised MacLaurin for his essay of 1740 on the tides. Unlike Daniel Bernoulli, whose essay of 1740 included tables of tides, MacLaurin dealt more in his essay of 1740 with the potential mathematical difficulties that might have to be faced in accounting for the causes of the tides. Like Bernoulli, MacLaurin based his essay on Newton's universal inverse-square law of attraction.

In his letter to MacLaurin Clairaut told MacLaurin that he admired in particular MacLaurin's methods of showing that the principle of the plumb line can hold at all points on the surface of a homogeneous figure shaped like an ellipsoid of revolution flattened at its poles which attracts according to the

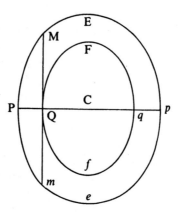

Figure 33. Clairaut's figure used for calculating the magnitude of the attraction at point M on the surface of a homogeneous ellipsoid of revolution which attracts according to the universal inverse-square law (Clairaut 1743: figure on p. 168).

universal inverse-square law and which revolves around its axis of symmetry and that when this principle does hold at all points on the surface of such a figure then the columns from the center of the figure to the surface of the figure all balance or weigh the same as well. For in his essay on the tides, MacLaurin achieved something that Clairaut had not done in his first paper on the theory of the earth's shape published in the *Philosophical Transactions*. Namely, MacLaurin proved his results geometrically without making any approximations. Hence his demonstrations held true for homogeneous figures shaped like ellipsoids of revolution flattened at their poles which attract in accordance with the universal inverse-square law, which revolve around their axes of symmetry, and whose ellipticities are *finite*. Clairaut instead had made first-order approximations which only held for homogeneous figures shaped like ellipsoids of revolution flattened at their poles which attract according to the universal inverse-square law, which revolve around their axes of symmetry, and whose ellipticities are infinitesimal.

Specifically, MacLaurin showed that the following is true (see Figures 33 and 34.) Let *PEpe* be a meridian at the surface of a homogeneous figure shaped like an ellipsoid of revolution whose ellipticity is finite and which attracts according to the universal inverse-square law. Let *C* be the center of the figure. Let *Pp* be the diameter of the figure which lies along the axis of symmetry of the figure. Let *M* be an arbitrary point on the meridian *PEpe*. Let *MQ* be the line segment which is perpendicular to *Pp* and whose endpoints are *M* and the point *Q* on *Pp*. Let *MR* be the line segment which is parallel to *Pp* and whose endpoints are *M* and the point *R* in the plane through *C* which is perpendicular to *Pp*. MacLaurin showed that the magnitude of the component of the attraction at *M* in the direction of *Q* produced by the figure in question was equal to the magnitude of the attraction at

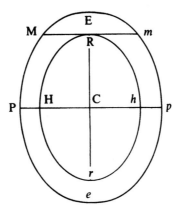

*Figure 34. Clairaut's figure used for calculating the magnitude of the attrac-
tion at point M on the surface of a homogeneous ellipsoid of revolution which
attracts according to the universal inverse-square law (Clairaut 1743: figure
on p. 170).*

R produced by a homogeneous figure shaped like an ellipsoid of revolution which
attracts in accordance with the universal inverse-square law, whose center is at C,
and which is geometrically similar to the figure in question, hence whose
ellipticity is the same as the ellipticity of the figure in question, where Hh is the
diameter of the figure which lies along the axis of symmetry of the figure, which is
the same as the axis of symmetry of the figure in question, whose density is the
same as the density of the figure in question, and where $HRhr$ is a meridian at
the surface of the figure which lies in the plane of the meridian $PEpe$. Similarly,
MacLaurin showed that the magnitude of the component of the attraction at M
in the direction of R produced by the figure in question was equal to the
magnitude of the attraction at Q produced by a homogeneous figure shaped like
an ellipsoid of revolution which attracts in accordance with the universal inverse-
square law, whose center is at C, and which is geometrically similar to the figure
in question, hence whose ellipticity is the same as the ellipticity of the figure in
question, where Qq is the diameter of the figure which lies along the axis of
symmetry of the figure, which is the same as the axis of symmetry of the figure in
question, whose density is the same as the density of the figure in question, and
where $QFqf$ is a meridian at the surface of the figure which lies in the plane of
the meridian $PEpe$.[31]

   To express this result another way, let the resultant of the attraction at any
point inside or on the surface of a homogeneous figure shaped like an ellipsoid of
revolution whose ellipticity is finite and which attracts according to the universal
inverse-square law be resolved into two components, one of which is perpendicular to the axis of symmetry of the figure and the other of which is parallel to the
axis of symmetry of the figure. Then the magnitude of the former component

varies directly as the distance of the point from the axis of symmetry of the figure, and the magnitude of the latter component varies directly as the distance of the point from the plane through the center of the figure which is perpendicular to the axis of symmetry of the figure. (The equivalence of these two ways of stating MacLaurin's results can be established using Corollary 3 to Proposition 91 of Book I of Newton's *Principia*, which I discussed in Chapter 1.)

In short, MacLaurin reduced the problem of calculating the exact values of the magnitudes of the resultants of the attraction at arbitrary points on the surfaces of homogeneous figures shaped like ellipsoids of revolution whose ellipticities are finite and which attract in accordance with the universal inverse-square law to the problem of calculating the exact values of the magnitudes of the resultants of the attraction at the poles and at the equators of such figures.

MacLaurin then used this result to demonstrate that it was possible to make a homogeneous figure shaped like an ellipsoid of revolution flattened at its poles whose ellipticity is finite and which attracts according to the universal inverse-square law revolve around its polar axis, the axis of symmetry of the figure, in such a way that (in other words, with a certain angular speed such that) the effective gravity at all points on the surface of the figure is perpendicular to the surface of the figure at those points. He found that this occurs (that is, that the principle of the plumb line holds at all points on the surface of the figure) when the equality

$$\frac{\text{Magnitude of Attraction at a Pole}}{\text{Magnitude of Effective Gravity at the Equator}} \quad (8.2.1)$$

$$= \frac{\text{Equatorial Radius}}{\text{Polar Radius}}$$

holds. [Equality (8.2.1) is the same as equalities (6.1.28) and (5.3.2).] In addition, MacLaurin proved that all columns from the center of the figure to the surface of the figure balance or weigh the same as well when equality (8.2.1) holds. In so doing he implicitly demonstrated that equality (6.1.28) entails equality (6.1.29).[32] (Thus MacLaurin solved the problem that Maupertuis had not even realized *was* a problem for homogeneous figures shaped like ellipsoids of revolution which are flattened at their poles, which attract according to the universal inverse-square law, which revolve around their axes of symmetry, and whose ellipticities are *finite*.)

In this same letter to MacLaurin of February 1714, Clairaut asked MacLaurin to tell him frankly what he thought of his second paper on the theory of the earth's shape published in the *Philosophical Transactions*. Clairaut was particularly concerned about the part of the paper that included results that contradicted some of Newton's statements in the *Principia* which were really nothing more than conjectures. Clairaut told MacLaurin that he "needed an opinion as reliable as" MacLaurin's "so as not to worry."[33]

In a reply to Clairaut's letter of February 1741 which has disappeared, MacLaurin mentioned his *A Treatise of Fluxions* (1742), which was then being published. The treatise was essentially finished by December of 1740, for MacLaurin wrote Stirling at that time to say that all but "the last three sheets" had been printed. MacLaurin added, however, that the printers were "very slow" in doing "the algebraic part"; that he had very little time during the winter; and that these factors together with the printing of "the figures" would delay the publication of the treatise until the spring of 1741, he thought. He told Stirling in the same letter that James Short was supposed to send him Clairaut's second paper on the theory of the earth's shape published in the *Philosophical Transactions* but that he still had not received the paper.[34] MacLaurin evidently informed Clairaut in his reply to Clairaut's letter of February 1741 that *A Treatise of Fluxions* had virtually been completed, for in a letter that he wrote to Euler dated 12 April 1741, Clairaut told Euler that MacLaurin's "*A Treatise of Fluxions* would soon appear. It has already been completely printed. The only thing left to do is to have the figures engraved and printed."[35] It also seems clear that MacLaurin could only have seen for the first time Clairaut's second paper on the theory of the earth's shape published in the *Philosophical Transactions* after he had already finished his own *A Treatise of Fluxions*.

Whatever MacLaurin told Clairaut, it sufficed to pique the French mathematician's curiosity. On 10 April 1741 Clairaut wrote MacLaurin again, asking him once more to look at his second paper on the theory of the earth's shape, if he had not already read it and formed an opinion about it. He reminded MacLaurin that he very much wanted to know what MacLaurin thought of it. Clairaut also wanted to know more about what *A Treatise of Fluxions* specifically contained and when it would appear. Clairaut asked MacLaurin if he would write to him about the treatise "in some detail."

In his letter to MacLaurin of April 1741, Clairaut mentioned MacLaurin's theorem that interrelates the magnitudes of the attractions produced by concentric, homogeneous figures which have the same density, which attract according to the universal inverse-square law, and which are shaped like ellipsoids of revolution, where the ellipsoids are confocal and have a common axis of symmetry, at a point situated on the common axis of symmetry of the figures or at a point located on any common axis of the figures through the common center of the figures which lies in the plane through the common center of the figures which is perpendicular to the common axis of symmetry of the figures, where the point is situated in the exteriors of all of the figures as well. According to this theorem, if concentric, homogeneous figures that have the same density and that attract according to the universal inverse-square law are shaped like ellipsoids of revolution, where the ellipsoids are confocal and have a common axis of symmetry, then the magnitudes of the attractions produced by the figures at any point situated on the common axis of symmetry of the figures or at any point located on a common axis of the figures through the common center of the figures which lies in the plane through the common center of the figures which is perpendicular to the common axis of symmetry of the figures, where the point is situated outside each figure, vary directly as the volumes of the figures. MacLaurin had evidently stated the theorem in his reply to Clairaut's letter of February 1741. Clairaut said that he saw the theorem's utility for determining the lengths of the axes of figures shaped like ellipsoids of revolution which attract in accordance with the universal inverse-square law and whose densities vary in their interiors. He asked MacLaurin to show him his proof of the theorem. He told MacLaurin that he imagined that in demonstrating the theorem MacLaurin had probably kept the same geometric style that he had used to prove the theorems which appeared in the essay on the tides.

Where his own theory of the shapes of planets was concerned, Clairaut said that he did not regret too much merely having produced a theory in his second paper published in the *Philosophical Transactions* which, like the theory in his first paper published there, was only valid to terms of first order [that is, a theory that only holds for stratified figures shaped like ellipsoids of revolution which are flattened at the poles, whose ellipticities are infinitesimal, which attract according to the universal inverse-square law, and which revolve around their axes of symmetry, because, he told MacLaurin, when the density is not uniform, it is not true in general that the figure of revolution representing a planet is even an ellipsoid. He gave as an example the case where all of the matter that the figure of revolution contains is assumed to be concentrated at its center. (Here Clairaut alluded to hypothesized central forces $f(r) = r^n$ of attraction and to the fact that such a force of attraction can never produce a homogeneous figure of revolution which is shaped like an ellipsoid and whose columns from center to surface all balance or weigh the same unless $n = 1$, as I mentioned in Chapters 1 and 3.)

Clairaut also wanted to know "what hypothesis" MacLaurin had made which resulted in a stratified figure shaped like an ellipsoid of revolution which is flattened at its poles, which attracts according to the universal inverse-square law, which revolves around its axis of symmetry, the ratio of whose equatorial and polar axes equals $\frac{179\frac{1}{4}}{178\frac{1}{4}}$, and along whose surface the magnitude of the effective gravity increases from the equator of the figure to a pole of the figure by a factor of $\frac{1}{220}$.[36] MacLaurin evidently had also communicated this finding to Clairaut in his reply now lost to Clairaut's letter of February 1741. It is obvious why the result troubled Clairaut: the ellipticity $\frac{1}{178\frac{1}{4}}$ of MacLaurin's figure is greater than $\frac{1}{230}$, and, at the same time, the factor $\frac{1}{220}$ that expresses the relative amount by which the magnitude of the effective gravity increases from the equator of the figure to a pole of the figure along the surface of the figure is also greater than $\frac{1}{230}$. The two numerical values together conflict with the main result concerning stratified figures that represent the earth which Clairaut arrived at in his second paper on the earth's shape published in the *Philosophical Transactions* (equation (6.2.4')). Now, the fraction $\frac{1}{178\frac{1}{4}}$ differs insignificantly (that is, it differs by terms of second order and higher orders) from the fraction $\frac{1}{177}$ which the members of the Lapland expedition had determined using their measurements of degrees of latitude in Lapland and in France to be the value of the earth's ellipticity. Specifically,

$$\frac{1}{177} = \frac{1}{178.5}\left(\frac{1}{1 - \frac{1}{119}}\right) \simeq \frac{1}{178.5}\left(1 + \frac{1}{119}\right) \simeq \frac{1}{178.5}.$$

Moreover, the fraction $\frac{1}{230}$ was close to the value of the relative increase in the magnitude of the effective gravity from the earth's equator to the earth's north pole along the earth's surface which had been deduced from experiments done with seconds pendulums at various latitudes. MacLaurin, in effect, intended to substantiate the conjectures that Newton had made in the *Principia* relating the shapes of inhomogeneous figures that represent the earth and the variations of the magnitude of the effective gravity with latitude along the surfaces of such figures.

Moreover, from what MacLaurin had told him thus far, Clairaut gathered that MacLaurin had convinced himself that his hypothesis about the way that the density varied, which MacLaurin had not yet described to Clairaut in detail, would allow "the planet to have the shape of an ellipsoid [of revolution] and remain in equilibrium." [37] That is, as far as Clairaut could determine, MacLaurin had worked here specifically with a stratified figure shaped like an ellipsoid of revolution which is flattened at its poles, whose ellipticity is infinitesimal, which attracts according to the universal inverse-square law, which revolves around its axis of symmetry, and which MacLaurin treated as a figure of equilibrium. Clairaut wanted to know how MacLaurin assured himself that the figure in question truly was a figure of equilibrium. We recall that Clairaut thought that the figures that he took to represent the earth in his second paper on the theory of the earth's shape published in the *Philosophical Transactions*, namely, stratified figures shaped like ellipsoids of revolution which are flattened at their poles, whose ellipticities are infinitesimal, which attract according to the universal inverse-square law, which revolve around their axes of symmetry, and which satisfy equation (6.2.4'), which means that the principle of the plumb line holds at all points on their surfaces, were figures of equilibrium. But MacLaurin's figure did not satisfy equation (6.2.4'). Thus it seemed to Clairaut that he and MacLaurin must have assumed a different hypothesis or different hypotheses, for example, about the geometry of the strata, or else at least one of them had to have made errors in doing calculations.

MacLaurin evidently replied to Clairaut's latest communication, in another letter that has also vanished, for the Frenchman wrote MacLaurin again, in September of 1741, to say how flattered and relieved he was that MacLaurin was pleased with his second paper on the theory of the earth's shape published in the *Philosophical Transactions*, even though the paper included results that conflicted with Newton's conjectures in the *Principia*. (As I mentioned, Clairaut's letters to MacLaurin and MacLaurin's letters to Stirling make clear that MacLaurin only saw for the first time Clairaut's second paper on the theory of the earth's shape published in the *Philosophical Transactions* after he had already completed his own *A Treatise of Fluxions*.)

Clairaut also told MacLaurin that he had learned in the meantime that he had angered the Reverend Patrick Murdoch, who found the introduction to Clairaut's second paper on the theory of the earth's shape published in the *Philosophical Transactions* to be presumptuous, disrespectful, and unwarranted and who accordingly, in Clairaut's opinion, completely misunderstood the aims of the two papers that he published in the *Philosophical Transactions*. Murdoch thought that Clairaut had underestimated Newton when Clairaut observed that Newton had not explained why he had chosen an ellipsoid of revolution as the shape of a flattened homogeneous figure of revolution which attracts according to the universal inverse-square law, which revolves around its axis of symmetry, and whose columns from center to surface all balance or weigh the same. MacLaurin's more positive attitude reassured Clairaut.

Clairaut asked MacLaurin if he believed Newton's assertions that the earth must be flatter than the homogeneous figure shaped like an ellipsoid of revolution which is flattened at its poles, which attracts in accordance with the universal inverse-square law, which revolves around its axis of symmetry, and whose columns from center to surface all balance or weigh the same, which Newton had originally taken to be his theoretical model of the earth, as well as denser at its center than at its surface, because the relative increase in the magnitude of the effective gravity from the earth's equator to the earth's north pole along the earth's surface was observed to be greater than the calculated relative increase in the magnitude of the effective gravity from the equator to a pole along the surface of the homogeneous figure shaped like an ellipsoid of revolution which is flattened at its poles, which attracts according to the universal inverse-square law, which revolves around its axis of symmetry, and whose columns from center to surface all balance or weigh the same, which Newton had originally taken to be his theoretical model of the earth. Clairaut also asked MacLaurin if he knew whether Newton had picked the ellipsoid of revolution by chance to be the shape of the flattened homogeneous figure of revolution in his theory of the earth's shape, and if Newton did not, then why did he fail to explain his choice.

Having asked MacLaurin in his letter of April 1741 to give details concerning what would appear in *A Treatise of Fluxions*, Clairaut was very surprised to learn in reply that the treatise had been composed since 1734. (In fact, MacLaurin's correspondence with Stirling during the years 1734–40 shows that MacLaurin learned more than a thing or two from Stirling which appear in the treatise. This makes such an early dating of the treatise doubtful. Indeed, George Berkeley's sagacious criticism of Newton's fluxions, which Berkeley published in *The Analyst* in 1734, is supposed to have been what induced MacLaurin to write *A Treatise of Fluxions* in the first place!)

Clairaut also informed MacLaurin that, no matter what he tried, he could not make his own theory of stratified figures shaped like ellipsoids of revolution which are flattened at their poles, whose ellipticities are infinitesimal, which attract according to the universal inverse-square law, which revolve around their axes of symmetry, and at whose surfaces the principle of the plumb line holds at all points and the calculations that were involved in developing the theory agree with MacLaurin's findings. We recall that the Edinburgh mathematician had talked earlier about a stratified figure shaped like an ellipsoid of revolution which is flattened at its poles, which attracts in accordance with the universal inverse-square law, whose ellipticity is $\frac{1}{178\frac{1}{4}}$, which revolves around its axis of symmetry, and along whose surface the magnitude of the effective gravity increases from the equator of the figure to a pole of the figure by a factor of $\frac{1}{220}$. Clairaut thought that his own approximate theory sufficed, because the shapes of his figures differed very little from a sphere, whatever the values of their ellipticities. But for a relative increase of $\frac{1}{220}$ of the magnitude of the effective gravity from the equator of such a figure to a pole of the figure along the surface of the figure, Clairaut always determined a value for the ellipticity of the figure which was much smaller than $\frac{1}{178\frac{1}{4}}$. Clairaut allowed for the possibility that he had made numerical errors.

On the other hand, Clairaut found MacLaurin's hypothesis, which Clairaut had asked MacLaurin to clarify in his letter to MacLaurin of April 1741 and which MacLaurin evidently explained in his reply now lost to the letter, to be improbable. In MacLaurin's theoretical model, a homogeneous spherical core that attracted in accordance with the universal inverse-square law, one of whose meridians is the circle *lmnr*, was surrounded by a layer that did not attract (therefore a layer whose density was essentially zero, or a vacuum, in other words). This layer that did not attract was itself surrounded by a dense, homogeneous outer layer that attracted according to the universal inverse-square law and whose inner and outer surfaces were geometrically similar ellipsoids of revolution flattened at their poles which had a common axis of symmetry and which were concentric with the spherical core (see Figure 35). The flattened ellipses *adbe* and *ADB* are meridians of these two flattened ellipsoids of revolution which lie in the same plane as the circle *lmnr*. The geometric similarity of the two flattened ellipsoids of revolution meant that they had the same ellipticities.

Thus the geometry of MacLaurin's strata in this theoretical model did differ from the geometry of Clairaut's strata. The ellipticity of MacLaurin's concentric, geometrically similar ellipsoids of revolution flattened at their poles which had a common axis of symmetry and which bounded the homogeneous outer layer of his figure did not equal the ellipticity of the homogeneous spherical core of the figure which was zero. This difference in the geometries of the strata could have caused Clairaut and MacLaurin to arrive at different results.

Moreover, MacLaurin evidently did have a figure of equilibrium in mind. That is, the central core and the outer layer consisted of fluid, just as MacLaurin had seemed to suggest in his earlier letter to Clairaut without going into details in that letter. But Clairaut could not understand how such a figure could possibly

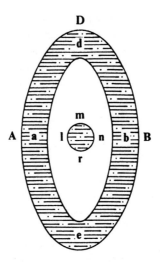

Figure 35. Meridians of a homogeneous outer fluid layer whose surfaces are geometrically similar, flattened ellipsoids of revolution, a layer whose density is zero, and a spherical fluid core of the same density as the outer fluid layer, where the outer layer and spherical core attract according to the universal inverse-square law (Clairaut 1741c: figure on p. 365).

maintain itself in equilibrium. Clairaut said that the inner surface of the outer layer of fluid would fall toward the center of the figure, since the spherical core whose center is the center of the figure attracted this inner surface of the outer layer, while the outer layer did not attract its inner surface, because it lay outside its inner surface.

Clairaut probably alluded here to Proposition 91, Corollary 3 of Book I of the *Principia*, which I discussed in Chapter 1 and which says, among other things, that a homogeneous figure shaped like an ellipsoid of revolution which attracts according to the universal inverse-square law attracts a point in its interior the same way that a homogeneous figure shaped like an ellipsoid of revolution which attracts in accordance with the universal inverse-square law and which is concentric with the other figure and geometrically similar to it, whose density is the same as the density of the other figure, and upon whose surface the point lies attracts the point, where both figures have a common axis of symmetry. Consequently the attraction produced by a homogeneous shell bounded by surfaces that are concentric, geometrically similar ellipsoids of revolution which have a common axis of symmetry is zero at points within the cavity that the shell surrounds. With nothing to counteract the attraction between the central core and the outer layer, Clairaut felt that these two parts of MacLaurin's figure would unite as a result of their mutual attraction. From what Clairaut said in this letter, MacLaurin apparently had his own doubts about the plausibility of the model too.[38]

Did MacLaurin arrive at his findings concerning the ratios of the axes of stratified figures shaped like ellipsoids of revolution which are flattened at their poles, which attract according to the universal inverse-square law, and which revolve around their axes of symmetry by utilizing the principle of the plumb line, as he had done for such homogeneous figures in his essay on the tides?. If he did, how could be have arrived at results that differed from Clairaut's findings? For Clairaut had employed the principle of the plumb line in his second paper on the theory of the earth's shape published in the *Philosophical Transactions* to arrive at the results in that paper which pertained to the shapes (that is, the ellipticities) of such stratified figures and which contradicted MacLaurin's findings. But if the geometry of MacLaurin's strata differed from the geometry of the strata in Clairaut's second paper on the theory of the earth's shape published in the *Philosophical Transactions*, as it did in the example just described, this could conceivably have caused the results which followed from an application of the principle of the plumb line to differ from Clairaut's findings.

In this same letter, Clairaut sketched an algebraic proof of the theorem that MacLaurin had evidently reported in his reply to Clairaut's letter to him of February 1741. We recall that Clairaut had asked MacLaurin in his letter to him of April 1741 to show him his demonstration of the theorem, which Clairaut guessed was geometric like the demonstrations that MacLaurin gave in his essay on the tides. But MacLaurin seems not to have complied with Clairaut's request. As I mentioned above, this theorem states that if concentric, homogeneous figures which have the same density and which attract according to the universal inverse-square law are shaped like confocal ellipsoids of revolution which have a common axis of symmetry, then the magnitudes of the attractions produced by the figures at any point situated on the common axis of symmetry of the figures or at any point located on a common axis of the figures through the common center of the figures which lies in the plane through the common center of the figures which is perpendicular to the common axis of symmetry of the figures, where the point is situated outside each figure, vary directly as the volumes of the figures.

Did MacLaurin use this theorem to determine the ellipticities of stratified figures shaped like ellipsoids of revolution which are flattened at their poles, which attract according to the universal inverse-square law, and which revolve around their axes of symmetry? MacLaurin evidently had not made himself very clear in his letters to Clairaut.

(One thing does seem certain, however. MacLaurin could not have used such a theorem to calculate the magnitude of the attraction at a pole of a heterogeneous figure shaped like an ellipsoid of revolution which is flattened at its poles, which attracts according to the universal inverse-square law, and which is made up either of strata that are concentric ellipsoids of revolution flattened at their poles which all have a common axis of symmetry and the same ellipticities or strata whose thicknesses are finite and whose inner and outer surfaces are concentric ellipsoids of revolution flattened at their poles which all have a common axis of

symmetry and the same ellipticities, because in either case the strata are geometrically similar and thus are not confocal.)

Clairaut realized that in developing theories of the shapes of the earth he and
MacLaurin could have arrived in at least three different ways at results that
disagreed:[39]

1. Clairaut produced an approximate theory, which only held to terms of first
order. Thus, for example, he had not tried to apply his theory to Jupiter, because
that planet was too flat. Its ellipticity was not infinitesimal. Clairaut acknowledged the superiority of MacLaurin's exact methods in this regard. For they
could be applied to Jupiter, whose ellipticity was finite.[40] Clairaut had not kept
track of the total error that accumulated as he made approximations. Hence he
had to consider the possibility that errors in making approximations had
multiplied more rapidly than he had thought, leading to a total error of the same
order as the infinitesimal ellipticity of the earth whose value he sought to
determine.

2. If the geometries of MacLaurin's strata and Clairaut's strata differed from
each other, as they did in fact in at least one case, that could have caused
MacLaurin and Clairaut to arrive at different results.

3. In determining ratios of axes (in other words, in determining ellipticites) of
stratified figures shaped like ellipsoids of revolution which are flattened at their
poles, which attract according to the universal inverse-square law, and which
revolve around their axes of symmetry, both MacLaurin and Clairaut had
supposed that the particular stratified figures that they treated were figures of
equilibrium. Thus Clairaut was concerned about the stability of MacLaurin's
figure that I described, two parts of which consisted of homogeneous fluid
separated by a vacuum. If something were fundamentally wrong with Mac-
Laurin's assumptions or Clairaut's assumptions about the equilibrium of heterogeneous fluid figures or both, that too could account for the differences between
Clairaut's findings and MacLaurin's findings.

Not until Clairaut received the letter from MacLaurin which Clairaut replied
to in September of 1741 did Clairaut really have a need to investigate deeply the
foundations of figures of equilibrium, and stratified figures of equilibrium in
particular, especially those that attract according to the universal inverse-square
law, in order to deal with the gaps in Newton's theory of the earth's shape in the
Principia. Bouguer had certainly paved the way for such a study in his Paris
Academy mémoire of 1734, although Bouguer himself did not examine in that
mémoire the case of homogeneous figures assumed to attract according to the
universal inverse-square law in order to determine whether such figures that
satisfy both the principles of balanced columns and the plumb line simultaneously could exist. In Chapter 6 I observed that in his two papers on the theory of the
earth's shape published in the Philosophical Transactions, Clairaut implicitly
showed that such figures do indeed exist. But were such homogeneous figures
truly figures of equilibrium? In his essay of 1740 on the tides, MacLaurin called

the homogeneous figures shaped like ellipsoids of revolution which are flattened at their poles, which attract according to the universal inverse-square law, whose ellipticities are finite, which revolve around their axes of symmetry, and which satisfy equality (8.2.1) figures of equilibrium.[41] But was it enough for a homogeneous figure to satisfy both the principles of balanced columns and the plumb line in order for the figure to be a figure of equilibrium? (I mentioned in Chapter 4 that thanks in part to an error that Bouguer made in his Paris Academy *mémoire* of 1734, Clairaut discovered that the answer to this question is no, for reasons that we shall learn in Chapter 9.) Moreover, in his Paris Academy *mémoire* of 1734 Bouguer did not investigate stratified fluid figures that attract according to the hypotheses that he assumed in establishing the conclusions that he came to about homogeneous fluid figures in that *mémoire*, much less stratified fluid figures that attract in accordance with the universal inverse-square law.

But until now Clairaut had had no reason to look into the matter carefully. His application in his second paper on the theory of the earth's shape published in the *Philosophical Transactions* of the principle of the plumb line at points on the surfaces of stratified figures shaped like ellipsoids of revolution which are flattened at their poles, which attract according to the universal inverse-square law, whose ellipticities are infinitesimal, and which revolve around their axes of symmetry, from which equation (6.2.4) followed, had seemed adequate until now to deal with the anomalies in Newton's theory of the earth's shape. To be sure, as a result of MacLaurin's findings that conflicted with his, Clairaut did have to think about the possibility that the geometry of the strata of the figures that he investigated in his second paper on the theory of the earth's shape published in the *Philosophical Transactions* was not general enough. But until September of 1741 Clairaut thought that the principle of the plumb line was all that he needed to use in studying the problem. His thoughts about stratified figures of equilibrium shaped like ellipsoids of revolution which are flattened at their poles, which attract according to the universal inverse-square law, whose ellipticities are infinitesimal, and which revolve around their axes of symmetry, which he introduced at the end of his second paper on the theory of the earth's shape published in the *Philosophical Transactions*, were really little more than idle speculations. MacLaurin's findings concerning such figures, which MacLaurin reported to Clairaut by letter and which included results concerning shapes of figures intended to represent the earth which contradicted Clairaut's findings and supported Newton's conjectures instead, changed all this. More than anything else, Clairaut's correspondence with MacLaurin, which the Paris Academy's contest of 1740 on the tides occasioned, forced the French mathematician to think seriously for the first time about the theory of figures of equilibrium, and the theory of stratified figures of equilibrium which attract according to the universal inverse-square law above all.

9

Clairaut's mature theory of the earth's shape (1741–1743): First substantial connections between the revival of mathematics in Paris and progress in mechanics there

9.1. Clairaut's principles of equilibrium: Partial differential calculus applied

In this chapter a figure of equilibrium will be a figure that happens to satisfy certain conditions that are *necessary* for figures of equilibrium to exist. [Toward the end of the eighteenth century mathematicians began to find that figures of equilibrium must fulfill even *more* necessary conditions than those that we shall consider in this chapter. For example, as I mentioned in Chapter 1, in order that a homogeneous figure shaped like a flattened ellipsoid of revolution which attracts according to the universal inverse-square law and which revolves around its axis of symmetry be a figure of equilibrium, the ratio $\omega^2/2\pi\rho$ cannot exceed a certain upper limit whose value is less than 1 and that Laplace was the first to determine, where ω is the angular speed at which the figure revolves around its axis of symmetry and ρ is the density of the figure. The discovery of other necessary conditions that figures of equilibrium must satisfy continued well into the twentieth century.] As I also stated in Chapter 1, the problem of determining conditions that are *sufficient for* figures of equilibrium to exist, hence the problem of determining whether or not a figure that satisfies conditions that are necessary for figures of equilibrium to exist is truly a figure of equilibrium, is outside the limits of the story told in these pages.

In April of 1743, Clairaut published his *Théorie de la figure de la terre, tirée des principes de l'hydrostatique.*[1] This work included few new results concerning the

426

earth's shape which had not already appeared in the first part of Clairaut's second paper on the theory of the earth's shape published in the *Philosophical Transactions*. Instead, Clairaut showed in his treatise of 1743 that the same results followed from hypotheses concerning the earth's internal structure which were more general than those that he had considered in the earlier paper. In addition, in his treatise Clairaut placed these hypotheses and the results that followed from them on new foundations. The conflict between the results that Clairaut arrived at in his second paper on the theory of the earth's shape published in the *Philosophical Transactions* and the findings that MacLaurin reported to Clairaut by letter induced Clairaut to study the question of foundations more carefully. Clairaut's additional research enabled him to recognize errors he had made in the second part of that second paper.

I shall refer to Clairaut's treatise of 1743 itself where its contents and substance are concerned. Clairaut's recently published correspondences with Euler and MacLaurin help fix the period during which Clairaut did his research and composed various parts of his treatise. I shall refer to these correspondences in order to determine when he solved certain problems and wrote the parts of his treatise which pertain to these problems.[2]

Clairaut dated his treatise back to the Paris Academy vacation from September to November of 1741. In the spring of 1742 he told Euler that he had "enjoyed considerable leisure" during that vacation to work on his book.[3] It has been noted that Clairaut spent much of the year 1741 preparing his *Eléments de géométrie* (1741) for publication.[4] It would therefore appear that he could not have begun to work on his treatise on the earth's shape any sooner than he actually did. Be that as it may, I have contended that Clairaut would have had no reason to begin work on his treatise any earlier in any case. Not until he read the letter from MacLaurin which he replied to in September 1741 did Clairaut have a real need to theorize about the earth's shape from the standpoint of the theory of stratified figures of equilibrium, and stratified figures of equilibrium which attract according to the universal inverse-square law in particular. Until September of 1741 Clairaut had had no real reason to deal with the theory of the earth's shape this way. MacLaurin's letter changed all that. It provided Clairaut with the motive for theorizing about the earth's shape from the standpoint of the theory of stratified figures of equilibrium. But in order to treat the theory of the earth's shape in this manner, Clairaut had to go much more deeply into the principles that the theory of stratified figures of equilibrium are based upon than he had in his brief, shallow argument that he presented at the end of his second paper on the theory of the earth's shape published in the *Philosophical Transactions*.

As we shall see, Clairaut did indeed arrive at the bases of much of his new theory of the earth's shape during the last four months of 1741. And yet his book only appeared in April of 1743. But the printer did not delay its publication. It took Clairaut a year to reach a point where he felt that the work was ready to be made public. In the course of examining Clairaut's theory, I shall explain the probable reasons that caused Clairaut to postpone the publication of his book.

At the beginning of his treatise Clairaut stated that the following principle must hold in order that a homogeneous fluid mass can be in equilibrium: a homogeneous fluid mass can be in equilibrium only if the "efforts" of all parts of the fluid in the homogeneous fluid mass which fill a channel whose shape is arbitrary which lies within the interior of the homogeneous fluid mass and traverses the whole mass, meaning that the channel joins two different points on the surface of the homogeneous fluid mass, mutually cancel each other. We shall say that this new principle of equilibrium holds for a homogeneous fluid mass whenever the "efforts" of all parts of the fluid in the homogeneous fluid mass which fill a channel whose shape is arbitrary which lies within the interior of the homogeneous fluid mass and traverses the whole mass, meaning that the channel joins two different points on the surface of the homogeneous fluid mass, mutually cancel each other. (We shall see shortly that a homogeneous fluid mass need not be in equilibrium in order for this to happen.) "Effort" here simply means the effort that Bouguer talked about in his Paris Academy *mémoire* of 1734. We recall that in that *mémoire* Bouguer called the weight of a homogeneous column of fluid whose length is infinitesimal the "effort" produced by effective gravity along the homogeneous column of fluid. The weight of a homogeneous column of fluid whose length is finite was then the sum of the efforts produced by effective gravity along the infinite number of homogeneous columns of fluid whose lengths are infinitesimal, all of which have the same density, and which make up the homogeneous column of fluid whose length is finite. Clairaut essentially generalized the notion of the sum of the efforts produced by effective gravity along the infinite number of straight channels whose lengths are infinitesimal, all of which are filled with homogeneous fluid of the same density, and which make up a straight channel filled with homogeneous fluid whose length is finite (essentially the weight of the homogeneous fluid that fills a straight channel whose length is finite, calculated or measured in the direction toward which the channel points) to include curved plane channels and skew channels filled with homogeneous fluid. Clairaut called the sum of the efforts produced by effective gravity along all of the infinitesimal parts that make up a straight channel, a curved plane channel, or a skew channel filled with homogeneous fluid the total effort produced by effective gravity along the channel filled with homogeneous fluid. Clairaut's new principle of equilibrium essentially states that in a homogeneous fluid mass in equilibrium the total effort produced by effective gravity along the kind of channel described at the beginning of this paragraph filled with fluid from the homogeneous fluid mass in equilibrium must be equal to zero. The new principle of equilibrium is not restricted to plane channels filled with homogeneous fluid; the principle applies as well to skew channels filled with homogeneous fluid. No one before Clairaut, including Huygens,[5] had ever utilized curved channels in theorizing about the earth's shape. We shall understand the likely reason why

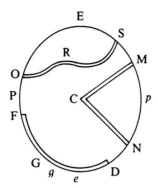

Figure 36. Figure used for showing that the principle of balanced columns and the plumb line are special cases of Clairaut's first new principle of equilibrium (Clairaut 1743: figure on p. 4).

shortly. And even though Clairaut had already introduced curved channels at the end of his second paper on the theory of the earth's shape published in the *Philosophical Transactions*, he used them there to reach conclusions that were false. Later in this chapter we shall discover why these conclusions were wrong.

We note that Clairaut did not specify that in order for a homogeneous fluid mass to be a figure of equilibrium, the mass must necessarily have the shape of a figure that revolves in such a way and that the force of attraction must act in such a way as to produce the symmetries that I mentioned in Chapters 1, 4, and 6. (That is, that a homogeneous fluid mass must be a figure of revolution which revolves around its axis of symmetry, in which case the axis of symmetry of the homogeneous fluid figure of revolution is also an axis of symmetry of the centrifugal force of rotation per unit of mass, and, moreover, that the axis of symmetry of the homogeneous fluid figure of revolution must also be an axis of symmetry of the attraction. These conditions did in fact always hold for the homogeneous fluid figures discussed in Chapters 1, 4, and 6.) We shall see later in this chapter that homogeneous fluid figures of equilibrium need not be such figures.

Clairaut demonstrated that this new principle of equilibrium included both the principles of balanced columns and the plumb line as special cases. When the principle of balanced columns holds for a homogeneous fluid mass, it follows that the total effort produced by effective gravity along any plane channel *MCN* (see Figure 36) which lies inside the mass, which is filled with fluid from the mass, which connects a point *M* on the surface of the mass to another point *N* on the surface of the mass, and which consists of two rectilinear channels *MC* and *NC*, respectively, which lie in the same plane through the center *C* of the mass, which meet at the center *C* of the mass, and which join *C* to the points *M* and *N*, respectively, must equal zero, where *PEpe* is the closed plane curve that lies at the surface of the homogeneous fluid mass in the plane of the channel MCN. For this reason the principle of balanced columns is a special case of Clairaut's new principle of equilibrium.

Moreover, by considering a plane channel FGD filled with homogeneous fluid which lies along the surface of the homogeneous fluid mass in the plane of the closed plane curve $PEpe$, whose density is the same as that of the homogeneous fluid mass, and whose ends are at the surface of the homogeneous fluid mass, and then allowing the length of channel FGD to become arbitrarily short, Clairaut showed that the principle of the plumb line also followed from his new principle of equilibrium. If the effective gravity at a point on the surface of the homogeneous fluid mass were not perpendicular to the surface at that point, then the effective gravity at that point would have a component tangent to the surface at the point. This component would produce a total effort that is not equal to zero along some very short, nearly rectilinear plane channel filled with homogeneous fluid which lies along the surface of the homogeneous fluid mass and which joins the point to a nearby point on the surface of the homogeneous fluid mass. But this would contradict the new principle of equilibrium. (In fact, we see that in deducing the principle of the plumb line as a special case of his new principle of equilibrium, Clairaut assumed that his new principle of equilibrium not only held for all plane or skew channels that lie inside the homogeneous fluid mass, that are filled with fluid from the homogeneous fluid mass, and that join two points on the surface of the homogeneous fluid mass, but that it also held for plane or skew channels filled with homogeneous fluid which lie *along the surface* of the homogeneous fluid mass, whose densities are the same as that of the homogeneous fluid mass, and whose ends are *at the surface* of the homogeneous fluid mass. From the derivation of Clairaut's first new principle of equilibrium which I shall present in a moment, we shall see why the principle can be applied to channels that lie at the surface of the homogeneous fluid mass, whose densities are the same as that of the homogeneous fluid mass, and whose ends are at the surface of the homogeneous fluid mass.[6])

Clairaut stated moreover that the following other principle must also hold: in order that a homogeneous fluid mass can be in equilibrium, it is necessary that the efforts of all parts of the fluid in the mass which fill a closed channel whose shape is arbitrary which lies within the interior of the mass mutually cancel each other. We shall say that this second new principle of equilibrium holds for a homogeneous fluid mass whenever the efforts of all parts of the fluid in the mass which fill a closed channel whose shape is arbitrary which lies within the interior of the homogeneous fluid mass mutually cancel each other. (We shall also see shortly that a homogeneous fluid mass need not be in equilibrium in order for this to occur.) Clairaut called the sum of the efforts produced by effective gravity along all of the infinitesimal parts that make up a closed channel filled with homogeneous fluid the total effort produced by effective gravity around the closed channel filled with homogeneous fluid. Clairaut's second new principle of equilibrium

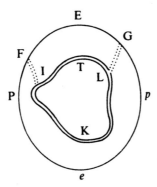

Figure 37. Figure used to derive Clairaut's second new principle of equilibrium from Clairaut's first new principle of equilibrium (Clairaut 1743: figure on p. 6).

essentially states that in a homogeneous fluid mass in equilibrium the total effort produced by effective gravity around the closed channel of fluid just described filled with fluid from the homogeneous fluid mass in equilibrium must be equal to zero. Like his first new principle of equilibrium, Clairaut's second new principle of equilibrium is not restricted to closed plane channels filled with homogeneous fluid; the principle also applies to closed skew channels filled with homogeneous fluid.

Clairaut established in the following way that his first new principle of equilibrium entails that his second new principle of equilibrium holds. Suppose that the first new principle of equilibrium holds for a homogeneous fluid mass. Then it follows that the total efforts produced by effective gravity along the two channels *FITLG* and *FIKLG* which lie inside the homogeneous fluid mass, which are filled with fluid from the mass, and which join a point *F* on the surface of the mass to another point *G* on the surface of the mass are both equal to zero. (Figure 37 is rather misleading. In the diagram the two channels of fluid *FITLG* and *FIKLG* appear to be plane channels that lie in the same plane, namely, the plane of the closed plane curve *PEpe* that lies at the surface of the homogeneous fluid mass. The closed channel of fluid *ITLK* also appears in the diagram to be a plane channel that lies in the plane of *PEpe*. In fact the closed channel of fluid *ITLK* need not be a closed channel that lies in the plane of *PEpe*. It does not even have to be a plane channel. It can either be a plane channel or a skew channel. Likewise, the two channels of fluid *FITLG* and *FIKLG* need not be channels that lie in the plane of *PEpe*. They do not even have to be plane channels. They can either be plane channels or skew channels.) Now, the total efforts produced by effective gravity along the two channels of fluid *ITL* and *IKL* will be the same, because the two channels of fluid *FITLG* and *FIKLG* both include the two channels of fluid *FI* and *LG*. That is, subtracting the total efforts produced by effective gravity along the two channels of fluid FI and LG from the total efforts produced by effective gravity along the two channels of fluid $FITLG$ and $FIKLG$, Clairaut inferred that the total efforts produced by effective gravity along the two channels of fluid ITL and IKL have to be the same, because the total efforts produced by effective gravity along the two channels of fluid $FITLG$ and $FIKLG$ are the same, namely, they are both equal to zero. Changing the sign of the total effort produced by effective gravity along one of the two channels of fluid ITL and IKL from positive to negative, Clairaut concluded that the total effort produced by effective gravity around the closed channel $ITLK$ that lies inside the homogeneous fluid mass and that is filled with fluid from the mass is equal to zero. In other words, Clairaut's second new principle of equilibrium also holds for the homogeneous fluid mass.[7] (We shall see shortly that Clairaut in fact assumed that when his second new principle of equilibrium held for a homogeneous fluid mass, this meant that it not only held for all closed plane or skew channels that lie entirely within the interior of the homogeneous fluid mass and that are filled with fluid from the homogeneous fluid mass, but that it also held for closed plane or skew channels of homogeneous fluid parts of which lie *along the surface* of the homogeneous fluid mass. If part of a closed plane or skew channel of homogeneous fluid lies along the surface of the homogeneous fluid mass, then the proof here still holds, because we recall that Clairaut in fact assumed that when his first new principle of equilibrium held for a homogeneous fluid mass, this meant that it also held for plane or skew channels filled with homogeneous fluid which lie along the surface of the homogeneous fluid mass, whose densities are the same as that of the homogeneous fluid mass, and whose ends are at the surface of the homogeneous fluid mass. In a moment we shall see how these two particular forms of the two principles are related.)

Clairaut also stated that his first new principle of equilibrium, which included the previously known principles of balanced columns and the plumb line as special cases, followed from the principle of solidification.[8] While this is indeed the case, Clairaut did not explain why this is true. It is worth taking a moment to examine how the various principles of equilibrium, old and new, are related, because the connections need to be made clear and must be understood before we proceed any further.

Clairaut affirmed that his second new principle of equilibrium also followed from the principle of solidification. He gave an alternate derivation of his second new principle of equilibrium in which he utilized a second form of the principle of solidification instead of his first new principle of equilibrium.[9] Clairaut's assertion here is plausible and his demonstration of it using the principle of solidification is sound or at least reasonable, insofar as Clairaut reasoned that a total effort that is not equal to zero produced by effective gravity around a closed plane or skew channel that lies within the interior of a homogeneous fluid mass in equilibrium and that is filled with fluid from the homogeneous fluid mass in equilibrium would produce a "perpetual current" of the homogeneous fluid in the closed channel.[10] But the existence of such a current would contradict the principle of solidification, because according to this principle solidifying all of the fluid that makes up a homogeneous fluid mass in equilibrium except the fluid in a closed channel that lies inside the homogeneous fluid mass in equilibrium leaves the fluid in the closed channel in equilibrium. A "perpetual current" of the homogeneous fluid in the closed channel, however, would obviously upset the state of equilibrium of the fluid in the closed channel. In other words, Clairaut interpreted a state of equilibrium of the fluid in the closed channel to be equivalent to a total effort produced by effective gravity around the closed channel which is equal to zero. This second derivation of Clairaut's second new principle of equilibrium also holds both for closed skew channels of homogeneous fluid $ITLK$ as well as for closed plane channels of homogeneous fluid $ITLK$.

Now, the second new principle of equilibrium alone does not suffice to make the first new principle of equilibrium hold. If, however, one or the other of the principles of balanced columns or the plumb line holds, together with the second new principle of equilibrium, then the first new principle of equilibrium follows as a consequence. This can be shown to be true for a plane channel or skew channel MHN that lies inside a homogeneous fluid mass, that is filled with fluid from the mass, and that joins two points M and N on the surface of the mass as follows.

If, say, the principle of balanced columns holds, then it follows that the total effort produced by effective gravity along the plane channel MCN that lies within the interior of the homogeneous fluid mass, that is filled with fluid from the mass, and that joins the point M to the point N is equal to zero, where C is the center of the mass and where $PEpe$ is the closed plane curve that lies at the surface of the mass in the plane of the channel of fluid MCN. But then the total effort produced by effective gravity along the channel of fluid MHN must be equal to zero as well, when the total efforts produced by effective gravity around all closed channels that lie inside the mass and that are filled with fluid from the mass are equal to zero. (The particular diagram labeled Figure 38 is also misleading. The channel of fluid MHN need not be a channel in the plane of the plane channel of fluid MCN. Nor does the channel of fluid MHN even have to be a plane channel. It can be a skew channel.)

Similarly, since the principle of the plumb line entails that the total effort produced by effective gravity along the plane channel of homogeneous fluid OES which lies along the surface of the homogeneous fluid mass be equal to zero, the total effort produced by effective gravity along the plane channel or skew channel ORS that lies inside the mass, that is filled with fluid from the mass, and that joins the points O and S on the surface of the mass must be equal to zero too, when the total efforts produced by effective gravity around all closed channels that lie within the interior of the mass and that are filled with fluid from the mass are equal to zero. (The particular diagram, Figure 38, is again misleading. The plane channel OES need not be part of $PEpe$. The plane channel OES can lie at the surface of the homogeneous fluid mass in a plane that is different from the plane of

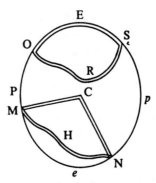

*Figure 38. Figure used to show that Clairaut's first new principle of equilib-
rium follows from Clairaut's second new principle of equilibrium together with
one or the other of the principles of balanced columns or the plumb line
(Clairaut 1743: figure on p. 8).*

PEpe. Moreover, the channel of fluid *ORS* need not be a channel that lies in the
plane of the plane channel *OES*. Nor does the channel of fluid *ORS* even have to
be a plane channel of fluid. It can be a skew channel of fluid. Here we see that
Clairaut in fact supposed that when his second new principle of equilibrium held
for a homogeneous fluid mass, this meant that it not only held for all closed plane
or skew channels that lie inside the mass and that are filled with fluid from the
mass, but that it held for closed plane or skew channels of homogeneous fluid
parts of which lie *along the surface* of the homogeneous fluid mass too.[11]

Furthermore, any application of the principle of solidification to a homogeneous fluid mass presupposes that *all* necessary conditions for equilibrium hold for
the homogeneous fluid mass, because in order for the principle of solidification to
hold when this principle is applied to a homogeneous fluid mass, the homogeneous fluid mass *has to be in equilibrium*. (The same is not true, for example of
Clairaut's second new principle of equilibrium. That is, as I noted previously, a
homogeneous fluid mass does not have to be in equilibrium in order for Clairaut's
second new principle of equilibrium to hold for the homogeneous fluid mass. This
statement will have been shown to be true by the end of this chapter. For the
moment it is enough to be aware of the following things. Shortly we shall see that
the effects of the centrifugal force produced when a mass of homogeneous fluid
revolves around an axis can be disregarded when trying to determine whether or
not Clairaut's second new principle of equilibrium holds for the homogeneous
rotating fluid mass; all that needs to be considered are the total efforts produced
by the attraction around closed channels that lie inside the homogeneous
rotating fluid mass and that are filled with fluid from the mass. But as I have
already mentioned in Chapter 6, the resultant of the attraction produced by a
mass that attracts according to the universal inverse-square law never produces a
total effort that is not equal to zero around a closed channel filled with

homogeneous fluid which lies within the interior of the mass, even when the mass is *not* in equilibrium. We shall discover later in this chapter why this is true. In short, Clairaut's second new principle of equilibrium does not require that a homogeneous mass of fluid be in equilibrium in order for the principle to hold for the homogeneous fluid mass. Today we would say that Clairaut's second new principle of equilibrium is a necessary condition for homogeneous figures of equilibrium to exist but that it is not a sufficient condition for homogeneous figures of equilibrium to exist.) This means in particular that if the principle of solidification holds for a homogeneous fluid mass, then both of the previously known necessary conditions for equilibrium, the principles of balanced columns and the plumb line, must hold as well for the homogeneous fluid mass, in addition to Clairaut's second new principle of equilibrium. But we have just seen that Clairaut's second new priniciple of equilibrium, together with either one of the two previously known principles of equilibrium, entail that Clairaut's first new principle of equilibrium holds as well. Thus from this chain of arguments Clairaut's first new principle of equilibrium can be observed to follow from the principle of solidification. Clairaut stated that this is the case, but he did not make clear why this is true.

That the first new principle of equilibrium applied to channels at the surface whose densities are the same as that of the homogeneous fluid mass and whose ends are at the surface holds follows from the first new principle of equilibrium applied to channels in the interior together with the second new principle of equilibrium. Here the homogeneous fluid mass in equilibrium consists of the homogeneous fluid mass $AHCFA$ and the homogeneous channel of fluid ADC that lies at the surface of $AHCFA$ and whose ends are at the surface of $AHCFA$ (see Figure 39). The densities of $AHCFA$ and the channel of fluid ADC are the same. The total mass of fluid in equilibrium consists of these two fluid masses together. (In other words, the homogeneous fluid figure of equilibrium consists of $AHCFA$ together with the channel of fluid ADC at the surface of $AHCFA$ and whose ends are at the surface of $AHCFA$.) If all but the closed channel $ABCDA$ of homogeneous fluid is then solidified, the closed channel $ABCDA$ of homogeneous fluid remains in equilibrium. Hence the total effort produced by effective gravity around the closed channel $ABCDA$ of homogeneous fluid is equal to zero; otherwise a "perpetual current" of the homogeneous fluid in this closed channel would be produced. But it follows from the first new principle of equilibrium applied to the channel ABC of homogeneous fluid in the interior of $AHCFA$ whose two ends are at the surface of $AHCFA$ that the total effort produced by effective gravity along the channel of fluid ABC is equal to zero. Hence the total effort produced by effective gravity along the channel ADC of homogeneous fluid at the surface of $AHCFA$ whose ends are at the surface of $AHCFA$ is also equal to zero.

Clairaut presented his new principles of equilibrium and their derivations in a mixed up way. The confusion may result from what one scholar has perceived to

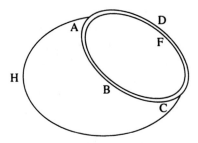

Figure 39. Figure used to show that Clairaut's first new principle of equilib-
rium holds for channels at the surface of a homogeneous fluid mass whose
densities are the same as the density of the mass when the principle holds for
channels in the interior of the mass filled with fluid from the mass.

be Clairaut's tendency to confound conditions for *figures of equilibrium* with
conditions for *forces*.[12] Clairaut also muddled the two conditions at the end of his
second paper on the theory of the earth's shape published in the *Philosophical
Transactions,* which caused Clairaut to make errors in analyzing the problem of
stratified figures of equilibrium in that paper. Once again, Clairaut may have
unknowingly jumbled conditions for forces and conditions for figures of equilibrium in that paper.

Clairaut evidently thought that his first new principle of equilibrium was the
strongest condition that must hold in order for homogeneous figures of equilibrium to exist, in the sense that he tried to show that in order for this principle of
equilibrium to hold for homogeneous fluid masses, all of the other principles of
equilibrium for homogeneous fluid masses (the principles of balanced columns
and the plumb line, in addition to Clairaut's second new principle of equilibrium)
must also hold. Clairaut eventually made more use of his second new principle of
equilibrium than he did his first new principle of equilibrium. Hence the second
new principle of equilibrium is the one of greater interest.

It is easy to show that when this second new principle of equilibrium holds for a
homogeneous fluid mass, then if one of the principles of balanced columns or the
plumb line holds as well for the homogeneous fluid mass, it follows directly as a
consequence that the other principle also holds for the homogeneous fluid mass.
(In order to demonstrate this, however, one must assume, as Clairaut in fact did
himself, that when this second new principle of equilibrium holds for a homogeneous fluid mass, this means that it not only holds for all closed plane or skew
channels that lie inside the homogeneous fluid mass and that are filled with fluid
from the homogeneous fluid mass, but that it also holds for closed plane or skew
channels of homogeneous fluid parts of which lie *at the surface* of the homogeneous fluid mass.[13]) In other words, the second new principle of equilibrium was
a condition that entailed that if both of the previously known principles of

equilibrium do not hold simultaneously for a homogeneous fluid mass, then neither of the previously known principles of equilibrium holds for the homogeneous fluid mass. (Desaguliers, for example, failed to realize this when, as we saw in Chapter 2, he gave an inconclusive demonstration that the columns from the center to the surface of a homogeneous elongated figure can never all balance or weigh the same. But if his proof were valid, it would have established that the principle of the plumb line cannot hold at the surface of a homogeneous elongated figure that attracts according to the universal inverse-square law, either, because, as I mentioned previously, this particular law of attraction permits Clairaut's second new principle of equilibrium to hold for homogeneous fluid masses. Later in this chapter we shall discover why. But Desaguliers did not recognize these connections, since he sometimes talked as if the principle of the plumb line could hold at the surfaces of homogeneous elongated figures of revolution which attract in accordance with the universal inverse-square law.) Thus the second new principle of equilibrium could be used to help identify hypotheses of attraction which allow one of the two previously known principles of equilibrium to hold for a homogeneous fluid mass but not the other, like some of the hypotheses of attraction which Bouguer investigated in his Paris Academy *mémoire* of 1734. (I use the qualification "help identify," because, as we shall see in a moment, although the second new principle of equilibrium is a *sufficient* condition for the two previously known principles of equilibrium to hold at the same time for a homogeneous fluid mass, the second new principle of equilibrium is not a *necessary* condition for this to happen.)

In arriving at these results, Clairaut may have taken his system of waters in equilibrium at the earth's surface together with waters in underground channels joined to the waters at the earth's surface, which he introduced at the end of his second paper on the theory of the earth's shape published in the *Philosophical Transactions*, as a model. For his new principles of equilibrium in effect explain why his intuitions about the equilibrium of such a system of waters, which he mentioned at the end of that paper, indeed turn out to be true. Clairaut's new principles of equilibrium explain why Clairaut found that columns of homogeneous fluid from the centers to the surfaces of the stratified figures discussed in that paper, where the homogeneous fluid columns in any such stratified figure all have the same density, all balance or weigh the same in any such stratified figure when the principle of the plumb line holds at all points on the surfaces of the stratified figures. (That his, Clairaut's new principles of equilibrium explain these intuitions and findings, once it is shown that the universal inverse-square law of attraction permits the second new principle of equilibrium to hold.)

In the discussion that immediately follows all homogeneous fluid masses will be homogeneous fluid figures of revolution which revolve in such a way and the forces of attraction will act in such a way as to produce the symmetries that I mentioned in Chapters 1, 4, and 6. (That is, a homogeneous fluid mass will be a figure of revolution which revolves around its axis of symmetry, in which case the axis of symmetry of the homogeneous fluid figure of revolution is also an axis of symmetry of the centrifugal force of rotation per unit of mass, and, moreover, the axis of symmetry of the homogeneous fluid figure of revolution will also be an axis of symmetry of the attraction. These conditions did in fact always hold for the homogeneous fluid figures discussed in Chapters 1, 4, and 6.) We recall that in such instances the problem becomes two dimensional. That is, it suffices to consider the effects of attraction and effective gravity at points in the plane of any meridian at the surface of such a homogeneous fluid figure of revolution, since these effects are exactly the same at corresponding points in different planes of meridians at the surface of such a homogeneous fluid figure of revolution because of the various symmetries.

We recall that Bouguer had concluded in his Paris Academy *mémoire* of 1734 that a central force of attraction whose magnitude at a point in a plane of a meridian at the surface of a homogeneous fluid figure of revolution which revolves around its axis of symmetry does not simply depend upon the distance of the point in the plane from the fixed center of force which produces the attraction and which is situated at the center of the homogeneous fluid figure of revolution but depends as well upon the angle that the direction of the attraction at the point in the plane, which is indicated by the line that passes through the point and the center of the homogeneous fluid figure of revolution, makes with some fixed line in the plane which passes through the center of the homogeneous fluid figure of revolution and which is chosen as an axis of reference would not allow the principles of balanced columns and the plumb line to hold at the same time, in which case such an hypothesis of attraction would not permit the homogeneous fluid figures of revolution which revolves around its axis of symmetry to exist in a state of equilibrium. Without referring to Bouguer, Clairaut easily showed, using the relevant example that appears in Bouguer's Paris Academy *mémoire* of 1734, that such an hypothesis of attraction would not allow his second new principle of equilibrium to hold.

That is, Clairaut considered a homogeneous fluid figure of revolution which revolves around its axis of symmetry and whose center is at C, and he let the closed plane curve *PEpe* that lies at the surface of the homogeneous fluid figure of revolution be one of the meridians of that figure. He assumed that C is a center of attraction, that the attraction at a point in the plane of *PEpe* is directed toward C, and that the magnitude of the attraction at the point does not simply depend upon the distance of the point from the center of attraction C but depends as well upon some other quantity, such as the angle that the line that passes through the point and C makes with the axis of symmetry of the homogeneous fluid figure of revolution chosen as an axis of reference. Next he considered the closed plane channel *abdc* which lies within the interior of the homogeneous fluid figure of revolution in the plane of *PEpe*, which is filled with fluid from the homogeneous fluid figure of revolution, and which consists of the two channels of fluid *ab* and *cd* that are shaped like circular arcs whose centers are at C and the two rectilinear

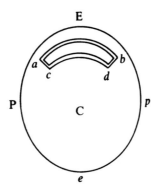

Figure 40. Figure used to show that a central force whose magnitude at points in a plane through the center of force depends upon the angles made with respect to some axis of reference through the center of force in the plane does not permit Clairaut's second new principle of equilibrium to hold (Clairaut 1743: figure on p. 29).

channels *ac* and *bd* that are directed toward *C*. As I mentioned, shortly we shall see that the effects of the centrifugal force produced when a mass of homogeneous fluid revolves around an axis can be ignored when trying to determine whether or not Clairaut's second new principle of equilibrium holds for the homogeneous fluid mass; all that needs to be considered is the total effort produced by the attraction around a closed channel that lies inside the homogeneous fluid mass and that is filled with fluid from the mass (see Figure 40). Then in the particular case in question the total efforts produced by the attraction along the two channels of fluid *ab* and *cd* that are shaped like circular arcs are equal to zero, because the attraction at each point on each of these two channels is perpendicular to the channel at the point on it. At the same time, the total efforts produced by attraction along the two rectilinear channels *ac* and *bd* are not equal in general, even though to each point on rectilinear channel *ac* there corresponds a point on rectilinear channel *bd* which is the same distance from the center of force *C* as the point on rectilinear channel *ac* which corresponds to it and vice versa and even though the homogeneous fluid in the two rectilinear channels have the same density, because the rectilinear channels *ac* and *bd* do not make the same angles with the axis of symmetry of the homogeneous fluid figure of revolution. Consequently the total effort produced by attraction around the closed plane channel *abdc* which lies inside the homogeneous fluid figure of revolution in the plane of *PEpe* and which is filled with fluid from the homogeneous fluid figure of revolution is not equal to zero in general. Hence this figure cannot be a figure of equilibrium.[14]

Thus Bouguer had come to the right conclusions, namely, that this homogeneous fluid figure of revolution which revolves around its axis of symmetry cannot be a figure of equilibrium. But he did so for the wrong reasons! Again without mentioning Bouguer, Clairaut demonstrated that a central force of attraction whose magnitude at a point in a plane of a meridian at the surface of a homogeneous fluid figure of revolution which revolves around its axis of symmetry does not simply depend upon the distance of the point in the plane from the fixed center of force which produces the attraction and which is situated at the center of the figure but depends as well upon the angle that the direction of the attraction at the point in the plane, which is indicated by the line that passes through the point and the center of the figure, makes with some fixed line in the plane which passes through the center of the figure and which is chosen as an axis of reference could be suitably defined so as to permit both of the principles of balanced columns and the plumb line to hold simultaneously.[15] Since the second new principle of equilibrium would not hold in this case, however, Clairaut thereby showed that this principle of equilibrium was truly independent of the two previously known priniciples of equilibrium.

Roughly speaking, if the magnitude of a central force of attraction is specified at all points on *one* column CM, where C is the center of the homogeneous fluid figure of revolution which revolves around its axis of symmetry, where the center of attraction is situated at C, and where M designates some particular point on a meridian at the surface of the homogeneous fluid figure of revolution which revolves around its axis of symmetry, and if it is stipulated moreover that the magnitude of the central force of attraction at a point in the plane of this meridian does not depend upon the angle that the line that passes through the point and the center of attraction, which is the line that indicates the direction of the attraction at the point, makes with some fixed line in the plane which passes through the center of attraction and which is chosen as an axis of reference, but depends only upon the distance r of the point from the center of attraction, then the magnitude of the central force of attraction at a point in the plane of the meridian in question has the form $f(r)$, where r stands for the distance of the point from the center of attraction, which is located at C, and this automatically fixes the magnitudes of the attraction at all points on *all* columns CM that join the center C of the homogeneous fluid figure of revolution to any other point M on the meridian in question. This severely limits the chances of having both of the principles of balanced columns and the plumb line hold at the same time. Introducing a variable θ that designates, say, the kinds of angles described previously, increases the possibilities of having both of these principles hold simultaneously instead of decreasing the chances. That is, it becomes possible to specify *individually* and *independently* the magnitudes $f(r, \theta)$ of a central force of attraction at points on each column CM that joins the center C of the homogeneous fluid figure of revolution to any other point M on the meridian in question, where r again stands for the distance of a point from the center of attraction, which is situated at C. This additional flexibility makes it easier for the two previously known principles of equilibrium to hold at the same time. (Today we would say that what Clairaut showed in essence is that while the principles of balanced columns and the plumb line are both necessary conditions that must be satisfied in order that homogeneous figures of equilibrium can exist, the two principles together do not constitute a condition that is sufficient for homogeneous figures of equilibrium to exist.)

In the "introduction" to his treatise of 1743, Clairaut stated that Bouguer's Paris Academy *mémoire* of 1734 is what had induced him to investigate conditions for equilibrium. Clairaut declared that Bouguer had made him wonder whether all of the conditions necessary for equilibrium had been determined.[16] Perhaps what Clairaut stated here is true, but it does seem more likely, based on the chronology of the events told so far in this story, that he first found the letters to him from MacLaurin, whom he neglected to mention at all in the "introduction," to be more troublesome than Bouguer's *mémoire* and that MacLaurin's letters, not Bouguer's *mémoire*, are what originally caused Clairaut to look into the question of conditions for equilibrium. At the same time, when Bouguer's work obviously did affect Clairaut's work in a very specific way, Clairaut did not mention that fact. Thus Clairaut failed to cite Bouguer's Paris Academy *mémoire* of 1734 as the source of the particular example above, although it is often easier to make progress when one already has the right example in hand to start with. It is not clear whether Clairaut deliberately concealed Bouguer's true influence or not. We must bear in mind that it was not customary to document sources at this time. If Clairaut did not clarify how Bouguer's work on figures of equilibrium specifically bore upon his own work on the same problem, it is also true that such disregard was a sign of the times.

Clairaut demonstrated that the question of whether his second new principle of equilibrium holds or not depends solely on the forces of attraction that act. That is, using the facts that the magnitudes of the centrifugal force per unit of mass at points that revolve around an axis at the same angular speed are the same at all such points which are the same distance from the axis and that the centrifugal force per unit of mass at such a point is perpendicular to the axis at the point, Clairaut claimed to show that the centrifugal force generated when a mass of homogeneous fluid revolves around an axis *never* produces a total effort around a closed channel that lies within the interior of the homogeneous rotating fluid mass and that is filled with fluid from the mass which is not equal to zero. (In fact, Clairaut did not really demonstrate this result, although it is in fact true. What he actually showed to be true is that the total efforts around closed plane channels filled with homogeneous fluid produced by the centrifugal force generated when a plane revolves around an axis that lies in the plane and the closed plane channels filled with homogeneous fluid all lie in that plane are equal to zero. That is, all that Clairaut really proved in detail is that when a plane revolves around an axis that lies in the plane, the centrifugal force generated *never* produces a total effort around a closed channel filled with homogeneous fluid which lies in that plane which is not equal to zero.) Hence in order to determine whether a mass of homogeneous fluid which revolves around an axis can exist in a state of relative equilibrium or not (more precisely, whether Clairaut's second new principle of equilibrium can hold for the homogeneous rotating fluid mass or not) when certain forces of attraction act, it is enough to determine whether a homogeneous fluid mass at absolute rest can exist in a state of equilibrium or not when those particular forces of attraction act. That is to say, the centrifugal force can be ignored when trying to determine whether or not Clairaut's second new principle of equilibrium holds for the homogeneous rotating fluid mass; it suffices to consider the total efforts produced by the attraction alone around closed channels that lie inside the mass and that are filled with fluid from the mass.[17]

Clairaut next demonstrated that if the law of attraction is such that the attraction has an axis of symmetry, then the total efforts produced by the attraction along two plane channels or skew channels filled with homogeneous fluid of the same density which lie at the surface of a figure of revolution whose axis of symmetry is the same as the axis of symmetry of the attraction and whose endpoints are situated on the same circles whose centers are both located on the axis of symmetry and which are both perpendicular to the axis of symmetry are equal. To do this he considered the two channels HI and KL filled with homogeneous fluid of the same density which lie at the surface of the figure of revolution $AFGB$. The endpoints I and L of the two channels are situated on the same circle $BILGR$ that is perpendicular to the axis of symmetry of the figure of revolution $AFGB$ and whose center is located on this axis of symmetry, and the endpoints H and K of the two channels are located on the same circle $AHKFQ$ that is perpendicular to the axis of symmetry of the figure of revolution $AFGB$ and whose center is situated on this axis of symmetry. Clairaut now imagined channel HI and the homogeneous fluid that fills it to be composed of an infinite number of infinitesimal homogeneous fluid right circular cylinders ("petits cylinders") Mm and, likewise, he imagined channel KL and the homogeneous fluid that fills it to be made up of an infinite number of infinitesimal homogeneous fluid right circular cylinders Nn, where all of the infinitesimal homogeneous fluid right circular cylinders have the same density. The points M and N are situated on the same circle $DMNE$ that is perpendicular to the axis of symmetry of the figure of revolution $AFGB$ and whose center is located on this axis of symmetry. This circle is perpendicular to the two planes of the meridians at the surface of the figure of revolution $AFGB$ which pass through M and N, respectively. The points m and n are also situated on the same circle $dmne$ that is perpendicular to the axis of symmetry of the figure of revolution $AFGB$ and whose center is also located on this axis of symmetry. This circle is perpendicular to the two planes of the meridians at the surface of the figure of revolution $AFGB$ which pass through the points r and s, respectively, where r and s are determined as follows (see Figure 41). The infinitesimal line segment Mr is an infinitesimal part of the plane curve formed where the plane of the meridian at the surface of the figure $AFGB$ which passes through M intersects the surface of $AFGB$. (In other words, Mr is just an infinitesimal part of the meridian at the surface of the figure of revolution

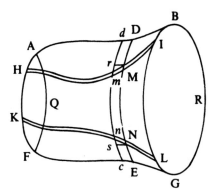

Figure 41. Figure of revolution and two channels at the surface whose end-points are situated on the same circles perpendicular to the axis of symmetry (Clairaut 1743: figure on p. 12).

AFGB which passes through *M*.) The infinitesimal line segment *rm* is perpendicular to *Mr*. In fact *rm* is just an infinitesimal part of the circle *dmne*. The infinitesimal line segment *Ns* is an infinitesimal part of the plane curve formed where the plane of the meridian at the surface of the figure of revolution *AFGB* which passes through *N* intersects the surface of *AFGB*. (In other words, *Ns* is just an infinitesimal part of the meridian at the surface of the figure of revolution *AFGB* which passes through *N*.) The infinitesimal line segment *sn* is perpendicular to *Ns*. In fact *sn* is just an infinitesimal part of the circle *dmne*. The distance between the two parallel circles *DMNE* and *dmne* is infinitesimal, since the lengths of the homogeneous fluid right circular cylinders *Mm* and *Nn* are infinitesimal. *Mr* is parallel to *Ns*, and *Mr* and *Ns* have the same length (namely, the distance between the two parallel circles *DMNE* and *dmne*).

Clairaut must now have reasoned in one of two possible ways. He may have taken the cross sections of the infinitesimal homogeneous fluid right circular cylinders *Mm* and *Nn* to have the same areas, in which case the volumes of the cylinders *Mm* and *Nn* are directly proportional to their lengths. Now, since cylinders *Mm* and *Nn* have the same density, the masses of the cylinders *Mm* and *Nn* are directly proportional to their volumes. But if the volumes of cylinders *Mm* and *Nn* are directly proportional to their lengths, then the masses of *Mm* and *Nn* are directly proportional to their lengths too, in which case the masses of *Mm* and *Nn* can be thought of as having units of length. (The actual value of the density of the infinitesimal homogeneous fluid right circular cylinders can be disregarded.)

Alternatively, since he actually treated *M*, *m*, *N*, and *n* like points, not like disks, Clairaut may have assumed instead that the lengths of the infinitesimal homogeneous fluid right circular cylinders *Mm* and *Nn* are at least one order of magnitude greater than the areas of their cross sections, so that the cylinders *Mm* and *Nn* are effectively one dimensional and thus can be imagined as having units of length. Then since Mm and Nn have the same density, their masses can be thought of as having units of length as well. (Again, the actual value of the density of the infinitesimal homogeneous fluid right circular cylinders can be ignored.)

Clairaut assumed that the magnitudes of the attraction at all points on the homogeneous fluid right circular cylinder Mm are the same, and he assumed that the directions of the attraction at all points on the cylinder are also the same. Likewise, he assumed that the magnitudes of the attraction at all points on the homogeneous fluid right circular cylinder Nn are the same, and he assumed that the directions of the attraction at all points on the cylinder are also the same. He made these assumptions because the lengths of cylinders are infinitesimal. Now, because of the symmetry of the attraction and the symmetry of the figure of revolution $AFGB$, the magnitudes of the attraction at M and at N are equal. Moreover, Clairaut stated that the attraction at M is directed toward r and that the attraction at N is directed toward s. Clairaut deduced that this statement is true from the fact that bodies placed at M and at N are not able to leave the surface of the figure of revolution $AFGB$. However, such a conclusion and the reason on which it is based are both rather dubious. The line that indicates the direction of the attraction at a point of the surface of a homogeneous fluid figure of revolution in equilibrium certainly lies in the plane of the meridian at the surface of the figure which passes through the point because of the symmetry of the attraction and the symmetry of the figure. But a body placed at a point of the surface of this figure does not leave the surface of the figure, either, yet the line that indicates the direction of the attraction at the point certainly is not tangent to the surface of the figure at the point, because the principle of the plumb line holds at all points of the surface of a homogeneous fluid figure of revolution in equilibrium. Thus Clairaut's last statement in fact need not be true. It is certainly true that because of the symmetry of the attraction and the symmetry of the figure of revolution $AFGB$, the line that indicates the direction of the attraction at M must lie in the plane of the meridian at the surface of the figure of revolution $AFGB$ which passes through M. As a result of the two symmetries, there can be no component of attraction at M whose direction is the same as the direction of rm, because there can be no component of attraction at M perpendicular to the plane of the meridian at the surface of the figure of revolution $AFGB$ which passes through M. But the attraction at M need not be directed toward r; it can be directed toward some other point in the plane of the meridian at the surface of the figure of revolution $AFGB$ which passes through M. Similarly, it is certainly true that because of the symmetry of the attraction and the symmetry of the figure of revolution $AFGB$ the line that indicates the direction of the attraction at N must lie in the plane of the meridian at the surface of the figure $AFGB$ which passes through N. Likewise, as a result of the two symmetries, there can be no component of attraction at N whose direction is the same as the direction of sn, because there can be no component of attraction at N perpendicular to the plane of the meridian at the surface of the figure of revolution $AFGB$ which passes through N. But the attraction at N need not be directed toward s; it can be directed toward some other point in the plane of the meridian at the surface of the figure of revolution $AFGB$ which passes through N. What Clairaut *should* have said is that the magnitudes of the *component* of the attraction at M directed toward r and the *component* of the attraction at N directed toward s are equal, which is clearly true because of the symmetry of the attraction and the symmetry of the figure of revolution $AFGB$. Moreover, as a result of the symmetry of the attraction and the symmetry of the figure of revolution $AFGB$, it is also clearly true that the angle between Mr and the line that indicates the direction of the attraction at M is the same as the angle between Ns and the line that specifies the direction of attraction at N.

It follows from the assumptions stated at the beginning of the preceding paragraph that the magnitudes of the components of the attraction directed toward m at all points on the homogeneous fluid right circular cylinder Mm are the same and that the magnitudes of the components of the attraction directed toward n at all points on the homogeneous fluid right circular cylinder Nn are also the same. Clairaut then stated that the magnitude of the component of the attraction at M directed toward m and the magnitude of the component of the attraction at N directed toward n are inversely proportional to the lengths of the infinitesimal homogeneous fluid right circular cylinder Mm and Nn. This follows from applying what Clairaut called "the theory of inclined planes" to the infinitesimal right triangles Mrm and Nsn. Now, for the reasons that I gave before, one can conclude in two different ways that the masses of the cylinders Mm and Nn are directly proportional to their lengths. Then the effort produced by the attraction along the cylinder Mm is just the constant value of the magnitudes of the components of the attraction directed toward m at all points on the cylinder, which equals the magnitude of the component of the attraction at M directed toward m, multiplied by the mass of the cylinder; the effort produced by the attraction along the infinitesimal homogeneous fluid right circular cylinder Nn is just the constant value of the magnitudes of the components of the attraction directed toward n at all points on the cylinder, which equals the magnitude of the component of the attraction at N directed toward n, multiplied by the mass of the cylinder; and these two products are the same because the magnitude of the component of the attraction at M directed toward m and the magnitude of the component of the attraction at N directed toward n are inversely proportional to the lengths of the infinitesimal homogeneous fluid right circular cylinders Mn and Nn and their masses are directly proportional to their lengths. Consequently the efforts produced by the attraction along the infinitesimal homogeneous fluid right circular cylinders Mm and Nn are equal. Hence the total efforts produced by the attraction along the channels HI and KL filled with homogeneous fluid of the same density are equal too. This is true because the total effort produced by the attraction along the channel HI filled with homogeneous fluid is just the sum of the efforts produced by the attraction along all of the infinitesimal homogeneous fluid right circular cylinders Mm that make up the channel HI filled with homogeneous fluid; the total effort produced by the attraction along the channel KL filled with homogeneous fluid is just the sum of the efforts produced by the attraction along all of the cylinders Nn that make up the channel KL filled with homogeneous fluid; and the densities of all of the cylinders Mm and Nn are the same.

Clairaut now used this result to demonstrate the following. Imagine that a homogeneous fluid figure of revolution revolves in such a way and that the forces of attraction act in such away as to produce the symmetries that I mentioned in Chapters 1, 4, and 6. (That is, the homogeneous fluid figure of revolution revolves around its axis of symmetry, in which case the axis of symmetry of the figure is also an axis of symmetry of the centrifugal force of rotation per unit of mass, and, moreover, the axis of symmetry of the figure is also an axis of symmetry of the attraction. These conditions did in fact always hold for the homogeneous fluid figures discussed in Chapters 1, 4, and 6.) Then according to what Clairaut showed, in order that this homogeneous fluid figure of revolution which revolves around its axis of symmetry can be a figure of equilibrium, the total efforts produced by the attraction around all closed channels that lie in the interior of the figure and that are filled with fluid from the figure, where the closed channels can be closed plane channels or closed skew channels, must be equal to zero. That is, according to what Clairaut demonstrated (above), the centrifugal force generated when a mass of homogeneous fluid revolves around an axis can be ignored when trying to determine whether or not Clairaut's second new principle of equilibrium holds for the homogeneous rotating fluid mass. It suffices to consider the total efforts produced by the attraction alone around closed channels that lie inside the mass and that are filled with fluid from the mass.

This condition can be restated another way. In order that this homogeneous fluid figure of revolution which revolves around its axis of symmetry can be a figure of equilibrium, the total efforts produced by the attraction along all channels, plane or skew, that lie inside the figure, that are filled with fluid from the figure, and that join the same two points H and I must be the same, for each pair of points H and I situated inside the figure.

Then consider any closed channel that lies within the interior of the homogeneous fluid figure of revolution and that is filled with fluid from the figure. This closed channel of fluid can be thought of as composed of two channels that lie inside the homogeneous fluid figure of revolution, that are filled with fluid from the figure, and that connect the same two points H and I. Let $AFGB$ be the figure of revolution whose axis of symmetry is the same as the axis of symmetry of the given homogeneous fluid figure of revolution and upon whose surface one of these two channels of fluid which join H and I lies. Then it follows from the preceding result that the total effort produced by the attraction along the channel formed where a plane of a meridian at the surface of the given homogeneous fluid figure of revolution, which is also a plane of a meridian at the surface of the figure of revolution $AFGB$, intersects the surface of the figure of revolution $AFGB$ and which is filled with fluid from the given homogeneous fluid figure of revolution (in other words, a channel along a meridian at the surface of the figure of revolution $AFGB$ which also lies in the particular plane of a meridian in question at the surface of the given homogeneous fluid figure of revolution and which is filled with fluid from the given homogeneous fluid figure of revolution) is the same as the total effort produced by the attraction along this channel filled with fluid from the given homogeneous fluid figure of revolution which joins H and I. Now we apply the same argument to the other channel of fluid which connects H and I. This time let $AFGB$ designate the figure of revolution whose axis of symmetry is the same as the axis of symmetry of the given homogeneous fluid figure of revolution and upon whose surface this other channel of fluid which joins H and I lies. Again it follows from the preceding result that the total effort produced by the attraction along the channel formed where the particular plane of a meridian in question at the surface of the given homogeneous fluid figure of revolution, which is also a plane of a meridian at the surface of this other figure of revolution $AFGB$, intersects the surface of this other figure of revolution $AFGB$ and which is filled with fluid from the given homogeneous fluid figure of revolution (in other words, a channel along a meridian at the surface of this other figure of revolution $AFGB$ which also lies in the particular plane of a meridian in question at the surface of the given homogeneous fluid figure of revolution and which is filled with fluid from the given homogeneous fluid figure of revolution) is the same as the total effort produced by the attraction along this other channel filled with fluid from the given homogeneous fluid figure of revolution which joins H and I. Moreover, the two channels in the plane of the meridian in question at the surface of the given homogeneous fluid figure of revolution have the same endpoints; hence these two channels together form a closed channel that lies within the interior of the given homogeneous fluid figure of revolution in the plane of the meridian in question at its surface. In short, we have produced a closed plane channel that lies inside the given homogeneous fluid figure of revolution in the plane of the meridian in question at its surface and that is filled with fluid from the given homogeneous fluid figure of revolution such that the total effort produced by the attraction around this closed channel of fluid is the same as the total effort produced by the attraction around the closed channel of fluid which we started with.

Thus in order to determine whether or not Clairaut's second new principle of equilibrium holds for a homogeneous fluid figure of revolution which revolves around its axis of symmetry, in which case the axis of symmetry is also an axis of symmetry of the centrifugal force of rotation per unit of mass, and, morevoer, whose axis of symmetry is also an axis of symmetry of the attraction, it is clearly enough to determine whether or not the total efforts produced by the attraction around all closed plane channels that lie within the interior of the homogeneous fluid figure of revolution in the plane of any meridian at the surface of the figure and that are filled with fluid from the figure are always equal to zero. Equivalently, in order to determine whether or not Clairaut's second new principle of equilibrium holds for a homogeneous fluid figure of revolution which revolves around its axis of symmetry, in which case the axis of symmetry is also an axis of symmetry of the centrifugal force of rotation per unit of mass, and, moreover, whose axis of symmetry is also an axis of symmetry of the attraction, it is clearly enough to determine whether or not the total efforts produced by attraction along any two channels that lie inside the homogeneous fluid figure of revolution in the plane of any meridian at the surface of the homogeneous fluid figure of revolution, that are filled with fluid from the figure, and that join the same two points are always the same, for each pair of points located inside the figure in the plane of that meridian at the surface of figure. Consequently in cases where homogeneous fluid masses and the attraction fulfill the condition stated, a problem in three dimensions reduces to a problem in two dimensions.[18]

Clairaut introduced in addition a third new principle of equilibrium for homogeneous masses of fluid. When the effective gravity is perpendicular to a surface at all of the points of a surface (in other words, when the principle of the plumb line holds at all of the points of the surface), Clairaut called the surface a level surface ("surface courbe de niveau"). He also defined a level layer ("couche de niveau") to be the space between two level surfaces. Clairaut then imagined a homogeneous fluid mass to be partitioned into an infinite number of level layers. Each point K inside the homogeneous fluid mass lies within the interior of one of the level layers or is a point of one of the two level surfaces that bound one of the level layers (see Figure 42). Here $PEpe$ and HKI are closed plane curves that lie in two different level surfaces and that also lie in the same plane, where it is assumed that the effective gravity at each point of one of these closed plane curves is perpendicular at the point to the level surface in which the closed plane curve lies. Moreover onl and qst are cross sections of two contiguous level layers which lie in the plane of $PEpe$ and HKI. (The homogeneous fluid mass need not be a figure of revolution.) Clairaut showed that in order for the homogeneous fluid mass to be in equilibrium, the thickness of the level layer at a point K of one of the two level surfaces that bound the level layer or in whose interior the point K lies must be inversely proportional to the magnitude of the effective gravity at the point K.[19] (Clairaut specified an infinity of such level layers. However, such a condition of course makes no sense, if each level layer has the finite thickness that Clairaut tacitly assumed in defining the notion of a level layer. The total number of such level layers in a three-dimensional figure whose size is finite must be finite. The true purpose of the stipulation that there be an infinite number of level layers is only to assure that all of the level layers be very thin, for reasons that we will understand in a moment, in which case the number of them will of course be very large. Moreover, as we shall also discover later in this chapter, level layers *cannot* in general be taken to be *infinitesimally thin* in Clairaut's theory of figures of equilibrium. That is, in Clairaut's theory of figures of equilibrium the level layers

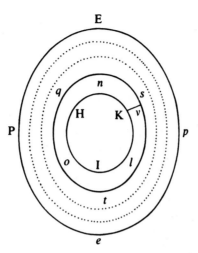

Figure 42. Cross sections of the level layers of a homogeneous mass of fluid (Clairaut 1743: figure on p. 41).

must in general *have finite thicknesses*, if the *densities* of the individually homogeneous level layers *are not all the same*. In other words, the level layers cannot in general be infinitesimally thin (that is to say, the level layers cannot be surfaces themselves) in Clairaut's theory of figures of equilibrium, if the densities of the individually homogeneous level layers are not all the same, unless special conditions are fulfilled. I shall discuss these conditions later in this chapter.)

To derive this third new principle of equilibrium, Clairaut reasoned as follows. He started with a core of homogeneous fluid which he assumed to be in equilibrium and in whose surface *HKI* lies. Then it appeared evident to Clairaut, as a result of what he called "the first principles of hydrostatics," that if a force is exerted at each point of the surface of this core of homogeneous fluid which is perpendicular to this surface at each point of the surface and whose magnitude is the same at all points of this surface, then the equilibrium of the core of homogeneous fluid would not be disturbed. Now, it turns out that if the inverse proportion that Clairaut stated as his third new principle of equilibrium holds, then surrounding this core of homogeneous fluid with a very thin level layer of homogeneous fluid whose density is the same as the density of the core of homogeneous fluid and whose cross section is *onl* would produce such a force whose magnitude is uniform at all points of the surface of the core of homogeneous fluid and which is orthogonal to this surface at all points of the surface. This can be seen as follows. First, the force would be perpendicular to the surface of the core of homogeneous fluid at all points of the surface because of what it means for a very thin layer of homogeneous fluid whose cross section is *onl* to be a "level layer." Second, the magnitude of the force at a point *K* of the surface of the core of homogeneous fluid would equal the thickness at the point K of the level layer of homogeneous fluid whose cross section in *onl* multiplied by the magnitude of the effective gravity at the point K, since the level layer is assumed to be so thin that the magnitudes of the effective gravity at all points of a line segment Kv that joins the two surfaces of the level layer and that is orthogonal to both surfaces of the level layer can be assumed to be the same and, at the same time, the effective gravity at all points of the line segment Kv can be assumed to have the same direction as the direction of the line segment Kv, in which case the magnitude of the force ("pression") at K exerted by the level layer of homogeneous fluid whose cross section is *onl* is simply the length of the line segment Kv, which is directly proportional to the mass of line segment Kv, multiplied by the magnitude of the effective gravity at K. (See Figure 42.)

[Here Clairaut imagined Kv to be a very short homogeneous fluid right circular cylinder ("petit cylindre") that traverses at the point K in question of the surface of the homogeneous core of fluid the level layer of homogeneous fluid whose cross section is *onl* and that is perpendicular to the two surfaces of the level layer. Again Clairaut must have reasoned in one of two possible ways. Since the fluid in the level layer whose cross section is *onl* is homogeneous, the masses of the very short homogeneous fluid right circular cylinders that make up this level layer are directly proportional to their volumes. Now, the cross sections of these very short cylinders can all be taken to have the same areas, and Clairaut may have made this assumption. In this case the volumes of the very short cylinders are directly proportional to their lengths. Then, since the masses of these cylinders are directly proportional to their volumes and, at the same time, the volumes of the cylinders are directly proportional to their lengths, it follows that the masses of the cylinders are also directly proportional to their lengths. Hence the masses of the very short homogeneous fluid right circular cylinders that make up the level layer can be thought of as having units of length. (The actual value of the density of the very short homogeneous fluid right circular cylinders that make up the level layer can be disregarded.)

Alternatively, since K in fact really designates a point and not a disk, Clairaut may have supposed instead that the area of a cross section of a very short homogeneous fluid right circular cylinder Kv is at least one order of magnitude smaller than its length (that is, at least one order of magnitude smaller than the thickness at K of the level layer of homogeneous fluid whose cross section is *onl*), in which case cylinder Kv is effectively one dimensional and thus can be thought of as having units of length. Then, since the fluid in the level layer whose cross section is *onl* is homogeneous, the mass of cylinder Kv can be imagined as having units of length too. (Again, the actual value of the density of the very short homogeneous fluid right circular cylinders that make up the level layer can be ignored.)

But if the inverse proportion that Clairaut stated as his third new principle of equilibrium holds, this product will have the same value for all points of the surface of the core of homogeneous fluid. In other words, since the level layer of homogeneous fluid whose cross section is *onl* is very thin, the magnitudes of the effective gravity at all points of a line segment *Kv* that traverses the level layer from its inner surface to its outer surface and that is perpendicular to both of these surfaces can be assumed to be the same, and, at the same time, the effective gravity at all points of the line segment *Kv* can be assumed to have the same direction as the direction of the line segment *Kv*. But then the effort produced by effective gravity along this line segment *Kv* is just the constant value of the magnitudes of the effective gravity at the points of the line segment *Kv* multiplied by the length of the line segment *Kv* since the masses of the line segments here have units of length. But this product, which is the same as the magnitude of the force ("pression") at *K* exerted by the level layer of homogeneous fluid whose cross section is *onl*, will have the same value for all points of the surface of the core of homogeneous fluid if the inverse proportion that Clairaut stated as his third new principle of equilibrium holds. Consequently the force exerted at each point *K* of the surface of the core of homogeneous fluid by the level layer of homogeneous fluid which surrounds the core of homogeneous fluid and whose cross section is *onl*, which is perpendicular to the surface of the core at each point *K* of its surface, will have the same magnitude at all points of the surface. Clairaut concluded from this that the equilibrium of the core of homogeneous fluid would not be disturbed, as a result of what he called "the first principles of hydrostatics."[20]

The masses of the core of homogeneous fluid and the level layer of homogeneous fluid which surrounds the core and whose cross section is *onl* can then be combined to form a new core of homogeneous fluid in a state of equilibrium. (The level layer of homogeneous fluid whose cross section is *onl* certainly cannot circulate around the original core of homogeneous fluid at the surface of this core, since the force acting on the level layer of homogeneous fluid whose cross section is *onl* is perpendicular to the level layer everywhere.) This argument can be repeated using a new level layer of homogeneous fluid whose cross section is *qst*. In fact the argument can be repeated over and over again to produce a homogeneous mass of fluid in equilibrium with as many level layers as desired.

Finally, since the original core of homogeneous fluid in equilibrium was chosen at will, it can be chosen to be arbitrarily small. In particular, it can be chosen to be "so small, that it be no more than an infinitely small particle, which cannot fail to be in equilibrium."[21] In other words, it is chosen to be a portion of fluid[22] which is smaller than any amount of fluid which could possibly disturb itself. This assures that the whole chain of reasoning does indeed have a starting point.

Clairaut also showed that this third new principle of equilibrium was not an independent principle of equilibrium – that in fact it followed from his second new principle of equilibrium. Namely, he considered two level surfaces that bound a homogeneous level layer of fluid whose cross section is *πσγλ*, and he considered a closed plane channel *ABDC* filled with this homogeneous fluid which lies in the plane of this cross section such that each of the channels of fluid

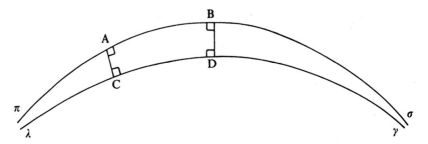

Figure 43. Figure used for deriving Clairaut's third new principle of equilibrium from Clairaut's second new principle of equilibrium .

AB and *CD* lies in one of the two level surfaces and *AC* and *BD* are rectilinear channels of fluid which are orthogonal to these same two level surfaces (see Figure 43). (Here we have evidence that Clairaut did in fact imagine his channels of fluid to be one dimensional. In other words, Clairaut's plane channels of fluid are really plane curves, and his skew channels of fluid are actually skew curves. We shall find more evidence that supports these statements shortly.) Then the principle of the plumb line holds at all points on the channels of fluid *AB* and *CD*. This statement follows from the meaning of level surfaces. But this means that the total efforts produced by effective gravity along the channels of fluid *AB* and *CD* must be equal to zero. Consequently the total efforts produced by effective gravity along the channels of fluid *AC* and *BD* will be the same, if the total effort produced by effective gravity around the closed channel of fluid *ABDC* is equal to zero. Thus the force that the transverse weight of the homogeneous level layer of fluid whose cross section is $\pi\sigma\gamma\lambda$ exerts upon the level surface that is its inner boundary will be uniform at all points of this level surface. In particular, this means that the thickness of the level layer at a point *K* of the inner surface of the level layer, which is just the length of the line segment *Kv* that joins the points *K* and *v* of the two level surfaces that bound the level layer and that traverses the layer at right angles to the two level surfaces that bound the level layer, must be inversely proportional to the magnitude of the effective gravity at *K* when the two level surfaces neighbor each other (that is, when the level layer is very thin), because in this case the magnitudes of the effective gravity all at points of the line segment *Kv* can be assumed to be the same and, at the same time, the effective gravity at all points of the line segment *Kv* can be assumed to have the same direction as the direction of the line segment *Kv*.[23] [Here we see again that Clairaut in fact assumed that when his second new principle of equilibrium held for a homogeneous fluid mass, this meant that it not only held for all closed plane or skew channels that lie entirely within the interior of the mass and that are filled with fluid from the mass, but that it also held for closed plane or skew channels of homogeneous fluid parts of which lie *along the surface* (in other words, lie *at the boundary*) of the mass.]

To apply his new principles of equilibrium, Clairaut often expressed them using mathematics. I turn to the mathematical representations of his new principles of equilibrium which he utilized. These mathematical representations do not appear anywhere in his second paper on the theory of the earth's shape published in the *Philosophical Transactions*. It is certain that he did not invent them before he learned of Fontaine's work involving the partial differential calculus which I discussed in Chapter 7. As we shall see, this can hardly be fortuitous. It seems unlikely that Clairaut *could* have thought up such mathematical representations before he learned late in 1738 how Fontaine used the partial differential calculus to integrate in finite terms or reduce to quadratures first-order ordinary "differential" equations of first degree and then subsequently undertook to do work in the same area himself early in 1739. Even more noteworthy is the lack of any evidence that Clairaut had even discovered the new principles of equilibrium *themselves* before he began to criticize Fontaine's application of the partial differential calculus to the integral calculus and began to try to improve upon Fontaine's work in this area early in 1739. In other words, it appears as if Fontaine's encounter with Clairaut, which had nothing whatever to do with principles of equilibrium and which was limited exclusively to matters having to do with the integral calculus, somehow functioned in some way in Clairaut's determining the new principles of equilibrium.

Clairaut first worked out mathematical representations of his new principles of equilibrium for the case of a homogeneous fluid figure of revolution which revolves in such a way and where the forces of attraction act in such a way as to produce the symmetries that I mentioned in Chapters 1, 4, and 6. (That is, the homogeneous fluid figure of revolution revolves around its axis of symmetry, in which case the axis of symmetry of the figure is also an axis of symmetry of the centrifugal force of rotation per unit of mass, and, moreover, the axis of symmetry of the figure is also an axis of symmetry of the attraction. These conditions did in fact always hold for the homogeneous fluid figures discussed in Chapters 1, 4, and 6.) In this case a problem in three dimensions reduces to a problem in two dimensions, as I have already explained. That is, it suffices to consider the effects of attraction and effective gravity at points in the plane of any meridian at the surface of such a homogeneous fluid figure of revolution, since these effects are exactly the same at corresponding points in planes of different meridians at the surface of such a homogeneous fluid figure of revolution because of the various symmetries. As we saw above, in order that Clairaut's second new principle of equilibrium hold in this case, it is enough that the total efforts produced by the attraction around all closed channels that lie inside the homogeneous fluid figure of revolution in a plane of a meridian at the surface of the figure and that are filled with fluid from the figure be equal to zero.

To express the principle of equilibrium that the total efforts produced by attraction around all closed channels that lie within the interior of the homogeneous fluid figure of revolution in a plane of a meridian at the surface of the figure and that are filled with fluid from the figure are equal to zero, where the figure revolves in such a way and where the forces of attraction act in such away as to produce the symmetries that I mentioned in Chapters 1, 4, and 6 – or "what amounts to the same thing, that the total efforts produced by attraction along all channels [that lie inside the homogeneous fluid figure of revolution in the plane of a meridian at the surface of the figure and that are filled with fluid from the figure, where the figure revolves in such a way and where the forces of attraction act in such a way as to produce the symmetries that I mentioned in Chapters 1, 4, and 6 and] that join the same two points O and N are the same" – Clairaut considered an arbitrary plane channel ON that lies within the interior of the homogeneous fluid figure of revolution in the plane of a meridian at the surface of the figure whose center is at C and that is filled with fluid from the figure. He let x and y, respectively, stand for rectangular coordinates in this plane of a meridian at the surface of the homogeneous fluid figure of revolution which are parallel and perpendicular, respectively, to the axis of symmetry of the figure, which the line segment CP is part of. PE represents one quarter of the meridian. (In other words, PE is the part of the meridian which lies in one quadrant of the plane of the meridian). Clairaut let S and s designate two points on the channel ON which are infinitesimally near each other, and he let $CH = x$, $HS = y$, $Sr = dx$, and $sr = dy$. (Hence $Ss^2 = dx^2 + dy^2$.) In effect Clairaut imagined the channel ON filled with homogeneous fluid to be composed of an infinite number of infinitesimal homogeneous fluid right circular cylinders Ss. (See Figure 44.)

Clairaut then resolved the attraction at each point S on the channel ON into the following two components: $P =$ the magnitude of the component of attraction at S which is perpendicular to the axis of symmetry of the homogeneous fluid figure of revolution, which the line segment CP is part of, and which lies in the plane of the meridian at the surface of the figure which includes the channel ON; and $Q =$ the magnitude of the component of attraction at S which is parallel to the axis of symmetry of the figure, which the line segment CP is part of, and which lies in the plane of the meridian at the surface of the figure which includes the channel ON. Thus the components of the attraction at S whose magnitudes are P and Q, respectively, lie in the plane of the meridian at the surface of the homogeneous fluid figure of revolution which includes the channel ON in the directions of the rectangular coordinates y and x, respectively. [There is no third component of the attraction at S which is perpendicular to the plane of the meridian at the surface of the figure which includes the channel ON, because the axis of symmetry of the figure, which the line segment CP is part of, is also an axis of symmetry of the attraction.] Next Clairaut determined the effort produced by the attraction along an infinitesimal homogeneous fluid right circular cylinder Ss. This he expressed in terms of the efforts along the cylinder Ss produced individually by the components of the attraction whose magnitudes are P and Q, respectively, which lie in the plane of the meridian at the surface of the homogeneous fluid figure of revolution which includes the channel ON in the directions of the rectangular

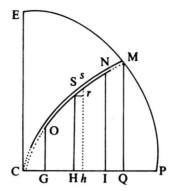

Figure 44. Meridian of a quadrant of a homogeneous fluid figure of revolution and interior channel expressed in two rectangular coordinates in the plane of the meridian (Clairaut 1743: figure on p. 36).

coordinates *y* and *x*, respectively. Clairaut assumed the values of *P* to be the same at all points on the infinitesimal homogeneous fluid right circular cylinder *Ss*, because the length of cylinder *Ss* is infinitesimal. For the same reason he also assumed the values of *Q* to be the same at all points on *Ss*.

Again Clairaut must have now reasoned in one of two possible ways. Since the channel *ON* is filled with fluid that is homogeneous, the masses of the infinitesimal homogeneous fluid right circular cylinders *Ss* that make up channel *ON* filled with homogeneous fluid are directly proportional to their volumes. Now, the cross sections of the cylinders *Ss* that make up channel *ON* can all be taken to have the same areas, and Clairaut may have made this assumption. In this case the volumes of cylinders *Ss* are directly proportional to their lengths. Then, since the masses of the cylinders *Ss* are directly proportional to their volumes and, at the same time, the volumes of the cylinders *Ss* are directly proportional to their lengths, it follows that the masses of the cylinders *Ss* are also directly proportional to their lengths. Hence the masses of these infinitesimal homogeneous fluid right circular cylinders *Ss* that make up channel *ON* filled with homogeneous fluid can in effect be thought of as having units of length. (The actual value of the density of the cylinders *Ss* that make up channel *ON* filled with homogeneous fluid can be disregarded.)

Alternatively, since *S* and *s* really stand for points, not disks, Clairaut may have assumed instead that the area of a cross section of an infinitesimal homogeneous fluid right circular cylinder *Ss* is at least one order of magnitude smaller than its length, in which case the cylinder *Ss* is effectively one dimensional and thus can be imagined as having units of length. Then, since the fluid in channel *ON* is homogeneous, the mass of cylinder *Ss* can also be thought of as having units of length. (Again, the actual value of the density of the cylinders *Ss* that make up channel *ON* filled with homogeneous fluid can be ignored.)

Clairaut first looked at the component of the attraction at S in the direction of the rectangular coordinate y whose magnitude is P. Clairaut determined the magnitude at S of the component of this component of the attraction whose direction is the same as the direction of the infinitesimal right circular cylinder Ss to be equal to

$$P \times \frac{rs}{Ss} = \frac{P \times rs}{Ss}.$$

Since the values of P are assumed to be the same at all points on the cylinder Ss, the values of the magnitudes of these components of the attraction are also the same at all points on Ss. Namely, they are equal to

$$P \times \frac{rs}{Ss} = \frac{P \times rs}{Ss}.$$

Now, we recall that the mass of the infinitesimal homogeneous fluid right circular cylinder Ss is just its length. Then the effort along cylinder Ss produced by the components of the attraction at all points on Ss in the direction of the rectangular coordinate y whose magnitudes all have the same value P is just the constant value at all points on Ss of the magnitudes of the components of these components of the attraction whose directions are the same as the direction of the infinitesimal homogeneous fluid right circular cylinder Ss, which is equal to

$$P \times \frac{rs}{Ss} = \frac{P \times rs}{Ss},$$

multiplied by the mass of the cylinder Ss. This product is

$$\frac{P \times rs}{Ss} \times Ss = P \times rs = P \, dy.$$

Thus $P \, dy$ is the effort along cylinder Ss produced by the components of the attraction at all points on cylinder Ss in the direction of the rectangular coordinate y whose magnitudes all have the same value P. Similarly, after next looking at the component of the attraction at S in the direction of the rectangular coordinate x whose magnitude is Q, Clairaut showed that $Q \, dx$ is the effort along the cylinder Ss produced by the components of the attraction at all points on cylinder Ss in the direction of the rectangular coordinate x whose magnitudes all have the same value Q. Then by composition of forces, Clairaut concluded that $P \, dy + Q \, dx$ is the effort produced by the attraction along the infinitesimal homogeneous fluid right circular cylinder Ss.[24]

Clairaut stated that in order to calculate the total effort produced by attraction along a straight channel or a curved plane channel ON that lies within the interior of the homogeneous fluid figure of revolution in the plane of a meridian at the surface of the figure and that is filled with fluid from the figure, when the total effort depends upon the particular plane channel that joins the two points O and N, the equation in x and y for the plane channel ON would ordinarily be solved for y as an expression in x, and then this expression for y in terms of x would be used to determine dy as an expression in x and dx. These two expressions for y and dy, respectively, would then be substituted for y and dy, respectively, in the expression $P(y,x)dy + Q(y,x)dx$, and by this means y and dy would be eliminated from the expression $P(y,x)dy + Q(y,x)dx$. The expression in x and dx which would result after y and dy are eliminated from the expression $Pdy + Qdx$ could then be integrated (which implicitly meant at the time that the fundamental theorem of the calculus could be utilized to antidifferentiate the expression) to find the total effort produced by the attraction along the plane channel ON filled with fluid from the homogeneous fluid figure of revolution, where the constant of integration would be chosen to make the integral equal to zero when $x = CG$, in which case the integral when $x = CI$ would be the total effort produced by the attraction along the plane channel ON filled with fluid from the homogeneous fluid figure of revolution. Here Clairaut explained, for the first time ever, how to calculate the total effort produced by attraction along a *curved* plane channel ON that lies inside a homogeneous fluid figure of revolution in the plane of a meridian at the surface of the figure and that is filled with fluid from the figure.[25] (Here we also have additional evidence that Clairaut really thought of his plane channels filled with homogeneous fluid as one dimensional. In other words, he actually imagined these channels to be plane curves. Moreover, Clairaut also assumed here that the equation of a plane curve can be solved explicitly for y in terms of x and, more precisely, that y can be expressed in a finite number of such terms in x. However, we remember from the discussion of the problem of orthogonal trajectories in Chapter 7 that what Clairaut assumed here in fact cannot be done for the equations of a large number of transcendental plane curves and that it can never be done for the equations of algebraic plane curves whose equations are algebraic equations of degree $\geqslant 5$ in y. Nor is it necessarily true that after eliminating y and dy, when this can be done, the resulting expression in x and dx can be antidifferentiated.)

However, in order for this calculation of the total effort produced by the attraction along the plane channel ON that lies within the interior of the homogeneous fluid figure of revolution in the plane of a meridian at the surface of the homogeneous fluid figure of revolution and that is filled with fluid from the figure to be independent of the particular plane channel that joins the two points O and N, which is what Clairaut's theory of homogeneous fluid figures of equilibrium required, Clairaut declared that "$Pdy + Qdx$ must be a complete differential"[26] in the sense that he and Fontaine had used the expression "complete differential" in their work on integrating in finite terms or reducing to quadratures first-order ordinary "differential" equations of first degree. Namely, an expression $\phi(y,x)$ must exist whose total differential is the differential $Pdy + Qdx$, where $\phi(y,x)$ is uniquely determined up to an additive constant.

Clairaut mentioned that in his second Paris Academy *mémoire* on integral calculus he had derived the condition

$$\frac{dP}{dx} = \frac{dQ}{dy} \qquad (7.3.4.6''')$$

that the differential $Pdy + Qdx$ must fulfill in order to be a complete differential, and he noted further that the symbols on the left-hand side and right-hand side of equality (7.3.4.6''') stood for partial differential coefficients.

Clairaut did not actually prove that a differential $Pdy + Qdx$ whose "line integral" is path independent is always a complete differential. In order to do that he would have had to show that the line integral of a differential $Pdy + Qdx$ whose line integral is path independent can actually *be used to construct* an antidifferential of the differential $Pdy + Qdx$ [that is, an expression $\phi(y, x)$ whose total differential is the differential $Pdy + Qdx$, where $\phi(y, x)$ is uniquely determined up to an additive constant], but Clairaut did not do this. Indeed, he did not use the fact that this can be done to produce his antidifferentials of complete differentials $Pdy + Qdx$ in his two Paris Academy *mémoires* on integral calculus but constructed these antidifferentials another way. It is quite plausible, nevertheless, that the mechanics problem, and, in particular, Clairaut's interpretation of the differential $Pdy + Qdx$ as the effort produced by the attraction along an infinitesimal homogeneous fluid right circular cylinder Ss, is what originally led Clairaut to introduce the line integral in his work and to associate the path independence of the line integral of the differential $Pdy + Qdx$ with the condition (7.3.4.6''') for the differential $Pdy + Qdx$ to be complete. [Once Clairaut had established the meaning of a "line integral"

$$\int_\gamma Pdy + Qdx,$$

of a differential $Pdy + Qdx$, whether he did this through the mechanics problem or not, he could have easily concluded that if an expression $\phi(y, x)$ exists whose total differential is the differential $Pdy + Qdx$, then for any plane curve γ that joins the two points $a = (y_0, x_0)$ and $b = (y_1, x_1)$, it is true that

$$\int_\gamma Pdy + Qdx = \int_\gamma d\phi = \phi(b) - \phi(a)$$

holds, independently of the particular plane curve γ that joins the two points $a = (y_0, x_0)$ and $b = (y_1, x_1)$ and that is the path of integration of the line integral.[27] This is the converse of what is stated in the first sentence of this paragraph, and when this converse is true, it is easy to demonstrate that it is true, as we have just seen. However, this converse is true if and only if the expression $\phi(y, x)$ whose total differential is the differential $Pdy + Qdx$ is a *single-valued function* of y and x. In order for this converse to be true the expression $\phi(y, x)$

cannot be a multiple-valued function of y and x. That is, it turns out that the total differentials of multiple-valued functions $\phi(y, x)$ can be complete differentials $P dy + Q dx$ whose line integrals are *not* path independent. In other words, the converse of what is stated in the first sentence of this paragraph need *not* be true. As we shall see shortly, there is evidence that Clairaut did not realize this.] Be that as it may, the geometric notion of line integrals and their path independence had not appeared at all in Fontaine's and Clairaut's formulations of the problem of integrating in finite terms or reducing to quadratures first-order ordinary "differential" equations of first degree in terms of complete differentials. The idea only materialized for the first time later in Clairaut's further work on the mechanics problem. Moreover, this idea is one of the rare exceptions to the trend of eighteenth-century algebraic analysis to develop independently of the geometric motifs that gave rise to it.[28]

Again we continue to assume that the homogeneous fluid figure of revolution revolves in such a way and that the forces of attraction act in such a way as to produce the symmetries that I mentioned in Chapters 1, 4, and 6. (That is, the homogeneous fluid figure of revolution revolves around its axis of symmetry, in which case the axis of symmetry of the homogeneous fluid figure of revolution is also an axis of symmetry of the centrifugal force of rotation per unit of mass, and, moreover, the axis of symmetry of the homogeneous fluid figure of revolution is also an axis of symmetry of the attraction. These conditions did in fact always hold for the homogeneous fluid figures discussed in Chapters 1, 4, and 6.) In this case a problem in three dimensions reduces to a problem in two dimensions, as I have already mentioned. If this homogeneous fluid figure of revolution is, moreover, a figure of equilibrium, then the following three principles of equilibrium all hold for this homogeneous fluid figure of revolution in equilibrium: (1) Clairaut's second new principle of equilibrium; (2) the principle of equilibrium that the total efforts produced by attraction around all closed channels that lie inside the homogeneous fluid figure of revolution in the plane of a meridian at the surface of the figure of revolution and that are filled with fluid from the figure are equal to zero, which follows from Clairaut's second new principle of equilibrium when the homogeneous fluid figure of revolution revolves around its axis of symmetry, in which case the axis of symmetry of the figure is also an axis of symmetry of the centrifugal force of rotation per unit of mass, and, moreover, when the axis of symmetry of the figure is also an axis of symmetry of the attraction; (3) and the principle of balanced columns. When the homogeneous fluid figure of revolution is a figure of equilibrium, an equation of the surface of the homogeneous fluid figure of revolution in equilibrium can be determined by simply finding an equation of any meridian at the surface of the homogeneous fluid figure of revolution in equilibrium. Clairaut showed how principle (1), his mathematical representation of principle (2), and principle (3) could be used together to derive an equation of a meridian at the surface of the homogeneous fluid figure of revolution in equilibrium.

Clairaut again let x and y, respectively, stand for rectangular coordinates in the plane of a meridian at the surface of a homogeneous fluid figure of revolution in equilibrium whose center is at C which are parallel and perpendicular, respectively, to the axis of symmetry of the homogeneous fluid figure of revolution in equilibrium, which the line segment CP is part of. PE again represents one quarter of the meridian. (In other words, PE is the part of the meridian which lies in one quadrant of the plane of the meridian. See Figure 44.) Clairaut again resolved the attraction at points S situated inside the homogeneous fluid figure of revolution in equilibrium in the plane of this meridian into components in the plane of this meridian which are perpendicular and parallel, respectively, to the axis of symmetry of the homogeneous fluid figure of revolution in equilibrium, which the line segment CP is part of, and whose magnitudes are P and Q, respectively. Thus the components of attraction at a point S in the plane of the meridian in question whose magnitudes are P and Q, respectively, lie in the plane of the meridian in question in the directions of the rectangular coordinates y and x, respectively. [Again there is no third component of the attraction at S which is perpendicular to the plane of the meridian in question because the axis of symmetry of the homogeneous fluid figure of revolution, which the line segment CP is part of, is also an axis of symmetry of the attraction.] The attraction produces a total effort equal to

$$\int_C^M P\,dy + Q\,dx \qquad (9.1.1)$$

along the straight channel or the curved channel $y = CSNM$ in the plane of the meridian in question which joins the center C of the homogeneous fluid figure of revolution in equilibrium to a point $M = (y, x)$ of the surface of the figure in the plane of the meridian in question [that is, to a point $M = (y, x)$ of the meridian in question] and which is filled with fluid from the figure, where $x = CQ$ and $y = QM$, and where the path of integration of the line integral (9.1.1) is the straight channel or the curved channel $y = CSNM$ in the plane of the meridian in question which joins C to M and which is filled with fluid from the homogeneous fluid figure of revolution in equilibrium. The constant of integration in the line integral (9.1.1) is chosen so that the value of the line integral (9.1.1) is equal to zero when $x = y = 0$. Clairaut stated that when $P\,dy + Q\,dx$ is a complete differential, this value of the line integral (9.1.1) will not depend on the straight channel or the particular choice of the curved channel $y = CSNM$ in the plane of the meridian in question which joins C to M, which is filled with fluid from the homogeneous fluid figure of revolution in equilibrium, and which is the path of integration of the line integral (9.1.1).

Now, I mentioned before that Clairaut demonstrated that when a plane revolves around an axis that lies in the plane, the centrifugal force generated *never* produces a total effort around a closed channel filled with homogeneous fluid which lies in the plane which is not equal to zero. In particular, then, the total effort produced by the centrifugal force per unit of mass around any closed channel that lies within the interior of a homogeneous fluid figure of revolution in equilibrium in the plane of any meridian at the surface of the homogeneous fluid figure of revolution in equilibrium which revolves around its axis of symmetry and that is filled with fluid from the homogeneous fluid figure of revolution in equilibrium is equal to zero. Now, channel $CQMC$ is one such closed channel in the plane of the meridian in question which is filled with fluid from the homogeneous fluid figure of revolution in equilibrium. But the total effort produced by the centrifugal force per unit of mass along the rectilinear part CQ of this closed channel of fluid is zero, because the magnitudes of the centrifugal force per unit of mass at all points of CQ are equal to zero, since CQ is part of the axis of symmetry of the figure, which is also the axis of rotation of the figure and which the line segment CP is part of. (The total effort produced by the centrifugal force per unit of mass along any rectilinear channel inside the homogeneous fluid figure of revolution in equilibrium which is parallel to the axis of symmetry of the figure and which is filled with fluid from the figure is also equal to zero, because although the magnitudes of the centrifugal force per unit of mass at all points on such a channel have the same value, and this value is not equal to zero, the directions of the centrifugal force per unit of mass at all points on such a channel are perpendicular to the channel.)

It follows that the total effort produced by the centrifugal force per unit of mass along the straight channel or the curved channel $\gamma = CSNM$ in the plane of the meridian in question which joins C to M and which is filled with fluid from the homogeneous fluid figure of revolution in equilibrium equals the total effort produced by the centrifugal force per unit of mass along the rectilinear channel QM in the plane of the meridian in question which is filled with fluid from the figure and that is perpendicular to the axis of symmetry of the figure, which is also its axis of rotation and which the line segment CP is part of. This equality can be expressed in the following way:

$$\int_C^M f\frac{y}{r}dy = \int_Q^M f\frac{y}{r}dy, \qquad (9.1.2)$$

where f designates the centrifugal force per unit of mass at a distance r from the axis of symmetry of the homogeneous fluid figure of revolution in equilibrium, where the path of integration of the integral on the left-hand side of equality (9.1.2) is the straight channel or the curved channel $\gamma = CSNM$ in the plane of the meridian in question which joins C to M and which is filled with fluid from the figure, and where the path of integration of the integral on the right-hand side of equality (9.1.2) is the rectilinear channel QM in the plane of the meridian in question which is filled with fluid from the figure and that is perpendicular to the figure's axis of symmetry.

The integral on the left-hand side of equality (9.1.2) represents the total effort produced by the centrifugal force per unit of mass along the straight channel or the curved channel $\gamma = CSNM$ in the plane of the meridian in question which joins C to M and which is filled with fluid from the homogeneous fluid figure of revolution in equilibrium. This is true because the centrifugal force per unit of mass at every point of the figure is perpendicular to the axis of symmetry of the figure, so that there is no component of the centrifugal force per unit of mass in the direction of the rectangular coordinate x at any point in the plane of the meridian in question. Consequently there is no term in dx in the integrand of the integral on the left-hand side of equality (9.1.2). In other words, the direction of the centrifugal force per unit of mass at each point on the straight channel or the curved channel $\gamma = CSNM$ in the plane of the meridian in question which joins C to M and which is filled with fluid from the homogeneous fluid figure of revolution in equilibrium is in the direction of the rectangular coordinate y. At the same time, the centrifugal force per unit of mass at each point on the straight channel or the curved channel $\gamma = CSNM$ in the plane of the meridian in question which joins C to M and which is filled with fluid from the figure varies directly as the distance of the point from the axis of symmetry of the figure.

The integral on the right-hand side of equality (9.1.2) represents the total effort produced by the centrifugal force per unit of mass along the rectilinear channel QM in the plane of the meridian in question which is filled with fluid from the homogeneous fluid figure of revolution in equilibrium. This is true because the direction of the centrifugal force per unit of mass at each point on the rectilinear channel QM in the plane of the meridian in question which is filled with fluid from the figure is in the direction of the rectangular coordinate y, since the rectilinear channel QM in the plane of the meridian in question is perpendicular to the axis of symmetry of the homogeneous fluid figure of revolution in equilibrium, which is also its axis of rotation and which the line segment CP is part of, and, moreover, as I noted, the centrifugal force per unit of mass at every point of the figure is also perpendicular to the axis of symmetry of the figure, so that there is no component of the centrifugal force per unit of mass in the direction of the rectangular coordinate x at any point in the plane of the meridian in question. At the same time, the centrifugal force per unit of mass at each point on the rectilinear channel QM in the plane of the meridian in question which is filled with fluid from the figure varies directly as the distance of the point from the axis of symmetry of the figure, which is also its axis of rotation and which the line segment CP is part of.

But the integral on the right-hand side of equality (9.1.2) equals $(f/(2r))y^2$, where $y = QM$. Therefore it follows from equality (9.1.2) that the total effort produced by the centrifugal force per unit of mass along the straight channel or any curved channel $\gamma = CSNM$ in the plane of the meridian in question which joins C to M and which is filled with fluid from the homogeneous fluid figure of revolution in equilibrium is simply $(f/(2r))y^2$, where $y = QM$.

Now, since the homogeneous fluid figure of revolution is a figure of equilibrium, it follows that (3), the principle of balanced columns, also must hold for the homogeneous fluid figure of revolution in equilibrium. This means in particular that the total effort produced by the effective gravity along the straight channel in the plane of the meridian in question which joins C to M and which is filled with fluid from the figure is independent of the particular point M of the surface of the figure in the plane of this meridian (in other words, is independent of the particular point M of this meridian). Moreover, according to principle (1), Clairaut's second new principle of equilibrium, which also must hold for the homogeneous fluid figure of revolution in equilibrium, the total effort produced by the effective gravity along the straight channel in the plane of the meridian in question which joins C to a particular point M of the surface of the homogeneous fluid figure of revolution in equilibrium in the plane of this meridian (in other words, to a particular point M of this meridian) and which is filled with fluid from the figure is the same as the total effort produced by the effective gravity along the curved channel $\gamma = CSNM$ in the plane of the meridian in question which joins C to that particular point M and which is filled with fluid from the figure.

From these two results together, which are results that follow from (3), the principle of balanced columns, and (1), Clairaut's second new principle of equilibrium, respectively, it follows that the total effort produced by the effective gravity along the straight channel or the curved channel $\gamma = CSNM$ in the plane of the meridian in question which joins C to a point M of the surface of the homogeneous fluid figure of revolution in equilibrium in the plane of the meridian in question (in other words, to a point M of the meridian in question) and which is filled with fluid from the figure is both independent of the particular point M of the surface of the figure in the plane of the meridian in question (in other words, is independent of the particular point M of the meridian in question) and independent of the straight channel or the particular curved channel $\gamma = CSNM$ in the plane of the meridian in question which joins C to a particular point M of the surface of the homogeneous fluid figure of revolution in equilibrium in the plane of the meridian in question (in other words, to a particular point M of the meridian in question) and which is filled with fluid from the figure. But the total effort produced by effective gravity along the straight channel or the curved channel $\gamma = CSNM$ in the plane of the meridian in question which joins C to a particular point $M = (y, x)$ of the surface of the figure in the plane of the meridian in question (in other words, to a particular point $M = (y, x)$ of the meridian in question) and which is filled with fluid from the figure is just the difference between the integrals (9.1.1) and (9.1.2), namely,

$$\int_C^M P\,dy + Q\,dx - \frac{f}{2r}y^2,$$

where the path of integration of the line integral in this expression is the straight channel or the curved channel $\gamma = CSNM$ in the plane of the meridian in question which joins C to the particular point $M = (y, x)$ of the surface of the figure in the plane of the meridian in question (in other words, to the particular point

$M = (y, x)$ of the meridian in question) and which is filled with fluid from the homogeneous fluid figure of revolution in equilibrium, and where $y = QM$.

Consequently it follows from the two results together, which are results that follow from (3), the principle of balanced columns, and (1), Clairaut's second new principle of equilibrium, respectively, that the equation

$$\int_C^M P\,dy + Q\,dx - \frac{f}{2r}y^2 = \text{a constant } A \text{ that is independent of } M \qquad (9.1.3)$$

in the two rectangular coordinates y and x, where the point $M = (y, x)$ of the surface of the homogeneous fluid figure of revolution in equilibrium in the plane of the meridian in question (in other words, where the point $M = (y, x)$ of the meridian in question) can be chosen arbitrarily, where the straight channel or the curved channel $\gamma = CSNM$ in the plane of the meridian in question which joins C to that particular point M, which is filled with fluid from the figure, and which is the path of integration of the line integral in this equation can also be chosen arbitrarily, and where $y = QM$ is an equation of the meridian in question at the surface of the figure.[29]

Clairaut also worked out a mathematical representation of his third new principle of equilibrium for a homogeneous fluid figure of revolution which revolves in such a way and where the forces of attraction act in such away as to produce the symmetries that I mentioned in Chapters 1, 4, and 6. (That is, the homogeneous fluid figure of revolution revolves around its axis of symmetry, in which case the axis of symmetry of the figure is also an axis of symmetry of the centrifugal force of rotation per unit of mass, and, moreover, the axis of symmetry of the figure is also an axis of symmetry of the attraction. These conditions did in fact always hold for the homogeneous fluid figures discussed in Chapters 1, 4, and 6.) In this case a problem in three dimensions reduces to a problem in two dimensions, as I have already explained.

Clairaut let PEp and $\Pi\varepsilon\pi$ designate halves of two plane curves formed where the plane of a meridian at the surface of such a homogeneous fluid figure of revolution intersects two different level surfaces with respect to effective gravity of the homogeneous fluid figure of revolution. The halves PEp and $\Pi\varepsilon\pi$ of these two plane curves lie in the plane of the meridian on the same side of the axis of symmetry of the homogeneous fluid figure of revolution, of which the line segment $\Pi Pp\pi$ is part. Clairaut called such plane curves level curves ("courbes de niveau"). He also assumed that the two different level surfaces of the homogeneous fluid figure of revolution are consecutive, neighboring level surfaces. In this case the level layer of the homogeneous fluid figure of revolution which the two level surfaces bound and whose cross section in the plane of the meridian in question is $\pi\varepsilon\Pi PEp$ is very thin. Clairaut let x and y stand for rectangular coordinates in the plane of the meridian in question which are parallel and perpendicular, respectively, to the axis of symmetry of the homogeneous fluid figure of revolution, of which the line segment $\Pi Pp\pi$ is part (see Figure 45). He let

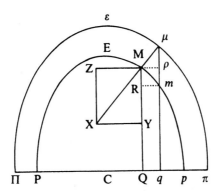

Figure 45. Figure used to work out a mathematical representation of Clairaut's third new principle of equilibrium for a homogeneous fluid figure of revolution (Clairaut 1743: figure on p. 44).

M designate a point of PEp, and he let $CQ = x$, $QM = y$, $Qq = Rm = dx$, and $MR = -dy$. He assumed that the effective gravity exerted a force whose magnitude is MX at each point $M = (y, x)$ of PEp. He let $MY = R$ and $MZ = Q$ symbolize the magnitudes at point M of the components of the effective gravity in the plane of the meridian in question in the directions of the rectangular coordinates y and x, respectively. (There is no third component of the effective gravity at M which is perpendicular to the plane of the meridian in question because of the conditions of symmetry.) Clairaut found that "since the force MX must be perpendicular [at the point M] to [the infinitesimal line segment] Mm [which is an infinitesimal part of PEp] by the conditions of the problem, the triangles MXY and MRm will be geometrically similar, from which the proportion

$$Q:R = -dy:dx \qquad (9.1.4)$$

follows."[30] Proportion (9.1.4) is nothing else than the equation that Maupertuis derived by converting into rectangular coordinates the equation that Bouguer had found to express the condition that the principle of the plumb line holds at all points of the surfaces of the homogeneous fluid figures of revolution which Bouguer had investigated in his Paris Academy *mémoire* of 1734. We remember that the axes of symmetry of the homogeneous fluid figures of revolution which Bouguer had examined in that *mémoire* were also the axes around which these homogeneous fluid figures of revolution revolved, in which case these axes were also axes of symmetry of the centrifugal force per unit of mass, and that these axes were also axes of symmetry of Bouguer's various hypotheses of attraction. Hence in each case that Bouguer considered, the three-dimensional problem of applying the principle of the plumb line at all points of the surface of a homogeneous fluid figure of revolution reduced to a problem in two dimensions as a result.[31] Clairaut rewrote the proportion (9.1.4) as the first-order ordinary "differential" equation of first degree

$$Rdy + Qdx = 0, \qquad (9.1.4')$$

which he called the general equation "for all level curves [in the particular plane of the meridian in question at the surface of the homogeneous fluid figure of revolution]."[32]

Clairaut next determined the characteristics of equation (9.1.4'), assuming that the magnitude MX of the force that the effective gravity exerts at each point $M = (y, x)$ of PEp is inversely proportional to the thickness $M\mu$ at $M = (y, x)$ of the very thin level layer of the homogeneous fluid figure of revolution whose cross section in the plane of the meridian in question is $\pi\varepsilon\Pi PEp$, which means that Clairaut's third new principle of equilibrium holds for the homogeneous fluid figure of revolution. We recall that this hypothesis entails that the force that the transverse weight of this very thin level layer of the homogeneous fluid figure of revolution exerts upon the level surface that is the inner boundary of this layer will be the same at all points of this level surface. To demonstrate the properties of equation (9.1.4') when he assumed that his third new principle of equilibrium holds for the homogeneous fluid figure of revolution, Clairaut again utilized his and Fontaine's work of 1738–40 in the integral calculus. Clairaut considered the differential $Rdy + Qdx$ on the left-hand side of equation (9.1.4') for any small values of dy and dx, and he let $\omega(y, x)$ designate an integrating factor for equation (9.1.4'), which means that $\omega(y, x)$ rendered the differential $Rdy + Qdx$ complete. Let $\phi(y, x)$ stand for an expression in y and x whose total differential is $\omega(Rdy + Qdx)$. Then

$$\phi(y, x) = a \qquad (9.1.5)$$

and

$$\phi(y, x) = a + da, \qquad (9.1.5')$$

respectively, say, where a and da designate real constants and da is very small, must be equations for the level curves in the particular plane of the meridian in question at the surface of the homogeneous fluid figure of revolution which PEp and $\Pi\varepsilon\pi$, respectively, are halves of, since

$$0 = d\phi = \omega(Rdy + Qdx) = Rdy + Qdx \qquad (9.1.4')$$

is the first-order ordinary "differential" equation of first degree of the level curves in the particular plane of the meridian in question at the surface of the homogeneous fluid figure of revolution. In particular, Clairaut let $M = (y, x)$ and $dx = Qq = Rm$ as he did previously. But instead of letting dy equal $-MR$ as he did before, Clairaut now let dy equal $\mu\rho$, in which case $M = (y, x)$ and $\mu = (y + dy, x + dx)$, respectively, are points of PEp and $\Pi\varepsilon\pi$, respectively. Then the points $M = (y, x)$

and $\mu = (y + dy, x + dx)$, respectively, satisfy the equations

$$\phi(y, x) = a \tag{9.1.5}$$

and

$$\phi(y + dy, x + dx) = a + da, \tag{9.1.5''}$$

respectively. If we subtract the left-hand side of equation (9.1.5) from the left-hand side of equation (9.1.5'') and, likewise, if we subtract the right-hand side of equation (9.1.5) from the right-hand side of equation (9.1.5''), we find that the points $M = (y, x)$ and $\mu = (y + dy, x + dx)$ together satisfy the equation

$$da = \phi(y + dy, x + dx) - \phi(y, x). \tag{9.1.6}$$

Now, inasmuch as dy and dx are very small, the right-hand side of equation (9.1.6) is equal to

$$d\phi = \omega R \, dy + \omega Q \, dx. \tag{9.1.7}$$

Then if we substitute the left-hand side of equation (9.1.7) for the right-hand side of equation (9.1.6), the equation

$$da = \omega R \, dy + \omega Q \, dx \tag{9.1.8}$$

results.

Now, since the length of the line segment $M\mu$ which joins points $M = (y, x)$ and $\mu = (y + dy, x + dx)$ is the thickness at $M = (y, x)$ of the very thin level layer of the homogeneous fluid figure of revolution whose cross section in the plane of the meridian in question is $\pi\varepsilon\Pi PEp$, this means that the line segment $M\mu$ is perpendicular to PEp and $\Pi\varepsilon\pi$, respectively, at $M = (y, x)$ and $\mu = (y + dy, x + dx)$, respectively. Thus $M\mu$, like MX, is perpendicular to the infinitesimal line segment Mm which is an infinitesimal part of PEp, and consequently μMX is a line segment. Therefore the equality

$$dy = \frac{R}{Q} dx \tag{9.1.9}$$

is true, because triangles $M\mu\rho$ and MXY are geometrically similar. Substituting the right-hand side of equality (9.1.9) for dy in equation (9.1.8) and then solving the equation that results for dx, we find that the equality

$$dx = \frac{da}{\omega} \left(\frac{Q}{R^2 + Q^2} \right) \tag{9.1.10}$$

is also true. Replacing dx in equality (9.1.9) by the right-hand side of equality (9.1.10), we find that the equality

$$dy = \frac{da}{\omega} \left(\frac{R}{R^2 + Q^2} \right) \tag{9.1.11}$$

is true too. Since the equality

$$M\mu = \sqrt{dx^2 + dy^2} \qquad (9.1.12)$$

is true as well, the thickness $M\mu$ at $M = (y, x)$ of the very thin level layer of the homogeneous fluid figure of revolution whose cross section in the plane of the meridian in question is $\pi\varepsilon\Pi PEp$ can be determined by replacing dx and dy, respectively, in equality (9.1.12) by the right-hand sides of equalities (9.1.10) and (9.1.11), respectively. If we do this we find that the thickness $M\mu$ at $M = (y, x)$ of the very thin level layer of the homogeneous fluid figure of revolution whose cross section in the plane of the meridian in question is $\pi\varepsilon\Pi PEp$ is

$$M\mu = \frac{da}{\omega\sqrt{R^2 + Q^2}}. \qquad (9.1.13)$$

Now,

$$MX = \sqrt{R^2 + Q^2} \qquad (9.1.14)$$

is the magnitude of the effective gravity at $M = (y, x)$. Moreover, because this level layer is very thin, Clairaut assumed that the magnitudes of the effective gravity at all points of the line segment $M\mu$ that joins the points $M = (y, x)$ and $\mu = (y + dy, x + dx)$ and that is perpendicular to the two level surfaces that bound the level layer are the same and, at the same time, that the directions of the effective gravity at all points of the line segment $M\mu$ are the same as the direction of the line segment $M\mu$.

Then it follows from equalities (9.1.13) and (9.1.14) that the product

$$M\mu \times MX = \frac{da}{\omega\sqrt{R^2 + Q^2}} \times \sqrt{R^2 + Q^2} = \frac{da}{\omega} \qquad (9.1.15)$$

is the effort produced by the effective gravity along the line segment $M\mu$, which is the same as the force that the transverse weight of the very thin level layer of the homogeneous fluid figure of revolution whose cross section in the plane of the meridian in question is $\pi\varepsilon\Pi PEp$ exerts at $M = (y, x)$. But the force that the transverse weight exerts upon the level surface that is the inner boundary of this very thin level layer is the same at all points of this level surface by the hypothesis, which means that da/ω on the right-hand side of equality (9.1.15) is a constant that is independent of the particular point $M = (y, x)$ of PEp. Finally, since da only depends upon the particular pair of level curves in the particular plane of the meridian in question at the surface of the homogeneous fluid figure of revolution of which PEp and $\Pi\varepsilon\pi$, respectively, are halves and does not depend upon the pair of points $M = (y, x)$ and $\mu = (y + dy, x + dx)$, respectively, of PEp and $\Pi\varepsilon\pi$, respectively, it follows that da is the same for all points $M = (y, x)$ of PEp too. From this Clairaut inferred that $\omega(y, x)$ is identically equal to a constant.

Clairaut stated his conclusion in the following way: "we said a moment ago that this weight ["poids"] must be proportional to da. [That is, the transverse weight of the level layer of the homogeneous fluid figure of revolution whose cross section in the plane of the meridian in question is $\pi\varepsilon\Pi PEp$, which at $M = (y, x)$ is the same as the effort produced by the effective gravity along the line segment $M\mu$, exerts a uniform force upon the level surface that is the inner boundary of the level layer, and, at the same time, da depends only upon the particular pair of level curves in the particular plane of the meridian in question at the surface of the homogeneous fluid figure of revolution which PEp and $\Pi\varepsilon\pi$, respectively, are halves of and does not depend upon the particular points $M = (y, x)$ and $\mu = (y + dy, x + dx)$, respectively, of PEp and $\Pi\varepsilon\pi$, respectively. Hence the equality

$$\frac{\text{the transverse weight of the level layer}}{da} = \text{a constant} \qquad (9.1.16)$$

is true at all points $M = (y, x)$ of PEp. But according to (9.1.15),

$$\text{the transverse weight of the level layer} = \frac{da}{\omega} \qquad (9.1.15')$$

is also true at all points $M = (y, x)$ of PEp. Substituting the right-hand side of equality (9.1.15′) for the numerator of the quotient on the left-hand side of equality (9.1.16), we find that

$$\frac{1}{\omega} = \frac{\dfrac{da}{\omega}}{da} = \frac{\text{the transverse weight of the level layer}}{da} = \text{a constant} \qquad (9.1.17)$$

is true at all points $M = (y, x)$ of PEp.] Therefore ω is a constant. Hence the differential $Rdy + Qdx$ needs no [integrating] factor in order to be complete. That is, $Rdy + Qdx$ must be the [total] differential of some function of x and y, in order for the weight ["pression"] of the [level] layer to be equal at all of its points and, consequently, in order for the [homogeneous fluid] spheroid to be in equilibrium."[33]

Once again the more algebraic or analytic notion of a complete differential $Rdy + Qdx$, which first arose in Fontaine's and Clairaut's work of 1738–40 in the integral calculus (that is, a differential $Rdy + Qdx$ such that an expression $\phi(y, x)$ exists whose total differential is the differential $Rdy + Qdx$, where the expression $\phi(y, x)$ is uniquely determined up to an additive constant) received an auxiliary, geometric interpretation *through* the mechanics problem. Namely, when $Rdy + Qdx$ is a complete differential, an integral of the first-order ordinary "differential" equation of first degree

$$Rdy + Qdx = 0 \qquad (9.1.4')$$

is geometrically just a "level curve" – that is, a plane curve formed where a plane of a meridian at the surface of a homogeneous fluid figure of revolution which revolves around its axis of symmetry, in which case the axis of symmetry of the figure is also an axis of symmetry of the centrifugal force of rotation per unit of mass, when, moreover, the axis of symmetry of the figure is also an axis of symmetry of the attraction, intersects a "level surface" with respect to effective gravity of the homogeneous fluid figure of revolution. Clairaut had already anticipated this geometric interpretation of an integral of equation (9.1.4') in 1740, in his second Paris Academy *mémoire* on integral calculus. In Chapter 7 I briefly discussed Clairaut's geometric result that appears in this *mémoire*, and I shall return to this geometric result in a moment.

Again we continue to assume that the homogeneous fluid figure of revolution revolves in such a way and that the forces of attraction act in such away as to produce the symmetries that I mentioned in Chapters 1, 4, and 6. (That is, the homogeneous fluid figure of revolution revolves around its axis of symmetry, in which case the axis of symmetry of the figure is also an axis of symmetry of the centrifugal force of rotation per unit of mass, and, moreover, the axis of symmetry of the figure is also an axis of symmetry of the attraction. These conditions did in fact always hold for the homogeneous fluid figures discussed in Chapters 1, 4, and 6.) In this case a problem in three dimensions reduces to a problem in two dimensions, as I have already mentioned. If this homogeneous fluid figure of revolution is, moreover, a figure of equilibrium, then the following two principles of equilibrium also both hold for this homogeneous fluid figure of revolution in equilibrium: (4), Clairaut's third new principle of equilibrium, and (5), the principle of the plumb line. As I have already noted, when the homogeneous fluid figure of revolution is a figure of equilibrium, an equation of the surface of the figure can be determined by simply finding an equation of any meridian at the surface of the figure. Clairaut showed how (5) and his mathematical representation of (4) could be used together to derive an equation of a meridian at the surface of the figure. The equations of the level curves formed where the plane of a meridian at the surface of the figure intersects the level surfaces with respect to effective gravity of the figure are all determined by setting the integral of the complete differential $Rdy + Qdx$ equal to constants, where R and Q, respectively, stand for the magnitudes of the components of the effective gravity in this plane in the directions of the rectangular coordinates y and x, respectively, in this plane at a point M of a level curve in this plane. In particular, since (5), the principle of the plumb line, holds at all points of the surface of a figure of equilibrium, the surface of the homogeneous fluid figure of revolution in equilibrium is itself a level surface. Consequently a meridian at the surface of the homogeneous fluid figure of revolution in equilibrium is itself a level curve. Hence an equation of a meridian at the surface of the figure must itself have the form

$$\int Rdy + Qdx = \text{a constant } A \tag{9.1.18}$$

in the two rectangular coordinates y and x in the plane of this meridian, where the left-hand side of equation (9.1.18) does not designate here a line integral of the complete differential $Rdy + Qdx$ but an antidifferential of the complete differential $Rdy + Qdx$ instead (that is, an expression $\phi(y, x)$ whose total differential is the complete differential $Rdy + Qdx$). But $R = P - (f/r)y$, where P stands for the magnitude of the component of the attraction in the plane of the meridian in question in the direction of the rectangular coordinate y in this plane at a point M of the meridian, and where f is the centrifugal force per unit of mass at a distance r from the axis of symmetry of the homogeneous fluid figure of revolution in equilibrium, which is also the axis of rotation of the homogeneous fluid figure of revolution in equilibrium and which the line segment $\Pi Pp\pi$ is part of. Hence

$$\int Rdy + Qdx = \int Pdy + Qdx - \int \frac{f}{r} y dy = \int Pdy + Qdx - \frac{f}{2r} y^2 \quad (9.1.19)$$

is true. Substituting the right-hand side of equality (9.1.19) for the expression on the left-hand side of equation (9.1.18), Clairaut arrived at the equation

$$\int Pdy + Qdx - \frac{f}{2r} y^2 = \text{a constant } A \quad (9.1.20)$$

as the equation of a meridian at the surface of the homogeneous fluid figure of revolution in equilibrium in the two rectangular coordinates y and x in the plane of this meridian. Now, the magnitude of the component of the centrifugal force per unit of mass in the direction of the rectangular coordinate x in the plane of the meridian in question is equal to zero at all points in this plane. Hence the magnitude of the component of the effective gravity in the plane of the meridian in question in the direction of the rectangular coordinate x in this plane at a point M of the meridian equals the magnitude of the component of the attraction in the plane of the meridian in question in the direction of the rectangular coordinate x in this plane at the point M of the meridian. In other words, Q in equation (9.1.20) is the same as Q in equation (9.1.3). Thus Clairaut found that the two equations (9.1.3) and (9.1.20) of a meridian at the surface of the homogeneous fluid figure of revolution in equilibrium, which he derived by making independent use of the mathematical representations of two different principles of equilibrium (principle of equilibrium 2, which follows from principle of equilibrium 1 when a homogeneous fluid figure of revolution revolves around its axis of symmetry, in which case the axis of symmetry of the homogeneous fluid figure of revolution is also an axis of symmetry of the centrifugal force of rotation per unit of mass, and, moreover, the axis of symmetry of the homogeneous fluid figure of revolution is also an axis of symmetry of the attraction, and principle of equilibrium 4), are the same.[34] [Note that in concluding that the two equations (9.1.3) and (9.1.20) are the same, Clairaut implicitly assumed here that the antidifferential of the differential $Pdy + Qdx$ that we know from the derivation of equation (9.1.3) is a complete differential [that is, an expression $\phi(y, x)$ whose total differential is the complete differential $Pdy + Qdx$] is the same as the line integral of the complete differential $Pdy + Qdx$. In making such an assumption Clairaut tacitly supposed that the line integral of the complete differential $Pdy + Qdx$ is path independent. As I mentioned, Clairaut made no use of the assumption that the antidifferential of a complete differential $Pdy + Qdx$ and the line integral of the complete differential $Pdy + Qdx$ are identically the same in producing his antidifferentials of complete differentials $Pdy + Qdx$ [that is, expressions $\phi(y, x)$ whose total differentials are complete differentials $Pdy + Qdx$] in his two Paris Academy *mémoires* on integral calculus, even though the equality of the two can be used to this end when the two are the same. Instead he constructed these antidifferentials of complete differentials $Pdy + Qdx$ by other means in those two *mémoires*. Perhaps he did so because he did not realize at the time that he wrote those *mémoires* that the antidifferential of a complete differential $Pdy + Qdx$ [that is, an expression $\phi(y, x)$ whose total differential is the complete differential $Pdy + Qdx$] and the line integral of $Pdy + Qdx$ are sometimes the same. I also mentioned, however, that the two are *not* always the same. That is, the line integral of $Pdy + Qdx$ is *not* always path independent. As it turns out, the line integral of $Pdy + Qdx$ is path independent if and only if the expression $\phi(y, x)$ whose total differential is the complete differential $Pdy + Qdx$ is a *single-valued function* of y and x. We will now examine evidence that shows that Clairaut did not realize this.]

It might appear to be a foregone conclusion that the two equations (9.1.3) and (9.1.20) must turn out to be the same. While trying out his theory of homogeneous figures of equilibrium, however, Clairaut stumbled upon some examples where the equation of a meridian at the surface of a homogeneous fluid figure of revolution which revolves around its axis of symmetry, in which case the axis of symmetry of the figure is also an axis of symmetry of the centrifugal force of rotation per unit of mass, when, moreover, the axis of symmetry of the figure is also an axis of symmetry of the attraction, which he derived by applying principle (1) in conjunction with principle (3) and his mathematical representation of principle (2) to the homogeneous fluid figure of revolution, where (2) follows from (1) when the homogeneous fluid figure of revolution fulfills the conditions of symmetry just specified, was not the same as the equation of a meridian at the surface of the same homogeneous fluid figure of revolution which he derived by applying (5) together with his mathematical representation of (4) to the homogeneous fluid figure of revolution. The equations differed from each other when Clairaut introduced hypotheses of attraction expressed in polar coordinates.

In the derivations that follow, PE again represents one quarter of a meridian at the surface of a homogeneous fluid figure of revolution whose center is at C. (In other words, PE is the part of the meridian which lies in one quadrant of the plane of the meridian.) M is an arbitrarily chosen point of PE, and CM is the straight channel that joins C to M and that is filled with fluid from the homogeneous fluid figure of revolution. ON stands for an arbitrarily chosen plane channel that lies

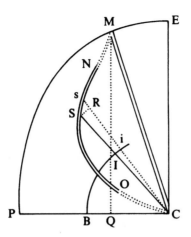

Figure 46. Meridian of a quadrant of a homogeneous fluid figure of revolution and interior channels expressed in polar coordinates in the plane of the meridian (Clairaut 1743: figure on p. 79).

within the interior of the homogeneous fluid figure of revolution in the plane of the meridian in question and that is filled with fluid from the homogeneous fluid figure of revolution. *ON* is part of an arbitrarily chosen plane channel *CONM* that lies inside the homogeneous fluid figure of revolution in the plane of the meridian in question, that joins *C* to *M*, and that is filled with fluid from the homogeneous fluid figure of revolution. When it was convenient to utilize polar coordinates, Clairaut let y = the length of the line segment *CS* that joins *C* to a point *S* inside the homogeneous fluid figure of revolution in the plane of the meridian in question, $CB = 1$, x = the radian measure of the angle that the line segment *CS* makes with the axis of symmetry of the homogeneous fluid figure of revolution, which the line segment *CP* is part of (in other words, the length of the circular arc *BI* that lies in the plane of the meridian in question), $dx = Ii$, $ydx = SR$, $dy = sR$ (see Figure 46), P = the component of the attraction at *S* which is directed toward *C*, and Q = the component of the attraction at *S* in the plane of the meridian in question which is perpendicular to *CS*. (Again, there is no component of attraction at *S* which is perpendicular to the plane of the meridian in question because of the conditions of symmetry.)

By the resolution and composition of forces, Clairaut concluded that $Pdy + Qydx$ is the expression in the polar coordinates y and x for the effort produced by attraction along the infinitesimal homogeneous fluid right circular cylinder *Ss* that is part of the plane channel *CONM* that is filled with fluid from the homogeneous fluid figure of revolution and that is made up of an infinite number of such infinitesimal homogeneous fluid right circular cylinders. From this he concluded that $Pdy + Qydx$ must be a complete differential, if a homogeneous fluid figure of revolution is to be a figure of equilibrium when such forces of attraction expressed in the polar coordinates y and x act. Consequently the equality

$$\frac{dP}{dx} = \frac{d(Qy)}{dy} \qquad (7.3.4.6''''')$$

must hold true. Accordingly, the equation

$$\int_C^M Pdy + Qydx - \frac{f}{2r}z^2 = \text{a constant } A \text{ that is independent of } M, \qquad (9.1.3')$$

in the polar coordinates y and x, where $M = (y, x)$ is the arbitrarily chosen point of PE, where $y = CM$, where $z = QM = CM \sin x = y \sin x$, where f is the centrifugal force per unit of mass at a distance r from the axis of symmetry of the homogeneous fluid figure of revolution, which is also the axis of rotation of the homogeneous fluid figure of revolution and which the line segment CP is part of, and where A is a constant that is independent of the particular point M of PE, should be an equation of the meridian in question at the surface of the homogeneous fluid figure of revolution, where the straight channel or any curved channel $CONM$ that lies in the plane of the meridian in question, that joins C to M, and that is filled with fluid from the homogeneous fluid figure of revolution can be taken as the path of integration γ of the line integral on the left-hand side of equation (9.1.3'). This result follows from applying principle (1) in conjunction with principle (3) and Clairaut's mathematical representation of (2) to the homogeneous fluid figure of revolution which revolves around its axis of symmetry, in which case the axis of symmetry of the homogeneous fluid figure of revolution is also an axis of symmetry of the centrifugal force of rotation per unit of mass, when, moreover, the axis of symmetry of the homogeneous fluid figure of revolution is also an axis of symmetry of the attraction, where (2) follows from (1) when the homogeneous fluid figure of revolution satisfies the conditions of symmetry just specified.[35]

Clairaut discovered, however, that when the differentials $Pdy + Qydx$ were complete, the equation (9.1.3') of a meridian at the surface of such a homogeneous fluid figure of revolution sometimes differed nevertheless from the equation

$$\int Pdy + Qydx - \frac{f}{2r}z^2 = \text{a constant } A \qquad (9.1.20')$$

in the polar coordinates y and x, where $M = (y, x)$ is the arbitrarily chosen point of PE, where $y = CM$, where $z = QM = CM \sin x = y \sin x$, where f is the centrifugal force per unit of mass at a distance r from the axis of symmetry of the homogeneous fluid figure of revolution, which is also the axis of rotation of the homogeneous fluid figure of revolution and which the line segment CP is part of, and where A is a constant that is independent of the particular point M of PE, of a meridian at the surface of the same homogeneous fluid figure of revolution arrived at by applying instead (4), expressed mathematically using the same complete differentials $Pdy + Qydx$, together with (5) to the homogeneous fluid figure of revolution, where the integral on the left-hand side of equation (9.1.20') is not in this case a line integral of the differential $Pdy + Qydx$ but an antidifferential of the differential $Pdy + Qydx$ instead (that is, an expression $\phi(y, x)$ whose total differential is the differential $Pdy + Qydx$). In particular, Clairaut observed that equations (9.1.3') and (9.1.20') were not always the same when he took the plane channel $CONM$ that lies inside the homogeneous fluid figure of revolution in the plane of the meridian in question, that joins C to M, and that is filled with fluid from the homogeneous fluid figure of revolution to be the straight channel CM, in which case the line integral

$$\int_C^M Pdy + Qydx$$

on the left-hand side of equation (9.1.3') is the total effort produced by the attraction along the straight channel CM that joins C to M and that is filled with fluid from the homogeneous figure of revolution.

Clairaut found that this "paradox," as he called it, arose, for example, when

$$P(y, x) \equiv Y(y) \quad \text{and} \quad Q(y, x) \equiv \frac{X(x)}{y},$$

where $Y(y)$ and $X(x)$, respectively, are arbitrary expressions in y and in x, respectively, or when

$$P(y,x) \equiv \frac{y}{\sqrt{x^2 + y^2}} \quad \text{and} \quad Q(y,x) \equiv \frac{x}{y\sqrt{x^2 + y^2}}.$$

But Clairaut thought that principles (1),(3), and his mathematical representation of (2) applied together to a homogeneous fluid figure of revolution which revolves around its axis of symmetry, in which case the axis of symmetry of the homogeneous fluid figure of revolution is also an axis of symmetry of the centrifugal force of rotation per unit of mass, when, moreover, the axis of symmetry of the homogeneous fluid figure of revolution is also an axis of symmetry of the attraction, should lead to the same equation of a meridian at the surface of the homogeneous fluid figure of revolution as the equation of a meridian at the surface of the homogeneous fluid figure of revolution arrived at by applying instead (5) together with his mathematical representation of (4) to the homogeneous fluid figure of revolution. After all, he had demonstrated that (3) and (5) applied to a homogeneous fluid mass entail each other, hence accord with the same shape of that homogeneous fluid mass, when (1) holds for the homogeneous fluid mass. He had also shown that when (1) holds for a homogeneous fluid mass, it follows that (4) holds too for the homogeneous fluid mass. [Recall that we have seen that it follows from the principle of solidification that if (1) holds for a homogeneous fluid mass, this means that (1) not only holds for all closed plane or skew channels that lie entirely within the interior of the homogeneous fluid mass and that are filled with fluid from the mass, but that (1) also holds for closed plane or skew channels of homogeneous fluid parts of which lie *at the surface* (that is, lie *at the boundary*) of the homogeneous fluid mass.] Moreover, he had demonstrated that when (1) holds for a homogeneous fluid mass, it follows that (2) holds as well for the homogeneous fluid mass when the mass is a homogeneous fluid figure of revolution which revolves around its axis of symmetry, in which case the axis of symmetry of the homogeneous fluid figure of revolution is also an axis of symmetry of the centrifugal force of rotation per unit of mass, and when, moreover, the axis of symmetry of the homogeneous fluid figure of revolution is also an axis of symmetry of the attraction. And Clairaut thought that the condition that the differential of the attraction that acts be complete, which in the case at issue is equality (7.3.4.6''''), was equivalent to principle (2).

To try to reconcile this difference of equations, Clairaut argued as follows. Normally the efforts produced by the attraction along circular arcs whose centers are at the center C of a homogeneous fluid figure of revolution, that lie inside the figure in the plane of a meridian at the surface of the figure, and that have the same radian measure approach zero as the radii of the circular arcs approach zero, because the lengths of the circular arcs approach zero as their radii approach zero and, at the same time, the values of Q at all points of the circular arcs remain finite and bounded as the radii approach zero. Clairaut noted however, that in the first example above, where $P(y, x) \equiv Y(y)$ and $Q(y, x) \equiv X(x)/y$, the values of Q increase without bound as y approaches zero. He said that this behavior of Q in this example produces the following condition. The efforts produced by the attraction along circular arcs whose centers are at $C = (0,0)$ of the homogeneous fluid figure of revolution, that lie in the interior of the figure in the plane of the meridian in question, that have the same radian measure, and whose radii approach zero remain finite and bounded and approach a finite limit as the radii of the circular arcs approach zero, even though the values of Q at points of the circular arcs approach infinity as y approaches zero. Moreover, these efforts produced by the attraction do not approach zero as a limit as the radii approach zero, even though the lengths of the circular arcs approach zero as their radii approach zero. Thus Clairaut contended that the total effort produced by the attraction along, say, the straight channel CM in the plane of the meridian in question which is filled with fluid from the homogeneous fluid figure of revolution is actually the total effort produced by the attraction along the straight channel ML in the plane of the meridian in question which is filled with fluid from the figure plus the total effort produced by the attraction along the circular arc AL in the plane of the meridian in question whose center is at $C = (0,0)$, as the radius AC of the circular arc AL becomes increasingly smaller and approaches zero (see Figure 47). What happens is that the total effort produced by the attraction along

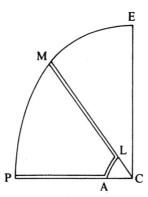

Figure 47. Meridian of a quadrant of a homogeneous fluid figure of revolution and interior channel in the plane of the meridian which is composed of a straight portion directed toward the center of the figure and a circular portion whose center is the center of the figure (Clairaut 1743: figure on p. 89).

the circular arc *AL* approaches a finite value as *AC* approaches zero, even though the values of *Q* at the points of the circular arc *AL* approach infinity as *AC* approaches zero. The value of the limit is finite essentially because these values of *Q* do not approach infinity any faster than the length of the circular arc *AL* approaches zero as *AC* approaches zero. Moreover, the value of the limit is not equal to zero, even though the length of the circular arc *AL* approaches zero as *AC* approaches zero, essentially because the length of the circular arc *AL* also does not approach zero any faster than the values of *Q* at the points of the circular arc *AL* approach infinity as *AC* approaches zero. By adding this finite, nonzero value, which is a kind of "residue," so to speak, to the total effort produced by attraction along the straight channel *CM* filled with fluid from the homogeneous fluid figure of revolution, where this total effort is computed the usual way, Clairaut arrived at an equation (9.1.3′) that was the same as the equation (9.1.20′) that he derived by applying (5) together with his mathematical representation of (4) to the homogeneous fluid figure of revolution. To summarize the reasoning that Clairaut used to account for the difference of equations in this first example, the unbounded values of *Q* in the neighborhood of the center *C* = (0,0) of the homogeneous fluid figure of revolution produce an effort at *C* = (0,0) whose value is finite and which is not equal to zero, where *C* = (0,0) is treated as an infinitesimal circular arc *AL* whose radius *AC* = 0. In this case the effort produced by the attraction along the infinitesimal circular arc *AL* will unknowingly be neglected when applying (1) and (3) together with Clairaut's mathematical representation of (2) to the homogeneous fluid figure of revolution in order to determine the equation of a meridian at the surface of the figure, unless the straight channel *CM* is thought of as having an infinitesimal circular arc *AL* at its end at *C* whose radius *AC* is equal to zero.

In the second example above, where

$$P(y,x) \equiv \frac{y}{x^2 + y^2} \quad \text{and} \quad Q(y,x) \equiv \frac{x}{y\sqrt{x^2 + y^2}}$$

the values of Q do not approach infinity as y approaches zero. Nevertheless, Clairaut found that the same device that he used to resolve the difference of equations in the first example also worked to resolve the difference of equations in this example.[36]

Clairaut intended here to resolve a puzzling discrepancy that he felt could not truly exist. At first sight the equation expressed in polar coordinates of a meridian at the surface of a homogeneous fluid figure of revolution which revolves around its axis of symmetry, in which case the axis of symmetry of the homogeneous fluid figure of revolution is also an axis of symmetry of the centrifugal force of rotation per unit of mass, when, moreover, the axis of symmetry of the homogeneous fluid figure of revolution is also an axis of symmetry of the attraction, which principles (1), (3), and Clairaut's mathematical representation of (2) determined when applied together to the homogeneous fluid figure of revolution did not always appear to be the same as the equation expressed in polar coordinates of a meridian at the surface of the same homogeneous fluid figure of revolution which principle (5) and Clairaut's mathematical representation of (4) determined when applied together to the homogeneous fluid figure of revolution. But Clairaut thought that the equation of a meridian at the surface of a homogeneous fluid figure of revolution which fulfills the conditions of symmetry just specified, derived using (1), (3), and his mathematical representation of (2), could not differ from the equation of a meridian at the surface of the homogeneous fluid figure of revolution derived using (5) and his mathematical representation of (4) for the following reasons. He had shown that (3) and (5) applied to a homogeneous fluid mass entail each other, hence accord with the same shape of that homogeneous fluid mass, when (1) holds for the homogeneous fluid mass. He had also demonstrated that when (1) holds for a homogeneous fluid mass, it follows that (4) holds too for the homogeneous fluid mass. Moreover, he had shown that when (1) holds for a homogeneous fluid mass, it follows that (2) holds as well for the homogeneous fluid mass when the homogeneous fluid mass is a homogeneous fluid figure of revolution which revolves around its axis of symmetry, in which case the axis of symmetry of the homogeneous fluid figure of revolution is also an axis of symmetry of the centrifugal force of rotation per unit of mass, and, moreover, when the axis of symmetry of the homogeneous fluid figure of revolution is also an axis of symmetry of the attraction. And Clairaut thought that the condition that the differential of the attraction that acts be complete, which in the case at issue is equality (7.3.4.6''''), was a condition that was equivalent to (2). That is, Clairaut had shown that if (1) holds for a homogeneous fluid mass, it follows that (2) also holds for the mass when the mass satisfies the conditions of symmetry just specified; he thought that the condition that the differential of the attraction that acts be complete, which in the case at issue is equality (7.3.4.6''''), was a condition that was equivalent to (2); he had shown that when (1) holds for a homogeneous fluid mass, it follows that (4) holds as well for the mass; and he had also demonstrated that if (1) holds for a homogeneous fluid mass, then (3) and (5) are equivalent and thus accord with the same shape of the homogeneous fluid mass.

Clairaut's ingenious physical argument does not really explain the contradictory results that Clairaut tried to make agree and that he thought that he had succeeded in doing. Not only can the physical argument be shown to be unsound because it turns out not to be uniquely determined, but the mathematical analysis that Clairaut used also has its problems. Clairaut guessed correctly that the singular point of the attraction at $C = (0,0)$ in the first of his two examples discussed previously caused the trouble in the example, but he did not truly understand the nature of the difficulties that the singular point creates. The measures that he proposed to take account of the singular point do not really resolve the problem that they were meant to resolve. From the way that he handled the second of his two examples, he appears to have guessed correctly that the behavior of the attraction at $C = (0,0)$ in that example also caused the problems in that example too. Again, however, the steps that he took to resolve the problem that the behavior of the attraction at $C = (0,0)$ in that example creates do not really do what they were intended to do. For one thing, neither the P nor the Q in the second of Clairaut's two examples even has a unique limit as (y,x) approaches $C = (0,0)$, so that the definitions of P and Q in this example cannot even be extended continuously to have *unique finite values* at $C = (0,0)$. But Clairaut failed to notice or perceive this. Clairaut confronted subtle matters in the realm of mathematical analysis which would only truly begin to be clarified later by other mathematicians.

It is true that principle (2) follows from (1) when the homogeneous fluid figure of revolution revolves around its axis of symmetry, in which case the axis of symmetry of the homogeneous fluid figure of revolution is also an axis of symmetry of the centrifugal force of rotation per unit of mass, and, moreover, when the axis of symmetry of the homogeneous fluid figure of revolution is also an axis of symmetry of the attraction. However, it turns out that when the homogeneous fluid figure of revolution fulfills the conditions of symmetry just specified, the condition that the differential of the attraction that acts be complete, which in the case at issue is equality (7.3.4.6'''') and which is equality (7.3.4.6''') when rectangular coordinates are employed, is *not* equivalent to (2). The condition that the differential of the attraction that acts be complete, which the equality (7.3.4.6'''') expresses in the case at issue and which the equality (7.3.4.6''') expresses when rectangular coordinates are employed, does indeed follow from (2), but (2) does not always follow from the condition that the differential of the attraction that acts be complete.

In fact a homogeneous fluid figure of revolution *cannot* be a figure of equilibrium when the kinds of laws of attraction expressed above in polar coordinates act. This is true because even when such a figure revolves around its axis of symmetry, in which case the axis of symmetry of the figure is also an axis of symmetry of the centrifugal force of rotation per unit of mass, and, moreover, even when the axis of symmetry of the figure is also an axis of symmetry of the attraction, it turns out that the total efforts produced by the attraction around closed plane channels that lie within the interior of such a homogeneous fluid figure of revolution in the plane of the meridian in question, that are filled with fluid from the figure, and that encircle the singular point of the attraction at $C = (0,0)$ produced by the kinds of laws of attraction expressed previously in polar coordinates are *not* in general equal to zero. Hence (2) does *not* hold in general for such a homogeneous fluid figure of revolution. Consequently (1), from which (2) follows when a homogeneous fluid figure of revolution satisfies the conditions of symmetry just specified, cannot hold in general for such a homogeneous fluid figure of revolution either when the kinds of laws of attraction expressed above in polar coordinates act.

The trouble is that although (2) follows from (1) when the homogeneous fluid figure of revolution revolves around its axis of symmetry, in which case the axis of symmetry of the homogeneous fluid figure of revolution is also an axis of symmetry of the centrifugal force of rotation per unit of mass, and, moreover, when the axis of symmetry of the homogeneous fluid figure of revolution is also an axis of symmetry of the attraction, the condition that the differential of the attraction that acts be complete, which the equality (7.3.4.6''''') expresses in the case at issue and which the equality (7.3.4.6''') expresses when rectangular coordinates are employed, is *not equivalent* to (2). Specifically, although the condition that the differential of the attraction that acts be complete does indeed follow from (2), the converse is not true in general. That is, the differential of the attraction that acts can be complete, yet (2) need not hold. And if (2) does not hold, then (1), from which (2) follows when the homogeneous fluid figure of revolution fulfills the conditions of symmetry just specified, cannot hold either.

The reason that the condition that the differential of the attraction that acts be complete, which the equality (7.3.4.6''''') expresses in the case at issue, and (2) are not equivalent is that when the positions of the points S that lie inside the homogeneous fluid figure of revolution in the plane of the meridian in question are expressed in the polar coordinates y and x, arg $S = x$ is a *multiple-valued function* of the positions of the points S. Two values of x which correspond to the same point S differ by $2n\pi$, where n is a positive or negative integer. Hence the polar coordinates y and x of the position of a point S are not unique. Both (y,x_1) and (y,x_2) can represent the same point S in the polar coordinates y and x, where $x_2 = x_1 + 2n\pi$ for some positive or negative integer n. Consequently an expression $\phi(y,x)$ whose total differential is a complete differential $Pdy + Qydx$ is *also* a *multiple-valued function* of the positions of the points S. Thus, for example, if the path of integration γ of the line integral

$$\int_\gamma Pdy + Qydx$$

from a point S to the same point S that lies inside the homogeneous fluid figure of revolution in the plane of the meridian in question and whose distance from $C = (0,0)$ is y is taken to be a circle whose center is at $C = (0,0)$ and whose radius is y, then

$$\int_\gamma Pdy + Qydx = \int_{S=(y,x)}^{S=(y,x+2\pi)} d\phi = \phi(y, x+2\pi) - \phi(y, x) \neq 0.$$

In short, the line integral of the complete differential $Pdy + Qydx$ whose path of integration γ is a *closed* path that *encircles* $C = (0,0)$ is *not* equal to zero.

But Clairaut did not understand that this is what caused the contradictory results that he thought that he had reconciled but in reality had not. Indeed, since the equality

$$\frac{d}{dx}\left(\frac{x}{x^2+y^2}\right) = \frac{d}{dy}\left(-\frac{y}{x^2+y^2}\right) \tag{7.3.4.6''''}$$

holds, Clairaut gave the differential

$$\frac{x}{x^2+y^2}dy - \frac{y}{x^2+y^2}dx$$

as an example of a complete differential in the rectangular coordinates y and x. Clairaut observed that the expression $\phi(y, x)$ whose total differential is the differential

$$\frac{x}{x^2+y^2}dy - \frac{y}{x^2+y^2}dx$$

is arc tan (y/x). However, Clairaut did not call the reader's attention to the fact that arc tan (y/x) is a multiple-valued function of y and x. He doubtlessly did not notice this fact, much less realize its significance. That is, he did not understand that because of this fact the condition that the differential of the attraction that acts be complete, which the equality (7.3.4.6''') expresses when rectangular coordinates are employed, and (2) are not equivalent. [It turns out that if the domain in which P and Q are defined is *simply connected*, then the equality

$$\frac{dP}{dx} = \frac{dQ}{dy} \tag{7.3.4.6'''}$$

holds at all points of this simply connected domain if and only if the expression $\phi(y, x)$ whose total differential is $Pdy + Qdx$ is a single-valued function of the rectangular coordinates y and x of points of the simply connected domain and the line integrals

$$\int_\gamma Pdy + Qdx$$

of the complete differential $Pdy + Qdx$ around all closed plane curves γ that lie within the simply connected domain are all equal to zero. However, in the preceding example the P and the Q,

$$P \equiv \frac{x}{x^2 + y^2} \quad \text{and} \quad Q \equiv -\frac{y}{x^2 + y^2},$$

do not even have unique limits as (y,x) approaches $C = (0,0)$, so that the definitions of P and Q in this example cannot even be extended continuously to have *unique finite values* at $C = (0,0)$, which in fact, as I noted previously, is also true of the P and the Q in the second of Clairaut's two examples in the polar coordinates y and x discussed before but which Clairaut did not notice or perceive. Hence in the preceding example the domain in which the equality (7.3.4.6''') holds, namely, the plane of the meridian in question without the point $C = (0,0)$, *is not simply connected.*[37]]

We recall that Clairaut derived principle (4) from (1) for a homogeneous fluid mass [although we remember that in order to do this he assumed that if (1) held for a homogeneous fluid mass, this meant that (1) not only held for all closed plane or skew channels that lie inside the homogeneous fluid mass and that are filled with fluid from the homogeneous fluid mass, but that (1) also held for closed plane or skew channels of homogeneous fluid parts of which lie *at the surface* of the homogeneous fluid mass]. But he did not demonstrate that (1) and (4) are strictly equivalent. Evidently the attraction must fulfill certain conditions in order to make (1) follow from (4). In particular, when the homogeneous fluid mass is a homogeneous fluid figure of revolution which revolves around its axis of symmetry, in which case the axis of symmetry of the homogeneous fluid figure of revolution is also an axis of symmetry of the centrifugal force of rotation per unit of mass, and, moreover, when the axis of symmetry of the homogeneous fluid figure of revolution is also an axis of symmetry of the attraction, the attraction must not have a singular point of one of the kinds illustrated in the examples in polar coordinates and rectangular coordinates discussed before, because (1) cannot even hold when the attraction has a singular point of one of these kinds, much less be made to follow from (4). That is, (1) entails (2) when the homogeneous fluid figure of revolution satisfies the conditions of symmetry just specified. Consequently (1) fails to hold if (2) fails to hold when the homogeneous fluid figure of revolution fulfills these conditions of symmetry. But (2) cannot hold when the homogeneous fluid figure of revolution satisfies these conditions of symmetry if the attraction has a singular point of one of the kinds illustrated in the examples in polar coordinates and rectangular coordinates. Therefore (1) cannot hold either when the homogeneous fluid figure of revolution fulfills these conditions of symmetry if the attraction has a singular point of one of these kinds.

Clairaut had evidently arrived at his new principles of equilibrium, as well as their mathematical representations, by the start of 1742. For he told Euler, in a letter dated 4 January 1742, that if y and x are rectangular coordinates in the plane of a meridian at the surface of a homogeneous fluid figure of revolution, where x is parallel to the axis of symmetry of the homogeneous fluid figure of revolution and y is perpendicular to the axis of symmetry of the homogeneous fluid figure of revolution, and if P and Q, respectively, are the components of the effective gravity in the plane of the meridian at the surface of the figure in the directions of y and x, respectively, then in order for the figure to be in equilibrium, it must be true that $Pdy + Qdx$ is an "integrable differential" ("différentielle intégrable"), by which he meant that the equality

$$\frac{dP}{dx} = \frac{dQ}{dy} \qquad (7.3.4.6''')$$

must hold, in which case he said that

$$\int Pdy + Qdx = \text{a constant } A \qquad (9.1.18')$$

is an equation of a meridian at the surface of the homogeneous fluid figure of revolution in equilibrium.[38]

Until now Clairaut had assumed that homogeneous fluid figures of equilibrium fulfilled the various conditions of symmetry which I have mentioned numerous times, namely, that they are homogeneous fluid figures of revolution which revolve around their axes of symmetry, in which case their axes of symmetry are also axes of symmetry of the centrifugal force per unit of mass, and, moreover, that their axes of symmetry are also axes of symmetry of the attraction. He said that he had made these assumptions because he was primarily concerned with "shapes of planets," in which case to have made other hypotheses would have given "a superfluous generality to the problem." Clairaut noted, however, that "there was only very little to add to the theory [of homogeneous fluid figures of equilibrium of the kind just specified]" in order to generalize this theory so that it includes homogeneous fluid figures of equilibrium which revolve around axes that are not axes of symmetry of the homogeneous fluid figures of equilibrium. Indeed the homogeneous fluid figures of equilibrium need not have axes of symmetry.

First let us consider a homogeneous fluid figure that revolves around an axis that is not an axis of symmetry of the homogeneous fluid figure (see Figure 48). Moreover, let us suppose that Clairaut's second new principle of equilibrium holds for this homogeneous fluid figure and that, consequently, Clairaut's third new principle of equilibrium also holds for this figure. Clairaut recalled that it

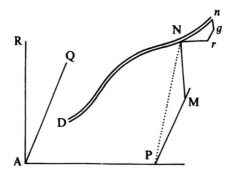

Figure 48. Figure used for determining efforts along channels when there is no symmetry (Clairaut 1743: figure on p. 98).

followed from his second new principle of equilibrium that the total effort produced by the effective gravity along a channel DN that lies within the interior of this homogeneous fluid figure and that is filled with fluid from the figure, where the channel DN can either be a plane channel or a skew channel, depends only upon the endpoints D and N of the channel. In other words, the total efforts produced by the effective gravity along channels that lie inside the homogeneous fluid figure, that join the same two points D and N, and that are filled with fluid from the figure are the same for all such channels. Clairaut showed that if P, Q, and R, respectively, are the components of the effective gravity in the directions of the rectangular coordinates $z = MN$, $y = PM$, and $x = AP$, respectively, which are mutually perpendicular, then $Pdz + Qdy + Rdx$ expresses the effort produced by the effective gravity along each of the infinite number of infinitesimal homogeneous fluid right circular cylinders Nn that make up the channel DN, where $dz = sn$, $dy = sr$, $dx = Nr$ (and consequently $Nn^2 = dz^2 + dy^2 + dx^2$), and, as I have said, where the channel DN can but need not lie in a plane. The principle that the total efforts produced by the effective gravity around arbitrary closed channels that lie inside the homogeneous fluid figure and that are filled with fluid from the figure, where the closed channels may either be closed plane channels or closed skew channels, are equal to zero, which is Clairaut's second new principle of equilibrium, is equivalent to the following statement: the total effort produced by the effective gravity along a channel DN that lies in the interior of the figure, that can either be a plane channel or a skew channel, and that is filled with fluid from the figure, which Clairaut represented by a line integral

$$\int_D^N Pdz + Qdy + Rdx,$$

where the path of integration of the line integral is the channel inside the homogeneous fluid figure which joins the two points D and N and which is filled with fluid from the figure, depends only upon the two points D and N and not upon the particular channel that joins these two points. (Here we have additional evidence that Clairaut thought of his plane and skew channels filled with homogeneous fluid as one dimensional. In other words, he actually imagined these channels of fluid to be plane curves and skew curves.) Clairaut affirmed that this condition entailed that the differential $Pdz + Qdy + Rdx$ must be a complete differential in z, y, and x. By this statement Clairaut meant that an expression $\phi(z, y, x)$ that is "algebraic or depends upon quadratures" exists whose total differential is the differential $Pdz + Qdy + Rdx$. Thus Clairaut again associated the more analytic or algebraic notion of a complete differential $Pdz + Qdy + Rdx$ with the more geometric idea of the path independence of the line integral of the differential $Pdz + Qdy + Rdx$. Furthermore, all evidence again points to Clairaut's having done so *through* the mechanics problem. Clairaut added that the same conclusion that the differential $Pdz + Qdy + Rdx$ must be a complete differential in z, y, and x could be reached by considering instead "the equality of the force that the transverse weight (pression) that a level layer" of the homogeneous fluid figure exerts at all points of the level surface that is its inner boundary, which, we remember, is a statement that is equivalent to Clairaut's third new principle of equilibrium. Hence the more analytic or algebraic notion of a complete differential $Pdz + Qdy + Rdx$ received another auxiliary, geometric interpretation. Namely, when $Pdz + Qdy + Rdx$ is a complete differential, an integral of the first-order "differential" equation of first degree

$$Pdz + Qdy + Rdx = 0$$

in the three rectangular coordinates z, y and x is geometrically just a level surface with respect to the effective gravity of the homogeneous fluid figure. In this case the axis that the homogeneous fluid figure revolves around is not an axis of symmetry of the homogeneous fluid figure. Once again, all evidence points to Clairaut's having arrived at this geometric interpretation of a complete differential $Pdz + Qdy + Rdx$ *through* the mechanics problem. Clairaut noted that in his second Paris Academy *mémoire* on integral calculus he had derived the conditions

$$\frac{dP}{dy} = \frac{dQ}{dz}, \quad \frac{dP}{dx} = \frac{dR}{dz}, \quad \text{and} \quad \frac{dQ}{dx} = \frac{dR}{dy} \qquad (7.3.4.29')$$

that the differential $Pdz + Qdy + Rdx$ in three variables z, y, and x must fulfill in order to be a complete differential.

Clairaut then stated that if the homogeneous fluid figure is a figure of equilibrium, in which case Clairaut's second and third new principles of equilibrium hold for the homogeneous fluid figure of equilibrium and the equalities

(7.3.4.29') hold, then the equation

$$\int_C^M Pdz + Qdy + Rdx - \frac{f}{2r}(y^2 + z^2)$$

$$= \text{a constant } A \text{ that is independent of } M \tag{9.1.3''}$$

in the three rectangular coordinates z, y, and x is an equation of the surface of the homogeneous fluid figure of equilibrium, where the path of integration of the line integral on the left-hand side of equation (9.1.3'') is any channel that joins the center C of the homogeneous fluid figure of equilibrium to an arbitrary point M of the surface of the figure and that is filled with fluid from the figure, where A is a constant that is independent of the particular point M chosen at the surface of the figure, where the figure revolves around its x-axis, and where f is the magnitude of the centrifugal force per unit of mass at a distance r from the axis of rotation (the x-axis).[39]

[Here Clairaut made a careless error, but fortunately it is one that is easy to rectify. Although Clairaut did not actually demonstrate that the following statement is true, evidently the total effort produced by the centrifugal force of rotation around an arbitrary closed channel, which may either be a plane channel or a skew channel, which lies in the interior of the homogeneous fluid figure of equilibrium, and which is filled with fluid from the figure is equal to zero. For it follows from such a result together with Clairaut's second new principle of equilibrium that the total effort produced by the *attraction* around an arbitrary closed channel, which may either be a plane channel or a skew channel, which lies inside the figure, and which is filled with fluid from the figure is equal to zero. Equivalently, for each pair of points situated inside the figure the total efforts produced by *attraction* along channels, which may be either plane channels or skew channels, which lie within the interior of the figure which are filled with fluid from the figure, and which join the same two points are the same for all channels that join these two points. In that case the differential $Pdz + Qdy + Rdx$, where P, Q, and R, respectively, are the magnitudes of the components of the *attraction* in the directions of the rectangular coordinates z, y, and x, respectively, is a complete differential, and the line integral

$$\int_C^M Pdz + Qdy + Rdx,$$

which represents the total effort produced by the *attraction* along a channel that joins the center C of the homogeneous fluid figure of equilibrium to a particular point M on the surface of the figure and that is filled with fluid from the figure, does not depend upon the particular channel that joins the center C of the homogeneous fluid figure of equilibrium to that point M of the surface of the figure, that is filled with fluid from the figure, and that is the path of integration of the line integral. For it should be *this* line integral that appears on the left-hand side of Clairaut's equation

$$\int_C^M Pdz + Qdy + Rdx - \frac{f}{2r}(y^2 + z^2)$$
$$= \text{a constant } A \text{ that is independent of } M, \qquad (9.1.3'')$$

not the line integral that represents the total effort produced by the effective gravity along a channel that joins the center C of the figure to a particular point M of the surface of the figure, that is filled with fluid from the figure, and whose path of integration is this channel. In fact this second line integral can be shown to · reduce to the expression

$$\int_C^M Pdz + Qdy + Rdx - \frac{f}{2r}(y^2 + z^2)$$

on the left-hand side of equation (9.1.3''), where P, Q, and R, respectively, are the magnitudes of the components of the attraction in the directions of the rectangular coordinates z, y, and x, respectively, not the magnitudes of the components of the effective gravity in the directions of the rectangular coordinates z, y, and x, respectively. We can easily see that this last expression just parallels the expression

$$\int_C^M Pdy + Qdx - \frac{f}{2r}y^2$$

on the left-hand side of equation (9.1.3) for the two-dimensional case (that is, the case where the homogeneous fluid figure of equilibrium is a figure of revolution which revolves around its axis of symmetry, in which case the axis of symmetry of the homogeneous fluid figure of revolution in equilibrium is also an axis of symmetry of the centrifugal force of rotation per unit of mass, and, moreover, where the axis of symmetry of the homogeneous fluid figure of revolution in equilibrium is also an axis of symmetry of the attraction).

In the three-dimensional case the total effort produced by the effective gravity along a channel that joins the center C of the homogeneous fluid figure of equilibrium to a particular point M of the surface of the figure and that is filled with fluid from the figure does not depend upon the particular channel that joins the center C of the figure to that point M of the surface of the figure and that is filled with fluid from the figure. This just follows from Clairaut's second new principle of equilibrium. Now, in particular the channel can be chosen to be the straight channel that joins the center C of the homogeneous fluid figure of equilibrium to the point M. But the homogeneous fluid figure is a figure of equilibrium, which means that the principle of balanced columns holds for the homogeneous fluid figure of equilibrium. But this means that the total efforts produced by the effective gravity along straight channels that join the center C of the figure to points M of the surface of the figure and which are filled with fluid from the figure are the same. The total efforts produced by the effective gravity along these straight channels do not depend upon the particular points M of the surface of the homogeneous fluid figure of equilibrium. Consequently it follows from these two results together (that is, from Clairaut's second new principle of equilibrium together with the principle of balanced columns) that the total effort produced by the effective gravity along a channel that joins the center C of the homogeneous fluid figure of equilibrium to a point M of the surface of the figure and that is filled with fluid from the figure neither depends upon the particular channel that joins the center C of the figure to a particular point M of the surface of the figure and that is filled with fluid from the figure nor depends upon the particular point M of the surface of the figure. But this is exactly what Clairaut's equation

$$\int_C^M Pdz + Qdy + Rdx - \frac{f}{2r}(y^2 + z^2)$$
$$= \text{a constant } A \text{ that is independent of } M, \tag{9.1.3''}$$

where the constant A is independent of the particular point M of the surface of the figure, expresses, when P, Q, and R, respectively, are the magnitudes of the components of the *attraction* in the directions of the rectangular coordinates z, y, and x, respectively, and where the path of integration of the line integral on the left-hand side of equation (9.1.3'') is any channel that joins the center C of the homogeneous fluid figure of equilibrium to an arbitrary point M of the surface of the figure and that is filled with fluid from the figure. We can easily see that equation (9.1.3'') just parallels equation

$$\int_C^M Pdy + Qdx - \frac{f}{2r} y^2 = \text{a constant } A \text{ that is independent of } M \tag{9.1.3}$$

for the two-dimensional case (that is, the case where the homogeneous fluid figure of equilibrium is a figure of revolution which revolves around its axis of symmetry, in which case the axis of symmetry of the homogeneous fluid figure of revolution in equilibrium is also an axis of symmetry of the centrifugal force of rotation per unit of mass, and, moreover, where the axis of symmetry of the homogeneous fluid figure of revolution in equilibrium is also an axis of symmetry of the attraction).

Now, since the homogeneous fluid figure is a figure of equilibrium, the principle of the plumb line holds at all of the points of the surface of the homogeneous fluid figure of equilibrium. Thus the surface of the figure is itself a level surface with respect to the effective gravity of the figure. Consequently, if we let P, Q, and R, respectively, stand for the magnitudes of the components of the effective gravity in the directions of the rectangular coordinates z, y, and x, respectively, instead of for the magnitudes of the components of the *attraction* in the directions of the rectangular coordinates z, y, and x, respectively, then, for the reasons given, the surface of the homogeneous fluid figure of equilibrium must be an integral of the first-order "differential" equation of first degree

$$Pdz + Qdy + Rdx = 0$$

in the three rectangular coordinates z, y and x, where the differential $Pdz + Qdy + Rdx$ is also a complete differential. In other words, the equation

$$\int Pdz + Qdy + Rdx = \text{a constant } A \qquad (9.1.18'')$$

in the three rectangular coordinates z, y, and x, where the integral on the left-hand side of equation (9.1.18″) does not designate a line integral of the complete differential $Pdz + Qdy + Rdx$ but an antidifferential of the complete differential $Pdz + Qdy + Rdx$ instead (that is, an expression $\phi(z, y, x)$ whose total differential is the complete differential $Pdz + Qdy + Rdx$), is an equation of the surface of the homogeneous fluid figure of equilibrium. We can easily see that equation (9.1.18″) just parallels equation

$$\int Rdy + Qdx = \text{a constant } A, \qquad (9.1.18)$$

for the two-dimensional case. Clairaut did not pay very close attention to what he was doing here, and as a result he mixed up effective gravity with attraction.]

Clairaut had already associated algebraic equations in three rectangular coordinates with surfaces in the treatise that he wrote on skew curves which was published in 1731. In doing so he became one of the first mathematicians ever to identify a single equation in three unknowns with a surface.[40] But in his treatise of 1731 he did not express surfaces as first-order "differential" equations of first degree

$$dz - \omega(z, x, y)\,dx - \theta(z, x, y)\,dy = 0 \qquad (7.3.5.4)$$

in three rectangular coordinates z, x, and y. For in order to investigate the properties and characteristics of surfaces expressed this way, Clairaut would have needed to use the partial differential calculus as a tool. Thus he would have had to introduce the partial differential calculus in his treatise of 1731, which he did not do. Nor does any evidence exist which suggests that he could have done so. There is no evidence that Clairaut began to write equations that involve the partial differential calculus before he learned late in 1738 of Fontaine's research in the partial differential calculus.

At the end of Chapter 7 I mentioned that Clairaut demonstrated in his second Paris Academy *mémoire* on integral calculus that if the first-order "differential" equation of first degree

$$dz - w(z, x, y)dx - \theta(z, x, y)dy = 0 \qquad (7.3.5.4)$$

in the three variables z, x, and y represents a surface, then Fontaine's "conditional

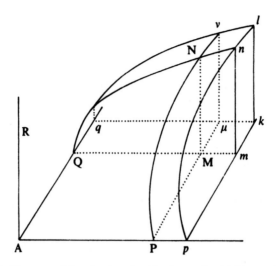

Figure 49. Figure used for showing that if a first-order "differential" equation
$dz - \omega(z, x, y) dx - \theta(z, x, y) dy = 0$ *represents a surface, then Fontaine's*
"conditional equation" must hold (Clairaut 1740b: figure on p. 309).

equation"

$$\frac{d\theta}{dx} + \omega\frac{d\theta}{dz} = \frac{d\omega}{dy} + \theta\frac{d\omega}{dz} \tag{7.3.5.5}$$

for the integrability of equation (7.3.5.4) must hold. Clairaut proved this result
in the following way. First he let the three variables z, x, and y be mutually
orthogonal rectangular coordinates. Then he utilized the idea that any two
infinitesimal arcs that lie in the surface and that begin at a common point of the
surface, whose rectangular coordinates are, say, (z_0, x_0, y_0), and whose points at
their other ends have the same values of x and y, respectively, should end at the
same point of the surface. In other words, the two points at the other ends of the
two infinitesimal arcs should have the same values of z as well in this case. [In
terms of functional relations, this amounts to assuming that the surface can be
expressed in the form $z = z(x, y)$ or that there are no multiple values of z for any
pair (x, y) of rectangular coordinates x and y such that a point whose rectangular
coordinates are (z, x, y) is a point of the surface.] In this way, Clairaut intuitively
characterized a surface. He then visualized plane curves formed where planes
perpendicular to the rectangular coordinate axes intersect the surface, and he
imagined certain infinitesimal arcs that lie in the surface and that these plane
curves determine (see Figure 49). Specifically, let AP stand for part of the x-axis;
let AQ designate part of the y-axis; and let AR stand for part of the z-axis. Thus if
Nvl is an infinitesimal arc in the surface whose points N and l, respectively, at its

two ends have coordinates (z, x, y) and $(z_1, x + dx, y + dy)$, respectively, and if Nnl' is an infinitesimal arc in the surface whose points N and l', respectively, at its two ends have coordinates (z, x, y) and $(z_2, x + dx, y + dy)$, respectively, then

$$z_1 \text{ must equal } z_2 \tag{9.1.21}$$

Clairaut aimed to express equality (9.1.21) assuming that the equation (7.3.5.4) represents the surface.

Since $dz = \theta(z, x, y) dy$ in this case is the equation of the plane curve formed where the plane perpendicular to the x-axis whose equation is $x = a$ constant intersects the surface, which PNv is part of, it follows that the coordinates of the point v at the end of the infinitesimal line segment Nv that is an infinitesimal part of the plane curve just specified and that is tangent to the surface at $N = (z, x, y)$ must be $(z + \theta(z, x, y) dy, x, y + dy)$. (Calling Nv a line segment involves making a linear approximation that only holds to terms of first order.) Consequently if the equation (7.3.5.4) represents the surface, in which case $dz = \omega(z, x, y + dy) dx$ is the equation of the plane curve formed where the plane perpendicular to the y-axis whose equation is $y + dy = a$ constant intersects the surface, which qvl is part of, and the infinitesimal line segment vl that is an infinitesimal part of the plane curve just specified is tangent to the surface at $v = (z + \theta(z, x, y) dy, x, y + dy)$, it follows that the coordinates of l should be

$$(z + \theta(z, x, y) dy + \omega(z + \theta(z, x, y) dy, x, y + dy) dx, x + dx, y + dy). \tag{9.1.22}$$

(Calling vl a line segment also involves making a linear approximation that only holds to terms of first order.)

Likewise, if the equation (7.3.5.4) represents the surface, in which case $dz = \omega(z, x, y) dx$ is the equation of the plane curve formed where the plane perpendicular to the y-axis whose equation is $y = a$ constant intersects the surface, which QNn is part of, it follows that the coordinates of the point n at the end of the infinitesimal line segment Nn that is an infinitesimal part of the plane curve just specified and that is tangent to the surface at $N = (z, x, y)$ must be $(z + \omega(z, x, y) dx, x + dx, y)$. (Again, calling Nn a line segment also involves making a linear approximation that only holds to terms of first order.) Therefore if the equation (7.3.5.4) represents the surface, in which case $dz = \theta(z, x + dx, y) dy$ is the equation of the plane curve formed where the plane perpendicular to the x-axis whose equation is $x + dx = $ constant intersects the surface, which pnl' is part of, and the infinitesimal line segment nl' that is an infinitesimal part of the plane curve just specified is tangent to the surface at $n = (z + \omega(z, x, y) dx, x + dx, y)$, it follows that the coordinates of l' should be

$$(z + \omega(z, x, y) dx + \theta(z + \omega(z, x, y) dx, x + dx, y) dy, x + dx, y + dy). \tag{9.1.23}$$

(Once again, calling nl' a line segment also involves making a linear approximation that only holds to terms of first order.)

Clairaut's hypotheses about the nature of surfaces require that l and l' coincide. Hence the first coordinates in (9.1.22) and in (9.1.23) must be the same, because the second and third coordinates, respectively, in (9.1.22) and in (9.1.23), respectively, are the same. In other words the equality

$$z + \theta(z, x, y)\, dy + \omega\, (z + \theta(z, x, y)\, dy, x, y + dy)\, dx$$
$$= z + \omega(z, x, y)\, dx + \theta(z + \omega(z, x, y)\, dx, x + dx, y)\, dy \qquad (9.1.24)$$

must hold.

Now, the expression

$$\theta(z + \omega(z, x, y)\, dx, x + dx, y) \qquad (9.1.25)$$

that appears in the third term on the right-hand side of equality (9.1.24) is equal to

$$\theta(z, x, y) + \frac{d\theta}{dx}(z, x, y)\, dx + \frac{d\theta}{dz}(z, x, y)\, \omega(z, x, y)\, dx \qquad (9.1.26)$$

(to terms of first order, which suffices here, since part of the reasoning is based on linear approximations that only hold to terms of first order anyway). Replacing the expression (9.1.25) that appears in the third term on the right-hand side of equality (9.1.24) by the expression (9.1.26), Clairaut found the expression on the right-hand side of equality (9.1.24) to be equal to the expression

$$z + \omega\, (z, x, y)dx + \left(\theta\, (z, x, y) + \frac{d\theta}{dx}(z, x, y)\, dx + \frac{d\theta}{dz}(z, x, y)\, \omega\, (z, x, y)\, dx \right) dy. \qquad (9.1.27)$$

Similarly, the expression

$$\omega(z + \theta(z, x, y)\, dy, x, y + dy) \qquad (9.1.28)$$

that appears in the third term on the left-hand side of equality (9.1.24) is equal to

$$\omega(z, x, y) + \frac{d\omega}{dy}(z, x, y)\, dy + \frac{d\omega}{dz}(z, x, y)\, \theta(z, x, y)\, dy \qquad (9.1.29)$$

(to terms of first order, as usual, of course). Replacing the expression (9.1.28) that appears in the third term on the left-hand side of equality (9.1.24) by the expression (9.1.29), Clairaut determined the expression on the left-hand side of equality (9.1.24) to be equal to the expression

$$z + \theta(z, x, y)\, dy + \left(\omega(z, x, y) + \frac{d\omega}{dy}(z, x, y)\, dy \right.$$
$$\left. + \frac{d\omega}{dz}(z, x, y)\, \theta(z, x, y)\, dy \right) dx. \qquad (9.1.30)$$

Equating expressions (9.1.27) and (9.1.30) and then canceling the terms

$$z + \omega(z, x, y)\,dx + \theta(z, x, y)\,dy$$

common to both expressions (9.1.27) and (9.1.30), Clairaut reduced equality (9.1.24) to the equality

$$\frac{d\theta}{dx}(z, x, y)\,dxdy + \frac{d\theta}{dz}(z, x, y)\,\omega(z, x, y)\,dxdy$$

(9.1.24')

$$= \frac{d\omega}{dy}(z, x, y)\,dydx + \frac{d\omega}{dz}(z, x, y)\,\theta(z, x, y)\,dydx.$$

Finally, dividing both sides of equality (9.1.24') by the factor $dxdy = dydx \neq 0$ that both sides of equality (9.1.24') have in common, Clairaut arrived at the equation

$$\frac{d\theta}{dx} + \omega\frac{d\theta}{dz} = \frac{d\omega}{dy} + \theta\frac{d\omega}{dz}. \qquad (9.1.24'')$$

But equation (9.1.24'') is the same as Fontaine's "conditional equation" (7.3.5.5) for the integrability of equation (7.3.5.4). Thus in this way Clairaut derived Fontaine's conditional equation (7.3.5.5) geometrically in his second Paris Academy *mémoire* on integral calculus by assuming that equation (7.3.5.4) represents a surface.[41]

[We recall that in his first Paris Academy *mémoire* on integral calculus, Clairaut used the condition (7.3.4.6) that he deduced in that *mémoire* from the equality (7.3.4.7) of mixed, second-order partial differential coefficients to arrive at equation (7.3.5.5). And we remember that in his second Paris Academy *mémoire* on integral calculus he demonstrated that equality (7.3.4.7) follows from Leibniz's inversion (7.1.1.1) of differentiation and integration. But the preceding geometrically based derivation of equation (7.3.5.5) does not involve condition (7.3.4.6) or equality (7.3.4.7) or Leibniz's inversion (7.1.1.1) of differentiation and integration. Hence the validity of this geometrically based derivation of equation (7.3.5.5) does not depend upon condition (7.3.4.6) or upon equality (7.3.4.7) or upon Leibniz's inversion (7.1.1.1) of differentiation and integration. In other words, it is not required that condition (7.3.4.6) or equality (7.3.4.7) or Leibniz's inversion (7.1.1.1) of differentiation and integration hold in order to use the preceding geometrically based argument to arrive at equation (7.3.5.5). That is, one need not assume that any of the following three conditions [condition (7.3.4.6), equality (7.3.4.7), or Leibniz's inversion (7.1.1.1) of differentiation and integration] holds in order to use the preceding geometrical based argument to arrive at equation (7.3.5.5). Indeed, as I noted at the end of Chapter 7, in his second Paris Academy *mémoire* on integral calculus Clairaut actually *deduced* condition (7.3.4.6) ($= (7.3.5.8')$) *as a special case of* the foregoing geometrically based derivation of equation (7.3.5.5), without having to assume that equality

(7.3.4.7) or Leibniz's inversion (7.1.1.1) of differentiation and integration holds.]

In his second Paris Academy *mémoire* on integral calculus Clairaut also proved that the statement that Fontaine's equation (7.3.5.5) is the conditional equation for the integrability of equation (7.3.5.4) is equivalent to the statement that the equation

$$N\frac{dP}{dx} - P\frac{dN}{dx} + M\frac{dN}{dz} - N\frac{dM}{dz} - M\frac{dP}{dy} + P\frac{dM}{dy} = 0 \qquad (7.3.4.37)$$

is the conditional equation for the integrability of the first-order "differential" equation of first degree

$$P(z, x, y)\, dz + M(z, x, y)\, dx + N(z, x, y)\, dy = 0 \qquad (7.3.4.28)$$

in the three variables z, x, and y. In Chapter 7 we saw that in his second Paris Academy *mémoire* on integral calculus Clairaut first showed that equation (7.3.4.37) is the conditional equation for the integrability of equation (7.3.4.28) by multiplying the differential on the left-hand side of (7.3.4.28) by an integrating factor $u(z, x, y)$ and then by applying the three conditions (7.3.4.29) to the resulting complete differential $uPdz + uMdx + uNdy$. The latter step gave rise to three equations, from which Clairaut eliminated $u(z, x, y)$ and its partial differential coefficients du/dz, du/dx, and du/dy. After he eliminated these quantities, the single equation (7.3.4.37) resulted. [In fact, he first demonstrated that equation (7.3.5.5) is the conditional equation for the integrability of (7.3.5.4) in his first Paris Academy *mémoire* on integral calculus essentially this same way.]

Now Clairaut showed that equation (7.3.4.37) is the conditional equation for the integrability of (7.3.4.28) another way. He remarked that if equation (7.3.4.28) is rewritten as

$$dz + \frac{M}{P}dx + \frac{N}{P}dy = 0, \qquad (9.1.31)$$

by dividing both sides of (7.3.4.28) by P (assuming that $P \neq 0$), and then if M/P is replaced by $-\omega$ and N/P is replaced by $-\theta$ in equation (9.1.31), then equation (7.3.5.4) results. At the same time, he noted that if ω is replaced by $-(M/P)$ and θ is replaced by $-(N/P)$ in equation (7.3.5.5), then equation (7.3.4.37) results. But equation (7.3.5.5) is the conditional equation for the integrability of equation (7.3.5.4). Hence it follows, as Clairaut observed, that equation (7.3.4.37) must be the conditional equation for the integrability of equation (7.3.4.28).

Conversely, Clairaut noted that if $P \equiv 1$ in equations (7.3.4.28) and (7.3.4.37), then (7.3.4.28) reduces to the equation

$$dz + M(z, x, y)\, dx + N(z, x, y)\, dy = 0, \qquad (9.1.32)$$

and (7.3.4.37) reduces to the equation

$$-\frac{dN}{dx} + M\frac{dN}{dz} = -\frac{dM}{dy} + N\frac{dM}{dz}. \tag{9.1.33}$$

But equation (7.3.4.37) is the conditional equation for the integrability of equation (7.3.4.28). Consequently equation (9.1.33) must be the conditional equation for the integrability of equation (9.1.32). Then if M is replaced by $-\omega$ and N is replaced by $-\theta$ in equation (9.1.32) and (9.1.33), respectively, the equations (7.3.5.4) and (7.3.5.5) result. Thus it follows that equation (7.3.5.5) must be the conditional equation for the integrability of equation (7.3.5.4).[42]

Clairaut was especially interested in geometrically characterizing the difference between first-order "differential" equations of first degree in three variables which are integrable and first-order "differential" equations of first degree in three variables which are not integrable. He believed that the former always represented surfaces and that the latter never represented surfaces. [As I mentioned in Chapter 7, however, he never actually demonstrated in his second Paris Academy *mémoire* on integral calculus that equation (7.3.5.5), which he derived by assuming that equation (7.3.5.4) represents a surface, is a *sufficient* condition for the integrability of equation (7.3.5.4) and not just a necessary condition for the integrability of equation (7.3.5.4). Equivalently, he did not prove in his second Paris Academy *mémoire* on integral calculus that equation (7.3.4.37) is a *sufficient* condition for the integrability of equation (7.3.4.28) and not just a necessary condition for the integrability of equation (7.3.4.28).] He demonstrated that his geometrically based derivation of equation (7.3.5.5) did not depend upon the particular functional relation assumed in representing the surface. That is, suppose that $x = x(z, y)$ or $y = y(z, x)$, instead of $z = z(x, y)$, were the right functional ways to represent other parts of the surface. (For example, consider the simple case of a rectangular parallelepiped. Then $z(x, y) = a$ constant a, $x(z, y) = a$ constant b, and $y(z, x) = a$ constant c express three of its faces that have a common vertex.) Then it is conceivable that Clairaut's reasoning applied to portions of the surface expressed these other ways could give rise to other "conditional equations," so that one conditional equation (7.3.5.5) alone did not suffice to specify the surface.

Clairaut made clear the advantage of expressing equation (7.3.5.5) in the equivalent form (7.3.4.37) for certain purposes. Namely, it follows from equation (7.3.4.37) that no other conditional equations exist. At the end of Chapter 7 I said that it was useful to rewrite equation (7.3.4.37) as

$$M\left(\frac{dN}{dz} - \frac{dP}{dy}\right) + N\left(\frac{dP}{dx} - \frac{dM}{dz}\right) + P\left(\frac{dM}{dy} - \frac{dN}{dx}\right) = 0. \tag{7.3.4.37'}$$

in order to determine certain properties of (7.3.4.37). [Namely, when (7.3.4.37) is rewritten as (7.3.4.37'), it can clearly be seen why (7.3.4.37) holds true when the three conditions (7.3.4.29) for the differential on the left-hand side of (7.3.4.28) to be complete are satisfied.] Equation (7.3.4.37) rewritten as (7.3.4.37′) displays other characteristics of (7.3.4.37) as well. Equation (7.3.4.37′) exhibits an inherent symmetry. Clairaut's arguments applied to portions of the surface expressed as $x = x(z, y)$ or $y = y(z, x)$ would lead to equations that can be arrived at by simply interchanging P and M and interchanging z and x in equation (7.3.4.37′) when $x = x(z, y)$ and by interchanging P and N and interchanging z and y in equation (7.3.4.37′) when $y = y(z, x)$. Such permutations, however, leave (7.3.4.37′) unchanged. [They cause the sign of the left-hand side of (7.3.4.37′) to change from positive to negative, but both sides of (7.3.4.37′) can then be multiplied or divided by -1.] Hence (7.3.4.37′) is intrinsic in the sense that it does not depend upon functional relations among the rectangluar coordinates used to express the surface. Consequently equation (7.3.4.37) is intrinsic in the same sense too. Thus (7.3.4.37) is not limited to surfaces whose dependent rectangular coordinates as functions of independent rectangular coordinates do not change. [Clairaut, in fact, did not rewrite (7.3.4.37) as (7.3.4.37′), but he came to these same conclusions by examining (7.3.4.37) itself.[43]]

We recall that Clairaut determined the equation of the surface of a homogeneous fluid figure of equilibrium which revolves around an axis that is not an axis of symmetry of the homogeneous fluid figure of equilibrium to be [44]

$$\int_C^M Pdz + Qdy + Rdx - \frac{f}{2r}(y^2 + z^2) = \text{a constant } A \text{ that is independent of } M.$$

$$(9.1.3'')$$

His interest in the problem of the shapes that homogeneous fluids in capillary tubes have appears to have been what induced him to generalize his theory of homogeneous fluid figures of equilibrium to include homogeneous fluid figures of equilibrium which revolve around axes that are not axes of symmetry of the homogeneous fluid figures of equilibrium. Indeed, he next spent a whole chapter of his treatise of 1743 discussing the theory of capillary action.[45]

Clairaut then turned to heterogeneous fluid figures consisting of individually homogeneous fluid strata, where two contiguous homogeneous fluid strata have different densities and where all of the individually homogeneous fluid strata have finite thicknesses. He introduced the first theory of equilibrium of such stratified fluid figures. He claimed to reduce the principles of equilibrium of such stratified fluid figures to the principles of equilibrium of homogeneous fluid figures. In particular, if a law of attraction permitted homogeneous fluid figures of equilibrium to exist, then, according to Clairaut, that law of attraction also allowed stratified fluid figures of the kind just described to exist in equilibrium. Once again, however, Clairaut presented his theory in a tangled, roundabout way. Here I unscramble the theory.

In effect, assuming that (1), the principle of the plumb line, holds at all points of the surface of a stratified fluid figure of the kind just described, which in fact must be true at all points of the surface of *any* fluid mass *in equilibrium*, and that (2), Clairaut's first new principle of equilibrium, holds for all plane or skew channels that lie within the interior of the stratified fluid figure, that are filled with fluid from the stratified fluid figure, and that join two different points of the surface of the figure (in other words, assuming that the total efforts produced by the effective gravity along all channels of the kind just described which are filled with fluid from the stratified fluid figure are equal to zero), Clairaut showed that it followed that (3) the surfaces between pairs of contiguous homogeneous fluid strata must all be "level surfaces" with respect to the effective gravity of the stratified fluid figure of the kind just described.

Clairaut's demonstration of the preceding statement is not complete, and I fill in the gaps. Although Clairaut did not specifically say so, the stratified fluid figure of the kind in question need not be a figure of revolution nor do the individually homogeneous strata have to be geometrically similar. That the surface of the stratified fluid figure is a "level surface" with respect to the effective gravity is true by hypothesis (1). Then consider the surface between any two contiguous homogeneous fluid strata, and let Q stand for a point of this surface. Suppose that the effective gravity at the point Q is not perpendicular to this surface at Q. Then the effective gravity at Q must have a component at Q which is tangent to this surface at Q. Consequently one can choose a closed plane curve $HKNQR$ in this surface such that the line that indicates the direction of this component of the effective gravity at Q, which is tangent to this surface at Q, lies in the plane of $HKNQR$ and is tangent to $HKNQR$ at Q. (Such a closed plane curve $HKNQR$ is not uniquely determined.) Next let O and S designate two different points of the surface of the stratified fluid figure, and let $PEpe$ stand for a closed plane curve in the surface of the stratified fluid figure which passes through O and S. (Given two different points O and S of the surface of the stratified fluid figure, such a closed plane curve $PEpe$ is also not uniquely determined.) The points O and S and the closed plane curve $PEpe$ can all be chosen so that the closed plane curves $PEpe$ and $HKNQR$ lie in the same plane, but the points O and S and the closed plane curve $PEpe$ need not be chosen this way. (Again, for this reason Figure 50 is misleading, like other diagrams that we have seen in Clairaut's treatise.) Let OQ designate a channel that joins O to Q, which can be a straight channel, a curved plane channel, or a skew channel; let SR designate a channel that joins S to another point R of the closed plane curve $HKNQR$, which can also be a straight channel, a curved plane channel, or a skew channel; and let QR designate the curved plane channel that coincides with the part of the closed plane curve $HKNQR$ which joins Q to R.

Then let us consider the channel $OQRS$ that is filled with fluid from the stratified fluid figure. This channel lies inside the stratified fluid figure; it is filled with fluid from the stratified fluid figure; and it joins two different points O and S of the surface of the figure. Then the total effort along the channel of fluid $OQRS$ produced by the effective gravity is equal to zero by hypothesis (2). Moreover, the

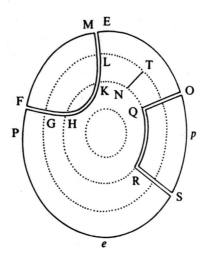

Figure 50. Clairaut's figure used for showing that when Clairaut's first new principle of equilibrium holds for channels inside a stratified fluid figure whose strata have finite widths, when the channels are filled with fluid from the figure, then the surfaces between pairs of contiguous homogeneous fluid strata must all be "level surfaces" with respect to the effective gravity if the principle of the plumb line holds at the surface of the figure and conversely (Clairaut 1743: figure on p. 129).

same thing would be true if the part *QR* of the channel of fluid *OQRS* were displaced in the plane of *HKNQR* by an infinitesimal amount from one side of the surface between the two contiguous homogeneous fluid strata to the other side of this surface.

Now, according to Clairaut, the total efforts produced by the effective gravity along the parts *OQ* and *SR* of the channel of fluid *OQRS* would not change appreciably as a result of such an infinitesimal displacement, presumably because the corresponding changes in the lengths of these two parts of the channel of fluid *OQRS* would be infinitesimal. Hence the total effort produced by the effective gravity along the part *QR* of the channel of fluid *OQRS* could not change perceptibly, either, as a result of such an infinitesimal displacement, because no matter how the channel of fluid *OQRS* is displaced inside the stratified fluid figure, the total effort produced by the effective gravity along the channel does not change. The total effort produced by the effective gravity along the channel of fluid *OQRS* remains constant, no matter how the channel of fluid *OQRS* is displaced inside the stratified fluid figure (namely, it is always equal to zero by hypothesis (2)).

Clairaut maintained, however, that the effort produced by the effective gravity along the part *QR* of the channel of fluid *OQRS* could not possibly change merely by an infinitesimal amount, after the part QR of the channel of fluid *OQRS* is displaced by an infinitesimal amount in the way just described, if there were components of the effective gravity at the points of QR which lie in the plane of $HKNQR$ and which are tangent to QR at those points. Clairaut said that the total effort produced by the effective gravity along the part QR of the channel of fluid $OQRS$ would be "very different" instead as a result of such an infinitesimal displacement of the part QR of the channel of fluid $OQRS$, if such components of the effective gravity at the points of QR existed.

Now, the point R of the closed plane curve $HKNQR$ was chosen arbitrarily. In particular, R can be chosen to be as close to Q as one wishes. Thus if we let R approach Q, we conclude that there can be no component of the effective gravity at Q in the direction in question. It follows that there can be no component of the effective gravity at Q which is tangent at Q to the surface between the two contiguous homogeneous fluid strata. In short, the effective gravity at Q must be perpendicular at Q to the surface between the two contiguous homogeneous fluid strata. But Q was taken to be an arbitrary point of the surface between the two contiguous homogeneous fluid strata. Hence the surface between the two strata must be a level surface with respect to the effective gravity. That is to say, (3) holds for the stratified fluid figure of the kind in question.[46]

[Clairaut did not call the reader's attention to the fact that if (2) holds for a stratified fluid figure of the kind in question, then it follows that (4), Clairaut's second new principle of equilibrium, also holds for all closed plane or skew channels that lie within the interior of the stratified fluid figure and that are filled with fluid from the stratified fluid figure (in other words, the total efforts produced by the effective gravity around all closed channels of the kind just described which are filled with fluid from the stratified fluid figure are equal to zero). This can easily be demonstrated to be true by applying the same method that Clairaut used to derive (4) from (2) in the case of homogeneous fluid figures.]

Apart from the gaps in Clairaut's demonstration which I have filled in, Clairaut reasoned in a particularly vague way at one point in the demonstration. He did not explain very intelligibly or convincingly why components of the effective gravity at the points of QR could not exist which lie in the plane of $HKNQR$, which are tangent to QR at those points, and which vary continuously across $HKNQR$ in the plane of $HKNQR$ so as to make the total effort produced by the effective gravity along the part QR of the channel of fluid $OQRS$ change by an infinitesimal amount when this part of the channel of fluid $OQRS$ is displaced in the manner described above. Why, for example, couldn't Clairaut's same arguments be used just as easily to prove that arbitrary surfaces that surround the inner surface that is a boundary of any particular homogeneous fluid stratum and that are surrounded by the outer surface that is also a boundary of this particular homogeneous fluid stratum are all level surfaces, which in fact such surfaces are not? Clairaut failed to make clear the essential, decisive part that assuming that the densities of two contiguous homogeneous fluid strata are *not the same*, in which case the difference in the densities of two contiguous homogeneous fluid strata is *not to equal zero*, plays in determining what he concluded to be the result of his demonstration, namely, that (3) holds for the stratified fluid figure of the kind in question. (In order to show why this assumption is particularly important in determining the result at issue, I will assume in order to simplify matters that the closed plane curves *PEpe* and *HKNQR* lie in the same plane and, moreover, that the parts *OQ* and *SR* of the channel of fluid *OQRS* are straight channels that also lie in this same plane, so that the channel of fluid *OQRS* is a plane channel that lies in this plane, which is how the channel of fluid *OQRS* appears in Figure 50.[47])

Clairaut next demonstrated that if hypotheses (1) and (3) hold for a stratified fluid figure of the kind in question and, moreover, if (4) holds for the figure as well, then (2) also holds for the figure. To do this Clairaut considered an arbitrary plane or skew channel *FGHKLM* that lies inside the stratified fluid figure, that is filled with fluid from the figure, and that joins two different points *F* and *M* of the surface of the figure. The closed plane curve *PEpe* lies in the surface of the figure. The closed plane curves *GLT, HKNQR*, etc., lie in the surfaces between different pairs of contiguous homogeneous fluid strata. (Here I allow for the possibility that, unlike the stratified fluid figure represented by Figure 50, the stratified fluid figure of the kind in question is made up of more than three individually homogeneous fluid strata that surround a core of homogeneous fluid concentric with the figure.) These closed plane curves will not in general all lie in the same plane. Indeed, it is not even essential that these closed curves be closed plane curves that lie in the surfaces between different pairs of contiguous homogeneous fluid strata. They can be closed skew curves that lie in the surfaces between different pairs of contiguous strata. All that really matters is that *G, H, K, L*, etc. designate the points where the channel *FGHKLM* intersects the surfaces between different pairs of contiguous strata. (Again, Figure 50 is misleading in this regard.)

Now, since hypothesis (4) is assumed to hold for the stratified fluid figure, then it follows of course a fortiori that the total efforts produced by the effective gravity around all closed plane or skew channels that lie wholly within the interior of an individual homogeneous fluid stratum and that are filled with fluid from that homogeneous fluid stratum are equal to zero. Clairaut used this fact together with (1) and (3) to deduce that the efforts produced by the effective gravity along the parts *FG* and *ML*, along the parts *GH* and *LK*, and so on, respectively, of channel *FGHKLM* must be equal and that the effort produced by the effective gravity along the part *HK* of channel *FGHKLM* must be equal to zero. Changing the sign of the efforts produced by the effective gravity along the parts *ML, LK*, etc. of channel *FGHKLM* from positive to negative, Clairaut concluded that the total effort produced by the effective gravity along the channel *FGHKLM* must be equal to zero. In other words, (2) holds for the stratified fluid figure of the kind in question.

[Once again we see, as we did when we examined Clairaut's theory of homogeneous fluid figures of equilibrium, that when Clairaut assumed that (4) held for a homogeneous fluid mass, which in the case in question is an individual homogeneous fluid stratum of a stratified fluid figure whose individually homogeneous fluid strata all have finite thicknesses, this meant in fact that he assumed that (4) not only held for all closed plane or skew channels that lie entirely *within the interior* of an individual stratum and that are filled with fluid from that stratum, but that it also held for all closed plane or skew channels that lie inside an individiual homogeneous fluid stratum and that are filled with fluid from that stratum but parts of which lie *at the surfaces* (in other words, lie *at the boundaries*) of that stratum.[48] On the other hand, one may argue that when (4) is assumed to hold for a stratified fluid figure of the kind in question, this means that (4) not only includes all closed plane or skew channels that lie entirely within the interior of an individual homogeneous fluid stratum and that are filled with fluid from that stratum, but that it also includes as special cases all closed plane or skew channels that lie inside an individual homogeneous fluid stratum and that are filled with fluid from that stratum but parts of which lie at the surfaces (in other words, lie at the boundaries) of that stratum. This conclusion can probably be made to follow from a judicious application of the principle of solidification.]

We note that the stratified fluid figure must be of the kind in question in order for the preceding statements and Clairaut's demonstrations of them to hold. That is, the individually homogeneous fluid strata that make up the stratified fluid figure *must* all have thicknesses that are *finite*. The statements and their proofs are not valid if the stratified fluid figure consists of individually homogeneous fluid strata that are *surfaces themselves,* in which case the strata are "infinitesimally" thin. Thus, for example, Clairaut's demonstrations cannot be applied to a stratified fluid figure whose density varies, say, *continuously* from its center to its surface, since in this case the individually homogeneous fluid strata that make up the stratified fluid figure are surfaces themselves. In other words, Clairaut's results hold for stratified fluid figures of equilibrium which are *made up of* homogeneous fluid figures of equilibrium (the individually homogeneous fluid strata whose thicknesses are finite), which a stratified fluid figure of equilibrium composed of individually homogeneous fluid strata that are surfaces themselves is not. This restriction is essentially what served as Clairaut's basis for claiming that he had reduced the principles of his theory of such stratified fluid figures of equilibrium to the principles of his theory of homogeneous fluid figures of equilibrium. I shall talk in the next section of this chapter about the limitations of the theory which such a condition entails.

Clairaut remarked that the preceding results could also be demonstrated "rather easily" provided his third new principle of equilibrium is assumed to hold for a homogeneous fluid figure. Hypothesizing that his third new principle of equilibrium holds for a homogeneous fluid figure, Clairaut again imagined such a figure to be partitioned into an infinite number of level layers with respect to effective gravity, which we remember actually means that the figure is partitioned into a very large number of very thin level layers with respect to effective gravity. Each point N inside the homogeneous fluid figures lies in the interior of one of the very thin level layers or is a point of one of the two level surfaces that bound one of the very thin level layers. Here $PEpe$, GLT, $HKNQR$, etc. are closed plane curves that lie in different level surfaces and that also lie in the same plane, where it is assumed that the effective gravity at each point of one of these closed plane curves is perpendicular at the point to the level surface in which the closed plane curve lies. [Here I take into account the assumption that the homogeneous fluid figure is made up of a very large number of very thin level layers with respect to effective gravity. Clairaut himself applied his argument to the homogeneous fluid figure represented by Figure 50, which is made up of only three very thin level layers that surround a core of fluid concentric with the homogeneous fluid figure. In his demonstration he kept the number of very thin level layers small because the same reasoning holds no matter how many thin level layers with respect to effective gravity make up a homogeneous fluid figure, so that the number of thin level layers might just as well be assumed to be a small number for purposes of illustrating the general argument.]

Then according to Clairaut's third new principle of equilibrium, the thickness NT of a very thin level layer at a point N of one of the two level surfaces that bound the very thin level layer or in whose interior the point N lies must be inversely proportional to the magnitude of the effective gravity at the point N. We recall that when this third new principle of equilibrium holds for a homogeneous fluid figure, it turns out that the transverse weight (pression) of a very thin level layer of the homogeneous fluid figure exerts a force upon the level surface that is the inner boundary of the very thin level layer which has the same magnitude at all of the points of this level surface and, moreover, which is perpendicular to this level surface at all of the points of the level surface.

Then changing the density of the fluid in such a very thin level layer without changing either the shape of the layer or the effective gravity at any of its points should permit the layer's transverse weight to continue to exert a force upon the surface that is the inner boundary of this layer which has the same magnitude at all of the points of this surface and, in addition, which is orthogonal to this surface at all of the points of the surface, where the magnitude of the force (pression) differs, of course, from the magnitude of the force exerted before the density of the fluid in the very thin layer was changed. In other words, the very thin layer should remain a level layer after changing the density of the fluid in it. In particular, if the very thin layer is part of a homogeneous fluid figure of equilibrium, then changing the density of the fluid in the layer should not disrupt the equilibrium of the fluid figure that the layer is part of and therefore should not cause the shape of this fluid figure to change. Thus Clairaut stated that when the conditions just mentioned held, "the equilibrium of a [fluid] planet [in equilibrium] will be preserved, whether all of the [very thin] homogeneous [level] layers [of the fluid planet] have the same density or different densities," provided, he said, "that the denser layers always lie underneath those that are less dense."[49] The last stipulation I will discuss later in this chapter.

In particular, it followed immediately from this result that if a stratified fluid figure of equilibrium of the kind in question which revolves around an axis and whose shape has to be determined is composed of very thin individually homogeneous strata that also happen to be level layers and that attract according to a law that is given independently of the shapes and densities of the individually homogeneous strata that make up the stratified fluid figure of equilibrium (which is true, for example, in case the figure is attracted by a force directed toward a fixed center of force whose magnitude at a point solely depends upon the distance of the point from the center of force), then the stratified fluid figure of equilibrium can be assumed to be homogeneous for purposes of finding its shape.[50]

As Clairaut realized himself, however, the preceding demonstration, which involves the use of his third new principle of equilibrium for homogeneous fluid figures, is only valid when the resultants of the attraction at the points of a stratified fluid figure do not themselves depend upon the shapes and the densities of the very thin individually homogeneous fluid strata that make up the stratified fluid figure and that are also level layers with respect to effective gravity. This condition holds, for example, if such a stratified fluid figure is attracted by a force directed toward a fixed center of force whose magnitude at a point only depends upon the distance of the point from the center of force.

In case fluid is assumed to attract according to the universal inverse-square law, however, the resultants of the attraction at the points of a stratified fluid figure themselves depend upon the shapes and the densities of the individually homogeneous fluid strata that make up the figure, and, conversely, the shapes of the individually homogenous fluid strata that make up the figure depend upon the resultants of the attraction at the points of the figure. In other words, in this case the resultants of the attraction at the points of a stratified fluid figure and the shapes of the individually homogeneous fluid strata that make up the figure are interdependent.

Indeed, changing the density of only a single very thin level layer that is part of a homogeneous fluid figure will be enough to cause the resultants of the attraction to change appreciably at the points of the fluid figure. This means that changing the density of this very thin level layer will cause the resultants of the attraction to change appreciably both at the points within this layer as well as at the points within the rest of what was originally a homogeneous fluid figure. This is true as long as the thickness of the homogeneous level layer whose density is changed is finite, no matter how thin the layer might be. But the perceptible changes in the resultants of the attraction at the points within what was originally a very thin homogeneous level layer that was part of what was initially a homogeneous fluid figure before the density of this layer was changed will also cause the effective gravity at the points within the layer to change appreciably, and the perceptible changes in the resultants of the attraction at the points within the rest of what was originally a homogeneous fluid figure will cause the effective gravity at the points within this part of what was originally a homogeneous fluid figure to change appreciably too.

Moreover, both the perceptible changes in the effective gravity at the points within what was originally a very thin homogeneous level layer that was part of what was initially a homogeneous fluid figure and the perceptible changes in the effective gravity at the points within the rest of what was originally a homogeneous fluid figure, both of which are caused by the change in the density of the very thin homogeneous layer, will also cause both the shape of the layer and the shape of the fluid figure that this layer is part of to change.

Consequently, because of all of these different changes, what was originally a very thin level layer with respect to effective gravity which was part of what was initially a homogeneous fluid figure is transformed, as a result of a change in the density of the layer, into a very thin homogeneous layer that will *not* in general be a level layer with respect to effective gravity. (In other words, the change in the shape of what was originally a very thin level layer with respect to effective gravity which was part of what was initially a homogeneous fluid figure, the changes in the effective gravity at the points of the very thin layer with its shape altered, and the changes in the effective gravity at the points of the rest of the fluid figure that this layer is part of, all of which are caused by the change in the density of the very thin homogeneous layer, will not in general be changes that make the resultants of the effective gravity at all of the points of the surfaces of the very thin homogeneous layer be perpendicular to these surfaces at the points of these surfaces.)

Moreover, as I just mentioned, the fluid figure that this very thin homogeneous layer is part of will not have the same shape as the original homogeneous fluid figure whose density differs from the density of this layer.

Consequently the preceding demonstration, which involves the application of Clairaut's third new principle of equilibrium for homogeneous fluid figures, does not hold in this particular case.[51]

9.2. *Applications of the new principles of equilibrium*

I shall present some of the ways that Clairaut tried to show in his treatise of 1743 that his second new principle of equilibrium holds for homogeneous fluid figures that revolve around axes that pass through them when it is hypothesized that these figures are attracted in certain specific ways. In each case the result that he stated is true. Sometimes, however, the diagrams that accompany certain demonstrations and the language that Clairaut used to describe various steps in these demonstrations and the intermediate results that follow from these steps suggest that in these instances Clairaut really tried to prove results that are less general than the results that he stated. (In other words, he actually tried to prove only special cases of the general results that he stated.) Some of the demonstrations have gaps, which I fill in. In one case in order to fill in the gap, one has to use implicitly the very result that Clairaut was then attempting to prove, so that although the result in question is in fact true, as it turns out, Clairaut's demonstration of it is circular.

In trying to demonstrate his first result, which I shall state in a moment, Clairaut does not appear to have assumed that the homogeneous fluid figures that he considered necessarily had to be homogeneous fluid figures of revolution. In particular, he does not appear to have assumed in trying to demonstrate this first result that a homogeneous fluid figure is necessarily a homogeneous fluid figure of revolution which revolves in such a way and that the forces of attraction act in such a way so as to produce the symmetries that I mentioned in Chapters 1, 4, and 6 (that is, that a homogeneous fluid figure is a homogeneous fluid figure of revolution which revolves around its axis of symmetry, in which case the axis of symmetry of the homogeneous fluid figure of revolution is also an axis of symmetry of the centrifugal force of rotation per unit of mass, and, moreover, that the axis of symmetry of the homogeneous fluid figure of revolution is also an axis of symmetry of the attraction, which are conditions, we recall, that did in fact always hold for the homogeneous fluid figures discussed in Chapters 1, 4, and 6.)

Clairaut first claimed to show that if such a homogeneous fluid figure is attracted by a force directed toward a fixed center of force situated inside the homogeneous fluid figure whose magnitude at a point solely depends upon the distance of the point from the fixed center of force, then his second new principle of equilibrium holds for arbitrary closed plane channels or skew channels that lie within the interior of the homogeneous fluid figure and that are filled with fluid from the homogeneous fluid figure. In fact what he actually tried to demonstrate is that if a homogeneous fluid figure is attracted by a force directed toward a fixed center of force located inside the homogeneous fluid figure whose magnitude at a point solely depends upon the distance of the point from the fixed center of force, then the total effort produced by the attraction around an arbitrary closed plane or skew channel that lies inside the homogeneous fluid figure and that is filled with fluid from the figure is equal to zero. I make this distinction because we remember that although Clairaut stated that the centrifugal force generated when a mass of homogeneous fluid revolves around an axis never produces a total effort around an arbitrary closed channel that lies in the interior of the mass and that is filled with fluid from the mass which is not equal to zero, which in fact is true, he never actually demonstrated that this result is true. Thus Clairaut only assumed implicitly here without proof that this result is true. We recall that all that he actually showed to be true is that the total efforts around closed plane channels filled with homogeneous fluid produced by the centrifugal force generated when a plane revolves around an axis that lies in the plane and the closed plane channels filled with homogeneous fluid all lie in that plane are equal to zero. That is, all that

Clairaut really proved in detail is that when a plane revolves around an axis that lies in the plane, the centrifugal force generated never produces a total effort that is not equal to zero around a closed channel filled with homogeneous fluid which lies in that plane.

Specifically, Clairaut claimed to show that if such a homogeneous fluid figure is attracted by a force directed toward a fixed center of force C situated inside the homogeneous fluid figure whose magnitude at a point solely depends upon the distance of the point from the fixed center of force, where *PEpe* designates a closed plane curve that lies at the surface of the homogeneous fluid figure, then the total efforts produced by the attraction around all closed plane or skew channels that lie within the interior of the homogeneous fluid figure and that are filled with fluid from the figure are equal to zero.

Now, in fact Clairaut did not demonstrate any such result as general as this, although the general result that he stated is nevertheless true. The reason that the general result is true is that the attraction produced by a fixed center of force whose magnitude at a point solely depends upon the distance of the point from the fixed center of force has a single-valued potential, in which case the attraction produced by the fixed center of force is a force per unit of mass which is conservative. Hence the work done by this attraction in moving a particle whose mass is 1 around any closed plane or skew curve is always equal to zero. But the total effort

$$\int_{\gamma} \mathbf{a} \cdot d\mathbf{s},$$

expressed here in terms of the vector calculus, produced by the attraction $\mathbf{a}$ around a closed plane or skew channel γ filled with homogeneous fluid cannot be distinguished in terms of units from the work done by the attraction $\mathbf{a}$ in moving a particle whose mass is 1 around γ. However, Clairaut did not realize that single-valued potentials of attraction can exist. This we recall from his treatment of the total efforts produced by attraction along plane channels filled with homogeneous fluid, where the lines that indicate the directions of the attraction at points in the plane of the plane channel lie in that plane, which we examined in the preceding section of this chapter. We remember that he expressed these efforts in terms of line integrals of differentials of attraction in two variables (two rectangular coordinates or polar coordinates), but that he failed to recognize the difference between complete differentials of attraction in two variables, whose integrals (that is, whose antidifferentials) can be multiple-valued functions of the positions of points when these positions are expressed in terms of the variables, and differentials of attraction in two variables whose line integrals are path independent. Surely then, Clairaut would not have realized that single-valued potentials of attraction can exist, much less have known the conditions that must hold in order for them to exist, in the more complex case of the total efforts produced by attraction along skew channels filled with homogeneous fluid, which are represented by line integrals of differentials in three variables, and where the attraction at points is no longer restricted to points that lie in a plane and where the lines that indicate the directions of the attraction at points are also not restricted to lie in a plane. In particular, Clairaut certainly did not know that if y, x, and z designate mutually perpendicular rectangular coordinates and if $P(y,x,z)$, $Q(y,x,z)$, and $R(y,x,z)$ are expressions in y, x, and z which are defined in a domain that is *simply connected* in three-dimensional space, then the line integral

$$\int_c^d \mathbf{a} \cdot d\mathbf{s} = \int_{(y_0, x_0, z_0)}^{(y,x,z)} P\,dy + Q\,dx + R\,dz$$

of the differential $P\,dy + Q\,dx + R\,dz$ in the three rectangular coordinates y, x, and z is independent of the path in the simply connected domain which joins the point $c \equiv (y_0, x_0, z_0)$ to the point $d \equiv (y, x, z)$ in the simply connected domain, in which case the line integral consequently defines a single-valued function of y, x, and z in the simply connected domain (the potential of $\mathbf{a}$), if and only if the differential $P\,dy + Q\,dx + R\,dz$ is a complete differential in the simply connected domain. For we recall that Clairaut established no result for plane paths and for differentials in two rectangular coordinates y and x in the plane which is analogous to the result just stated, nor did he even realize that there is such a result in this case which is comparable to the result just stated. Indeed, we remember that Clairaut mistook complete differentials of attraction in two variables (two rectangular coordinates or polar coordinates) whose integrals (that is, whose antidifferentials) can be multiple-valued functions of the two variables for differentials of attraction in the two variables whose line integrals are path independent. Thus it is hard to imagine how Clairaut, who did not know these things, could possibly have proven the general result stated before which he claimed to demonstrate. This general result can be shown to be true by using Stokes's theorem, but this theorem dates back to the nineteenth century.

As I have mentioned several times already, Clairaut's diagrams often do not help in trying to determine exactly what it was that Clairaut was trying to prove. The diagrams frequently only add to the uncertainty and confuse matters even further instead of help clarify them. In Figure 51 which accompanies Clairaut's supposed demonstration of the result stated before, the closed channel that lies within the interior of the homogeneous fluid figure and that is filled with fluid from the homogeneous fluid figure, which is designated by $MBNA$, in fact appears to be a closed plane channel that lies in the plane of $PEpe$. Moreover, the fixed center of force C also appears to be situated in the plane of $PEpe$. If Clairaut stated that the general result in question is true, which it is, it nevertheless seems likely that Clairaut only tried to demonstrate this result for this special case and this case alone, for when Clairaut really meant to talk about skew channels filled with homogeneous fluid, he had the means to make this clear in his diagrams. Namely, when he demonstrated results that he specifically meant to apply to skew channels filled with homogeneous fluid, he used diagrams that show skew

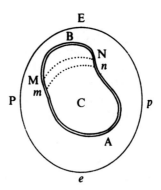

Figure 51. Clairaut's figure used to show that if a homogeneous fluid figure is attracted by a central force, where the fixed center of force is situated inside the figure, then Clairaut's second new principle of equilibrium holds for all closed channels inside the figure which are filled with fluid from the figure (Clairaut 1743: figure on p. 17).

channels filled with homogeneous fluid, as in the case of Figure 41, in the preceding section of this chapter. Furthermore, as we shall see in a moment, there is additional evidence that supports my conjecture that Clairaut only meant to treat the special case.

Clairaut claimed to demonstrate the result in question for the case shown in Figure 51 as follows. He imagined the closed plane channel of homogeneous fluid *MBNA* to be composed of two branches *BMA* and *BNA*. He noted that an infinity of circular arcs *MN*, *mn*, etc. whose centers are all located at *C* partition the two branches *BMA* and *BNA* of the closed plane channel of homogeneous fluid *MBNA* into the same number of infinitesimal homogeneous fluid right circular "cylinders" ("cylindres") *Mm*, *Nn*, etc., where all of the infinitesimal homogeneous fluid right circular cylinders have the same density. In fact, the number of homogeneous fluid right circular cylinders in each branch is infinite, and the cylinders in one branch are in a one-to-one correspondence with the cylinders in the other branch.

Clairaut then stated that since the infinitesimal homogeneous fluid right circular cylinders *Mm* and *Nn* have the same height ("hauteur") and since the magnitudes of the attraction at *M* and at *N* are equal because *M* and *N* are the same distance from the center of force *C*, it follows that the efforts produced by the attraction along the cylinders *Mm* and *Nn* are equal. Now, it is true that if the closed channel *MBNA* is a plane channel that lies in the plane of *PEpe* and if the center of force *C* is also situated in the plane of *PEpe*, then the two cylinders *Mm* and *Nn* and the two circular arcs *MN* and *mn* whose centers are located at *C* all lie in the plane of *PEpe* and, moreover, the distance between the two circular arcs *MN* and *mn* is constant. As we shall see in a moment, this distance can be thought of as the height of both cylinders Mm and Nn in this particular case. Then since this distance can be thought of as the height of both cylinders and, moreover, since both cylinders turn out to have the same height in this particular case, it seems likely that Clairaut was only talking about this particular case in his demonstration of the result in question. But Clairaut did not really explain why the efforts produced by the attraction along the two infinitesimal homogeneous fluid right circular cylinders Mm and Nn must be equal if the cylinders have the same height. Indeed, as I shall show in a moment, in arriving at this conclusion Clairaut in all probability implicitly used the very result that he was trying to demonstrate. By the same token, we can see that Clairaut certainly did not prove that the total effort produced by the attraction due to a fixed center of force C located inside a homogeneous fluid figure around any arbitrary closed plane or skew channel $MBNA$ that lies within the interior of the homogeneous fluid figure and that is filled with fluid from the figure is equal to zero, since he does not really appear even to have proven the result correctly in the special case in question.

However, if it is *already assumed or known at the start* that the total efforts produced by the attraction due to the fixed center of force C around all closed plane channels $MBNA$ that lie in the interior of the homogeneous fluid figure, that are filled with fluid from the figure, and that lie in the plane of $PEpe$, where the fixed center of force C is also situated inside the figure in the plane of $PEpe$, *then* it can be shown that the efforts produced by the attraction along the two infinitesimal homogeneous fluid right circular cylinders Mm and Nn must be equal if the two cylinders have the same height. This can be done in the following way. The infinitesimal homogeneous fluid right circular cylinder Mm can be thought of as a side of the infinitesimal closed plane curve Mrm in the plane of $PEpe$, where r is the point where the line through C and M intersects the circular arc mn, where the infinitesimal line segment Mr is consequently an infinitesimal part of the line through C and M, and where rm is part of the circular arc mn. Now we consider Mrm to be a closed plane channel in the plane of $PEpe$ which is filled with fluid from the homogeneous fluid figure. Then the total effort produced by the attraction due to the fixed center of force C around Mrm is equal to zero by hypothesis. But the effort produced by the attraction due to the fixed center of force C along rm is equal to zero, because the attraction produced by the fixed center of force C is perpendicular to rm at all points of rm. Hence the effort produced by the attraction due to the fixed center of force C along Mm is the same as the effort produced by the attraction due to the fixed center of force C along Mr. Similarly, the infinitesimal homogeneous fluid right circular cylinder Nn can be imagined to be a side of the infinitesimal closed plane curve Nsn in the plane of $PEpe$, where s is the point where the line through C and N intersects the circular arc mn, where the infinitesimal line segment Ns is consequently an infinitesimal part of the line through C and N, and where sn is part of the circular arc mn. Now we consider Nsn to be a closed plane channel in the plane of $PEpe$ which is filled with fluid from the homogeneous fluid figure. Then the total effort produced by the attraction due to the fixed center of force C around Nsn is again equal to zero by hypothesis. But the effort produced by the attraction due to the fixed center of force C along sn is equal to zero, because the attraction produced by the fixed center of force C is perpendicular to sn at all points of sn. Hence the effort produced by the attraction due to the fixed center of force C along Nn is the same as the effort produced by the attraction due to the fixed center of force C along Ns. Now, Mr is the distance between the circular arcs MN and mn at M; Ns is the distance between the circular arcs MN and mn at N; and Mr and Ns are equal, because the distance between the circular arcs MN and mn is constant. Mr must be what Clairaut meant by the height of Mm, and Ns must be what he meant by the height of Nn. But M and N are the same distance from the fixed center of force C; m and n are the same distance from the fixed center of force C; and Mr and Ns are both parts of lines of attraction through the fixed center of force C. Hence the efforts produced by the attraction due to the fixed center of force C along Mr and Ns are clearly the same. Therefore the efforts produced by the attraction due to the fixed center of force C along Mm and Nn must be the same too, since the effort produced by the attraction due to the fixed center of force C along Mm is the same as the effort produced by the attraction due to the fixed center of force C along Mr, and the effort produced by the attraction due to the fixed center of force C along Nn is the same as the effort produced by the attraction due to the fixed center of force C along Ns. This explains what I meant when I said above that Clairaut very likely implicitly used the very result that he was trying to demonstrate.

In fact, Clairaut could have correctly demonstrated the result in question using the same kinds of arguments that he had employed earlier in his treatise of 1743 to prove some of the results that we examined in the preceding section of this chapter. Namely, the infinitesimal homogeneous fluid right circular cylinder Mm can be thought of as the hypotenuse of an infinitesimal right triangle Mrm in the plane of $PEpe$, where the infinitesimal line segment Mr is an infinitesimal part of the line through C and M and where the infinitesimal line segment rm is perpendicular to Mr. Similarly, the infinitesimal homogeneous fluid right circular cylinder Nn can be imagined to be the hypotenuse of an infinitesimal right triangle Nsn in the plane of $PEpe$, where the infinitesimal line segment Ns is an infinitesimal part of the line through C and N and where the infinitesimal line segment sn is perpendicular to Ns. More precisely, as I have already noted above, if, as we have assumed, the closed channel $MBNA$ is a plane channel that lies in the plane of $PEpe$ and if the fixed center of force C is also situated in the plane of $PEpe$, then the two infinitesimal homogeneous fluid right circular cylinders Mm and Nn and the two circular arcs MN and mn whose centers are located at C all lie in the plane of $PEpe$. Then in this case the distance between the two circular arcs MN and mn is constant. The infinitesimal right triangles Mrm and Nsn in the plane of $PEpe$ can therefore be imagined as follows. In exactly the same way that I argued before, the infinitesimal homogeneous fluid right circular cylinder Mm

can be thought of as a side of the infinitesimal closed plane curve *Mrm* in the plane of *PEpe*, where *r* is the point where the line through *C* and *M* intersects the circular arc *mn*, where the infinitesimal line segment *Mr* is consequently an infinitesimal part of the line through *C* and *M*, and where *rm* is part of the circular arc *mn*. Now, in addition, *Mr* is clearly perpendicular to *mn*, and since the circular arc *rm* is infinitesimal, *rm* can be treated as a rectilinear segment. Thus *Mrm* can be thought of as an infinitesimal right triangle. Similarly, the infinitesimal homogeneous fluid right circular cylinder *Nn* can be imagined to be a side of the infinitesimal closed plane curve *Nsn* in the plane of *PEpe*, where *s* is the point where the line through *C* and *N* intersects the circular arc *mn*, where the infinitesimal line segment *Ns* is consequently an infinitesimal part of the line through *C* and *N*, and where *sn* is part of the circular arc *mn*. Again, *Ns* is clearly perpendicular to *mn*, and since the circular arc *sn* is infinitesimal, *sn* can also be treated as a rectilinear segment. Hence *Nsn* can also be thought of as an infinitesimal right triangle. Moreover, the lengths of *Mr* and *Ns* are equal, because both lengths are the distance between the two circular arcs *MN* and *mn* whose centers are located at *C*, and this distance is constant.

For the same reasons that I explained in the preceding section of this chapter, one can now reason in one of two possible ways. One can take the cross sections of the infinitesimal homogeneous fluid right circular cylinders *Mm* and *Nn* to have the same areas, in which case the volumes of the cylinders are directly proportional to their lengths. Now, since the cylinders *Mm* and *Nn* have the same density, the masses of the cylinders are directly proportional to their volumes. But if the volumes of cylinders *Mm* and *Nn* are directly proportional to their lengths, then the masses of the cylinders are directly proportional to their lengths too, in which case the masses of the cylinders can be thought of as having units of length. (The actual value of the density of the infinitesimal homogeneous fluid right circular cylinders can be disregarded.)

Alternatively, since Clairaut actually treated *M*, *m*, *N*, and *n* like points, not like disks, one can assume instead that the lengths of the infinitesimal homogeneous fluid right circular cylinders *Mm* and *Nn* are at least one order of magnitude greater than the areas of their cross sections, so that the cylinders are effectively one dimensional and thus can be imagined as having units of length. Then since cylinders *Mm* and *Nn* have the same density, their masses can be thought of as having units of length as well. (Again the actual value of the density of the infinitesimal homogeneous fluid right circular cylinders can be ignored.)

Now, we can assume that the magnitudes of the attraction at all points on the homogeneous fluid right circular cylinder *Mm* are the same, and we can assume that the directions of the attraction at all points on the homogeneous fluid right circular cylinder *Mm* are also the same. Likewise, we can assume that the magnitudes of the attraction at all points on the homogeneous fluid right circular cylinder *Nn* are the same, and we can assume that the directions of the attraction at all points on *Nn* are also the same. We can make these assumptions because the lengths of the cylinders Mm and Nn are infinitesimal. But the magnitudes of the attraction at M and at N are equal, because M and N are the same distance from the center of force C. Moreover, the attraction at M is directed toward r (that is, the attraction at M is directed toward C), and the attraction at N is directed toward s (that is, the attraction at N is directed toward C), because the line through C and M indicates the direction of the attraction at M, and the line through C and N specifies the direction of the attraction at N.

It follows from the assumptions stated in the preceding paragraph that the magnitudes of the components of the attraction directed toward m at all points on the homogeneous fluid right circular cylinder Mm are the same and that the magnitudes of the components of the attraction directed toward n at all points on the homogeneous fluid right circular cylinder Nn are also the same. But the magnitude of the component of the attraction at M directed toward m and the magnitude of the component of the attraction at N directed toward n are inversely proportional to the lengths of the cylinders Mm and Nn. This follows from applying what Clairaut called "the theory of inclined planes" to the infinitesimal right triangles Mrm and Nsn. Furthermore, for the same reasons that I explained in the preceding section of this chapter and that I also stated above, one can conclude in two different ways that the masses of the cylinders Mm and Nn are directly proportional to their lengths. Then the effort produced by the attraction along cylinder Mm is just the constant value of the magnitudes of the components of the attraction directed toward m at all points on cylinder Mm, which equals the magnitude of the component of the attraction at M directed toward m, multiplied by the mass of the cylinder Mm; the effort produced by the attraction along the infinitesimal homogeneous fluid right circular cylinder Nn is just the constant value of the magnitudes of the components of the attraction directed toward n at all points on cylinder Nn, which equals the magnitude of the component of the attraction at N directed toward n, multiplied by the mass of cylinder Nn; and these two products are the same because the magnitude of the component of the attraction at M directed toward m and the magnitude of the component of the attraction at N directed toward n are inversely proportional to the lengths of the infinitesimal homogeneous fluid right circular cylinders Mm and Nn and the masses of the cylinders are directly proportional to their lengths. Consequently the efforts produced by the attraction along the infinitesimal homogeneous fluid right circular cylinders Mm and Nn are equal. Hence the total efforts produced by the attraction along the two branches BMA and BNA of the closed channel of homogeneous fluid $MBNA$ are equal too. This is true because the total effort produced by the attraction along the branch BMA is just the sum of the efforts produced by the attraction along all of the infinitesimal homogeneous fluid right circular cylinders Mm that make up the branch BMA; the total effort produced by the attraction along the branch BNA is just the sum of the efforts produced by the attraction along all of the infinitesimal homogeneous fluid right circular cylinders Nn that make up the branch BNA; and the densities

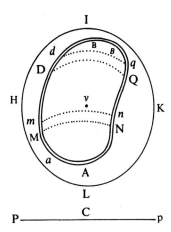

Figure 52. Clairaut's figure used to show that Clairaut's second new principle of equilibrium holds for all closed channels inside a homogeneous fluid figure of revolution shaped like a ring which is attracted by multiple centers of force, when the channels are filled with fluid from the figure (Clairaut 1743: figure on p. 19).

of all of the cylinders Mm and Nn are the same. Then if the sign of the effort produced by the attraction along one of the two branches BMA and BNA is changed from positive to negative, it follows that the total effort produced by the attraction around the closed channel of homogeneous fluid $MBNA$ is equal to zero.[52]

In his *Discours sur les différentes figures des astres* (1732) Maupertuis had determined the meridians at the surfaces of homogeneous fluid figures of revolution shaped like rings, with an eye to finding the shapes of Saturn's rings. With Maupertuis's end in view and the particular hypotheses of attraction which Maupertuis had made to this end also in view, Clairaut next considered a torrent of homogeneous fluid which revolves around an axis Pp. The torrent was a homogeneous fluid figure of revolution whose axis of symmetry was the axis Pp (see Figure 52). Clairaut took C to be a fixed center of force situated outside the torrent. He assumed that the magnitude of the attraction exerted by this fixed center of force C at a point depended solely upon the distance of the point from the fixed center of force C. He let $HIKL$ designate a meridian at the surface of the torrent, and he let v stand for a fixed center of force located inside the torrent in the plane of the meridian $HIKL$ which he assumed only attracts points that are also situated in the plane of the meridian $HIKL$. He hypothesized that the magnitude of the attraction exerted by this fixed center of force v at a point in the plane of the meridian $HIKL$ depended solely upon the distance of the point from the fixed center of force v. [In assuming that v only attracts points in the plane of the meridian *HIKL*, Clairaut made the same hypotheses that Maupertuis had made in determining the meridians at the surfaces of homogeneous fluid figures of revolution shaped like rings in his *Discours sur les différentes figures des astres* (1732).] Clairaut also assumed that *C* was situated in the plane of the meridian *HIKL*. In fact, he implicitly assumed that *C* was located on the axis of symmetry *Pp* of the torrent and that, moreover, the fixed centers of force *v* in all of the planes of the meridians *HIKL* at the surface of the torrent formed a circle whose center was located at *C* and that was perpendicular to *Pp*, so that *Pp* was an axis of symmetry of the set of fixed centers of force composed of the fixed center of force *C* and the circle made up of the fixed centers of force *v*, in which case *Pp* was also an axis of symmetry of the attraction produced by the fixed center of force *C* and the circle made up of the fixed centers of force *v*. Moreover, he supposed that each fixed center of force *v* was situated on the line through *C* and the center of the cross section of the torrent which is bounded by the meridian *HIKL*, in which case *C* was located at the center of the torrent.

[As a result of these assumptions the axis through *C* and *v* should be an axis of symmetry of the meridian *HIKL*, which we note is in fact true in Clairaut's diagram (Figure 52) as well as in Maupertuis's diagram (see Figure 53), where the meridian is *AQaPD* and where *C* and *v* are interchanged.]

Clairaut needed to make these assumptions in order for the axis of symmetry *Pp* of the torrent to be an axis of symmetry of the attraction too. He needed this to be the case for the following reason. Clairaut stated that it sufficed to show that the total effort produced by the attraction around any closed channel that lies within the interior of the torrent in the plane of a meridian *HIKL* at the surface of the torrent and that is filled with fluid from the torrent is equal to zero. But we remember that in order for it be enough to show that this is true, a homogeneous fluid figure of revolution must revolve in such a way and the forces of attraction must act in such a way so as to produce the symmetries that I mentioned in Chapters 1, 4, and 6. (That is, the homogeneous fluid figure of revolution must revolve around its axis of symmetry, in which case the axis of symmetry of the homogeneous fluid figure of revolution is also an axis of symmetry of the centrifugal force of rotation per unit of mass, and, moreover, the axis of symmetry of the homogeneous fluid figure of revolution must also be an axis of symmetry of the attraction. These conditions, we recall, did in fact always hold for the homogeneous fluid figures discussed in Chapters 1, 4, and 6.) Moreover, Clairaut evidently must have also implicitly assumed that the magnitude of the attraction exerted by a fixed center of force *v* at points in the plane of the meridian *HIKL* varied with distance from *v* the same way for each meridian *HIKL*. It is necessary to make such an assumption in order for the torrent to be a homogeneous fluid figure of revolution whose axis of symmetry *Pp* is also an axis of symmetry of the attraction.

In order to prove that the total effort produced by the attraction around any closed channel that lies within the interior of the torrent in the plane of a meridian

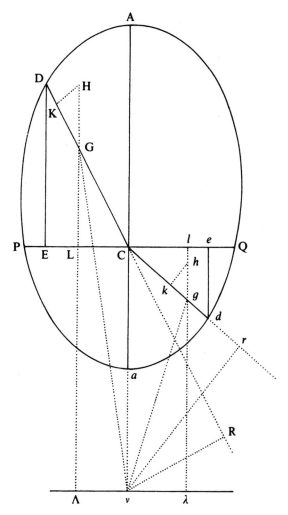

Figure 53. Maupertuis's figure used to determine the meridian of a homogeneous fluid figure of revolution shaped like a ring which is attracted by multiple centers of force (Maupertuis 1732b: figure on p. 63).

HIKL at the surface of the torrent and that is filled with fluid from the torrent is equal to zero, Clairaut simply applied the preceding result involving one fixed center of force to the total efforts produced by the attractions due to *C* and *v* individually around the closed channel *DMANQ* that lies within the interior of the torrent in the plane of a meridian *HIKL* at the surface of the torrent and that is filled with fluid from the torrent. (Here it is important to recall that we only really gave a complete demonstration that this result is true for closed channels that lie in the interior of a homogeneous fluid figure in the plane of a closed plane curve *PEpe* at the surface of the homogeneous fluid figure, where the fixed center of force *C* which attracts the channel is located inside the homogeneous fluid figure and is also situated in this plane. But it is easy to see that this result is also true when the fixed center of force *C* is situated outside the homogeneous fluid figure in this plane. The proof of this is exactly the same as the proof of the result in the case that the fixed center of force *C* is located inside the homogeneous fluid figure in this plane. We should also remember that Clairaut's own proof of the result in this special case is circular, although, as I demonstrated before, the result in this special case is indeed true. And the problem here does satisfy all of the conditions of the special case that Clairaut tried to prove of the general result that he stated.) Clairaut asserted that since the total efforts produced by the attractions due to *C* and *v* individually around the closed channel *DMANQ* are equal to zero, it follows that the total effort produced by the attraction due to *C* and *v* together around the closed channel *DMANQ* is equal to zero. Here Clairaut tacitly made use of a superposition principle. (We must again keep in mind, however, that in trying to demonstrate that the total effort produced by the attraction due to a single fixed center of force *C* around a closed plane channel that lies within the interior of a homogeneous fluid figure in the plane of a closed plane curve *PEpe* at the surface of the homogeneous fluid figure and that is filled with fluid from the homogeneous fluid figure is equal to zero, where the fixed center of force *C* is situated inside the homogeneous fluid figure in the plane of *PEpe*, Clairaut implicitly seems to have used the very result that he was trying to demonstrate. But as I showed, the result in this particular case is true, for it can be proved in a way that is not circular.) Clairaut added that it was easy to see that the same result must hold if the plane of each meridian *HIKL* is assumed to have the same arbitrary number of fixed centers of force instead of only two fixed centers of force *C* and *v* (provided of course that the fixed centers of force are distributed in a way that makes the axis of symmetry *Pp* of the torrent be an axis of symmetry of the attraction as well.)[53] This must be true by the superposition principle.

Clairaut next asserted that if the fixed center of force *v* in the plane of each meridian *HIKL* at the surface of the torrent is assumed to act in all directions and not just attract points that are also situated in the plane of the meridian *HIKL*, where Clairaut now stated clearly that all of the fixed centers of force *v* in all of the planes of the meridians *HIKL* do indeed form "a circle that serves, so to speak, as the center of the [torrent of homogeneous fluid shaped like a] ring [anneau]," then it is easy to demonstrate that the result above still holds. This time, however, Clairaut apparently did not assume that the magnitude of the attraction exerted by a fixed center of force *v* in the plane of the meridian *HIKL* varied with distance from *v* the same way for each such fixed center of force *v*. For Clairaut now considered "an arbitrary closed plane or skew channel that lies within the interior of the [torrent of homogeneous fluid shaped like a] ring [and that is filled with fluid from the torrent]," instead of an arbitrary closed plane channel that lies inside the torrent in the plane of a meridian *HIKL* and that is filled with fluid from the torrent. However, if Clairaut had again assumed that the magnitude of the attraction exerted by a fixed center of force *v* in the plane of the meridian *HIKL* varied with distance from *v* the same way for each such fixed center of force *v*, then the axis of symmetry *Pp* of the torrent would again be an axis of symmetry of the attraction too, in which case it would again be enough to show that the total efforts produced by the attraction around arbitrary closed plane channels that lie inside the torrent in the plane of a meridian *HIKL* and that are filled with fluid from the torrent are always equal to zero. In other words, in this case it would again suffice just to show that the total effort produced by the attraction around any closed channel that lies within the interior of the torrent in the plane of a meridian *HIKL* at the surface of the torrent and that is filled with fluid from the torrent is equal to zero. It would not be necessary to consider arbitrary closed plane or skew channels that lie inside the torrent and that are filled with fluid from the torrent. Thus Clairaut must have imagined the fixed centers of force *C* and *v* to attract in such a way that the axis *Pp* is not an axis of symmetry of the attraction, which can only be the case if the magnitude of the attraction exerted by a fixed center of force *v* in the plane of the meridian *HIKL* does not vary with distance from *v* the same way for each such fixed center of force (What Clairaut neglected to mention, however, is that in this case the torrent will not in general be a figure of revolution and, in particular, *Pp* will not in general be an axis of symmetry of the torrent. That is, the assumption that the magnitude of the attraction exerted by a fixed center of force *v* in the plane of the meridian *HIKL* varies with distance from *v* the same way for each such fixed center of force *v* is also necessary in order for the torrent even to be a homogeneous fluid figure of revolution whose axis of symmetry is the axis *Pp*.) In any case, Clairaut showed that the result above still holds when the hypothesis about the way that the fixed centers of force *v* attract is modified in the way that I mentioned before by simply stating that since the total effort produced by the attraction due to each fixed center of force *v* individually around any closed plane or skew channel that lies within the interior of the torrent and that is filled with fluid from the torrent is equal to zero, according to the first result stated earlier for a single fixed center of force, it follows that the total effort produced by the attraction due to the circle made up of the fixed centers of force *v* around the closed channel is also equal to zero. In other words, since the result holds for each fixed center of force individually, it follows that the result also holds for all of the fixed centers of force together. Here Clairaut again tacitly made use of a superposition principle. (Again we must bear in mind, however, that Clairaut did not actually demonstrate the general result for a single fixed center of force which he first stated, although the general result is true, as I mentioned previously. Indeed, as we saw, he did not really demonstrate satisfactorily this result even for the special case, which I discussed earlier. We also remember that Clairaut never actually demonstrated that centrifugal force never produces a total effort that is not equal to zero around an arbitrary closed plane or skew channel filled with homogeneous fluid which revolves around an axis, although the result is in fact true, so that in order to show that Clairaut's second new principle of equilibrium holds for arbitrary plane or skew channels that lie within the interior of a homogeneous fluid figure and that are filled with fluid from the figure, it is enough to consider the total efforts produced by the attraction around such closed channels and to show that these total efforts are always equal to zero. In other words, the effects of centrifugal force can be ignored.)

Clairaut added that it would not be more difficult to show that the same result holds for any homogeneous fluid figure whose surface is a surface of revolution ("sphéroïde") which revolves around its axis of symmetry, which includes any homogeneous fluid figure shaped like a ring ("anneau") which revolves around its axis of symmetry, if such a homogeneous fluid figure of revolution is attracted by a three-dimensional, solid nucleus that is situated inside the homogeneous fluid figure of revolution, whose surface is a surface of revolution, which includes a solid nucleus shaped like a ring, and whose center is located at the center of the homogeneous fluid figure of revolution. Here Clairaut evidently meant that it sufficed to think of such a three-dimensional, solid nucleus as composed of fixed centers of force, from which it followed that the result in question holds for the homogeneous fluid figure of revolution which surrounds the three-dimensional, solid nucleus by the superposition principle. (Here again, however, we must not forget that Clairaut did not actually demonstrate the general result for a single fixed center of force.) Clairaut probably assumed that the homogeneous fluid figure is some kind of figure of revolution and that the solid nucleus inside the homogeneous fluid figure which attracts the homogeneous fluid figure is also some kind of figure of revolution in order to ensure that conditions be fulfilled which entail that the homogeneous fluid figure be a figure of revolution which revolves in such a way and the forces of attraction act in such a way so as to produce the symmetries that I mentioned in Chapters 1, 4, and 6. (That is, the homogeneous fluid figure is a figure of revolution which revolves around its axis of symmetry, in which case the axis of symmetry of the homogeneous fluid figure of revolution is also an axis of symmetry of the centrifugal force of rotation per unit of mass, and, moreover, the axis of symmetry of the homogeneous fluid figure of revolution is also an axis of symmetry of the attraction. These conditions, we recall, did in fact always hold for the homogeneous fluid figures discussed in Chapters 1, 4, and 6.[54])

We note that such homogeneous fluid figures of revolution include as special cases homogeneous fluid figures of revolution which are attracted by three-dimensional, solid nuclei that are situated inside the figures, that are figures of revolution, and whose centers are located at the centers of the homogeneous fluid figures of revolution. In making the statement that begins the preceding paragraph, Clairaut probably had in mind such homogeneous fluid figures of revolution which are attracted by such three-dimensional, solid nuclei situated inside them, for, as I mentioned in Chapter 3, Cartesians like Samuel Koenig thought that such three-dimensional, solid nuclei that attract could be used to "save phenomena." Clairaut very likely meant to show that his result also held for homogeneous fluid figures of revolution which are attracted by such three-dimensional, solid nuclei located inside the homogeneous fluid figures of revolution at the centers of these figures.

In his treatise of 1743 Clairaut made the preceding attempts to demonstrate the result in question for the figures in question and for the hypotheses of attraction in question (that is, essentially attempts to demonstrate that his second new principle of equilibrium holds for the figures in question and for the hypotheses of attraction in question) *before* he introduced the mathematics that he used to represent mathematically his second and third new principles of equilibrium, two principles which must hold in order for a homogeneous fluid figure to be a figure of equilibrium, and that he also used to express the conditions on the forces of attraction which act which must be fulfilled if the second and third new principles of equilibrium are to hold – namely, complete differentials of attraction in rectangular coordinates and differentials of attraction in rectangular coordinates whose line integrals are path independent. The fact that he did not use this mathematics to demonstrate the result in question for the figures in question and for the hypotheses of attraction in question is additional evidence that in practice Clairaut did not know how to use this mathematics to demonstrate the result in question for the figures in question and for the hypotheses of attraction in question, which is a hypothesis that I proposed at the beginning of this section of this chapter.

In all of the examples just mentioned which Clairaut considered, the resultants of the attraction can be determined independently of the shapes of the homogeneous fluid figures which the hypotheses of attraction in question determine and for which Clairaut's second new principle of equilibrium holds. In each case where the homogeneous fluid figure is a figure of revolution and the axis of symmetry of the figure is also an axis of symmetry of the attraction, the components of the attraction at a point in the plane of a meridian at the surface of the figure whose magnitudes are P and Q, respectively, and which lie in the plane of the meridian in the directions of the rectangular coordinates y and x, respectively, can be determined without having to know the shape of the figure which any one of the hypotheses of attraction in question determine and for which Clairaut's second new principle of equilibrium holds.

Now, homogeneous fluid figures of revolution in equilibrium are figures for which Clairaut's second and third new principles of equilibrium necessarily hold. Using his mathematical representations of his second and third new principles of equilibrium in two rectangular coordinates and the equations

$$\int_c^M Pdy + Qdx - \frac{f}{2r}y^2 = \text{a constant } A \text{ that is independent of } M \quad (9.1.3)$$

and

$$\int Pdy + Qdx - \frac{f}{2r} y^2 = \text{a constant } A \qquad (9.1.20)$$

in two rectangular coordinates of the meridians at the surfaces of homogeneous fluid figures of revolution in equilibrium, which are valid when the figures revolve in such a way and when the forces of attraction act in such a way so as to produce the symmetries that I mentioned in Chapters 1, 4, and 6 (that is, the homogeneous fluid figures of revolution in equilibrium revolve around their axes of symmetry, in which case the axes of symmetry of the homogeneous fluid figures of revolution in equilibrium are also axes of symmetry of the centrifugal force of rotation per unit of mass, and, moreover, the axes of symmetry of the homogeneous fluid figures of revolution in equilibrium are also axes of symmetry of the attraction, which are conditions, we recall, that did in fact always hold for the homogeneous fluid figures discussed in Chapters 1, 4, and 6) and where equations (9.1.3) and (9.1.20) are equations that follow from his mathematical representations of his second and third new principles of equilibrium in two rectangular coordinates when these conditions of symmetry are fulfilled, as we saw in the preceding section of this chapter, Clairaut derived the equations of the meridians at the surfaces of homogeneous fluid figures of revolution in equilibrium, which are figures that his second and third new principles of equilibrium necessarily hold for, when the homogeneous fluid figures of revolution in equilibrium revolve in such a way and when they are attracted by one or more fixed, discrete centers of force whose positions are given, whose magnitudes at a point depend solely upon the distances of the point from the fixed centers of force, and which are distributed in such a way that the fixed centers of force together attract the homogeneous fluid figures of revolution in equilibrium so as to produce the symmetries that I mentioned in Chapters 1, 4, and 6. (That is, the homogeneous fluid figures of revolution in equilibrium revolve around their axes of symmetry, in which case the axes of symmetry of the homogeneous fluid figures of revolution in equilibrium are also axes of symmetry of the centrifugal force of rotation per unit of mass, and, moreover, the axes of symmetry of the homogeneous fluid figures of revolution in equilibrium are also axes of symmetry of the attraction. These conditions, we recall, did in fact always hold for the homogeneous fluid figures discussed in Chapters 1, 4, and 6.)

In deriving these equations Clairaut reproduced Huygens's and Hermann's earlier findings concerning the shapes of homogeneous fluid flattened figures of revolution, which I mentioned in Chapters 1 and 3, as well as Maupertuis's results concerning the shapes of homogeneous fluid flattened figures of revolution and homogeneous fluid figures shaped like rings which Maupertuis published in the *Discours sur les différentes figures des astres* (1732). These results included Maupertuis's equation for the meridians at the surfaces of homogeneous fluid figures shaped like rings,[55] from which the shapes of the homogeneous fluid figures can be determined since homogeneous fluid figures shaped like rings are figures of revolution.

Clairaut also found the equations of the meridians at the surfaces of homogeneous fluid figures of revolution in equilibrium which are attracted by three-dimensional, solid nuclei that are situated inside the homogeneous fluid figures of revolution in equilibrium, that are figures of revolution, and whose centers are located at the centers of the homogeneous fluid figures of revolution in equilibrium, where the nuclei can either be homogeneous or be composed of individually homogeneous layers such that the densities of any two contiguous layers differ, when the conditions of symmetry mentioned before are satisfied.[56] Once again, Clairaut probably had in mind here homogeneous fluid figures of revolution in equilibrium which have three-dimensional, solid nuclei situated inside them, where the nuclei are figures of revolution, where the centers of the nuclei are located at the centers of the homogeneous fluid figures of revolution in equilibrium, and where the nuclei attract the homogeneous fluid figures of revolution in equilibrium. As I mentioned in Chapter 3 and mentioned again above, Cartesians like Samuel Koenig thought that such three-dimensional, solid nuclei that attract could be used to "save phenomena," and Clairaut showed here how to find the equations of the meridians at the surfaces of homogeneous fluid figures of revolution in equilibrium which are attracted by such three-dimensional, solid nuclei situated inside the homogeneous fluid figures of revolution in equilibrium at the centers of these figures.

In addition, Clairaut determined the equation of a meridian of a homogeneous fluid figure of revolution in equilibrium when the magnitude of the component of the force per unit of mass at a point which is perpendicular to the x-axis only depends upon the distance of the point from the x-axis, in which case $P(y, x) = P(y)$, and the magnitude of the component of the force per unit of mass at a point which is perpendicular to the y-axis only depends upon the distance of the point from the y-axis, so that $Q(y, x) = Q(x)$. As a special case he derived Daniel Bernoulli's equation of a meridian at the free surface of a mass of homogeneous fluid contained in a cylindrical vase that revolves around the axis through the centers of the two circles at the two ends of the vase, which appears in Bernoulli's *Hydrodynamica*, published in 1738. In this case there is a vertical component of force per unit of mass whose magnitude at a particle of water is the same for all particles of water, namely, the earth's uniform terrestrial gravity, and there is also a horizontal component of force per unit of mass whose magnitude at a particle of water only depends upon the distance of the particle of water from the axis of rotation, namely, the centrifugal force per unit of mass.[57] (We note that in these examples, Clairaut applied his theory of homogeneous fluid figures of equilibrium to homogeneous fluid figures of revolution which are not in general shaped like ellipsoids.)

Since stratified fluid figures of revolution in equilibrium can be treated as homogeneous fluid figures for purposes of determining the shapes of the stratified fluid figures of revolution in equilibrium, when the resultants of the attraction at points do not depend upon the shapes of the stratified fluid figures of revolution in equilibrium, which is true in the cases just mentioned, Clairaut also found by the same means, without having to do any additional work, the equations of the shapes of such stratified fluid figures of revolution in equilibrium when the figures are assumed to be attracted in accordance with any of the hypotheses of attraction stated above and when the conditions of symmetry mentioned above are fulfilled.

Clairaut also employed his new principles of equilibrium to analyze Bouguer's thought-provoking Paris Academy *mémoire* of 1734 and, in particular, to uncover other errors in it. We recall that Bouguer had investigated hypotheses of attraction which tacitly fulfilled conditions of symmetry which ensured that all homogeneous fluid figures considered were figures of revolution, where the figures revolve around their axes of symmetry. We remember that according to the ways that Bouguer hypothesized that homogeneous fluid figures of revolution are attracted, the lines that indicate the directions of the attraction at all points in the plane of a meridian $PEpe$ at the surface of a homogeneous fluid figure of revolution were always assumed to lie in that plane and were always assumed to be perpendicular to a given closed plane curve $KLkl$ that also lies in that plane. Moreover, the variation of the magnitude of the attraction at points in the plane was fixed by specifying this variation at the points situated on each individual line in the plane which is perpendicular to this directrix $KLkl$ and which indicates the direction of the attraction at the points in the plane which are located on this line.

First Clairaut remarked that it seemed to him that if a homogeneous fluid figure of revolution were to be attracted in accordance with any such hypothesis of attraction, then in order to be a figure of equilibrium the homogeneous fluid figure of revolution had to have a solid nucleus that is a figure of revolution whose center is situated at the center of the homogeneous fluid figure of revolution and whose meridian in the plane of $KLkl$ is the smallest involute $RSrs$ of the evolute of $KLkl$ (that is, the smallest oval plane curve $RSrs$ that has the same evolute as $KLkl$) (see Figure 54). As Clairaut expressed the matter: "one must assume that the [homogeneous fluid] spheroid has a solid nucleus inside it which is the smallest $RSrs$ of the ovals that have the same evolute as the oval $KLkl$ [in other words, $RSrs$ is the smallest of the involutes of the plane curve that is the evolute of $KLkl$]." Clairaut believed this to be true for the following reason. If there were no such solid nucleus, which he interpreted to mean that there is such a nucleus but that this nucleus is composed of fluid instead, then the directions of the attraction at the particles of the fluid nucleus which are located in the plane of $PEpe$ would also have to be perpendicular to the directrix $KLkl$ in that plane. Hence the changes in the directions of the attraction at two particles of fluid ("particules de fluide") n and v of the fluid nucleus which are situated in the plane of $PEpe$ infinitely near each other on opposite sides of the axis Pp would be discontinuous

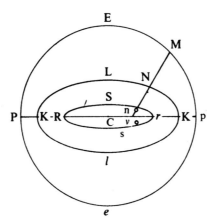

Figure 54. Clairaut's figure used to show that if a homogeneous fluid figure of revolution is attracted in accordance with any of Bouguer's hypotheses, then the figure must have a solid nucleus in order to be a figure of equilibrium (Clairaut 1743: figure on p. 64).

instead of continuous. This would be true because the angle between the lines that indicate the directions of attraction at n and at v is finite. But Clairaut did not believe such a phenomenon to be "natural." He found the possibility of such a phenomenon to be "shocking" ("choquant"). Here it is not clear whether Clairaut meant shocking in the sense of surprising or startling or shocking in the sense that he could not imagine how discontinuous changes in the directions of the attraction at particles of a fluid nucleus which are situated in the plane of $PEpe$ infinitely near each other on opposite sides of the axis Pp would not produce shocks that would upset a state of equilibrium of the fluid in a fluid nucleus. In any case, he believed that a homogeneous fluid figure of revolution which is assumed to be attracted in accordance with any of the hypotheses of attraction in question had to have a solid nucleus of the kind just described whose center is located at the center of the homogeneous fluid figure of revolution in order for the homogeneous fluid figure of revolution to be a figure of equilibrium. (Clairaut did not state whether the points of the solid nucleus are also assumed to be attracted in accordance with such a hypothesis of attraction. If the points of the nucleus are assumed to be attracted in accordance with such a hypothesis of attraction, the "unnatural," discontinuous changes in the directions of the attraction at points in the nucleus which are situated in the plane of $PEpe$ infinitely near each other on opposite sides of the axis Pp would presumably have no effect on the nucleus, since the nucleus is solid. Outside such a solid nucleus the directions of the attraction at particles of the homogeneous fluid figure of revolution which are situated in the plane of $PEpe$ infinitely near each other on opposite sides of the axis Pp change continuously. We can recognize this to be true by studying Figure 54. Moreover, Clairaut did not explain why the closed plane curve $RSrs$ in the plane of $KLkl$ has to have the same evolute as $KLkl$. In a moment we shall see that if the nucleus is solid, then $RSrs$ in fact *cannot* in general have the same evolute as $KLkl$, if the homogeneous fluid that surrounds the solid nucleus is to be in equilibrium, and we shall understand why this is true. If $RSrs$ does have the same evolute as $KLkl$, in which case $RSrs$ and $Klkl$ are involutes of the same plane curve, and if the points of the solid nucleus are also assumed to be attracted in accordance with one of the hypotheses of attraction in question, then the lines that indicate the directions of attraction at points n and v will be perpendicular to $RSrs$, because the lines that indicate the directions of attraction at points n and v are perpendicular to $KLkl$ and any two involutes of the same plane curve are parallel. But there is nothing in Clairaut's reasoning that requires that the lines that indicate the directions of attraction at points n and v must be perpendicular to $RSrs$. What must be true is that if the nucleus is solid and if the homogeneous fluid that surrounds the solid nucleus is to be in equilibrium, then the principle of the plumb line must hold at all points of the surface of the solid nucleus. Now, if $RSrs$ does have the same evolute as $KLkl$, which means that $RSrs$ and $KLkl$ are involutes of the same plane curve, then the lines that indicate the directions of the attraction at the points of $RSrs$ will all be perpendicular to $RSrs$ at the points of $RSrs$, because these lines are all perpendicular to $KLkl$, and any two involutes of the same plane curve are parallel. However, it is the effective gravity, not the attraction, that must be perpendicular to $RSrs$ at the points of $RSrs$ in order for the principle of the plumb line to hold at all points of the surface of the solid nucleus. But if the figure of revolution revolves around its axis of symmetry, then the effective gravity will not in general be perpendicular to $RSrs$ at the points of $RSrs$ if the attraction is perpendicular to $RSrs$ at the points of $RSrs$. This is true for the same reasons that I explained in Chapter 2 in discussing Mairan's Paris Academy *mémoire* of 1720. Thus in fact if the homogeneous fluid that surrounds the solid nucleus is to be in equilibrium when the figure of revolution made up of the solid nucleus and the homogeneous fluid that surrounds the solid nucleus revolves around its axis of symmetry, the closed plane curve $RSrs$ in the plane of $KLkl$ must *not* in general have the same evolute as $KLkl$. In taking $RSrs$ to have the same evolute as $KLkl$, Clairaut may have made an error that is similar to the kind of mistake that Mairan made.) Be that as it may, Clairaut stated, whether one assumes that such a solid nucleus does exist or whether one assumes instead that the homogeneous figure of revolution is made up entirely of fluid, the calculations to determine whether the homogeneous figure of revolution can exist in a state of equilibrium are the same in both cases.[58]

Using the mathematical representations of his new principles of equilibrium expressed in terms of complete differentials in two rectangular coordinates and differentials in two rectangular coordinates whose line integrals are path independent, Clairaut then showed that the fluid in the homogeneous figure of revolution which revolves around its axis of symmetry can exist in a state of equilibrium whenever the magnitudes of the attraction at points in the plane of a meridian *PEpe* at the surface of the figure only depend upon the distances of the points from the directrix *KLkl* in that plane and do not vary in any other way.[59] We remember that Bouguer had concluded in his Paris Academy *mémoire* of 1734 that in order that the principles of balanced columns and the plumb line applied separately to a homogeneous fluid figure of revolution which revolves around its axis of symmetry determine the same shape of the figure, in case the attraction is constant in magnitude, the lines that indicate the directions of the attraction at points in the plane of a meridian *PEpe* at the surface of the homogeneous fluid figure of revolution all had to intersect at the center of the figure. In other words, the directrix *KLkl* in that plane must be a circle. And Clairaut noted that if the problem were expressed using differentials in two rectangular coordinates, Bouguer's conclusion did appear to be correct at first sight. That is, at first sight the relevant differential in two rectangular coordinates only seemed to be complete when the conclusion that Bouguer had reached was true.[60] But then Clairaut showed, however, that a more careful analysis of the relevant differential in two rectangular coordinates revealed that in order for the shapes determined by the principles of balanced columns and the plumb line applied individually to a homogeneous fluid figure of revolution which revolves around its axis of symmetry to be the same, no such restriction concerning the directrix *KLkl* was required when the magnitude of the attraction was assumed to be constant. In other words, the directrix *KLkl* did not have to be a circle when the magnitude of the attraction was assumed to be constant. To demonstrate that this is true Clairaut cleverly rewrote the relevant differential in two rectangular coordinates a different way, from which it could be seen that Bouguer's conclusion was not necessary in order for this differential to be complete. Clairaut showed that the directrix *KLkl* could be any closed curve in the plane of *PEpe* and the relevant differential in two rectangular coordinates would always be complete.[61] Indeed, he showed that the magnitude of the attraction did not even have to be constant in order for this to be true. When the magnitudes of the attraction at points in the plane of *PEpe* were assumed to depend solely upon the distances of the points from the directrix *KLkl* in that plane, he showed that the relevant differential in two rectangular coordinates also turned out to be complete. [In order to demonstrate that this is true, Clairaut made use of what is called "the coefficient lemma for total differentials" in Engelsman (1984) pp. 148–149.] In other words, in general as long as the magnitudes of attraction at points in the plane of *PEpe* solely depended upon the distances of the points from the directrix *KLkl* in that plane, the relevant differential in two rectangular coordinates was complete, which meant that Clairaut's new principles of equilibrium held for the figure, whatever the shape of the directrix *KLkl*. But we recall from the preceding section of this chapter that when Clairaut's second new principle of equilibrium holds for a homogeneous fluid figure, then this ensures that either both of the principles of balanced columns and the plumb line also hold for the homogeneous fluid figure, or else that neither of these two principles holds for the homogeneous fluid figure. Hence in order for the principles of balanced columns and the plumb line applied separately to a homogeneous fluid figure of revolution which revolves around its axis of symmetry to determine the same shape of the figure, it was enough that the magnitudes of attraction at points in the plane of *PEpe* depend solely upon the distances of the points from the directrix *KLkl* in that plane, whatever the shape of the directrix *KLkl* (where of course the directrices *KLkl* in the planes of different meridians *PEpe* at the surface of the homogeneous fluid figure of revolution which revolves around its axis of symmetry are all assumed to be congruent and can be gotten by revolving any one of them around the axis of symmetry of the figure and where of course the magnitudes of the attraction at points in the plane of a meridian *PEpe* at the surface of the homogeneous fluid figure of revolution which revolves around its axis of symmetry are assumed to depend the same way upon the distances of the points from the directrix *KLkl* in that plane for all meridians *PEpe* at the surface of the homogeneous fluid figure of revolution which revolves around its axis of symmetry, because the conditions of symmetry mentioned before which concern the figure and the attraction require that these hypotheses be stipulated).[62]

Clairaut also tried to prove the same results a second, more geometric way using his new principles of equilibrium. His second demonstration of these results did not involve having to show that certain differentials in two rectangular coordinates are complete. We shall discover, however, that his argument has a lacuna. We recall that in order to demonstrate that Clairaut's second new principle of equilibrium holds for a homogeneous fluid figure of revolution which revolves in such a way and when the forces of attraction act in such a way so as to produce the symmetries that I mentioned in Chapters 1, 4, and 6 (that is, the homogeneous fluid figure of revolution revolves around its axis of symmetry, in which case the axis of symmetry of the homogeneous fluid figure of revolution is also an axis of symmetry of the centrifugal force of rotation per unit of mass, and, moreover, the axis of symmetry of the homogeneous fluid figure of revolution is also an axis of symmetry of the attraction, which are conditions, we recall, that did in fact always hold for the homogeneous fluid figures discussed in Chapters 1, 4, and 6), it is enough to show that the total effort produced by the attraction along a plane channel that lies within the interior of the homogeneous fluid figure of revolution in the plane of a meridian *PEpe* at the surface of the figure, that is filled with homogeneous fluid from the figure, and that joins the same two points *O* and *M* must be independent of the particular plane channel that joins *O* to *M*. Clairaut applied this result to two conveniently chosen plane channels that lie in the interior of the figure in the plane of *PEpe*, that are filled with homogeneous fluid from the figure, and that join *O* to *M*: (1) channel *OSM* and (2) a suitably chosen second channel *OTZM* (namely, a plane channel whose part *TZM* is perpendicular to the directrix *KLk* in the plane of *PEpe* and whose

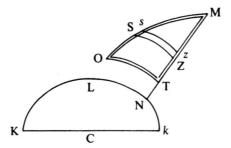

Figure 55. Clairaut's figure used to show that if a homogeneous fluid figure of revolution is attracted in accordance with any of Bouguer's hypotheses, then the fluid in the figure can exist in a state of equilibrium whenever the magnitudes of the attraction at points in the plane of a meridian only depend upon the distances of the points from Bouguer's directrix in that plane and do not vary in any other way, whatever the shape of the directrix (Clairaut 1743: figure on p. 69).

part *OT* has the same evolute as *KLk*). (See Figure 55.) Now, since the lines that indicate the directions of the attraction at all points in the plane of *PEpe* are all perpendicular to *KLk*, this means in particular that the lines that indicate the directions of the attraction at the points on the channel *OT* are all perpendicular to *KLk*. But the channel *OT* has the same evolute as *KLk*, which means that the channel *OT* and *KLk* are involutes of the same plane curve, and any two involutes of the same plane curve are parallel. Therefore the lines that indicate the directions of the attraction at the points on the channel *OT* are all perpendicular to the channel *OT* at the points on the channel *OT*, since the lines that indicate the directions of the attraction at the points on the channel *OT* are all perpendicular to *KLk*, and the channel *OT* is parallel to *KLk*. Hence the total effort produced by the attraction along the channel *OT* is equal to zero. Consequently the total effort produced by the attraction along the channel *OSM* must be equal to the total effort produced by the attraction along the channel *TZM*. Or "what amounts to the same thing," Clairaut said, if *SZ* and *sz* are any two plane curves in the plane of *PEpe* which intersect the channels *OSM* and *TZM*, which are infinitesimally near each other, and which have the same evolute as the channel *OT* (which means that the two plane curves *SZ* and *sz* have the same evolute as *KLk*), then the efforts along the two infinitesimal homogeneous fluid right circular cylinders ("deux petits cylindres") *Ss* and *Zz* must be the same. (Here Clairaut tried to reverse the reasoning that he had used several times before to demonstrate in his treatise of 1743 that the efforts produced by attraction along a certain pair of channels filled with homogeneous fluid of the same density are equal. We recall that he would imagine two channels filled with homogeneous fluid of the same density to be partitioned into an infinite number of infinitesimal homogeneous fluid right circular cylinders, where the cylinders that make up one of the two channels are in a one-to-one correspondence with the cylinders that make up the other channel; he would show that the efforts produced by the attraction along two corresponding infinitesimal homogeneous fluid right circular cylinders are equal; and from this result it followed that the total efforts produced by the attraction along the two channels filled with homogeneous fluid of the same density are the same. In fact, however, Clairaut provided no real basis for concluding that the efforts produced by the attraction along the two cylinders Ss and Zz must be equal from the fact that the total efforts produced by the attraction along the channels OSM and TZM are the same. That the efforts produced by the attraction along the two cylinders Ss and Zz are the same is certainly a sufficient condition for the total efforts produced by the attraction along the channels OSM and TZM to be equal, but this condition is not a necessary condition for the total efforts produced by the attraction along the channels OSM and TZM to be equal. The efforts produced by the attraction along two corresponding infinitesimal homogeneous fluid right circular cylinders Ss and Zz can be different, yet the total efforts produced by the attraction along the channels OSM and TZM can still be the same. This is the lacuna in Clairaut's argument which I alluded to earlier.) But this can only be true, Clairaut concluded, if the magnitude of the attraction at S is equal to the magnitude of the attraction at Z. Now, any two involutes of the same plane curve are parallel. In particular, SZ is parallel to KLk, since SZ and KLk have the same evolute, so that SZ and KLk are involutes of the same plane curve. Therefore the channel TZM must be perpendicular to SZ at Z, since the channel TZM is perpendicular to KLk, and SZ is parallel to KLk. But the fact that any two involutes of the same plane curve are parallel also means that any two involutes of the same plane curve are equidistant from each other. In other words, the segments of lines that are perpendicular to any two involutes of the same plane curve which are intercepted by the two involutes always have the same length. But since the channel TZM is perpendicular to KLk and is also perpendicular to SZ at Z, it clearly follows that NZ is a segment of a line that is perpendicular to the two involutes SZ and KLk of the same plane curve which is intercepted by the two involutes. Then since all such segments must have the same length, it follows that the distance of S from KLk must be the same as the length of the segment NZ, which is just the distance of Z from KLk. Finally, this fact together with the result that the magnitude of the attraction at S is equal to the magnitude of the attraction at Z means that the magnitudes of the attraction at points in the plane of $PEpe$ must be the same at points that are the same distance from the directrix KLk in that plane. Thus did Clairaut try to demonstrate, this time more geometrically, that the only thing required for his second new principle of equilibrium to hold for a homogeneous fluid figure of revolution which revolves around its axis of symmetry, when the figure and the attraction fulfill the conditions of symmetry mentioned, is that the magnitudes of the attraction be the same at points in the plane of $PEpe$ which are equidistant from the directrix *KLk* in that plane, whatever the shape of the directrix *KLk*.[63] Clairaut tried to establish this result by utilizing curved plane channels. The result had slipped through Bouguer's analysis partly because Bouguer had only employed straight channels. Because he chose his plane channels advantageously, Clairaut did not have to determine the values of any line integrals of the complete differential in two rectangular coordinates which in general represent the total efforts produced by the attraction along plane channels that lie within the interior of the homogeneous fluid figure of revolution in the plane of a meridian *PEpe* at the surface of the homogeneous fluid figure of revolution, that are filled with homogeneous fluid from the homogeneous fluid figure of revolution, and that join the same two points *O* and *M*. (We remember that the values of these line integrals are the same for all plane channels that join the same two points *O* and *M*.)

Assuming that homogeneous fluid figures of revolution which revolve around their axes of symmetry and for which Clairaut's second and third new principles of equilibrium hold when these figures are attracted in accordance with any of the hypotheses of attraction in question are, moreover, figures of equilibrium, Clairaut also derived the equations of meridians at the surfaces of these homogeneous fluid figures of revolution in equilibrium when these figures are attracted in accordance with any such hypothesis of attraction and when the figures and the attraction fulfill the conditions of symmetry mentioned before.[64] Probably with Bouguer's error mentioned previously in mind, Clairaut showed how without paying careful attention one could be misled into falsely concluding, when applying Bouguer's methods for determining the equations of meridians at the surfaces of homogeneous fluid figures of revolution which revolve around their axes of symmetry and for which both the generalized principle of balanced columns and the principle of the plumb line hold, which appear in Bouguer's Paris Academy *mémoire* of 1734, that hypotheses of attraction where the magnitudes of the attraction are constant at points in planes of meridians at the surfaces of homogeneous fluid figures of revolution and where the directrices in planes of meridians at the surfaces of homogeneous fluid figures of revolution are not circles in fact do not permit homogeneous fluid figures of revolution which revolve around their axes of symmetry and for which both the generalized principle of balanced columns and the principle of the plumb line hold to exist. Assuming that such a homogeneous fluid figure of revolution which revolves around its axis of symmetry is attracted in accordance with such a hypothesis of attraction, Clairaut showed how the generalized principle of balanced columns and the principle of the plumb line applied individually to such a figure could indeed lead to different equations of meridians at the surface of the figure instead of to the same equation, if one did not pay careful attention.[65] In other words, Clairaut located Bouguer's error and explained it in detail.

When he turned to the problem of finding the shapes of stratified fluid figures in relative equilibrium, Clairaut disregarded from the outset all hypotheses of attraction where the resultants of the attraction at points do not themselves depend upon the shapes of the stratified fluid figures. He only considered hypotheses of attraction where the resultants of the attraction at points depend themselves upon the shapes of the stratified fluid figures. The universal inverse-square law of attraction or any other universal law of attraction, which means that all of the parts of a stratified fluid figure mutually attract each other, where the line through two parts of a stratified fluid figure indicates the directions of the attractions at the two parts of the figure which the two parts exert on each other and where the magnitudes of the attractions at two parts of a stratified fluid figure which the two parts exert on each other are given by an arbitrary function $f(r)$, if r is the distance between the two parts (so that in the case of the universal inverse-square law, $f(r) = k/r^2$, where k is a constant), are examples of hypotheses of attraction where this is true. Clairaut did this for reasons that I explained at the end of the preceding section of this chapter, namely, if the resultants of the attraction at points are already specified and known at the start, then all of the individually homogeneous strata all of which must be "level layers" according to Clairaut's theory of figures of equilibrium, where the densities of any two contiguous strata differ, can be replaced by strata whose densities are all the same. When the densities of the strata change, the strata remain level layers and retain the same shapes that they originally had. Such modifications consequently give rise to a homogeneous fluid figure of equilibrium whose shape is the same as that of the stratified fluid figure of equilibrium. Hence the problem reduces to determining the shapes of homogeneous fluid figures of equilibrium in these instances.

Clairaut maintained that stratified fluid figures that are made up of finite numbers of individually homogeneous strata whose thicknesses are all finite and that attract according to the universal inverse-square law or any other universal law of attraction, which means that all of the parts of a stratified fluid figure mutually attract each other, where the line through two parts of a stratified fluid figure indicates the directions of the attractions at the two parts of the stratified fluid figure which the two parts exert on each other and where the magnitudes of the attractions at two parts of a stratified fluid figure which the two parts exert on each other are given by an arbitrary function $f(r)$, if r is the distance between the two parts of the stratified fluid figure (so that in the case of the universal inverse-square law, $f(r) = k/r^2$, where k is a constant), can exist in a state of equilibrium. In order to prove that this is true Clairaut stated that it was enough to show that the resultants of the attraction at points due to a stratified fluid figure of this kind which attracts in accordance with the universal inverse-square law or any other universal law of attraction permit his second new principle of equilibrium to hold for all closed plane or skew channels that lie within the interior of the stratified fluid figure and that are filled with homogeneous fluid (in other words, the total efforts produced by the effective gravity around all closed plane or skew channels that lie in the interior of such a stratified fluid figure that attracts according to the universal inverse-square law or any other universal law of attraction and that are filled with homogeneous fluid are equal to zero).

In order to try to demonstrate that this is true Clairaut simply treated each particle or element of such a stratified fluid figure that attracts according to the universal inverse-square law or any other universal law of attraction[66] as a center of force whose attraction produces a total effort equal to zero around a closed plane or skew channel that lies within the interior of the stratified fluid figure and that is filled with homogeneous fluid. He did so because the attraction due to any particular particle of the stratified fluid figure is everywhere directed toward that particle and, moreover, the magnitude of the attraction due to that particle at a point solely depends upon the distance of the point from that particle. The magnitude of the attraction due to the particle at a point is proportional to $1/r^2$ or, more generally, is given by an arbitrary function $f(r)$, where r is the distance of the point from the particle. (In other words, the only real difference between these particles of fluid treated as centers of force which attract and the fixed centers of force which attract in the case of hypothesized central forces of attraction is that the particles of fluid attract each other, which the fixed centers of force are not assumed to do, as well as attract all other matter.) By applying a superposition principle, Clairaut concluded that the total effort produced by the attraction due to the stratified fluid figure of the kind in question which attracts in accordance with the universal inverse-square law or any other universal law of attraction around a closed plane or skew channel that in lies the interior of the stratified fluid figure and that is filled with homogeneous fluid is equal to the sum of the total efforts produced by the attractions due to the individual particles that make up the stratified fluid figure around the closed channel filled with homo-geneous fluid. But the total effort produced by the attraction due to each individual particle around the closed channel filled with homogeneous fluid is equal to zero. (We recall, however, that Clairaut only really *stated* that this result is true; he never *proved* that the result is true.) Hence the sum of the total efforts produced by the attractions due to the individual particles that make up the stratified fluid figure around the closed channel filled with homogeneous fluid is also equal to zero by the superposition principle. Consequently the total effort produced by the attraction due the stratified fluid figure around the closed channel filled with homogeneous fluid is equal to zero.

Thus did Clairaut "demonstrate" that the total efforts produced by the attraction due to a stratified fluid figure of the kind in question which attracts according to the universal inverse-square law or any other universal law of attraction around all closed plane or skew channels that lie inside the stratified fluid figure and that are filled with homogeneous fluid are equal to zero. In other words, Clairaut "showed" that his second new principle of equilibrium holds for all closed plane or skew channels that lie within the interior of a stratified fluid figure of the kind in question which attracts in accordance with the universal inverse-square law or any other universal law of attraction and that are filled with homogeneous fluid. (We remember that in order to determine whether or not Clairaut's second new principle of equilibrium holds for all closed plane or skew channels that lie in the interior of a fluid figure and that are filled with homogeneous fluid, the effects of centrifugal force of rotation can be ignored. That is, we recall that Clairaut correctly stated in general and demonstrated in one special case that the centrifugal force generated when a closed plane or skew channel filled with homogeneous fluid revolves around an axis never produces a total effort around the closed channel filled with homogeneous fluid which is not equal to zero. Consequently whether or not the effective gravity produces total efforts around closed plane or skew channels that lie inside a fluid figure and that are filled with homogeneous fluid which are equal to zero solely depends upon the forces of attraction which act.) From this result Clairaut claimed that it followed that if stratified fluid figures of the kind in question attract according to the universal inverse-square law or any other universal law of attraction, then (1) these stratified fluid figures can exist in a state of equilibrium and, moreover, that in the case of stratified fluid figures of the kind in question which attract according to the universal inverse-square law or any other universal law of attraction (2) the principles of his theory of equilibrium of stratified fluid figures reduce to the principles of his theory of equilibrium of homogeneous fluid figures.[67]

In fact, Clairaut really *only* "proved" that the total efforts produced by the attraction due to a stratified fluid figure of the kind in question which attracts according to the universal inverse-square law or any other universal law of attraction around all closed plane or skew channels that lie within the interior of the stratified fluid figure and that are filled with homogeneous fluid are equal to zero (in other words, that his second new principle of equilibrium holds for all closed plane or skew channels that lie inside a stratified fluid figure of the kind in question which attracts in accordance with the universal inverse-square law or any other universal law of attraction and that are filled with homogeneous fluid). Now, this result in reality is only a *necessary* condition even for a *homogeneous* fluid figure that attracts according to the universal inverse-square law or any other universal law of attraction to exist in a state of equilibrium. It is not a *sufficient* condition for a homogeneous fluid figure or a stratified fluid figure of the kind in question which attracts in accordance with the universal inverse-square law or any other universal law of attraction to exist in a state of equilibrium. In other words, a homogeneous fluid figure or a stratified fluid figure of the kind in question which attracts in accordance with the universal inverse-square law or any other universal law of attraction need not even be a figure of equilibrium in order for this condition to hold for the figure. This is true for the following reason. From Clairaut's arguments we can conclude just as well that at no instant does *any* homogeneous fluid mass or stratified fluid mass of the kind in question which attracts according to the universal inverse-square law or any other universal law of attraction ever produce a total effort that is not equal to zero around a closed plane or skew channel that lies within the interior of the homogeneous fluid mass or the stratified fluid mass of the kind in question which attracts according to the universal inverse-square law or any other universal law of attraction and that is filled with homogeneous fluid, whether the homogeneous fluid mass or the stratified fluid mass of the kind in question which attracts according to the universal inverse-square law or any other universal law of attraction is in a state of equilibrium or not. What Clairaut really assumed is that he had at the start a stratified fluid mass of the kind in question which attracts according to the universal inverse-square law or any other universal law of attraction and which has a certain shape, and then he went on to claim to show that the attraction due to the stratified fluid mass does nothing to cause that shape to change. In fact Clairaut proved no such thing. Time nowhere enters into Clairaut's demonstration. Thus Clairaut's reasoning can be applied just as well *at any instant to any* fluid mass that attracts according to the universal inverse-square law or any other universal law of attraction, for at each instant such a fluid mass has some specific shape, whether the fluid mass is in a state of equilibrium or not. And it follows from his reasoning that at no instant does such a fluid mass ever produce a total effort that is not equal to zero around a closed plane or skew channel that lies inside the fluid mass and that is filled with homogeneous fluid, whether the fluid mass is in a state of equilibrium at that instant or not. Clairaut's arguments do not require that the particles that make up a fluid mass that attracts in accordance with the universal inverse-square law or any other universal law of attraction endure in a state of equilibrium over time in order for the total efforts produced by the attraction due to the fluid mass at any instant around all closed plane or skew channels that lie within the interior of the fluid mass and that are filled with homogeneous fluid to be equal to zero at that instant.

The conclusion that the condition that the total efforts produced by the attraction around all closed plane or skew channels that lie within the interior of a fluid figure and that are filled with homogeneous fluid are always equal to zero (in other words, that Clairaut's second new principle of equilibrium holds for all closed plane or skew channels that lie inside a fluid figure and that are filled with homogeneous fluid) is a necessary condition but not a sufficient condition even for a homogeneous fluid figure to exist in a state of equilibrium can also be reached another way. Namely, we recall that Clairaut showed in two different ways that the total effort produced by the effective gravity around a closed plane or skew channel that lies within the interior of a homogeneous fluid figure of equilibrium and that is filled with fluid from the figure must be equal to zero. One of the two ways that he demonstrated that this result is true was to note that this result follows from the principle of solidification applied to a homogeneous fluid figure of equilibrium. But the principle of solidification is itself a necessary condition but not a sufficient condition for a homogeneous fluid figure to exist in a state of equilibrium. Hence any result deduced from the principle of solidification applied to a homogeneous fluid figure is a necessary condition but not a sufficient condition for the homogeneous fluid figure to exist in a state of equilibrium. (Attraction can be substituted for effective gravity in this argument because, as I just mentioned, in order to determine whether or not Clairaut's second new principle of equilibrium holds for all closed plane or skew channels that lie within the interior of a fluid figure and that are filled with homogeneous fluid (in other words, whether or not the total efforts produced by the effective gravity around all closed plane or skew channels that lie within the interior of the fluid figure and that are filled with homogeneous fluid are equal to zero), the effects of centrifugal force of rotation can be ignored.)

Moreover, although Clairaut "showed" that his second new principle of equilibrium holds for all closed plane or skew channels that lie within the interior of a stratified fluid figure of the kind in question which attracts according to the universal inverse-square law or any other universal law of attraction and that are filled with homogeneous fluid (in other words, the total efforts produced by the effective gravity around all closed plane or skew channels that lie within the interior of a stratified fluid figure of the kind in question which attracts according to the universal inverse-square law or any other universal law of attraction and that are filled with homogeneous fluid are equal to zero), he did not state, much less demonstrate, whether or not his second new principle of equilibrium can hold for all closed plane or skew channels that lie within the interior of a stratified fluid figure of the kind in question and that are filled with homogeneous fluid (in other words, whether or not the total efforts produced by the effective gravity around all closed plane or skew channels that lie within the interior of a stratified fluid figure of the kind in question and that are filled with homogeneous fluid are equal to zero), if the stratified fluid figure of the kind in question is assumed to attract in a way that makes the resultants of the attraction depend upon the shape of the stratified fluid figure but the hypothesis about the way that the stratified fluid figure attracts differs from the universal inverse-square law and any other universal law of attraction. In fact, Clairaut's second new principle of equilibrium need not hold for all closed plane or skew channels that are filled with homogeneous fluid and that lie within the interior of a stratified fluid figure of the kind in question which is assumed to attract in a way that makes the resultants of the attraction depend upon the shape of the stratified fluid figure but the hypothesis about the way that the figure attracts differs from the universal inverse-square law and any other universal law of attraction (in other words, the total efforts produced by the effective gravity around all closed plane or skew channels that are filled with homogeneous fluid and that lie within the interior of a stratified fluid figure of the kind in question which is assumed to attract in a way that makes the resultants of the attraction depend upon the shape of the figure but the hypothesis about the way that the figure attracts differs from the universal inverse-square law and any other universal law of attraction need not all be equal to zero).

Furthermore, Clairaut did not state, much less demonstrate, whether or not his second new principle of equilibrium can hold for all closed plane or skew channels that are filled with homogeneous fluid and that lie within the interior of a stratified fluid figure that is not composed of a finite number of individually homogeneous strata whose thicknesses are all finite, but whose individually homogeneous strata are surfaces instead, so that the strata are "infinitesimally" thin and infinite in number, and that attracts according to the universal inverse-square law or any other universal law of attraction (in other words, whether or not the total efforts produced by the effective gravity around all closed plane or skew channels that lie inside such a stratified fluid figure that attracts according to the universal inverse-square law or any other universal law of attraction and that are filled with homogeneous fluid are equal to zero). This kind of stratified fluid figure that attracts according to the universal inverse-square law or any other universal law of attraction includes stratified fluid figures whose densities vary, say, continuously from their centers to their surfaces, since in this case the individually homogeneous fluid strata that make up the figures are surfaces themselves, and that attract according to the universal inverse-square law or any other universal law of attraction. [We shall see below that Clairaut's second new principle of equilibrium does in fact hold for all closed plane or skew channels that lie within the interior of a stratified fluid figure of the kind in question which attracts according to the universal inverse-square law or any other universal law of attraction and that are filled with homogeneous fluid (in other words, the total efforts produced by the effective gravity around all closed plane or skew channels that lie inside a stratified fluid figure of the kind in question which attracts according to the universal inverse-square law or any other universal law of attraction and that are filled with homogeneous fluid are in fact equal to zero).]

In addition, it turns out, as we shall see in a moment, that the condition that the total efforts produced by the effective gravity around all closed plane or skew channels that lie within the interior of a fluid figure and that are filled with homogeneous fluid are equal to zero (in other words, that Clairaut's second new principle of equilibrium holds for all closed plane or skew channels that lie inside a fluid figure and that are filled with homogeneous fluid) is not even a condition that in general must hold in order for a heterogeneous fluid figure to exist in a state of equilibrium. This is the case in general, for example, for stratified fluid figures of equilibrium which are not composed of finite numbers of individually homogeneous strata whose thicknesses are all finite, but whose individually homogeneous strata are surfaces instead, so that the strata are "infinitesimally" thin and infinite in number. These stratified fluid figures of equilibrium include figures whose densities vary, say, continuously from their centers to their surfaces, since in this case the individually homogeneous fluid strata that make up the figures are surfaces themselves. When the strata that make up a stratified fluid figure of equilibrium are surfaces themselves, it turns out that it is not true in general that the total efforts produced by the attraction due to the stratified fluid figure of equilibrium around closed plane or skew channels that lie in the interior of the figure and that are filled with homogeneous fluid are equal to zero. That is, Clairaut's second new principle of equilibrium does not in general hold for all closed plane or skew channels that lie inside a stratified fluid figure of equilibrium whose strata are surfaces themselves and that are filled with homogeneous fluid (in other words, the total efforts produced by the effective gravity around all closed plane or skew channels that lie within the interior of a stratified fluid figure of equilibrium whose strata are surfaces themselves and that are filled with homogeneous fluid are not in general equal to zero). (As I just remarked, however, Clairaut's second new principle of equilibrium does always hold for all closed plane or skew channels that lie inside a stratified fluid figure of equilibrium which attracts according to the universal inverse-square law or any other universal law of attraction and that are filled with homogeneous fluid (in other words, the total efforts produced by the effective gravity around all closed plane or skew channels that lie within the interior of such a stratified fluid figure of equilibrium which attracts according to the universal inverse-square law or any other universal law of attraction and that are filled with homogeneous fluid are equal to zero). This exception to the preceding statement is important for reasons that we shall discover. Moreover, as I also mentioned, in order to determine whether or not Clairaut's second new principle of equilibrium holds for all closed plane or skew channels that lie within the interior of a fluid figure and that are filled with homogeneous fluid (in other words, whether or not the total efforts produced by the effective gravity around all closed plane or skew channels that lie inside the fluid figure and that are filled with homogeneous fluid are equal to zero), the effects of centrifugal force of rotation can be ignored.)

In particular, although Clairaut "demonstrated" that his second new principle of equilibrium holds for all closed plane or skew channels that are filled with homogeneous fluid and that lie within the interior of a stratified fluid figure that is composed of a finite number of individually homogeneous strata whose thicknesses are all finite and that attracts according to the universal inverse-square law or any other universal law of attraction (in other words, the total efforts produced by the effective gravity around all closed plane or skew channels that lie inside such a stratified fluid figure that attracts according to the universal inverse-square law or any other universal law of attraction and that are filled with homogeneous fluid are equal to zero), it does not follow from this result that a stratified fluid figure whose individually homogeneous strata are surfaces themselves and that attracts according to the universal inverse-square law or any other universal law of attraction can exist in a state of equilibrium. Nor does it really even follow from this result that a stratified fluid figure that is made up of a finite number of individually homogeneous strata whose thicknesses are all finite and that attracts according to the universal inverse-square law or any other universal law of attraction can exist in a state of equilibrium. In other words, simply because Clairaut's second new principle of equilibrium holds for all closed plane or skew channels that are filled with homogeneous fluid and that lie within the interior of a stratified fluid figure that is composed of a finite number of individually homogeneous strata whose thicknesses are all finite and that attracts according to the universal inverse-square law or any other universal law of attraction (in other words, simply because the total efforts produced by the effective gravity around all closed plane or skew channels that lie within the interior of a stratified fluid figure that is made up of a finite number of individually homogeneous strata whose thicknesses are all finite and that attracts according to the universal inverse-square law or any other universal law of attraction and that are filled with homogeneous fluid are equal to zero), one cannot really conclude that statement (1) above holds for such a stratified fluid figure that attracts according to the universal inverse-square law or any other universal law of attraction.

In fact, as we shall see, however, it follows from Clairaut's theory of equilibrium of stratified fluid figures that are composed of finite numbers of individually homogeneous strata whose thicknesses are all finite, which we examined in the preceding section of this chapter, that statement (1) does hold for stratified fluid figures that are made up of finite numbers of individually homogeneous strata whose thicknesses are all finite and that attract according to the universal inverse-square law or any other universal law of attraction. Moreover, we shall see below that statement (1) also holds for stratified fluid figures whose individually homogeneous strata are surfaces themselves and that attract according to the universal inverse-square law or any other universal law of attraction. We shall discover later in this chapter that Clairaut in fact made use of the second result, although he did not really explain anywhere in his treatise of 1743 why it turns out that this result is true.

Indeed, we already saw at the end of the preceding section of this chapter that Clairaut's theory of stratified fluid figures of equilibrium did not generally apply to stratified fluid figures of equilibrium whose strata are surfaces themselves. We saw that Clairaut only really showed that his theory of stratified fluid figures of equilibrium could be applied to stratified fluid figures of equilibrium which are *made up of* homogeneous fluid figures of equilibrium (the individually homogeneous fluid strata whose thicknesses are finite), which a stratified fluid figure of equilibrium composed of individually homogeneous fluid strata that are surfaces themselves is not. This restriction is essentially what served as Clairaut's basis for claiming that statement (2) above holds for such stratified fluid figures of equilibrium (that is, that he had reduced the principles of his theory of such stratified fluid figures of equilibrium to the principles of his theory of homogeneous fluid figures of equilibrium). I shall call attention to this reduction again in a moment.

In a work published in 1752, D'Alembert contended correctly that Clairaut's theory of heterogeneous fluid figures of equilibrium had a restricted validity. He showed that Clairaut's theory of heterogeneous fluid figures of equilibrium could not be a general theory. D'Alembert emphasized that Clairaut's theory could not be applied to all heterogeneous figures of equilibrium. D'Alembert observed that the true condition necessary for a heterogeneous fluid figure to exist in a state of equilibrium is that the total efforts produced by the effective gravity around all closed plane or skew channels that lie within the interior of a heterogeneous fluid figure and that are filled with fluid from the heterogeneous fluid figure must be equal to zero. Such a condition follows naturally, for example, from the principle of solidification applied to a heterogeneous fluid figure of equilibrium. Of course, the closed channels that lie within the interior of such a fluid figure and that are filled with fluid from the figure will generally be filled with fluid whose density varies, not homogeneous fluid, since the fluid figure that the channels lie inside is heterogeneous, not homogeneous. This will certainly be true, for example, if the density of a heterogeneous fluid figure of equilibrium varies continuously from the center of the figure to its surface.

Thus D'Alembert stated that the condition necessary for a heterogeneous fluid mass to exist in a state of equilibrium, which I express here in terms of the vector calculus, is, in general, that

$$\int_\gamma P\mathbf{F}\cdot d\mathbf{s} = 0$$

around all closed plane or skew channels γ that lie inside the heterogeneous fluid mass and that are filled with fluid from the heterogeneous fluid mass, where P stands for density and $\mathbf{F}$ symbolizes the effective gravity. (Unlike the masses of channels filled with homogeneous fluid of the same density, the masses of channels that lie within the interior of the heterogeneous fluid mass and that are filled with fluid from the heterogeneous fluid mass are not in general proportional to their lengths. Hence in the heterogeneous fluid mass the one-dimensional quantity $d\mathbf{s}$ has units of length but not units of mass. Instead the one-dimensional quantity $Pd\mathbf{s}$ has units of mass. Then, since the effective gravity $\mathbf{F}$ is the force per unit of mass, it follows that

$$\int_\Gamma \mathbf{F}\cdot Pd\mathbf{s} = \int_\Gamma P\mathbf{F}\cdot d\mathbf{s}$$

is in general the total effort produced by the effective gravity $\mathbf{F}$ along a plane or skew channel Γ that lies inside a heterogeneous fluid mass whose variable density is P, that is filled with fluid from the heterogeneous fluid mass, and that joins two different points. Moreover, it follows that $\int_\gamma P\mathbf{F}\cdot d\mathbf{s}$ is in general the total effort produced by the effective gravity $\mathbf{F}$ around a closed plane or skew channel γ that lies inside a heterogeneous fluid mass whose variable density is P and that is filled with fluid from the heterogeneous fluid mass.) At the same time, it will *not* be true in general that $\int_\gamma \mathbf{F}\cdot d\mathbf{s} = 0$ around arbitrary closed plane or skew channels γ that lie inside a heterogeneous fluid figure of equilibrium and that are filled with homogeneous fluid. In other words, it is *not* true in general that Clairaut's second new principle of equilibrium holds for all closed plane or skew channels that lie within the interior of a heterogeneous fluid figure of equilibrium and that are filled with homogeneous fluid (that is, the toal efforts produced by the effective gravity **F** around all closed plane or skew channels that lie within the interior of a heterogeneous fluid figure of equilibrium and that are filled with homogeneous fluid are not in general equal to zero).

Now, in the case of a stratified fluid figure that is made up of a finite number of individually homogeneous strata whose thicknesses are all finite, Clairaut in fact stated in his theory of equilibrium of such figures, which we examined at the end of the preceding section of this chapter, the condition that amounts to $\int_\gamma P\mathbf{F}\cdot d\mathbf{s} = 0$ around all closed plane or skew channels γ that lie within the interior of such a stratified fluid figure of equilibrium and that are filled with fluid from the stratified fluid figure of equilibrium, where P stands for density and **F** symbolizes the effective gravity. Namely, we recall that he observed that if a stratified fluid figure is composed of a finite number of individually homogeneous strata whose thicknesses are all *finite*, then P is *constant* within each individual stratum, so that if $\int_\gamma P\mathbf{F}\cdot d\mathbf{s} = 0$ is true around all closed plane or skew channels γ that lie inside the stratified fluid figure and that are filled with fluid from the figure (in other words, as Clairaut actually expressed the matter, if his second new principle of equilibrium holds for all closed plane or skew channels γ that lie within the interior of the stratified fluid figure and that are filled with fluid from the figure), then the condition $\int_\gamma P\mathbf{F}\cdot d\mathbf{s} = 0$ is reduced to $\int_\gamma \mathbf{F}\cdot d\mathbf{s} = 0$ for all closed plane or skew channels γ that lie entirely *within the interior* of a single homogeneous fluid stratum and that are filled with fluid from that stratum. Moreover, we have seen that it follows from the principle of solidification that in this case it is also true that $\int_\gamma \mathbf{F}\cdot d\mathbf{s} = 0$ around all closed plane or skew channels γ that lie inside a single homogeneous fluid stratum and that are filled with fluid from that stratum but parts of which *lie at the surfaces that bound* that homogeneous fluid stratum. Using these results Clairaut showed in effect in one of his demonstrations that we examined at the end of the preceding section of this chapter that if the two surfaces that bound each homogeneous fluid stratum within a stratified fluid figure of the kind in question are "level surfaces" of the effective gravity **F**, then it follows that the total effort $\int_\Gamma P\mathbf{F}\cdot d\mathbf{s}$ produced by the effective gravity **F** along an arbitrary plane of skew channel Γ that lies inside the stratified fluid figure, that is filled with fluid from the figure, and that joins two different points of the surface of the figure will be equal to zero. In other words, as Clairaut actually expressed the matter, it follows that his first new principle of equilibrium holds for all plane or skew channels Γ that lie inside the stratified fluid figure, that are filled with fluid from the figure, and that join two different points of the surface of the figure.

In addition, we saw at the end of the preceding section of this chapter that Clairaut also demonstrated that if his first new principle of equilibrium is assumed to hold for all plane or skew channels Γ that lie within the interior of a stratified fluid figure of the kind in question, that are filled with fluid from the figure, and that join two different points of the surface of the figure (in other words, assuming that the total efforts $\int_\Gamma PF \cdot ds$ produced by the effective gravity F along all plane or skew channels Γ that lie inside a stratified fluid figure of the kind in question, that are filled with fluid from the figure, and that join two different points of the surface of the figure are equal to zero), from which it follows that the total efforts $\int_\gamma PF \cdot ds$ produced by the effective gravity F around all closed plane or skew channels γ that lie in the interior of the stratified fluid figure and that are filled with fluid from the figure will be equal to zero (in other words, Clairaut's second new principle of equilibrium holds for all closed plane or skew channels γ that lie inside the figure and that are filled with fluid from the figure), then if the principle of the plumb line is assumed to hold at all points of the surface of the figure, then the surfaces between pairs of contiguous homogeneous fluid strata must all be "level surfaces" of the effective gravity F.

The two preceding results are the ones that Clairaut essentially used as the basis for claiming that statement (2) holds for stratified fluid figures of equilibrium which are composed of finite numbers of individually homogeneous strata whose thicknesses are all finite (that is, that he had reduced the principles of his theory of stratified fluid figures of equilibrium which are made up of finite numbers of individually homogeneous strata whose thicknesses are all finite to the principles of his theory of homogeneous figures of equilibrium). In particular, this reduction holds for stratified fluid figures of equilibrium which are composed of finite numbers of individually homogeneous strata whose thicknesses are all finite and which attract according to the universal inverse-square law or any other universal law of attraction. Thus statement (1) does indeed hold for stratified fluid figures that are made up of finite numbers of individually homogeneous strata whose thicknesses are all finite and that attract according to the universal inverse-square law or any other universal law of attraction. At the same time, we see that in Clairaut's theory of stratified fluid figures of equilibrium which are composed of finite numbers of individually homogeneous strata whose thicknesses are all finite, the condition that the total efforts $\int_\gamma PF \cdot ds$ produced by the effective gravity F around all closed plane or skew channels γ that lie inside such a stratified fluid figure of equilibrium and that are filled with fluid from the stratified fluid figure of equilibrium are equal to zero, where P stands for density (in other words, the condition that Clairaut's second new principle of equilibrium holds for all closed plane or skew channels γ that lie inside such a stratified fluid figure of equilibrium and that are filled with fluid from the figure), and the condition that the two surfaces that bound each homogeneous fluid stratum within such a stratified fluid figure of equilibrium are "level surfaces" of the effective gravity F *are interrelated*. In particular, since the total efforts $\int_\gamma PF \cdot ds$ produced by the effective gravity F around all closed plane or skew channels γ that lie within the interior of a stratified fluid figure of equilibrium which is made up of a finite number of strata whose thicknesses are all finite and that are filled with fluid from the stratified fluid figure of equilibrium are equal to zero (in other words, since Clairaut's second new principle of equilibrium holds for all closed plane or skew channels γ that lie inside a stratified fluid figure of equilibrium which is composed of a finite number of strata whose thicknesses are all finite and that are filled with fluid from the stratified fluid figure of equilibrium), it follows that the two surfaces that bound each homogeneous fluid stratum within the stratified fluid figure of equilibrium are "level surfaces" of the effective gravity F.

Naturally, it does *not* follow from Clairaut's theory of such stratified fluid figures of equilibrium that a surface of constant density within a heterogeneous fluid figure of equilibrium whose density P varies *continuously* must in general be a "level surface" of the effective gravity F. On the contrary, D'Alembert also showed that surfaces of constant density within a heterogeneous fluid figure of equilibrium whose density P varies continuously will *not* ordinarily be "level surfaces" of the effective gravity F. Thus D'Alembert demonstrated that surfaces of constant density within heterogeneous fluid figures of revolution in equilibrium are not in general level surfaces of the effective gravity F, yet the total efforts $\int_\gamma PF \cdot ds$ produced by the effective gravity F around all closed plane or skew channels γ that lie inside a heterogeneous fluid figure of equilibrium and that are filled with fluid from the heterogeneous fluid figure of equilibrium must nevertheless be equal to zero. In other words, the two conditons that are interrelated in Clairaut's theory of stratified fluid figures of equilibrium which are composed of finite numbers of individually homogeneous strata whose thicknesses are all finite are not in general interrelated in heterogeneous fluid figures of equilibrium. Indeed, one of the two conditions always holds for heterogeneous fluid figures of equilibrium, and the other condition does not in general hold for heterogeneous fluid figures of equilibrium. In short, Clairaut's theory of stratified fluid figures of equilibrium which are made up of finite numbers of individually homogeneous strata whose thicknesses are all finite cannot be generalized to cover all heterogeneous fluid figures of equilibrium.

Now, as I explained, Clairaut also "demonstrated" by applying a superposition principle to individual particles of fluid which attract, where the attraction due to a particle of fluid at all points is assumed to be directed toward the particle of fluid and where the magnitude of the attraction due to the particle of fluid at a point is assumed to vary directly as $1/r^2$ or, more generally, is given by an arbitrary function $f(r)$, where r is the distance of the point from the particle of fluid, that for any stratified fluid figure that is composed of a finite number of individually homogeneous strata whose thicknesses are all finite and that attracts according to the universal inverse-square law or any other universal law of attraction, the total effort $\int_\gamma F \cdot ds$ produced by the effective gravity F around any closed plane or skew channel γ that lies inside the stratified fluid figure and that is filled with homogeneous fluid is equal to zero. (We recall that Clairaut correctly stated in general and demonstrated in one special case that the centrifugal force generated when a closed plane or skew channel filled with homogeneous fluid revolves around an axis never produces a total effort around the closed channel filled with homogeneous fluid which is not equal to zero. Thus in order to determine whether or not Clairaut's second new principle of equilibrium holds for all closed plane or skew channels that lie within the interior of a fluid figure and that are filled with homogeneous fluid (that is, whether or not the total efforts produced by the effective gravity around all closed plane or skew channels that lie within the interior of the fluid figure and that are filled with homogeneous fluid are equal to zero), the effects of centrifugal force of rotation can be ignored. In other words, whether or not the effective gravity produces total efforts around these closed plane or skew channels filled with homogeneous fluid which are equal to zero solely depends upon the forces of attraction which act.)

But it is easy to see that it follows from the superposition principle applied to the kind of particles of fluid just described that this result not only holds for stratified fluid figures of the kind in question which attract in accordance with the universal inverse-square law or any other universal law of attraction, but that the result also holds for stratified fluid figures whose individually homogeneous strata are themselves "infinitesimally" thin surfaces (for example, whose densities P vary continuously from their centers to their surfaces) and that attract according to the universal inverse-square law or any other universal law of attraction. Indeed, it is also easy to see that it follows from the superposition principle applied to the kind of particles of fluid just described that when *any* homogeneous or heterogeneous fluid figure attracts according to the universal inverse-square law or any other universal law of attraction, the effective gravity F always produces total efforts $\int_\gamma F \cdot ds$ that are equal to zero around all closed plane or skew channels γ that lie within the interior of the homogeneous or heterogeneous fluid figure and that are filled with homogeneous fluid. This is true because the resultants of the attraction due to a homogeneous or heterogeneous fluid figure that attracts according to the universal inverse-square law or any other universal law of attraction always produces total efforts that are equal to zero around all closed plane or skew channels γ that lie inside the homogeneous or heterogeneous fluid figure and that are filled with homogeneous fluid and, at the same time, the centrifugal force generated when such a homogeneous or heterogeneous fluid figure revolves around an axis always produces total efforts that are equal to zero around closed plane or skew channels γ that lie within the interior of the homogeneous or heterogeneous fluid figure and that are filled with homogeneous fluid.

Now, in his "proof" that attraction directed toward a fixed center of force or toward a particle of fluid whose magnitude at a point depends solely upon the distance of the point from the fixed center of force or from the particle of fluid produces total efforts that are equal to zero around closed plane or skew channels, Clairaut made use of the assumption that the closed plane or skew channels were filled with *homogeneous* fluid, not fluid whose density *varies*. Clairaut's "proof" does *not* apply to closed plane or skew channels that are filled with fluid whose density *varies*. But we can consider a closed plane or skew channel that lies within the interior of a heterogeneous fluid figure that attracts according to the universal inverse-square law or any other universal law of attraction and that is filled with homogeneous fluid to be so thin that allowing the density of the fluid in the closed plane or skew channel to vary will not cause the effective gravity F due to the heterogeneous fluid figure and the closed plane or skew channel of fluid together to change appreciably. (In other words, we suppose that the closed plane or skew channel is "infinitesimally" thin.) In particular, we can assume that the closed plane or skew channel lies within the interior of a heterogeneous fluid figure whose density P varies continuously and that attracts according to the universal inverse-square law or any other universal law of attraction and that the closed plane or skew channel is filled with fluid from the heterogeneous fluid figure. Then since the variation of the density of the fluid in the closed plane or skew channel does not cause the effective gravity F to change to all intents and purposes, it follows that $\int_\gamma F \cdot ds$ is still equal to zero around all closed plane or skew channels γ that lie inside the heterogeneous fluid figure that attracts according to the universal inverse-square law or any other universal law of attraction and whose density P varies continuously and that are now filled with fluid from the heterogeneous fluid figure instead of homogeneous fluid. But in his treatise of 1743 Clairaut did not even state that this result holds, much less demonstrate that the result is true or explain why the result is true.

It is also easy to see by applying the same reasoning that the same result $\int_\gamma F \cdot ds = 0$ must equally be true around all closed plane or skew channels γ that lie inside a stratified fluid figure that attracts according to the universal inverse-square law or any other universal law of attraction and that is composed of a finite number of individually homogeneous strata whose thicknesses are all finite and that are filled with fluid from the stratified fluid figure. That is, $\int_\gamma F \cdot ds$ *will* necessarily be equal to zero around *all* closed plane or skew channels γ that lie within the interior of a stratified fluid figure that attracts according to the universal inverse-square law or any other universal law of attraction and that is made up of a finite number of individually homogeneous strata whose thicknesses are all finite and that are filled with fluid from the stratified fluid figure. Thus in the case of a closed plane or skew channel γ that lies within the interior of the stratified fluid figure that attracts according to the universal inverse-square law and that is composed of a finite number of individually homogeneous strata whose thicknesses are all finite and that is filled with fluid from the stratified fluid figure but that does not lie entirely within the interior of a single homogeneous fluid stratum, so that the closed plane or skew channel γ is consequently filled with fluid whose density varies instead of homogeneous fluid, it is nevertheless true that $\int_\gamma F \cdot ds$ is still equal to zero around the closed plane or skew channel γ.

Again, in his treatise of 1743 Clairaut did not call attention to this fact, much less explain why it is true.

Now, it turns out that one of Clairaut's demonstrations that we examined at the end of the preceding section of this chapter, which applies to a stratified fluid figure composed of a finite number of individually homogeneous strata whose thicknesses are all finite, can be suitably modified so as to show that if the density P of a heterogeneous fluid figure varies *continuously* and if $\int_\gamma \mathbf{F} \cdot d\mathbf{s}$ is never equal to zero around a closed plane or skew channel γ that lies within the interior of the heterogeneous fluid figure whose density P varies continuously and that is filled with fluid from the heterogeneous fluid figure (in other words, $\int_\gamma \mathbf{F} \cdot d\mathbf{s}$ is equal to zero around all closed plane or skew channels γ that lie inside the heterogeneous fluid figure whose density P varies continuously and that are filled with fluid from the heterogeneous fluid figure), which is true, for example, for all heterogeneous fluid figures that attract according to the universal inverse-square law or any other universal law of attraction and whose densities P vary continuously, as I just explained, then the surfaces of constant density within the heterogeneous fluid figure must also be "level surfaces" of the effective gravity $\mathbf{F}$. In other words, in the event that $\int_\gamma \mathbf{F} \cdot d\mathbf{s} = 0$ does happen to be true around all closed plane or skew channels γ that lie inside a heterogeneous fluid figure whose density P varies continuously and that are filled with fluid from the heterogeneous fluid figure, then it turns out that the surfaces of constant density within the heterogeneous fluid figure will all be level surfaces of the effective gravity $\mathbf{F}$. In particular, the surfaces of constant density within a heterogeneous fluid figure of equilibrium which attracts according to the universal inverse-square law or any other universal law of attraction and whose density P varies continuously *will* all be level surfaces of the effective gravity $\mathbf{F}$. We shall understand the significance of this result later in this chapter. We shall see that in his treatise of 1743 Clairaut made important use of this result, although, as I mentioned, in his treatise of 1743 he did not even specifically state that such a result holds, much less demonstrate that the result is true or explain why the result is true.

Consequently, although D'Alembert showed that statement (2) is *not* true in general for heterogeneous fluid figures of equilibrium, statement (1) nevertheless turns out to be true *both* for stratified fluid figures that are composed of finite numbers of individually homogeneous strata whose thicknesses are all finite and that attract according to the universal inverse-square law or any other universal law of attraction *and* for stratified fluid figures whose densities P vary continuously from their centers to their surfaces and that attract according to the universal inverse-square law or any other universal law of attraction. In particular, the surfaces of constant density within stratified fluid figures of equilibrium whose densities P vary continuously from their centers to their surfaces and which attract according to the universal inverse-square law or any other universal law of attraction are always level surfaces of the effective gravity $\mathbf{F}$. As we shall see later in this chapter, these results turn out to be indispensable to Clairaut's attempts in his treatise of 1743 to determine the shapes of stratified fluid figures of equilibrium which attract according to the universal inverse-square law.

9.3. Clairaut's and Euler's debate

Clairaut in effect showed that there were more necessary conditions that had to be fulfilled in order for fluid figures to exist in a state of equilibrium than had previously been known. We shall now see that Clairaut's research gave rise to other kinds of questions that have to do with the equilibrium of fluids.

In Chapter 7 we saw that during the period that Clairaut challenged Fontaine because he did not like Fontaine's particular way of using the partial differential calculus to integrate or reduce to quadratures first-order ordinary "differential" equations of first degree, Clairaut first wrote Euler specifically in order to try to learn whether mathematicians outside France had already done what Clairaut, Bouguer, and Fontaine thought that they had been the first to achieve. In the correspondence with Euler which ensued, Clairaut outlined his theory of figures of equilibrium. He did this before he published his treatise in 1743. What Euler read in Clairaut's letters caused him to question the theory's adequacy at first. These doubts brought about quite an interesting discussion between the two mathematicians.

Euler evidently found Clairaut's letter of 4 January 1742, in which Clairaut first mentioned his new theory of figures of equilibrium, not to be very illuminating. Indeed, Clairaut simply stated in that letter that not only were fluid figures of equilibrium ruled out when the principles of balanced columns and the plumb line failed to be consistent with each other and consequently failed to hold at the same time, as Bouguer had shown, but, in addition, in order for a fluid figure of revolution to be a figure of equilibrium, the differential $Pdy + Qdx$ had to be a complete differential (that is, condition (7.3.4.6''') $(dP/dx = dQ/dy)$ had to hold) as well, where x and y, respectively, stand for rectangular coordinates in the plane of a meridian at the surface of the fluid figure of revolution which are parallel and perpendicular, respectively, to the axis of symmetry of the fluid figure of revolution and where $P(y, x)$ and $Q(y, x)$, respectively, designate the magnitudes of the components of the effective gravity at a point in the plane of the meridian in question which lie in the plane of the meridian in question in the directions of the rectangular coordinates y and x, respectively, in which case equation (9.1.18) was the equation of the meridian, he said. But Clairaut did not make clear at all in the letter what mechanical principle or principles the complete differential $Pdy + Qdx$ expressed.[68] Moreover, although Clairaut did not say so in his letter, we recall that the fluid figure of revolution in equilibrium here is a homogeneous fluid figure of revolution and that the figure is assumed to revolve around its axis of symmetry, in which case the axis of symmetry of the homogeneous fluid figure of revolution is also an axis of symmetry of the centrifugal force per unit of mass, and that the axis of symmetry of the homogeneous fluid figure of revolution is also assumed to be an axis of symmetry of the attraction.

Euler replied sometime during the weeks that followed:

I cannot yet understand how it could be possible for any hypothesis not to permit some state of equilibrium [to exist] and to allow instead a perpetual motion that tends toward a state of rest yet that never reaches such a state. Such a situation appears to me to be too paradoxical to be accepted without proofs drawn from the first principles of hydrostatics which are as clear as the day itself. The principle that you use is undoubtedly the following one: a fluid mass can only be in equilibrium in case each molecule [*molécule*] [of fluid] be equally pressed [*comprimée*] at all parts.[69]

Euler told Clairaut that he had recently received some letters from Henri Kuhn, and Euler reported to Clairaut what Kuhn had written in the letters. In these letters Kuhn maintained that the surfaces of the land and of the seas were so irregular that no local observations whatever, including those of surveyors, could be used to determine the earth's shape, no matter how accurate the observations might be. Kuhn told Euler that he didn't doubt the accuracy of the observations made; Kuhn simply considered them to be useless for purposes of determining the earth's shape. He maintained that the surfaces of the land and of the seas were so unequal and uneven in comparison with measurements made with a surveyor's spirit level that one place on land or at sea can be several German miles higher than another spot on land or at sea. For example, Kuhn claimed that Lapland was five German miles higher than France in relation to measurements made with a surveyor's spirit level. As a result he contended that the earth could be spherical or elongated at its poles according to measurements made with a surveyor's spirit level yet still have a degree of latitude in Lapland that was much larger than the degree of latitude in France. In order to prove his statement that Lapland was five German miles higher than France in relation to measurements made with a surveyor's spirit level, Kuhn cited observations of the leveling out of rivers. From calculations he did based on these observations, he concluded that from its source to the Caspian Sea four hundred German miles away, the Volga River drops six German miles. Moreover, from the mouths of rivers in the seas he believed that the surface of the seas was as unequal and as uneven as the surface of the land and that this imperceptible inequality greatly exceeded the height of the highest mountains, although no one until now had paid any attention to this inequality. As a result, Euler said, Kuhn also rendered doubtful the method of surveying or measuring heights above the true level of the land and of the seas by means of barometers, which Kuhn only believed to be good when the heights rise suddenly. But if the heights rise imperceptibly, then, according to Kuhn, the earth's atmosphere adapts to the earth's surface, and as a result the earth's atmosphere does not make some decrease in its weight be noticed, although one actually goes up a distance much greater than the distance that one goes in climbing the highest mountain. Kuhn also categorically denied that the seas existed in a state of equilibrium. This assertion brought Bouguer's Paris Academy *mémoire* of 1734 to Euler's mind and the possibility that the law that governed the earth's attraction might not permit both of the principles of balanced columns and the plumb line to accord with each other, in which case the law that governed the earth's attraction would not allow both principles to hold at the same time, which would consequently make it impossible for the earth ever to have existed in a state of equilibrium.[70]

Henri Kuhn, professor of mathematics in Dantzig, had recently won a prize from the Bordeaux Academy for his *Méditations sur l'origine des fontaines, l'eau des puits, et autres problèmes qui ont du rapport à ce sujet. Ouvrage qui a remporté le prix au jugement de l'Académie Royale des Belles-Lettres, Sciences et Arts de Bordeaux* (Bordeaux: P. Brun, 1741). Although he originally defended Jacques Cassini's elongated earth against Newton's flattened earth, Kuhn ultimately concluded from the alternating rise and fall of the tides that it was unlikely that the earth's true shape would ever be found by using methods that depended upon assumptions made about the condition of the seas.[71]

Euler's reply made Clairaut realize that he had not explained his theory of figures of equilibrium very clearly in his letter of January 1742. Late in March of 1742 Clairaut wrote Euler again. He restated that a fluid figure failed to exist in a state of a equilibrium when the principles of balanced columns and the plumb line were incompatible, thus failed to hold at the same time, as Bouguer had shown. But now he added that it was not enough for these two principles to hold at the same time in order for a fluid figure to exist in a state of equilibrium. "Much more" [*bien plus*] was necessary in order for a fluid mass to exist in a state of equilibrium, he said. Clairaut now plainly stated his first and third new principles of equilibrium, and he mentioned that it followed from each of these two principles of equilibrium individually that the condition that the differential $Pdy + Qdx$ be complete must hold. In other words, he remarked that this condition expressed each of these two particular principles of equilibrium separately. He also mentioned that he had seen Kuhn's *Méditations sur l'origine des fontaines* and that it did not appear to him to merit any attention. Clairaut said that he found that the work included nothing precise or definite or specific where surveying and observations of leveling out were concerned. In addition, Kuhn did not appear to Clairaut to be any more conversant with theory than he was with practice. For example, Clairaut told Euler, Kuhn claimed in the work in question that the earth is elongated at the poles, because at the equator the water is saltier, hence more dense there.[72]

Although Clairaut had found additional conditions that had to be satisfied in order for a fluid mass to exist in a state of equilibrium, Euler was still not convinced that Clairaut had not overlooked or skirted fundamental issues in the realm of first principles of mechanics which he believed should be looked into. Euler answered Clairaut's latest communication in April of 1742. In his letter he said he agreed that both of the principles of balanced columns and the plumb line had to hold at the same time in order for a fluid mass to exist in a state of equilibrium. But then he added:

Here are my difficulties: when a system of bodies is not in equilibrium, it follows that motion necessarily occurs, but this motion always tends toward the state of equilibrium. This rule seems general to me, which is why I cannot imagine how a motion can take place when there is no state of equilibrium. For this reason people have tried until now to demonstrate that perpetual motion is impossible. But in the case that I am talking about [where the principles of balanced columns and the plumb line do not accord with each other and therefore cannot hold at the same time], it seems to me that perpetual motion cannot be denied. Why couldn't such a machine be invented, by using various forces like gravity, the spring, magnetism, etc., where there is no state of equilibrium either? Such a machine would in fact be a perfect perpetual motion. Then the [magnitudes of these] forces should continually increase, because all parts of such a machine are ceaselessly prompted, and [the magnitudes of] these forces should increase infinitely, unless resistance and friction limit them. These are the difficulties that I do not know how to resolve and that I would like you to think about.[73]

Euler called Clairaut's attention to Orfiré's perpetual motion machine, publicized in *Acta Eruditorum* for 1715, which no one had yet disproved. For purposes of demonstrating perpetual motion, Orfiré (whose true name was J. E. E. Bessler) had constructed a machine that he said continually sought a state of equilibrium but could never find one. Euler thought that Orfiré might have constructed a machine that had no state of equilibrium.[74]

We recall that in his Paris Academy *mémoire* of 1734, Bouguer had himself imagined, if rather vaguely, a kind of perpetual oscillation of homogeneous fluid with no limiting state (that is, no particular figure with a well-determined shape could be taken as a figure toward which the oscillating fluid tended). This no doubt explains why Bouguer's Paris Academy *mémoire* of 1734 came to Euler's mind. But several other, more recent developments might have affected Euler's outlook as well. In 1739, Euler created the first theory of an undamped, simple harmonic oscillator that is driven by a periodic (sinusoidal) force. In particular, in developing this theory, he also produced the first theory of resonance. Because finite forces could be used to produce motions (specifically, oscillations) whose amplitudes increase infinitely, the theory of resonance seemed to Euler "to allow the invention of perpetual motion."[75] (The fact that finite forces could be used to produce motions (specifically, oscillations) whose amplitudes increase infinitely, which Euler discovered, in effect invalidates the statement that Euler made in the passage cited, where Euler implies that were it not for the forces of resistance and friction which dampen the motions of the parts of a machine, the magnitudes of forces like gravity, the spring, magnetism, etc. should increase infinitely, if these forces produce motions of bodies which go on ceaselessly. Indeed, even a free, undamped simple harmonic oscillator, like a simple pendulum in a vacuum whose oscillations have a small amplitude, disproves the statement in question which Euler made.) Moreover, in his *Scientia navalis*, which he completed in 1738 but which was not published until 1749, Euler brought up the problem of stability in one of the chapters on hydrostatics (Chapter III).[76] In addition, unlike MacLaurin and Daniel Bernoulli, who proposed static theories of equilibrium of the tides in the essays that they submitted to the Paris Academy when they competed for the Academy's prize of 1740, Euler put forward a dynamical theory of the tides in the essay that he submitted to the Academy in order to try to win the same contest. Euler hypothesized that the restoring force of the water is proportional to the elevation of the free surface. He thus reduced the problem of the tides to a problem in nonlinear oscillations, in which an elastic ocean, rather than a fluid one, is attracted by the sun and moon.[77] Although the theory has since been described as not having been very successful,[78] it still helps to give an idea of the kinds of problems in mechanics which were on Euler's mind in the years around 1740.

Clairaut responded at the end of May 1742. He admitted that he did not understand Euler's worries. He thought Euler's argument that no motion exists which does not tend toward a state of equilibrium to be inappropriate and irrelevant to the problem in question. He saw no connections between demonstrations meant to refute the existence of perpetual motion and the manner of showing the fluids cannot exist in states of equilibrium when attraction is assumed to act in certain ways. He maintained that in order to construct a perpetual motion machine, the forces of nature would have to be increased to offset the continual losses of motion caused by friction and resistance. But whatever kind of forces act on a system of bodies, where the forces act continuously instead of by impulses ("*forces acceleratrices*"), like gravity, the spring, magnetism, and so on, it was known, he added, that the machines that hold the bodies in the system together can only transfer forces from some bodies in the system to other bodies in the system, to give certain bodies in the system a quantity of force that other bodies in the system lose, and can never change the total quantity of forces. Consequently, Clairaut affirmed, in a resistant medium and whenever parts of a machine rub against something, which inevitably happens in the case of a man-made machine, the forces continually diminish, so that in the end the system reaches a state in which all of its parts are all in equilibrium. This situation must always be the case, Clairaut stated categorically. Although he had never heard of Orfiré before, Clairaut nevertheless felt confident, without actually having to look at a detailed description of Orfiré's alleged perpetual motion machine, that like all of the many so-called perpetual motion machines that were continually submitted to the Paris Academy for approval, Orfiré's perpetual motion machine was quackery or fakery (*imposture*).[79]

Clairaut turned to the issue of the substructure of a fluid, whose problems Euler originally raised, and he gave the following lecture:

As for fluids, that is a whole other question. If they are simply regarded as a mass ["*amas*"] of little bodies, it is certain that whatever hypotheses of attraction are assumed or however attraction acts, one could imagine one way or even an infinite number of different ways to arrange these little bodies so that they will all be at rest, and these arrangements would not require that either Huygens's law [the principle of the plumb line] or Newton's law [the principle of balanced columns] holds. For example, in the same way that a pile of cannonballs supports itself in the form of a pyramid, if a mass of water is considered to be a collection [*assemblage*] of little cannonballs, the mind could conceive these little cannon-balls to be put together in such ways that they support themselves without having to be level [*sans niveau*]. But one sees clearly that the equilibrium that would depend upon how one would have arranged these corpuscles to his liking is not the equilibrium at issue. To answer the question how the true equilibrium of fluids comes about, the nature of these bodies [that is, fluids] would have to be understood, which I believe to be a very difficult matter. But without going back to first principles, one can, it seems to me, make progress in studying the theory of fluids by starting with the truths of experience which the laws that these bodies [that is, fluids] obey give us without showing us their nature. We know that if a piston presses on one part of a fluid enclosed in a vessel, that same pressure [*pression*] is transferred to all parts of the fluid in the vessel.[80] Granting that to be true, I assume that the [fluid] body having this property be moved to a place where the attraction acts according to any law, and I determine whether this law allows each particle of fluid [*particule du fluide*] to be pressed equally on all sides.[81] For example, let us imagine a planet [shaped like a figure of revolution] that attracts in a way that makes all bodies tend toward a fixed center. Instead of assuming that the magnitude of the attraction at a point solely depends upon the distance of the point from this fixed center, however, let us hypothesize instead that the magnitude of the attraction at a point is given as a function of the distance of the point from the fixed center and the angle that the radius [that is, the line segment that joins the fixed center to the point] makes with the axis [of symmetry which the planet revolves around]. I say that if a vase filled with water or any other fluid that we know is transported to this planet, all of the particles [*particules*] of this fluid will not be equally pressed on all sides. I cannot say what would happen to the fluid, because I do not know what the fluid quality is due to. Perhaps the fluid would oscillate incessantly; perhaps it would become solid. All that I can be certain of is that it would no longer have the property that our fluids have, of making themselves level of their own accord, of making bodies of the same density float, etc.[82]

After receiving Clairaut's reply, Euler raised no more questions of the kind that he had been asking his French correspondent until then. The two mathematicians were at cross purposes. In the course of their discussion various issues got mixed together: whether perpetual motion is present or absent in nature, whether the forces of nature are conservative or non-conservative, and whether states of equilibrium are stable or unstable, not to mention the question whether states of equilibrium can fail to exist altogether in nature.[83]

Perhaps with Euler's reservations in mind, Clairaut concluded the "Introduc-tion" to his treatise of 1743 in the following way. He stated that it was easy to see how a solid body can conserve of its own accord its rotational motion. He said that in order to understand how this can happen it was enough to take a look at a stick that has two weights attached to it. One gives impulses to these two weights which point in opposite directions and whose magnitudes are inversely proportional to the masses of the two weights. Then according to Clairaut, from the same principles that determine that this stick with two weights attached to it will turn incessantly around its center of gravity [*centre de gravité*] one easily realizes that if one gives one suitable impulse to any solid body, the body will turn ceaselessly around a line passing through its center of gravity. But when the body is fluid, Clairaut stated, the same thing is no longer true. He considered the particles that make up a fluid body, and he said that each particle of fluid, which is completely detached from the other particles of fluid, seems to want to move separately and on its own, around the point toward which the attraction pushes it, like a planet that follows an orbit around a central body in accordance with an impulse given to the planet and the force that pushes the planet toward the central body. In other words, the particles that make up a fluid body, which are unconnected, do not force each other to move the same way. Why, then, do the particles that make up a rotating fluid figure of equilibrium "circulate all together,"[84] instead of in some way that causes the motions of individual particles of fluid to disturb each other mutually? How can a rotating fluid mass even attain a relative state of equilibrium? How can a mass of fluid, assumed to be made up of separate, unconnected particles, ever reach a state of rotation in which the individual particles do not obstruct or interfere with each other's motions? Clairaut tacitly suggested here that such interference would produce a state of disequilibrium. It was easier to imagine a body assumed to be solid, whose parts mutually constrain one another to rotate together as rotation begins, to exist in a state of equilibrium (in other words, to rotate in a way that allows the motion of the body to be conserved). But what was there to prevent the paths of individual particles of fluid from crisscrossing and in so doing cause the rotating fluid mass to fail to arrive at a state of equilibrium? One could always postulate that some sort of "subtle matter" exists which "carries along all parts of the [fluid] planet and guides them as if it pushed them in circular pipes," so that the subtle matter makes all of the particles of fluid revolve around the same axis of rotation with the same angular speed. But in Clairaut's opinion anyone who resorts to such an expedient only dodges the issue, because the rotation of the presumably fluid subtle matter must then be explained. In other words, such a device only transfers the problem to another level without resolving it and thereby begs the question, Clairaut observed.

Clairaut dealt with the problem as follows. To understand how a state of equilibrium once attained could *endure*, he declared that it sufficed to imagine the problem from the standpoint of an observer rotating with the fluid mass. The frame of reference of such an observer is a frame of reference attached to the rotating fluid mass. Such a frame of reference is an accelerated frame of reference. Viewed from this accelerated frame of reference, the rotating fluid mass appears to be at rest. At the same time, the "disorderly" centrifugal force of rotation simply behaves when viewed from this accelerated frame of reference like a force that appears to accelerate bodies when these bodies are observed from an inertial frame of reference. Viewed from the accelerated frame of reference, the centrifugal force of rotation appears to push the particles of fluid which make up the rotating fluid mass away from the axis of rotation of the rotating fluid mass, and the magnitudes of this force per unit of mass at the particles of fluid which make up the rotating fluid mass vary directly as the distances of the particles of fluid from the axis of rotation of the rotating fluid mass. In this way Clairaut reduced the problem of explaining how a fluid mass can exist in a state of relative equilibrium when it rotates to explaining how a fluid mass that is at rest at one moment when certain forces act on it (the forces of attraction and a force that acts in an inertial frame of reference like centrifugal force of rotation) can remain at rest as these same forces continue to act on it. Clairaut had also utilized accelerated coordinate systems in other works that he wrote around this time (for example, in his Paris Academy of Sciences *mémoire* of 1742 on dynamics, published in 1745[85]). In fact he was among the first mathematicians to employ such coordinate systems.[86] His reasoning where figures of equilibrium are concerned was all right, as far as it went, but his analysis was not complete. It did not really explain how fluid masses can exist in states of absolute (hydrostatic) equilibrium.[87]

As to the question how a rotating fluid mass reaches a state of relative equilibrium in the first place, Clairaut responded that although the preceding analysis did not explain "how the planets got their shapes of their own accord, won't it suffice for us to know how the planets are able to preserve their shapes?"[88] Having asked this rhetorical question, Clairaut let Euler's questions about how equilibrium is approached, which involved dynamical issues, alone. He ultimately avoided the whole matter in his treatise of 1743.

Later, in justifying having limited his investigation this way, which Clairaut did in answering certain critics, he appealed to the work of his illustrious predecessors. He stressed that no one – Newton, Huygens, nor anyone else – had ever "tried to calculate the infinite number of oscillations that must have occurred before the [fluid] mass reached a permanent state [of equilibrium]." (Here Clairaut tacitly *assumed* without proof that homogeneous fluid figures of equilibrium shaped like flattened ellipsoids of revolution whose ellipticities are infinitesimal, which revolve around their axes of symmetry, and which attract according to the universal inverse-square law are *stable* figures of equilibrium. We now know that this need not be true if the ellipticities of the homogeneous fluid figures of equilibrium just described are *finite*.[89])

By the time that he and Clairaut discussed these matters concerning the equilibrium of fluids in their correspondence in 1742, Euler had already begun to make great progress in the study of the dynamics of continuous systems. Clairaut, it seems, never became an adept in continuum mechanics. At least, little evidence can be found that supports the conclusion that Clairaut ever became skilled in continuum mechanics. In his Paris Academy *mémoire* of 1742 on dynamics, he did solve some problems that have to do with constrained motions of bodies. But he never solved problems in elasticity, fluid dynamics, or in the mechanics of any continuous medium which involve many degrees of freedom (for example, the motion of an extended, rigid body, which is a problem in 6 degrees of freedom). Years later, for example, in a letter to his close friend Daniel Bernoulli, Clairaut admitted that he never truly understood the problem of a vibrating string, the modes of vibration, and so forth, [90] which had led to a protracted controversy that began in 1746 and that Bernoulli had participated in along with D'Alembert and Euler.

If Clairaut's reasons for treating figures of equilibrium as problems in statics hardly persuaded Euler to put dynamical questions out of mind, Clairaut's thoughts about the substructure of a fluid may have nevertheless helped to influence the way that Euler subsequently viewed fluids. In the first of three important papers on fluid mechanics which Euler wrote and which were published in the Berlin Academy's *Mémoires* for 1755, Euler argued that no corpuscular model of a fluid would ever lead to an understanding of the state of equilibrium of a fluid mass. Euler considered lattices of corpuscles, in which corpuscles are in contact with each other. He then allowed each corpuscle to become infinitesimally small and the number of corpuscles to become infinitely large. He concluded that in order for the mass of corpuscles to be in a state of equilibrium, it was necessary that an infinite number of forces act from all sides of the mass of corpuscles, so that if one of these forces was missing or suddenly disappeared, a state of equilibrium of the mass of corpuscles would be upset. Euler noted as well that the corpuscles could also be imagined to be arranged so that all of the forces required for equilibrium become equal among themselves, which, Euler stated, would represent exactly the case of a fluid. However, not only would this be "morally impossible, so to speak," Euler said, in the case of a mass of corpuscles, but, in addition, the forces required in order for the mass of corpuscles to be in equilibrium would not fail to become unequal among themselves, and, in fact, to differ greatly from each other, as soon as the mass of corpuscles underwent the slightest change. At the same time, the equality of these forces in a mass of fluid in equilibrium necessarily endures instead, whatever changes the fluid mass undergoes. In other words, a mass of corpuscles cannot be visualized or imagined in any state of equilibrium besides an unstable one, whereas a fluid mass can exist in a stable state of equilibrium.[91] Euler ultimately concluded from these thoughts that fluidity cannot be explained by means of "a mass of solid corpuscles" [*"un amas de corpuscules solides"*], even if the solid corpuscles are assumed to be infinitesimally small, wholly detached from one another, and the number of solid corpuscles is assumed to be infinitely large. Moreover, Euler thought it highly doubtful that an internal motion of any kind could make up for this deficiency.[92] Euler's notion of a "particle of fluid" simply became a mathematical point in a continuum of matter.[93] While various considerations probably caused Euler to adopt this viewpoint, his discussion with Clairaut in his correspondence with the French mathematician in 1742 is undoubtedly among them. The treatments of masses of corpuscles by the two mathematicians resemble each other too closely to allow one to conclude that the dialogue between the two had no part in leading Euler to imagine a fluid as a continuum of matter.[94] By the same token, when Clairaut talked about "particles of fluid" in his treatise of 1743, he too evidently meant in fact mathematical points in a continuum of matter.

9.4. *Toward solving the problem of the earth's shape*

As we shall now see, Clairaut treated the problem of the earth's shape as a problem in figures of equilibrium in the following way. He began by assuming the basic shape of a fluid figure that attracts according to the universal inverse-square law which he sought – namely, a fluid figure shaped like an ellipsoid of revolution which is flattened at its poles, which revolves around its axis of symmetry (its polar axis), and whose ellipticity is infinitesimal. He then determined for a given rate of rotation (that is, for a given ratio of the magnitude of the centrifugal force per unit of mass to the magnitude of the effective gravity at the equator of the figure) what ellipticity would enable this fluid figure to fulfill conditions that are *necessary* in order for the fluid figure to exist in a state of equilibrium. He implicitly assumed that conditions that are *sufficient* in order for the fluid figure to exist in a state of equilibrium are satisfied. He did not actually demonstrate this to be true. He did not prove that a fluid mass that attracts according to the universal inverse-square law and that rotates at a certain rate *would* eventually assume a specific shape that satisfied the conditions necessary for the fluid mass to exist in a state of equilibrium. He just showed that *if* a rotating fluid mass that attracts according to the universal inverse-square law had a certain shape, *then* the mass would satisfy the conditions necessary for the fluid mass to exist in a state of equilibrium. As it turns out, the fluid figures that attract according to the universal inverse-square law – shapes he assumed beforehand – namely, fluid figures shaped like ellipsoids of revolution flattened at their poles, which revolve around their axes of symmetry, and whose ellipticities are infinitesimal – we know happen to satisfy some condition or conditions that are sufficient in order for the fluid figures to exist in states of equilibrium. In other words, Clairaut assumed at the beginning that his fluid figures that attract in accordance with the universal inverse-square law had shapes that we now know happen to fulfill some condition or conditions that are *sufficient* in order for the fluid figures to exist in states of equilibrium. (As I mentioned in Chapter 1 and at the beginning of this chapter, sufficient conditions for equilibrium go beyond the limits of the story told in these pages. What these conditions are is a question that would only begin to be answered by mathematicians much later, which is why I do not discuss them here.) Clairaut in effect found among these fluid figures the ones which also happen to satisfy the various conditions *necessary* in order for the fluid figures to exist in states of equilibrium. Thus despite the qualifications that Clairaut stipulated which we examined in the preceding section of this chapter, which limited the kind of investigation that he would undertake, he did nevertheless introduce enough conditions that determined a problem that can be solved – namely, to find the ellipticity of a stratified fluid figure of equilibrium shaped like an ellipsoid of revolution flattened at its poles which attracts according to the universal inverse-square law, which revolves around its axis of symmetry, and whose ellipticity is infinitesimal. Indeed, modern researchers have succeeded in generalizing Clairaut's problem,[95] and I now turn to that problem.

When a fluid mass attracts in accordance with the universal inverse-square law, the resultant of the attraction of the fluid mass, which tries to preserve a state of equilibrium of the mass when the mass is in a state of equilibrium, depends itself upon the shape of the mass. Consequently, although a stratified fluid figure of equilibrium shaped like an ellipsoid of revolution flattened at its poles which attracts according to the universal inverse-square law, which revolves around its axis of symmetry, and whose ellipticity is infinitesimal fulfills the conditions of symmetry mentioned several times earlier in this chapter, Clairaut could not use the mathematical representations of his new principles of equilibrium for homogeneous fluid figures of equilibrium which satisfy these conditions of symmetry, which are expressed in terms of complete differentials in two rectangular coordinates and differentials in two rectangular coordinates whose integrals are path independent, in order to find the equation of a meridian at the surface of such a stratified fluid figure of equilibrium by suitably modifying these mathematical representations. The reason is that in order to utilize these mathematical representations, the magnitudes of the components of the attraction $P(y, x)$ and $Q(y, x)$, respectively, at points (y, x) in the plane of the meridian in question which lie in the plane of the meridian in question in the directions of the rectangular coordinates y and x, respectively, in the plane of the meridian in question must be known at the start. But in the case of the stratified fluid figure of equilibrium in question they are not known at the start, because they depend upon the unknown shape of the stratified fluid figure of equilibrium.

In order to deal with this situation, Clairaut resorted to the kind of mathematics which he had employed in his two papers on the theory of the earth's shape published in the *Philosophical Transactions*. Namely, he utilized a great deal of geometry in addition to analysis, which included many approximations. In resorting to such mathematics, he confined himself to solving the problem to terms of first order. First he calculated the infinitesimal ellipticity δ of a figure composed of two parts. (See Figure 56.) He assumed that the figure was shaped like an ellipsoid of revolution and that it attracted according to the universal inverse-square law. One of the poles of the figure was situated at P; PME is the part of a meridian at the surface of the figure which lies in one quadrant; and the figure revolved around its polar axis, which was its axis of symmetry. The surface of the figure was the outer surface of a homogeneous fluid layer in equilibrium whose density was 1. This fluid layer surrounded a solid body shaped like an ellipsoid of revolution which was made up of an infinite number of concentric,

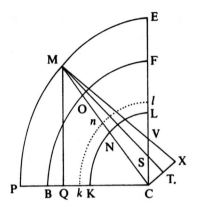

Figure 56. Meridian of a quadrant of a stratified ellipsoid of revolution whose ellipticity is infinitesimal surrounded by a homogeneous fluid layer in equilibrium whose surface is also an ellipsoid of revolution whose ellipticity is infinitesimal, where the whole figure attracts according to the universal inverse-square law (Clairaut 1743: figure on p. 210).

individually homogeneous ellipsoids of revolution whose ellipticities were infinitesimal and could either be positive or negative, meaning that the ellipsoids of revolution could either be flattened or elongated at their poles, and all of which had a common axis of symmetry, namely, the polar axis of the figure. Hence the strata of the solid part of the figure were in effect surfaces whose infinitesimal ellipticities $\rho(r)$ were given at the start, where $\rho(r)$ can be positive or negative. Here C designates the center of the figure; $\alpha \equiv$ the ellipticity of the outer surface of the solid body shaped like an ellipsoid of revolution; $\delta \equiv$ the ellipticity of the figure (in other words, $\delta \equiv$ the ellipticity of the ellipsoid of revolution which is the outer surface of the homogeneous fluid layer in equilibrium which surrounds the solid body); $CP = 1$ (normalization), $a \equiv CB$ is the polar radius of the solid body (hence $a < 1$); $R(r)$ and $\rho(r)$, respectively, are the density and the ellipticity, respectively, of the solid stratum whose polar radius equals r, where r is measured along CB (in which case $\rho(a) = \alpha$); and ϕ is the ratio of the magnitude of the centrifugal force per unit of mass to the magnitude of the effective gravity at the equator E of the outermost surface, the surface of the figure.[96] In short, the ellipticities of the solid strata were permitted to vary. Moreover, these infinitesimal ellipticities $\rho(r)$ were allowed to vary as a function of distance r along the polar radius CB of the solid body in an arbitrary way. In particular, these ellipticities did not necessarily have to be positive numbers. They could be negative numbers. Furthermore, Clairaut allowed the densities of the strata that make up the solid body to vary as an arbitrary nonnegative function $R(r)$ of distance r along the polar radius CB of the solid body too. Clairaut ceased to use the infinite series (6.2.1) to express the way that density varies as a function of r

which he had introduced in his second paper on the theory of the earth's shape published in the *Philosophical Transactions*. (We recall that he had already stopped using infinite series of monomials in two variables to express functions of two variables in his Paris Academy *mémoire* of 1740 on integral calculus, possibly to avoid the question whether such series can represent arbitrary functions of two variables.) As we shall see, sometimes he assumed $R(r)$ to be differentiable; at other times he thought of $R(r)$ as a step function. Here Clairaut undoubtedly allowed the ellipticities of the ellipsoids of revolution which are the strata of the solid body to vary in order to take into account the possibility that Newton imagined the density to decrease from the earth's center to the earth's surface in a way that differed from the one that accorded with the geometry of the strata in the figure that Clairaut had studied in his second paper on the theory of the earth's shape published in the *Philosophical Transactions* (namely, in that paper the ellipticities of the strata are assumed to be constant, in which case the ellipsoids of revolution are geometrically similar.) Moreover, Clairaut also very likely permitted the ellipticities of the strata to vary in order to generalize the geometry of the strata to include the strata in MacLaurin's figure whose ellipticity was $\frac{1}{178\frac{1}{4}}$ and whose effective gravity increased in magnitude from the equator of the figure to a pole of the figure by a factor of $\frac{1}{220}$.

Such a problem resembles to a certain extent the one that Daniel Bernoulli formulated and solved in the essay on the tides which earned Bernoulli a share of the Paris Academy's prize of 1740, although, of course, the conditions given in Clairaut's problem and in Bernoulli's problem differ. (Clairaut presupposed the stratified solid body that is surrounded by a homogeneous fluid layer in equilibrium to be shaped like an ellipsoid of revolution, not like Bernoulli's sphere. Moreover, Bernoulli's homogeneous outer fluid layer in equilibrium was shaped like a meniscus cap. The surface of the homogeneous outer fluid layer in equilibrium and the surface of the stratified solid body that the fluid layer surrounds *met* at the equator of Bernoulli's figure shaped like an elongated ellipsoid of revolution. But along *both* axes of Clairaut's figure the homogeneous fluid layer in equilibrium had a finite depth at the surface of the stratified solid part underlying it. Furthermore, the attractions of the sun and moon have no part in Clairaut's problem, nor does the earth's revolution around the sun or around the center of gravity of the earth–moon system, while all four of these phenomena are involved in Bernoulli's problem. Clairaut wanted to find the ellipticity of the outer surface of the homogeneous fluid layer in equilibrium when the shape of the layer was produced solely by the combined effects of the attraction of the stratified solid body shaped like an ellipsoid of revolution, the attraction of the homogeneous fluid layer in equilibrium which surrounds the stratified solid body, and the rotation of the stratified solid body and the homogeneous fluid layer in equilibrium around their common axis of rotation.)

In order to show that the homogeneous fluid layer satisfies the relevant conditions necessary for the fluid layer to exist in a state of equilibrium, Clairaut only needed to demonstrate that the principle of the plumb line held at the outer surface of the homogeneous fluid layer. We recall that Clairaut "proved" that the effective gravity due to a rotating figure that attracts according to the universal inverse-square law or any other universal law of attraction always produces total efforts equal to zero around closed plane or skew channels that lie within the interior of the figure and that are filled with homogeneous fluid, which is to say that Clairaut's second new principle of equilibrium holds for such a figure, so that if the principle of the plumb line could be shown to hold at the outer surface of the homogeneous fluid layer, this would be enough to assure that the homogeneous fluid layer fulfilled the relevant conditions necessary for the homogeneous fluid layer to exist in a state of equilibrium. (We recall that Clairaut "showed" that centrifugal force of rotation never produces a total effort around a closed, revolving plane or skew channel filled with homogeneous fluid which is not equal to zero. Whether effective gravity produces total efforts around closed, revolving plane or skew channels filled with homogeneous fluid which are equal to zero or not depends solely on the forces of attraction which act.) As a result, Clairaut could calculate much the same way that he had in his second paper on the theory of the earth's shape published in the *Philosophical Transactions*, when he applied in that paper the principle of the plumb line at the surface of what he thought at the time was a stratified fluid figure of equilibrium shaped like a flattened ellipsoid of revolution which attracted according to the universal inverse-square law, which revolved around its axis of symmetry, whose ellipticity was infinitesimal, and which was made up of concentric, individually homogeneous, flattened ellipsoids of revolution all of which had a common axis of symmetry, namely, the polar axis of the figure, and whose infinitesimal ellipticities were all equal, namely, equal to the ellipticity of the figure.

(Unlike the "infinitesimal" thicknesses of the solid strata, which are surfaces, the thickness of the outer, fluid layer is finite. This is the only feature of the problem which differentiates it from the problem that Clairaut tried to solve in his second paper on the theory of the earth's shape published in the *Philosophical Transactions*. However, as I mentioned in Chapter 6, the calculations of magnitudes of attraction which Clairaut did in his second paper on the theory of the earth's shape published in the *Philosophical Transactions* can be generalized to cover stratified figures shaped like ellipsoids of revolution which attract in accordance with the universal inverse-square law, which revolve around their axes of symmetry, whose ellipticities are infinitesimal, and whose densities vary from their centers to their surfaces as step functions. But in such cases the figures consist of finite numbers of strata whose thicknesses are finite, where two contiguous homogeneous strata in such a figure have different densities.[97] Hence the figure at issue, which has an outer, homogeneous fluid layer whose thickness is finite, does not call for the use of any methods for calculating magnitudes of attractions which Clairaut had not already utilized before.)

Clairaut found that if the ratio ϕ of the magnitude of the centrifugal force per unit of mass to the magnitude of the effective gravity at the equator E of the figure is infinitesimal (that is, $\phi^2 \ll \phi$), then the principle of the plumb line held at the outer surface of such an outer, homogeneous fluid layer if the outer surface of the homogeneous fluid layer is a particular ellipsoid of revolution. More precisely, he showed that the principle of the plumb line held to terms of first order in this case (that is, the principle of the plumb line held after neglecting terms in δ^n, $n \geqslant 2$, where δ is the ellipticity of the outer surface of the outer, homogeneous fluid layer, as well as discarding all other terms of the same order as δ^n, $n \geqslant 2$, such as ρ^n, $n \geqslant 2$, where $\rho = \rho(r)$, $0 \leqslant r \leqslant a$, is the ellipticity of an underlying, solid stratum, and ϕ^n, $n \geqslant 2$). In other words, Clairaut demonstrated that such an outer, homogeneous fluid layer satisfied the relevant conditions necessary for the fluid layer to exist in a state of equilibrium in this instance.

(The equation expressing the principle of the plumb line which Clairaut wrote in his treatise of 1743 differed from the equation that he wrote in his second paper on the theory of the earth's shape published in the *Philosophical Transactions*. The equation that Clairaut wrote in his second paper on the theory of the earth's shape published in the *Philosophical Transactions* held *exactly*, even though he only used it at the time to arrive at results that held good to terms of first order. By the same token, the equation also held true for figures that are not almost spherical – for example, stratified figures shaped like ellipsoids of revolution flattened at their poles which attract according to the universal inverse-square law, which revolve around their axes of symmetry, and whose ellipticities are finite – although Clairaut did not point this fact out in his second paper on the theory of the earth's shape published in the *Philosophical Transactions*. However, unlike the equation that Clairaut wrote in that paper, the equation expressing the principle of the plumb line which Clairaut wrote in his treatise of 1743 *only holds itself to terms of first order*. That is, Clairaut must have used the assumption that the figures shaped like ellipsoids of revolution to which he applied the principle of the plumb line were all nearly spherical, in which case their ellipticities were infinitesimal, in order to derive the equation. But in his treatise of 1743 Clairaut did not mention the fact that he used this assumption to arrive at the equation, much less show how he found the equation. Furthermore, in applying this equation, Clairaut wrote equalities which also turn out to hold good only to terms of first order, although, once again, Clairaut neglected to call attention to this fact.[98])

As we shall now see, the infinitesimal ellipticity δ of the ellipsoid of revolution which is the outer surface of the homogeneous fluid layer that fulfills the relevant conditions necessary for the fluid layer to exist in a state of equilibrium could either be positive or negative, depending upon the ways that the densities of the homogeneous fluid layer and the solid strata and the ellipticities of the solid strata and δ are interrelated. Using the equation for the ellipticity δ

which resulted from applying the principle of the plumb line at the outer surface of the outer, homogeneous fluid layer, Clairaut showed that heterogeneous flattened figures of the kind just described, which were flatter than the corresponding homogeneous fluid figure of equilibrium shaped like an ellipsoid of revolution flattened at its poles which attracts according to the universal inverse-square law, which revolves around its axis of symmetry, and whose ellipticity is infinitesimal (that is, the homogeneous fluid figure of equilibrium shaped like an ellipsoid of revolution flattened at its poles which attracts in accordance with the universal inverse-square law, which revolves around its axis of symmetry, whose ellipticity is infinitesimal, and which has the same value of ϕ as heterogeneous flattened figures of the type in question), could exist, when the solid strata all had the same density, one that was less than the density of the surrounding homogeneous fluid layer that satisfies the relevant conditions necessary for the fluid layer to exist in a state of equilibrium, and, in addition, all of the solid strata had the same ellipticity as the outer surface of the homogeneous fluid layer.[99] Moreover, he demonstrated that by specifying that the solid strata all have the same density, one that was greater than the density of the surrounding homogeneous fluid layer that fulfills the relevant conditions necessary for the fluid layer to exist in a state of equilibrium, the ellipticity of the outer surface of the homogeneous fluid layer could be made greater than the ellipticity of the corresponding homogeneous fluid figure of equilibrium shaped like an ellipsoid of revolution flattened at its poles which attracts according to the universal inverse-square law, which revolves around its axis of symmetry, and whose ellipticity is infinitesimal (that is, the homogeneous fluid figure of equilibrium shaped like an ellipsoid of revolution flattened at its poles which attracts in accordance with the universal inverse-square law, which revolves around its axis of symmetry, whose ellipticity is infinitesimal, and which has the same value of ϕ as heterogeneous flattened figures of the type in question), when the ellipticity of the homogeneous, solid part underlying the homogeneous fluid layer that fulfills the relevant conditions necessary for the fluid layer to exist in a state of equilibrium was suitably chosen. Similarly, by appropriately picking this ellipticity in a different way, while keeping the densities the same as those just indicated, the ellipticity of the outer surface of the homogeneous fluid layer could be made positive but less than the ellipticity of the corresponding homogeneous fluid figure of equilibrium shaped like an ellipsoid of revolution flattened at its poles which attracts according to the universal inverse-square law, which revolves around its axis of symmetry, and whose ellipticity is infinitesimal.[100]

From these findings Clairaut concluded that in order to make a rotating, homogeneous figure of revolution shaped like a flattened ellipsoid of revolution which attracts according to the universal inverse-square law, which revolves around its axis of symmetry, and whose ellipticity is infinitesimal even flatter, it did not suffice simply to make it denser at the center, although Newton had seemed to believe that it did.[101] In a figure made up of a homogeneous solid nucleus shaped like a flattened ellipsoid of revolution surrounded by a homogeneous fluid layer in equilibrium which is less dense than the nucleus and whose outer surface was also a flattened ellipsoid of revolution, where the two flattened ellipsoids of revolution have the same axis of symmetry, namely, the polar axis of the figure, the outer surface of the homogeneous fluid layer could either be made flatter or less flat than the corresponding homogeneous fluid figure of equilibrium shaped like an ellipsoid of revolution flattened at its poles which attracts according to the universal inverse-square law, which revolves around its axis of symmetry, and whose ellipticity is infinitesimal (that is, the homogeneous fluid figure of equilibrium shaped like an ellipsoid of revolution flattened at its poles which attracts in accordance with the universal inverse-square law, which revolves around its axis of symmetry, whose ellipticity is infinitesimal, and which has the same value of ϕ as heterogeneous flattened figures of the type in question), depending upon how the ellipticities of the homogeneous solid nucleus and the outer surface of the homogeneous fluid layer were chosen relative to each other. Consequently, greater density at the center than at the surface of a figure shaped like a flattened ellipsoid of revolution which attracts according to the universal inverse-square law, which revolves around its axis of symmetry, and whose ellipticity is infinitesimal did not mean that the figure necessarily had to be flatter than the corresponding homogeneous fluid figure of equilibrium shaped like an ellipsoid of revolution flattened at its poles which attracts according to the universal inverse-square law, which revolves around its axis of symmetry, and whose ellipticity is infinitesimal (that is, the homogeneous figure of equilibrium shaped like an ellipsoid of revolution flattened at its poles which attracts in accordance with the universal inverse-square law, which revolves around its axis of symmetry, whose ellipticity is infinitesimal, and which has the same value of ϕ as heterogeneous flattened figures of the type in question), contrary to what Newton had suggested to be true in the case of the earth. At the same time, greater density at the center than at the surface of a figure shaped like a flattened ellipsoid of revolution which attracts according to the universal inverse-square law, which revolves around its axis of symmetry, and whose ellipticity is infinitesimal could make the figure be flatter than the corresponding homogeneous fluid figure of equilibrium shaped like an ellipsoid of revolution flattened at its poles which attracts according to the universal inverse-square law, which revolves around its axis of symmetry, and whose ellipticity is infinitesimal, provided the ellipticities of the solid homogeneous nucleus and the outer surface of the homogeneous fluid layer that surrounds the nucleus were suitably interrelated. But Jupiter is flatter than it would be if it were homogeneous. As a result Clairaut deduced that it was totally unnecessary for Newton to have introduced entirely different internal structures for the earth and for Jupiter.[102] Clairaut had already implied as much in his second paper on the theory of the earth's shape published in the *Philosophical Transactions*, but he only truly displayed for the first time the calculations which led him to conclude this in his treatise of 1743. (Moreover, we must not forget that the calculations that led Clairaut to come to such conclusions are only valid to terms of first order, whereas Jupiter's ellipticity *is finite, not infinitesimal.*)

Clairaut also showed that by assuming that all of the strata of the solid nucleus have the same density, in which case the nucleus is homogeneous, and that this density differs from the density of the homogeneous fluid layer that surrounds the nucleus and that fulfills the relevant conditions necessary for the fluid layer to exist in a state of equilibrium, and, moreover, that the strata of the nucleus all have the same suitably chosen negative ellipticity α, in which case the nucleus is shaped like an ellipsoid of revolution elongated at its poles whose ellipticity is infinitesimal, then the ellipticity δ of the outer surface of the homogeneous fluid layer that fulfills the relevant conditions necessary for the fluid layer to exist in a state of equilibrium could also be made negative . That is, in this case the outer surface of the homogeneous outer fluid layer is an ellipsoid of revolution elongated at its poles whose ellipticity is infinitesimal, and, moreover, the homogeneous outer fluid layer would fulfill the relevant conditions necessary for the fluid layer to exist in a state of equilibrium.[103]

By allowing the homogeneous outer fluid layer to become "infinitesimally thin," in which case $\alpha = \delta$ and $a = 1$, Clairaut arrived at the equation for the ellipticity δ of a solid stratified figure shaped like an ellipsoid of revolution which attracts according to the universal inverse-square law, which revolves around its axis of symmetry, whose ellipticity is infinitesimal, and at whose outer surface the principle of the plumb line holds, where the figure is composed of an infinite number of concentric, individually homogeneous ellipsoids of revolution whose ellipticities are infinitesimal and all of which have a common axis of symmetry, namely, the polar axis of the figure. Because the ellipticities $\rho(r)$ of the strata could be either positive or negative, the figure could be shaped like a flattened ellipsoid of revolution or an elongated ellipsoid of revolution. In other words, the ellipticity δ that satisfies the equation could be negative, which means in particular that the principle of the plumb line could hold at the surface of a solid stratified figure shaped like an elongated ellipsoid of revolution which attracts according to the universal inverse-square law, which revolves around its axis of symmetry, and whose ellipticity is infinitesimal.[104]

Clairaut showed that this equation included as a special case the equation for the ellipticity δ of the flattened stratified figure that he had treated in his second paper on the theory of the earth's shape published in the *Philosophical Transactions*, which we examined in Chapter 6. We remember that Clairaut had restricted the densities of the strata of this figure to vary continuously as

$$R(r) = fr^p + gr^q + \text{etc.} \tag{6.2.1}$$

and that he had specified the ellipticities $\rho(r)$ of strata of this figure all to be the same, namely, equal to δ, where δ was positive.[105] (We must not forget, however, that in his second paper on the theory of the earth's shape published in the *Philosophical Transactions*, Clairaut did *not* assume his stratified figure to be solid. He supposed the stratified figure to be a *fluid figure of equilibrium*. But the attraction at the outer surface of a stratified figure assumed to attract according to the universal inverse-square law is the same, whether the figure is assumed to be solid or fluid. It makes no difference for purposes of computing attraction what the state of matter of the figure is. The attraction only depends upon the shapes of the strata.)

Clairaut inferred from the equation for the general case that if the densities of the strata continually decrease from the center to the surface of a solid stratified figure shaped like an ellipsoid of revolution flattened at its poles which attracts according to the universal inverse-square law, which revolves around its axis of symmetry, whose ellipticity is infinitesimal, and at whose outer surface the principle of the plumb line holds, then the figure again will be less flat than the corresponding homogeneous fluid figure of equilibrium shaped like an ellipsoid of revolution flattened at its poles which attracts according to the universal inverse-square law, which revolves around its axis of symmetry, and whose ellipticity is infinitesimal (that is, the homogeneous fluid figure of equilibrium shaped like an ellipsoid of revolution flattened at its poles which attracts in accordance with the universal inverse-square law, which revolves around its axis of symmetry (its polar axis), whose ellipticity is infinitesimal, and which has the same value of ϕ as the solid stratified figure), unless the ellipticities of the strata also decrease from the center of the figure to its surface and, moreover, do so faster than $1/r^2$ decreases from the center of the figure to its surface.[106] Again, Clairaut had already alluded to such a result in his second paper on the theory of the earth's shape published in the *Philosophical Transactions*, but he had not demonstrated it there. As we shall see, Clairaut tried to utilize this particular result in his effort to determine the origins of the conflict between his findings and those of MacLaurin concerning the extent of the flattening of stratified figures shaped like flattened ellipsoids of revolution which attract according to the universal inverse-square law, which revolve around their axes of symmetry, and whose ellipticities are infinitesimal.

When the densities of the strata varied as

$$R(r) = fr^p + gr^q \qquad (6.2.2)$$

and the ellipticities of the strata were all assumed to be the same, Clairaut could reproduce the hypotheses in a number of the examples which appeared in MacLaurin's *A Treatise of Fluxions*. However, in each instance, Clairaut found, using his formula for the ellipticity δ, a value of the ellipticity that differed from the one determined by the author of *A Treatise of Fluxions*. In particular, when $\phi = \frac{1}{288}$, MacLaurin, unlike Clairaut, always arrived at a value for the ellipticity δ which was greater than $\frac{1}{230}$ instead of less than $\frac{1}{230}$.[107] Thus did Clairaut illustrate in his treatise of 1743 the problem that had originally induced him to look into the problem of stratified figures of equilibrium shaped like flattened ellipsoids of revolution which attract according to the universal inverse-square law, which revolve around their axes of symmetry, and whose ellipticities are infinitesimal more deeply and to investigate the principles that underlie the theory of figures of equilibrium.

Clairaut determined as well how the magnitude of the effective gravity varied from the equator to a pole along the surface of a solid stratified figure shaped like an ellipsoid of revolution which attracts according to the universal inverse-square law, which revolves around its axis of symmetry, whose ellipticity is infinitesimal, and at whose surface the principle of the plumb line holds. Here the figure in question is made up of strata that are concentric, individually homogeneous ellipsoids of revolution all of which have a common axis of symmetry, namely, the polar axis of the figure, whose ellipticities are infinitesimal and are not constant but vary in an arbitrary way instead, and whose nonnegative densities also vary from the center of the figure to its surface in an arbitrary way. He arrived at exactly the same results that he had found when he had stipulated conditions that were more restricted in his second paper on the theory of the earth's shape published in the *Philosophical Transactions*.

First he showed that if the ratio ϕ of the magnitude of the centrifugal force per unit of mass to the magnitude of the effective gravity at the equator of the solid stratified figure is infinitesimal (that is, $\phi^2 \ll \phi$), then the increase in the magnitude of the effective gravity with latitude from the equator to a pole along the surface of the figure varies directly as the square of the sine of the latitude. Only this time he proved that this result is true whether the principle of the plumb line holds at the surface of the figure or not. That is, he showed that this result held whatever the rate of rotation of the figure, provided ϕ is infinitesimal. Nor did the figure even have to be flattened; it could be elongated instead and the result will still hold.[108] (Clairaut really demonstrated this result holds true for the *components* of the effective gravity at points N at the surface of the figure which are directed toward the center C of the figure, not for the *total* or the *resultants of* the effective gravity at the points N. The result happens to turn out to hold true for the components of the effective gravity in question for essentially the same reasons that I explained in Chapter 6. However, when the ellipticities of the strata that make up the figure are all infinitesimal, the two magnitudes of effective gravity at a point N at the surface of the figure differ by terms of second order and higher orders at most. That is, the two magnitudes of effective gravity are equal to terms of first order.)

Second, Clairaut combined this result with the equation for the ellipticity δ mentioned above, an equation which was valid when the principle of the plumb line held at the surface of the solid, stratified figure, which, as I explained, did not have to be a flattened figure. The figure could be elongated; its ellipticity δ could be negative. Having put the result and the equation for δ together, Clairaut found that the equation

$$\frac{P - \Pi}{\Pi} = \tfrac{5}{2}\phi - \delta \qquad (9.4.1)$$

followed, where $P =$ the magnitude of the attraction at a pole of the stratified figure, $\Pi =$ the magnitude of the effective gravity at the equator of the stratified figure, $\delta =$ the ellipticity of the stratified figure, where δ can be positive (for flattened stratified figures) or negative (for elongated stratified figures), depending upon the given conditions, namely, the densities and the ellipticities of the solid strata, and $\phi =$ the ratio of the magnitude of the centrifugal force per unit of mass to the magnitude of the effective gravity at the equator of the stratified figure.

Now,

$$\epsilon = \tfrac{5}{4}\phi \tag{9.4.2}$$

is the infinitesimal ellipticity ϵ of the homogeneous fluid figure of equilibrium shaped like an ellipsoid of revolution flattened at its poles which attracts according to the universal inverse-square law, which revolves around its axis of symmetry, and whose infinitesimal ratio of the magnitude of the centrifugal force per unit of mass to the magnitude of the effective gravity at its equator is ϕ. Equation (9.4.2) is the equation that I mentioned in Chapter 6 which Clairaut could have written in his first paper on the theory of the earth's shape published in the *Philosophical Transactions* but did not. It is the same equation as Newton's equation (1.1) to terms of first order. [Equation (9.4.2) is also the same as equation (5.3.4) to terms of first order.]

Using equation (9.4.2), Clairaut replaced $\tfrac{5}{2}\phi$ by 2ϵ in equation (9.4.1), and he thereby arrived at the equation

$$\frac{P - \Pi}{\Pi} = 2\epsilon - \delta, \tag{9.4.1'}$$

where $\epsilon =$ the infinitesimal ellipticity of the homogeneous fluid figure of equilibrium shaped like an ellipsoid of revolution flattened at its poles which attracts in accordance with the universal inverse-square law, which revolves around its axis of symmetry, and whose infinitesimal value of ϕ is the same as the value of ϕ of the stratified figure. But equation (9.4.1) is the same as equation (6.2.4). In short, Clairaut discovered that the results that he arrived at in his second paper on the theory of the earth's shape published in the *Philosophical Transactions* which he found using the principle of the plumb line and which I discussed in Chapter 6 held true for a solid stratified figure shaped like an ellipsoid of revolution which attracts in accordance with the universal inverse-square law, which revolves around its axis of symmetry, whose ellipticity is infinitesimal, at whose surface the principle of the plumb line holds, and whose strata are individually homogeneous, concentric ellipsoids of revolution all of which have a common axis of symmetry, namely, the polar axis of the figure, and whose ellipticities are infinitesimal and can vary in an *arbitrary* manner and whose densities can also vary in an *arbitrary* way. It even turns out that the ellipticities of the strata do *not* have to be *positive*. That is, the same results hold for solid stratified figures shaped like ellipsoids of revolution which attract according to the universal inverse-square law, which revolve around their axes of symmetry, at whose surfaces the principle of the plumb line holds, and whose ellipticities are infinitesimal but negative, which means that the figures are *elongated* at their poles.[109] Thus Clairaut essentially found that equation (6.2.4) still held for a wider range of hypotheses concerning the ellipticities and internal structure of stratified figures shaped like ellipsoids of revolution which attract according to the universal inverse-square law, which revolve around their axes of symmetry, whose ellipticities are infinitesimal, and at whose surfaces the principle of the plumb line holds.

The search for an internal structure that would be consistent with Newton's conjectures about the degree of flattening of heterogeneous figures which attract in accordance with the universal inverse-square law had originally led Clairaut to investigate stratified figures shaped like ellipsoids of revolution flattened at their poles which attract according to the universal inverse-square law, which revolve around their axes of symmetry, and whose ellipticities are infinitesimal. Clairaut now explained in his treatise of 1743 that five years earlier, in 1738, at the time that he wrote his second paper on the theory of the earth's shape published in the *Philosophical Transactions*, he had not yet examined the second edition of Newton's *Principia* but only the more recent, third edition.[110] Having subsequently found that the results that he arrived at in that second paper could be extended to cover an even broader range of stratified figures shaped like flattened ellipsoids of revolution which attract in accordance with the universal inverse-square law, which revolve around their axes of symmetry, whose ellipticities are infinitesimal, and at whose surfaces the principle of the plumb line holds (namely, figures made up of concentric, individually homogeneous ellipsoids of revolution such that all of the ellipsoids which make up such a figure have a common axis of symmetry, namely, the polar axis of the figure, whose infinitesimal ellipticities need not be uniform, and whose nonnegative densities can vary in arbitrary ways), Clairaut looked at the second edition of the *Principia*. He said that he found evidence in that edition that Newton had made the same mistake that David Gregory made.[111] We recall that in his second paper on the theory of the earth's shape, Clairaut had mentioned Gregory's error, and now he stated that Gregory had very likely simply copied Newton's mistake.

Clearly with Jacques Cassini's work of 1718 on the earth's shape in mind, as well as possibly Mairan's Paris Academy *mémoire* of 1720 too, Clairaut applied his findings to stratified figures shaped like elongated ellipsoids of revolution which attract according to the universal inverse-square law, which revolve around their axes of symmetry, whose ellipticities are infinitesimal, and at whose surfaces the principle of the plumb line holds. Substituting the value $\frac{1}{230}$ for ϵ and Cassini's value $-\frac{1}{93}$ for δ into equation (9.4.1'), which meant tacitly assuming that the principle of the plumb line holds at the surface of Cassini's figure, a condition that Clairaut believed was observed to hold at the earth's surface, Kuhn's doubts, which I mentioned in the preceding section of this chapter, notwithstanding, Clairaut concluded that extraordinarily large errors would have had to be made in carrying out the experiments with seconds pendulums at different latitudes, in order for Cassini's value of δ to be correct, because the value of $(P - \Pi)/\Pi$ which had been deduced from experiments made with seconds pendulums at different latitudes differed appreciably from (specifically, was appreciably smaller than) the value of $(P - \Pi)/\Pi$ which equation (9.4.1') gave, when the values of ϵ and δ in it were specified to be those just mentioned.

(Mairan, of course, had based all of his arguments in his Paris Academy *mémoire* of 1720 upon hypotheses of attraction which differed completely from Clairaut's hypotheses of attraction. We also note that in order to apply Clairaut's reasoning to Cassini's figure shaped like an elongated ellipsoid of revolution which is assumed to attract according to the universal inverse-square law and whose ellipticity is infinitesimal, the figure *cannot* be assumed to be *homogeneous*. The figure must be assumed to be heterogeneous instead. In particular, in order to apply Clairaut's theory to the figure, the figure *must be* assumed to be *stratified* – that is, that its density varies from its center to its surface. This is true because if the principle of the plumb line holds at all points at the surface of a homogeneous figure shaped like an ellipsoid of revolution which attracts according to the universal inverse-square law, then all of the columns from the center of the figure to the surface of the figure must balance or weigh the same as well, because Clairaut's second new principle of equilibrium necessarily holds for such a figure. However, as I observed in note 40 of Chapter 2, the principle of the plumb line cannot hold at all points at the surface of a homogeneous figure shaped like an elongated ellipsoid of revolution which attracts according to the universal inverse-square law, in which case the columns from the center to the surface of a homogeneous figure shaped like an elongated ellipsoid of revolution which attracts according to the universal inverse-square law cannot all balance or weigh the same either. Instead such a homogeneous figure that satisfies those particular conditions necessary for the figure to exist in equilibrium which Clairaut knew, like the two conditions just mentioned, must be shaped like a flattened ellipsoid of revolution whose infinitesimal ellipticity ϵ is determined by the equation (9.4.2), $\epsilon = \frac{5}{4}\phi$, where ϕ is the ratio of the magnitude of the centrifugal force per unit of mass to the magnitude of the effective gravity at the equator of the figure. And when ϕ has the earth's value $\frac{1}{288}$, then $\epsilon = \frac{1}{230}$. In other words, a figure shaped like an elongated ellipsoid of revolution which attracts according to the universal inverse-square law, whose value of ϕ is $\frac{1}{288}$, and which satisfies equation (9.4.1), which means that the principle of the plumb line holds at its surface, cannot possibly be homogeneous.[112])

We recall that in his second paper on the theory of the earth's shape published in the *Philosophical Transactions*, Clairaut believed that he had satisfactorily treated the problem of a stratified fluid figure of equilibrium shaped like a flattened ellipsoid of revolution which attracts according to the universal inverse-square law, which revolves around its axis of symmetry, whose ellipticity is infinitesimal, and *all* of whose concentric, individually homogeneous strata are flattened ellipsoids of revolution whose ellipticities are infinitesimal and all of which have a common axis of symmetry, namely, the polar axis of the figure. In fact he had not done what he thought he had, and in a moment we shall understand why he had not.

In order to deal adequately with such stratified fluid figures of equilibrium, Clairaut had to confront new mathematical obstacles. In the preceding problem, the densities and ellipticities of the interior, solid strata were all given at the start. Their values are what in part fixed the ellipticity of the outer surface of the outer, homogeneous fluid layer, which could be determined by applying the principle of the plumb line at the outer surface of the outer, homogeneous fluid layer. In a figure that is entirely made up of fluid and that is in equilibrium, however, the ellipticities of the interior, homogeneous fluid strata are, like the ellipticity of the outer surface of the figure, themselves unknown at the beginning. In his second paper on the theory of the earth's shape published in the *Philosophical Transactions*, Clairaut had assigned the ellipticities of the strata of such fluid figures of equilibrium beforehand, namely, he had assumed them all to be equal. But he had since discovered that it was a mistake to have done this.

Clairaut utilized the following principle of equilibrium to formulate the problem mathematically. In the foregoing problem, the principle of the plumb line applied at the outer surface of the outer, homogeneous fluid layer determined the value of the ellipticity of that outer surface, which enabled that value to be found. In the problem in question, the infinitesimally thin strata are all surfaces exactly like the outer surface in the other problem. (See Figure 57.) Thus Clairaut stated: "By the principles established in Part I [of the *Théorie de la figure de la terre, tirée des principes de l'hydrostatique*].... the [fluid] spheroid cannot be in equilibrium unless all forces that act upon an arbitrary particle [*particule*] N in the interior of the spheroid having been combined into a single resultant force, this particle be pulled in a direction perpendicular at N to the layer FNG which has the same density as this particle."[113] In other words, the infinitesimally thin strata, or surfaces of constant density, are also the "level surfaces" of the fluid figure with respect to the effective gravity. (Recall, however, that I explained in Section 2 of this chapter that Clairaut in fact never actually demonstrated that this is the case or explained why this is the case, although it does in fact turn out be true, as I showed. He really only proved that the result holds true for the surfaces of strata whose thicknesses are finite.)

To handle the new problem, Clairaut once again turned to the kinds of approximate, geometric arguments that he had first employed in his second paper on the theory of the earth's shape published in the *Philosophical Transactions* to express the principle of the plumb line applied at the outer surface of a stratified figure shaped like a flattened ellipsoid of revolution which attracts in accordance with the universal inverse-square law, which revolves around its axis of symmetry, and whose ellipticity is infinitesimal. He now applied these arguments and this principle at *all* surfaces of constant density. Having originally conceived the

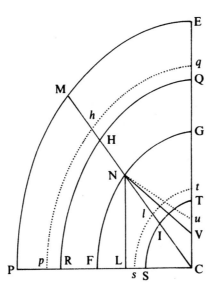

Figure 57. Clairaut's figure used to illustrate his assertion that the infinite number of infinitesimally thin strata, or surfaces of constant density, of a stratified fluid figure that attracts according to the universal inverse-square law are also the "level surfaces" of the fluid figure with respect to the effective gravity (Clairaut 1743: figure on p. 266).

problem in terms of surfaces of constant density which are flattened ellipsoids of revolution whose ellipticities are infinitesimal, where the densities and the ellipticities of the ellipsoids of revolution both normally varied continuously as functions of distance from the center of the fluid figure to its surface along its polar radius, Clairaut was led to express the problem by means of ordinary "differential" equations. (Hence, in effect, Clairaut usually assumed more than simply that the densities $R(r)$ and ellipticities $\rho(r)$ of the strata varied continuously with r. Namely, he assumed that the differential coefficients dR/dr, $d\rho/dr$, and $d^2\rho/dr^2$ exist.)

As we saw in Sections 1 and 2 of this chapter, however, Clairaut's theory of stratified fluid figures of equilibrium could only really be applied to figures that consist of strata whose thicknesses were all *finite*, in which case the densities $R(r)$ of the figures vary discontinuously as step functions $R(r)$ of r. Indeed, as I indicated in Sections 1 and 2 of this chapter, this is essentially what enabled Clairaut to reduce this theory to his theory of homogeneous fluid figures of equilibrium. In particular, the surfaces of strata whose thicknesses are finite are level surfaces of the effective gravity. From Clairaut's theory it does *not* follow in general, however, that a surface of constant density within a heterogeneous fluid figure in equilibrium whose density varies continuously must be a level surface of the effective gravity. On the contrary, surfaces of constant density will *not* ordinarily be level surfaces of the effective gravity in this case. As I mentioned in Section 2 of this chapter, in 1752, D'Alembert showed that Clairaut's theory could not be general. D'Alembert proved that surfaces of constant density within a heterogeneous fluid figure of equilibrium whose density varies continuously need not in general be level surfaces of the effective gravity. D'Alembert observed that the true principle of equilibrium for such figures was that $\int_\gamma P\mathbf{F}\cdot d\mathbf{s}$ be equal to zero around all closed plane or skew channels γ that lie inside a heterogeneous fluid figure of equilibrium and that are filled with fluid from the heterogeneous fluid figure of equilibrium, where P designates the density and $\mathbf{F}$ stands for the effective gravity. In other words, as we saw in Section 2 of this chapter, this means that the total "effort" produced by the effective gravity around all closed plane or skew channels that lie within the interiors of such figures and that are filled with fluid from such figures must be equal to zero. Such a principle follows naturally, for example, from the principle of soldification applied to heterogeneous fluid figures of equilibrium. Of course, the closed channels that lie within the interiors of such figures will generally be filled with *heterogeneous* fluid, *not homogeneous* fluid, since the figures that the channels lie inside are heterogeneous not homogeneous. This will certainly be true, for example, if the density of a heterogeneous fluid figure of equilibrium varies *continuously* from the center of the figure to its surface.

As I mentioned in Section 2 of this chapter, however, it turns out that surfaces of constant density within heterogeneous fluid figures of equilibrium *will* be level surfaces of effective gravity if $\int_\gamma \mathbf{F}\cdot d\mathbf{s}$ is equal to zero around all closed plane or skew channels γ that lie inside the heterogeneous fluid figure of equilibrium and that are filled with fluid from the heterogeneous fluid figure of equilibrium. But, as we saw in Section 2 of this chapter, it is also true that $\int_\gamma \mathbf{F}\cdot d\mathbf{s}$ *is* always equal to zero around closed plane or skew channels γ that lie inside a heterogeneous fluid figure of equilibrium that attracts according to the universal inverse-square law of attraction or any other universal law of attraction and that are filled with fluid from the heterogeneous fluid figure of equilibrium, although Clairaut himself in fact never actually demonstrated that this is true or explained why this is true.[114] In any case Clairaut's theory *can* consequently be applied to the problem at issue.

In determining the ellipticity of a stratified fluid figure of equilibrium shaped like a flattened ellipsoid of revolution which attracts according to the universal inverse-square law, which revolves around its axis of symmetry, and whose ellipticity is infinitesimal, Clairaut assumed that the strata were all concentric, individually homogeneous ellipsoids of revolution whose ellipticities were infinitesimal and all of which had a common axis of symmetry, namely, the polar axis of the figure. He then derived equations that were consistent with this hypothesis and with the principles of equilibrium that underlay his theory of stratified fluid figures of equilibrium. Applying the principle of equilibrium quoted above to points N within the interior of the stratified fluid figure (see Figure 57) and, as usual, disregarding terms of second and higher orders in $\rho(r)$ and ϕ (since $\rho(r)^2 \ll \rho(r)$ and $\phi^2 \ll (\phi)$, Clairaut arrived at the equation

$$5r^2\rho \int_{r=0}^{r=r} Rrrdr = \int_{r=0}^{r=r} Rd(r^5\rho) + Fr^5 + \tfrac{5}{2}A\phi r^5 - r^5 \int_{r=0}^{r=r} Rd\rho, \quad (9.4.3)$$

where P is a pole of the stratified fluid figure of equilibrium shaped like a flattened ellipsoid of revolution which attracts according to the universal inverse-square law, which revolves around its axis of symmetry, and whose ellipticity is infinitesimal; PME is the part of a meridian at the surface of the figure which lies in one quadrant; the length of the polar radius $CP = 1$ (normalization); the length of $CF = r$ along the polar radius; $\rho = \rho(r)$ is the ellipticity of the level surface, hence the ellipticity of the surface whose density is constant, through N, which intercepts the polar radius CP a distance $CF = r$ from the center C of the figure; $R(r) =$ the density of the level surface through N;

$$F \equiv \int_{r=0}^{r=1} Rd\rho; \quad A \equiv \int_{r=0}^{r=1} Rrrdr; \quad (9.4.4)$$

and $\phi =$ the ratio of the magnitude of the centrifugal force per unit of mass to the magnitude of the effectivity gravity at the equator E of the figure.[115] When conditions were such that the equation (9.4.3) had solutions, stratified figures shaped like flattened ellipsoids of revolution which attracted in accordance with the universal inverse-square law, which revolved around their axes of symmetry, whose ellipticities were infinitesimal, and which satisfied these conditions were figures which fulfilled the relevant conditions necessary for the figures to exist in a state of equilibrium. Equation (9.4.3) is an integro-differential equation. Clairaut's theory of stratified fluid figures of equilibrium led to this kind of equation when the figures attract according to the universal inverse-square law because in this particular case the resultant attraction that acts upon the fluid figure, causing it to take on the shape that it does, itself depends upon that shape.

The calculations that Clairaut did in applying the principle of the plumb line at a point inside the fluid figure closely resemble those that he had done in applying the principle of the plumb line at points at the surfaces of figures with the following difference. A particle in the interior of the fluid figure is also attracted by those layers within whose cavities the particle lies. We recall that according to Proposition 91, Corollary 3 of Book I of the *Principia*, the attraction due to a homogeneous layer of matter which attracts according to the universal inverse-square law and which is bounded by two concentric, geometrically similar ellipsoids of revolution, hence two ellipsoids of revolution whose ellipticities are the *same*, both of which have a common axis of symmetry is *zero* at points within the cavity that the layer surrounds. However, the attraction due to a homogeneous layer of matter which attracts in accordance with the universal inverse-square law is *not* zero within the cavity that the layer surrounds, if the layer is bounded by two concentric ellipsoids of revolution which have a common axis of symmetry but whose ellipticities *differ* (in other words, the two concentric ellipsoids of revolution are *not* geometrically similar).

If r is chosen equal to 1, in which case the point N lies at the outer surface of the fluid figure and not inside the fluid figure, the integro-differential equation (9.4.3) reduces to the equation that Clairaut had derived for the ellipticity $\rho(1) = \delta$ of a solid stratified figure shaped like a flattened ellipsoid of revolution which attracts according to the universal inverse-square law, which revolves around its axis of symmetry, whose ellipticity is infinitesimal, and at whose outer surface the principle of the plumb line holds. But the reason why this occurs is obvious. If a stratified fluid figure of equilibrium shaped like a flattened ellipsoid of revolution which attracts in accordance with the universal inverse-square law, which revolves around its axis of symmetry, and whose ellipticity is infinitesimal is solidified, a solid stratified figure shaped like a flattened ellipsoid of revolution which attracts according to the universal inverse-square law, which revolves around its axis of symmetry, whose ellipticity is infinitesimal, and at whose surface the principle of the plumb line holds results. (Again, the result in question in this case does not depend upon the states of the strata; only the geometry of the strata matters.)

As we shall soon see, Clairaut spent a great deal more time trying to solve the integro-differential equation (9.4.3) than he did deriving the equation. Taking differentials of the integro-differential equation (9.4.3) twice, the second time "in supposing dr constant,"[116] Clairaut arrived at the equation[117]

$$\frac{d^2\rho}{dr^2} = \left(\frac{6\rho}{rr} - \frac{2\rho Rr}{\int_{r=0}^{r=r} Rrrdr}\right) - \frac{2Rrr}{\int_{r=0}^{r=r} Rrrdr}\frac{d\rho}{dr}. \tag{9.4.5}$$

When the densities $R = R(r)$ of the strata (that is, the surfaces of constant density) are specified beforehand, equation (9.4.5) is a second-order, linear ordinary "differential" equation with coefficients that are not constant. In this case the solutions of equation (9.4.5) are the ellipticities $\rho = \rho(r)$ of the strata, where $\rho(1) =$ the ellipticity δ of the figure.[118] Using the changes of variables

$$\rho \equiv c^{\int udr}\left(\text{that is, } \frac{d\rho}{du} \equiv u\rho\right), \quad \text{and} \quad t \equiv u + \left(\frac{Rrr}{\int_{r=0}^{r=r} Rrrdr}\right), \tag{9.4.6}$$

where c is Euler's notation for e, Clairaut converted the second-order ordinary "differential" equation (9.4.5) into the first-order ordinary "differential" equation[119]

$$\frac{dt}{dr} + tt = \frac{rr\frac{dR}{dr}}{\int_{r=0}^{r=r} Rrrdr} + \frac{6}{rr}. \tag{9.4.7}$$

It is possible, although not absolutely certain, that Clairaut learned of the first of

the two transformations in (9.4.6.) from his good friend Daniel Bernoulli. At least
years later Clairaut mentioned in a letter to Bernoulli that Bernoulli had known
the transformation before he had and that, as far as he knew, Bernoulli had been
the first to construct "an infinity of equations" of the kind produced by employing
this transformation.[120]

Clairaut's first-order ordinary "differential" equation (9.4.7) is an example of
"the famous equation"[121]

$$dy + y^2 \, dx = X(x) \, dx. \tag{9.4.8}$$

This is a Riccati equation, named after the Italian count who first investigated the
first-order ordinary "differential" equation

$$\frac{dy}{dx} + ay^2 = bx^n, \tag{9.4.9}$$

where a and b are constants, while trying to reduce second-order ordinary "differential" equations to first-order ordinary "differential" equations.[122] Riccati
put forward his Riccati equation (9.4.9) and his particular use of it in Volume 8 of
the supplements to *Acta Eruditorum*, published in 1724. Using the first of the two
transformations in (9.4.6.), Bernoulli succeeded in constructing solutions of "an
infinite number" of Riccati equations (9.4.9) for those values of n for which a
solution can be expressed in finite terms. This is what Clairaut meant by "an
infinity of equations" in his reference to Bernoulli mentioned in the preceding
paragraph. Bernoulli published these solutions in *Acta Eruditorum* for 1725. But
as far as Clairaut knew, no one "had yet been able to separate the variables in
general" in equation (9.4.8),[123] so that Riccati equations generally could not be
integrated in finite terms or reduced to quadratures by separating variables. In
particular, this was true of equation (9.4.7). Nor could Clairaut apply to equation
(9.4.7) his or Fontaine's methods of 1738–41 for integrating in finite terms or
reducing to quadratures first-order ordinary "differential" equations of first
degree any more successfully.[124] As I mentioned in Chapter 7, only in 1841 was it
shown by Joseph Liouville that the Riccati equation (9.4.9) cannot in general
even be reduced to quadratures, much less be integrated in finite terms.

Clairaut first confined himself to solving equation (9.4.7) when the densities of
the strata varied as $R(r) = r^n$. In these cases the change of variables $w = 1/r$
transforms the Riccati equation (9.4.7) into an equation that is homogeneous in
its variables. Consequently the variables in this equation can be separated, and
the equation can be integrated as a result. Integrating this equation and then
inverting the changes of variables $w = 1/r$ and (9.4.6) in order to return to the
original variables, Clairaut found that the expression

$$\rho(r) = \delta a r^{-n-\frac{5}{2}-q} + \delta r^{-n-\frac{5}{2}+q}, \tag{9.4.10}$$

where $q \equiv \sqrt{nn + 3n + \frac{25}{4}}$, satisfies the second-order ordinary "differential" equation (9.4.5) in these instances.[125] In order for the density to decrease from the

center to the surface of the stratified fluid figure of equilibrium shaped like a flattened ellipsoid of revolution which attracts according to the universal inverse-square law, which revolves around its axis of symmetry, and whose ellipticity is infinitesimal, which Clairaut maintained must be the case if a stratified fluid figure is even to exist in a state of equilibrium, n must be negative in $R(r) = r^n$. (In fact, in order for the integral A in (9.4.4) even to exist, it must be true that $0 < m < 3$, if $n = -m$.[126])

Here Clairaut encountered a difficulty that took him some time to resolve. The solution (9.4.10) of the second-order ordinary "differential" equation (9.4.5) included two arbitrary constants a and δ, which resulted from integrating the equation (9.4.5) twice. Now, all solutions of the integro-differential equation (9.4.3) also satisfy the second-order ordinary "differential" equation (9.4.5), but the converse is not true. That is, not all solutions of (9.4.5) are solutions of (9.4.3), even though the second-order ordinary "differential" equation (9.4.5) is arrived at by differentiating the integro-differential equation (9.4.3) twice. In expressions (9.4.10) that satisfy both the second-order ordinary "differential" equation (9.4.5) and the integro-differential equation (9.4.3), the constants of integration a and δ must fulfill relations determined by the integro-differential equation (9.4.3) and the equation arrived at by differentiating the integro-differential equation (9.4.3) once. The constants of integration a and δ can be replaced by a pair of initial conditions on $\rho(r)$, say,

$$\rho(1) \quad \text{and} \quad \left.\frac{d\rho}{dr}\right|_{r=1}, \tag{9.4.11}$$

for such initial conditions (9.4.11) can be expressed in terms of the two constants of integration a and δ and vice versa. Ordinarily, the integro-differential equation (9.4.3) examined at $r = 1$ should determine an equation in the two initial conditions (9.4.11), which can be converted into an equation in a and δ. Then, when the integro-differential equation (9.4.3) is differentiated once, the equation that results when examined at $r = 1$ should give rise to a second, independent equation in the two initial conditions (9.4.11), which can be transformed accordingly into a second, independent equation in a and δ. But two independent, consistent equations in two initial conditions (9.4.11), or, alternatively, in two constants of integration a and δ instead, should fix the two initial conditions (9.4.11), or, alternatively, the two constants of integration a and δ, uniquely. Clairaut found that when the integro-differential equation (9.4.3) is differentiated once, the resulting equation examined at $r = 1$ determined an equation in the two initial conditions (9.4.11) and, consequently, an equation in the two constants of integration a and δ.[127] However, as Clairaut discovered, the integro-differential equation (9.4.3) examined at $r = 1$ did *not* give rise to a second, independent equation in the two initial conditions (9.4.11), or, alternatively, a second, independent equation in the two constants of integration a and δ, needed to fix the two initial conditions (9.4.11), or, alternatively, the two constants of integration a and

δ, uniquely. It sufficed to choose one of the two constants of integration a or δ arbitrarily, then pick the other constant of integration so that the single equation relating the two constants of integration a and δ is fulfilled. When this is done, Clairaut said, any solution (9.4.10) of the second-order ordinary "differential" equation (9.4.5) which also happens to satisfy the equation obtained by differentiating the integro-differential equation (9.4.3) once cannot fail to satisfy (9.4.3) too.[128] Clairaut inferred that instead of being specified uniquely, the ellipticity $\rho(1) = \delta$ of the figure in question could be chosen arbitrarily! In short, the conditions of Clairaut's theory of stratified figures of equilibrium did not appear to determine uniquely the ellipticity of a stratified figure of equilibrium shaped like a flattened ellipsoid of revolution which attracts in accordance with the universal inverse-square law, which revolves around its axis of symmetry, and whose ellipticity is infinitesimal. Yet it seemed obvious to Clairaut that specifying the ratio ϕ of the magnitude of the centrifugal force per unit of mass to the magnitude of the effective gravity at the equator of a stratified figure of equilibrium shaped like a flattened ellipsoid of revolution which attracts according to the universal inverse-square law, which revolves around its axis of symmetry, and whose ellipticity is infinitesimal ought to be enough to fix the shape of the figure uniquely and, in particular, fix the ellipticity of the figure.[129]

Eventually Clairaut's oversight occurred to him. Namely, he finally realized or remembered that the mathematics only represented the mechanics when ϕ and $\rho(r)$, $0 \leqslant r \leqslant 1$, were infinitesimal. The whole physical argument leading to the integro-differential equation (9.4.3) only holds true when ϕ and $\rho(r)$, $0 \leqslant r \leqslant 1$, are infinitesimal. All solutions $\rho(r)$ of the integro-differential equation (9.4.3) which do not remain infinitesimal are physically extraneous solutions of the equation. In order for $\rho(r)$ in expression (9.4.10) to stay infinitesimal when ϕ is infinitesimal and n is negative, Clairaut contended that the first term on the right-hand side of expression (9.4.10) had to vanish. This statement is true for the following reasons. The exponent appearing in the first term on the right-hand side of expression (9.4.10) is always a negative quantity. As a result, if the first term on the right-hand side of expression (9.4.10) did not vanish, it would approach infinity as r approaches zero. At the same time the exponent appearing in the second term on the right-hand side of expression (9.4.10) is always a positive quantity. Consequently that term approaches zero as r approaches zero. Therefore if the first term on the right-hand side of expression (9.4.10) did not vanish, $\rho(r)$ [that is, expression (9.4.10)] would increase without bound as r approaches zero, instead of remaining infinitesimal. Hence the first term on the right-hand side of expression (9.4.10) had to be made to disappear. But choosing the constant δ to be equal to zero would make both terms on the right-hand side of expression (9.4.10) be equal to zero, not just the first term. Consequently, the constant a that only appears in the first term on the right-hand side of expression (9.4.10) should be equal to zero. In other words, the mechanics did not permit the constant a to be chosen arbitrarily. Substituting zero for a in the equation in a and δ then uniquely fixed the value of δ. Replacing δ by this value in the two terms on the right-hand side of expression (9.4.10) and replacing a by zero in the first term on the right-hand side of expression (9.4.10), Clairaut arrived at the expression

$$\rho(r) = \frac{5\phi}{2q - 1 - 2n} r^{q-n-(5/2)}, \qquad (9.4.10')$$

as the solution of the integro-differential equation (9.4.3),

where[130]

$$q \equiv \sqrt{nn + 3n + \frac{25}{4}}.$$

The converse problem of solving the integro-differential equation (9.4.3) for the densities $R(r)$ of strata, when the ellipticities $\rho(r)$ of the strata are specified at the start, turned out to be a good deal easier to handle. In this case the second-order ordinary "differential" equation (9.4.5) could be readily solved for $R(r)$ using logarithmic integration.[131] Solving this problem, however, was more an academic exercise than anything else. It did not answer any questions that had real interest to Clairaut.

As far as Clairaut was concerned the principal problem remained the one where the value of $\rho(1) =$ the ellipticity δ of the figure is not known beforehand but must be found instead. But, as we have seen, the integro-differential equation (9.4.3) could not in general be solved for $\rho(r)$ when the density varied as an arbitrary differentiable function $R(r)$ of r along the polar radius from the center to the surface of a stratified figure of equilibrium shaped like a flattened ellipsoid of revolution which attracts in accordance with the universal inverse-square law, which revolves around its axis of symmetry, and whose ellipticity is infinitesimal. The inextricable difficulties that the Riccati equation (9.4.7) presented stood in the way of solving the second-order ordinary "differential" equation (9.4.5), which ultimately kept the integro-differential equation (9.4.3) from being solved.

Clairaut did succeed, however, in dealing with the problem when the number of strata was *finite*, in which case the thickness of each stratum was also finite and the density varied as a step function $R(r)$ of r along the polar radius from the center to the surface of the stratified figure of equilibrium shaped like a flattened ellipsoid of revolution which attracts according to the universal inverse-square law, which revolves around its axis of symmetry, and whose ellipticity is infinitesimal. Concerning stratified figures of equilibrium which attract according to the universal inverse-square law, we recall from the first and second sections of this chapter that the surfaces of constant density are level surfaces of the effective gravity when the density varies continuously and that the surfaces between contiguous strata are also level surfaces of the effective gravity when the thicknesses of the strata are finite. Redoing for the second case the calculations that had led him to the integro-differential equation (9.4.3), Clairaut arrived this time at a finite system of equations in a finite number of unknowns instead of equation (9.4.3). The unknowns were the ellipticities of the flattened ellipsoids of revolution which bound the strata all of whose thicknesses were finite, whose number was finite, and where, as we saw in Sections 1 and 2 of this chapter, the surfaces of the strata were all level surfaces of the effective gravity. Hence in this case the problem reduced to solving a finite system of equations in a finite number of unknowns.[132]

After analyzing this system of equations, Clairaut came to what was probably the most important conclusion of his whole investigation of stratified figures of equilibrium shaped like flattened ellipsoids of revolution which attract in accordance with the universal inverse-square law, which revolve around their axes of symmetry, and whose ellipticities are infinitesimal. It was the result that he ultimately attempted to use to account for the conflict between his findings and MacLaurin's. Assuming that "the densities of the layers always decrease from center to surface," which "the laws of hydrostatics required," Clairaut stated that it turned out that the "ellipticities of the surfaces that bound the layers must increase from the center to the surface" of a stratified figure of equilibrium shaped like a flattened ellipsoid of revolution which attracts in accordance with the universal inverse-square law, which revolves around its axis of symmetry, whose ellipticity is infinitesimal, and whose strata all have finite thicknesses.[133] Clairaut concluded that he had erred in his second paper on the theory of the earth's shape published in the *Philosophical Transactions* when he assumed that the stratified figure that he treated in that paper, which was shaped like a flattened ellipsoid of revolution which attracted according to the universal inverse-square law, which revolved around its axis of symmetry, whose ellipticity was infinitesimal, and which was composed of concentric, flattened ellipsoids of revolution all of which had a common axis of symmetry, namely, the polar axis of the figure, and whose infinitesimal ellipticities were all equal to the ellipticity of the figure, was a figure of equilibrium.[134] (On the other hand, we shall see in a moment why such a conclusion does not truly follow from the result quoted.) Clairaut also realized, however, that the calculations that he did in his second paper on the theory of the earth's shape published in the *Philosophical Transactions* which only involved his application of the principle of the plumb line at the surfaces of stratified figures and the results that followed from these calculations were perfectly valid provided the stratified figures that he treated in that paper were assumed to be solid, not fluid.[135]

Clairaut tried to apply this result to the stratified figures shaped like ellipsoids of revolution flattened at their poles which attracted in accordance with the . universal inverse-square law, which revolved around their axes of symmetry, and whose ellipticities were infinitesimal which appeared in MacLaurin's *A Treatise of Fluxions*. In fact, Clairaut did not truly establish an analogous result when densities of strata vary as arbitrary differentiable functions $R(r)$ of r, because he confronted obstacles in handling the Riccati equation (9.4.7) which he could not surmount. He declared that in the examples that MacLaurin gave in which density varied as

$$R(r) = fr^p + gr^q, \tag{6.2.2}$$

where the values of f, g, p, and q were specified and in which the infinitesimal ellipticities $\rho(r)$ of the strata were all assumed to be equal, the preceding result explained the significantly different values of the ellipticities of the figures in question which he and MacLaurin found. Namely, Clairaut asserted that MacLaurin had erred in treating his stratified figures shaped like flattened ellipsoids of revolution which attracted in accordance with the universal inverse-square law, which revolved around their axes of symmetry, and whose ellipticities were infinitesimal as fluid figures of equilibrium. As we shall see in the next section of this chapter, MacLaurin did this implicitly in balancing columns from the centers of the figures to their surfaces which lie along the polar and equatorial axes of the figures. Clairaut maintained that these figures were not figures of equilibrium, because the ellipticities of the strata did not increase from the centers of the figures to their surfaces in these particular examples. Such a conclusion, however, does not truly follow from Clairaut's arguments. As I say, Clairaut could not really establish such a result when the densities of strata varied as differentiable functions $R(r)$ of r in general. But in the first two of three examples from *A Treatise of Fluxions* which Clairaut analyzed:

$$(1) \ R(r) = fr \quad \text{and} \quad (2) \ R(r) = fr^3, \tag{9.4.12}$$

the densities of the strata *increase* as differentiable functions $R(r)$ of r from the centers to the surfaces of the figures. Although Clairaut failed to recognize the following fact, or if not at least he neglected to point this fact out, he could have ruled out the two examples in (9.4.12) without even having to consider the ellipticities of the strata, because the densities of the strata in the figures in these two examples do not fulfill a condition that Clairaut thought followed from the laws of hydrostatics, namely, that the density must *decrease* from the center to the surface of a stratified figure of equilibrium. In the third example that Clairaut examined in MacLaurin's treatise:

$$(3) \ R(r) = \frac{n}{n-1} - r, \tag{9.4.13}$$

$R(r)$ is greater than zero on the closed interval $[0, 1]$, because when r is an element of $[0, 1]$, then $0 \leqslant r \leqslant 1 < n/(n-1)$. Moreover, $R(r)$ decreases on $[0, 1]$, since r increases on $[0, 1]$. But Clairaut could not solve the Riccati equation (9.4.7) that corresponds to this differentiable variation $R(r)$ of r in density. Consequently Clairaut did not truly have the means to rule out this example on the grounds that he had maintained.[136] In the one case of a differentiable function $R(r)$ of r which decreases on $[0, 1]$ where he could integrate the Riccati equation (9.4.7), namely, when $R(r) = r^n$, $n < 0$, the solution (9.4.10') of the integro-differential equation (9.4.3) *does increase* on the interval $[0, 1]$,[137] but that does not prove that the solution $\rho(r)$ of (9.4.3) which corresponds to an *arbitrary* differentiable function $R(r)$ of r which decreases on $[0, 1]$ *always* increases on $[0, 1]$.

(It is possible that Clairaut simply assumed that the following was true. As long as the number of equations derived for a stratified figure shaped like a flattened ellipsoid of revolution which attracts according to the universal inverse-square law, which revolves around its axis of symmetry, whose ellipticity is infinitesimal, and whose strata all have finite thicknesses remained finite, the conclusion that he drew for such a finite system of equations held true, regardless of how many equations there were in the system and how many unknowns the system of equations included. Since the system of equations and the number of unknowns stayed finite, no matter how thin the strata might become, where dr, say, stands for the length of the widest part of the thickest stratum, and no matter how large the number N of strata might become, the conclusion arrived at for such a system continued to hold good. Clairaut may have just assumed that the same result also proved true in the limit, as dr approached zero and N approached infinity.

Be that as it may, we have already studied another aspect of Clairaut's theory of figures of equilibrium which shows that this type of reasoning is not sound in general when applied to his theory. Namely, we have seen that Clairaut's theory of heterogeneous fluid figuers of equilibrium always holds for fluid figures composed of a finite number of individually homogeneous fluid strata whose thicknesses are all finite, no matter how thin each stratum might be. But Clairaut's theory of heterogeneous fluid figures of equilibrium no longer holds in general when the individually homogeneous fluid strata are "infinitesimally thin" (in other words, when the strata are surfaces themselves). In short, the theory breaks down in the limit, as dr approaches zero and N approaches infinity. Specifically, we recall from the discussion in Section 2 of this chapter that as long as the thicknesses of the individually homogeneous fluid strata that make up a stratified fluid figure of equilibrium are finite, no matter how thin the strata are, the two conditions:

$$\int_\gamma P\mathbf{F}\cdot d\mathbf{s} = 0 \tag{1}$$

around all closed plane or skew channels that lie within the interior of the stratified fluid figure of equilibrium and that are filled with fluid from the figure; and (2) the surfaces of the individually homogeneous fluid strata are all level surfaces of the effective gravity are *interrelated*. However, when the thicknesses of the individually homogeneous fluid strata that make up a stratified fluid figure of equilibrium approach zero, in which case the individually homogeneous fluid strata are surfaces themselves, condition (1) must hold once again, but the strata, which are surfaces of constant density, are *not* in general level surfaces of the effective gravity in this case. In other words, condition (2) does not hold in general in this case.)

Clairaut tried to account for the conflict between his findings and MacLaurin's in another kind of case as well. Each time that MacLaurin assumed that the earth was filled from its center to its surface with matter that attracts according to the universal inverse-square law, he found that when the ellipticity was greater than $\frac{1}{230}$, the relative increase in the magnitude of the effective gravity from equator to pole turned out to be less than $\frac{1}{230}$,[138] contrary to Newton's conjectures. In order to "save Newton" and the observations of the day, MacLaurin consequently resorted to the model that he had described to Clairaut by letter, whose reasonableness and plausibility Clairaut had questioned in his reply to MacLaurin in September of 1741. We recall that MacLaurin imagined a homogeneous, spherical, fluid nucleus in equilibrium which was surrounded by a hollow shell. The shell was surrounded by a homogeneous outer layer of fluid in equilibrium whose density was equal to the density of the fluid nucleus and whose inner and outer surfaces were ellipsoids of revolution flattened at the poles which were concentric with the fluid nucleus, which had a common axis of symmetry (the polar axis of the figure), and whose infinitesimal ellipticities were the same. Thus the concentric ellipsoids of revolution were geometrically similar. The fluid nucleus and outer fluid layer were both assumed to attract in accordance with the universal inverse-square law and to revolve at the same rate around the axis of symmetry of the figure. Clairaut could only conceive of two possibilities: (1) either the hollow part was truly empty space, in which case Clairaut was certain that the nucleus and the outer layer would attract each other and unite to form a homogeneous fluid figure of equilibrium shaped like a flattened ellipsoid of revolution which attracts according to the universal inverse-square law, which revolves around its axis of symmetry, and whose ellipticity is $\frac{1}{230}$; or else (2) if the hollow cavity is assumed to be filled with a dense fluid that does not attract, whose purpose is to keep the nucleus and the outer layer from coming together, such a layer was, according to Clairaut's theories, simply a layer whose density is equal to zero, because in Clairaut's theories density is a measure that only quantifies matter that attracts.[139] Neither Clairaut's theory of the shapes of solid figures which attract according to the universal inverse-square law and at whose outer surfaces the principle of the plumb line holds nor his theory of the shapes of fluid figures of equilibrium which attract in accordance with the universal inverse-square law included anything that presupposed that *all* individually homogeneous strata in a stratified figure shaped like a flattened ellipsoid of revolution which attracts according to the universal inverse-square law, which revolves around its axis of symmetry, and whose ellipticity is infinitesimal had to have densities that were positive. Some strata could have densities that were equal to zero. Therefore Clairaut applied his theory of stratified figures of equilibrium shaped like flattened ellipsoids of revolution which attract in accordance with the universal inverse-square law, which revolve around their axes of symmetry, whose ellipticities are infinitesimal, and whose individually homogeneous strata all have finite thicknesses to MacLaurin's figure (the spherical nucleus is simply a homogeneous figure shaped like an ellipsoid of revolution whose ellipticity is zero). All that truly mattered was the geometry of the figure. From Clairaut's standpoint MacLaurin's model could not possibly exist in a state of equilibrium, because the ellipticities of the flattened ellipsoids of revolution which bound the individually homogeneous strata in figures of equilibrium, when the strata all have finite thicknesses, must increase from the center of the figure to its surface, in order for the strata, each of which is a layer whose density is constant, to be level layers of effective gravity. But the inner and outer surfaces of the outer layer of MacLaurin's model were geometrically similar flattened ellipsoids of revolution. They both had the same ellipticity. In other words, the ellipticities of the surfaces that bounded the individually homogeneous strata in MacLaurin's figure did not increase from the center of the figure to its surface. Even in this particular case, however, Clairaut's conclusion did not truly follow from the result that he had been able to demonstrate, because the densities of the layers in MacLaurin's model do not decrease from the center of the figure to its surface. Once again Clairaut could have reasoned that this figure could not possibly exist in a state of equilibrium simply because its density does not decrease from its center to its surface. The outer layer has the same density as the spherical nucleus, while the density of the layer that lies between these two parts of the figure is zero. However, Clairaut did not rule out the figure on these grounds. Perhaps he did not notice that the figure did not satisfy what he had considered to be a condition that followed from the laws of hydrostatics. In any case, he failed to mention this fact. In the final analysis, Clairaut only had one real piece of evidence to support his belief that MacLaurin's stratified figures which attracted according to the universal inverse-square law were not figures of equilibrium. Moreover, this evidence only applied to MacLaurin's stratified figures whose densities did not decrease from the centers of the figures to their surfaces. Namely, such figures, but only such figures among all those that MacLaurin discussed, did not satisfy a condition that Clairaut thought that the laws of hydrostatics required that all stratified figures of equilibrium must fulfill: that the densities of the individually homogeneous strata within stratified figures of equilibrium must decrease from the centers of the figures to their surfaces. (And as we shall see in the next section of in this chapter, even *that* turns out not to be real evidence that MacLaurin's stratified figures could not be figures of equilibrium.)

Likewise, Clairaut did not truly establish in his treatise of 1743 that the stratified figures that he dealt with in his second paper on the theory of the earth's shape published in the *Philosophical Transactions*, whose densities vary as:

$$R(r) = fr^p + gr^q, \tag{6.2.2}$$

were not figures of equilibrium, because the densities of the individually homogeneous strata that made up the figures discussed in that paper varied continuously from the centers of the figures to their surfaces. Thus the numbers of the individually homogeneous strata that made up the figures treated in that paper were not finite, and the individually homogeneous strata that made up the figures discussed in that paper were not strata whose thicknesses were finite. Again, however, according to Clairaut such figures cannot be figures of equilibrium if the densities of the individually homogeneous strata do not decrease from the centers of the figures to their surfaces.

Given the geometry of the strata in MacLaurin's figures, Clairaut believed that MacLaurin should have treated the last model in question as a solid figure at whose outer surface the principle of the plumb line holds. But in that event Clairaut's findings that concerned such figures bore upon MacLaurin's model, because the geometry of that model coincided with the geometry of the figures that appeared in Clairaut's theory of the shapes of solid figures.[140] Hence equation (9.4.1') could be applied to MacLaurin's model, and this, Clairaut thought, disproved MacLaurin's assertion that $(P - \Pi)/\Pi > \frac{1}{230}$ and $\delta > \frac{1}{230}$ both held at the same time for this model, when $\epsilon = \frac{1}{230}$.[141]

In 1752, D'Alembert claimed that MacLaurin had first introduced the notion of level surfaces,[142] and we can now judge the significance of such an assertion. D'Alembert stated his opinion during a period when he and Clairaut were arch rivals and showed hostility to each other,[143] and this fact probably should be taken into account. In any case, even if what D'Alembert said were true, its importance would still remain questionable. For MacLaurin never truly made use of the idea of level surfaces of effective gravity in his *A Treatise of Fluxions*. Indeed, the fact that he did not could have been what caused him to make mistakes. MacLaurin no more utilized level surfaces of effective gravity in *A Treatise of Fluxions* than did Mairan employ level surfaces of attraction in his Paris Academy *mémoire* of 1720. Mairan, if we remember, had already introduced the concept of level surfaces of attraction from another standpoint in the section on the problem of "orthogonal trajectories" appearing in that *mémoire*, but he could not really put the idea to work there. Level surfaces of effective gravity served no important purpose in MacLaurin's *A Treatise of Fluxions*, whereas they formed part of the foundations of Clairaut's theory of heterogeneous figures of equilibrium published in 1743.

9.5. Creation and application of the theory of figures of equilibrium: Problems of chronology

It is not an easy matter to order and date Clairaut's efforts to solve the various problems that he had to confront in determining stratified figures of equilibrium which attract according to the universal inverse-square law; nor is it an easy matter to determine the total amount of time that it took him to establish all of the results that he thought that he needed in order to publish. As early as January of 1742, Clairaut appears, at least at first sight, to have already arrived at everything that he eventually used to try to explain why he and MacLaurin arrived at conflicting results. And yet it took Clairaut almost another whole year to reach the point where he felt ready to make his work public. Here I try to account for the delay, using hints revealed in Clairaut's correspondence.

In a letter dated 4 January 1742, Clairaut told Euler:

By means of my theory I recognized that I made a mistake in [my second paper on the theory of the earth's shape published in] the *Philosophical Transactions* when I determined the earth's shape, supposing that the earth [is a fluid figure of equilibrium that is shaped like a flattened ellipsoid of revolution which attracts according to the universal inverse-square law, which revolves around its axis of symmetry, whose ellipticity is infinitesimal, and which] is made up of geometrically similar [concentric, individually homogeneous] layers that are [flattened] ellipsoids [of revolution], where the densities [of two contiguous layers] differ. Bouguer's [Paris Academy] *mémoire* [of 1734] led me astray, because I contented myself with seeing to it that the columns [from the center to the surface of the figure] were in equilibrium and that effective gravity was perpendicular to the surface [of the figure]. Having examined the problem more carefully, I found that the layers cannot be geometrically similar but must be ellipsoids whose flattenings [that is, whose positive ellipticities] increase from the center to the surface [of the figure], because the denser parts lie nearer to the center [of the figure].[144]

Now, if Clairaut had truly made such a discovery and had demonstrated it by 4 January 1742, he could have only proven it for the *finite* system of equations which expressed the problem in cases where the numbers of strata are *finite* and each stratum has *finite* thickness. (In the passage just quoted, Clairaut described the strata ambiguously. First he termed them layers (*couches*), but then he called them flattened ellipsoids, which are surfaces.) In Clairaut's second paper on the theory of the earth's shape published in the *Philosophical Transactions*, however, the densities varied as *differentiable* functions

$$R(r) = fr^p + gr^q \qquad (6.2.2)$$

of *r*, in which case a generally unsolvable integro-differential equation (9.4.3) represented the problem of the equilibrium of stratified fluid figures that are shaped like flattened ellipsoids of revolution which attract according to the universal inverse-square law, which revolve around their axes of symmetry, whose ellipticities are infinitesimal, and whose densities vary this way, not a finite system of equations in a finite number of unknowns. Clairaut admitted in the same letter to Euler that he had failed to solve a first-order ordinary "differential" equation which he showed Euler in the letter and which he said an integro-differential equation that also appears in the letter reduced to.[145] Here Clairaut referred to the mathematical representation of the problem of stratified fluid figures of equilibrium which are shaped like flattened ellipsoids of revolution which attract according to the universal inverse-square law, which revolve around their axes of symmetry, and whose ellipticities are infinitesimal in cases where the densities of the figures vary as differentiable functions *R(r) of r* from the centers of the figures to their surfaces along their polar radii. In fact, the integro-differential equation that Clairaut communicated to Euler in January of 1742 differs from the one examined above which appears in Clairaut's treatise of 1743 [integro-differential equation (9.4.3)]. Nor is the corresponding first-order ordinary "differential" equation reported in the same letter the Riccati equation (9.4.8) that appears in the treatise of 1743.[146] Whether the equations transmitted to Euler in January of 1742 include typographical errors or evidence instead mistakes that Clairaut had made at that time makes little difference. Clairaut did not make progress in solving the problem of finding stratified fluid figures of equilibrium which are shaped like flattened ellipsoids of revolution which attract according to the universal inverse-square law, which revolve around their axes of symmetry, and whose ellipticities are infinitesimal in cases where the densities vary as differentiable functions $R(r)$ of r any further by late July of 1742. In a letter to Euler dated 25 July 1742, in which the correct integro-differential equation (9.4.3) appears for the first time, Clairaut asked Euler whether he could do anything with this equation.[147] By the same token, the remarks in Clairaut's two letters which concern results established and difficulties that still frustrated Clairaut in his attempt to deal with equation (9.4.3) collectively point to ambiguities in Clairaut's letter of January 1742 regarding what Clairaut had actually shown to be true at that time and what he had not. I shall return to this particular problem in a moment.

Then in a letter to Euler dated 14 September 1742, Clairaut stated that his integro-differential equation (9.4.3) could be reduced, by means of differentiation and changes of variables, to a Riccati equation. Here he reported for the first time the two changes of variables (9.4.6), as well as the correct first-order ordinary "differential" equation which results when these changes of variables are applied in conjunction with differentiation to the integro-differential equation (9.4.3), namely, the Riccati equation (9.4.7). He also acknowledged that he had still not found a way out of his impasse. He doubted that a method for constructing Riccati equations by means of tractrices, which Euler had evidently told Clairaut about earlier, would serve his purposes, but he asked Euler to show him the method anyway.[148]

But if Clairaut had at least demonstrated by 4 January 1742 the result mentioned in the passage quoted, which holds for stratified fluid figures of equilibrium which are shaped like flattened ellipsoids of revolution which attract according to the universal inverse-square law, which revolve around their axes of symmetry, whose ellipticities are infinitesimal, and which consist of finite numbers of concentric, individually homogeneous strata, where each stratum has finite thickness, then why did it take Clairaut so long to give his work to the publisher? What deterred him from having it printed immediately? Did the problems with the Riccati equation (9.4.7) cause him to postpone publication? In fact, Clairaut *never* overcame the obstacles that he faced in trying to deal with this equation, but this did not keep him from ultimately publishing his work. Instead, Clairaut must not yet have entirely convinced himself that he had truly determined what caused him to arrive at results that did not accord with MacLaurin's. He would have to take more time to try to understand why his and MacLaurin's findings did not agree with each other, so that he could firmly resolve the problem of their conflicting results once and for all. How could this be? Well, for one thing, as of January 1742, *A Treatise of Fluxions*, which included some of the ideas that MacLaurin had discussed with Clairaut in correspondence, had not yet been published. Unitl Clairaut saw the printed work, he could not be completely sure just what methods MacLaurin had employed and what conclusions he had ultimately come to.

Around October of 1742, Clairaut received a copy of the finally published *A Treatise of Fluxions* from MacLaurin.[149] It seems likely that Clairaut only learned how MacLaurin actually did his calculations after reading it. Clairaut had already guessed from reading MacLaurin's letters that the Scottish mathematician worked with what he thought were stratified fluid figures of equilibrium. But as *A Treatise of Fluxions* makes clear, MacLaurin did not apply the principle of the plumb line to what he thought were such figures. Instead MacLaurin balanced, meaning that he equated the weights of, columns from the centers to the surfaces along the polar and equatorial axes of his stratified figures shaped like flattened ellipsoids of revolution which attract according to the universal inverse-square law, which revolve around their axes of symmetry, and whose ellipticities are infinitesimal. (Here is where he made use of his theorem for calculating the magnitudes of the attraction at points along the axes of stratified figures shaped like flattened ellipsoids of revolution which attract in accordance with the universal inverse-square law and which are made up of strata that are confocal ellipsoids of revolution, hence strata whose ellipticities differ.) But Clairaut probably only found this out for certain after receiving his copy of MacLaurin's treatise. At least, Clairaut only referred specifically for the first time to MacLaurin's equating the weights of columns after the copy of the treatise arrived. This Clairaut did in his own treatise, published the next year, in 1743, in connection with MacLaurin's examples, which I mentioned before, of stratified figures shaped like flattened ellipsoids of revolution which attract according to the universal inverse-square law, which revolve around their axes of symmetry, whose strata all have the same infinitesimal ellipticity, and whose densities vary continuously as

$$R(r) = fr^p + gr^q \qquad (6.2.2)$$

for particular choices of f, g, p, and q,[150] as well as MacLaurin's model composed of a homogeneous, spherical, fluid nucleus surrounded by a hollow shell which is itself surrounded by a homogeneous fluid layer whose thickness is finite, whose inner and outer surfaces are concentric, flattened, geometrically similar ellipsoids of revolution whose infinitesimal ellipticities are consequently the same, which have a common axis of symmetry, namely, the polar axis of the figure, and whose

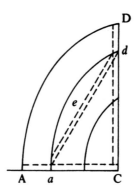

Figure 58. Clairaut's figure used to show that the homogeneous outer fluid layer of one of MacLaurin's figures that attracts according to the universal inverse-square law cannot be a figure of equilibrium (Clairaut 1742f: figure on p. 380).

density is the same as the density of the spherical nucleus, where the figure again attracts in accordance with the universal inverse-square law and revolves around its axis of symmetry.[151]

In the last of these figures (Figure 58), MacLaurin equated the weights of columns *Aa* and *Dd*, which lie in the plane of a merdian at the surface of the figure of revolution and which traverse the outer layer of the figure, in order to calculate the ellipticity of the figure. I already mentioned what seemed to Clairaut to be the defects of this model, which he pointed out in his letter to MacLaurin of September 1741 and which he repeated in his treatise of 1743. MacLaurin had implicitly treated the figure as a fluid figure of equilibrium, and Clairaut believed that his theory of stratified fluid figures of equilibrium which are shaped like flattened ellipsoids of revolution which attract according to the universal inverse-square law, which revolve around their axes of symmetry, and whose ellipticities are infinitesimal revealed that the figure could not possibly be a figure of equilibrium.

In addition, in a letter to MacLaurin dated 19 November 1742, in which Clairaut acknowledged having received the copy of *A Treatise of Fluxions*, Clairaut explained in simple terms why he thought that MacLaurin's equating the weights of columns *Aa* and *Dd* could not lead to the true value of the ellipticity of this figure of revolution when MacLaurin assumed, as he had in his treatise, that the figure attracts according to the universal inverse-square law, that it revolves around its axis of symmetry, and that the inner and outer surfaces of the outer layer are concentric, flattened, geometrically similar ellipsoids of revolution whose inifinitesimal ellipticities are consequently the same and which have a common axis of symmetry, namely, the polar axis of the figure. Namely, Clairaut noted that if the columns *Aa* and *Dd*, which lie in the plane of a meridian at the surface of the figure of revolution, were joined by a rectilinear channel *aed* which lies in the same plane of a meridian as the columns *Aa* and *Dd* and which is filled with homogeneous fluid whose density is the same as the density of the fluid whose parts make up the outer layer of the figure, then the homogeneous fluid that fills the plane channel *AaedD* that lies in the plane of the meridian in question and joins two points at the surface of the figure would be in equilibrium. In other words, the total effort along the plane channel *AaedD* produced by the effective gravity would be equal to zero. (This, we recall, is true, provided the principle of the plumb line holds at the surface of the figure, because the effective gravity due to a rotating figure that attracts according to the universal inverse-square law always produces a total effort equal to zero around closed plane or skew channels that lie within the interior of the figure and that are filled with homogeneous fluid, which is to say that Clairaut's second new principle of equilibrium holds for such a figure. In other words, Clairaut's second new principle of equilibrium holds for a stratified fluid figure that attracts according to the universal inverse-square law. But in treating his figure as a fluid figure of equilibrium, MacLaurin tacitly assumed that the principle of the plumb line does hold at its outer surface. That the total effort along the plane channel *AaedD* produced by the effective gravity would be equal to zero also follows from Clairaut's first new principle of equilibrium applied to the stratified fluid figure assumed to be in equilibrium.) The inner surface of the outer layer of MacLaurin's figure includes the curved plane channel *ad* that lies in the plane of the meridian in question. The outer surface of the outer layer of MacLaurin's figure includes the curved plane channel *AD* that lies in the plane of the meridian in question. In MacLaurin's figure the two surfaces are geometrically similar; hence their ellipticities are the same.

Clairaut then replaced the inner surface of the outer layer of the figure by a level surface of the effective gravity which included points *a* and *d*, hence a surface whose ellipticity was less than the ellipticity of the outer surface, not equal to it. (Strictly speaking, however, Clairaut did not truly demonstrate that the ellipticity of the level surface of the effective gravity which points *a* and *d* are situated on must be less than the ellipticity of the outer surface of the outer layer of the figure, for the reasons that I mentioned before. Namely, the densities of the strata in this figure do not decrease from the center of the figure to its surface. To put the matter another way, Clairaut did not really prove that the surface which MacLaurin chose to be the inner surface of the outer layer, whose ellipticity was the same as the ellipiticity of the outer surface of the outer layer, could not be a level surface of the effective gravity. Clairaut did not truly establish that this must be the case because the densities of the strata in MacLaurin's figure do not decrease from the center of the figure to its surface, so that Clairaut could not really apply the chief result that followed from his theory of stratified fluid figures of equilibrium which are shaped like flattened ellipsoids of revolution which attract according to the universal inverse-square law, which revolve around their axes of symmetry, and whose ellipticities are infinitesimal to MacLaurin's figure.) Clairaut then considered a plane channel *aed* that lies on this level surface of the effective gravity in the plane of the merdian in question, which was also filled with homogeneous fluid whose density was the same as the density of the fluid whose parts made up the outer layer of the figure. Once again, the homogeneous fluid that fills the plane channel *AaedD* that lies in the plane of the meridian in question and connects two points at the surface of the figure would be in equilibrium. In other words, the total effort along plane channel *AaedD* produced by the effective gravity is also zero. But the effort produced by the effective gravity along the plane channel *aed* that lies on the level surface of the effective gravity must be equal to zero, because the effective gravity at all points on a level surface of the effective gravity is perpendicular to that surface at those points. In particular, the effective gravity at all points along the plane channel *aed* must be perpendicular to the plane channel *aed* at all of its points, from which it follows that the effort along the plane channel *aed* produced by the effective gravity must be equal to zero. Hence the efforts along columns *Aa* and *Dd* produced by the effective gravity must be the same. That is, columns *Aa* and *Dd* must balance or weigh the same in this altered figure.

In that case, however, columns *Aa* and *Dd* could not possibly balance or weigh the same in MacLaurin's figure, because a plane channel *aed* could also be chosen to lie in the plane of the meridian in question on the surface that MacLaurin chose to be the inner surface of the outer layer of the figure. But this inner surface is not a level surface of the effective gravity (although again, strictly speaking, there really is no basis for drawing such a conclusion from Clairaut's theory, because the densities of the layers in MacLaurin's figure do not decrease from the center of the figure to its surface), so that the effort produced by the effective gravity along the plane channel *aed* that lies on this inner surface does not equal zero, and, at the same time, the effort produced by the effective gravity along the plane channel *AaedD* in the plane of the meridian in question, which again joins two points at the surface of the figure, would be equal to zero, because the homogeneous fluid that fills this channel would, once again, be in equilibrium. Consequently the efforts along columns *Aa* and *Dd* produced by the effective gravity could not be the same in MacLaurin's figure. In other words, columns *Aa* and *Dd* could not balance or have the same weight in MacLaurin's figure.[152]

In his treatise of 1743, Clairaut mentioned that MacLaurin had not been completely unaware that his model had problems. Thus at one point in *A Treatise of Fluxions*, MacLaurin acknowledged, to his credit in Clairaut's opinion, that equating the weights of columns *Aa* and *Dd* had no real foundation. Consequently, Clairaut emphasized, MacLaurin did not himself insist that equating the weights of columns *Aa* and *Dd* was a valid procedure but granted instead that his approach could appear to be based upon what Clairaut termed an "arbitrary supposition."[153]

Clairaut may have hesitated to publish his treatise too soon for other reasons as well. We must not forget that in developing his theory of stratified fluid figures of equilibrium shaped like flattened ellipsoids of revolution which attract according to the universal inverse-square law, which revolve around their axes of symmetry, and whose ellipticities are infinitesimal in order to derive his integro-differential equation (9.4.3), Clairaut relied on approximations, as he had done in deriving equations that appear in his two papers on the theory of the earth's shape published in the *Philosophical Transactions*. Moreover, not only did he frequently make use of approximations, but, as I indicated earlier, these approximations often concealed the orders to which they held true, particularly when Clairaut employed geometry instead of analysis. Furthermore, even in those parts of his theory where it is clear, as a result of his having used infinite series, that terms of second order and higher orders have been discarded, Clairaut did not keep track of the errors that accumulated in making such approximations. For these two reasons Clairaut could not logically rule out the possibility that his total error was of the same order as the infinitesimal ellipticity δ that he sought. If these orders were equal, that would render Clairaut's whole theory meaningless. Thus Clairaut thought that the problem of the order of the total error too could account in part for the fact that his findings and MacLaurin's did not agree with each other. For MacLaurin demonstrated his results rigorously and exactly and made no use of approximations. Hence his results should have held true exactly, not just approximately, provided his theory involved no errors.

Upon receiving his copy of *A Treatise of Fluxions*, Clairaut told its author that he found some of the theorems that the work included to be useful for dealing with the problem of the errors that accumulated in making his approximations. In his letter to MacLaurin dated 19 November 1742, Clairaut said that he profited from some of his Scottish colleague's propositions. They enabled Clairaut to determine the degree of accuracy of his own method.[154] Clairaut must have alluded here to the fact that in his treatise of 1743 he used MacLaurin's geometric, rigorously derived theorems to determine the ellipticities of homogeneous fluid figures of equilibrium shaped like flattened ellipsoids of revolution which attract according to the universal inverse-square law, which revolve around their axes of symmetry, and whose ellipticities are infinitesimal.

As I mentioned in Chapter 6, Clairaut did not actually determine in his first paper on the theory of the earth's shape published in the *Philosophical Transactions* the ellipticity of a homogeneous fluid figure shaped like a flattened ellipsoid of revolution which attracts in accordance with the universal inverse-square law, which revolves around its axis of symmetry, whose ellipticity is infinitesimal, and whose columns from center to surface all balance or weigh the same, which is a figure of equilibrium according to the theory of homogeneous fluid figures of equilibrium in Clairaut's treatise of 1743, although this ellipticity can be found by combining the calculations that do appear in Clairaut's first paper on the theory of the earth's shape published in the *Philosophical Transactions*. I also noted in Chapter 6 that Clairaut's equation for the ellipticities of stratified figures shaped like flattened ellipsoids of revolution which attract according to the universal inverse-square law, which revolve around their axes of symmetry, whose ellipticites are infinitesimal, and at whose surfaces the principle of the plumb line holds, which appears in his second paper on the theory of the earth's shape published in the *Philosophical Transactions*, reduces in the special case where all strata are assumed to have the same density to the equation that he did not write but could have written in his first paper on the theory of the earth's shape published in the *Philosophical Transactions* for the ellipticity of a homogeneous fluid figure shaped like a flattened ellipsoid of revolution which attracts according to the universal inverse-square law, which revolves around its axis of symmetry, whose ellipticity is infinitesimal, and whose columns from center to surface all balance or weigh the same. That is, the same homogeneous figures shaped like flattened ellipsoids of revolution which attract in accordance with the universal inverse-square law, which revolve around their axes of symmetry, and whose ellipticities are infinitesimal satisfy both the principles of the plumb line and balanced columns simultaneously. Now we understand why this is true. Homogeneous figures shaped like flattened ellipsoids of revolution which attract according to the universal inverse-square law, which revolve around their axes of symmetry, whose ellipticities are infinitesimal, and at whose surfaces the principle of the plumb holds are, once again, figures that satisfy all of the conditions then known which were necessary for homogeneous fluid figures of equilibrium to exist, according to the theory of homogeneous fluid figures of equilibrium in Clairaut's treatise of 1743. (That is, the effective gravity due to a rotating figure that attracts according to the universal inverse-square law always produces a total effort equal to zero around closed plane or skew channels that lie within the interior of the rotating figure and that are filled with homogeneous fluid, which is to say that Clairaut's second new principle of equilibrium holds for such a rotating figure that attracts according to the universal inverse-square law, in which case when such a rotating figure that attracts in accordance with the universal inverse-square law is homogeneous, each of the principles of the plumb line and balanced columns entails the other.)

In using MacLaurin's geometric, rigorously derived theorems to determine the ellipticities of homogeneous fluid figures of equilibrium shaped like flattened ellipsoids of revolution which attract according to the universal inverse-square law, which revolve around their axes of symmetry, and whose ellipticities are infinitesimal, Clairaut mentioned in his treatise of 1743 that he followed the exposition that appears in *A Treatise of Fluxions*.[155] In fact, as I indicated in Chapter 8, MacLaurin had already demonstrated these same results for homogeneous figures shaped like flattened ellipsoids of revolution which attract according to the universal inverse-square law, which revolve around their axes of symmetry, and whose ellipticities are finite for the first time in his essay on the tides which earned him a share of the Paris Academy's prize of 1740. Clairaut, however, may not have felt that he needed to make a close study of the problem of errors until he finally received *A Treatise of Fluxions*, which he found to include numerous results that contradicted his own. He reproduced MacLaurin's geometric results for homogeneous figures shaped like flattened ellipsoids of revolution which attract according to the universal inverse-square law, which revolve around their axes of symmetry, and whose ellipticities are finite and then converted them into equivalent statements in continental analysis. MacLaurin's exactly stated theorems furnished Clairaut with a means of estimating the total error that he made in doing the approximate calculation in his first paper on the theory of the earth's shape published in the *Philosophical Transactions*. Using MacLaurin's theorems, Clairaut showed that in having neglected terms of second order and higher orders in the theory of a homogeneous fluid figure shaped like a flattened ellipsoid of revolution which attracts in accordance with the universal inverse-square law, which revolves around its axis of symmetry, whose ellipticity is infinitesimal, and whose columns from center to surface all balance or weigh the same, which appeared in his first paper on the theory of the earth's shape published in the *Philosophical Transactions*, he had not made appreciable errors. Thus he confirmed that those earlier calculations were adequate.

We recall from Chapter 8 that MacLaurin had found that equality (8.2.1) results when the principle of the plumb line holds at the surface of a homogeneous figure shaped like a flattened ellipsoid of revolution which attracts according to the universal inverse-square law, which revolves around its axis of symmetry, and whose ellipticity is finite. We also remember from Chapter 8 that MacLaurin had also discovered that when equality (8.2.1) holds for this same figure, all of the columns from the center of the figure to its surface balance or weigh the same as well. But if one of the principles of the plumb line or balanced columns holds for a homogeneous figure shaped like a flattened ellipsoid of revolution which attracts according to the universal inverse-square law, which revolves around its axis of symmetry, and whose ellipticity is finite, it follows from Clairaut's theory of homogeneous fluid figures of equilibrium appearing in Clairaut's treatise of 1743 that the other principle holds necessarily as a result, which explains MacLaurin's findings.

Using MacLaurin's geometric way of determining the exact values of the magnitudes of the attractions at the pole and at the equator of such a figure, translating the results into analysis, and then making use of equality (8.2.1), Clairaut expressed the ellipticity δ of the figure as an infinite series consisting of terms of all orders in the ratio ϕ of the magnitude of the centrifugal force per unit of mass to the magnitude of the effective gravity at the equator of the figure[156]:

$$\delta = \tfrac{5}{4}\phi + \tfrac{5}{224}\phi^2 - \tfrac{135}{6272}\phi^3 + \dots \qquad (9.5.0)$$

He maintained that from the expansion of δ in an infinite series in ϕ it could be easily seen that the first term $\tfrac{5}{4}\phi$ of the infinite series did indeed approximate δ closely enough in the case of the earth.[157] In other words, the equation

$$\delta = \tfrac{5}{4}\phi \qquad (9.5.1)$$

for the ellipticity δ held true to terms of first order when the ellipticity δ and ratio

ϕ were infinitesimal (which means that $\delta^2 \ll \delta$ and $\phi^2 \ll \phi$). Equation (9.5.1) is the same as equations (5.3.4) and (9.4.2). It is precisely the equation for the infinitesimal ellipticity δ which Clairaut could have written in his first paper on the theory of the earth's shape published in the *Philosophical Transactions* but did not. (In the two examples that Newton had talked about, the earth and Jupiter, the infinite series that Clairaut derived for the ellipticity of a planet that is assumed to be a homogeneous figure of equilibrium shaped like a flattened ellipsoid of revolution which attracts according to the universal inverse-square law and which revolves around its axis of symmetry converges in both cases. However, the infinite series need not converge for flatter bodies. Specifically, the ellipticity δ must be less than $\sqrt{2} - 1$ in order for the series to converge.[158] Moreover, in the case of Jupiter, the series cannot be truncated after the first term, because δ and ϕ are *not* infinitesimal in the case of Jupiter, which means that $\delta^2 \ll \delta$ and $\phi^2 \ll \phi$ are *not* true in the case of Jupiter.) But Clairaut did not discuss the convergence of infinite series. His failure to do so was a sign of the times. By the same token, Clairaut did not submit the errors that he made in discarding terms to a truly rigorous analysis. Nevertheless, his use of infinite series was a step in the direction of such an analysis.

In his treatise of 1743 Clairaut also derived the equation

$$\delta = \tfrac{5}{4} \phi \tag{9.5.1}$$

another way. That is, for the first time he also arrived at this equation by using calculations that involved approximations like the ones that he had done in his two papers on the theory of the earth's shape published in the *Philosophical Transactions*. We recall that in his treatise of 1743 Clairaut calculated the infinitesimal ellipticity δ of a figure composed of two parts (see Figure 56). He assumed the figure to be shaped like an ellipsoid of revolution. One of the poles of the figure was situated at P, and the figure revolved around its polar axis, which was its axis of symmetry. The surface of the figure was the outer surface of a homogeneous fluid layer in equilibrium whose density was 1. This fluid layer surrounded a solid body shaped like an ellipsoid of revolution which was made up of an infinite number of concentric, individually homogeneous ellipsoids of revolution whose ellipticities were infinitesimal and all of which had a common axis of symmetry, namely, the axis of symmetry of the figure. Hence the strata of the solid part of the figure were in effect surfaces whose infinitesimal ellipticities $\rho(r)$ were given at the start. Here C designates the center of the figure; $\alpha \equiv$ the ellipticity of the outer surface of the solid body shaped like an ellipsoid of revolution; $\delta \equiv$ the ellipticity of the ellipsoid of revolution which is the outer surface of the homogeneous fluid layer in equilibrium which surrounds the solid body and whose density was 1; $CP = 1$ (normalization); $a \equiv CB$ is the polar radius of the solid body (hence $a < 1$); $R(r)$ and $\rho(r)$ are, respectively, the density and the ellipticity of the solid stratum whose polar radius equals r, where r is measured along CB [in which case $\rho(a) = \alpha$]; and ϕ is the ratio of the magnitude of the centrifugal force per unit of mass to the magnitude of the effective gravity at the equator E of the outermost surface, the surface of the figure. By assuming that all of the strata that make up the solid part had the same ellipticity α and had the same density as the outer fluid layer (namely, 1), in which case the entire figure is homogeneous, and letting $\alpha = \delta$, Clairaut showed that the equation for δ reduced to equation (9.5.1).[159]

While deriving the "$\sin^2$ law" of the variation of the increase in the magnitude of the effective gravity with latitude along the surface of a homogeneous fluid figure of equilibrium shaped like a flattened ellipsoid of revolution which attracts according to the universal inverse-square law, which revolves around its axis of symmetry, and whose ellipticity is infinitesimal, Clairaut employed MacLaurin's rigorously precise, geometric arguments to prove that the magnitude of the resultant of the effective gravity at points at the surface of a homogeneous fluid figure of equilibrium shaped like a flattened ellipsoid of revolution in equilibrium which attracts according to the universal inverse-square law, which revolves around its axis of symmetry, and whose ellipticity is finite varies directly as the length of the rectilinear segment that joins a point at the surface to the equatorial axis and that is perpendicular to the surface at that point of the surface. (MacLaurin in fact had already demonstrated this result; Clairaut simply followed him here.[160] Moreover, Stirling had also alluded to such a result in his paper on the earth's shape published in the *Philosophical Transactions* for 1735. It was another result that did not appear as part of Newton's theory of the earth's shape.[161])

Clairaut, however, could not make use of MacLaurin's geometrical theorems that appeared in *A Treatise of Fluxions* which pertained to stratified figures shaped like ellipsoids of revolution which attract according to the universal inverse-square law and whose ellipticities are finite to estimate his errors in making approximations in his theories of the shapes of stratified figures. As I indicated earlier, MacLaurin had already mentioned some of his theorems that bore upon stratified figures shaped like ellipsoids of revolution which attract according to the universal inverse-square law and whose ellipticities are finite in his correspondence with Clairaut.

The main theorem is the following one: Consider two homogeneous bodies that have the same densities and whose outer surfaces are confocal ellipsoids of revolution which have the same axis of symmetry. Since the two ellipsoids of revolution are confocal, they must be concentric. Moreover, because the ellipsoids of revolution are confocal, their axes that are not axes of symmetry must lie in the same plane, namely, the plane that is perpendicular to the common axis of symmetry of the two ellipsoids of revolution and that includes the common center and the two common foci of the two ellipsoids of revolution. In fact, these two axes can be chosen to lie along the same line, namely, the line that includes the two common foci of the two ellipsoids of revolution. The curves formed where planes that include the two common foci of the two ellipsoids of revolution intersect one of the two ellipsoids of revolution will be ellipses. But the ellipses formed where different planes that include the two common foci of the two ellipsoids of revolution intersect this ellipsoid of revolution will not be the same. The ellipses will differ from one another, because, as I just pointed out, the line that includes the common foci of the two ellipsoids of revolution is not an axis of symmetry of the two ellipsoids of revolution. That is, if the ellipses formed where planes that include the two common foci of the two ellipsoids of revolution intersect one of the two ellipsoids were identical, then these ellipses could all be obtained by revolving any one of them around the line that includes the two common foci of the two ellipsoids of revolution. But this contradicts the fact that this line is not an axis of symmetry of the two ellipsoids of revolution. Where the plane that includes *both* the two common foci of the two ellipsoids of revolution *and* the common axis of symmetry of the two ellipsoids intersects one of the ellipsoids of revolution, the curve formed will just be an ellipse that generates that ellipsoid by revolving this ellipse around the common axis of symmetry of the two ellipsoids. Where the plane that includes the two common foci of the two ellipsoids of revolution and which is perpendicular to the common axis of symmetry of the two ellipsoids intersects one of the ellipsoids, the curve formed will be a *circle*, because of what it means to be an axis of symmetry. A circle is an ellipse whose ellipticity is zero. It is a limiting case of an ellipse. Its two "foci" coincide, at the center of the circle.

Then MacLaurin's principal theorem can be stated as follows: at any point on either of the two common axes of the two confocal ellipsoids of revolution, where in this case the ellipticities of the two confocal ellipsoids of revolution can be finite, the magnitudes of the attractions at that point produced by the two homogeneous figures that have the same densities and whose outer surfaces are these confocal ellipsoids of revolution, where the two figures are assumed to attract in accordance with the universal inverse-square law, will have the same ratio as the volumes of the two homogeneous figures whose outer surfaces are these two confocal ellipsoids of revolution, provided the point lies in the exteriors of both figures.[162] From this theorem it follows immediately as a corollary that if a homogeneous figure shaped like a flattened ellipsoid of revolution attracts according to the universal inverse-square law, then the magnitude of the attraction due to the figure at a point along its polar axis or along the line that includes its two foci, where the point is assumed to lie in the exterior of the figure in either case, will equal the magnitude of the attraction at the same point produced by a homogeneous figure upon whose surface the point lies and which also attracts in accordance with the universal inverse-square law, whose mass is the same as the mass of the other figure, and which is shaped like a flattened ellipsoid of revolution which has the same two foci as the other figure, where the two figures have a common axis of symmetry. The ellipticities of the two flattened ellipsoids of revolution can be finite.

Since the flattened ellipse formed where *any* plane that includes the polar axis of a flattened ellipsoid of revolution which attracts according to the universal inverse-square law, which is the axis of symmetry of the ellipsoid of revolution in this particular case, intersects that flattened ellipsoid of revolution can be thought of as the ellipse that generates the flattened ellipsoid of revolution by revolving the ellipse around its polar axis, and the two foci of this particular ellipse can also be regarded as the foci of the flattened ellipsoid of revolution because of the symmetry, it is clear that the preceding result and its corollary can be used to determine the magnitude of the attraction at any point that is both situated in the plane of the equator of the homogeneous figure shaped like a flattened ellipsoid of revolution and in the exterior of the figure, where the figure attracts in accordance with the universal inverse-square law and its ellipticity can be finite. Moreover, it is easy to see how the same result and its corollary could also be used to calculate the magnitudes of the attraction at points along the polar axis of a stratified figure shaped like a flattened ellipsoid of revolution which lie in the exterior of the figure, as well as at points along a line that lies in the plane of the equator of the figure and that includes the center of the figure and, again, which are located in the exterior of the figure, when the strata of the figure are concentric, flattened, individually homogeneous confocal ellipsoids of revolution all of which have a common axis of symmetry, namely, the axis of symmetry of the figure, and whose densities differ, where the figure is assumed to attract according to the universal inverse-square law. Again, the ellipticity of the figure can be finite.[163]

But MacLaurin could not generalize his results in a way that would allow him to calculate the magnitudes of the attraction at points located at the surface of the kind of stratified figure just described but not situated at one of its poles or along its equator. In determining the ellipticities of stratified figures shaped like flattened ellipsoids of revolution which attract according to the universal inverse-square law and which revolve around their axes of symmetry, MacLaurin equated the weights of columns from the centers to the surfaces of the figures which lie along their axes. Hence his theorems sufficed for his purposes. As Clairaut suggested in his letter to MacLaurin dated 10 April 1741, MacLaurin could use his theorems to calculate the magnitudes of the attraction at points along these columns and thereby compute the weights of the columns. There were evidently ways to calculate the magnitudes of the attraction at points along the axes in the interior of a stratified figure shaped like a flattened ellipsoid of revolution which attracts according to the universal inverse-square law and whose strata are concentric, flattened confocal ellipsoids of revolution which have a common axis of symmetry, namely, the axis of symmetry of the figure.

But Clairaut, on the other hand, applied the principle of the plumb line instead to find the ellipticities of stratified figures shaped like flattened ellipsoids of revolution which attract in accordance with the universal inverse-square law, which revolve around their axes of symmetry, and whose ellipticities are infinitesimal. Consequently, he needed to know the magnitudes of the attraction at *all* points at the surfaces of solid, stratified figures shaped like flattened ellipsoids of revolution which attract according to the universal inverse-square law and whose ellipticities are infinitesimal. He needed to know as well the magnitudes of the attraction at *all* points located on level surfaces with respect to effective gravity of stratified figures of equilibrium shaped like flattened ellipsoids of revolution which attract according to the universal inverse-square law, which revolve around their axes of symmetry, and whose ellipticities are infinitesimal, which meant at the points that made up each stratum when the thicknesses of the strata were infinitesimal (that is, when the strata were surfaces themselves, namely, the level surfaces with respect to effective gravity of the stratified figures of equilibrium, which are concentric, flattened, individually homogeneous ellipsoids of revolution all of which have a common axis of symmetry, namely, the axis of symmetry of the figure, and whose densities differ). In both cases, then, he had to determine the magnitudes of the attraction at points at the surfaces of stratified figures shaped like flattened ellipsoids of revolution which attract in accordance with the universal inverse-square law and whose ellipticities are infinitesimal, the majority of which were not situated at poles or along equators of these figures. As a result, unlike MacLaurin's theorems for determining the magnitudes of the attraction at all points at the surfaces of *homogeneous* figures of equilibrium shaped like flattened ellipsoids of revolution which attract according to the universal inverse-square law and whose ellipticities are finite, MacLaurin's theorems for calculating the magnitudes of the attraction produced by the type of stratified figure described would not serve Clairaut's purposes.

(When the strata of figures were not concentric, flattened *confocal* ellipsoids of revolution which had a common axis of symmetry, MacLaurin himself could not use the particular theorem stated earlier to equate the weights of columns from the centers to the surfaces of such figures which lie along their axes. In order to compute the weights of such columns, the magnitudes of the attraction at all points along the columns have to be determined. MacLaurin, however, could not use the theorem in question to calculate these magnitudes of the attraction in this case. For example, concentric ellipsoids of revolution which have a common axis of symmetry and have the same ellipticities, hence are geometrically similar ellipsoids of revolution, are *not* confocal ellipsoids of revolution. But in the examples that Clairaut cited from *A Treatise of Fluxions*, which I examined above, the strata *were* concentric, flattened ellipsoids of revolution which were geometrically similar and had common axes of symmetry. Hence MacLaurin could not apply the theorem in question to these examples. MacLaurin had other theorems for calculating the magnitudes of the attraction at points along the axes of stratified figures composed of such strata.[164])

Because of this Clairaut had no choice but to avail himself of the kinds of mathematical arguments, including the inevitable approximations without estimates of errors, which he had first employed in his two papers on the theory of the earth's shape published in the *Philosophical Transactions.*

(Methods for generalizing MacLaurin's theorems so that the magnitudes of the attraction produced by a stratified figure shaped like a flattened ellipsoid of revolution which attracts according to the universal inverse-square law, whose ellipticity is finite, whose strata are flattened, individually homogeneous, confocal ellipsoids of revolution all of which have a common axis of symmetry, namely, the polar axis of the figure, and whose densities differ could be calculated at points in the exterior of the figure but *not* located on the polar axis of the figure or in the plane of the equator of the figure were not produced until the 1780s, when Legendre and Laplace introduced harmonic analysis while working with "Legendre series" and "Laplace functions.")

Clairaut must have more or less accepted on faith alone that his approximations held good to terms of first order for stratified figures shaped like flattened ellipsoids of revolution which attract in accordance with the universal inverse-square law, which revolve around their axes of symmetry, and whose ellipticities are infinitesimal, perhaps because he had verified them to hold true to terms of first order for homogeneous figures shaped like flattened ellipsoids of revolution which attract according to the universal inverse-square law, which revolve around their axes of symmetry, and whose ellipticities are infinitesimal.

In his letter to MacLaurin of November 1742, in which he acknowledged receipt of MacLaurin's treatise, Clairaut indicated that MacLaurin's theorems had helped him to assure himself that he had not made a nonnegligible, cumulative error in discarding terms of second order and higher orders, so that something else caused his findings to fail to accord with MacLaurin's.[165] Shortly afterward, Clairaut wrote Euler, on 3 December 1742. He told Euler in this letter that "not having paid attention to Monsieur MacLaurin's principles at first," he checked his own first-order approximations instead for nonnegligible errors but could not find any.[166] In alluding to MacLaurin's principles, Clairaut doubtlessly meant the principle of balanced columns, whose application by MacLaurin more than likely only became clear to Clairaut after reading *A Treatise of Fluxions.*

Leaving specific principles and calculations aside, Clairaut already knew by September of 1741 that MacLaurin treated stratified figures that the Scottish mathematician assumed were fluid figures of equilibrium. Consequently unanswered questions remain concerning what, apart from the problem of estimating the order of the errors in making approximations, could have caused Clairaut to put off publishing his treatise for so long. The solution perhaps can be found in Clairaut's letter to Euler dated 3 December 1742, where Clairaut explained that he finally felt "obliged to demonstrate a proposition whose truth" he "had only suspected at first." This was his proposition that if the density decreases from the center to the surface of a stratified fluid figure of equilibrium shaped like a flattened ellipsoid of revolution which attracts according to the universal inverse-square law, which revolves around its axis of symmetry, and whose ellipticity is infinitesimal, a condition that Clairaut thought that the laws of hydrostatics required that a stratified fluid figure shaped like a flattened ellipsoid of revolution which attracts according to the universal inverse-square law, which revolves around its axis of symmetry, and whose ellipticity is infinitesimal must fulfill in order that the stratified fluid figure be a figure of equilibrium, then the strata must become increasingly flatter from the center of the figure to its surface and the figure must always be less flat than it would be if it were homogeneous.[167] It is possible that Clairaut simply *conjectured* that this result was true when he wrote his letter to Euler early in January of 1742. He might not have actually demonstrated the crucial result, which he believed the application of his theory of stratified figures of equilibrium to stratified figures shaped like flattened ellipsoids of revolution which attract according to the universal inverse-square law, which revolve around their axes of symmetry, and whose ellipticities are infinitesimal depended upon, until late in 1742.

What examples did Clairaut have in mind in his letter to Euler dated 3 December 1742? Clairaut never did solve his integro-differential equation (9.4.3) in generality, because he still could not solve the equivalent Riccati equation (9.4.7) except in one special case. Hence he could not have had in mind examples in which densities vary as differentiable functions $R(r)$ of r. Instead he must have meant stratified figures shaped like flattened ellipsoids of revolution which attract in accordance with the universal inverse-square law, which revolve around their axes of symmetry, whose ellipticities are infinitesimal, and which are composed of finite numbers of concentric, individually homogeneous strata, where each stratum has finite thickness, in which case the densities of the stratified figures vary as step functions $R(r)$ of r. If Clairaut did not demonstrate until late 1742 that the critical result holds true in these particular cases, this fact would help account for the ambiguity of the statements that appear in his letter to Euler of January 1742, which I mentioned earlier. In that letter, Clairaut did not distinguish results that held true in cases where $R(r)$ is a differentiable function of r from results that proved true in cases where $R(r)$ is a step function of r. On the contrary, he spoke as if the former followed from the latter, which is not true. (The confusion caused by Clairaut's failure at this time to distinguish between the two kinds of variations in density remains, in fact, to a certain extent in the published treatise. I explained earlier how Clairaut could have conceivably reasoned that the former followed from the latter.)

In his letter to Euler dated 3 December 1742, Clairaut called Euler's attention to the other obstacles that he had to confront in developing his theory of stratified figures of equilibrium shaped like flattened ellipsoids of revolution which attract according to the universal inverse-square law, which revolve around their axes of symmetry, and whose ellipticities are infinitesimal. These obstacles held him back, retarded his progress, and also caused him to delay the publication of his treatise. "If it took me some time to examine Monsieur MacLaurin's propositions." Clairaut told Euler, "that examination did not approach the trouble that the problem that had led me to" the integro-differential equation (9.4.3) "gave me," he said.[168] Although he had found this equation "a long time ago," Clairaut continued, the need to stipulate two constants of integration for the integrable case $R(r) = r^n$, when the conditions of the problem only specified one constant of integration determined by the ratio ϕ of the magnitude of the centrifugal force per unit of mass to the magnitude of the effective gravity at the equator, alarmed and dismayed him for some time. He did not know if he had made errors, so that his solution of the integro-differential equation (9.4.3) was incorrect, or whether his solution of (9.4.3) was correct and he had simply overlooked something.[169]

Clairaut verified that Euler's construction of Riccati equations through the motions of tractices, which Clairaut had asked Euler to show him when he wrote Euler in September of 1742 and which Euler subsequently did, would not serve his purposes in the general case.[170] In the 1730s, Euler had also employed his theory of "modular equations," which I discussed in Chapter 7, to construct Riccati equations by means of rectifications.[171] Even earlier in the 1730s, Euler also had utilized a method that brings to mind Nicole's use of integrands of integrals expanded in infinite series and integration of these infinite series term by term to solve problems of trajectories, which I also examined in Chapter 7, in order to construct Riccati equations through rectifications.[172] As Euler must have realized himself, however, neither method would have helped Clairaut solve the problem that he was working on.

The fact that Clairaut found such constructions "useless" for his purposes indicated a trend. After 1740, problems in continuum mechanics largely determined the directions of subsequent research in differential equations. Clairaut's negative reply to Euler signified this tendency. Clairaut's failure to bring Euler's constructions of Riccati equations to bear in a useful way upon the problem of solving the integro-differential equation (9.4.3) in the context of stratified figures of revolution in equilibrium which attract according to the universal inverse-square, which revolve around their axes of symmetry, and whose ellipticities are infinitesimal is one of the earliest examples of this trend.

We remember that in his letter to Euler dated 3 December 1742, Clairaut said that "if it took me some time to examine Monsieur MacLaurin's propositions, that examination did not approach the trouble that the problem that had led me to" the integro-differential equation (9.4.3) "gave me." This assertion is an understatement. For Clairaut *never* solved the general integro-differential equation (9.4.3). Consequently he could not truely use the equation to explain once and for all why his values of the ellipticities of stratified figures of equilibrium shaped like flattened ellipsoids of revolution which attract in accordance with the universal inverse-square law, which revolve around their axes of symmetry, and whose ellipticities are infinitesimal and MacLaurin's values of the ellipticities of the same figures did not agree with each other, when the densities of the stratified figures of equilibrium varied as differentiable functions $R(r)$ of r. That is, what

Clairaut could not do was use the integro-differential equation (9.4.3) to prove conclusively that MacLaurin's stratified figures shaped like flattened ellipsoids of revolution which attract according to the universal inverse-square law, which revolve around their axes of symmetry, and whose ellipticities are infinitesimal could not be figures of equilibrium, when the densities of the figures varied as differentiable functions $R(r)$ of r. He could not do so because he could not use the integro-differential equation (9.4.3) to show that the ellipticities of the strata could not vary the way that MacLaurin had chosen them to vary if the stratified figures are really figures of equilibrium. By the same token, as I have already mentioned, since the density varied as a differentiable function

$$R(r) = fr^p + gr^q \qquad (6.2.2)$$

of r in the stratified figure shaped like a flattened ellipsoid of revolution which attracts according to the universal inverse-square law, which revolves around its axis of symmetry, and whose ellipticity is infinitesimal which he studied in his second paper on the theory of the earth's shape published in the *Philosophical Transactions*, Clairaut did not truly establish that this figure cannot be a figure of equilibrium either, even though the ellipticities of the strata that make up the figure do not increase from the center of the figure to its surface but remain constant instead. He did not establish this result because he was only able to demonstrate that the ellipticities of the strata must increase from the center to the surface of a stratified figure of equilibrium shaped like a flattened ellipsoid of revolution which attracts according to the universal inverse-square law, which revolves around its axis of symmetry, and whose ellipticity is infinitesimal when the number of strata is finite and the thicknesses of the strata are all finite, which is not true of the strata that make up the figure that he examined in his second paper on the theory of the earth's shape published in the *Philosophical Transactions*.

Finally, what about Clairaut's assertion that the densities of the strata that make up a stratified fluid figure of equilibrium shaped like a flattened ellipsoid of revolution which attracts according to the universal inverse-square law, which revolves around its axis of symmetry, and whose ellipticity is infinitesimal must decrease from the center of such a figure to its surface "as the laws of hydrostatics require," which, as we have seen, is a condition that some of MacLaurin's figures whose ellipticities are infinitesimal do not fulfill? It was customary at this time to believe that the laws of hydrostatics did necessitate such a decrease. Desaguliers, for example, thought that these laws did in his articles published in the *Philosophical Transactions* of 1725. As proof Desaguliers noted that after oil and water are mixed together, the two fluids separate into two layers. The layer of oil, which is the denser of the two fluids, lies beneath the layer of water.[173] Daniel Bernoulli, moreover, assumed the same thing in his essay on the tides which earned him a share of the Paris Academy's prize of 1740. As evidence he cited observations similar to Desaguliers's, namely, that water is found at the earth's surface instead of other fluids because water is less dense than other fluids.[174] And we know that Clairaut read Bernoulli's essay very, very carefully.

(In fact, it turns out that Clairaut's assumption that density decreases from center to surface is *not* needed to ensure that stratified figures of equilibrium made of incompressible fluid and shaped like flattened ellipsoids of revolution which attract according to the universal inverse-square law, which revolve around their axes of symmetry, and whose ellipticities are infinitesimal *exist*, an assumption that D'Alembert was the first to dispute,[175] but that it is needed to assure that the figures are in a state of *stable* equilibrium.)

9.6. Remaining difficulties that caused Clairaut to cease working on the problem of the earth's shape

Clairaut now tried to draw further conclusions about the shapes of stratified figures of revolution in equilibrium which attract according to the universal inverse-square law, which revolve around their axes of symmetry, and whose ellipticities are infinitesimal from his findings that concerned shapes of solid stratified figures of revolution which attract in accordance with the universal inverse-square law, which revolve around their axes of symmetry, whose ellipticities are infinitesimal, and at whose outer surfaces the principle of the plumb line holds. In a fluid figure of equilibrium shaped like a flattened ellipsoid of revolution which attracts according to the universal inverse-square law, which revolves around its axis of symmetry, whose ellipticity is infinitesimal, and which consists of a finite number of concentric, individually homogeneous strata, each of which has finite thickness, whose bounding surfaces are all concentric, flattened ellipsoids of revolution whose ellipticities are infinitesimal and all of which have a common axis of symmetry, namely, the polar axis of the figure, the ellipticities of the surfaces that bound the strata must increase from the center of the figure to the surface of the figure (assuming that the densities of the strata decrease from the center of the figure to the surface of the figure). But then the ellipticities of the surfaces that bound the strata a fortiori do not decrease from the center of the figure to the surface of the figure faster than the inverse squares of the distances of the surfaces that bound the strata from the center of the figure, measured along the polar axis of the figure, do.

Applying his findings that concern solid, stratified figures shaped like flattened ellipsoids of revolution which attract in accordance with the universal inverse-square law, which revolve around their axes of symmetry, whose ellipticities are infinitesimal, and at whose outer surfaces the principle of the plumb line holds, Clairaut concluded that a stratified fluid figure of equilibrium shaped like a flattened ellipsoid of revolution which attracts according to the universal inverse-square law, which revolves around its axis of symmetry, and whose ellipticity is infinitesimal must not be flatter than the corresponding homogeneous fluid figure of equilibrium shaped like a flattened ellipsoid of revolution which attracts in accordance with the universal inverse-square law, which revolves around its axis of symmetry, and whose ellipticity is infinitesimal (that is, the homogeneous fluid figure of equilibrium whose ratio ϕ of the magnitude of the centrifugal force per unit of mass to the magnitude of the effective gravity at its equator is the same as the ratio ϕ of the magnitude of the centrifugal force per unit of mass to the magnitude of the effective gravity at the equator of the stratified fluid figure of equilibrium) but must be less flat instead.[176] Clairaut was able to deduce this from his results which pertained to the shapes of solid stratified figures shaped like flattened ellipsoids of revolution which attract according to the universal inverse-square law, which revolve around their axes of symmetry, and whose ellipticities are infinitesimal because these results really only depended upon the assumption that the principle of the plumb line held at the outer surfaces of such solid figures. The assumption that the figures were in a solid state was not used in arriving at the results. But the principle of the plumb line *also* holds at the surface of a stratified fluid figure of equilibrium shaped like a flattened ellipsoid of revolution which attracts according to the universal inverse-square law, which revolves around its axis of symmetry, and whose ellipticity is infinitesimal. Indeed, this condition is a condition necessary for a stratified fluid figure to exist in a state of equilibrium.

Clairaut viewed the homogeneous fluid figure of equilibrium shaped like a flattened ellipsoid of revolution which attracts according to the universal inverse-square law, which revolves around its axis of symmetry, and whose ellipticity is infinitesimal as one of two limiting cases of the way that the densities of the bounding surfaces of concentric, individually homogeneous strata can decrease from the center of a fluid figure of revolution in equilibrium to its surface when the value of ϕ has been specified. In this particular figure the density of the figure decreases from the center of the figure to its surface the most slowly. Namely, the density doesn't decrease at all but remains constant instead. When the value of ϕ is $\frac{1}{288}$, the ellipticity δ of the figure equals $\frac{1}{230}$.

Clairaut treated the fluid figure of revolution in equilibrium produced when attraction acts like a central force, where the center of force is situated at the center of the figure, where the attraction at all points is directed toward the center of force, and where the magnitudes of attraction at all points vary inversely as the squares of the distances of the points from the center of force, as the other limiting case. In this figure the density of matter that attracts decreases the fastest from the center of the figure to its surface, because all of the matter that attracts is concentrated at the center of force. Hence the density jumps from a finite value to zero abruptly. (As I have already indicated, however, this last limit is really little more than suggestive; it is not a conclusion that can be drawn rigorously from Clairaut's theory. I explained in Chapter 6 why this last limit does not truly follow from his theory.[177]) When the value of ϕ is $\frac{1}{288}$, the ellipticity δ of this figure equals $\frac{1}{576}$.

Hence Clairaut concluded that if δ is the ellipticity of a fluid figure of equilibrium shaped like a flattened ellipsoid of revolution which attracts according to the universal inverse-square law, which revolves around its axis of symmetry, whose ellipticity is infinitesimal, and which consists of a finite number of concentric, individually homogeneous strata, each of which has finite thickness, whose bounding surfaces are all concentric, flattened ellipsoids of revolution whose ellipticities are infinitesimal and all of which have a common axis of symmetry, namely, the polar axis of the figure, then

$$\tfrac{1}{576} \leqslant \delta \leqslant \tfrac{1}{230} \qquad (9.6.1)$$

when $\phi = \tfrac{1}{288}$.[178]

These conclusions that Clairaut drew from his theory of figures of revolution in equilibrium which attract according to the universal inverse-square law, which revolve around their axes of symmetry, and whose ellipticities are infinitesimal, together with the equation

$$\frac{P - \Pi}{\Pi} = 2\epsilon - \delta, \qquad (9.4.1')$$

required that the magnitude of the effective gravity increase more from the equator to a pole along the surface of a stratified fluid figure of equilibrium shaped like a flattened ellipsoid of revolution which attracts according to the universal inverse-square law, which revolves around its axis of symmetry, and whose ellipticity is infinitesimal than it does along the surface of the corresponding homogeneous fluid figure of equilibrium shaped like a flattened ellipsoid of revolution which attracts in accordance with the universal inverse-square law, which revolves around its axis of symmetry, and whose ellipticity is infinitesimal, where two such figures correspond when their values of the ratio ϕ are the same.[179] And the greater increase did agree with the terrestrial observations of the day.

At the same time, equation (9.4.1') applies to *any* stratified figure shaped like a flattened ellipsoid of revolution which attracts according to the universal inverse-square law, which revolves around its axis of symmetry, whose ellipticity is infinitesimal, and which is made up of an infinite number of concentric, individually homogeneous strata that are flattened ellipsoids of revolution whose ellipticities are all infinitesimal and all of which have a common axis of symmetry, namely, the polar axis of the figure, or which is composed of a finite number of concentric, individually homogeneous strata whose thicknesses are all finite and which are all bounded by surfaces that are concentric, flattened ellipsoids of revolution whose ellipticities are all infinitesimal and all of which have a common axis of symmetry, namely, the polar axis of the figure, when the principle of the plumb line holds at the outer surface of the figure. That is, in deriving equation (9.4.1'), Clairaut did not use the assumption that the figure was solid. All that truly mattered was the hypothesis that the principle of the plumb line held at the outer surface of the figure. The densities and ellipticities of the concentric, individually homogeneous strata can vary in any manner, whether the strata are surfaces or layers whose thicknesses are finite. Hence without even having to assume that the stratified figure shaped like a flattened ellipsoid of revolution which attracts according to the universal inverse-square law, which revolves around its axis of symmetry, and whose ellipticity is infinitesimal is a figure of equilibrium, it followed from equation (9.4.1') that $\delta < \frac{1}{230}$ when $\epsilon = \frac{1}{230}$, because the earth's value of $(P - \Pi)/\Pi$ had been inferred from experiments with seconds pendulums carried out at different latitudes to be greater than $\frac{1}{230}$. Clairaut found nothing that would harmonize a mathematical theory of shapes of heterogeneous figures of revolution with Newton's "intuitions."[180]

Nor could Clairaut reconcile his mathematical theory with the observations of the day which seemed to confirm Newton's conjectures. Clairaut suspected that the degree of latitude in Lapland had not been measured precisely enough. He noted that if the difference between the degrees of latitude measured in Paris and in Tornéo, Lapland, were reduced by a mere 60 *toises* (1 *toise* $\simeq$ 6.5 feet), the earth's ellipticity would have come out to be $\frac{1}{230}$ instead of $\frac{1}{177}$. Clairaut felt certain that 60 *toises* easily fell within the margin of error that had to be allowed for in making all of the astronomical observations and geographical measurements at the two different locations, in Lapland and in France, from which the extent of the earth's flattening had been deduced.[181] While the degrees of latitude measured in France and in Lapland did, in Clairaut's opinion, suffice to confirm Newton's deduction that the earth was flattened at its poles, Clairaut also thought that more work would have to be done in order to determine accurately the extent of the earth's flattening by direct measurement. He believed that it was still necessary to compare degrees of latitude "as far apart as possible" in order to find the true value of the flattening,[182] namely, degrees of latitude measured in Lapland and in Peru. He hoped that such a comparison would prove his suspicion that errors had been made in carrying out the astronomical observations made and geographical measurements made in Lapland and that such a finding would confirm his theory.

Nevertheless, despite all of his efforts to reconcile theory with observations, Clairaut ultimately concluded that the universal inverse-square law of attraction could never be tested in the terrestrial realm. Even before he published his treatise on the earth's shape, in 1743, he began to doubt that the earth would ever serve as a suitable ground for testing the universal inverse-square law of attraction. And he took this uncertainty into account in writing his treatise, in making allowances for other hypotheses of attraction. Namely, he also determined conditions that other hypotheses of attraction had to fulfill in order for them to lead to results and conclusions that also agreed with the observations of the day.

For example, we recall that in his Paris Academy *mémoire* of 1734 Bouguer had investigated hypotheses of attraction which tacitly fulfilled conditions of symmetry which ensured that all homogeneous fluid figures considered were figures of revolution, where the figures revolve around their axes of symmetry. We remember that according to the ways that Bouguer hypothesized that homogeneous fluid figures of revolution are attracted, the lines that indicate the directions of the attraction at all points in the plane of a meridian at the surface of a homogeneous fluid figure of revolution were always assumed to lie in that plane and were always assumed to be perpendicular to a given closed plane curve that also lies in that plane. Moreover, the variation of the magnitudes of the attraction at points in the plane was fixed by specifying this variation at the points situated on each individual line in the plane which is perpendicular to this closed plane curve and which indicates the direction of the attraction at the points in the plane which are located on this line. Furthermore, the closed plane curves in different planes of meridians at the surface of the homogeneous fluid figure of revolution which revolves around its axis of symmetry are all assumed to be congruent and can be derived by revolving any one of them around the axis of symmetry of the homogeneous fluid figure of revolution. In addition, the magnitudes of the attraction at points in the plane of a meridian at the surface of the homogeneous fluid figure of revolution which revolves around its axis of symmetry are assumed to depend the same way upon the distances of the points from the closed plane curve in that plane for all meridians at the surface of the homogeneous fluid figure of revolution which revolves around its axis of symmetry. Among these hypotheses of attraction Clairaut found one of the kind that allowed his second new principle of equilibrium to hold – that is, one where the magnitudes of the attraction at points in the plane of a meridian are assumed to depend solely upon the distances of points from the closed curve in the plane of the meridian – and which would permit a homogenous fluid figure shaped like a figure of revolution to exist whose ellipticity is $\frac{1}{177}$ when the ratio ϕ of the magnitude of the centrifugal force per unit of mass to the magnitude of the effective gravity at its equator is $\frac{1}{288}$ and whose relative increase in the effective gravity along its surface from its equator to one of its poles is $\frac{10}{2025}$, which is the value of this relative increase which the "$\sin^2$ law" of the variation of the increase in the effective gravity with latitude along the earth's surface together with the results of the experiments with seconds pendulums made at different terrestrial latitudes required. Moreover, according to Clairaut's theory of homogeneous fluid figures of equilibrium, the homogeneous figure shaped like a figure of revolution would be a fluid figure of equilibrium, except for a solid nucleus which surrounds the point at its center. I explained in Section 2 of this chapter how Clairaut reasoned that there must be such a solid nucleus in order for the mass of fluid which surrounds it to exist in a state of equilibrium when attraction is assumed to act according to such hypotheses. In short, Clairaut discovered hypotheses of attraction among those that Bouguer had examined which allowed homogeneous fluid figures of revolution to exist in a state of equilibrium and which accorded with both the measured degrees of latitude at the earth's surface and the results of the experiments made with seconds pendulums at the earth's surface within the limits of observational errors.[183] The shapes of these figures of course differed from the shapes of homogeneous fluid figures of revolution in equilibrium assumed to attract according to the universal inverse-square law, which revolve around their axes of symmetry, and whose ellipticities are infinitesimal.

Clairaut also remarked that certain hypotheses of attraction could immediately be ruled out once and for all – for example, all hypothesized central forces of attraction. In 1690, Huygens had already suggested that when the ratio ϕ of the magnitude of the centrifugal force per unit of mass to the magnitude of the attraction at the equator was infinitesimal, changing the magnitude of a central force (that is, changing the value of n in the function $f(r) = kr^n$ that expressed the magnitude of a central force as a function of distance r from the center of force, where k is a positive constant and where n is positive, negative, or zero) would produce negligible variations in the ellipticity of a homogeneous fluid figure of revolution at whose center a central force of attraction acts and which satisfies the principles of the plumb line and balanced columns. In particular, when $\phi = \frac{1}{289}$, the ellipticity of the figure would always be very close to $\frac{1}{578}$, whatever the value of n.[184] In his *Discours* published in 1732, Maupertuis noted that the value $\sqrt{289}/\sqrt{288}$ of the ratio of the equatorial axis to the polar axis appearing in Hermann's *Phoronomia* (1716), which Hermann found by assuming that attraction acts like a central force, where the center of force is situated at the center of the figure, where the attraction at all points is directed toward the center of force, and where the magnitudes of the attraction at all points vary directly as the distances of the points from the center of force, "did approach very closely" Huygens's value $\frac{578}{577}$ of the same ratio. Huygens had arrived at this value of the ratio by assuming that attraction acts like a central force, where the center of force is situated at the center of the figure, where the attraction at all points is directed toward the center of force, and where the magnitudes of the attraction at all points are the same.[185]

Clairaut verified that the conclusion that others had come to by induction was indeed true. Specifically, he showed that if ϵ is the ellipticity of a homogeneous fluid figure of revolution in equilibrium when attraction acts like a central force, where the center of force is situated at the center of the figure, where the attraction at all points is directed toward the center of force, and where the magnitudes of the attraction at all points vary directly as a function $f(r) = kr^n$ of the distances r of the points from the center of force, where k is a positive constant and where n is positive, negative, or zero, then

$$\epsilon \simeq \tfrac{1}{2}\phi, \qquad (9.6.2)$$

where ϕ here is the ratio of the magnitude of the centrifugal force per unit of mass to the magnitude of the effective gravity at the equator of the figure. This result does not depend upon the way that the magnitudes of the attraction at points vary with the distances of points from the center of force. The only condition that must be fulfilled is that the figure rotate "slowly." That is, equality (9.6.2) holds provided $\phi^2 \ll \phi$ and $\epsilon^2 \ll \epsilon$. From this finding Clairaut concluded that whatever the correct value of the earth's ellipticity, whether it was $\frac{1}{177}$ or not, it "differed so much" from $\frac{1}{2}\phi = \frac{1}{576}$, the value of $\frac{1}{2}\phi$ when ϕ has the earth's value $\frac{1}{288}$, that terrestrial attraction could not possibly act like a force directed toward the earth's center whose magnitudes at points depend solely upon the distances of points from its center.[186] In other words, observational errors considerably larger than ones that it was reasonable to assume might have been made in Lapland would have had to be made in order for the earth to have a value for its ellipticity as small as $\frac{1}{576}$.

In Chapter 5 we saw that Maupertuis appeared to demonstrate in his Paris Academy *mémoire* of 1734 that the theory of a homogeneous figure of revolution which revolves around its axis of symmetry and whose columns from its center to its surface all balance or weigh the same when a central force of attraction acts, where the center of force is situated at the center of the figure, where the attraction due to the center of force at all points is directed toward the center of force, and where the magnitudes of the attraction at points vary inversely as the squares of the distances of points from the center of force, work better when applied to Jupiter than did Newton's theory of a homogeneous figure shaped like a flattened ellipsoid of revolution which attracts according to the universal inverse-square law, which revolves around its axis of symmetry, and whose columns from its center to its surface all balance or weigh the same. Specifically, we recall that in his Paris Academy *mémoire* of 1734, Maupertuis showed how to calculate the ratio of the magnitude of the centrifugal force per unit of mass to the magnitude of the attraction at the equator of a spherical celestial body assumed to revolve around its polar axis at a constant rate (that is, at a constant angular speed) using the celestial body's radius, its observed period of revolution around its polar axis, the observed period of a satellite assumed to revolve at a constant speed around the celestial body in a circle whose center is located at the center of the celestial body, and the radius of the satellite's orbit relative to the radius of the celestial body around which it revolves (that is, the ratio of the radius of the satellite's orbit to the radius of the celestial body around which it revolves). He also used the fact that the magnitude of the centripetal acceleration of a celestial body moving in a circle at a constant speed varies inversely as the square of the radius of the circle. This fact is a consequence of Kepler's third law of motion together with Newton's expression for the magnitude of the centripetal acceleration of a body moving in a circle at a constant speed (which was the same as Huygens's expression for the magnitude of the centrifugal force per unit of mass of a body moving in a circle at a constant speed.) The centripetal acceleration of such a body is directed toward the center of the circle that is the body's path of motion, and Maupertuis implicitly assumed that the centripetal acceleration of the revolving body was caused by the attraction due to a center of force situated at the center of the circle that the body moves along, where the magnitudes of the attraction due to this center of force at points varied inversely as the squares of the distances of the points from this center of force. In other words, he assumed that the magnitudes of the attraction due to the center of force at points vary with the distances of points from the center of force the same way that the magnitudes of the centripetal accelerations of bodies that revolve around the center of force in circles at constant speeds vary with the distances of the bodies from the center of force. This is a natural assumption to make, if it is hypothesized that the central force of attraction *causes* the centripetal acceleration of the body revolving in a circle whose center is situated at the center of force. Thus he treated the satellite as if it were attracted by a fixed center of force situated at the center of the celestial body around which it revolves, where the magnitudes of the attraction due to this center of force at points vary inversely as the squares of the distances of the points from this center of force. Applying his calculations to Jupiter, Maupertuis determined the ratio of the magnitude of the centrifugal force per unit of mass to the magnitude of the attraction at Jupiter's equator to be $\frac{1}{7.48}$. Taking this to be the value of the ratio of the magnitude of the centrifugal force per unit of mass to the magnitude of the attraction at the equator of a figure in his theory of a homogeneous figure of revolution which revolves around its axis of symmetry and which is attracted by a single, fixed center of force located at the center of the figure, where the attraction due to the center of force at all points is directed toward the center of force, where the magnitudes of the attraction at all points vary solely as a function of the distances of points from the center of force, and where the columns from the center of such a figure to its surface all balance or weigh the same, Maupertuis found that when the magnitudes of the attraction at all points vary inversely as the squares of the distances of the points from the center of force, the value δ of Jupiter's ellipticity is $\frac{1}{14.96}$, which is close to the value $\frac{1}{15}$ of the extent of Jupiter's flattening observed by Gian-Domenico Cassini, in 1691, and confirmed by Philippe de La Hire. (I mentioned in Chapter 5, that the two values $\frac{1}{14.96}$ and $\frac{1}{15}$ are equal to terms of first order.) Maupertuis's value also lay closer to the value

$$\frac{1}{12\frac{11}{48}}$$

of the extent of Jupiter's flattening observed by the English astronomer James Pound, in 1719, than did the value

$$\frac{1}{9\frac{1}{3}}$$

of the extent of Jupiter's flattening which Newton determined by assuming that Jupiter is a homogeneous fluid figure shaped like a flattened ellipsoid of revolution which attracts according to the universal inverse-square law, which revolves around its axis of symmetry, whose ellipticity is infinitesimal, and whose columns from center to surface all balance or weigh the same. We also know from Clairaut's theory of homogeneous fluid figures of equilibrium that if a homogeneous fluid figure of equilibrium is assumed to be attracted by a fixed center of force situated at the center of the figure, then changing the figure into a stratified figure by changing the constant density into a density that varies will not change the ellipticity of the figure.

Clairaut came to entirely different conclusions about Jupiter than Maupertuis did, however. Doing exactly the same kind of calculations that Maupertuis had done for Jupiter, Clairaut found that the ratio of the magnitude of the centrifugal force per unit of mass to the magnitude of the attraction at Jupiter's equator is $\frac{1}{11.615}$ instead of $\frac{1}{7.48}$. Then assuming that a center of force situated at the center of Jupiter attracts points of Jupiter inversely as the squares of the distances of points of Jupiter from the center of force, where the attraction due to the center of force at all points is directed toward the center of force, and assuming that Jupiter is a homogeneous figure of equilibrium, Clairaut used the ratio $\frac{1}{11.615}$ of the magnitude of the centrifugal force per unit of mass to the magnitude of the attraction at Jupiter's equator to determine the ellipticity δ of Jupiter to be $\frac{1}{23.23}$ when such hypotheses are made,[187] a value of Jupiter's ellipticity which differed considerably from the value $\frac{1}{14.96}$ that Maupertuis had determined in making the same assumptions. The value $\frac{1}{23.23}$ was not nearly as close to the observed values of Jupiter's ellipticity as the value $\frac{1}{14.96}$ was.

On the other hand, treating Jupiter as a homogeneous figure of equilibrium shaped like a flattened ellipsoid of revolution which attracts according to the universal inverse-square law, which revolves around its axis of symmetry, and whose ellipticity is finite, Clairaut found that the infinite series (9.5.0) for the ellipticity δ of such a figure, which he had derived using the results that MacLaurin expressed geometrically in *A Treatise of Fluxions*, gave $\frac{1}{9.05}$ as the value of Jupiter's ellipticity δ.[188] Although he doubtlessly used this infinite series to determine the value $\frac{1}{9.05}$ of Jupiter's ellipticity δ, this particular infinite series is not the infinite series for Jupiter's ellipticity δ which he published. Instead he published a meaningless infinite series for Jupiter's ellipticity δ, which he found by making a mistake.[189] The value

$$\frac{1}{9\frac{1}{3}}$$

of Jupiter's ellipticity δ which Newton found by assuming that Jupiter is a homogeneous fluid figure shaped like a flattened ellipsoid of revolution which attracts according to the universal inverse-square law, which revolves around its axis of symmetry, whose ellipticity is infinitesimal, and whose columns from center to surface all balance or weigh the same is not equal to $\frac{1}{9.05}$ to terms of first order. That is, these two values of Jupiter's ellipticity δ differ by terms of first order, not just by terms of second order and higher orders.[190] But this is to be expected, since Jupiter's ellipticity δ was in fact observed *not* to be infinitesimal. (Clairaut's calculations also prove that Jupiter cannot be a *homogeneous* figure of equilibrium shaped like a flattened ellipsoid of revolution which attracts according to the universal inverse-square law, which revolves around its axis of symmetry, and whose ellipticity is finite, because the value $\frac{1}{9.05}$ of Jupiter's ellipticity which follows from assuming that Jupiter is such a homogeneous figure of equilibrium is too far from the observed values.)

Clairaut viewed the two values $\frac{1}{23.23}$ and $\frac{1}{9.05}$ as the limiting values of the ellipticity δ that Jupiter can have if it is a stratified figure of revolution in equilibrium which attracts according to the universal inverse-square law and which revolves around its axis of symmetry. Thus with help from MacLaurin's *A Treatise of Fluxions*, Clairaut concluded that

$$\frac{1}{23.23} \leqslant \delta \leqslant \frac{1}{9.05}. \tag{9.6.3}$$

(Once again, however, the lower limit $\frac{1}{23.23}$ is really little more than suggestive; it is not a conclusion that can be drawn rigorously from Clairaut's theory. I explained in Chapter 6 why this limit does not truly follow from his theory.) He stated that the values $\frac{1}{13}$ and $\frac{1}{10}$ of Jupiter's ellipticity observed by Cassini and Pound, respectively, agreed with (9.6.3).[191]

Now, (9.6.3) does not truly follow from Clairaut's theory of stratified fluid figures of equilibrium. The upper limit $\frac{1}{9.05}$ of Jupiter's ellipticity δ is valid, because that is the ellipticity that Jupiter would have if it were a *homogeneous fluid* figure of equilibrium shaped like a flattened ellipsoid of revolution which attracts according to the universal inverse-square law, which revolves around its axis of symmetry, and whose ellipticity is finite. That follows directly from Clairaut's conversion of the results of MacLaurin's theory of such homogeneous fluid figures of revolution in equilibrium whose ellipticities are finite into analysis. By the same token, if Jupiter is assumed to be a figure of equilibrium shaped like a flattened ellipsoid of revolution which attracts according to the universal inverse-square law, which revolves around its axis of symmetry, and whose ellipticity δ is finite but less than this upper limit, where the last condition was in fact observed to hold, then Jupiter cannot be homogeneous. To apply Clairaut's theory of stratified fluid figures of equilibrium to Jupiter one must then assume that Jupiter is a stratified figure of equilibrium shaped like a flattened ellipsoid of revolution which attracts according to the universal inverse-square law and which revolves around its axis of symmetry. However, Clairaut only showed this theory to be good when the ellipticities of such stratified figures of equilibrium are *infinitesimal*. But the number $\frac{1}{23.23}$ is *not* infinitesimal. Clairaut must have concluded that (9.6.3) held for stratified fluid figures of revolution in equilibrium which attract in accordance with the universal inverse-square law, which revolve around their axes of symmetry, and whose ellipticities are *finite* by analogy with the result (9.6.1) that he found to hold for his theory of stratified fluid figures of revolution in equilibrium which attract according to the universal inverse-square law, which revolve around their axes of symmetry, and whose ellipticities are infinitesimal applied to the earth. (The upper limit $\frac{1}{230}$ and lower limit $\frac{1}{576}$ of the earth's ellipticity which he deduced by applying this theory to the earth are both infinitesimal.)

The fact that Maupertuis and Clairaut found such different values of the ratio ϕ of the magnitude of the centrifugal force per unit of mass to the magnitude of the attraction at Jupiter's equator, when in fact both had made the same assumptions and had done the same kind of calculations to arrive at their values of the ratio ϕ, can be explained by the fact that in doing the same calculations they used different values of the ratio of the radius of Jupiter's fourth satellite, the satellite that is a farthest from Jupiter among Jupiter's four observable satellites, to Jupiter's radius. Maupertuis used the value of the ratio which Cassini had found, and Clairaut used the value of the ratio which Pound had determined.[192] It is worth noting that both Maupertuis's and Clairaut's determinations of the value of Jupiter's ellipticity δ when Jupiter is assumed to be a homogeneous figure of revolution which is attracted by a fixed center of force located at the center of the figure, where the attraction due to the center of force at all points is directed toward the center of force, where the magnitudes of the attraction at points are assumed to vary inversely as the squares of the distances of points from the center of force, which revolves around its axis of symmetry, and whose columns from center to surface all balance or weigh the same are consistent with

$$\delta \simeq \tfrac{1}{2}\,\phi. \tag{9.6.2'}$$

That is,

$$\frac{1}{14.96} = \frac{1}{2}\left(\frac{1}{7.48}\right) \quad \text{and} \quad \frac{1}{23.23} = \frac{1}{2}\left(\frac{1}{11.615}\right).$$

But (9.6.2') was derived assuming that δ and ϕ are *infinitesimal*, and Jupiter's ellipticity δ was observed at the time *not* to be infinitesimal.[193]

Clairaut also noted that the earth was unique among the planets inasmuch as it required that certain conditions be fulfilled which the other planets did not. Unlike the other planets along whose surfaces the variations in the magnitudes of the effective gravity from their equators to their poles cannot be measured directly or be found by any indirect means, the variation of the magnitude of the effective gravity with latitude along the earth's surface could be determined. Consequently in the case of the earth, not only did the earth's true shape (that is, the correct value of the earth's ellipticity) have to follow from an assumed hypothesis of attraction in order for that hypothesis to be considered the possible law that governs the earth's attraction, but the hypothesis of attraction had to be consistent with the observed variation in the increase in the magnitude of the effective gravity with latitude along the earth's surface as well.[194] Now, by means of experiments made with seconds pendulums, the magnitudes of the effective gravity at all latitudes along the earth's surface could, in principle, be found. Furthermore, by measuring degrees of latitude at many different latitudes, the earth's shape could also, in principle, be determined (assuming, of course, that the earth is a figure of revolution). From these measurements, the way that the centrifugal force per unit of mass affects effective gravity at the earth's surface, which causes the effective gravity at the surface of the rotating earth to differ from the attraction at its surface, could be found. (Here Clairaut implicitly assumed either that the principle of the plumb line held at the earth's surface when the earth rotates, so that the effective gravity at all points at the surface of the rotating earth is perpendicular to the surface of the earth at those points, or else if the principle of the plumb line does not hold at the earth's surface when the earth rotates, then at least the directions of the effective gravity at all points at the surface of the rotating earth can be determined. The centrifugal force per unit of mass is everywhere perpendicular to the axis of rotation. If the directions of the effective gravity at points on the earth's surface when the earth rotates are not known and cannot be found, then it is impossible to determine the directions of the attraction at points on the earth's surface.) Hence the attraction at all points on the surface of an earth that did not rotate could, in principle, be found empirically. (Assuming that the principle of the plumb line holds at the surface of the earth when the earth rotates, the principle of the plumb line would not hold at the surface of the earth if the earth did not rotate, assuming that the earth does not change its shape when it rotates. In other words, the directions of the attraction at points on the earth's surface could not be perpendicular to the earth's surface at those points.)

But Clairaut now emphasized that the values of the magnitude of attraction at the points on the surface of the stationary earth did not uniquely determine the law of attraction in the earth's interior or in its exterior. The magnitude of attraction, he noted, could vary in the interior and in the exterior of a stationary figure in different ways and still accord with a given set of values of the magnitude of the attraction at the points on the surface of the figure.[195] Thus, for example, as Koenig had suggested, modified Cartesian hypotheses of attraction, where centrally located, three-dimensional nuclei attract, instead of centrally located points, might also "save" all of the phenomena,[196] just as Clairaut had shown some of Bouguer's hypotheses of attraction to do.

Clairaut may have taken certain works produced as a result of events that went on in Paris at this time into consideration here, too. In the years 1734–40, Malebranche's followers attempted to reconcile vortices with Newtonian attraction. Mairan had been a leading Malebranchist in the 1710s and 1720s. Other members of the Malebranche circle headed the later effort. One of Malebranche's partisans, Joseph Privat, Abbé de Molières, actively took part in the movement, in the later 1730s.[197] Another member of the group, Etienne-Simon, Abbé de Gamaches, published his *Astronomie physique* in 1740. Gamaches argued in that work that Newton's use of totally different internal structures to account for observed anomalies in the cases of the earth and Jupiter, respectively, appeared no less ad hoc than did tinkering with the features of vortices in order to "save" phenomena.[198] Although Malebranche's supporters failed in their endeavors, Clairaut may have nevertheless granted the truth of Gamaches's criticism. The Malebranchist astronomer had indeed perceived a disadvantage or inconvenience of Newton's theory – the very trouble that had begun to bother Clairaut himself in 1738.

Other works of the period on the earth's shape may have also influenced Clairaut's outlook and expectations. In his *Dissertatio de telluris figurâ* (1739) and in his *Dissertatio de inaequalitate gravitatis indiversis terrae locis* (1740), Roger-Joseph Boscovich stressed the uncertainty of theories of gravity and suggested that the earth's interior might be highly irregular. Nor did Boscovich think that it was possible to determine the earth's shape solely by measuring degrees of latitude, because he did not believe the earth to be regularly shaped. (If that were true, Clairaut's results concerning solid, stratified figures shaped like flattened ellipsoids of revolution which attract according to the universal inverse-square law, which revolve around their axes of symmetry, whose ellipticities are infinitesimal, and at whose surfaces the principle of the plumb line holds would not even be useful, much less his findings that pertained to stratified figures of equilibrium shaped like flattened ellipsoids of revolution which attract in accordance with the universal inverse-square law, which revolve around their axes of symmetry, and whose ellipticities are infinitesimal, because in both instances Clairaut assumed the earth to be regularly shaped and to have a highly regular internal structure.) Boscovich even contrived an ad hoc law of attraction which would permit the "$\sin^2$ law" of the variation in the increase in the magnitude of the effective gravity from the equator to a pole to hold along the surface of a homogeneous figure shaped like an ellipsoid of revolution which was *elongated* at its poles and at whose surface the principle of the plumb line holds.[199] Now, the Paris Academy actually received a copy of *Dissertatio de telluris figurâ* in 1739,[200] which increases the likelihood that Paris Academicians at that time knew about the work and perhaps paid attention to it. The work possibly even affected the points of view of some Paris Academicians in the years that followed. Clairaut in particular may have taken Boscovich's words to heart. We know that the French mathematician came to admire the Jesuit *savant* greatly.[201]

By the time Clairaut wrote his appendix to the French edition of Newton's *Principia*, edited by his friend the Marquise du Chatelet, he had concluded that a theory of the earth's shape which was founded upon the universal inverse-square law of attraction was impossible to prove, because too many auxiliary hypotheses concerning the earth's internal structure had to be introduced which could not be verified independently of the theory. The same difficulties troubled theories of the tides, when the ebb and flow of the tides occurred at the surface of an earth that was assumed to be heterogeneous and to attract according to the universal inverse-square law. Not only did the variation of the density of the earth's interior have to be known at the start, Clairaut observed, but in addition it made a difference whether the earth was assumed to be solid or fluid, in case the density of its interior was not constant, he noted as well.[202]

Clairaut may have had in mind Daniel Bernoulli's equilibrium theory of the tides, which earned Bernoulli a share of the Paris Academy's prize of 1740. We recall that in that essay, Bernoulli had applied the principle of balanced columns to a figure that consisted of a core made up of concentric, solid, individually homogeneous strata that are all spheres, which means that the strata are all surfaces, hence are all "infinitesimally" thin, and which is surrounded by an outer, homogeneous fluid layer whose thickness was finite and whose outer surface was an ellipsoid of revolution elongated at its poles whose center was situated at the common center of the strata and whose ellipticity was infinitesimal. Bernoulli assumed that the figure attracted itself in accordance with the universal inverse-square law. He also assumed that the figure is attracted by the sun and moon according to the universal inverse-square law and that the figure revolves around the sun and around the center of gravity of the system consisting of the earth and moon. In applying the principle of balanced columns to this figure, he implicitly assumed that the entire figure exists in a state of equilibrium.

In a work that he published in 1747, D'Alembert maintained that Bernoulli had made an error.[203] The alleged mistake resembled the kind that Clairaut tried to show that MacLaurin had made, although Clairaut's theory sometimes lacked the bases that Clairaut needed to demonstrate conclusively that MacLaurin had definitely made an error. It turns out that in claiming that Bernoulli had made a mistake, D'Alembert essentially reasoned as Clairaut had done in the case of MacLaurin, meaning that, in particular, D'Alembert supported his argument using these same bases that Clairaut's theory lacked. We remember that Bernoulli had assumed the earth's solid part to be composed of an infinite number of concentric, individually homogeneous strata that were all spheres, which means that they were all surfaces, hence whose thicknesses were all "infinitesimal," and that attract according to the universal inverse-square law. This solid part of the earth was surrounded by a layer of homogeneous fluid whose thickness was finite, which was concentric with the spherical strata, whose outer surface was an ellipsoid of revolution elongated at its poles whose ellipticity is infinitesimal, and which attracts according to the universal inverse-square law. Then in order to account for the phenomena of the tides, Bernoulli made the densities of these individually homogeneous strata that made up the solid part of the earth and the density of the outer, homogeneous fluid layer decrease from the earth's center to the surface of the outer, homogeneous fluid layer. But according to Clairaut's theory of stratified figures of equilibrium shaped like flattened ellipsoids of revolution which attract according to the universal inverse-square law, which revolve around their axes of symmetry, and whose ellipticities are infinitesimal, the strata that make up the earth's solid part could not be the geometrically similar, spherical ones that Bernoulli had assumed. According to Clairaut's theory, the strata that make up the earth's solid part must be concentric, flattened, individually homogeneous ellipsoids of revolution which all have a common axis of symmetry, namely, the axis of symmetry of the figure, and whose infinitesimal ellipticities increase from the center of the earth to the surface of the outer, homogeneous fluid layer. In other words, the surfaces that make up the earth's solid part cannot be the spheres that Bernoulli had assumed these surfaces to be.

In fact this conclusion does not truly follow from Clairaut's theory. One trouble is that Clairaut was only able to prove the result in case the stratified fluid figure of equilibrium shaped like a flattened ellipsoid of revolution which attracts according to the universal inverse-square law, which revolves around its axis of symmetry, and whose ellipticity is infinitesimal is made up of a finite number of strata whose thicknesses are all finite, not made up of strata that include an infinite number of strata that are surfaces, hence whose thicknesses are infinitesimal. In other words, D'Alembert's conclusion that Bernoulli's stratified figure was not a figure of equilibrium did not follow from Clairaut's theory in part for the same reason that Clairaut's conclusion that some of MacLaurin's stratified figures were not figures of equilibrium did not follow from Clairaut's theory.

Another problem is that integro-differential equation (9.4.3), which holds when the strata of the fluid figure of equilibrium are surfaces, and the corresponding finite system of equations, which holds when the strata of the fluid figure of equilibrium have finite thicknesses, are actually only valid when the stratified fluid figure of equilibrium attracts itself *alone* according to the universal inverse-square law and is not attracted by another body. And the equations that result in each of the two cases depend upon the assumption that the stratified fluid figure of equilibrium is shaped like a figure of revolution flattened at its poles. But in Bernoulli's figure the outer surface of the outer, homogeneous fluid layer is an ellipsoid of revolution elongated at its poles, not an ellipsoid of revolution flattened at its poles which it must be in Clairaut's theory.

[If there were no outer, homogeneous fluid layer whose surface is an ellipsoid of revolution elongated at its poles, whose thickness is finite, and which attracts according to the universal inverse-square law, then one could conclude that the spherical, solid part of Bernoulli's earth, which attracts according to the universal inverse-square law, which revolves around its polar axis, and which is made up of an infinite number of concentric, individually homogeneous strata that are spheres, which means that the strata are all surfaces, hence are all infinitesimally thin, cannot be a figure of equilibrium for the following elementary reason. If the spherical, solid part of Bernoulli's earth does not move and there is no outer, homogeneous fluid layer whose surface is an ellipsoid of revolution elongated at its poles, whose thickness is finite, and which attracts according to the universal inverse-square law, the attraction at all points on the surface of the spherical, solid part of Bernoulli's earth is perpendicular to the surface of the spherical, solid part of Bernoulli's earth at those points. Consequently, if the spherical, solid part of Bernoulli's earth revolves around its polar axis, the effective gravity at the surface of the spherical, solid part of Bernoulli's earth cannot possibly be perpendicular to the surface of the spherical, solid part of Bernoulli's earth at all points on the surface for the reasons that I explained in Chapter 2. In other words, the principle of the plumb line cannot hold at all points on the surface of the rotating, spherical, solid part of Bernoulli's earth. This in fact is even true if the number of concentric, individually homogeneous spherical strata that make up the spherical, solid part of Bernoulli's earth is finite, if the thicknesses of all of the spherical strata are finite, and if the densities of the spherical strata decrease from the center of the spherical, solid part of Bernoulli's earth to its surface. The principle of the plumb line is, however, a condition that must necessarily hold at the surface of any figure of equilibrium. Hence the rotating, spherical, solid part of Bernoulli's earth cannot be a figure of equilibrium in this case. Then one could say that perhaps Bernoulli could neglect the earth's flattening at its poles caused by centrifugal force of the earth's rotation in calculating the heights of tides, as he in fact did, but this still did not permit him to treat the rotating, solid part of the earth, which he assumed to be spherical instead of flattened at its poles, as a figure of equilibrium. But the presence of the outer, homogeneous fluid layer whose surface is an ellipsoid of revolution elongated at its poles, whose thickness is finite, and which attracts according to the universal inverse-square law invalidates or nullifies the preceding argument.]

[We recall that the densities of MacLaurin's stratified figures shaped like flattened ellipsoids of revolution which attract according to the universal inverse-square law, which revolve around their axes of symmetry, and whose ellipticities are infinitesimal did not *always* decrease from the centers of the figures to their surfaces. And when they did not, Clairaut's theory of stratified fluid figures of equilibrium shaped like flattened ellipsoids of revolution which attract according to the universal inverse-square law, which revolve around their axes of symmetry, and whose ellipticities are infinitesimal did not apply to such figures, although Clairaut knowingly tried even in these instances to use his theory to explain why his values and MacLaurin's values of the ellipticities of such figures did not agree with each other. Based on what Clairaut believed, Clairaut should have simply noted, which he did not do, that such figures did not satisfy a condition that he thought was a basic law of hydrostatics: that the densities of stratified figures of equilibrium must *always* decrease from the centers of the figures to their surfaces. However, as I mentioned at the end of the preceding section of this chapter, even the failure of this condition to hold is not enough to assure that stratified fluid figures are not figures of equilibrium.]

In his appendix to the French edition of the *Principia*, Clairaut restricted himself to reproducing Bernoulli's calculations of the heights of the tides in the special case that the earth's solid part is assumed to be homogeneous and to have the same density as the homogeneous fluid at its surface. This is the example that Newton had limited himself to in the *Principia* – with good reason, as it turned out. Because such a figure was homogeneous throughout, Bernoulli's calculations, which followed in part from applying the principle of balanced columns, were valid. In other words, in this particular case and this one alone, Bernoulli could treat the earth and its seas as a figure of equilibrium and consequently apply the principle of balanced columns to the figure. (The homogeneous figure of equilibrium is an *elongated* one, and this may appear paradoxical at first sight. We must not forget, however, that this figure not only attracts itself, but that the sun and moon are assumed to attract it as well. What cannot exist are elongated figures of equilibrium which attract according to the universal inverse-sqaure law and which are attracted by their own forces of attraction *alone*.) Clairaut even "forgave" Newton for having disregarded all but this simplest case![204]

In order to write his appendix to the French edition of the *Principia*, Clairaut had to review Newton's work very carefully. Clairaut won the Toulouse Academy's prize for the year 1750 by submitting an essay that was one by-product of this endeavor.[205] In the essay Clairaut put forward a new theory of the earth's shape. In the introduction to the essay he emphasized that theories of the earth's shape and theories of phenomena like the tides and the precession of the equinoxes not only depended upon the assumption that arbitrary hypotheses about the earth's internal structure which could not be verified held true, but that testing such theories depended as well upon a capability to make exceedingly critical observations. He maintained that it was all too easy for the magnitude of the cumulative errors made in measuring quantities to reach the same order as the magnitudes of the quantities being measured.[206]

In his new theory of the earth's shape, Clairaut again assumed that the earth is shaped like an ellipsoid of revolution flattened at its poles which attracts according to the universal inverse-square law, which revolves around its axis of symmetry, and whose ellipticity is infinitesimal. He also assumed that the earth's interior included a homogeneous nucleus whose center was located at the earth's center and whose density differed from the matter, also assumed to be homogeneous, which surrounded it. Unlike his theory of 1743, however, the nucleus was a figure of revolution *not* shaped like an ellipsoid of revolution. Clairaut postulated that the ordinates of the points along any meridian at the surface of the nucleus be larger than the ordinates of the corresponding points on a circle concentric with that meridian or on an ellipse concentric with that meridian by amounts proportional to the cubes of the cosines of the latitudes of the points.[207] Using such an hypothesized internal structure for the earth, Clairaut finally succeeded in reconciling the universal inverse-square law of attraction with the observations of the day. That is, by choosing the shape of the nucleus suitably, as well as the densities of the homogeneous nucleus and the homogeneous layer of matter which surrounded the nucleus, Clairaut showed that values of the ellipticity δ could be produced which were greater than $\frac{1}{230}$, when the value of $(P - \Pi)/\Pi$ was greater than $\frac{1}{230}$.

In order to arrive at such results, however, Clairaut of course had to abandon the hypothesis that the earth is a figure of equilibrium. In fact, he abandoned the assumption that the earth was once entirely composed of fluid. Clairaut evidently did not think that the earth could have been at one time, say, a fluid figure that was not in equilibrium. Such a belief would be consistent with Euler's failure, after repeated attempts, to get him to address the dynamical questions of approach to equilibrium, stability, and so on. We saw in Section 9.3 that Clairaut was totally unable to deal with such issues. As I also indicated earlier in this chapter, Clairaut's theories of homogeneous and stratified fluid figures of equilibrium are only based on conditions that are *necessary* for equilibrium. The theories are not based on any conditions that are *sufficient* for equilibrium.

Moreover, in order to achieve his goal, Clairaut equated the weights of columns from the center of the figure to its surface which lay along its axes. That is, he treated the two columns from center to surface as two columns *of fluid in equilibrium*. Having pointed out himself that the figure that represented the earth in his new theory could not possibly have been a body that once consisted entirely of fluid in equilibrium, or for that matter that, as far as he was concerned, the body could not even "have been [entirely] fluid originally,"[208] much less a fluid figure of equilibrium, this was a rather strange way to proceed. Indeed, he used here essentially the same method that MacLaurin had employed. But I mentioned earlier that when the strata in a figure that is shaped like a flattened ellipsoid of revolution which attracts according to the universal inverse-square law, which revolves around its axis of symmetry, and whose ellipticity is infinitesimal are all shaped like concentric, flattened ellipsoids of revolution whose ellipticities are infinitesimal and all of which have a common axis of symmetry, namely, the axis of symmetry of the figure, MacLaurin could not make δ be greater than $\frac{1}{230}$ and $(P - \Pi)/\Pi$ be greater than $\frac{1}{230}$ at the same time when he balanced columns from the center of such a figure to its surface along its axes, except for the figure made up of a homogeneous spherical nucleus that was surrounded by a stratum that did not attract, whose thickness was finite and whose outer surface was an ellipsoid of revolution flattened at its poles, which was itself surrounded by a homogeneous stratum whose thickness was finite, whose outer surface was also an ellipsoid of revolution flattened at its poles, and whose density was the same as the density of the nucleus, where the outer surface of the stratum that did not attract and the outer surface of the whole figure were concentric with the center of the nucleus, had a common axis of symmetry, namely, the axis of symmetry of the figure, and were geometrically similar (that is, had the same ellipticity). But Clairaut thought that he had ruled out this particular figure on grounds that it was not a figure of equilibrium, although Clairaut could not truly rule the figure out for the reasons that he imagined, because the density of this figure did not decrease from the center of the figure to its surface. He could have observed that *because* the density of the figure did not decrease from the center of the figure to its surface, the figure failed to fulfill a condition that Clairaut believed to be a fundamental law of hydrostatics, but Clairaut did not mention this. Now we find Clairaut himself balancing columns from center to surface along the axes of figures that he believed could not be figures of equilibrium.

This change in strategy (that is, not treating the earth as a figure of equilibrium or even a body once composed entirely of fluid) surprised some of Clairaut's faithful correspondents, like the Genevan mathematician Gabriel Cramer.[209] In fact Clairaut simply introduced the new hypothesis in an ad hoc fashion. The whole problem had become for him by this time little more than an academic exercise – one to be resolved by whatever means served the purpose. Indeed, by allowing the nucleus to be heterogeneous, by assuming, for example, that a homogeneous spherical figure lies inside the nucleus, whose density differs from the rest of the nucleus which surrounds it, Clairaut observed that the number of parameters that could be varied increased. These parameters could be chosen in an infinite number of ways, and by choosing them appropriately, he demonstrated that *any* proposed combination of a value of the ellipticity δ which was greater than $\frac{1}{230}$ and a value of the relative increase $(P - \Pi)/\Pi$ of the effective gravity with latitude which was greater than $\frac{1}{230}$ could be produced. However, Clairaut remarked, all this did was prove that hypotheses about the earth's internal structure could always be found which accorded with a particular postulated combination of a value the ellipticity δ which was greater than $\frac{1}{230}$ and a value of the relative increase $(P - \Pi)/\Pi$ of the effective gravity with latitude which was greater than $\frac{1}{230}$. But this did not help resolve the problem of determining the true structure of the earth's interior.[210]

With this seemingly impenetrable problem of determining the true structure of the earth's interior in mind, Clairaut ceased to investigate terrestrial problems as · soon as he had completed his *Théorie de la figure de la terre, tirée des principes de l'hydrostatique*, which finally appeared in 1743. In the years that followed the publication of this work he turned to the heavens instead as a new scene of action for testing the universal inverse-square law of attraction. Since the enormous distances between celestial bodies were larger than their mean diameters by several orders of magnitude, Clairaut reasoned that the motions of these bodies relative to their centers of mass could be disregarded. Hence the bodies could be thought of as spherical bodies, which meant that they could ultimately be treated as points, because a homogeneous spherical body or a spherical body composed of concentric, spherical strata which attracts according to the universal inverse-square law attracts bodies in its exterior like an inverse-square central force. (In fact such spherical bodies attract bodies in their exteriors as if the whole masses of the spherical bodies were concentrated at their centers.) In other words, Clairaut thought that celestial bodies would obey the laws that govern the way that mass-points attract each other. In particular, Clairaut chose the anomalous motion of the moon's apogee, another problem that Newton had handled unsatisfactorily in the *Principia*, which remained unresolved, as the new phenomenon for testing the universal inverse-square law of attraction.[211] As Clairaut expressed the matter, in the celestial realm, "the data are so simple that geometry has no need for auxiliary hypotheses."[212]

IO

Epilogue: Fontaine's and Clairaut's advances in the partial differential calculus revisited, or the virtues of interrelated developments in mathematics and science, and the fall of "normal" science

The preceding text is in part a story about how mathematical science progressed in Paris during the first half of the eighteenth century after the wars that had isolated France from scientific developments that took place elsewhere in Europe came to an end, and about how foreign influences contributed to this development of mathematical science in Paris when the borders that separated France from her neighbors were reopened after the wars. It is also in part a story about the way a leading scientific institution in Paris, the Paris Academy of Sciences, operated during the first half of the eighteenth century.

Despite the strategic location of Chapter 7, the "internal" part of the story is not about applied mathematics. What we call today applied mathematics did not exist during the eighteenth century,[1] and even if it had, the internal part of the story just told would not be about applied mathematics. Instead, the internal part of the story just recounted began with unrelated developments in mechanics and in mathematics. As the story unfolded, further advances in the mechanics and in the mathematics became interconnected to their mutual advantage. Both mathematics and mechanics progressed thanks to the interaction. Indeed, it is conceivable that the improvements in both domains might not have occurred without the interplay. The internal part of the story traced the origins and development of a new theory of figures of equilibrium. At a critical juncture in the story, certain new, contemporaneous advances in mathematics intervened and subsequently played a major part in the further development of the new theory of figures of equilibrium. Conversely, the steps that had to be taken in order to develop the physical theory to which the mathematics gave rise caused the mathematics itself to

620

develop further as well. The story illustrates a purpose that mathematics serves in the growth of physical sciences which few physicists who are not mathematically inclined would ever emphasize. This use would simply never occur to most physicists, or else they would deny that mathematics has such a function.

Nor do mathematicians always recognize what is involved. They tend to simplify matters from a different point of view. Thus Todhunter, for example, did not bother to look into the history of the partial differential calculus. As a result he did not realize that the partial differential calculus that Clairaut used to create his theory of figures of equilibrium of 1743 was itself brand new. Todhunter simply ignored the mathematics, because he assumed that the mathematics already existed. He treated the mathematics as if it had been available all along, just ready and waiting to be applied.

The way that Todhunter proceeded points up a difficulty that has afflicted the history of physical sciences at least since the nineteenth century and that continues to cause trouble today. Historians of physics avoid the mathematics of physics, while historians of mathematics focus primarily on technical mathematics and exclude physical contexts. Consequently advances in eighteenth-century mathematics and mechanics which are historically interrelated have been artificially disunited by historians of the period, who attempt to separate with the benefit of hindsight what they view as the "purely physical" parts of problems from the "purely geometrical" or "purely analytical" aspects. In retrospect the historians of mathematics disregard the mechanics and manufacture so-called developments in pure analysis.

One effect of this is that the story related in the preceding pages has never been properly told. Instead, the idea of "normal science" materialized, along with the accompanying notion of "puzzle solving" within "paradigms."[2] The "mopping up" of Newton's *Principia* has acted as the paradigm of a "paradigm." Associated with puzzle solving is a tacit supposition, which in the extreme becomes a feeling of inevitability, that puzzles will be solved, thereby strengthening the paradigm, until a point is reached where "anomalies" remain which the paradigm cannot be made to explain. These leftover anomalies are important enough to endanger the well-being of the paradigm. They can cause the paradigm to break down, or they can cause people to ignore the paradigm thenceforth and to shift their interests elsewhere, namely, toward a different and, more often than not, new paradigm that can be made to resolve the anomalies.

I do not wish to argue that a distinction between "revolutionary" science and "nonrevolutionary" science cannot be made. It would doubtlessly be foolish to try to defend such a thesis, when, for example, revolutions in mathematics, like non-Euclidean geometries, have been observed to occur.[3] I do maintain, however, that the usual way of describing the spread of Newtonianism, and the mopping up of the *Principia* in particular, leaves much to be desired. Standard accounts of this process excessively simplify events. The oversimplification is convenient, since it is invariably taken as a, if not *the*, model or doctrinal example of normal science.

But this fact simply illustrates that the idea of normal science, as it is ordinarily conceived, has some problems. Whether the range that a paradigm covers can be expanded or not can depend on factors that one usually does not call attention to. For this reason there is a tendency to distinguish nonrevolutionary science or normal science from revolutionary science in a manner that seems too simplistic, at least to me.

Physicists have often treated mathematics as a convenient language for expressing "physical" ideas that they say come to mind independently of the mathematics. In principle one could, they believe, substitute another language for the mathematics to express these ideas, although the representation of the ideas in the other language could turn out to be more awkward and cumbersome than if mathematics were employed to express the ideas instead. In my opinion, "normal science" as it is usually defined, is an idea that is readily and eagerly upheld by historians of science and philosophers of science who, following the physicists, view mathematics as purely utilitarian instead of an essential component of the science into which it enters. Like the physicists, these historians and philosophers treat mathematics as a language that could in theory be eliminated from science by replacing it by some other language. I think that historians and philosophers who believe this often do not appreciate, perhaps simply as a result of ignorance, how mathematics works in science. Mathematics need not simply serve as a vehicle for expressing scientific developments. Developments in mathematics can act to *shape* science. In the opinion of many historians and philosophers, broadening the range that paradigms take in, by means of solving puzzles, characterizes nonrevolutionary science. In thinking this way, these historians and philosophers either presuppose conditions or neglect matters that they should not. For one thing, they usually assume implicitly that if the task of solving puzzles can be carried out at all, a sufficient amount of mathematics exists for doing this, which is a rather gratuitous supposition to make at the start. The mathematics for doing so may *not* exist, and it *need not ever* be invented. There is certainly *nothing* to assure that the contrary will happen. Even physicists who do attach an importance to the "elegance" of mathematics sometimes forget that the physics *cannot even be done* without that "elegant" mathematics.

The usual way that the story told in this book has been sketched makes that story exemplify normal science. As I have recounted the story, however, it evidences all of the points I have just indicated. If my version of the story can be believed, then we have to conclude that mathematics can appear in a physical science as *much more* than just a convenient language for expressing the physical science. "Physical" ideas are not as independent of mathematics as they are customarily thought to be. It is hard to see how the story as I have told it illustrates "normal science" in the usual sense of the term. It is impossible for me to imagine this story without the episode from the history of mathematics. In my view, the idea of solving puzzles within a Newtonian framework does not alone do justice to the story's complexity. It simply does not capture some basic elements that are involved in the story's unfolding. Newton's theory of the earth's shape, which was based on the universal inverse-square law of attraction, did not accord with terrestrial observations. It is true that this anomaly in today's parlance is what originally induced Clairaut to investigate the problem of the degrees of flattening of stratified figures shaped like ellipsoids of revolution flattened at their poles which attract according to the universal inverse-square law, which revolve around their axes of symmetry, and whose ellipticities are infinitesimal. One could view an "inevitable" solution of problems involving the universal inverse-square law of attraction which remained after the *Principia* was published as the theme that unifies the parts of the story. Such a standpoint, however, minimizes developments that are critical parts of the story, yet which cannot easily be included under the heading "Newtonianism." To ignore these developments in recounting the story is to oversimplify the chain of events that make up the story. If one overlooks Bouguer's important contributions to mechanics and Fontaine's new mathematics, neither of which had its origins in Newton's work, he literally misses half of the story.

Nor should we forget that Clairaut's mature theory of 1743 of figures of equilibrium provided foundations for results that contradicted Newton's speculations. Newtonianism sheds no light whatever upon the insuperable *mathematical* difficulies that Clairaut had to face in developing his theory in order to try to explain why his findings and MacLaurin's did not agree with each other. Both Clairaut and MacLaurin, after all, were "Newtonians," but their conclusions differed. Indeed, their aims and objectives conflicted. (That fact in itself, however, does not undermine the notion of normal science and the idea of working within a paradigm which is so fundamental to that notion.) As we have seen, although Clairaut introduced mathematical tools to deal with the problem of the earth's shape, which resulted in mathematical problems whose solutions more than likely could have answered the question why his findings and MacLaurin's differed, Clairaut did not truly clear up the mystery, at least not entirely, because he could not solve the mathematical problems. The kinds of mathematical obstacles that Clairaut had to confront in dealing with terrestrial problems did not facilitate his endeavor to verify the universal inverse-square law of attraction. The law was only conclusively established later, after problems having to do with *heavenly* phenomena, *not earthly* phenomena, were solved.

Clairaut and MacLaurin both worked on *problems* that Newton did not solve in the *Principia*. But what, then, is the paradigm that Clairaut worked within? Robert Hooke had to explain the first law of motion twice to Newton before Newton finally understood it.[4] Hooke also arrived at the universal inverse-square law of gravitation independently of Newton. (It is also true that Hooke could not and did not deduce mathematically any natural phenomena from the law.[5]) The physical meaning and mathematical formulation of the second law of motion that we know today are not the ones in the *Principia*. The ones that we know today were introduced by Euler in 1750, and the mathematical formulation of the second law of motion that we know today probably owes more to Galileo's time-squared law of falling bodies together with the continental differential calculus than to anything that Newton ever wrote.[6] Newton interpreted centrifugal force in terms of the third law of motion, but during the eighteenth century other mathematicians interpreted centrifugal force differently, in terms of Cartesian ideas, "fictitious forces," and so forth.[7] In other words, some parts of what we ordinarily think of as the Newtonian paradigm were not really due to Newton; some parts changed during the eighteenth century; and some parts were ambiguously interpreted in ways that conflicted, and the differences in these interpretations were not well understood in the eighteenth century. It is thus hard to maintain that in trying to solve problems that Newton had left unsolved in the *Principia*, Newton's successors strictly "worked within" a framework that Newton had clearly formulated.

It is often said that Hooke's role in introducing ideas into mechanics like the first law of motion and the universal inverse-square law of gravitation is unimportant or inconsequential, because Hooke could not make the ideas mathematical. In other words, Hooke's ideas came cheaply, and creating a mathematical theory out of such ideas, as Newton tried to do, is what is important. But the same reasoning can be applied to Newton. He did not have the mathematics required to solve certain problems in the *Principia* which he consequently left unsolved, like the problem of the earth's shape. And when he could not solve a problem mathematically he resorted to conjectures and "intuitions" which were not so different from the kind that Hooke employed. Making ideas mathematical is a relative notion. A theory is "more or less" mathematical. If Newton's *Principia* is far more mathematical than anything Hooke ever wrote or ever could have written, Clairaut's mature theory of 1743 of the earth's shape was also a good deal more mathematical than Newton's theory of the earth's shape in the *Principia*. Some of Newton's conjectures about the earth in the *Principia* could only be substantiated mathematically later in an ad hoc fashion, which is what Clairaut did in his essay that won the Toulouse Academy's prize for 1750. And in his treatise of 1743 Clairaut went as far as he could go with his theory of the earth's shape until his progress was stopped by mathematical problems that he could not solve.

Finally, the discoveries that Bouguer made concerning figures of equilibrium, which Clairaut exploited to fullest advantage in his treatise of 1743, owed nothing to Newton. Nor did the mathematics that Clairaut used to develop his theory of figures of equilibrium in that treatise owe anything to Newton.

To imagine what Newton might have done, had he invented the partial differential calculus, which he did not do, or if other mathematicians who did participate in the creation of the partial differential calculus, like Leibniz and Johann I Bernoulli, had also perfected this calculus when Newton was in the prime of life, which they did not do, is to speculate idly and pointlessly. In contrast to this futility, one can probably reason with some assurance that Einstein might not have gone as far as he did in General Relativity, had he not had Riemannian geometry, the absolute differential calculus of Gregorio Ricci-Curbastro and Tullio Levi-Civita, later renamed tensor analysis by Einstein, and, in particular, the Ricci tensor at his disposal.[8] As I said earlier, it probably would be foolish, and perhaps pointless, to argue that scientific revolutions do not take place. It is possible, however, that what determines whether a scientific revolution is needed or not, and whether it can occur if it is needed, may depend in part upon whether suitable mathematics is invented or not. If such mathematics is created, it may postpone a need for a scientific revolution. If it is not invented, then unsolvable "anomalies" persist, which call for a scientific revolution to clear them up. Once again, however, whether scientific revolutions can come about or not may depend in part upon whether suitable mathematics is created or not. If these suggestions are true, then scientific revolutions too involve the issues in mathematics which are central to the internal part of the story told in the preceding pages.

If Clairaut failed to resolve conclusively all of the problems that had first induced him to investigate stratified figures of equilibrium shaped like ellipsoids of revolution flattened at their poles which attract in accordance with the universal inverse-square law, which revolve around their axes of symmetry, and whose ellipticities are infinitesimal, it is also true that his research cannot be judged to have been a vain attempt. For in the course of carrying out his research, Clairaut laid new foundations for theories of the state of equilibrium of fluid masses, which other mathematicians, like Euler, would find useful later.

In the internal part of my story I have not argued that any of the events that occurred were inevitable. On the contrary, chance played a major part in the internal part of the story. Nearly simultaneous, but originally unrelated developments in mechanics and in mathematics, respectively, were united. These developments individually cannot be traced back to any common source that might have assured that they would converge at some time in the future. Nothing in the prior histories of the mechanics and the mathematics foreshadows what happened later.

Normal science connotes a routine, namely, puzzle solving. The events involved in the internal part of the preceding story form anything but a routine. Fortunately, some of the earliest methods for integrating or reducing to quadratures first-order ordinary "differential" equations of first degree and first-order "differential" equations of first degree in three variables, which involved use of the partial differential calculus, just happened to be developed at a time when it had recently been made clear that the correct foundations of the theory of figures of equilibrium were not truly known. The new mathematics, the partial differential calculus, turned out to be exactly what was needed to establish the foundations that had been missing, thereby enabling people to understand what they had not really understood until then. The idea of a complete differential $P(y, x)\,dy + Q(y, x)\,dx$ in two variables y and x, meaning a differential for which an expression $\phi(y, x)$ exists whose total differential is the differential $P\,dy + Q\,dx$, first arose in connection with the methods of solving the problems in integral calculus just mentioned. The idea was extended to a differential $P(y, x, z) dy + Q(y, x, z) dx + R(y, x, z) dz$ in three variables y, x, and z, which means that an expression $\phi(y, x, z)$ exists whose total differential is the differential $Pdy + Qdx + Rdz$. Clairaut found this idea to be just what he needed to help him formulate mathematically a new theory of figures of equilibrium. At the same time, in developing the theory, Clairaut infused the new mathematical analysis with geometric meaning. The property that a differential $Pdy + Qdx + Rdz$ be complete could be expressed through "line integrals" of the differential $Pdy + Qdx + Rdz$. In all probability, Clairaut came to this conclusion while he was applying arguments that involved infinitesimals to channels that lie within the interior of a homogeneous fluid mass in equilibrium and that are filled with fluid from the homogeneous fluid mass in equilibrium, when forces of attraction act upon the fluid mass which permit all of the conditions then known which were necessary for the fluid mass to exist in a state of equilibrium to hold. The notion of integration along curved plane and skew paths very likely *grew out of* the problem in mechanics. Rectilinear columns or channels had customarily been applied to figures thought to be homogeneous figures of equilibrium before Clairaut developed his theory of figures of equilibrium. The representation of resultant "effort" (that is, weight) by means of "line integrals" became essential when calculations had to be carried out along curved plane or skew paths instead of along rectilinear ones. The tools of synthetic geometry (proportions and the like) would not do when paths were curved. Clairaut doubtlessly never even imagined the idea of a "line integral" until he had to calculate the "effort" along a curved plane or skew channel produced by forces of attraction and centrifugal force of rotation which act upon the fluid in the channel.[9] Clairaut here was neither acting like Thomas S. Kuhn's mathematical physicist, who tends to take the physics problem as conceptually fixed and to develop powerful mathematical techniques for application to it, nor was he acting like Kuhn's theoretical physicist, who thinks more physically, adapting the conception of his problem to the often more limited mathematical tools at his disposal.[10] Clairaut falls between the two classes of physicist, or else he straddles them, for the mathematics and the mechanics interacted in his work. The mathematics and the mechanics *mutually shaped each other*.

Clairaut's reasons for introducing line integrals have analogues in twentieth-century physics. The calculations involved in Werner Heisenberg's creation of quantum mechanics required that Heisenberg reinvent matrix algebra. Richard Feynman's reformulation of quantum mechanics required that Feynman invent the Feynman integral.

Clairaut began to talk about the problem of stratafied figures of equilibrium shaped like ellipsoids of revolution flattened at their poles which attract according to the universal inverse-square law, which revolve around their axes of symmetry, and whose ellipticities are infinitesimal, without having thought much about it, in his second paper on the theory of the earth's shape published in the *Philosophical Transactions*. At the time he had no formal mathematics to handle the problem. Having proceeded incautiously and precipitately, Clairaut came to a number of erroneous conclusions in the paper concerning such figures of equilibrium. He only began to develop the mathematics that ultimately transformed his vague, ambiguous reflections into a coherent theory, which enabled him to find and correct the mistakes that he made in his second paper on the theory of the earth's shape published in the *Philosophical Transactions*, during his controversy with Fontaine about methods of integrating or reducing to quadratures first-order ordinary "differential" equations of first degree. Fontaine unintentionally touched off what Clairaut tried to turn into a contest, with Fontaine cast for the part of the unwilling participant, nearly two months after Clairaut had sent his second paper on the theory of the earth's shape published in the *Philosophical Transactions* to the Royal Society. I have suggested that the mathematics and mechanics subsequently interacted to their mutual benefit. Clairaut's argument with Fontaine lasted from November of 1738 until the summer of 1741. Is it any wonder that Clairaut did not introduce the line integral into the partial differential calculus that was at bottom the topic of their debate until some time after September of 1741, when, as a result of having corresponded with MacLaurin, Clairaut finally began to work earnestly on the problem of stratified fluid figures of equilibrium shaped like ellipsoids of revolution flattened at their poles which attract according to the universal inverse-square law, which revolve around their axes of symmetry, and whose ellipticities are infinitesimal? How can the developments in mathematics and mechanics in question be anything but closely interrelated? The principle of the total effort equal to zero around closed channels that lie within the interior of a homogeneous fluid mass in equilibrium and that are filled with fluid from the homogeneous fluid mass in equilibrium did not itself depend upon arguments that involved infinitesimals. Instead, this principle of equilibrium preceded its analytic expression, in coming directly from mechanics (the principle of solidification applied to arbitrarily shaped, closed plane or skew channels that lie inside a homogeneous fluid mass in equilibrium and that are filled with fluid from the homogeneous fluid mass in equilibrium). Arguments that involved infinitesimals then received a new interpretation *through* the problem in mechanics, when the principles of equilibrium were expressed in terms of the infinitesimal calculus. Clairaut's geometric interpretations of complete differentials, as opposed to the algebraic or analytic interpretation of complete differentials which first made its appearance during Clairaut's and Fontaine's controversy, emerged in the course of Clairaut's investigation of the problem in mechanics. Separate developments in mathematics and in mechanics, respectively, mutually affected the two domains, and the two domains advanced as a result of the mutual influences. But Clairaut's geometric interpretations of complete differentials also contradicted a tendency of mathematicians during the eighteenth century to try to purge the infinitesimal calculus of its original geometric foundations.[11] As I mentioned in Chapter 7, this trend only began to be reversed in France toward the end of the eighteenth century.

Indeed, it is reasonable to maintain that even more is true. "Conditional equations" like

$$\frac{dP}{dx} = \frac{dQ}{dy} \qquad (7.3.4.6''')$$

only really attracted attention *after and as a result of* having appeared in Clairaut's theory of 1743 of figures of equilibrium. Had these equations never gotten beyond the confines of Clairaut's and Fontaine's polemic, they might have gone unnoticed for years, much as the literature on the problem of orthogonal trajectories, published in *Acta Eruditorum* for 1717–21, had. Nor can ignorance of the literature published in *Acta Eruditorum* during these years simply be attributed to the fact that Latin was gradually losing its place as the universal language of scholars. But Clairaut's quick application of such equations as (7.3.4.6''') to the theory of figures of equilibrium made it practically impossible for these equations to be overlooked and forgotten. Clairaut's treatise of 1743 on the earth's shape, not his Paris Academy *mémoires* of 1739 and 1740 on integral calculus, soon prompted Euler to begin to look for and find ways to make widespread use of these "conditional equations" in hydrostatics and in hydrodynamics.[12]

We recall that Fontaine had published *nothing* as yet concerning the integral calculus. As I indicated in Chapter 7, there is good evidence that Clairaut's aggressive behavior toward Fontaine in 1739–41, ostensibly brought on by Clairaut's dislike of Fontaine's particular method of integrating or reducing to quadratures first-order ordinary "differential" equations of first degree using the partial differential calculus, which Clairaut showed in his Paris Academy *mémoire* of 1739 on integral calculus, bothered Fontaine during the summer of 1741. Clairaut's conduct is very likely one of the factors that caused Fontaine to put off publishing his work on integral calculus.

One consequence was that Euler did not learn until years later just how crucial a part Fontaine had played in introducing the partial differential calculus into France. To be sure, as I showed in Chapter 7, one could have guessed what Fontaine had done by paying very careful attention to the references to Fontaine which appear in Clairaut's printed Paris Academy *mémoires* of 1739 and 1740 on integral calculus. But then when most individuals who are not historians read, they do not ordinarily look for the kinds of facts and details that historians do. We recall from Chapter 7 how much Euler admired Fontaine's "fluxio-differential" method for solving the problem of tautochrones in resistant media. But in the Berlin Academy's *Mémoires* for 1748, published in 1750, Euler falsely attributed the "conditional equation" for the integrability of the first-order "differential" equation $Q(x, y, z) dx + R(x, y, z) dy + S(x, y, z) dz = 0$ of first degree in three variables x, y, and z to Clairaut and D'Alembert.[13] Startled by the erroneous reference to him in the Berlin Academy's *Mémoires*, D'Alembert wrote his friend Maupertuis, then the Berlin Academy's president, a letter in 1750 to be published in the Academy's *Mémoires* which rectified the mistake. D'Alembert emphasized that the equation in question was Fontaine's, and he stated his opinion that it was time that its author publish it along with numerous other interesting results.[14] D'Alembert's communication, which appeared in the Berlin Academy's *Mémoires* for 1749, published in 1751, enlightened Euler and brought Fontaine back to his mind. In 1751, Euler told the French astronomer Jérôme de Lalande, then visiting Berlin, that "if there is something yet to be discovered which we lack even a vague idea of, Fontaine is the one who will find it."[15] And we remember from Chapter 7 that in 1752, Euler wrote that Fontaine's "whole-symbol" notation for partial differential coefficients greatly "facilitated calculation."

But memories can be short, and even Euler's tributes were not enough to assure Fontaine immortality. Thus, for example, even the dictum: "admit infinitesimals as a hypothesis, study the calculus as practiced, and the faith will surely come to you," which is ordinarily ascribed to D'Alembert, has been traced to Fontaine.[16] Fontaine's part in the history of the partial differential calculus, which was to have introduced this mathematics into France, was quickly forgotten. Why? We have only to read France's most objective late-eighteenth-century mathematician-historian Jacques-Antoine-Joseph Cousin to find part of the answer to the question, I think.

Cousin had a unique reputation as a humble, modest individual for a Frenchman, even among foreigners like the Swedish astronomer Anders-Johann Lexell, who portrayed French men of science of the time unfavorably on the whole, but who called Cousin "a very sensible and moderate man."[17] In his unpublished "Eloge" of Cousin, which he drafted after the French mathematician-historian died, Jean-Baptiste-Joseph Delambre, secretary of the Paris Academy of Sciences reorganized as the French Institute during the French Revolution, characterized the late mathematician-historian as someone who had tried to "render to each individual exactly what belonged to him – something that is perhaps a little too rare."[18] Much more recently, one scholar has paid homage to Cousin for his "earnest and finally successful attempt to determine the history of partial differential equations."[19] In Cousin's opinion, Clairaut's Paris Academy *mémoires* of 1739 and 1740 on integral calculus added little to Fontaine's work. Clairaut's achievement instead was to have applied the results of the mathematical controversy toward solving a notoriously difficult problem in mechanics.[20]

Cousin exaggerated in suggesting that Clairaut had produced little, if anything new in integral calculus. Fontaine unquestionably introduced the partial differential calculus into France, but Clairaut used this calculus to devise a method for integrating or reducing to quadratures first-order ordinary "differential" equations of first degree which differed from Fontaine's method and which was in the long run the more practical of the two methods, in the sense that it was the easier one to apply. (It is still taught today to students of "advanced calculus.") Moreover, the novelty of Clairaut's geometric interpretation of Fontaine's "conditional equation" (7.3.3.111) for the integrability of the first-order "differential" equation (7.3.3.97) of first degree in three variables, which appears in Clairaut's Paris Academy *mémoire* of 1740 on integral calculus and involves surfaces, cannot be denied. In addition, we must not forget that when Fontaine and Clairaut talked about the integrals of complete differentials $Adx + Bdy$, $Adx + Bdy + Cdz$, and so on, as they argued about methods òf integrating or reducing to quadratures first-order "differential" equations of first degree, they meant and utilized the analytic or algebraic notion of such integrals exclusively (that is, expressions $\phi(x,y)$, $\phi(x,y,z)$, and so on, respectively, whose total differentials are, respectively, the differentials $Adx + Bdy$, $Adx + Bdy + Cdz$, and so on). But in his treatise of 1743 on the earth's shape, Clairaut geometrically interpreted constant values of the integral in the sense just described of a complete differential in three variables as "level surfaces" of the effective gravity. Moreover, in the same work, Clairaut introduced the geometric idea of the "line integral" of a differential in two or three variables, and he identified such differentials whose line integrals were path independent with differentials that were complete in the analytic or algebraic sense.

In assessing Cousin's conclusions, we must take into account the fact that Cousin was one of Fontaine's followers and supporters. He cared much more for Fontaine than for either Clairaut or D'Alembert, Fontaine's two distinguished colleagues in the Paris Academy.[21] Consequently, Cousin's feelings more than likely interfered with his powers of discrimination, and, as a result, he overstated Fontaine's importance, in trying to redress the balance by giving Fontaine the due that others had failed to give him. In the advanced lectures that he gave at the Collège de France, where he taught late in the eighteenth century, Cousin tried to perpetuate the memory of his former mentor Fontaine, in recounting anecdotes about the extraordinary exploits in mathematics which Fontaine brought off during his earliest years in Paris.[22] But the efforts that Cousin made on behalf of his late teacher, who died in 1771, were to no avail. Why? Cousin must have known the answer to the question himself. For he taught the works of Clairaut and D'Alembert, not those of Fontaine, because Clairaut and D'Alembert had produced important applications of the new mathematical analysis. In particular, these two mathematicians put the new analysis to use to solve problems in celestial mechanics. The works of Clairaut and D'Alembert provided Cousin with the substance for the lectures that he gave on physics and astronomy at the Collège de France for more than thirty years.[23] Cousin had to admit himself that the applications of analysis found in the works of Clairaut and D'Alembert attracted all of the attention, not the analysis itself which had been Fontaine's forte. Fontaine took no interest at all in the problem in mechanics which Clairaut had taken steps toward solving thanks to the mathematics that Fontaine had introduced into Paris. Lalande conveys Fontaine's apathy in a delightful anecdote:

it is astonishing that Fontaine did not know what the equator and meridian were. When the earth's shape was the subject of dispute at the Paris Academy in 1734, Fontaine led one of his colleagues to the [Jardin des] Tuileries. The colleague then proceeded to explain the problem to Fontaine in the sand. Fontaine told his colleague when he had finished: "If that is all there is to it, you have only been looking at elementary questions for the past month." Fontaine's genius for analysis showed him beyond the problem in question.[24]

In attributing the major French achievements of the late 1730s in the partial differential calculus to Fontaine and the important, immediate physical applications of them to Clairaut, Cousin indicated, perhaps not unawares, key factors that would eventually cause Fontaine to sink into oblivion.

Moreover, as I mentioned in Chapter 7, Fontaine *published nothing* concerning integral calculus until his complete works, which the publishers brought out in 1764, when Fontaine was sixty years old. By that time, however, the partial differential calculus and the calculus of variations had advanced considerably in the hands of Euler, D'Alembert, and Lagrange. As a result, the knowledgeable reader found little in the complete works of Fontaine which struck him as particularly original or surprising. Moreover, what the complete works did include that was new and unique often appeared not to be very useful, because Fontaine continued to tackle problems that he defined in extreme generality, as he had done with first-order ordinary "differential" equations of first degree, instead of starting with special cases.

When Clairaut died in 1765, Diderot wrote an obituary of the famous French mathematician. Although Diderot did no mathematics, meaning that he did not produce any mathematics, himself, the editor of the *Encyclopédie* by no means ignored the subject completely. In fact he has been judged to have been a fairly competent amateur.[25] The great *philosophe* knew enough about the mathematics scene in Paris to be able to make intelligent statements about it. In his obituary of Clairaut, Diderot wrote that the recently deceased, "very great geometer" was "nearly on the same level as Euler, Fontaine, the Bernoullis, and D'Alembert. He had less genius than Fontaine, but more soundness and steadiness, and less insight and perception than D'Alembert." Diderot figuratively termed Fontaine "the wheelwright, who sought to perfect the plow," while Clairaut and D'Alembert "stuck to laboring with the plow as it was."[26] Diderot made these remarks after having broken with D'Alembert and after having renounced mathematics. Diderot now viewed mathematics as an exhausted, fruitless, sterile discipline, in which obsession with abstraction as a guiding principle had led to what he judged to be an oversimplification of real phenomena and, as a result, had led, according to Diderot, to the field's destruction. By the time Diderot developed this new attitude toward mathematics he had converted to experimental sciences like chemistry and the life sciences. These he considered to be the sciences of the future. Even so, his new interests did not alter his sensibilities so much that he could no longer recognize "who was who" in mathematics.

Yet in 1784, the Library of the Paris Academy of Sciences did not even possess one copy of Fontaine's complete works, either in the first edition (1764) or in the second (1770). Indeed, Fontaine's name does not even appear in the Library's inventory for 1784.[27] This is utterly astonishing. Practically every other eighteenth-century Paris Academician of standing, including some who were much less renowned at the time for their work, is cited in the inventory, usually along with at least one major work as well.

Dirty French politics may have also contributed to Fontaine's downfall. The year 1765 was a bad one for D'Alembert. He fell seriously ill. Clairaut also died that year, and D'Alembert had extreme difficulty obtaining the position of pensioner in the Paris Academy which Clairaut vacated when he died, which D'Alembert should have automatically gotten, and which in D'Alembert's case was long overdue. D'Alembert was rightfully entitled to occupy the position based on seniority. But even after the Paris Academy chose D'Alembert to fill the position, the king tried to block the decision, because D'Alembert was notoriously antiaristocratic. D'Alembert had to use his reputation abroad as power to force the crown to decide in his favor.

Moreover, in his complete works, published the year before, Fontaine began by stating a principle of dynamics which he said he had first introduced in 1739 for resolving problems of motions of bodies.[28] D'Alembert's adversaries among the journalists, of whom there were many, interpreted this principle to be the same as the one known as "D'Alembert's principle," which D'Alembert presented in his *Traité de dynamique*, published in 1743. For example, writers for the *Mercure de France* and the *Journal de Trévoux* did this in their reviews of Fontaine's complete works, in 1765. The reviewers for the *Journal de Trévoux* called Fontaine's principle a principle that "the famous Monsieur D'Alembert makes continual use of in his *Traité de dynamique*" and in various works that D'Alembert wrote after he published the *Traité de dynamique*.[29] In fact, Fontaine's principle as it appears in the Paris Academy's unpublished proceedings for December 1739 and February 1740 is not stated very precisely.[30] Moreover, Fontaine never actually claimed in his complete works that he had inspired D'Alembert with the idea behind D'Alembert's principle or that he had been D'Alembert's source. Condorcet, who wrote Fontaine's obituary for the Paris Academy, believed that some meddlesome journalists only meant to put Fontaine on bad terms with D'Alembert. Condorcet found Fontaine's principle, as it was stated in the complete works of 1764, to be "metaphysical and vague," whereas he found D'Alembert's principle to be "geometric and precise."[31]

The behavior of the reviewers for the *Journal de Trévoux* is easy enough to explain. There was no love lost between the Jesuits and D'Alembert, because the French mathematician was also a *philosophe,* which meant, among other things, that he was anticlerical. In 1765, *Sur la destruction des Jésuites en France, par un auteur désintéressé*, which D'Alembert wrote, anonymously appeared. In the book, D'Alembert gave a concise account of the main facts, events, and political and moral reasons that had led to the official suppression of the Jesuits as a religious order in France in 1762. Voltaire saw the publication of the work through. He had it printed in Geneva, but everyone knew who had actually written it. D'Alembert did not even try to hide the fact that he wrote the book; in his correspondence he talked about having written it. In the manner that his friend Maupertuis, now deceased, had guilefully ridiculed the advocates of an elongated earth in the *Examen désintéressé* anonymously published in 1738, D'Alembert also underhandedly attacked the Jansenists, the enemies of the Jesuits, in his book on the suppression of the Jesuits. And it is more than a little ironic that the French "ex-Jesuits" cited D'Alembert's treatise in their campaign against the Jansenists. But the French ex-Jesuits remained anti*philosophe*, as did the *Journal de Trévoux*, which was taken out of their hands after the religious order was suppressed. And it is hard not to imagine that some French ex-Jesuits were pleased with the review of Fontaine's complete works, published at D'Alembert's expense in what had been their journal.

Be that as it may, Fontaine angered D'Alembert by not publicly denying the journalists' attribution of D'Alembert's principle to Fontaine. D'Alembert had good reason to be furious. If the reviewers of Fontaine's complete works had misunderstood Fontaine's principle, why, then, did Fontaine fail to call attention to that fact? After all, as I noted in Chapter 7, in the *Encyclopédie* D'Alembert praised Fontaine highly for his solution of the problem of the tautochrones in resistant media published in the Paris Academy's *Mémoires* for 1734. And as I also mentioned, D'Alembert had voluntarily corrected Euler's mistaken impression, which Euler had expressed publicly in the Berlin Academy's *Mémoires*, that D'Alembert had originated Fontaine's conditional equation (7.3.3.111). And D'Alembert pointed this error out with only the best intentions in mind. This was one of several grievances that D'Alembert spoke about in a letter that he drafted and originally intended to send to the *Mecure de France*, in which he replied to that journal's review of Fontaine's complete works.[32]

But for some reason Fontaine kept quiet. Why? It is very difficult to say with certainty. Perhaps he had not forgotten having been mistreated by Clairaut in Clairaut's Paris Academy *mémoires* during the 1730s. He could have kept silent as a way of taking revenge on D'Alembert for this earlier persecution. If Fontaine said nothing for these reasons, he must have done so subconsciously, because D'Alembert, who only entered the Paris Academy in 1741, obviously had absolutely nothing whatever to do with Clairaut's high-handed treatment of Fontaine in Clairaut's Paris Academy *mémoires* in the 1730s.

Then in 1767–68, Fontaine became involved in an unpleasant, protracted quarrel with D'Alembert's disciple Lagrange, then a member of the Berlin Academy, about the tautochrones and the calculus of variations. For example, after having stated in a Paris Academy *mémoire* of 1767 that Lagrange "went astray" ("s'était égaré") in his first work on the calculus of variations in a way that Fontaine did not specify, Fontaine simply rewrote Lagrange's first work in the "d," "δ" calculus using his own "·," "d" notation and thereby insinuated that Lagrange's "d," "δ" calculus was just his own fluxio-differential method in disguise. In other words, as Fontaine saw the matter, if Lagrange derived new results, he had essentially used what amounts to Fontaine's own method to do it.[33] Lagrange understandably got furious with Fontaine. Fontaine conducted himself in a way that was not in keeping with his character. Lagrange's anger waxed, as D'Alembert, who sought redress for the injustices that he felt Fontaine and the journalists had done him, stirred up his follower's wrath and goaded him into striking back.[34]

Moreover, in his article "Equation" appearing in the *Encyclopédie*, D'Alembert had spoken very favorably of Fontaine's Paris Academy *mémoire* of 1747 on the theory of equations,[35] while mentioning that it rested upon three unproven conjectures.[36] In the heat of the controversy about the tautochrones and the calculus of variations, Lagrange called D'Alembert's attention to the fact that his observations concerning Fontaine's theory of equations were entirely relevant and that, in particular, one of Fontaine's three conjectures included a defect that could not be fixed. D'Alembert's resentful protégé also notified his Parisian supporter that the version of the theory of equations appearing in Fontaine's collected works published in 1764 took none of D'Alembert's remarks into consideration.[37] It seems that Fontaine paid absolutely no attention whatever to D'Alembert's criticisms. (Had Fontaine simply not known about D'Alembert's article in the *Encyclopédie*?) D'Alembert felt slighted, and this only added to his own list of complaints. In 1769, D'Alembert took revenge, when he stated in the Paris Academy *Mémoires* of 1769, published in 1772, that: "It is according to this principle of the equality of $z + dz + dz$ with $z + \partial z + \delta z$, which I provided the late Monsieur Clairaut with the idea of, that Monsieur Clairaut derived the conditional equation for a differential equation in three variables to have a possible integral."[38] In other words, D'Alembert now insinuated that he had led Clairaut to the conditional equation (7.3.3.111), not Fontaine!

By this time Condorcet had broken with Fontaine, who had been the first member of the Paris Academy to back Condorcet's efforts to join the Academy.[39] Condorcet transferred his allegiance to D'Alembert instead, who greatly influenced the young nobleman from that time onward. D'Alembert was probably delighted to have a young aristocrat court him, when in principle he should have been the one who paid homage to members of the nobility, which in practice he always refused to do. D'Alembert, Lagrange, and Condorcet were a mutual admiration society to which Fontaine did not belong. This society formed a powerful coalition, and the combined efforts of its three members did much to blacken the image and sully the reputation of the elderly Fontaine, an outsider. The sixty-four-year-old mathematician simply could not stand up to the formidable alliance of younger mathematicians who opposed him and who, ultimately, helped lead to his ruin.

What caused Fontaine to behave indiscreetly and impudently toward Lagrange during the later 1760s? Did the elderly mathematician simply envy younger, talented mathematicians? The evidence suggests that the exact opposite is true. We know that in 1763–65 Fontaine held Lagrange in high esteem.[40] In 1763 or 1764, when he was twenty-one or twenty-two years old, the brilliant Jean-Jacques de Marguerie came to Paris. De Marguerie's mathematical talent amazed Fontaine so much that he offered to share his quarters in Paris with De Marguerie, who took advantage of the generous offer.[41] Fontaine began to tutor De Marguerie. The Comte de Roquefeuil, who heard about De Marguerie's ability, asked Fontaine what he thought of the young man. Fontaine replied: "He is at least as strong in [mathematical] analysis as I am."[42] In 1774, Lagrange congratulated De Marguerie for having inherited the genius of his late mentor Fontaine, who, had died in 1771.[43] (Evidently Lagrange rated Fontaine highly, despite having resorted at D'Alembert's suggestion to some malicious tactics to get even with him.) But De Marguerie joined the navy and became a naval officer. In 1779 he was killed in action at the age of thirty-seven, but not before he wrote five remarkable *mémoires* which are included in the first and only volume of *Mémoires de l'Académie Royale de Marine* (Brest, 1773).[44] Moreover, Alexandre-Théophile Vandermonde became interested in mathematics because he met Fontaine and found him to be so stimulating.[45]

Thus if Fontaine did not feel that young, talented mathematicians threatened him as he grew older, then what did happen to him? Did memories of the controversies with Clairaut during the 1730s haunt him and cause his behavior to change? Did he now regret having waited so long to publish the work on integral calculus which he had done at the end of the 1730s and at the beginning of the 1740s? What had been truly original work in 1738 scarcely impressed mathematicians in 1764.

One scholar has noted that even the great Daniel Bernoulli suffered from this kind of dawdling. Bernoulli failed to publish some of his ideas when he should have. Later Euler, D'Alembert, and Lagrange rediscovered them and developed them further, whereupon, in a vain attempt to reestablish the brilliant work he had done in younger years, Bernoulli quickly wrote up that work and published it, although by this time his ideas appeared old and primitive instead of fresh and original. Bernoulli came to regret his procrastination. Had he shaped and published ideas while they were still new, he would have earned a greater name in the history of mechanics than in fact he has.[46] In later years Fontaine could have conceivably come to reproach himself for his slowness too, and this could have caused him to become bitter.

As I mentioned in Chapter 7 and in this chapter, in the *Encyclopédie* D'Alembert praised Fontaine highly for his solution of the problem of the tautochrones in resistant media published in the Paris Academy's *Mémoires* for

1734. In 1767 Lagrange formulated problems of tautochrones in resistant media of a new, more general kind. D'Alembert first solved these problems "by generalizing and simplifying Monsieur Fontaine's method," which included replacing the notation "·" and "d," respectively, which Fontaine had used in 1734, by "d" and "δ," respectively. Then D'Alembert found "another solution" of the same problems which was "much simpler and more general than the first solution [that is, the solution that D'Alembert had based on Fontaine's method]."[47] Did D'Alembert anger Fontaine by "meddling with" his fluxio-differential method in order to solve more advanced problems in tautochrones and then by claiming that these problems could be handled and solved more easily using other means? As I mentioned, it was Lagrange's formulation of and solving of the more advanced problems in tautochrones which in part seems to have caused Fontaine to turn against Lagrange.

Perhaps Fontaine's hesitation and procrastination for decades suddenly made him sour. Or Lagrange and D'Alembert may have irritated Fontaine, because they "dared" to touch his "sacred" problems of tautochrones and improved upon his methods for dealing with such problems. Be that as it may, we have very little hard evidence to go on. It is known that when he died in 1771, he suffered from a terrible disease with painful, debilitating symptoms, which he neglected to have treated during the early stages of the illness.[48] If he was afflicted with some kind of chronic illness, this could have caused his personality to change during the years before he died, leading him to behave badly toward mathematicians like D'Alembert and Lagrange, whom he had admired and with whom he had been on good terms. In any case, it seems inconceivable that an individual as upright as Cousin would have worshipped and devoted himself to a villain. In fact, Fontaine was not a villain. Instead he was a victim throughout much of his career. But a mix-up over Fontaine's character can be explained easily enough. Fontaine was a misfit. To use an expression that the French are wont to say, Fontaine simply "did not fit the mold" ("il n'était pas dans le moule").

Setting politics and personal problems to one side, the factors mentioned earlier, however, probably contribute at least as much, if not more to the lack of attention paid to Fontaine's work of the 1730s for more than two hundred years. Today, just as in Fontaine's time, mathematics rarely, if ever, finds its way outside a small group of practitioners and reaches a wider audience unless it proves itself to be useful for solving problems in some other field.

Notes

1. Isaac Newton's theory of a flattened earth (1687, 1713, 1726)

1. Contrary to what some French colleagues believe, the third edition of the *Principia* (1726) was available in Paris during the 1730s. Newton sent six copies of the third edition to the Paris Academy of Sciences, along with a covering letter to the Academy's Secrétaire Perpétuel Fontenelle, in 1726 (see Cohen (1964: 75–76); Cohen (1978: 284); and Letters No. 1491: Newton to Fontenelle, ? June 1726 and No. 1492: Fontenelle to Newton, 3 July 1726, in Newton (1977: 349)). I refer to the theory that appears in the critical third edition of the *Principia:* Newton (1972: vol. 2, 592–610).

2. In the first edition of the *Principia* (1687), Newton made use of Picard's value of a degree of latitude. In the second edition (1713), Newton utilized instead the average value of a degree of latitude measured by Paris Academy astronomers along the piece of the meridian through Paris from Paris to Collioure, France, which Gian-Domenico Cassini had announced in 1701. These astronomers, however, subsequently determined that the value that they had found was too large. Hence in the second edition of the *Principia* Newton calculated a value of the radius of a spherical earth which was also too great. When the astronomers completed measuring the part of the meridian through Paris from Dunkerque, France to Collioure, in 1718, Newton used their conclusions to calculate the average degree of latitude along this portion of the meridian through Paris. He discovered that it differed from Picard's degree by one *toise* ($\simeq$ 6.5 feet), which was a meaningless quantity from the standpoint of observation. (That is, taking into account errors in making measurements, the values of the two degrees could not be distinguished.) This value of a degree of latitude is the one that appears in the third edition of the *Principia* (1726). In principle, the successive values of average degrees of latitude only caused estimates of the earth's *size* to change from one edition of the *Principia* to the next, not its *shape*.

3. In the first edition of the *Principia* (1687), Newton introduced without showing the calculation $1/290\frac{4}{5}$ as the ratio of the magnitude of the centrifugal force per unit of mass to the magnitude of the attraction at the equator, and he actually used the value 1/290 in developing his theory. Huygens first put forward $1/289 = 1/17^2$ as the value of this ratio in Huygens (1690: 146). Concerning the way that Huygens determined this value, see Aiton (1972: 83–84) and Costabel (1988: 98). It is possible that Newton borrowed this value of the ratio from Huygens, for it is the value that appears in the second edition of the *Principia* (1713) and in the third edition (1726). The difference between 1/290 and 1/289 is negligible for the following reasons: each of the two values of the ratio is an "infinitesimal" number ε (that is, ε is nonzero in each case, and $\varepsilon^2 \ll \varepsilon$ is true of each), while the two values differ by terms of second or higher order. In other words,

$$\frac{1}{290} = \frac{1}{289+1} = \frac{1}{289}\left(\frac{1}{1+(1/289)}\right) \simeq \frac{1}{289}\left(1-\frac{1}{289}\right) \simeq \frac{1}{289},$$

because

$$\left(\frac{1}{289}\right)^2 \ll \frac{1}{289}.$$

Newton demonstrated in Propositions 70 to 74 of Book I of the *Principia* that a homogeneous spherical body whose radius is R attracts like a $1/r^2$ central force in its exterior ($R \leqslant r$) and like an r central force in its interior ($r < R$), if the body attracts according to the universal inverse-square law. The sequence of Newton's ideas appearing in Propositions 70 to 74 is a subtle one, not readily

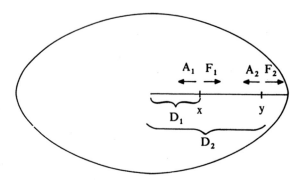

Figure 59. Meridian of a homogeneous figure shaped like an ellipsoid of revolution which attracts in accordance with the universal inverse-square law and which revolves around its axis of symmetry (its polar axis).

grasped, which has been handled unskillfully in the past. For a recent treatment that greatly clarifies Newton's chain of arguments, calls attention to an inapplicable justification of a correct geometric statement appearing in Newton's proof of Proposition 71 and supplies a valid proof of that statement, and which adds the missing element that would have enabled Newton to show that the magnitude and direction of the attraction due to a homogeneous spherical body which attracts according to the universal inverse-square law are the same as it would produce at points in its exterior if all of its mass were concentrated at its center, see Weinstock (1984). It turns out that that the earth could *not be spherical*, if it attracts in accordance with either Newton's or Huygens's hypotheses of attraction. The reason is simple enough: in both cases, attraction is *perpendicular* to the surface of a sphere at all points on the sphere's surface. But effective gravity is *itself* perpendicular to the earth's surface at all points on its surface, because waters at the earth's surface are observed to find their own levels (the principle of the plumb line). In order for this condition regarding effective gravity to hold at the earth's surface if the earth were spherical, however, the earth could not rotate. For if a spherical earth rotated, the attraction at the surface, which is perpendicular to the surface at all points on the surface, and the centrifugal force per unit of mass at the surface would combine in this instance to produce a resultant, which is the effective gravity at the surface, that would *not* be perpendicular to the earth's surface at all points on its surface. But the earth does rotate. Hence it must not be spherical. John Keill reasoned this way in 1698, but his argument was only published for the first time in Desaguliers (1725b: 247-249), where Keill's words are faithfully reproduced.

4. For an analysis of Alexis-Claude Clairaut's proof of this fact, which Clairaut gave in 1743 and in which he assumed the earth to be a homogeneous fluid figure of equilibrium shaped like an ellipsoid of revolution flattened at its poles which attracts according to the universal inverse-square law, which revolves around its axis of symmetry (its polar axis), and whose ellipticity is infinitesimal, see Greenberg (1988).

5. See note 40 of Chapter 2. See as well Lamb (1932: 697-699, 702) and Jardetzky (1958: 13-14).

6. The easiest way to see this is as follows. Let A_1 stand for the attraction at point x on the equatorial axis; let F_1 designate the magnitude of the centrifugal force per unit of mass at x; and let D_1 denote the distance of x from the center of the figure. Let A_2 specify the attraction at point y on the equatorial axis; let F_2 symbolize the magnitude of the centrifugal force per unit of mass at y; and let D_2 designate the distance of y from the center of the figure. (See Figure 59.)

Then

$$\frac{A_1}{A_2} = \frac{D_1}{D_2} = \frac{F_1}{F_2}. \quad \text{But} \quad \frac{A_1}{A_2} = \frac{F_1}{F_2}$$

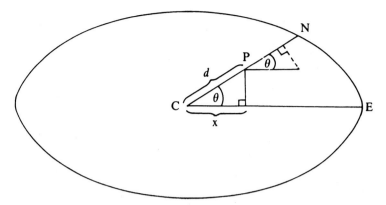

Figure 60. Meridian of a homogeneous figure shaped like a flattened ellipsoid of revolution which attracts in accordance with the universal inverse-square law and which revolves around its axis of symmetry (its polar axis).

means that $F_1 A_2 = A_1 F_2$. Hence $A_1 A_2 - F_1 A_2 = A_1 A_2 - A_1 F_2$, or $(A_1 - F_1)A_2 = (A_2 - F_2)A_1$, in which case

$$\frac{A_1 - F_1}{A_2 - F_2} = \frac{A_1}{A_2} = \frac{D_1}{D_2}.$$

It is obvious that the same is true along the polar axis, since the magnitude of the centrifugal force per unit of mass along the polar axis is zero, because the figure revolves around that axis.

7. In view of note 6 of Chapter 1, in order to show that the attraction and the centrifugal force per unit of mass at points along *any* column from the center C of a homogeneous body shaped like a flattened ellipsoid of revolution to an arbitrary point N at its surface determine the weight of the column in *exactly the same way* that the attraction and the centrifugal force per unit of mass at points along two columns from the center of the body to its surface which lie along its polar and equatorial axes determine the weights of these two columns, Newton would have had to establish that (1) the magnitudes of the components of the centrifugal force per unit of mass at points along an arbitrary column CN in the direction of CN vary directly as the distances of the points from the center C of the body; (2) the magnitudes of the components of the attraction at points along column CN in the direction of CN vary directly as the distances of the points from the center C of the body. Now, (1) is easy to demonstrate. (See Figure 60.) Let d denote the distance from a point P along column CN to C. The magnitude of the centrifugal force per unit of mass at point P is proportional to $x = d\cos\theta$. Therefore the magnitude of the component of centrifugal force per unit of mass at P in the direction of CN is proportional to $x\cos\theta = d\cos^2\theta$. Then let $F_1 \equiv$ the magnitude of the component of the centrifugal force per unit of mass at the point P_1 on CN in the direction of CN, and let $F_2 \equiv$ the magnitude of the component of the centrifugal force per unit of mass at the point P_2 on CN in the direction of CN. Then

$$\frac{F_1}{F_2} = \frac{d_1 \cos^2\theta}{d_2 \cos^2\theta} = \frac{d_1}{d_2}.$$

Moreover, as it turns out, (2) is concealed in Newton's statement of Corollary 3 to Proposition 91 of Book I of the *Principia* and in the results that Newton used to prove this corollary (see note 11 of Chapter 1). We also note that Newton treated the *component* of the effective gravity *directed toward the center of the figure* at a point on the surface of his homogeneous figure shaped like an ellipsoid of revolution flattened at its poles which attracts according to the universal inverse-square law and which revolves around its axis of symmetry as the *total*, or *resultant of*, effective gravity at the point. But in general the magnitudes of these two specifications of effective gravity at the point are not the same, because the total effective gravity at points on the surface of a figure of equilibrium, which in fact Newton's figure turns out to be, is perpendicular to the surface of the figure at those points, because the principle of the plumb line holds at the points on the surface of a figure of equilibrium. (See note 3 of Chapter 1.) However, for a homogeneous figure of equilibrium shaped like an ellipsoid of revolution flattened at its poles which attracts according to the universal inverse-square law, which revolves around its axis of symmetry, and which is nearly spherical, meaning that the figure is shaped like a flattened ellipsoid of revolution whose ellipticity ε is infinitesimal (that is, ε is positive, but $\varepsilon^2 \ll \varepsilon$), the total effective gravity at a point on the surface of the figure and the component of effective gravity at the point directed toward the center of the figure nearly coincide, and the magnitudes of the two specifications of effective gravity at the point are the same to all intents and purposes. They differ by terms of second order and higher orders in ε and all other infinitesimal quantities of the same order as ε. In other words, the two values are equal to terms of first order. However, Newton did not make clear whether he only meant his theory to apply to homogeneous figures of equilibrium shaped like ellipsoids of revolution flattened at their poles which attract according to the universal inverse-square law, which revolve around their axes of symmetry, and whose ellipticities are *infinitesimal and not finite*. For example, he tried to apply his theory to Jupiter, but Jupiter's ellipticity is *not infinitesimal*.

8. In stating without proof that the "sin² law" of the variation in the increase in the magnitude of the effective gravity with latitude followed from his result that the magnitudes of the effective gravity at points at the surface of his homogeneous figure varied inversely as the distances of the points from the center of the figure, Newton again evidently assumed that effective gravity at a point on the surface of the figure is directed toward the center of the figure (see note 63 of Chapter 5 for Maupertuis's defective demonstration of the law interpreted this way when the ellipticity ε of the homogeneous figure is infinitesimal, and my corrections of his mistakes), whereas in fact the resultant of effective gravity at a point on the surface of a figure of equilibrium, which Newton's homogeneous figure turns out to be, is perpendicular to the surface of the figure at the point, because the principle of the plumb line holds at the points on the surface of a figure of equilibrium. (See note 3 of this chapter.) However, Newton's assumption again involves no appreciable error, at least for practical purposes, when the figure is nearly spherical, which means that its ellipticity ε is infinitesimal. (See the end of note 7 of this chapter.) Clairaut (1743: 188–191) showed later that the sin² law of the variation in the increase in the magnitude of the effective gravity with latitude at the surface of a homogeneous fluid figure of equilibrium shaped like an ellipsoid of revolution flattened at its poles which attracts according to the universal inverse-square law and which revolves around its axis of symmetry *does* actually hold true for the *total,* or *resultant of,* effective gravity itself when the ellipticity of the figure is infinitesimal.

9. For a proof of the geometric statement, see Kellogg (1953: 22).

10. The quoted passages are from Weinstock (1984: 887). For a remarkably illuminating exposition of Newton's obscure proof of Proposition 72 to Book I of the *Principia*, see Weinstock (1984: 887–888). This exposition fills in the roughly 90% of the proof left out by Newton, thereby making the proposition and its *Principia* proof comprehensible – perhaps for the first time on a printed page.

11. For example, Robert Weinstock thinks it likely that Newton understood these concepts and could have stated and proved the version of Corollary 3 to Proposition 72 of Book I of the *Principia* which includes the various results mentioned above. Weinstock concludes this from the way that Newton did prove Proposition 72 of Book I of the *Principia*. I am grateful to Weinstock for stating and demonstrating in his letters to me these various results using methods that have their origins in Newton's arguments in Proposition 72. Weinstock's proofs extend his exposition of Proposition 72, which appears in Weinstock (1984: 887–888), to cover generalizations of Corollary 3 of Proposition 72 which Weinstock formulates more precisely than the Corollary 3 to Proposition 72 which Newton stated.

12. A proof of this result follows, which makes use of the integral calculus: (See Figure 61.) Let $A(r) \equiv$ the magnitude of the component of the attraction at p in the direction of the column CA on which p lies. Let $F(r) \equiv$ the magnitude of the component of the centrifugal force per unit of mass at p in the direction of the column CA on which p lies. Then Newton deduced that

$$A(r) \alpha r \qquad\qquad (I)$$

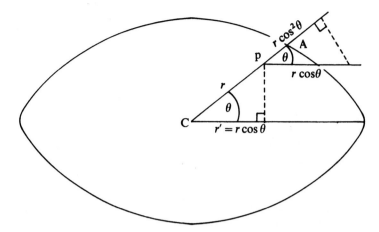

Figure 61. Meridian of a homogeneous figure shaped like a flattened ellipsoid of revolution which attracts in accordance with the universal inverse-square law, which revolves around its axis of symmetry (its polar axis), and all of whose columns from center to surface balance or weigh the same.

from Corollary 3 to Proposition 91 of Book I of the *Principia*. It is also easy to see that

$$F(r) \alpha r.\tag{II}$$

Namely, for any particular value of θ, the magnitude of the centrifugal force per unit of mass at $p = Lr' = Lr\cos\theta$, where L is a constant. Then the magnitude of the component of the centrifugal force per unit of mass at p in the direction of $CA = Lr\cos^2\theta$. For a constant value of θ, $L\cos^2\theta$ is a constant $\equiv k$. Hence (II) is true. Now, (I) means that

$$A(r) = Kr\tag{I'}$$

for some constant K, and (II) means that

$$F(r) = kr.\tag{II'}$$

(In the particular problem in question, $k < K$ and

$$\left(\frac{k}{K}\right)^2 \ll \frac{k}{K}.)$$

Let $|CA|$ denote the length of the column $CA \equiv R$, and let δ stand for the uniform density of the figure. Then the weight of column CA equals

$$\int_{r=0}^{r=R} (A(r) - F(r))\delta \, dr = \int_{r=0}^{r=R} \delta(K-k)r \, dr = \frac{\delta(K-k)R^2}{2} = \text{a constant } M$$

that is independent of the particular column from the center of the figure to the surface of the figure, because all columns from the center of the figure to the surface of the figure are assumed to balance or weigh the same. Thus the magnitude of the component of the effective gravity at A in the direction of column

$$CA = A(R) - F(R) = (K-k)R = \frac{2M}{\delta}\frac{1}{R}\alpha\frac{1}{R}.$$

Here Newton treated the *component* of effective gravity at a point on the surface of the figure *directed toward the center of the figure* as the *total*, or *resultant of*, effective gravity at the point. But the two specifications of effective gravity differ in general, because the total effective gravity at points on the surface of a figure of equilibrium, which in fact Newton's homogeneous figure turns out to be, as I mentioned in the text above, is perpendicular to the surface at those points, because the principle of the plumb line holds at the points on the surface of a figure of equilibrium. (See note 3 of this chapter.) However, for a homogeneous fluid figure of equilibrium shaped like an ellipsoid of revolution flattened at its poles which attracts according to the universal inverse-square law, which revolves around its axis of symmetry, and which is nearly spherical, meaning that the figure is shaped like a flattened ellipsoid of revolution whose ellipticity ε is infinitesimal (that is, ε is positive, but $\varepsilon^2 \ll \varepsilon$), the total effective gravity at a point on the surface of the figure and the component of effective gravity at the point directed toward the center of the figure almost coincide, and the magnitudes of the two specifications of effective gravity at the point are the same to all intents and purposes. They differ by terms of second order and higher orders in ε and all other infinitesimal quantities of the same order as ε. That is, the two values are equal to terms of first order. (See the end of note 7 of Chapter 1.)

13. In a page-long annotation in Newton (1974: 226–227), Derek Whiteside expresses in modern notation the integrals that Newton probably used to compute these values of attraction. This is the *only* place in *The Mathematical Papers of Isaac Newton* where Whiteside refers to Newton's theory of the earth's shape. Whiteside mentions that in his *Tractatus de Quadratura Curvarum* (1704) Newton only published for the first time the analysis (that is, the conic integrals) that underlay this geometric corollary.

14. Maupertuis (1731a: fol. 2v). Except where otherwise stated, all translations from French to English are mine.

15. Johann I Bernoulli (1731a: fols. 3v–4r).

16. For a study of the *Principia's* mathematics in which the style is characterized as neither classical geometry nor infinitesimal calculus, see De Gandt (1986).

17. As late as 1750, even Alexis-Claude Clairaut, who probably advanced the study of nonplane curves (that is, skew curves) and two-dimensional surfaces more than any other French mathematician of his era, specifically called attention to problems that still made studying surfaces that are not generated by conic sections difficult, which Clairaut did in connection with the problem of the earth's shape. See Clairaut (1750: 8). I thank Craig B. Waff for providing me with a photocopy of the copy of this rare work, which is in the Widener Library, Harvard University.

18. Van Maanen (1987: 33).

2. The state of the problem of the earth's shape in the 1720s

1. Jacques Cassini (1718: 237–245).
2. Mairan (1720).
3. Huygens (1690: 149); and Mairan (1720: 247).
4. Mairan (1720: 231–232).
5. Huygens (1690: 154) stated the principle of the plumb line.
6. Mairan (1720: 232–233 and [Mairan's] Figure 1).
7. Ibid., pp. 239, 248, and 264.
8. Ibid., pp. 238–240 and Figure 4.
9. Ibid., pp. 240–242 and Figure 5.
10. Ibid., pp. 243–245 and Figure 7.
11. Ibid., p. 246.
12. See note 3 of Chapter 1.
13. Mairan (1722: 570). Hartsoëker, who was no mathematician, believed, for example, that the planets floated in equilibrium with the surrounding fluid, without need of gravity or centrifugal force either. Leibniz told him that he and Newton, respectively, had proved the contrary, but in different ways, in *Acta Eruditorum* and in the *Principia*, respectively. See Aiton (1985: 337).
14. Mairan (1720: 253–255 and Figure 8).
15. Ibid., p. 255.
16. Ibid., p. 256.
17. Ibid., p. 257 and Figure 9.
18. Ibid., pp. 257–258, 262, and Figure 10.
19. Ibid., p. 262.

20. Ibid., p. 264.
21. Ibid., p. 265.
22. Ibid., p. 265.
23. Ibid., p. 252.
24. Concerning this French positivistic tradition, see Popkin (1979: Chapters 5 and 7); Hahn (1971: 31–34); Marsak (1959: 20–40); and Lenoble (1943, 1971).
25. Mairan (1720: 231).
26. Ibid., pp. 267–270 and Figure 12.
27. Ibid., pp. 270–272 and Figure 13.
28. Ibid., p. 266.
29. Ibid., p. 268 and Figure 12.
30. Ibid., pp. 269–270 and Figure 12.
31. Ibid., p. 270 and Figure 12.
32. Ibid., pp. 270–272 and Figure 13.
33. Ibid., p. 275.
34. In particular see Lafuente and Peset (1984: 237); and Lafuente and Delgado (1984: 21–22). Beeson (1985: 194) calls "Cartesian theory" the basis of Mairan's paper, too.
35. Mairan (1720: 251).
36. Guerlac (1977: 486); and Guerlac (1981: 65).
37. Fontenelle (1732: 92). Years earlier, in December of 1700, Fontenelle wrote Leibniz that Cassini had given up many years ago the Keplerian ellipses as the orbits of planets. Fontenelle mentioned that Cassini preferred instead an "ellipse" whose product of the "focal distances" is constant; he noted that Philippe de La Hire did not appear to favor any particular curve; and he stated that he himself suspected that planets would eventually be shown to move within certain limits, but that they would not be found to follow any curve that was regular and exact. See Aiton (1985: 248–249).
38. Desaguliers (1725c).
39. Ibid., pp. 278–279.
40. Laplace (1784: 125–128, §12) found the limiting value of the angular velocity for which a homogeneous fluid figure of equilibrium shaped like a flattened ellipsoid of revolution which attracts according to the universal inverse-square law is possible; he gave numerical results for the case of the earth; and he proved for the first time that a homogeneous fluid figure shaped like an elongated ellipsoid of revolution which attracts in accordance with the universal inverse-square law is not a possible figure of relative equilibrium. Laplace virtually reproduced this discussion in the first edition of his *Mécanique Céleste*, Volume III, Chapter III. See the second edition: Laplace (1829: vol. II, Chapter III). See as well Lamb (1932: 700–702) and Jardetzky (1958: 28–30) for modern proofs that if a homogeneous fluid figure shaped like an ellipsoid of revolution which attracts according to the universal inverse-square law is a figure of equilibrium, the figure must be flattened at its poles. The pressure must be constant at the surface of a homogeneous fluid figure of equilibrium which attracts according to the universal inverse-square law. Lamb and Jardetzky show that pressure can be constant at the surface of a homogeneous fluid figure shaped like an ellipsoid of revolution which attracts according to the universal inverse-square law if the ellipsoid of revolution is flattened at its poles but that pressure cannot be constant at the surface of the figure if the ellipsoid of revolution is elongated at its poles. But a level surface within a fluid mass in equilibrium, meaning a surface within the fluid mass along which effective gravity is perpendicular to the surface at all of its points, must always be a surface of constant pressure [see Truesdell (1954: LVI)]. Consequently the surface of a homogeneous fluid figure shaped like an elongated ellipsoid of revolution which attracts according to the universal inverse-square law can never be a level surface because it can never be a surface of constant pressure. In other words, the principle of the plumb line cannot hold at every point at the surface of a homogeneous fluid figure shaped like an elongated ellipsoid of revolution which attracts according to the universal inverse-square law. Hence such a homogeneous fluid figure cannot possibly be a figure of equilibrium. (This, in fact, is how Laplace showed in the works mentioned that a homogeneous fluid figure shaped like an elongated ellipsoid of revolution which attracts according to the universal inverse-square law cannot be a figure of equilibrium. Namely, he showed directly that the surface of such a homogeneous fluid figure cannot be a level surface of the effective gravity.) In Chapter 9 we will discover why as a result of this fact, the columns from the center to the surface of a homogeneous

figure shaped like an elongated ellipsoid of revolution which attracts according to the universal inverse-square law cannot all balance or weigh the same either. But the principles of the plumb line and balanced columns are two conditions that a homogeneous figure must necessarily fulfill in order to exist in a state of equilibrium. Hence a homogeneous figure shaped like an elongated ellipsoid of revolution which attracts according to the universal inverse-square law can never be a figure of equilibrium. Both Maupertuis (1732e: 343) and Clairaut [(1743: 265) and Newton (1966: vol. II, 267)] had *assumed* that if a homogeneous fluid mass at rest which attracts according to the universal inverse-square law is a figure of equilibrium, the mass must be spherical. In July of 1769, Lagrange wrote D'Alembert: "I am almost convinced that a homogeneous fluid [mass at rest], all of whose parts attract each other according to any law that [only] depends on the distance [between two parts], cannot be in equilibrium unless it forms a spherical mass. Nevertheless, it seems impossible to me to find a general demonstration." [See Lagrange (1769a: 139).] In fact it was not shown until the twentieth century, by A. Lyapunov (1918) and, independently, by T. Carleman (1919), that a fluid mass in equilibrium which attracts in accordance with the universal inverse-square law and which is at rest must be spherical [see Jardetzky (1958: 9–12, 28) and Lamb (1932: 698)]. From the fact that it took so long to demonstrate this result we get an inkling of just how difficult problems of figures of equilibrium are.

41. See the end of note 7 of Chapter 1.
42. See note 8 of Chapter 1.
43. Desaguliers [1725c: 295]
44. Ibid., p. 285.
45. Ibid., p. 295.
46. Ibid., p. 295.
47. Ibid., p. 295.
48. Ibid., p. 296.
49. For example,

$$\frac{g - (c - (l + m + n))}{g - c} = 1 + \frac{l + m + n}{g - c} \tag{I}$$

and

$$\frac{g - (c + 1 - (l + m))}{g - (c + 1)} = 1 + \frac{l + m}{g - (c + 1)}. \tag{II}$$

From equalities (I) and (II) it follows that in order to have

$$\frac{g - (c + 1 - (l + m))}{g - (c + 1)} < \frac{g - (c - (l + m + n))}{g - c}. \tag{III}$$

it must be true that

$$\frac{l + m}{g - (c + 1)} < \frac{l + m + n}{g - c}. \tag{IV}$$

Now, inequality (IV) reduces to

$$\frac{l + m + n + nc}{n} < g. \tag{V}$$

But for given positive values of c, l, m, and n, a value of g can always be chosen so small so that

$$0 < g \leqslant \frac{l + m + nc}{n}, \tag{VI}$$

in which case inequality (V) does not hold.

50. Ibid., p. 286.
51. Ibid., p. 288.
52. Ibid., p. 296.
53. Clairaut showed (1809: 120–122 and Figure 12) that the attraction at a point E at the equator of a homogeneous figure shaped like a flattened ellipsoid of revolution which attracts according to the

universal inverse-square law, whose density is 1, and whose ellipticity is infinitesimal equals

$$\tfrac{2}{3}(2\pi)r + \tfrac{6}{15}(2\pi)ra$$

to terms of first order in the ellipticity. Here $AC = r$, $DE = ar$, and

$$\text{ellipticity} \equiv \frac{(CE - AC)}{AC} = \frac{r + ar - r}{r} = a,$$

where $a^2 \ll a$ (see Figure 16). Now, the volume of Clairaut's figure shaped like a flattened ellipsoid of revolution equals

$$\tfrac{4}{3}\pi r(r + ar)^2 = \tfrac{4}{3}\pi r^3(1 + a)^2 = \tfrac{4}{3}\pi r^3(1 + 2a + a^2).$$

To terms of first order in the ellipticity a, the volume equals

$$\tfrac{4}{3}\pi r^3(1 + 2a) = \tfrac{4}{3}\pi(r(1 + 2a)^{1/3})^3.$$

That is, to terms of first order in a, the figure has the same volume as a spherical figure whose radius is

$$r(1 + 2a)^{1/3}.$$

We observe that r and a can be chosen in various ways that will make this expression have the same value. In other words, a homogeneous spherical figure of given radius and whose density is 1 does not uniquely determine a figure shaped like a flattened ellipsoid of revolution whose density is 1, whose ellipticity is infinitesimal, and whose volume is the same as the volume of the spherical figure. The attraction at the surface of a homogeneous spherical figure which attracts according to the universal inverse-square law, whose density is 1, and whose radius is

$$r(1 + 2a)^{1/3}$$

equals

$$\tfrac{4}{3}\pi r(1 + 2a)^{1/3}.$$

To terms of first order in a, this attraction equals

$$\tfrac{2}{3}(2\pi)r(1 + \tfrac{1}{3}(2a)) = \tfrac{2}{3}(2\pi)r + \tfrac{4}{9}(2\pi)ra.$$

And we note that

$$\tfrac{2}{3}(2\pi)r + \tfrac{6}{15}(2\pi)ra < \tfrac{2}{3}(2\pi)r + \tfrac{4}{9}(2\pi)ra.$$

That is, the attraction at the equator of the figure shaped like a flattened ellipsoid of revolution is less than the attraction at the surface of the spherical figure. Now, in Daniel Bernoulli (1740: 144–146 and the figure on p: 145), Bernoulli showed that the attraction at a point G at the equator of a homogeneous figure shaped like an elongated ellipsoid of revolution which attracts in accordance with the universal inverse-square law, whose density is 1, and whose ellipticity is infinitesimal equals:

$$\tfrac{2}{3}(2\pi)d + \tfrac{4}{15}(2\pi)d\left(\frac{\beta}{d}\right)$$

to terms of first order in the ellipticity. Here $GC = d$ and $bB = \beta$. (See Figure 31.) According to Clairaut's definition of ellipticity, the ellipticity of Bernoulli's elongated spheroid $\equiv$

$$\frac{GC - CB}{CB} = \frac{d - (d + \beta)}{(d + \beta)} = -\frac{\beta}{d + \beta}.$$

However, when the ellipticity is infinitesimal, as it is in the case of Bernoulli's ellipsoid, then

$$\left(\frac{\beta}{d + \beta}\right)^2 \ll \frac{\beta}{d + \beta}.$$

Hence in this instance

$$\frac{\beta}{d} = \frac{\beta}{d + \beta - \beta} = \frac{\beta}{d + \beta}\left(\frac{1}{1 - (\beta/(d + \beta))}\right)$$

equals

$$\frac{\beta}{d+\beta}\left(1+\frac{\beta}{d+\beta}\right)=\frac{\beta}{d+\beta}$$

to terms of first order in the ellipticity. In other words, $-\beta/d$ equals the ellipticity $-\beta/(d+\beta)$ to terms of first order in the ellipticity, or $-\beta/d$ can be taken to be the ellipticity of the figure shaped like an elongated ellipsoid of revolution, when its ellipticity is infinitesimal. Now, the volume of Bernoulli's figure shaped like an elongated ellipsoid of revolution equals

$$\tfrac{4}{3}\pi d^2(d+\beta)=\tfrac{4}{3}\pi d^3\left(1+\frac{\beta}{d}\right)=\tfrac{4}{3}\pi\left(d\left(1+\frac{\beta}{d}\right)^{1/3}\right)^3.$$

That is, Bernoulli's figure has the same volume as a spherical figure whose radius is

$$d\sqrt[3]{1+\frac{\beta}{d}}.$$

Again we observe that d and β can be chosen in different ways that will make

$$d\sqrt[3]{1+\frac{\beta}{d}}$$

have the same value. In other words, a spherical figure of given radius and whose density is 1 does not uniquely determine a figure shaped like an ellipsoid of revolution whose density is 1, whose ellipticity is infinitesimal, and whose volume is the same as the volume of the spherical figure. To make

$$r\sqrt[3]{1+2a}=d\sqrt[3]{1+\frac{\beta}{d}},$$

let us take

$$r=d,$$

and let us choose a and β so that

$$2a=\frac{\beta}{d}.$$

Then these choices determine a figure shaped like a flattened ellipsoid of revolution, a spherical figure, and a figure shaped like an elongated ellipsoid of revolution whose volumes are the same to terms of first order in the ellipticities of the two figures shaped like ellipsoids of revolution. Moreover, the attraction at the equator of the figure shaped like an elongated ellipsoid of revolution equals

$$\tfrac{2}{3}(2\pi)d+\tfrac{4}{15}(2\pi)d\left(\frac{\beta}{d}\right)=\tfrac{2}{3}(2\pi)r+\tfrac{4}{15}(2\pi)r(2a)=\tfrac{2}{3}(2\pi)r+\tfrac{8}{15}(2\pi)ra.$$

And we note that

$$\tfrac{2}{3}(2\pi)r+\tfrac{4}{9}(2\pi)ra<\tfrac{2}{3}(2\pi)r+\tfrac{8}{15}(2\pi)ra.$$

That is, the attraction at the surface of the spherical figure is indeed less than the attraction at the equator of the figure shaped like an elongated ellipsoid of revolution. However, we note that the average value of the attractions at the equators of the figure shaped like a flattened ellipsoid of revolution and the figure shaped like an elongated ellipsoid of revolution is

$$\tfrac{2}{3}(2\pi)r+\tfrac{7}{15}(2\pi)ra.$$

Since

$$\tfrac{4}{9}\neq\tfrac{7}{15},$$

the attraction at the surface of the spherical figure whose volume is the same as the volumes of the two figures shaped like ellipsoids of revolution chosen in the manner described above to terms of first order in the ellipticities of the two figures shaped like ellipsoids of revolution is not exactly the *average* value of the attractions at the equators of these two figures shaped like ellipsoids of revolution, contrary to what Desaguliers assumed, although it is obviously quite close to the average value. To be more precise, $4/9 = 20/45$ and $7/15 = 21/45$, hence

$$\tfrac{2}{3}(2\pi)r + \tfrac{7}{15}(2\pi)ra = \tfrac{2}{3}(2\pi)r + \tfrac{4}{9}(2\pi)ra + \tfrac{1}{45}(2\pi)ra.$$

Whereas $a^2 \ll a$, it is questionable whether $(\tfrac{1}{45})^2$ is $\ll \tfrac{1}{45}$. It is a borderline case. Hence it is debatable whether $\tfrac{1}{45}a$ is a quantity of second order, and consequently the same question about the term $\tfrac{1}{45}(2\pi)ra$ arises. If a is sufficiently small then $\tfrac{1}{45}(2\pi)ra$ is a quantity of second order. If $\tfrac{1}{45}(2\pi)ra$ is treated as a term of second order, then

$$\tfrac{2}{3}(2\pi)r + \tfrac{7}{15}(2\pi)ra$$

and

$$\tfrac{2}{3}(2\pi)r + \tfrac{4}{9}(2\pi)ra$$

can be considered equal to terms of first order.
54. It is always true that $0 < g$, $0 < s$, and $0 < c$.
   1. Suppose that $c \leqslant s$. Then $0 \leqslant s - c$; therefore $g \leqslant g + s - c$. But $g - s - (c + 2) < g - (c + 1) < g$. Hence

   $$g - s - (c + 2) < g - (c + 1) < g + s - c. \tag{I}$$

   2. Suppose that $s < c$. Then $0 < c - s$; therefore $g + s - c = g - (c - s) < g$. In order to have $g - (c - s) \leqslant g - (c + 1)$, $c + 1 \leqslant c - s$ would have to be true. But $c + 1 \leqslant c - s$ reduces to $s + 1 \leqslant 0$, which is impossible because $0 < s$. Hence

   $$g - (c + 1) < g - (c - s) = g + s - c. \tag{II}$$

   In order to have $g - (c + 1) \leqslant g - s - (c + 2) = g - (s + c + 2)$, $s + c + 2 \leqslant c + 1$ would have to be true. But $s + c + 2 \leqslant c + 1$ reduces to $s + 1 \leqslant 0$, which is impossible since $0 < s$. Consequently

   $$g - s - (c + 2) < g - (c + 1). \tag{III}$$

   From inequalities (II) and (III) it follows that

   $$g - s - (c + 2) < g - (c + 1) < g + s - c. \tag{I}$$

55. Desaguliers (1725c: 297).
56. Ibid., p. 297.
57. Ibid., p. 297.
58. Ibid., p. 297.
59. Ibid., p. 297.
60. Ibid., p. 282.
61. Ibid., pp. 281–282.
62. Ibid., pp. 283–284.
63. Heilbron (1983: 102–103) finds Desagulier's rebuttal unconvincing for various other reasons.
64. Mairan (1720: 248); also Mairan (1722). In the second and third editions of the *Principia*, Newton observed that the heat in the Torrid Zone could have caused seconds pendulums to lengthen slightly (see, for example, Newton (1972: vol. 2, 607). However, he stated immediately afterward that the total difference in the lengths of isochronal pendulums in different climates could not be ascribed to or accounted for by means of differences in temperature (607–608). Nevertheless, as I mentioned in Chapter 1, Newton did note that if the observed difference in the lengths of seconds pendulums in Paris and in Cayenne were reduced slightly to allow for the lengthening of metallic pendulums in the Torrid Zone caused by the heat there, then the observed increase in the magnitude of effective gravity with latitude, corrected in this way, would agree closely with the increase in the lengths of seconds pendulums with latitude which Newton's theory of homogeneous figures shaped like ellipsoids of revolution flattened at their poles which attract according to the universal inverse-square law, which revolve around their axes of symmetry, whose ellipticities are infinitesimal, and whose columns from center to surface all balance or weigh the same required when the theory was applied to the earth [Newton (1972: vol. 2, 609–610); and Costabel (1988: 112–113)]. Now, Desaguliers, we recall, thought that he had deduced properties of

Cassini's figure shaped like an elongated ellipsoid of revolution whose ellipticity is infinitesimal which conflicted with the observed variation of the magnitude of the effective gravity with latitude along the earth's surface, assuming that Cassini's elongated figure attracts according to the universal inverse-square law. But Newton himself had done this same thing even earlier, in the second edition of the *Principia*. Newton left his discussion of the conflict out of the third edition of the *Principia*. But Mairan specifically referred to the second edition of the *Principia* in his note in the margin, yet he seemed totally unaware that Newton had tried to show in this edition that Cassini's figure shaped like an elongated ellipsoid of revolution whose ellipticity is infinitesimal could not be reconciled with the observed variation of the magnitude of the effective gravity with latitude along the earth's surface, when Cassini's figure was assumed to attract in accordance with the universal inverse-square law. In fact, in order for Newton to have established that Cassini's figure shaped like an elongated ellipsoid of revolution whose ellipticity is infinitesimal contradicted the observed variation of the magnitude of the effective gravity with latitude along the earth's surface, assuming that Cassini's figure attracts according to the universal inverse-square law, he would have had to be able to apply his theory of a homogeneous figure shaped like an ellipsoid of revolution which attracts according to the universal inverse-square law, which revolves around its axis of symmetry, whose ellipticity is infinitesimal, and whose columns from center to surface all balance or weigh the same to Cassini's figure. However, the columns from the center to the surface of a homogeneous figure shaped like an elongated ellipsoid of revolution which attracts in accordance with the universal inverse-square law can *never* all balance or weigh the same, because such a figure fails to fulfill two necessary conditions that any homogeneous figure must satisfy in order to be a figure of equilibrium – namely, the principle of balanced columns and the principle of the plumb line (see note 40 of Chapter 2).

65. Delisle (1720a: fols. 1r–2r); and Louville (1720: fol. 2v).
66. See note 3 of Chapter 1. Dating Mairan's familiarity with the contents of the *Principia* is not an easy task. Guerlac (1977: 481, 485) conjectured that Mairan first came across Newton's treatise during his four-year sojourn in Paris from 1698 until 1702, at which time he frequented Malebranche and the members of Malebranche's circle, all of whom were acquainted with the *Principia*. In 1702 Mairan went back to Beziers, his home town. He returned to Paris in 1718, when he joined the Paris Academy of Sciences. In his *Dissertation sur les variations du baromètre*, which won the Bordeaux Academy's prize of 1715, Mairan mentioned in a note the theories of a flattened earth appearing in Newton's *Principia* and in Huygen's *Discours de la cause de la pesanteur* (1690). Guerlac (1981: 66, note 64) observed, however, that Mairan could have easily obtained the reference to the *Principia* from Huygens's *Discours*, the second of which Mairan had doubtlessly read. But Guerlac thought it less certain whether Mairan had actually seen the *Principia* at this time. Guerlac (1977: 482) believed that Mairan did "display some acquaintance with the *Principia*" in his *Dissertation sur la glace*, which won the Bordeaux Academy's prize of 1716.
67. Maupertuis (1731a: fol. 2v); and Johann I Bernoulli (1731: fols 3r–3v).
68. Mairan (1743: 381).
69. Brown (1975: 70); and Brown (1976: 168).
70. Maupertuis (1732b: 8–9); and Maupertuis (1733a: 154).
71. Brown (1975: 75); and Brown (1976: 172).
72. Desaguliers (1725a); also Desaguliers (1725b: 254–255).
73. See Greenberg (1983); and Greenberg (1984a). In Nordmann (1966: 78), Nordmann gives as his earliest French reference to Desaguliers's criticism of Cassini's measurements the Royal Geographer Bourguignon d'Anville's *Proposition d'une mesure de la terre* (1735). Maupertuis did not really discuss Desaguliers's three articles published in the *Philosophical Transactions* at all until 1741. This he did in an appendix to his (1741) entitled "Examen des trois dissertations de Monsieur Desaguliers sur la figure de la terre." In this appendix, Maupertuis trumped up deceitfully arguments against Desaguliers and "attacked" his three articles.
74. For example, Lafuente and Peset (1984: 237–239); and Lafuente and Delgado (1984: 24).

3. The revival of geodesy in Paris (1733–1735)

1. See Greenberg (1983) and Greenberg (1984a).
2. The published version is Maupertuis (1733a).

3. Beeson (1985: 196) and Beeson (1992: 100) gives reasons for thinking that Joncourt was the author.
4. Maupertuis (1732b) and Maupertuis (1732a). Hereafter when I refer to the latter of the two works, I shall refer to its translation into English: Maupertuis (1809).
5. Maupertuis (1733a: 156).
6. Maupertuis (1735: 98–105; diagram on p. 99).
7. Works that muddle the issues include Boss (1972: Chapter 14); Paolo Cassini (1975); Salomon-Bayet (1975: 192); Taton (1978: 488, 493); and Taton (1988: 117). Nordmann (1966: 93–95) thought wrongly that the results of the expedition to Lapland marked a triumph of Newton's principles over Descartes's and, moreover, that the members of the Paris Academy on the whole viewed these results as such a triumph and that they challenged the findings of the members of the expedition because they could not accept such a victory. At one of the roundtable discussions held during the Maupertuis conference in Paris in 1973, Merleau-Ponty (1975) doubted assertions made at the conference that the members of the expeditions to Peru and to Lapland carried out "crucial experiments." It turns out that Merleau-Ponty had good reasons to be skeptical. Beeson (1985: 191–193, 203) noted that until 1713, astronomers in Paris emphasized that both Newtonian and Cartesian world systems entailed that the earth is flattened at its poles, so that an earth found to be flattened at its poles would not confirm one of the two world systems more than the other. But then Beeson (1985: 195–196) went on to argue in a way that makes no sense in view of this. Namely, Beeson claimed that the problem of the earth's shape was later "elevated into a question capable of distinguishing between rival Newtonian and Cartesian cosmologies. More than that, almost uniquely among such matters of controversy between the two [world] systems, it was a question capable of resolution by immediate experimental means." Specifically, according to Beeson, it became possible to "envisage" an "acid test of the two [world] systems," by "obtaining a conclusive and incontrovertible observational determination of the shape of the Earth." Unlike Beeson, however, I shall maintain that it never became possible to do any such experiment and that the problem of deciding between rival Newtonian and Cartesian cosmologies was not solved in the terrestrial realm. Instead, Newton's universal inverse-square law of attraction was ultimately and only verified by solving problems in celestial mechanics.
8. Lafuente and Peset (1984); and Lafuente and Delgado (1984).
9. Other members of the Malebranche Group who sought to reconcile Descartes and Newton include Joseph Privat, abbé de Molières, and Etienne-Simon, abbé de Gamaches. Works that highlight the influence of Descartes upon the members of the group include Brunet (1931, 1970); and Aiton (1972). Studies of the influence of Newton upon the members of the group include Mouy (1938); Guerlac (1977); Guerlac (1979); and Guerlac (1981). I shall discuss one of the works of Gamaches in Chapter 9.
10. See Roche (1969: 711).
11. See Greenberg (1983); and Greenberg (1984a).
12. Koenig (1737: 109).
13. See Greenberg (1986). Those historians whom I refer to as wishful thinkers are named in this article, as well their works in which this wishful thinking appears. At least Beeson (1992: 79–82, 88–89] summarizes in words some of the results of Maupertuis's mathematics published in the *Discours*.
14. Concerning Coste's translation of Locke's work and Leibniz's use of the translation, see Aiton (1985: 277–278, 341); and Dubois (1986: 36).
15. For a discussion of des Maizeaux's *Recueil* and the reference to the *Principia* as an "intractable book," see Gillispie (1958: 429–430).
16. Maupertuis (1731a: fol. 2v). Nordmann (1966: 93–95) says that Maupertuis and Clairaut were attacked by colleagues in Paris because the two French mathematicians confirmed by measurements a shape for the earth theoretically deduced by an Englishman and a Dutchman.
17. Maupertuis (1738a); revised and enlarged as Maupertuis (1741).
18. Maupertuis (1738a: 49–50).
19. Desaguliers (1725b).
20. Maupertuis (1738a: 49–50).
21. Ibid., pp. 64–73.
22. Voltaire to Maupertuis, *circa* 10 January 1738, in Voltaire (1954: 14); also appearing in Voltaire (1969: 457). In the *Discours préliminaire de l'Encyclopédie*, D'Alembert praised Maupertuis for his "courage." D'Alembert's eulogy is quoted in Brunet (1931, 1970: 203, note 2).

23. *Journal des Savants*, June 1742, pp. 335–336. In fact the first edition *was* reviewed in the *Journal des Savants*, April 1733, pp. 206–217.
24. La Beaumelle (1856: 32–34).
25. Maupertuis (1734b: 64). For information about Roberval's explanation of gravity by attraction, see Koyré (1968: 59, note 2; 233, note 1).
26. Fontenelle (1734: 94).
27. Fontenelle (1732a: 93).
28. La Beaumelle (1856: 54–56).
29. For recent studies that shed much light on the controversies in Paris after the members of the Lapland expedition returned, see Terrall (1992) and Beeson (1992: 116–134).

4. Pierre Bouguer and the theory of homogeneous figures of equilibrium

1. Bouguer (1734).
2. Bouguer (1733: 96–98). For the early history of this technique in the partial differential calculus, see Engelsman (1984) and (1986).
3. Maheu (1966: 206) definitely exaggerated when he maintained that Bouguer had *no* direct interest whatever in ordinary and partial differential equations.
4. For additional discussion of the overall improvement and some of the reasons for it, see Greenberg (1986).
5. Bouguer (1734: 22, and Figure 1). Alexis-Claude Clairaut later coined the term directrix (*directrice*) for this oval curve, in Clairaut (1743: 65).
6. Maupertuis (1731b: fol. 1r).
7. See, for example, *L'Académie Royale des Sciences, Registres des Procès-Verbaux*, 20 August 1721, p. 237r; 23 August 1721, p. 239r; 3 September 1721, p. 243v; 6 September 1721, p. 256r; 1 February 1724, p. 51r; 23 August 1724, pp. 291v–295v; 30 August 1724, pp. 297r–302v; 21 July 1725, p. 167v; 9 February 1726, pp. 49v–51v; 4 September 1726, p. 275r; 7 September 1726, p. 277v; 26 April 1727, p. 153r; 23 August 1727, pp. 294r–294v; Mairan (1721: especially p. 105); Mairan (1724); Grandjean de Fouchy (1758: 129–130); and Morère (1965: 352–354).
8. Bouguer (1734: 25). Both Todhunter (1873, 1962: §55) and Truesdell (1954: XV, XX) attribute the generalization of Newton's principle of balanced columns to Huygens. Huygens (1690: 156) had stated that Newton's principle of balanced columns held for a channel of *any* form within a homogeneous figure of equilibrium provided each end of the channel touches the surface of the figure.
9. Bouguer (1734: 27, and Figure 2).
10. Ibid., p. 28. In Figure 62, *m* is a point on the meridian through *M* and infinitesimally near *M*. From the diagram it is clear that Angle $MmN = \alpha =$ Angle GMT, because the two sides forming one angle are perpendicular to the two sides forming the other angle, which follows from the way that the diagram is constructed. Moreover, Angle $MmO =$ Angle PMT, because the two sides forming one angle are perpendicular to the two sides forming the other angle, which again follows from the way that the diagram is constructed. Furthermore, Angle $PMT =$ Angle GST, because MP and GS are parallel. Hence Angle $MmO = \beta =$ Angle GST. If we now introduce the point Q such that GQ is perpendicular to ST (that is, perpendicular to MT), and if we let $|XY|$ stand for the length of a line segment XY, then it is easy to see from the diagram that

$$\frac{|MO|}{|MN|} = \frac{\dfrac{|MO|}{|Mm|}}{\dfrac{|MN|}{|Mm|}} = \frac{\sin \beta}{\sin \alpha}$$

$$= \frac{\dfrac{|QG|}{|GS|}}{\dfrac{|QG|}{|GM|}} = \frac{|GM|}{|GS|}.$$

11. Ibid., p. 29. Concerning the differences between "differential" equations and "differential equations," see Bos (1974–75).

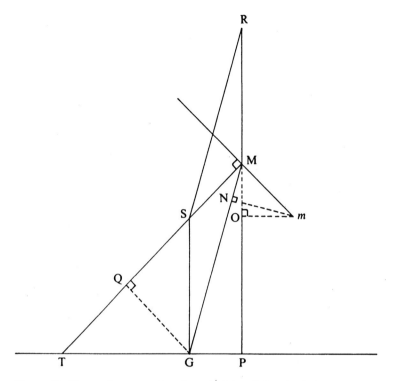

Figure 62. Figure for expressing the principle of the plumb line (see Figure 13).

12. Bouguer (1734: 28).
13. Ibid., pp. 30–31, and Figure 2.
14. Ibid., p. 31.
15. Ibid., p. 37, and Figure 3.
16. Ibid., p. 38, and Figure 4.
17. Ibid., p. 39.
18. See Aiton (1972: 219–228); Beeson (1985: 198, note 17); and Beeson (1992:103, note 39).

5. Maupertuis

1. Maupertuis (1734b).
2. Brunet (1952: 22–23). Taton (1978: 488–489) suggests that the idea of a second expedition, to Lapland to measure a degree of latitude there, first arose during the talks between Maupertuis, Clairaut, and Johann I Bernoulli in Basel in November of 1734. I have discussed the origins of the first expedition, to Peru, in Greenberg (1983) and in Greenberg (1984a). Johann I Bernoulli's Paris Academy prize essay of 1734 on the causes of the inclinations of the planetary orbits to the plane of the ecliptic is discussed in Brunet (1931, 1970: 272–293); in Aiton (1972: 228–235); and in Shea (1988: 84–89).
3. Maupertuis (1734b: 57).
4. Ibid., p. 60. In Figure 63, Angle $MdD = \alpha =$ Angle SDT, since the two line segments that form one of the angles are perpendicular to the two line segments that form the other angle, because of the way that the diagram is constructed. Consequently Triangle DST is similar to Triangle dMD. Let

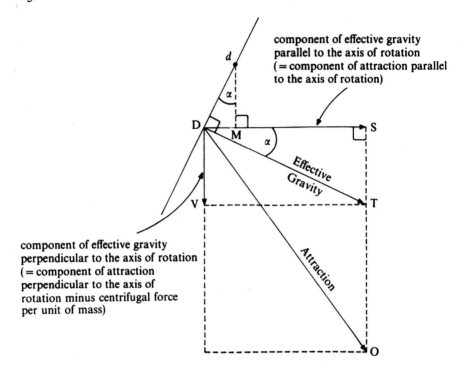

Figure 63. Figure for expressing the principle of the plumb line in rectangular coordinates.

$|AB|$ stand for the length of a line segment AB. Since $|DV| = |ST|$, it follows that

$$\frac{|DV|}{|DS|} = \tan\alpha = \frac{|DM|}{|dM|}. \tag{I}$$

Let $\vec{AB}$ designate the vector that lies along line segment AB. Let $\hat{\imath}$ and $\hat{\jmath}$, respectively, symbolize unit vectors that are parallel to and perpendicular to, respectively, the axis of rotation. We can write $\vec{Dd} = |DM|\hat{\imath} + |dM|\hat{\jmath}$ and $\vec{DT} = |DS|\hat{\imath} - |DV|\hat{\jmath}$. Then Maupertuis' condition

$$\frac{|DV|}{|DS|} = \frac{|DM|}{|dM|} \tag{I}$$

means that

$$\begin{aligned} 0 &= |DM||DS| - |dM||DV| \\ &= (|DM|\hat{\imath} + |dM|\hat{\jmath}) \cdot (|DS|\hat{\imath} - |DV|\hat{\jmath}) = \vec{Dd} \cdot \vec{DT}. \end{aligned} \tag{II}$$

That is, Maupertuis's condition is just the condition that $\vec{Dd}$ and $\vec{DT}$ be orthogonal, when the two vectors are expressed using orthonormal vectors $\hat{\imath}$ and $\hat{\jmath}$. Clairaut made early use of rectangular coordinates in his treatise of 1731 on skew curves: Clairaut (1731b).

5. Maupertuis (1734 b: 63).
6. Maupertuis (1731b: fol. 1r).
7. Rappaport (1981: 237, note 45) mentions that in addition to Mairan and Bouguer, two other individuals entered the Paris Academy directly as associate members during the eighteenth century: Exupère-Joseph Bertin in 1744 and Jean Darcet in 1784. Bertin was thirty-one years old

when he joined the Academy, and Darcet was fifty-eight years old when he entered the Academy. Age again could have been a factor that the Academy took into account in admitting these two individuals as associate members.

8. Maupertuis (1732c: fol. 1v).
9. Maupertuis (1733b: fol. 1v).
10. Maupertuis (1738a: 82).
11. Maupertuis (1731b: fol. 1r).
12. Maupertuis (1731c: 297). Maupertuis did not refer specifically to Guisnée here, but Clairaut did while describing Maupertuis's Paris Academy *mémoire* to the Genevan mathematician Gabriel Cramer in a letter in March of 1732 [see Clairaut (1732a)].
13. Bouguer (1732: esp. p. 5 concerning logarithms).
14. Maupertuis (1732d). For an account of Bouguer's and Maupertuis's Paris Academy *mémoires* on curves of pursuit, see Clarinval (1957: 26–29). Clarinval observes that Maupertuis could not really solve his generalization of the problem (p. 29). See as well Bernhart (1959).
15. Maupertuis (1734b: 56). In the original version of this *mémoire*, which Maupertuis read before the Paris Academy, "integral calculus" appears in place of "inverse method of tangents." See Maupertuis (1734a: 238v). Later, in Maupertuis (1738a: 54), Maupertuis said that "in Huygens's time the calculus of tangents was altogether lacking or else Huygens did not know it well enough" and that balancing columns enabled one to "dispense with the inverse method of tangents."
16. Mairan (1720: 253).
17. See Greenberg (1986).
18. Ibid.
19. In his anonymously published, "disinterested" account of the controversy over the earth's shape, published in 1738, Maupertuis wrote: "among the *mémoires* of the Paris Academy of Sciences for 1734, there are two that show much affinity in the determination of the earth's shape by the laws of hydrostatics. The first one is by Monsieur Bouguer and is filled with excellent things." See Maupertuis (1738a: 62). The second *mémoire* that Maupertuis alluded to was, of course, his own: Maupertuis (1734b).
20. Aiton (1972: 219–221).
21. La Condamine to Boscovich, 21 March 1770, in Boscovich (1912: 300–302).
22. Condorcet (1774: 99).
23. Ibid., pp. 93, 108–109.
24. Weiss (1825: 513–514).
25. In his biography of his friend Maupertuis, La Beaumelle said the following with regard to the two Paris Academy *mémoires* on curves of pursuit: "Would this be the origin of the estrangement between Messieurs Bouguer and Maupertuis, not to say the enmity that Monsieur Bouguer never stopped showing toward Monsieur Maupertuis ever since this moment? At least, no other cause can be found. Besides, we know that jealousy was Monsieur Bouguer's weak point and that it sometimes made him deviate from his naturally gentle nature." See La Beaumelle (1856: 22). At least two facts contradict La Beaumelle's account of the antagonism between Maupertuis and Bouguer:
    a. La Beaumelle evidently did not know that Maupertuis had already begun to show a dislike for Bouguer even *before* the two Paris Academy *mémoires* of 1732 on curves of pursuit – for example, in Maupertuis's letters to Johann I Bernoulli in 1731.
    b. It was *Maupertuis's jealousy* of Bouguer which very likely caused Maupertuis to write the kind of *mémoire* that he did – one that intentionally but covertly ridiculed Bouguer.
26. Maupertuis (1740a: 29r); and Maupertuis (1740b).
27. *L'Académie Royale des Sciences, Registres des Procès-Verbaux*, 4 July 1742, p. 310.
28. Maupertuis (1731d: 86r); and Maupertuis (1731e: 107–109).
29. Saurin (1721). Portions of two manuscripts of this paper are preserved at the Archives of the Paris Academy of Sciences, "Dossiers des Séances," Pochette: 1721. The published version of the paper appears as: Saurin (1722a). See as well Louville (1722a). The published version of this paper appears as: Louville (1722b). Saurin's reply to Louville, read to the Paris Academy, is Saurin (1722b). The manuscript of this reply is preserved at the Archives of the Paris Academy of Sciences, "Dossiers des Séances," Pochette: 1722.
30. The quotation concerning Fontenelle's censoring Varignon appears in the *Journal des Savants*, December 1725, p. 728.

31. Clairaut (1734a: 198).
32. Fontaine (1735: 17r–17v).
33. Clairaut (1734a: 198).
34. Fontaine (1734a). This is the published version of Fontaine (1735).
35. Buffon (1748); and the published version: Buffon (1745).
36. *Académie Royale des Sciences, Registres des Procès-Verbaux*, 27 May 1701, pp. 183r–183v.
37. Costabel (1983: 43–44).
38. *Académie Royale des Sciences, Registres des Procès-Verbaux*, 9 January 1706, pp. 1r–4r. For discussions of the "Rolle affair," see Costabel (1966), Blay (1986), and Mancosu (1989).
39. Mahoney (1975: 117).
40. In naming Jean Hellot as the only Paris Academician to be promoted directly to pensioner from assistant member during the eighteenth century, Rappaport (1981: 237, note 45) overlooked Saurin, because Rappaport took the year 1716 as the first year that she studied. Hellot was already forty-nine years old when he joined the Academy in 1735, which could explain his unusual promotion in 1743.
41. Schier (1941: 3–58).
42. Bibliothèque National (Paris), Fonds Français, Nouvelles Acquisitions, ms 5148, Claude II Bourdelin's minutes for the Paris Academy's meeting on 23 June 1705, p. 74; also *Académie Royale des Sciences, Registres des Procès-Verbaux*, 23 June 1705, p. 211r; 27 June 1705, p. 215v; 1 July 1705, p. 217v; 18 July 1705, pp. 243r–243v; and 24 July 1705, p. 294v.
43. Concerning Parent's contributions to mechanics and the strength of materials, see Truesdell (1960: 109–114).
44. Briggs (1974: 319).
45. L'Hôpital's recommending Parent to Leibniz is discussed in Aiton (1985: 244).
46. I thank John Pappas for having informed me of this fact.
47. Delisle (1717a: fols. 1r–1v); Delisle (1717b: fols. 1r–2v); Delisle (1717c: fol. 1r–2r); Delisle (1724a: fols. 4r); Delisle (1724b: fols. 2r–3r); Delisle (1724c: fols. 1r–2r); Delisle (1724d: fols. 1r–4v); Delisle (1724e: fol. 1r–2r); Delisle (1724f: fols. 1r–lv); and Delisle (1725: fols. 3r–4r).
48. See Greenberg (1983) and Greenberg (1984a). Delisle's criticisms of Cassini's measurements and Delisle's idea of measuring degrees of longitude appear in Delisle (1716). This manuscript and Delisle (1720b) are identical.
49. See Greenberg (1983: 243–246, 256); also Delisle (1734b: fols. 1r–1v). The latter is Delisle's copy of his letter to Fontenelle. The original is Delisle (1734a).
50. See Greenberg (1983: 255–256).
51. There is no mistaking which edition of the *Principia* Maupertuis used here. It is clear from internal evidence that Maupertuis gave an exposition of the theory of the earth's shape appearing in the third edition. The evidence for this is the following: In the third edition, Newton (1972: vol. 2, 599) gave the ratio of the densities of the earth and Jupiter as $400/(94\frac{1}{2})$ and the value of Jupiter's ellipticity as approximately $1/(9\frac{1}{3})$. But these are precisely the values of these quantities appearing in Maupertuis's *mémoire* (1734b: 96). The values for the earth's ellipticity given in the first edition of the *Principia* $((3/689) = (1/229\frac{2}{3}))$ and in the second and third editions $(1/229)$ differ insignificantly from each other, because both are "infinitesimals" ε (which is to say that they satisfy $\varepsilon^2 \ll \varepsilon$), while the two values differ from each other by terms of second or higher order (that is,

$$1/(229\tfrac{2}{3}) = 1/(229 + \tfrac{2}{3}) = (1/229)(1/(1 + (\tfrac{2}{3}/299))) \approx (1/229)(1 - (\tfrac{2}{3}/229)) \approx 1/229.$$

This turns out to be true essentially because of the negligible difference described in note 3 of Chapter 1, and for the same reason as the negligible differences described in note 14 of Chapter 6. By contrast, the values of Jupiter's ellipticity given in the first, second, and third editions of the *Principia* differ considerably from each other (namely, $1/(39\frac{3}{5})$, $1/8$, and $1/(9\frac{1}{3})$, respectively). The differences between them are not negligible, because, for one thing, these values are not "infinitesimal" (meaning that if we call each one ε, none satisfies $\varepsilon^2 \ll \varepsilon$), while the differences between the values include terms that are finite and/or terms of first order. [For example,

$$1/(9\tfrac{1}{3}) = 1/(8 + 1\tfrac{1}{3}) = (1/8)(1/(1 + (1\tfrac{1}{3}/8))) = (1/8)(1 - (1\tfrac{1}{3}/8) + (1\tfrac{1}{3}/8)^2 - \cdots).$$

But the term $(1/8)(1\frac{1}{3}/8)$, although infinitesimal, is of first order, not second or higher order, hence it cannot be neglected.]
52. Maupertuis (1733c). Maupertuis stated specifically in this version of his *mémoire* that he was

expounding parts of the *second* edition of the *Principia* [Maupertuis (1733c: 42r)], but he did not make this clear in the published version of the *mémoire*: Maupertuis (1732e). Moreover, in the published version of the *mémoire*, Section XIV of Book I of the *Principia* was omitted.

53. Maupertuis (1734b: 71–72). Maupertuis reasoned as follows. Suppose that Jupiter is a homogeneous spherical figure whose radius is r. If R is the radius of the orbit of one of Jupiter's satellites, where we assume the orbit to be a circle whose center is situated at Jupiter's center and we assume the satellite to revolve around Jupiter at a constant speed, and if τ is the period of that satellite (in other words, the amount of time it takes that satellite to revolve around Jupiter once), then R/τ^2 is the centripetal acceleration of the satellite as it revolves around Jupiter. The centripetal acceleration of the satellite is directed toward the center that the satellite revolves around. The magnitude of the centripetal acceleration of the satellite varies inversely as the square of the distance R of the satellite from the point that the satellite revolves around. This last result follows from Kepler's third law of motion together with Newton's expression for the magnitude of the centripetal acceleration of a body moving in a circle at a constant speed (which was the same as Huygens's expression for the magnitude of the centrifugal force per unit of mass of a body moving in a circle at a constant speed) [see Cohen (1960: 172–173, or 1988: 46–47, note 5) for a demonstration]. From this it follows that if the centripetal acceleration of the satellite is assumed to be caused by a central force that attracts the satellite, where the center of force that produces the attraction is situated at the point around which the satellite revolves, namely, the center of Jupiter, then if A designates the magnitude of the attraction produced by this center of force at Jupiter's surface and B stands for the magnitude of the attraction of the satellite produced by the center of force, then

$$\frac{B}{A} = \frac{r^2}{R^2}. \tag{I}$$

[Newton demonstrated that a homogeneous spherical figure that attracts according to the universal inverse-square law and whose radius is r attracts like a $1/r^2$ central force in its exterior $r \leqslant \underline{r}$ (see note 3 of Chapter 1), but Jupiter does not have to be assumed to attract according to the universal inverse-square law in order for equation (I) to hold.] But

$$B = \frac{R}{\tau^2} \tag{II}$$

is true, since the centripetal acceleration of the satellite is assumed to be caused by a central force that attracts the satellite, and B is the magnitude of the attraction of the satellite produced by the center of force. Consequently it follows from substituting the right-hand side of (II) for the numerator of the left-hand side of (I) that

$$\frac{R/\tau^2}{A} = \frac{r^2}{R^2} \tag{III}$$

is true, and (III) can be rewritten as

$$A = \frac{R^3}{\tau^2 r^2}. \tag{IV}$$

At the same time, the magnitude of the centripetal acceleration of rotation at Jupiter's equator is

$$\frac{r}{T^2},$$

where T is the length of Jupiter's day. Hence it follows from (IV) that the ratio

$$\frac{r/T^2}{A}$$

of the magnitude of the centripetal acceleration of rotation to the magnitude of the attraction at Jupiter's equator is

$$\frac{r/T^2}{A} = \frac{r/T^2}{R^3/\tau^2 r^2} = \left(\frac{r}{R}\right)^3 \frac{\tau^2}{T^2}, \tag{V}$$

where r/R, τ, and T were all measurable quantities. Maupertuis talked about "the centripetal or centrifugal force (*la force centripete ou centrifuge*)" of a body moving in a circle (p. 71). In other words, he used the adjectives "centripetal" and "centrifugal" interchangeably, which, of course, we do not do today. It turns out that there were conflicting interpretations of centrifugal force during the early decades of the eighteenth century, and, as a consequence, conflicting justifications of the equality of centripetal and centrifugal force. See Bertoloni Meli (1990). However, we now know that the magnitude of the centripetal acceleration of a body moving in a circle at a constant angular speed in an inertial frame of reference equals the magnitude of the centrifugal force per unit of mass of the body measured in the frame of reference attached to the moving body.

If Maupertuis had assumed that Jupiter is not a spherical figure but a figure shaped like an ellipsoid of revolution flattened at its poles whose ellipticity is infinitesimal (which in fact Jupiter's ellipticity *is not*), the ratio of the magnitude of the centrifugal force per unit of mass to the magnitude of the attraction at Jupiter's equator would have had to turn out to be the same. To see this, let $P + \delta$ designate the magnitude of the attraction at the equator of a homogeneous figure shaped like an ellipsoid of revolution flattened at its poles which attracts according to the universal inverse-square law, which revolves around its axis of symmetry, and whose ellipticity is infinitesimal, where δ stands for the difference between this value of the magnitude of the attraction and the magnitude of the attraction P at the surface of a homogeneous spherical figure which attracts according to the universal inverse-square law, whose density is the same as the density of the homogeneous figure shaped like a flattened ellipsoid of revolution, and whose radius is the same as the equatorial radius of the homogeneous figure shaped like a flattened ellipsoid of revolution. Let F symbolize the magnitude of the centrifugal force per unit of mass of the figure shaped like a flattened ellipsoid of revolution at a distance from its axis of rotation equal to its equatorial radius. Let us suppose that the two figures rotate with the same angular speed, and let us make as well the reasonable assumption that δ/P in infinitesimal (that is,

$$\left(\frac{\delta}{P}\right)^2 \ll \frac{\delta}{P}.$$

Then

$$\frac{F}{P+\delta} = \frac{F}{P}\left(\frac{1}{1+(\delta/P)}\right) \simeq \frac{F}{P}\left(1 - \frac{\delta}{P}\right) = \frac{F}{P} - \frac{\delta}{P}\frac{F}{P} \simeq \frac{F}{P},$$

where the first approximation holds to terms of first order because δ/P is infinitesimal and where the second approximation also holds to terms of first order because F/P is also infinitesimal [that is, $(F/P)^2 \ll F/P$], since little flattening means "slow" rotation, in which case $(\delta/P)(F/P)$ is a term of second order and can be neglected. Hence to terms of first order the ratios of the magnitudes of the centrifugal force per unit of mass to the magnitudes of the attraction at the equators of the two rotating figures are the same.

I mentioned previously that if Jupiter were a homogeneous spherical figure whose radius is r and which is assumed to attract according to the universal inverse-square law, then Jupiter would attract like a $1/\underline{r}^2$ central force in its exterior $r \leqslant \underline{r}$, where r is Jupiter's radius (see note 3 of Chapter 1). But Jupiter is *not* shaped like a sphere, so that if Jupiter is assumed to attract in accordance with the universal inverse-square law, it does *not* attract like a $1/\underline{r}^2$ central force in its exterior $r \leqslant \underline{r}$. However, Clairaut (1743; 196) maintained that if Jupiter is assumed to attract according to the universal inverse-square law, the centripetal acceleration of the observable satellite that is furthest from Jupiter (Jupiter's fourth satellite), which is caused by Jupiter's attraction of the satellite, *is* very nearly the same as it would be if Jupiter were spherical. In other words, if Jupiter is assumed to attract in accordance with the universal inverse-square law, it appears to its fourth satellite to attract like a $1/\underline{r}^2$ central force.

54. Newton's calculation of the ratio of Jupiter's density to the Earth's density appears in Newton (1972: vol. 2, 577–584). Essentially Newton showed the following. Assume that the Earth and Jupiter are homogeneous spherical figures that attract according to the universal inverse-square law and that the orbit of the Earth's moon and the orbit of one of Jupiter's satellites, respectively, are circles whose centers are located at the centers of the Earth and Jupiter, respectively. Let A designate the magnitude of the attraction at Jupiter's surface; let B stand for the magnitude of the attraction at the Earth's surface; let C symbolize the centripetal acceleration of Jupiter's satellite;

and let F stand for the centripetal acceleration of the Earth's moon. Then, once again, since the centripetal acceleration of a satellite is assumed to be caused by the force that attracts the satellite, it follows that

$$\frac{A}{B} = \frac{C \times (R^2/r^2)}{F \times (D^2/u^2)},$$
(I)

where R is the radius of the orbit of Jupiter's satellite, r is Jupiter's radius, D is the radius of the orbit of the Earth's moon, and u is the Earth's radius. This is again true because Jupiter attracts like a $1/r^2$ central force in its exterior $r \leqslant \underline{r}$, since Jupiter is assumed to be a homogeneous spherical figure, and, at the same time, the earth attracts like a $1/\underline{r}^2$ central force in its exterior $u \leqslant \underline{r}$, because the Earth is assumed to be a homogeneous spherical figure. (See note 3 of Chapter 1.) In particular, in case R were equal to D, it would follow from (I) that

$$\frac{A}{B} = H \times \left(\frac{u}{r}\right)^2,$$
(II)

where H is the ratio of the centripetal accelerations of satellites that revolve around Jupiter and the Earth, respectively, at equal distances D from them. If X were the centripetal acceleration of a satellite revolving around Jupiter at a distance equal to D from Jupiter, then

$$\frac{R/\tau^2}{X} = \frac{D^2}{R^2},$$
(III)

where τ is the period of Jupiter's satellite whose orbit has a radius equal to $R \neq D$, since Jupiter's attraction of the two satellites are assumed to cause the centripetal accelerations of the two satellites, and Jupiter attracts like a $1/r^2$ central force in its exterior $r \leqslant \underline{r}$, because Jupiter is assumed to be a homogeneous spherical figure. (See note 3 of Chapter 1.) But (III) can be rewritten as follows:

$$X = \frac{R^2}{\tau^2 D^2}.$$
(IV)

Therefore if T is the period of the Earth's moon, it follows from the definition of H together with (IV) that

$$H \equiv \frac{X}{D/T^2} = \left(\frac{R}{D}\right)^3 \left(\frac{T}{\tau}\right)^2.$$
(V)

Now, it follows from Proposition 72 of Book I of the *Principia* that the magnitudes of the attractions at the surfaces of homogeneous spherical figures that attract according to the universal inverse-square law are proportional to the radii of the figures. [For a remarkably illuminating exposition of Newton's obscure, nearly incomprehensible proof of this proposition, see Weinstock (1984: 887–888)]. Moreover, Newton implied without explanation in Corollary 3 to Proposition 8 of Book III of the *Principia* that the magnitude of the attraction at the surface of such a homogeneous spherical figure is also proportional to the density of the figure. In other words, he evidently assumed that if density of the homogeneous spherical figure is multiplied by a fixed factor, the magnitude of the attraction produced by the homogeneous spherical figure at any point is also multiplied by that same fixed factor. Newton may have reasoned that this was the case using arguments like those that appear in Weinstock (1984: 888). Be that as it may, he did not actually demonstrate that this result is true. It follows from putting the two results together that the magnitudes of the attraction at the surfaces of homogeneous spherical figures that attract in accordance with the universal inverse-square law are proportional to the products of the radii and the densities of the figures. Consequently

$$\frac{A}{B} = \frac{r\delta}{u\sigma},$$
(VI)

where δ is Jupiter's density and σ is the Earth's density. Then if (II), (V), and (VI) are combined, it follows that

$$\frac{r\delta}{u\sigma} = \frac{r\delta}{u\sigma}\frac{A}{B} = H\left(\frac{u}{r}\right)^2 = \left(\frac{R}{D}\right)^3\left(\frac{T}{\tau}\right)^2\left(\frac{u}{r}\right)^2,$$
(VII)

and from (VII) it follows that

$$\frac{\delta}{\sigma} = \left(\frac{R}{r}\right)^3 \left(\frac{u}{D}\right)^3 \left(\frac{T}{\tau}\right)^2. \tag{VIII}$$

Finally, (VIII) expresses the ratio of Jupiter's density to the Earth's density in terms of R/r, u/D, T, and τ, which were all measurable quantities. If we compare Newton's derivation of this result with Maupertuis's calculations in note 53 of Chapter 5, we see that Newton carried out the same kinds of calculations that Maupertuis did later.

55. Maupertuis (1809: 523). These remarks concerning Maupertuis's determination of shapes to be "more mathematical than physical," which appear in a "Scholium" that ends the English translation of Maupertuis's original paper in Latin, apparently do not appear in the original version. Maupertuis sent his remarks to the Royal Society after he sent his paper, but the remarks were evidently received too late to be included in the paper published in the *Philosophical Transactions* [see Brown (1975: 77, and 1976: 173)].

56. In Maupertuis (1732b: 49), Maupertuis said: "what appears here concerning shapes of planets and suns must not pass for exact determinations," because of the differences between his calculations and the conclusions that they led to and Newton's calculations and results.

57. Hankins (1976).

58. Newton (1972: vol. 2, 599–600). In fact, in the third edition of the *Principia*, taking the irregular refraction of light into account, Newton changed $1/9\frac{1}{3}$ to $1/10\frac{1}{6}$ (pp. 599–600).

59. Maupertuis (1734b: 98–99).

60. Ibid., pp. 99–100.

61. Logically speaking one would think that a "spheroid" means any surface that is nearly spherical, but in fact it does not mean this. Agnew (1962: 249) and James and James (1968: 127) say that a "spheroid" means *any* ellipsoid of revolution. Its ellipticity can be infinitesimal *or finite*. This modern definition of a "spheroid" in fact accords with what Clairaut already said in 1737 about the term "spheroid": "as we intend to apply our discoveries to the spheroid of the earth, which all agree to be very little different from a sphere, our computations must be adapted to those spheroids which have the smallest difference between the two axes" [Clairaut (1809: 120)]. The term "spheroid" meant the same thing to Clairaut in 1743 as it did in 1737, for in 1743 he wrote: "... if ... one supposes ... that the spheroid differs very little from a sphere (si ... on suppose ... que le sphéroïde diffère très-peu d'une sphère, ...)" [Clairaut (1743: 190–191)]. In reproducing MacLaurin's results that held for homogeneous figures shaped like ellipsoids of revolution flattened at their poles which attract according to the universal inverse-square law, which revolve around their axes of symmetry, and whose ellipticities are *finite*, Clairaut said that these figures were also shaped like spheroids (see Clairaut [1743: 168–174]). However, Clairaut also applied the term "spheroids" to figures of revolution which were *not* shaped like ellipsoids. In talking about the figure that appears in Corollaries 2 and 3 to Proposition 91 of Book I of the *Principia*, he said that it was shaped like an "elliptic spheroid" [Clairaut (1809: 119, 122)]. The need for an adjective "elliptic" means that Clairaut imagined that "spheroids" could be surfaces of revolution which were *not* ellipsoids or revolution. In 1750 he talked about finding "... spheroids whose attractions do not require formulas much more complicated than [the formulas for attractions produced by figures shaped like] ellipsoids [of revolution] [Clairaut (1750: 9)]." Here we see Clairaut use the term "spheroid" to mean a certain figure, not the shape of the figure (that is, not the surface of the figure) as we do today. In Clairaut (1743: 23), Clairaut also used the term "spheroid" in one paragraph to mean *both* a certain figure as *well as* the shape of that figure (that is, the surface of the figure). Moreover, throughout Mairan (1720), Desaguliers (1725c), Maupertuis (1732a), and Maupertuis (1732b), Mairan, Desaguliers, and Maupertuis also applied the term "spheroid" to figures of revolution which were not shaped like ellipsoids. In short, the term "spheroid" evidently did not have exactly the same meaning in the eighteenth century that it has now.

62. Maupertuis (1734b: 91)

63. Maupertuis (1734b: 98) tried to establish the result as follows:
AMR is a quadrant of a circle, PGA a quadrant of an ellipse. From the properties of an ellipse, $|RP|/|MG| = |CR|/|EM|$. Now, the tangent to the circle at M is perpendicular to MD. When the ellipse approaches the circle – in other words, $P \to R - DG$ approaches a direction parallel to the tangent to the circle at M. Hence DG approaches a direction perpendicular to DM, as $P \to R$. In

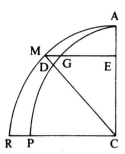

Figure 64. Quadrants of an ellipse and inscribed circle (Maupertuis 1734b: figure on p. 98).

other words, $\angle MDG \rightarrow$ a right angle, as the ellipse approaches a circle. In this case, $\Delta DMG \sim \Delta EMC$, and so $|MG|/|MD| = |MC|/|ME|$. Eliminating $|MG|$ from the two equalities and using $|MC| = |CR|$ gives $|MD|/|RP| = |ME|^2/|CR|^2 = \sin^2 \angle ACM$, or $|MD| = |RP| \sin^2 \angle ACM$. For a given ellipsoid, $|RP|$ is a constant K. Maupertuis now represents the Magnitude of the Effective Gravity at A by $|CD|$ and the Magnitude of the Effective Gravity at D by $|CA|$, and from this he concludes that the Magnitude of the Effective Gravity at D – the magnitude of the Effective Gravity at $A = |CA| - |CD| = |CM| - |CD| = |MD| = K \sin^2 \angle ACM$. Maupertuis's demonstration contains a defect, however. Taking Magnitude of the Effective Gravity at A to be $|CD|$ makes no sense, because the Magnitude of the Effective Gravity at A is a constant, while $|CD|$ varies with D. What *is* true, however, is that (the Magnitude of the Effective Gravity at D/the Magnitude of the Effective Gravity at $A) = (|CA|/|CD|)$, because the magnitude of the effective gravity at the surface of the homogeneous ellipsoid of revolution in equilibrium varies inversely with distance to the center. Now, when the ellipsoid is nearly a sphere, $|MD|/|CA|$ is infinitesimally small $((|MD|/|CA|)^2 \ll |MD|/|CA|)$, for all D, and so terms of second and higher order in $|MD|/|CA|$ can be neglected. Thus

the Magnitude of the Effective Gravity at D – the Magnitude of the Effective gravity at A

= ((the Magnitude of the Effective Gravity at D/the Magnitude of the Effective Gravity at A) − 1) × the Magnitude of the Effective Gravity at A

= $((|CA|/|CD|) - 1)$ × the Magnitude of the Effective Gravity at A

= $((|CM|/|CD|) - 1)$ × the Magnitude of the Effective Gravity at A

= $((|CM| - |CD|)/|CD|)$ × the Magnitude of the Effective Gravity at A

= $(|MD|/|CD|)$ × the Magnitude of the Effective Gravity at A = (the Magnitude of the Effective Gravity at A/$|CD|$) × $|MD|$

= (the Magnitude of the Effective Gravity at A/$(|CM| - |MD|))$ × $|MD|$

= (the Magnitude of the Effective Gravity at A/$(|CA| - |MD|))$ × $|MD|$

= (the Magnitude of the Effective Gravity at A/$|CA|)(1/(1 - (|MD|/|CA|)))$ × $|MD|$

= (the Magnitude of the Effective Gravity at A/$|CA|)(1 + (|MD|/|CA|))$ × $|MD|$

= (the Magnitude of the Effective Gravity at $A)((|MD|/|CA|) + (|MD|/|CA|)^2)$

= (the Magnitude of the Effective Gravity at $A)(|MD|/|CA|)$

= (the Magnitude of the Effective Gravity at A/$|CA|)$ × $|MD|$, where (the Magnitude of the Effective Gravity at A/$|CA|)$ is a constant L.

Thus the Magnitude of the Effective Gravity at D – the Magnitude of the Effective Gravity at

$$A = L|MD| = LK\sin^2 \angle ACM.$$

Hence the same is also true of the *relative* increase in the magnitude of the effective gravity with latitude:

$$\frac{\text{the Magnitude of the Effective gravity at } D - \text{the Magnitude of the Effective Gravity t } A}{\text{the Magnitude of the Effective Gravity at } A}$$

$$= L'K\sin^2 \angle ACM,$$

where

$$L' \equiv \frac{L}{\text{the Magnitude of the Effective Gravity at } A}$$

Following Newton, Maupertuis took the resultant of effective gravity at a point on the surface to be directed toward the center, when, in fact, effective gravity at a point on the surface of a figure of equilibrium is perpendicular to the surface at that point (see notes 3 and 8 of Chapter 1). However, when the ellipsoid is nearly spherical, the two directions nearly coincide, and the total effective gravity at the point and the component of effective gravity at the point directed toward the center are equal to terms of first order (see the end of note 7 of Chapter 1 and note 8 of Chapter 1).

6. Clairaut's first theories of the earth's shape

1. Clairaut (1737a). Hereafter I refer to the English translation: Clairaut (1809).
2. Clairaut (1737b: fol. 253).
3. As I mentioned at the end of note 7 of Chapter 1, Newton gives the impression that the theorem holds for the resultants of effective gravity at points on the surface of the figure, whereas in fact it is only true for the components of effective gravity at points on the surface of the figure which are directed toward the center of the figure. However, in case the figure is nearly spherical, the resultant of effective gravity at a point on its surface and the component of effective gravity in question at the point coincide for all practical purposes.
4. Maupertuis (1733c: 46v).
5. Clairaut (1809: 120, and Figure 12). Newton did not make clear that it is the component of effective gravity at a point on the surface directed toward the center of the figure, not the resultant of effective gravity at the point, which varies inversely as the distance of the point from the center of the figure (see note 3 of Chapter 6 and the end of note 7 of Chapter 1). Thus Clairaut stated the result with the precision that Newton's statement of it in the *Principia* lacked, although for bodies that are nearly spherical, the resultant of effective gravity at a point on the surface and the component of effective gravity in question at the point coincide for all practical purposes.
6. Ibid., p. 120, and Figure 12.
7. Clairaut (1737e).
8. Clairaut (1809: 122–123).
9. Ibid., p. 123, and Figure 12.
10. Ibid., p. 121, and Figure 12.
11. This statement can be demonstrated as follows: The integral

$$2\pi \int_{\sqrt{2ru-u^2}}^{(1+\alpha)\sqrt{2ru-u^2}} \frac{\rho u}{(u^2 + \rho^2)^{3/2}} \, d\rho \tag{I}$$

expresses the magnitude of the attraction at the pole A due to the annulus whose center is at P formed by revolving NM around the polar axis Aa, where $u = AP, r = AC = DC, \alpha$ is the ellipticity

$$\frac{EC - AC}{AC}, \quad \frac{PN^2}{(r + \alpha r)^2} + \frac{(r - u)^2}{r^2} = 1$$

is the equation of the ellipse in Figure 16 which generates the ellipsoid of revolution when the ellipse is revolved around the polar axis Aa, and $MP \leqslant \rho \leqslant NP$. Letting $w \equiv \sqrt{2ru - u^2}$ and making the change of variables $\rho = w + E$, $d\rho = dE$, integral (I) becomes

$$2\pi \int_{w}^{w+\alpha w} \frac{\rho u}{(u^2 + \rho^2)^{3/2}} \, d\rho = 2\pi \int_{0}^{\alpha w} \frac{(w + E)u}{(u^2 + (w + E)^2)^{3/2}} \, dE. \tag{II}$$

The integral on the right-hand side of (II) can be written as

$$2\pi \int_{0}^{\alpha w} \frac{w(1 + (E/w))u}{(u^2 + w^2(1 + (E/w)^2)^{3/2}} \, dE. \tag{III}$$

Making use of the mean value theorem, integral (III) can be written as

$$2\pi \int_{0}^{\alpha w} \left(\frac{wu}{(u^2 + w^2)^{3/2}} + \frac{\partial F}{\partial x}(u, w, x) \bigg|_{x = \xi} \left(\frac{E}{w} \right) \right) dE, \tag{IV}$$

where

$$F(u, w, x) \equiv \frac{w(1 + x)u}{(u^2 + w^2(1 + x)^2)^{3/2}},$$

for some value ξ where $0 \leqslant \xi \leqslant (E/w)$. For a fixed value of u, the value of w is also fixed. ξ, of course, still depends upon E for each fixed u and w. Since $(E/w) \leqslant \alpha$, however, we know that $0 \leqslant \xi \leqslant \alpha$, for each value of u where $0 \leqslant u \leqslant 2r$. It follows from (III) and (IV) that

$$\left| \int_{0}^{\alpha w} \frac{w(1 + (E/w))u}{(u^2 + w^2(1 + (E/w))^2)^{3/2}} \, dE - \int_{0}^{\alpha w} \frac{wu}{(u^2 + w^2)^{3/2}} \, dE \right| \leqslant \int_{0}^{\alpha w} \left| \frac{\partial F}{\partial x}(u, w, x) \right|_{x = \xi(E)} \left(\frac{E}{w} \right) dE. \tag{V}$$

Now,

$$\frac{\partial F}{\partial x}(u, w, x) = \frac{wu(u^2 + w^2(1 + x)^2)^{3/2} - w(1 + x)u(\frac{3}{2}(u^2 + w^2(1 + x)^2)^{1/2})(2w^2(1 + x))}{(u^2 + w^2(1 + x)^2)^3}. \tag{VI}$$

The numerator of the right-hand side of (VI) can be written as

$$((u^2 + w^2(1 + x)^2)^{3/2} - 3w^2(1 + x)^2(u^2 + w^2(1 + x)^2)^{1/2}) wu$$

$$= (u^2 + w^2(1 + x)^2 - 3w^2(1 + x)^2)(u^2 + w^2(1 + x)^2)^{1/2} wu$$

$$= (u^2 - 2w^2(1 + x)^2)(u^2 + w^2(1 + x)^2)^{1/2} wu.$$

Consequently the right-hand side of (VI) can be written as

$$\frac{(u^2 - 2w^2(1 + x)^2)(u^2 + w^2(1 + x)^2)^{1/2}}{(u^2 + w^2(1 + x)^2)^{3/2}} \frac{wu}{(u^2 + w^2(1 + x)^2)^{3/2}},$$

which equals

$$\frac{u^2 - 2w^2(1 + x)^2}{u^2 + w^2(1 + x)^2} \frac{wu}{(u^2 + w^2(1 + x)^2)^{3/2}}.$$

Thus

$$\frac{\partial F}{\partial x}(u, w, x) \bigg|_{x = \xi} = \frac{u^2 - 2w^2(1 + \xi)^2}{u^2 + w^2(1 + \xi)^2} \frac{wu}{(u^2 + w^2(1 + \xi)^2)^{3/2}}. \tag{VII}$$

Since

$$|u^2 - 2w^2(1 + \xi)^2| \leqslant u^2 + 2w^2(1 + \xi)^2,$$

it follows that

$$\left| \frac{u^2 - 2w^2(1 + \xi)^2}{u^2 + w^2(1 + \xi)^2} \right| \leqslant \frac{u^2 + 2w^2(1 + \xi)^2}{u^2 + w^2(1 + \xi)^2}$$

$$= \frac{u^2 + w^2(1 + \xi)^2}{u^2 + w^2(1 + \xi)^2} + \frac{w^2(1 + \xi)^2}{u^2 + w^2(1 + \xi)^2} = 1 + \frac{w^2(1 + \xi)^2}{u^2 + w^2(1 + \xi)^2}$$

$$\leqslant 1 + \frac{w^2(1 + \xi)^2}{w^2(1 + \xi)^2} = 1 + 1 = 2. \tag{VIII}$$

Hence it follows from (VII) and (VIII) that

$$\int_0^{aw} \left| \frac{\partial F}{\partial x}(u,w,x) \right|_{x=\xi(E)} \left(\frac{E}{w}\right) dE$$

$$\leqslant \int_0^{aw} \frac{2wu}{(u^2+w^2(1+\xi(E))^2)^{3/2}} \left(\frac{E}{w}\right) dE$$

$$= \int_0^{aw} \frac{2uE}{(u^2+w^2(1+\xi(E))^2)^{3/2}} dE$$

$$\leqslant \int_0^{aw} \frac{2uE}{(u^2+w^2)^{3/2}} dE = \frac{uw^2}{(u^2+w^2)^{3/2}} \alpha^2. \tag{IX}$$

From (V) and (IX) we conclude that

$$\left| \int_0^{aw} \frac{w(1+(E/w))u}{(u^2+w^2(1+(E/w))^2)^{3/2}} dE - \int_0^{aw} \frac{wu}{(u^2+w^2)^{3/2}} dE \right| \leqslant \frac{uw^2}{(u^2+w^2)^{3/2}} \alpha^2. \tag{X}$$

Since

$$\int_0^{aw} \frac{wu}{(u^2+w^2)^{3/2}} dE = \frac{uw^2}{(u^2+w^2)^{3/2}} \alpha, \tag{XI}$$

it follows from (X) and (XI) that

$$\int_0^{aw} \frac{w(1+(E/w))u}{(u^2+w^2(1+(E/w))^2)^{3/2}} dE = \frac{uw^2}{(u^2+w^2)^{3/2}} \alpha + f(u,w), \tag{XII}$$

where

$$|f(u,w)| \leqslant \frac{uw^2}{(u^2+w^2)^{3/2}} \alpha^2$$

for all $0 \leqslant u \leqslant 2r$. In particular, if $\alpha^2 \ll \alpha$, then

$$\frac{uw^2}{(u^2+w^2)^{3/2}} \alpha^2 \ll \frac{uw^2}{(u^2+w^2)^{3/2}} \alpha.$$

Thus, in this case, $f(u,w)$ can be neglected in (XII), and the magnitude of the attraction

$$2\pi \int_0^{aw} \frac{w(1+(E/w))u}{(u^2+w^2(1+(E/w))^2)^{3/2}} dE \tag{III}$$

at the pole A due to the annulus whose center is at P formed by revolving NM around the polar axis Aa has the approximate value

$$\frac{2\pi w(\alpha w)u}{(u^2+w^2)^{3/2}}$$

to terms of first order in α, which effectively says to treat the annulus, whose surface area is $2\pi w(\alpha w)$ to terms of first order in α, as if it were a circle in the plane of the annulus, whose center is at P, whose radius is $w = PM$, and each of whose points attracts the pole A at a distance $\sqrt{u^2+w^2}$ from it.

12. For an equation that holds for finite ellipticities δ and finite ratios ϕ of the magnitudes of the centrifugal force per unit of mass to the magnitudes of the attraction at the equators, see Clairaut (1743: 188). This equation expresses δ as a power series in ϕ. In Greenberg (1988: 230) I explain why this power series only converges to finite ellipticities δ which are less than $\sqrt{2^{1/2}} - 1$. Jupiter's ellipticity δ, for example, happens to be less than this positive number, so that the power series represents δ in the particular case of Jupiter.

13. In 1743, MacLaurin's English colleague Thomas Simpson published his *Mathematical Dissertations on a Variety of Physical and Analytical Subjects*. In his first essay, entitled "A Mathematical Dissertation on the Figure of the Earth," Simpson employed analysis to find expressions for the

magnitudes of the attraction at the surfaces of homogeneous figures shaped like ellipsoids of revolution which attract according to the universal inverse-square law and whose ellipticities are finite which held true exactly. Clairaut, as far as I know, had no contact with Simpson before 1743, the year that Clairaut published his mature work on the theory of the earth's shape, which I will examine in Chapter 9. The mathematical analyst's problem was to evaluate the nonelementary integrals that expressed the magnitudes of the attraction of figures shaped like ellipsoids which attract according to the universal inverse-square law when the ellipticities of the figures were finite, in which case first-order approximations cannot be made. Simpson utilized infinite series to do this. But his methods lacked the simplicity and elegance of MacLaurin's geometry. Lagrange proclaimed later that although Simpson did determine "a purely analytic solution of the problem," Simpson's use of infinite series, which included infinite series that did not always even converge, left much to be desired. Not only was Simpson's solution "long and complicated," Lagrange observed, but it was "indirect and not rigorous," whereas MacLaurin's solution was direct and rigorous. See Lagrange (1773: 121–122); also Legendre (1785: 411–412).

14. Clairaut found

$$f = \tfrac{8}{15} pam \tag{6.1.30}$$

to be the approximate value of the magnitude of the centrifugal force per unit of mass at the equator of a homogeneous figure shaped like an ellipsoid of revolution flattened at its poles which attracts according to the universal inverse-square law, which revolves around its axis of symmetry, and whose ellipticity m is infinitesimal, when equalities (6.1.28) and (6.1.29) hold, where a is the polar radius of the figure, and where $c \equiv 2\pi$ should appear in place of p. He also found

$$\tfrac{2}{3} pa + \tfrac{6}{15} pam \tag{6.1.16}$$

to be the approximate value of the magnitude of the attraction at the equator of this figure, where, again, c should appear in place of p. If we divide expression (6.1.30) by expression (6.1.16), the ratio

$$\phi \equiv \frac{\tfrac{8}{15} pam}{\tfrac{2}{3} pa + \tfrac{6}{15} pam} = \tfrac{4}{5} m \left(\frac{1}{1 + \tfrac{3}{5} m} \right) \simeq \tfrac{4}{5} m \left(1 - \tfrac{3}{5} m \right) \simeq \tfrac{4}{5} m$$

results. In other words,

$$\phi = \tfrac{4}{5} m \tag{I}$$

is true when terms of second order and higher orders in the infinitesimal ellipticity m are neglected. (We note that the presence of the spurious p in place of c does not affect the calculation, since the p's cancel out, just as the c's would have.) In other words, the ratio m/ϕ equals 5/4, a constant, for all homogeneous figures shaped like ellipsoids of revolution flattened at their poles which attract according to the universal inverse-square law, which revolve around their axes of symmetry, and whose ellipticities m are infinitesimal, when equalities (6.1.28) and (6.1.29) hold. Now, equation (I) can be written instead as

$$m = \tfrac{5}{4} \phi. \tag{I'}$$

At the same time, Newton's equation

$$\frac{\delta}{1/100} = \frac{\phi}{4/505} \tag{1.1}$$

can be rewritten as

$$\delta = \frac{1/100}{4/505} \phi. \tag{5.3.3}$$

If we replace δ by m in equation (5.3.3), equation (5.3.3) can be rewritten as

$$m = \frac{1/100}{4/505} \phi. \tag{II}$$

But equation

$$m = \tfrac{5}{4} \phi \tag{I'}$$

for the ellipticity m differs from Newton's equation

$$m = \frac{1/100}{4/505}\phi \tag{II}$$

for the ellipticity m by terms of second order and higher orders. That is,

$$m = \frac{1/100}{4/505}\phi = \frac{505}{400}\phi = \left(\frac{5}{4} + \frac{1}{80}\right)\phi = \frac{5}{4}\phi + \frac{1}{80}\phi. \tag{II}$$

But $1/80$ is infinitesimal, since

$$\left(\frac{1}{80}\right)^2 = \frac{1}{6400} \ll \frac{1}{80}.$$

Moreover, ϕ is infinitesimal, since $\phi^2 \ll \phi$. Consequently, since both ϕ and $\frac{1}{80}$ are infinitesimal, the term $\left(\frac{1}{80}\right)\phi$ is a term of second order; hence it can be discarded in equation (II). Finally, if δ is substituted for m in equation (I'), equation (I') can be rewritten as

$$\delta = \tfrac{5}{4}\phi. \tag{5.3.4}$$

If F designates the magnitude of the centrifugal force per unit of mass at the equator and A stands for the magnitude of the attraction there, Newton had calculated the ratio $\phi \equiv F/A$ in the third edition of the *Principia* (see note 3 of Chapter 1). Clairaut worked instead with the ratio of the magnitude of the centrifugal force per unit of mass to the magnitude of the effective gravity at the equator:

$$\phi \equiv \frac{F}{A-F} = \frac{F}{A}\left(\frac{1}{1-(F/A)}\right) = \frac{1}{289}\left(\frac{1}{1-(1/289)}\right) = \frac{1}{289}\frac{289}{288} = \frac{1}{288}.$$

Newton found a value $m = \frac{1}{229}$ for the earth's ellipticity (see note 51 of Chapter 5) by substituting $\phi = \frac{1}{289}$ into his equation

$$m = \frac{1/100}{4/505}\phi, \tag{II}$$

whereas a value $m = \frac{1}{230}$ for the earth's ellipticity is found by substituting $\phi = \frac{1}{288}$ into the equation

$$m = \tfrac{5}{4}\phi. \tag{I'}$$

Now, both Newton's equation

$$m = \frac{1/100}{4/505}\phi \tag{II}$$

and the equation

$$m = \tfrac{5}{4}\phi \tag{I'}$$

are only accurate to terms of first order and, as we have seen above, the first equation cannot be distinguished from the second to terms of first order. At the same time, the two ratios

$$\phi = \frac{F}{A} = \frac{1}{289} \quad \text{and} \quad \phi = \frac{F}{A-F} = \frac{1}{288}$$

differ insignificantly from each other, because F/A is also infinitesimal

$$\left(\left(\frac{F}{A}\right)^2 \ll \frac{F}{A}\right),$$

so that

$$\frac{F}{A-F} = \frac{F}{A}\left(\frac{1}{1-(F/A)}\right) \simeq \frac{F}{A}\left(1 + \frac{F}{A}\right) = \frac{F}{A} + \left(\frac{F}{A}\right)^2 \simeq \frac{F}{A}.$$

In other words,

$$\frac{F}{A - F} = \frac{F}{A}$$

to terms of first order. Consequently Newton and Clairaut inevitably determined values of the earth's ellipticity m which differ negligibly from each other. It is easy to check that the two values really do differ insignificantly from each other: $m = \frac{1}{229}$ and $m = \frac{1}{230}$ are both infinitesimal (that is, $m^2 \ll m$ in both cases), and, at the same time, the two numbers differ by terms of second order and higher orders. That is,

$$\frac{1}{229} = \frac{1}{230 - 1} = \frac{1}{230}\left(\frac{1}{1 - \frac{1}{230}}\right) \simeq \frac{1}{230}\left(1 + \frac{1}{230}\right) \simeq \frac{1}{230}.$$

It is also easy to show that Clairaut's expressions for the approximate values of the magnitudes of the attraction at the poles and equators of homogeneous figures shaped like flattened ellipsoids of revolution which attract according to the universal inverse-square law and whose ellipticities are infinitesimal accord with the results that Newton simply stated numerically. Clairaut found

$$\tfrac{2}{3} pa + \tfrac{8}{15} pam, \tag{6.1.15}$$

where, again, c should appear in place of p, to be the approximate value of the magnitude of the attraction at a pole of such a figure. If we divide expression (6.1.15) by expression (6.1.16) we find that

$$\frac{\tfrac{2}{3} pa + \tfrac{8}{15} pam}{\tfrac{2}{3} pa + \tfrac{6}{15} pam} = \frac{1 + \tfrac{4}{5} m}{1 + \tfrac{3}{5} m}$$

$$= \frac{1}{1 + \tfrac{3}{5} m} + \tfrac{4}{5} m\left(\frac{1}{1 + \tfrac{3}{5} m}\right) \simeq 1 - \tfrac{3}{5} m + \tfrac{4}{5} m(1 - \tfrac{3}{5} m) \tag{III}$$

$$\simeq 1 - \tfrac{3}{5} m + \tfrac{4}{5} m = 1 + \tfrac{1}{5} m.$$

(Again we observe that the ps, which should be replaced by cs, cancel out, so that the presence of the ps does not affect the calculation.) Now, Newton had stated without proof that when $m = \frac{1}{100}$, the ratio of the magnitude of the attraction at the pole to the magnitude of the attraction at the equator equals $\frac{501}{500}$. But if we substitute $\frac{1}{100}$ for m in $1 + \tfrac{1}{5} m$ on the right-hand side of equation (III), we find that $1 + \tfrac{1}{5}\frac{1}{100} = 1 + \frac{1}{500} = \frac{501}{500}$. Thus Clairaut's expressions for the magnitudes of the attraction accord with the results that Newton stated in numerical form.

15. Clairaut (1738a).
16. Clairaut (1737c).
17. Clairaut (1737d: 80–81).
18. Clairaut (1738b).
19. Ibid., p. 277.
20. Ibid., p. 278. See as well Newton (1972: vol. 2, 600, 607).
21. Clairaut (1738b: 279).
22. Ibid., pp. 281–282.
23. In practice, the density $D(r)$ varied continuously with r from the center of the figure to the surface of the figure in Clairaut's theory. (Indeed, $D(r)$ in fact always has the form of a differentiable function with respect to r in Clairaut's theory.) Now, the densities of an infinite number of individually homogeneous ellipsoids of revolution can also vary as step functions $D(r)$ of r. But Clairaut's results really only depend upon the existence of certain integrals. These integrals also exist when $D(r)$ is a step function of r. Thus $D(r)$ could just as well be taken to be a step function of r in Clairaut's theory. This means that Clairaut's theory and the results that follow from it can be extended to include a figure made up of a *finite* number of individually homogeneous strata, each of *finite* thickness, whose inner and outer surfaces are all concentric ellipsoids of revolution flattened at their poles, all of which have a common axis of symmetry (the axis of symmetry of the figure, which is the polar axis of the figure), and whose infinitesimal ellipticities are all the same. [See Kopal (1960: 7).]

24. Clairaut [1738a: fol. 28]; and Clairaut (1738b: Figure 2).
25. Clairaut (1738a: fol. 28).
26. Clairaut (1738b: Figures 2 and 3).
27. Ibid., pp. 285–286, and Figure 5.
28. I give one example of a geometric approximation that conceals the order of magnitude of the error. Clairaut (1738b: 286–287, and Figure 5) calculated

$$\frac{2cfne^{1+p}}{5+p} + \frac{2cgne^{1+q}}{5+q} \tag{I}$$

to be the approximate value $|CI|$ of the magnitude of the component of the attraction at N in the direction of CX perpendicular to CN due to the stratified figure shaped like a flattened ellipsoid of revolution. Here $e =$ the polar radius, $c = 2\pi$, the radius C_{uv} bisects the line segment Rr perpendicular to CN at H, line segment NX is perpendicular to the surface of the figure at N, and $n = |HY|/|CH|$ (see Figure 20). Since NX is perpendicular to the surface of the figure at N, angle vNX is a right angle. Clairaut says that because Nv is very small and angle vNC is almost a right angle, it follows that Nv and CX are almost parallel, since angle NCX is a right angle. Since angle vNX is also a right angle, it follows that angle NXC is almost a right angle. Because angle vNC and angle NCX are almost the same angles (right angles) and also angle vNX and angle NXC are almost the same angles (right angles), it follows that triangle vNC and triangle XCN are almost similar. (In fact, because they share NC, the two triangles are almost congruent.) It follows that angle NCv and angle CNX are nearly equal, because they are corresponding angles of triangles that are nearly similar. Hence the two angles have nearly the same sines. But $n =$ the sine of angle NCv, and $|CX|/|CN| =$ the sine of angle CNX. Thus Clairaut replaced n by $|CX|/|CN|$ in the expression (I) above for the magnitude of the attraction $|CI|$.
29. Clairaut (1738b: 287–288, and Figure 5). X is the point where the perpendicular to the line segment CN at C intersects the perpendicular to the surface of the figure at N. $|NI|$ is the magnitude of the resultant of the attraction at N; $|CN|$ is the magnitude of the component of the attraction at N in the direction of CN; and $|CI|$ is the magnitude of the component of the attraction at N in the direction of line segment CX perpendicular to CN. (See Figure 20.) Clairaut took $|CN|$ to be

$$\frac{2cfe^{1+p}}{3+p} + \frac{2cge^{1+q}}{3+q}, \tag{I}$$

which is the magnitude of the resultant of the attraction at a point N at the surface of a stratified figure shaped like an ellipsoid of revolution whose ellipticity is zero (in other words, a stratified spherical figure) and whose strata have the same densities as the strata of the stratified figure shaped like a flattened ellipsoid of revolution "by expunging what may be here expunged" (p. 287). This can be explained as follows: the true value of $|CN|$ often has the form:

$$\frac{2cfe^{1+p}}{3+p} + \frac{2cge^{1+q}}{3+q} + \varepsilon, \tag{II}$$

where

$$\xi \equiv \frac{\varepsilon}{(2cfe^{1+p}/(3+p)) + (2cge^{1+q}/(3+q))}$$

is infinitesimal ($\xi^2 \ll \xi$). [This statement I now verify. Clairaut's value of $|CN|$ is

$$\frac{2cfe^{1+p}}{3+p} + \frac{2cge^{1+q}}{3+q} + \frac{(2p-2)cf\lambda e^{1+p}}{(3+p)(5+p)} + \frac{(2q-2)cg\lambda e^{1+q}}{(3+q)(5+q)} + \frac{8cf\alpha e^{1+p}}{(3+p)(5+p)} + \frac{8cg\alpha e^{1+q}}{(3+q)(5+q)}$$

[Clairaut (1738b: 284)]. Hence the ε in (II) is

$$\varepsilon = \frac{(2p-2)cf\lambda e^{1+p}}{(3+p)(5+p)} + \frac{(2q-2)cg\lambda e^{1+q}}{(3+q)(5+q)} + \frac{8cf\alpha e^{1+p}}{(3+p)(5+p)} + \frac{8cg\alpha e^{1+q}}{(3+q)(5+q)},$$

and

$$\xi \equiv \left(\frac{((2p-2)cfe^{1+p}/(3+p)(5+p)) + ((2q-2)cge^{1+q}/(3+q)(5+q))}{(2cfe^{1+p}/(3+p)) + (2cge^{1+q}/(3+q))} \right) \lambda$$

$$+ \left(\frac{(8cfe^{1+p}/(3+p)(5+p)) + (8cge^{1+q}/(3+q)(5+q))}{(2cfe^{1+p}/(3+p)) + (2cge^{1+q}/(3+q))} \right) \alpha. \tag{III}$$

We recall that Clairaut took $p \geqslant 0$ and $q \geqslant 0$ in practice. Assume $f > 0$ and $g > 0$. Then we look for least upper bounds of the coefficients of λ and α in expression (III) for ξ. First we find the smallest positive number N for which

$$\left| \frac{2p-2}{(3+p)(5+p)} \right| < \left| \frac{2N}{3+p} \right| \tag{IV}$$

holds for all $0 \leqslant p$. Now, (IV) simplifies to

$$\left| \frac{2p-2}{5+p} \right| < 2N,$$

or

$$\left| \frac{p-1}{p+5} \right| < N, \tag{V}$$

since $0 < 3 + p$. But $0 \leqslant p_1 < p_2 \Rightarrow 6p_1 < 6p_2 \Rightarrow 5p_1 - p_2 < 5p_2 - p_1 \Rightarrow p_1 p_2 + 5p_1 - p_2 - 5$
$< p_1 p_2 + 5p_2 - p_1 - 5 \Rightarrow (p_1 - 1)(p_2 + 5) < (p_2 - 1)(p_1 + 5) \Rightarrow$

$$\frac{p_1 - 1}{p_1 + 5} < \frac{p_2 - 1}{p_2 + 5}.$$

Hence $(p-1)/(p+5)$ is a monotonically increasing function of $0 \leqslant p$. Moreover, for $0 < p$,

$$\frac{p-1}{p+5} = \frac{1-(1/p)}{1+(5/p)} \to 1$$

as $p \to \infty$. In particular, if $1 \leqslant p$, then

$$\left| \frac{p-1}{p+5} \right| = \frac{p-1}{p+5} < 1.$$

Moreover,

$$0 \leqslant p_1 < p_2 \Rightarrow \frac{p_1 - 1}{p_1 + 5} < \frac{p_2 - 1}{p_2 + 5} \Rightarrow \frac{1 - p_1}{p_1 + 5} > \frac{1 - p_2}{p_2 + 5} \Rightarrow \frac{1-p}{p+5}$$

has its maximum at $p = 0$ when $0 \leqslant p < 1$. So if $0 \leqslant p < 1$, then

$$\left| \frac{p-1}{p+5} \right| = \frac{1-p}{p+5} \leqslant \frac{1-0}{0+5} = \frac{1}{5}.$$

Therefore $N = 1$ is the smallest positive number for which (IV) holds for all $0 \leqslant p$. If q is substituted for p in (IV), it is easy to see that 1 must be the smallest upper bound for the absolute value of the coefficient of λ in the expression (III) for ξ when $0 \leqslant p$ and $0 \leqslant q$. Now we find the smallest positive number M for which

$$\left| \frac{8}{(3+p)(5+p)} \right| < \left| \frac{2M}{3+p} \right| \tag{VI}$$

holds for all $0 \leqslant p$. Inequality (VI) reduces to $|\frac{8}{5+p}| < 2M$, or

$$\left| \frac{4}{5+P} \right| < M, \tag{VII}$$

since $0 < 3 + p$. But

$$\left|\frac{4}{5+p}\right| = \frac{4}{5+p} \to 0$$

monotonically as $0 \leqslant p \to \infty$. Hence $M = \frac{4}{3}$ is the smallest positive number for which (VI) holds for all $0 \leqslant p$. If p is replaced by q in (VI), it is also easy to see that $\frac{4}{3}$ must be the least upper bound for the absolute value of the coefficient of α in the expression (III) for ξ when $0 \leqslant p$ and $0 \leqslant q$. Hence

$$|\xi| < |\lambda| + \tfrac{4}{3}|\alpha| \tag{VIII}$$

when $0 \leqslant p$ and $0 \leqslant q$. But

$$0 \leqslant \lambda \equiv \frac{CN - BC}{BC} \leqslant \frac{CE - BC}{BC} \equiv \alpha,$$

and from this and (VIII) it follows that

$$\xi < \alpha + \tfrac{4}{3}\alpha = \tfrac{9}{3}\alpha \tag{IX}$$

when $0 \leqslant p$ and $0 \leqslant q$. Finally, when α is a sufficiently small infinitesimal, $(9/5)\alpha$ is also infinitesimal. If $fg < 0$, then (I) need not be the dominant portion in (II). For example, (I) can equal 0 and $\varepsilon \neq 0$. I thank Robert Weinstock for pointing this out to me.]

Hence

$$\frac{|CI|}{|CN|} = \frac{|CI|}{(2cfe^{1+p}/(3+p)) + (2cge^{1+q}/(3+q)) + \varepsilon}$$

$$= \frac{|CI|}{(2cfe^{1+p}/(3+p)) + (2cge^{1+q}/(3+q))}\left(\frac{1}{1+\xi}\right) \tag{X}$$

$$= \frac{|CI|}{(2cfe^{1+p}/(3+p)) + (2cge^{1+q}/(3+q))}(1-\xi),$$

because $\xi^2 \ll \xi$. But

$$|CI| = \left(\frac{2cfe^{1+p}}{5+p} \times \frac{|CX|}{|CN|}\right) + \left(\frac{2cge^{1+q}}{5+q} \times \frac{|CX|}{|CN|}\right)$$

$$= \left(\frac{2cfe^{1+p}}{5+p} + \frac{2cge^{1+q}}{5+q}\right) \times \frac{|CX|}{|CN|}.$$

Hence

$$\frac{|CI|}{(2cfe^{1+p}/(3+p)) + (2cge^{1+q}/(3+q))} = \left(\frac{(2cfe^{1+p}/(5+p)) + (2cge^{1+q}/(5+q))}{(2cfe^{1+p}/(3+p)) + (2cge^{1+q}/(3+q))}\right) \times \frac{|CX|}{|CN|} < \frac{|CX|}{|CN|}.$$

Therefore

$$\frac{|CI|}{(2cfe^{1+p}/(3+p)) + (2cge^{1+q}/(3+q))}$$

is infinitesimal, because $|CX|/|CN|$ is infinitesimal. Thus

$$\frac{|CI|}{(2cfe^{1+p}/(3+p)) + (2cge^{1+q}/(3+q))} \times \xi$$

is a term of second order in (X). Consequently

$$\frac{|CI|}{|CN|} = \frac{|CI|}{(2cfe^{1+p}/(3+p)) + (2cge^{1+q}/(3+q))} \tag{XII}$$

to terms of first order. Then (I) follows from equating the denominators of the quotients on the left-hand side and right-hand side of (XII). Finally, (XII) together with (XI), which is the value of

$|CI|$ calculated in note 28 of Chapter 6, leads to

$$\frac{|CI|}{|CX|} = \frac{|CI|}{|CN|}\frac{|CN|}{|CX|} = \frac{(2cfe^{1+p}/(5+p)) + (2cge^{1+q}/(5+q))}{(2cfe^{1+p}/(3+p)) + (2cge^{1+q}/(3+q))}, \qquad \text{(XIII)}$$

where $c = 2\pi$ and $e = $ the polar radius. Equation (XIII) expresses $|CI|$ through $|CX|$. In other words, (XIII) determines the location of the point I along CX. In this way (XIII) determines the direction of the resultant of the attraction NI at N. But Clairaut did not really explain why point I is such that NI indicates the direction of the total or the resultant of the attraction at N.

30. Clairaut (1738b: 288).
31. Ibid., p. 288.
32. See Figure 20. Since CN is perpendicular to CX,

$$\text{Effective Gravity at } N = - \text{ Effective Gravity} (N, CN) \, \widehat{CN}$$
$$+ \text{ Effective Gravity } (N, CX) \, \widehat{CX}. \qquad \text{(I)}$$

If the principle of the plumb line holds at N, then Effective Gravity at N is parallel to NX, because NX is perpendicular to the surface of the figure at N. In other words

$$\text{Effective Gravity at } N = \lambda NX \qquad \text{(II)}$$

for some real value of $\lambda \neq 0$. Writing

$$NX = -|CN|\widehat{CN} + |CX|\widehat{CX}, \qquad \text{(III)}$$

combining (I), (II), and (III), and setting the corresponding scalar components on both sides of the resulting equation equal to each other, we find that

$$\text{Effective Gravity } (N, CN) = \lambda |CN| \qquad \text{(IV)}$$

and

$$\text{Effective Gravity } (N, CX) = \lambda |CX|. \qquad \text{(V)}$$

Dividing the left-hand side and the right-hand side, respectively, of (V) by the left-hand side and the right-hand side, respectively, of (IV), we find that

$$\frac{\text{Effective Gravity } (N, CX)}{\text{Effective Gravity } (N, CN)} = \frac{|CX|}{|CN|}. \qquad (6.2.3)$$

33. Let $|CN| = $ the magnitude of the component of the attraction at N in the direction of CN, $|CI| = $ the magnitude of the component of the attraction at N in the direction of CX, $\phi_N = $ the magnitude of the resultant of the centrifugal force per unit of mass at N, $\phi_N^{CN} = $ the magnitude of the component of the centrifugal force per unit of mass at N in the direction of CN, and $\phi_N^{CX} = $ the magnitude of the component of the centrifugal force per unit of mass at N in the direction of CX. From Figure 65

we can see that

$$\text{Effective Gravity } (N, CN) = |CN| - \phi_N^{CN}$$

and

$$\text{Effective Gravity } (N, CX) = |CI| + \phi_N^{CX}.$$

Clairaut took as $|CI|$ the value calculated in note 28 of Chapter 6. He took $|CN|$ to be

$$\frac{2cfe^{1+p}}{3+p} + \frac{2cge^{1+q}}{3+q}$$

for essentially the same reasons that I explained in note 29 of Chapter 6.

Clairaut calculated ϕ_N^{CX} to be

$$\frac{\phi \, |CX|}{2\alpha \, |CE|},$$

where α is the ellipticity of the figure and ϕ is the magnitude of the centrifugal force per unit of

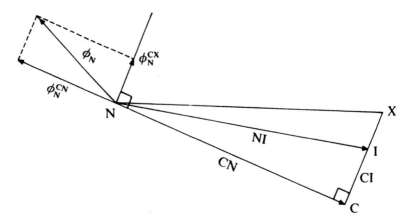

Figure 65. Figure used for expressing the principle of the plumb line to terms of first order.

mass at the equator of the figure. Applying the principle of the plumb line, we find that

$$\frac{|CX|}{|CN|} = \frac{\text{Effective Gravity}(N,CX)}{\text{Effective Gravity }(N,CN)} = \frac{\phi_N^{CX} + |CI|}{|CN| - \phi_N^{CN}} =$$

$$\frac{\left((\phi/2\alpha)(|CX|/|CE|)\right) + \left((2cfe^{1+p}/(5+p)) \times (|CX|/|CN|)\right) + \left((2cge^{1+q}/(5+q)) \times (|CX|/|CN|)\right)}{\left(2cfe^{1+p}/(3+p)\right) + \left(2cge^{1+q}/(3+q)\right) - \phi_N^{CN}},$$

(I)

where $c = 2\pi$ and $e = $ the polar radius. Letting

$$\gamma \equiv \frac{\phi_N^{CN}}{\left(2cfe^{1+p}/(3+p)\right) + \left(2cge^{1+q}/(3+q)\right)},$$

Equation (I) can be written as

$$\frac{|CX|}{|CN|} = \frac{\left((\phi/2\alpha)(|CX|/|CE|)\right) + \left(2cfe^{1+p}/(5+p)\right) \times \left((|CX|/|CN|)\right) + \left(2cge^{1+q}/(5+q)\right) \times \left((|CX|/|CN|)\right)}{\left(2cfe^{1+p}/(3+p)\right) + \left(2cge^{1+q}/(3+q)\right)} \left(\frac{1}{1-\gamma}\right).$$

(II)

But the right-hand side of (II) equals

$$\frac{\left((\phi/2\alpha)(|CX|/|CE|)\right) + \left((2cfe^{1+p}/(5+p)) \times (|CX|/|CN|)\right) + \left((2cge^{1+q}/(5+q)) \times (|CX|/|CN|)\right)}{\left(2cfe^{1+p}/(3+p)\right) + \left(2cge^{1+q}/(3+q)\right)}(1+\gamma)$$

(III)

to terms of first order, because γ, a ratio of a magnitude of centrifugal force per unit of mass to a magnitude of attraction, is infinitesimal ($\gamma^2 \ll \gamma$) at all points N at the surface of the figure. Finally, since $|CX|/|CE|$ and $|CX|/|CN|$ are also infinitesimal

$$\left(\left(\frac{|CX|}{|CE|}\right)^2 \ll \frac{|CX|}{|CE|}, \left(\frac{|CX|}{|CN|}\right)^2 \ll \frac{|CX|}{|CN|}\right), \frac{|CX|}{|CE|} \times \gamma \quad \text{and} \quad \frac{|CX|}{|CN|} \times \gamma$$

are both terms of second order in (III). Hence

$$\frac{|CX|}{|CN|} = \frac{\left((\phi/2\alpha)(|CX|/|CE|)\right) + \left((2cfe^{1+p}/(5+p)) \times (|CX|/|CN|)\right) + \left((2cge^{1+q}/(5+q)) \times (|CX|/|CN|)\right)}{\left(2cfe^{1+p}/(3+p)\right) + \left(2cge^{1+q}/(3+q)\right)}$$

(IV)

to terms of first order. This explains why Clairaut expressed the principle of the plumb line as

$$\frac{|CN|}{|CX|} = \frac{\left(2cfe^{1+p}/(3+p)\right) + \left(2cge^{1+q}/(3+q)\right)}{\left((\phi/2\alpha)(|CX|/|CE|)\right) + \left((2cfe^{1+p}/(5+p)) \times (|CX|/|CN|)\right) + \left((2cge^{1+q}/(3+q)) \times (|CX|/|CN|)\right)},$$

(V)

which is simply (IV) inverted [Clairaut (1738b: 289)].

34. Independence of the point N at the surface of the figure follows in part from Clairaut's statement that "CN and CE may be assumed as the same on this occasion" in Colson's words [Clairaut (1738b: 289)] in the ratios $|CX|/|CE|$ and $|CX|/|CN|$ (see Figure 20). The reason for taking the two ratios to be the same can be understood as follows: Let $|CE| = |CN| + \delta$, where $0 \leqslant \delta \leqslant |CE| - |CB|$. Since $|CB| \leqslant |CN|$ it follows that

$$0 \leqslant \frac{\delta}{|CN|} \leqslant \frac{\delta}{|CB|} \leqslant \frac{|CE| - |CB|}{|CB|} \equiv \alpha.$$

Now, $|CX|/|CN| =$ the tangent of angle CNX, and its value is approximately equal to zero when the ellipticity α of the figure is infinitesimal, because in that case the radius CN and the perpendicular NX to the surface of the figure at N almost coincide (in Figure 20 CX is perpendicular to CN). Consequently

$$\frac{|CX|}{|CN|} \text{ is infinitesimal} \left(\left(\frac{|CX|}{|CN|}\right)^2 \ll \frac{|CX|}{|CN|}\right)$$

in this case. But since $\delta/|CN| \leqslant \alpha$, $\delta/|CN|$ is also infinitesimal $((\delta/|CN|)^2 \ll \delta/|CN|)$ when α is infinitesimal. Consequently

$$\frac{|CX|}{|CE|} = \frac{|CX|}{|CN| + \delta} = \frac{|CX|}{|CN|}\left(\frac{1}{1 + (\delta/CN)}\right) = \frac{|CX|}{|CN|}\left(1 - \left(\frac{\delta}{CN}\right)\right) = \frac{|CX|}{|CN|}$$

to terms of first order, since $(|CX|/|CN|) \times (\delta/|CN|)$ too is a term of second order. That is, $|CX|/|CE|$ and $|CX|/|CN|$ are equal to terms of first order. Substituting $|CN|$ for $|CE|$ in equation (V) in note 33 of Chapter 6, we find that

$$\phi = \frac{8cfe^{1+p}\alpha}{(3+p) \times (5+p)} + \frac{8cge^{1+q}\alpha}{(3+q) \times (5+q)}$$

(I)

is the relation sought, where $c = 2\pi$ and $e =$ the polar radius.

35. In the case of a homogeneous figure shaped like a flattened ellipsoid of revolution whose density is $1, f + g = 1$ and $p = q = 0$ in

$$D(r) = fr^p + gr^q.$$

(6.2.2)

Setting $p = q = 0$ and $f + g = 1$ in equation (I) in note 34 of Chapter 6, this equation reduces to

$$\phi = \tfrac{8}{15}ce\alpha.$$

(A)

But equation (A) is exactly the same as the equation

$$f = \tfrac{8}{15}pam$$

(6.1.30)

in Clairaut's paper of 1737, when the spurious p appearing in that equation is replaced by c (see note 14 of Chapter 6).

36. Clairaut (1738b: 292). The equation is

$$\frac{2cfe^{1+p}}{3+p} + \frac{(2p-2)cfe^{1+p}\alpha}{(3+p) \times (5+p)} + \frac{2cge^{1+q}}{3+q} + \frac{(2q-2)cge^{1+q}\alpha}{(3+q) \times (5+q)}$$
$$= \frac{8cfme^{1+p}\alpha}{(3+p) \times (5+p)} + \frac{8cgme^{1+q}\alpha}{(3+q) \times (5+q)},$$

(I)

where $c \equiv 2\pi$, $e \equiv$ the polar radius, $\alpha \equiv$ the ellipticity, and $1/m$ is the ratio of the magnitude of the centrifugal force per unit of mass to the magnitude of the effective gravity at the equator.

37. In the case of a homogeneous figure shaped like a flattened ellipsoid of revolution, $p = q = 0$ in

$$D(r) = fr^p + gr^q.$$

(6.2.2)

If we substitute $p = q = 0$ into equation (I) in note 36 of Chapter 6, this equation reduces to

$$\tfrac{2}{3}cfe - \tfrac{2}{15}cfe\alpha + \tfrac{2}{3}cge - \tfrac{2}{15}cge\alpha = \tfrac{8}{15}cfme\alpha + \tfrac{8}{15}cgme\alpha, \tag{A}$$

which simplifies further to

$$\tfrac{2}{3}c(f+g)e - \tfrac{2}{15}c(f+g)e\alpha = \tfrac{8}{15}c(f+g)me\alpha, \tag{B}$$

which reduces to

$$\tfrac{2}{3} - \tfrac{2}{15}\alpha = \tfrac{8}{15}m\alpha. \tag{C}$$

Consequently

$$\alpha = \frac{\tfrac{2}{3}(1/m)}{\tfrac{2}{15}(1/m) + \tfrac{8}{15}} = \frac{\tfrac{5}{4}(1/m)}{1 + \tfrac{1}{4}(1/m)} \simeq \frac{5}{4}\frac{1}{m}\left(1 - \frac{1}{4}\frac{1}{m}\right) \simeq \frac{5}{4}\frac{1}{m} \tag{D}$$

to terms of first order, when $1/m$ is infinitesimal $((1/m)^2 \ll 1/m)$. But (D) is the equality that Clairaut could have written in his paper of 1737 but did not (see note 14 of Chapter 6). That is, (D) is the same as equality (5.3.4). We observe that, unlike the result in note 35 of Chapter 6, the density $f + g$ of the homogeneous figure shaped like a flattened ellipsoid of revolution *cancels out* in going from (A) to (C). That is, the result here is independent of the particular density of the homogeneous figure. And we note that equality (5.3.4) also does not depend upon the density of the homogeneous figure shaped like a flattened ellipsoid of revolution either.

38. Consider a homogeneous spherical figure whose density is ρ and which attracts according to the universal inverse-square law. Then the resultant of attraction at points on the surface of the figure is perpendicular to the surface at those points. If R is the radius of the figure, then the magnitude of the attraction at points on the surface of the figure is proportional to

$$\frac{\rho\frac{4}{3}\pi R^3}{R^2}. \tag{I}$$

Let us now consider a spherical figure whose radius is R, whose density $\rho(r)$ varies directly as the distance r from the center of the figure (in other words, the figure consists of an infinite number of concentric, individually homogeneous spheres), and which attracts in accordance with the universal inverse-square law. Then the resultant of the attraction at points on the surface of such a figure will again be perpendicular to the surface at those points, and the magnitude of the attraction at points on the surface of the figure will be proportional to

$$\frac{\int_0^R \rho(r)4\pi r^2 dr}{R^2}. \tag{II}$$

It is easy to see from expressions (I) and (II) that any variation in density $\rho(r)$ which happens to have ρ as its average value (that is,

$$\int_0^R \rho(r)4\pi r^2 dr = \rho\frac{4}{3}\pi R^3$$

is true) will result in a stratified spherical figure whose radius is R and which produces an attraction at points on its surface whose magnitude will be the same as the magnitude of the attraction at points on the surface of the homogeneous spherical figure whose radius is R and whose density is ρ. In other words, the density, hence the internal structure, of a figure whose shape is given can be varied in different ways that will produce the same values of the magnitude of the attraction at points on its surface.

39. Equation (6.2.7) is

$$\frac{P' - P}{P} = \frac{\big((5-p)f\alpha/(3+p) \times (5+p)\big) + \big((5-q)g\alpha/((3+q) \times (5+q))\big)}{\big((p-1)f\alpha/(3+p) \times (5+p)\big) + \big(f/(3+p)\big) + \big(g/(3+q)\big) + \big((q-1)g\alpha/((3+q) \times (5 \times q))\big)}$$

[Clairaut (1738b: 293)]. It is clear that given values of $(P' - P)/P$ and α in this single equation do not uniquely fix the values that f, g, p, and q can have. Consequently the given values of $(P' - P)/P$ and α do not uniquely determine the densities of the strata in the interior of the figure.

40. Clairaut (1738b: 293–294).

41. Clairaut (1738b: 295). In a passage appearing in the first and second editions of the *Principia*, which was left out of the third edition, Newton explained in words how a spherical nucleus within a figure shaped like an ellipsoid of revolution flattened at its poles which attracts according to the universal inverse-square law and which revolves around its axis of symmetry, where the homogeneous matter that makes up the nucleus is denser than the homogeneous matter that surrounds the nucleus, would cause the effective gravity to increase more from the equator to a pole along the surface of this figure than along the surface of a homogeneous figure which has the same shape as this figure, which attracts according to the universal inverse-square law, which revolves around its axis of symmetry, and whose columns from center to surface all balance or weigh the same. Namely, Newton said that the ratio of the magnitude of the effective gravity at a pole to the magnitude of the effective gravity at the equator of the homogeneous figure whose columns from center to surface all balance or weigh the same equals the ratio of the distance of the equator from the center of the figure to the distance of a pole from the center of the figure. Now, increasing the density of a spherical nucleus inside this figure shaped like a flattened ellipsoid of revolution, which is concentric with the flattened ellipsoid of revolution which is the surface of the figure, increases the magnitude of the effective gravity both at the pole and at the equator of the figure. But the ratio of the magnitude of the additional effective gravity at the pole to the magnitude of the additional effective gravity at the equator of the figure equals the ratio of the square of the distance of the equator from the center of the figure to the square of the distance of the pole from the center of the figure. Consequently, since the figure is flattened at the poles, in which case its equatorial radius is greater than its polar radius, the ratio of the magnitude of the resultant of the effective gravity at the equator of the figure to the magnitude of the resultant of the effective gravity at a pole of the figure is *less* than it is for the homogeneous figure. Newton's verbal reasoning can be elaborated in detail and his conclusions shown to be true using mathematics as follows:

Let G_P designate the magnitude of the effective gravity at a pole P of a homogeneous figure shaped like a flattened ellipsoid of revolution whose center is C, which attracts according to the universal inverse-square law, and whose columns from center to surface all balance or weigh the same. Let G_E stand for the magnitude of the effective gravity at a point E at the equator of this same figure. Then

$$\frac{G_E}{G_P} = \frac{|PC|}{|EC|} \tag{I}$$

(where $|PC| < |EC|$). Let σ designate the density of the homogeneous figure whose columns from center to surface all balance or weigh the same. Suppose the density of a spherical nucleus of this figure whose center is C and whose radius is R (where $R < |PC|$) increases to $\sigma + \beta$, $0 < \beta$. Then it is clear that the additional attraction at P and at E will come from a homogeneous spherical figure whose center is C, whose radius equals R, and whose density is β. Let A_P stand for the magnitude of the attraction at P due to this homogeneous spherical figure, and let A_E designate the magnitude of the attraction at E due to this homogeneous spherical figure. Let G_P^* designate the magnitude of the effective gravity at the pole P of the figure shaped like a flattened ellipsoid of revolution with the spherical nucleus inside it whose density has increased, and let G_E^* stand for the magnitude of the effective gravity at the point E at the equator of the same figure. Then

$$G_P^* = G_P + A_P, \quad \text{and} \quad G_E^* = G_E + A_E. \tag{II}$$

From (II) it follows that

$$\frac{G_E^*}{G_P^*} = \frac{G_E + A_E}{G_P + A_P}. \tag{III}$$

But

$$\frac{A_E}{A_P} = \frac{|PC|^2}{|EC|^2}, \tag{IV}$$

follows from the universal inverse-square law. At the same time

$$\frac{|PC|^2}{|EC|^2} < \frac{|PC|}{|EC|} \tag{V}$$

because $|PC|/|EC| < 1$. From (IV),(V), and (I) it follows that

$$\frac{A_E}{A_P} < \frac{G_E}{G_P}. \tag{VI}$$

Therefore

$$\frac{A_E}{G_E} < \frac{A_P}{G_P}, \tag{VI'}$$

which means that

$$1 + \frac{A_E}{G_E} < 1 + \frac{A_P}{G_P} \tag{VI''}$$

too. Consequently, from (III) and (VI'') it follows that

$$\frac{G_E^*}{G_P^*} = \frac{G_E + A_E}{G_P + A_P} = \frac{G_E}{G_P}\left(\frac{1 + (A_E/G_E)}{1 + (A_P/G_P)}\right) < \frac{G_E}{G_P}.$$

42. From Newton's theory of a homogeneous figure shaped like an ellipsoid of revolution flattened at its poles which attracts according to the universal inverse-square law, which revolves around its axis of symmetry, whose ellipticity ϵ is infinitesimal, and whose columns from center to surface all balance or weigh the same, it follows that if P = the magnitude of the attraction at a pole, Π = the magnitude of the effective gravity at the equator, A = the polar radius, and B = the equatorial radius, then $P = k/A$ and $\Pi = k/B$, for some constant k, because the magnitude of the effective gravity at points on the surface of the homogeneous figure shaped like an ellipsoid of revolution flattened at its poles which attracts according to the universal inverse-square law, which revolves around its axis of symmetry, whose ellipticity ϵ is infinitesimal, and whose columns from center to surface all balance or weigh the same varies inversely as the distances of the points from the center of the figure. (See the end of note 7 of Chapter 1.) Hence $P/\Pi = B/A$, from which we conclude immediately that $(P - \Pi)/\Pi = (P/\Pi) - 1 = (B/A) - 1 = (B - A)/A \equiv$ the ellipticity ϵ.

43. Maupertuis (1732b: 49). Maupertuis considered this limiting case after stating the qualification that I cited in note 56 of Chapter 5.

44. Clairaut (1738b: 296-297).

45. In fact, this argument is merely tentative and suggestive. The homogeneous figure of revolution found by balancing columns from center to surface is not even shaped like an ellipsoid unless $n = 1$, when a central force $f(r) = r^n$ attracts the figure, where r is the distance from a center of force situated at the center of the figure. Nor can Clairaut's formal expression

$$fr^p + gr^q + \text{etc.} \tag{6.2.1}$$

for the density $D(r)$ really be extended to include hypothesized central forces of attraction (that is, $D(r) = 0$ everwhere except at the center of the figure, where $D(r)$ has a finite value) as a limiting case. Indeed, even more is true. While Clairaut's theory can be generalized to cover densities $D(r)$ that vary as step functions of r (see note 23 of Chapter 6), it cannot be extended to include densities $D(r)$ that are zero except on a set of measure zero. Clairaut must have realized the limitations of his "example" himself, for he told Colin MacLaurin in a letter dated 10 April 1741 that when the density is not uniform, the figure of revolution is not in general shaped like an ellipsoid, "which is easy to see by supposing all of the density to be concentrated at the center" [Clairaut to MacLaurin, 10 April 1741, in MacLaurin (1982: 347-354, on p. 348)].

46. Clairaut (1738b: 297).

47. Ibid., p. 300.

48. Ibid., p. 297.

49. Ibid., p. 301.

50. Ibid., p. 301. Clairaut said "more convincing" (*plus convaincante*) in his manuscript in French [see Clairaut (1738a: fol. 47)].
51. Clairaut (1738b: 301). In Clairaut's manuscript in French, the important passage is "elle ne tiendra en equilibre qu'en vertu de ce que sa superficie est la même que celle qu'elle aurait si tout doit fluide" [see Clairaut (1738a: fol. 481)].
52. Clairaut (1738b: 303).
53. Ibid., p. 304. Clairaut possessed a copy of Stevin (1634). [See Clairaut (1765: 15, No. 205). I thank Craig B. Waff for having provided me with a photocopy of this inventory of Clairaut's library.] Newton's own treatment of hydrostatics in the *Principia* is so terse that it obscures his whole approach [see Shapiro (1974–1975: 241–242)]. For a recent history of the principle of solidification, see Casey (1992a).
54. Clairaut (1738b: 304).
55. Todhunter (1873 or 1962).

7. Interlude I

1. Much of my account of the early development of the partial differential calculus outside France relies heavily on Engelsman (1984), whose mathematical notation I also adopt. The particularly noteworthy developments that make up the early history are summarized in Engelsman (1986).
2. Engelsman (1984: 61, 75).
3. For example, a polynomial equation of fifth degree or higher in a, with coefficients in x and y, is not of this form, because the general quintic equation is not algebraically solvable.
4. "Variable-parameter equation" is an expression coined by Engelsman (1984).
5. Ibid., p. 70.
6. These accusations and charges of plagiarism are analyzed in Feigenbaum (1985).
7. Engelsman (1984: 78).
8. Ibid., p. 78.
9. Ibid., pp. 89–90.
10. Ibid., p. 107.
11. Ibid., p. 115.
12. Ibid., pp. 79, 95–96, 120–122.
13. Ibid., pp. 79, 86–87.
14. See Greenberg (1981: 284; 1982: 1–3; and 1986).
15. *Académie Royale des Sciences, Registre des Procès-Verbaux*, 15 July 1711. Pp. 302v–314r; Grandjean de Fouchy (1744: 66); and Johann I Bernoulli (1702).
16. Montmort (1703a: 1r and 1704: 1r–1v) and Johann I Bernoulli [1704]. Concerning the part that Montmort played in the diffusion of *De quadratura curvarum* in France, see Greenberg (1982: 2).
17. The early correspondence is published in Montmort (1713, 1980: 283–414). As the title of Montmort's book might suggest, the correspondence included in it mainly pertains to probability theory. The bulk of Montmort's correspondence with Nikolaus I Bernoulli, including the part that involves integral calculus, remains unpublished. It is located at the Universitätsbibliothek, Basel, Bernoulli Archives, in mss LIa 21^2 and LIa 22^2.
18. Montmort (1703b, c) and Nicole (1703).
19. Montmort (1712).
20. Engelsman (1984: 69–70).
21. Johann I Bernoulli (1702a: 292–293 and 1702b: 300–301).
22. Nicole (1715: 56–61). As the title of this *mémoire* suggests, Nicole dealt with oblique trajectories in it as well.
23. Engelsman (1984: 71–73).
24. Montmort (1715a: 195); and *L'Académie Royale des Sciences, Registres de Procès-Verbaux*, 26 June 1715, p. 149r.
25. Montmort (1715b: 81).
26. Nicole (1717).
27. Nikolaus I Bernoulli (1716: 237r–238v).
28. Montmort (1716a).
29. Montmort (1716b: 87).
30. Taylor (1716c: 95). Montmort also informed Nikolaus I Bernoulli of Taylor's challenge in Montmort (1716a).

31. The published portion of Montmort's letters to Taylor appears in Taylor (1793: 81-150).
32. Nicole (1718).
33. In effect, Nicole worked with subtangents instead of tangents.
34. Johann I Bernoulli (1718a: 2v-3r).
35. Montmort (1718a: 1r).
36. Johann I Bernoulli (1718a: 2v-3r).
37. Nicole (1725).
38. Nicole (1715: 54-56, esp. 56).
39. Engelsman (1984: 79-80, 87).
40. The Archives of the Paris Academy of Sciences have a manuscript that includes a construction of the solution to the Leibniz–Bernoulli challenge (the plane curves orthogonal to the generalized brachistochrone cycloids) by a certain de Brochenu, accompanied by a covering letter sent from Grenoble and dated 30 November 1719. The same Archives also have an anonymous referee's report on de Brochenu's manuscript. The referee mentions that the solution to the problem appears without a demonstration in the article by Nikolaus II Bernoulli in *Acta Eruditorum* for 1718; he notes that de Brochenu's solution and Bernoulli's solution are the same; and, apart from a few remarks that he made about superfluous steps in the construction, he praised the solution as the work of a truly skillful mathematician. [See the *Archives de L'Académie Royale des Sciences* (Paris), "Dossiers des Séances," Pochette: November 1719. The anonymous referee's handwriting closely resembles the handwriting in a letter from Pierre Varignon, dated 18 May [1721?], located in the *Archives de L'Académie Royale des Sciences* (Paris), "Dossier Varignon."] No mention of de Brochenu's solution is made in the Paris Academy's *Registres des Procès-Verbaux*, nor is the referee's report recorded there, which is rather unusual and without an apparent explanation.
41. For Castel's comments about Nicole's solution of the problem of finding the plane curves orthogonal to the generalized brachistochrone cycloids, see Castel (1730: 104 or 1968a: 32). For Clairaut's corrections of errors in Reyneau's *Analyse démontrée*, see Clairaut (1728 or 1968). For D'Alembert's correction of other mistakes that remained in the second edition of Reyneau's treatise, see *L'Académie Royale des Sciences, Registres des Procès-Verbaux*, 29 July 1739, pp. 145r-146r. For Castel's critical remarks about Volume 2 of the first edition of Reyneau's treatise, see Castel (1735: xcviii). In a letter to Dodart dated October 1725, Montesquieu, who then studied at the Oratorian *collège* at Juilly, wrote that he was surprised by what he called the "clarity" of Reyneau's treatise. (Reyneau was an Oratorian.) Montesquieu told Dodart that he had read " a large part of Father Reyneau's *Analyse demontrée*" and that he had "hardly ever got stuck anywhere." [See Desgraves (1986: 99).] It is doubtful that Montesquieu understood a word of Reyneau's treatise!
42. Concerning the solutions to the problems of orthogonal trajectories worked out by early-eighteenth-century Italian mathematicians, see Pepe (1986: 201). The letter from Montmort to Taylor is Montmort (1718e). The quotation from it appears in Feigenbaum (1992: 387). For additional evidence that the level of integral calculus in Paris remained low in general throughout most of the first three decades of the eighteenth century, see Greenberg (1982: 1-3) and Greenberg (1986: 59-61, 66-67).
43. Montmort (1709: 1v).
44. Bernoulli rebutted Fontenelle in Johann I Bernoulli (1728: 1v-2v).
45. Johann I Bernoulli (1717: 3v-4v).
46. Aiton (1972: 166-168).
47. Saulmon (1717: 3).
48. Daniel Bernoulli (1968: 278).
49. Marsak (1959: 33).
50. Lagrange (1774a).
51. Delorme (1972: 61). For Johann I Bernoulli's diplomatic criticism of Fontenelle for belaboring in his *Eléments de la géométrie de l'infini* (1727) difficulties that Bernoulli thought were more imagined than real, see Delorme (1957: 343-347). The correspondence between Bernoulli and Fontenelle concerning Fontenelle's *Eléments de la géometrie de l'infini* has been discussed more recently in Blay (1990), where, once again, Bernoulli's "relatively little enthusiasm for Fontenelle's theory from the point of view of its mathematical presentation" (p. 95) is mentioned. Couder (1961: 43-44) calls Fontenelle's book "his only prolix, tedious, and boring" work. Fontenelle was more concerned with what we would now call the foundations of the infinitesimal calculus than he

was with the infinitesimal calculus itself. For accounts of Fontenelle's aims in trying to construct an arithmetic of infinitely large numbers, see McMillan (1970), Blay (1989b, c). Both McMillan and Blay view Fontenelle as a precursor of Georg Cantor. In my opinion they are wrong. In fact, Fontenelle's rules for treating the infinitely large numbers of different orders which he tried to define simply parallel the rules of the day for treating infinitely small numbers of different orders. The latter numbers lacked foundations, and Fontenelle thought that he had provided the missing foundations when in fact he had not. In defining infinitely large numbers of different orders and the rules for handling them, he really only imagined a correlate with the rules of the day for handling infinitely small numbers of different orders. He tried to transpose or transfer the rules for the latter numbers to what he believed he had defined to be infinitely large numbers of different orders. Moreover, Fontenelle's scheme of infinitely large numbers of different orders and its arithmetic is internally inconsistent.

52. Johann I Bernoulli (1730a: 4r-4v).
53. Johann I Bernoulli (1721: 5v and 1723: 5r).
54. See Mahoney (1975: 118) and Boyer (1946: 166).
55. Johann I Bernoulli (1724: 14-15). Some early-eighteenth-century mathematicians in the Paris Academy who advocated the differential calculus may have continued to employ this calculus awkwardly, unskillfully, and laboriously as late as the 1720s in part because they continued well into the 1720s to worry about and to be preoccupied with the sometimes specious arguments against the differential calculus which the algebraist Michel Rolle (see note 38 of Chapter 5) had put forward at the beginning of the eighteenth century (see Greenberg (1986: 59, note 2)).
56. La Beaumelle (1856: 14).
57. Johann I Bernoulli (1730b: 3r-3v).
58. The relations between Maupertuis and Bernoulli are discussed at length in Brown (1975 and 1976)., and in Beeson (1992: 67-87).
59. McClellan (1981: 560, note 4; 561).
60. See Greenberg (1982: 1-5 and 1986).
61. The most recent study of Varignon's mathematics can be found in Blay (1992). Costabel (1966: 11-12) described Varignon as a competent, indefatigable mathematician instead of a talented one. He called Varignon's writings "prolix," and he noted that Varignon treated the subjects that he discussed in his works "in excessive detail." Although he found Varignon's work to be "extremely technical," he also found it to be formal and not deep and often to involve generalization for its own sake.
62. For discussions of Varignon's efforts to treat the direct and inverse problems of central forces, see Aiton (1964 and 1989: 51-52); Hankins (1967: 60-65); and De Gandt (1987: 292, 296-298). For a very detailed discussion of Varignon's use of the differential calculus to create a general formalism composed of algorithms for solving problems in motions of projectiles in resistant media and problems in the motions of orbiting bodies which are governed by central forces which Newton had formulated and solved in the *Principia* using other means, see Blay (1985-1986, 1988, 1989a, and 1992). Blay (1988: 618) does not exaggerate when he states that Varignon wrote some "very long, and perhaps too long ("trés longs, et peut-être trop longs")" Paris Academy *mémoires* concerning these problems. Concerning Varignon's failure to learn integral calculus, see Aiton (1964: 98 and 1986: 142).
63. Robinet (1978: 169) mentions that Varignon lacked either the interest or the ability to make his work comprehensible to a general audience. Although Varignon's published works are long-winded and confusing, his mathematics students at the Collège Mazarin nevertheless had a good opinion of him. One of his former students translated from Latin into French and edited Varignon's elementary mathematics courses, which were then posthumously published in 1731 [Jean-Baptiste Cochet, *Eléments de mathématiques de Monsieur Varignon* (Paris: Brunet, 1731)]. See Costabel (1966: 11-12, 28 and 1976: 585). At the Collège Mazarin D'Alembert was taught mathematics by a Professor Caron, who led him through Varignon's published lectures [see Hankins (1970: 19)].
64. Taton (1958); Torlais (1961: 29-34); and Gough (1975: 328).
65. Maupertuis (1731b: 3r).
66. Maupertuis (1730a: 235).
67. Maupertuis (1731e).
68. Johann I Bernoulli (1731b: 1r-2r and 1731c: 4r). In March of 1732 Clairaut told his friend the

Genevan mathematician Gabriel Cramer that Maupertuis had read a paper to the Paris Academy in the spring of 1731 in which he revealed what Clairaut judged to be "on many occasions a very ingenious and happy method for integrating or separating the differentials in equations" [Clairaut (1732a: 212)].
69. Maupertuis (1731d: 83v).
70. In Nikolaus I Bernoulli (1718a; 259v), Bernoulli told Montmort that he had heard that Craige had published a new book on integral calculus which Montmort had "apparently" seen. In this letter Nikolaus I Bernoulli asked Montmort to tell him what was in Craige's book. In fact Montmort probably had not yet seen Craige's book when he received Nikolaus I Bernoulli's letter, because in Montmort (1718b: 4v), Montmort told Johann I Bernoulli that he had heard from someone in England that Craige had recently published a book on integral calculus entitled *De Calculo fluentiun* in which Craige showed, among other things, how to separate the variables in certain kinds of first-order ordinary "differential" equations of first degree in two finite variables. In this letter Montmort told Johann I Bernoulli that he "was surprised" to hear that Craige claimed to be able to separate the variables in equations of one particular type that he treated. Montmort added that he would tell Johann I Bernoulli more about Craige's book when he received the copy of the book which he was waiting for. In Montmort (1718d: 203v), Montmort told Nikolaus I Bernoulli that he had received a copy of Craige's book and that he would write a review of the book for Nikolaus I Bernoulli. In Montmort (1718c: 1r–2v), Montmort told Johann I Bernoulli that he had received a copy of Craige's book, and Montmort reviewed the book in this letter. In particular, he gave all of Craige's examples of his method of separating the variables in certain types of first-order ordinary "differential" equations of first degree in two finite variables. He added that Taylor had recently informed him by letter that Craige had made an error in separating the variables in equations of the kind in question. In Nikolaus I Bernoulli (1718b: 261r), Nikolaus I Bernoulli told Montmort that he had read the letter concerning Craige's book which Montmort had sent to Johann I Bernoulli [Montmort (1718b)], and Nikolaus I Bernoulli said that he was "as surprised as" Montmort was that Craige claimed to be able to separate the variables in equations of the type at issue. Nikolaus I Bernoulli told Montmort that if Craige had managed to do that, then he knew how to do something that the Bernoullis had believed to be impossible. In Maupertuis (1731f: 2r–2v), Maupertuis asked Johann I Bernoulli about Craige's attempt to separate the variables in equations of the kind in question. In Johann I Bernoulli (1731c: 3r–3v), Johann I Bernoulli answered Maupertuis's question. He told Maupertuis that Craige had indeed made a serious error and that he thought that it was not possible in general to separate the variables in equations of the type at issue; Bernoulli pointed out Craige's error to Maupertuis; and Bernoulli added that he had explained Craige's error to Montmort in a letter that he wrote to Montmort "shortly before his [Montmort's] death." In fact, no such letter from Johann I Bernoulli to Montmort sent "shortly before his [Montmort's] death," in which Bernoulli talks about Craige's error, can be found in the Johann I Bernoulli–Montmort correspondence located at the Universitätsbibliothek, Basel, Bernoulli Archives.
71. Maupertuis (1731d).
72. Maupertuis (1731d: 82v–83r).
73. Maupertuis (1730b: 2r). In this letter Maupertuis told Bernoulli that he had not read the copy of James Stirling's *Methodus differentialis, sive Tractatus de summatione et interpolatione serierum infinitarum* (1731) which Stirling had sent him, because he was repelled by all of the infinite series that the work included.
74. Johann I Bernoulli (1726), which reappears in Johann I Bernoulli (1742b: Tome III, 108–124, esp. 115–116, §IX).
75. Maupertuis (1731d: 83r–83v).
76. Concerning Leibniz's universal characteristic, see Aiton (1985: 91–98).
77. Whiteside (1961: 262).
78. Bos (1974–1975: 5–7).
79. Maupertuis (1736: 61v).
80. Schier (1941: 4–5).
81. Desautels (1956: Ch. IV: "La physique Cartésienne et le Newtonianisme"); and Desautels (1971: 114).
82. Castel plainly expressed his preference for British mathematics in a letter to Woolhouse, dated 1726 [Castel (1726)], a letter that Castel's biographer Donald Schier (see note 80 of Chapter 7) did not see.

83. Schier (1941: 89).
84. Dainville (1964: 58–60 or 1978: 385–387).
85. The experimental approach to the teaching of mathematics, as it was developed in the 1740s by such researchers in mathematics as Clairaut, Simpson, and MacLaurin, is discussed in Boyer (1968).
86. Castel (1728a: 556).
87. Dainville (1964: 59 or 1978: 386).
88. Schiner (1941: 18, 120).
89. Castel (1735: xcviii). See as well note 41 of Chapter 7.
90. Schier (1941: 20–22, 123).
91. Castel (1732 or 1968d).
92. Castel (1735) and Stone (1735).
93. Review of Stone (1735), published in the *Journal des Savants*, June 1735, pp. 325–338.
94. Mairan (1736: 1r).
95. Johann I Bernoulli (1742a).
96. Johann I Bernoulli (1730b: 4r).
97. Johann I Bernoulli (1730b: 5v–6r); and Castel (1722: 992–994 or 1968b: 251–252).
98. Castel (1728a: 4).
99. Castel (1728a: 252–253).
100. Desgraves (1986: 156–158).
101. Johann I Bernoulli (1730c: 81). How much of Bernoulli's solution is due to Bernoulli and how much of it he owed to Euler is a debatable question whose answer is difficult to determine precisely because of Bernoulli's and Euler's differing accounts of the matter. At the end of August 1730 Bernoulli wrote Maupertuis to say that he had discussed the problem of tautochrones with his son after Maupertuis visited Basel. (This cannot be a reference to Bernoulli's oldest son Nikolaus II, because this son died in Saint Petersburg in 1726. Nor can it be a reference to Bernoulli's third son Johann II, because he was not a mathematician. Consequently Johann must have meant his second son Daniel, who was then in Saint Petersburg.) Johann also told Maupertuis that he had solved the general problem of the tautochrones in media that resist the motions of a body as the square of the body's speed and that Euler's solution to the problem which they had both seen while Maupertuis was still in Basel was only a particular case and not the easiest to construct [see Johann I Bernoulli (1730a: 2v–4r)]. Many years later Euler told Lagrange that Bernoulli only found his solution to the tautochrones after Euler had already communicated his own solution to him and that Bernoulli never claimed to be the first to discover the solution [see Euler (1768: 462)]. Euler first published his solution in Euler (1729). He republished it in Euler (1736a). The republished version can be found in Euler (1912: vol. 2, pp. 357–376, "Propositio 81" and "Propositio 82"). Thus it is difficult determine with certainty who influenced whom.
102. Maupertuis (1733d: 2r).
103. See Greenberg (1981, 1982, and 1984b).
104. Bos (1974–1975) is the definitive study of the Leibnizian differential. It includes an examination of the distinction between "independent variables" and "dependent variables." The making of this distinction accompanied the emergence of the notion of "functions of independent variables." This notion grew out of problems connected specifically with differential equations, not finite equations.
105. Ibid., pp. 5–9.
106. Engelsman (1984: 35).
107. Lagrange (1790?: 68). I thank Ivor Grattan-Guinness for having brought this manuscript to my attention.
108. Fontaine (1764a: 2–3).
109. *L'Académie Royale des Sciences, Registres des Procès-Verbaux*, 7 May 1732, p. 170r; and Fontenelle (1732b). Fontaine also published for the first time in Fontaine (1764b: 6) his solution to the problem of finding the path of motion of "least action." See as well Fontaine (1770: 6). Fontaine used his fluxio-differential method to arrive at his published solution of this problem too.
110. Fontaine (1764a: 3). Figure 2 appears on "Page 22, Plate 1," located at the end of Fontaine (1764b).
111. Woodhouse (1810: 85–86) or Woodhouse (1964: 85–86). Had Fontaine had an opportunity to

examine Volume 2 of papers published by the Turin Academy, which includes Lagrange's first paper on the calculus of variations, in which Lagrange made use of two differential operators "d" and "δ", before his own collected works were published in 1764? It seems unlikely. Lagrange's patron D'Alembert received a copy of Volume 2 of the Turin Academy's papers by 1762 [see D'Alembert (1762)]. But Lagrange's paper on the calculus of variations included in the volume never came up as a subject for discussion between D'Alembert and Lagrange until March of 1768 [see D'Alembert (1768)]. But this merely suggests that Fontaine did not see Lagrange's first paper on the calculus of variations before 1764. It does not prove that he did not. It is not yet possible to declare with certainty that Fontaine, who was not at all on the same intimate terms with Lagrange that D'Alembert was, did not see Lagrange's first paper on the calculus of variations before his own complete works were published in 1764. (On pp. 633–4 of Chapter 10 we shall find evidence that Fontaine did not see Lagrange's first paper on the calculus of variations before 1764.) I raise the question because we shall see later in this chapter that Fontaine did in fact modify one piece of his work which appeared in the *L'Académie Royale des Sciences, Registres des Procès-Verbaux* for the 1730s, using simplifications of this piece of work which someone else had found, and then Fontaine published the simplified version in his complete works without acknowledging the author of the simplifications.

112. Jakob Bernoulli (1697: Figure V in Table IV, 212) or Jakob Bernoulli (1744: 769–770). I have quoted the English translation that appears in Woodhouse (1810: 4) and in Woodhouse (1964: 4). The phrase within brackets is Woodhouse's.

113. Fontaine (1764a: 1).

114. Fontaine (1734b).

115. Lagrange (1760–1761a: 173–174 and 1867a: 336) introduced two first-order differential operators d and δ and explained that "δZ will express a difference in Z which will not be the same as dZ, but which will however be formed by the same rules, so that by having an arbitrary equation $dZ = mdx$ one will have $\delta Z = m\delta x$ as well."

116. Fontaine (1734b: 378 "Exemple IV", figure on p. 371).
Fontaine solved

$$\frac{\mp (3/n)\gamma + \lambda}{2\gamma} = \mp \frac{1}{n} \qquad (7.3.3.34)$$

for ϕ and γ by what amounts to the following: equation (7.3.3.34) reduces to

$$\lambda = \pm \frac{1}{n}\gamma. \qquad (I)$$

Here I replace the awkward notation "$\cdot$" for differentials by the standard symbol "d." In this notation (I) appears as

$$\frac{d^2\phi}{dx^2} = \pm \frac{1}{n}\frac{d\phi}{dx}, \qquad (II)$$

or

$$\frac{d^2\phi}{dx^2} \mp \frac{1}{n}\frac{d\phi}{dx} = 0. \qquad (III)$$

But (III) can be rewritten as

$$\frac{d}{dx}\left(\frac{d\phi}{dx} \mp \frac{1}{n}\phi\right) = 0, \qquad (IV)$$

and (IV) means that

$$\frac{d\phi}{dx} \mp \frac{1}{n}\phi = \text{a constant } k. \qquad (V)$$

Now, (V) can be rewritten as

$$\pm n\frac{d\phi}{dx} - \phi = \pm nk, \qquad (VI)$$

then as

$$\pm n\frac{d\phi}{dx} = \phi \pm nk, \tag{VII}$$

and finally as

$$\pm \frac{1}{n} dx = \frac{d\phi}{\phi \pm nk} = d\ln(\phi \pm nk). \tag{VIII}$$

It follows from (VIII) that

$$\pm \frac{x}{n} = \ln(\phi \pm nk) + \ln R = \ln(R\phi \pm nRk), \tag{IX}$$

where R is a positive constant. Then from (IX) it follows that

$$R\phi \pm nRk = e^{\pm x/n}. \tag{X}$$

But $\phi(0, X) = 0$ and (X) require that

$$\pm nRk = 1. \tag{XI}$$

Now, (XI) means that (X) can be written as

$$R\phi + 1 = e^{\pm x/n}, \tag{XII}$$

in which case

$$\phi = \frac{e^{\pm x/n} - 1}{R}. \tag{XIII}$$

But $\phi(X, X) = 1$ and (X) require that

$$R \pm nRk = e^{\pm X/n}. \tag{XIV}$$

Then

$$R = e^{\pm X/n} - 1 \tag{XV}$$

follows from (XI) and (XIV). Finally, from (XIII) and (XIV), it follows that

$$\phi = \frac{e^{\pm x/n} - 1}{e^{\pm X/n} - 1} \tag{XVI}$$

and

$$\gamma = \frac{d\phi}{dx} = \frac{\pm(1/n)e^{\pm x/n}}{e^{\pm X/n} - 1}. \tag{XVII}$$

Fontaine then integrated equation

$$\frac{p\gamma + q\phi}{2\gamma} = p \tag{7.3.3.33}$$

by what amounts to the following: Equation (7.3.3.33) reduces to

$$\frac{q}{p} = \frac{\gamma}{\phi}. \tag{XVIII}$$

Now, (XVI)–(XVIII) together mean that

$$\frac{q}{p} = \frac{\gamma}{\phi} = \frac{\pm(1/n)e^{\pm x/n}}{e^{\pm x/n} - 1}. \tag{XIX}$$

But

$$q = \frac{dp}{dx}. \tag{XX}$$

From (XIX) and (XX) it follows that

$$\frac{d}{dx} \ln p = \frac{1}{p}\frac{dp}{dx} = \frac{q}{p} = \frac{\gamma}{\phi} = \frac{\pm (1/n)\,e^{\pm x/n}}{e^{\pm x/n} - 1}. \qquad \text{(XXI)}$$

From (XXI) it follows that

$$d \ln p = \frac{\pm (1/n)\,e^{\pm x/n}}{e^{\pm x/n} - 1}\, dx = d \ln (e^{\pm x/n} - 1). \qquad \text{(XXII)}$$

But (XXII) means that

$$\ln (e^{\pm x/n} - 1) = \ln p + \ln m = \ln mp, \qquad \text{(XXIII)}$$

where m is a positive constant. From (XXIII) it follows that

$$mp = e^{\pm x/n} - 1. \qquad \text{(XXIV)}$$

But

$$p = \frac{dz}{dx}. \qquad \text{(XXV)}$$

If (XXIV) and (XXV) are combined, the equation

$$m\frac{dz}{dx} = e^{\pm x/n} - 1 \qquad \text{(XXVI)}$$

results. Integrating equation (XXVI), we arrive at the equation

$$mz = \pm n\,e^{\pm x/n} - x + c, \qquad \text{(XXVII)}$$

where c is a constant. If $z(0) = 0$, then it follows from (XXVII) that

$$0 = \pm n + c, \qquad \text{(XXVIII)}$$

which means that

$$c = \mp n$$

in (XXVII). Therefore

$$mz = \pm n\,e^{\pm x/n} - x \mp n, \qquad \text{(XXIX)}$$

and thus

$$z(x) = \frac{\pm n\,e^{\pm x/n} - x \mp n}{m}. \qquad \text{(XXX)}$$

117. Hence the methods of Jakob Bernoulli, Johann I Bernoulli, and Brook Taylor, applied to the problem of the brachistochrone in a resistant medium, had been subject to certain inherent limitations: "they do not extend to cases, in which the differential function expressing the maximum should depend on a quantity, not given except under the form of a "differential" equation, and that not integrable; for instance, they will not solve the case of the curve of the quickest descent, in a resisting medium, the descending body being solicited by any forces whatever" [Woodhouse (1810: 30)].
118. Johann I Bernoulli (1734: 2r).
119. Euler (1742a: 113–114).
120. D'Alembert (1765b: 946 or 1967a: 946).
121. The important use that Euler was to make of such coefficients is a principal theme of Bos (1974–1975).
122. The distinction made between these two kinds of equations is another of the major topics in Bos (1974–1975).
123. The definitive study of the Leibnizian differential calculus of a single differential operator is Bos (1974–1975). Concerning Euler, see pp. 66–77 of this work.
124. Bos (1974–1975: 59–62) says that Leibniz attempted to introduce quantities ddx that were not

equal to zero but that were fundamentally incompatible with his calculus of a *single* differential operation "*d*"; hence Leibniz did not succeed as a result.

125. Let us note the following: while $f(x, X) = g(x, X)$ for all x and X is obviously sufficient to ensure that $\int_0^X f(x, X)\,dx = \int_0^X g(x, X)\,dx$ for all X, it is not a necessary condition for the equality of the two integrals. As a simple illustration of this, consider two functions

$$f(x, X) = F(X)\left\{x - \frac{X}{2}\right\}, \quad g(x, X) = -F(X)\left\{x - \frac{X}{2}\right\},$$

where $F(X) \not\equiv 0$ is any function of X. Then

$$\int_0^x f(x, X)\,dx = \int_0^x g(x, X)\,dx = 0, \quad \text{while } f(x, X) \not\equiv g(x, X).$$

126. Fontaine (1734b: 378 "Première remarque").
127. See, for example, Clairaut (1740a: 68-69).
128. Euler (1756-1757), republished in Euler (1954: 133-168, esp. 137, including note 1); see Clifford Truesdell's English translation of the passage (1960: LXIII). Truesdell (p. LXII) dates this passage to the year 1752.
129. Fontaine (1734b: 379 "Seconde remarque").
130. Nicole (1737).
131. Fontaine (1737).
132. Fontaine (1764b: first page of "Table des Mémoires contenus dans ce Volume," in the section entitled "Le Calcul intégral. Première méthode.").
133. Engelsman (1984: 138-139). It is easy to show that if $f(x, p)$ is a homogeneous expression of degree n in the two finite variables x and p, then provided the integrals

$$\phi(x, p) \equiv \int_0^x f(x^*, p)\,dx^* \qquad (7.3.3.72)$$

and

$$\int_p^x f(x^*, p)\,dx^*$$

exist, these two integrals are the only integrals of $f(x, p)$ which are homogeneous expressions in x and p, and their degree of homogeneity in x and p is $n + 1$. Moreover, the first of the two integrals is the only one that has a fixed lower limit of integration, and, more particularly, a lower limit of integration that does not depend upon the value of the parameter p. Thus, for example, even though $f(x, p) \equiv 1/x$ and $f(x, p) \equiv p/x$, respectively, are homogeneous expressions of degree -1 and 0 in x and p, respectively, neither

$$\ln x = \int_1^x \frac{1}{x^*}\,dx^* \quad \text{nor} \quad p \ln x = \int_1^x \frac{p}{x^*}\,dx^*$$

is a homogeneous expression in x and p.
134. Engelsman (1984: 98).
135. Ibid., pp. 95-96.
136. Ibid., pp. 87, 116.
137. Ibid., pp. 120-122.
138. Ibid., pp. 116-119.
139. Ibid., p. 100-109.
140. Concerning Nikolaus's withholding his unpublished work in partial differential calculus from his correspondents, see ibid., p. 95.
141. Mairan's inquiries about Nikolaus appear in Mairan (1724: 4r-4v).
142. Engelsman (1984: 88-90).
143. Ibid., pp. 90-91.
144. Ibid., p. 100.
145. Ibid., pp. 110-111.

146. Ibid., pp. 79–87.
147. Johann I Bernoulli (1730: 80); reappearing in Johann I Bernoulli (1742b: vol. 3, 173–197, on pp. 175–176.)
148. Fontaine (1737: 139r).
149. Nikolaus II Bernoulli (1721: 322).
150. Ibid., pp. 319–320. The "homogeneous function theorem" itself does not appear in any of the articles in *Acta Eruditorum* on the problem of orthogonal trajectories, but this literature does include information that could have helped an Euler or a Fontaine discover the theorem. Namely, in a supplement to his construction of 1717 of the plane curves orthogonal to the generalized brachistochrone cycloids, published in *Acta Eruditorum* for 1718, Jakob Hermann gave

$$-\frac{da}{a} = \frac{(1+p^2)\,dy}{y+p\int pdy}$$

where

$$\int pdy \equiv \int_0^y p(y^*,a)dy^*,$$

as a general equation for orthogonal trajectories. In their correspondence with Montmort, both Johann I Bernoulli and Nikolaus I Bernoulli maintained that this equation was erroneous. They pointed out, for example, that it would not even work for the simplest examples, like a family of straight lines that intersect at the same point. Hermann's equation did not give circles centered at the point of intersection as the orthogonal trajectories [see Johann I Bernoulli (1718b: 1v–2r) and Nikolaus I Bernoulli (1718: 260v–261r)]. In the supplements to Nikolaus II Bernoulli's article in the *Acta* for 1720, the Bernoullis illustrated the mistakes in Hermann's equation by taking $p = p(y,a)$ to be a similar function, meaning a homogeneous function, of degree zero in y and a, and then they showed how Hermann's equation produced conflicts or inconsistencies in dimensions. In fact, the Bernoullis did this precisely on those pages of volume 7 of the supplements to the *Acta* which Johann I Bernoulli cited in his Paris Academy *mémoire* of 1730 on tautochrones in resistant media (1730c) in connection with similar (that is, homogeneous) functions, namely, Nikolaus II Bernoulli (1721: 322–324). Now, if Hermann's equation above and the variable-parameter equation

$$-da = \frac{(1+p^2)dy}{pq}$$

are combined, an equation

$$\int pdy = \frac{1}{p}y + qa$$

results. If someone searched for the source of Hermann's error, found it, and combined Hermann's equation corrected with the variable-parameter equation, a new equation

$$\int pdy = py + qa$$

might have resulted. But when $p = p(y,a)$ is a homogenous expression of degree zero in y and a, this last equation just expresses the homogeneous function theorem for homogeneous functions

$$\int_0^y p(y^*,a)dy^*$$

of degree one in y and a. (See as well note 133 of Chapter 7.)
151. Concerning Euler's familiarity with Nikolaus II Bernoulli's article in *Acta Eruditorum* for 1720, see Engelsman (1984: 135).
152. Ibid., pp. 96, 137–138.
153. Ibid., pp. 136–137.
154. Ibid., pp. 110–111.
155. Ibid., pp. 90–91.

156. Ibid., p. 91. This is Englesman's description, not Bernoulli's.
157. For a discussion of Euler's "higher-order modular equations" and how Euler planned to use them, see ibid., pp. 142–145, 150–156.
158. Euler (1736a: vol. 2, "Propositio 14"), reappearing in Euler (1912: vol. 2, 43–45, "Propositio 14").
159. In the very first letter he ever wrote Euler, in May of 1738, Maupertuis (1738b) told Euler that he had read Euler's *Mechanica*. Maupertuis more than likely only asserted this in order to attract Euler's attention, flatter him, and thereby entice him to reply. The trick worked, as it had already in the case of Johann I Bernoulli. It is unlikely that Maupertuis ever read the *Mechanica*, much less understood its contents. Pierre Costabel, coeditor of the Euler–Maupertuis correspondence, told me privately that Maupertuis simply was not a good enough mathematician to follow the *Mechanica*. *None* of Maupertuis's five letters to Euler include *any* mathematics whatever.
160. A short review of Euler's *Mechanica* did appear nevertheless in the *Journal de Trévoux* for the year 1737 – and during the month of July no less, the month that Fontaine read his paper on problems of trajectories to the Paris Academy! (See the *Journal de Trévoux*, July 1737, pp. 1318–1321; reprinted in facsimile as the *Journal de Trévoux*, Volume 37 (Geneva: Slatkine Reprints: 1968), pp. 335–336. Full length, detailed reviews of the *Mechanica* only appeared for the first time in the *Journal de Trévoux* for 1740. [See the review of Volume 1 in the *Journal de Trévoux*, May 1740, pp. 816–834; reprinted in facsimile as the *Journal de Trévoux*, Volume 40 (Geneva: Slatkine Reprints, 1968), pp. 214–219. A review of Volume 2 is published in the *Journal de Trévoux*, July 1740, pp. 1407–1422; reprinted in facsimile as the *Journal de Trévoux*, Volume 40 (Geneva: Slatkine Reprints, 1968), pp. 364–367.] Clairaut's ignorance of the contents of the *Mechanica* until Euler brought it to his attention late in 1740 I shall discuss in detail later.
161. There is evidence that Fontaine indeed had not *thoroughly* read Nikolaus II Bernoulli's article in the *Acta* for 1720 or its supplements. In 1733, Bouguer made use of "differentiation under the integral sign" – that is, the reversal (7.1.1.1) of differentiation and integration [see Bouguer (1733: 96–98)]. Fontaine then stated the result in generality. But neither Bouguer, nor Fontaine, nor Clairaut seems to have known during the 1730s that this technique was first published in explicit form in Nikolaus II Bernoulli's article in the *Acta* for 1720. Clairaut thought that Bouguer had been the first to apply the technique, but Clairaut's friend Daniel Bernoulli informed him that, according to Euler, he was mistaken. This revelation troubled and embarrassed Clairaut. His sudden realization of his profound ignorance of mathematical precedents abroad induced him, more than anything else, to write Euler for the first time, in September of 1740. In his first letter to the Swiss mathematician, Clairaut asked him if he knew of anyone else who had ever inverted differentiation and integration before. Clairaut (1739: 431) had used the procedure himself in a Paris Academy *mémoire* of 1739 on integral calculus without mentioning any earlier applications of it by mathematicians outside France. In his publications until then he had failed to show the slightest awareness of such applications, and he was anxious about possible oversights concerning precedents. Euler replied that the result in question had been published in explicit form in the literature in *Acta Eruditorum* on orthogonal trajectories [see Clairaut (1740a) and Euler (1740a)]. In fact, the general result appears in Nikolaus II Bernoulli (1721: 307–308), some fourteen pages before the particular page in this supplement to the *Acta* cited by Johann I Bernoulli in his Paris Academy *mémoire* of 1730 on tautochrones in resistant media [Johann I Bernoulli (1730c)]. Euler's announcement caused Clairaut to search for and find this prior application of the technique in question, which he subsequently cited in a second Paris Academy *mémoire*, of 1740, on integral calculus [Clairaut (1740b: 296)].
162. Thus, for example, Weinstock (1952, 1974: vii) intentionally avoids "the vague, mechanical 'δ method'... throughout." Weinstock alludes here to Lagrange's original calculus of variations consisting of two differential operators "d" and "δ", which to all intents and purposes virtually cannot be distinguished from Fontaine's calculus of operators "$\cdot$" and "d".
163. Synge and Griffith (1959: 443–444).
164. *L'Académie Royale des Sciences, Registres des Procès-Verbaux*, 26 November 1738, p. 190r; 29 November 1738, p. 191r.
165. Boucharlat (1816: 180); and Condorcet (1771: 105).
166. Boucharlat (1816: 179–180).
167. The Castel–Fontenelle correspondence is published in Fontenelle (1766: 141–172).
168. Castel (1728b). For Castel's use of the words "routine" and "trial and error" to describe the

infinitesimal calculus done by L'Hôpital and other mathematicians, see Castel (1728c: esp. 1238) or Castel (1968c: esp. 299). Castel used the words "routine" and "trial and error" even earlier, to describe the infinitesimal calculus done by L'Hôpital and other mathematicians, in Castel (1728b: 157), which is a letter to Fontenelle in which Castel mentioned that he would have liked Fontenelle to go further into the metaphysics of the infinite in his *Eléments de la géométrie de l'infini* (1727) than Fontenelle actually did.

169. Dainville (1964: 59 or 1978: 386).
170. Boucharlat (1816: 180 and Condorcet 1771: 106).
171. Maupertuis and Nicole (1737) and Fontenelle (1731).
172. Condorcet (1771: 105–106).
173. Clairaut and Nicole (1739). The manuscript of this report was sold to a private buyer. The sale is listed in the "Ficher Charavay," 45, CIC-CLE. No. 26805 (89) [*Salle des Manuscrits, Bibliothèque Nationale* (Paris)].
174. Clairaut (1739: 435).
175. Clairaut (1739: 434).
176. If $Adp + Bdx + Cdy$ is the differential of a homogeneous function ϕ of degree e in x, y, p, then A, B, C must themselves be homogeneous of degree $e - 1$. This is easy to see in terms of modern definitions as follows: For $\lambda \neq 0$,

$$\frac{\partial \phi}{\partial p}(\lambda p, \lambda x, \lambda y) = \lim_{\Delta p \to 0} \frac{\phi(\lambda p + \Delta p, \lambda x, \lambda y) - \phi(\lambda p, \lambda x, \lambda y)}{\Delta p}$$

$$= \lim_{\Delta p \to 0} \frac{\phi(\lambda(p + (\Delta p/\lambda)), \lambda x, \lambda y) - \phi(\lambda p, \lambda x, \lambda y)}{\lambda(\Delta p/\lambda)} = \lim_{\Delta p \to 0} \frac{\lambda^e}{\lambda} \frac{\phi(p + (\Delta p/\lambda), x, y) - \phi(p, x, y)}{(\Delta p/\lambda)}$$

$$= \lambda^{e-1} \lim_{\Delta p/\lambda \to 0} \frac{\phi(p + (\Delta p/\lambda), x, y) - \phi(p, x, y)}{(\Delta p/\lambda)} = \lambda^{e-1} \frac{\partial \phi}{\partial p}(p, x, y). \quad \text{But } A \equiv \frac{\partial \phi}{\partial p}.$$

The same argument applied to each of the other variables shows that $B \equiv \partial \phi/\partial x, C \equiv \partial \phi/\partial y$ are also homogeneous of degree $e - 1$.

The converse of this result can be stated in general as follows: if A, B, C, D, E, etc. are homogeneous expressions of degree e in the variables x, y, p, q, etc. such that the differential

$$Adx + Bdy + Cdp + Ddq + \text{etc.}$$

is integrable, then there exists an expression F that is homogeneous of degree $e + 1$ in x, y, p, q, etc. such that

$$dF = Adx + Bdy + Cdp + Ddq + \text{etc.}$$

This expression is

$$F = \frac{Ax + By + Cp + Dq + \text{etc.}}{e + 1}.$$

Clairaut (1740b: 322–323) in fact proved this converse of the homogeneous function theorem in general, as we shall discover in the text below. We shall also see, however, that the converse of the homogeneous function theorem cannot hold if A, B, C, D, E, etc. are homogeneous expressions of degree $e = -1$ in the variables x, y, p, q, etc. This is true because, as we shall see, the derivation of the expression

$$F = \frac{Ax + By + Cp + Dq + \text{etc.}}{e + 1}$$

does not hold if $e + 1 = 0$.

177. In the summer of 1741 Fontaine sent a substantial manuscript on integral calculus to the Paris Academy in a sealed envelope (*pli cacheté*). I have used a verbal description of Fontaine's special technique for dealing with this particular case, which is included in the referees' report on Fontaine's manuscript written by D'Alembert and Jean-Paul de Gua de Malves, to reconstruct

the technique [see D'Alembert and De Gua (1742: 16)]. I have not been able to find the contents of the *pli cacheté* itself. Fontaine's letter to Mairan, dated 15 June 1741, in which Fontaine told the Paris Academy's secretary that he was sending the *pli cacheté* to him immediately, has been sold. The sale is listed in the "Fichier Charavay," 74, FLO-FOR. No. 236 (191) [*Salle des Manuscrits, Bibliothèque Nationale* (Paris)].

178. Clairaut (1739: 434–435).
179. Clairaut (1739: 436).
180. The equation

$$x = \sqrt{p^2 - y^2}$$

(equation (7.3.3.102') when $k = 0$) is the equation of a one-parameter family of plane curves parametrized by p which lie in the same plane. This family of curves is a solution of the equation

$$dx + \frac{y}{x}dy - \frac{p}{x}dp = 0$$

(equation (7.3.3.126'). On the other hand, the equation

$$x^2 + y^2 - p^2 = k$$

is not an equation of a one-parameter family of plane curves parametrized by p which lie in the same plane, although for constant values of k, the equation is a solution of the equation

$$2xdx + 2ydy - 2pdp = 0,$$

which (except for division by $2x$) is the same as equation

$$dx + \frac{y}{x}dy - \frac{p}{x}dp = 0.$$

Since the nonsingular solution of equation (7.3.3.96) in the text is *already* a one-parameter family of plane curves which lie in the same plane, it stands to reason, or one could guess, that a one-parameter family of plane curves which lie in the same plane could not possibly be the general solution of an integrable equation (7.3.3.97), which is an equation more general than equation (7.3.3.96), and that, instead, a two-parameter object of some sort, in the parameters p and k, must be the general solution of equation (7.3.3.97). Clairaut (1740b: 308–311) seems to have been the first person to interpret the integral or an integrable equation (7.3.3.97) as an integral *surface* in rectangular coordinates x, y and p. This naturally suggests the idea that perhaps the general solution of equation (7.3.3.97) is a one-parameter family of *surfaces*.
181. Fontaine (1764b: 15).
182. Johann I Bernoulli (1739: 1v).
183. This the way that Hans Freudenthal characterized these developments, in a letter to me dated 4 February 1980.
184. The date can be found on the covering letter sent with the paper [Clairaut (1738c)].
185. *L'Académie Royale des Sciences, Registres des Procès-Verbaux,* 4 March 1739, p. 47r.
186. Clairaut (1739).
187. Ibid., p. 433.
188. Ibid., pp. 427–428.
189. Ibid., pp. 428–430.
190. Ibid., p. 434.
191. The expression $\gamma(x, y)$ found by using Clairaut's method such that

$$d\gamma = \frac{dx + \alpha\,dy}{x + \alpha y}, \tag{7.3.4.18}$$

when α is a homogeneous expression of degree zero in x and y such that $x + \alpha y \neq 0$, cannot be a homogeneous expression in x and y. For if γ were a homogeneous expression in x and y, say of degree e, then according to the homogeneous function theorem applied to γ, it would have to be true that

$$e\gamma = \frac{x + \alpha y}{x + \alpha y} = 1,$$

which is impossible. (If $e = 0$, the preceding equation reduces to $0 = 1$, which is false; if $e \neq 0$, γ must be the constant $1/e$ in the preceding equation, which is also false.) In other words, no such homogeneous expression γ in x and y exists. This can also be understood by using the results in note 176 of Chapter 7. If

$$A dx + B dy \equiv \frac{dx + \alpha dy}{x + \alpha y}$$

then A and B are homogeneous expression of degree $e = -1$ in x and y, in which case the converse of the homogeneous function theorem cannot hold. In integrating the equation

$$dx + \alpha dy = 0,$$

when α is a homogeneous expression of degree zero in x and y such that $x + \alpha y \neq 0$, Clairaut found the expression $\gamma(x, y)$ such that

$$d\gamma = \frac{dx + \alpha dy}{x + \alpha y} \tag{7.3.4.18}$$

as follows.
Let

$$\gamma \equiv \frac{l\phi}{e}.$$

From

$$d\gamma = d\left(\frac{l\phi}{e}\right) = \frac{d\phi}{e\phi} = \frac{dx + \alpha dy}{x + \alpha y}, \tag{I}$$

Clairaut deduced that

$$\gamma \equiv \frac{l\phi}{e} = \int \frac{dx}{x + \alpha y} + [y], \tag{II}$$

where

$$\frac{d}{dy}[y] = \frac{\alpha}{x + \alpha y} - \frac{d}{dy} \int \frac{dx}{x + \alpha y}. \tag{III}$$

Clairaut stated without explanation that the right-hand side of (III) is homogeneous of degree -1 in y. This can be understood as follows: If $d\gamma = A dx + B dy$, then

$$\gamma(x, y) = \int_{x_0}^{x} A(x^*, y)\, dx^* + [y], \quad \text{and}$$

$$B(x, y) = \frac{d\gamma}{dy}(x, y) = \frac{d}{dy}\left(\int_{x_0}^{x} A(x^*, y)\, dx^* + [y]\right)$$

$$= \int_{x_0}^{x} \frac{dA}{dy}(x^*, y)\, dx^* + \frac{d}{dy}[y] = \int_{x_0}^{x} \frac{dB}{dx^*}(x^*, y)\, dx^* + \frac{d}{dy}[y]$$

$$= B(x, y) - B(x_0, y) + \frac{d}{dy}[y],$$

which means that

$$\frac{d}{dy}[y] = B(x_0, y).$$

In the case in question,

$$B(x_0, y) = \frac{\alpha(x_0, y)}{x_0 + \alpha(x_0, y)\, y}.$$

If we let $x_0 = 0$, then

$$\frac{d}{dy}[y] = B(0, y) = \frac{1}{y},$$

which explains Clairaut's statement. Then

$$\gamma(x, y) \equiv \int_0^x \frac{1}{x^* + \alpha y} dx^* + \ln y$$

satisfies

$$d\gamma = \frac{dx + \alpha\, dy}{x + \alpha y}.$$

This can be verified as follows:

$$\frac{d\gamma}{dx} = \frac{1}{x + \alpha y}.$$

$$\frac{d\gamma}{dy} = \int_0^x \frac{d}{dy}\left(\frac{1}{x^* + \alpha y}\right) dx^* + \frac{d}{dy} \ln y$$

$$= \int_0^x \frac{d}{dx^*}\left(\frac{\alpha}{x^* + \alpha y}\right) dx^* + \frac{1}{y}$$

$$= \frac{\alpha}{x + \alpha y} - \frac{\alpha(0, y)}{0 + \alpha(0, y) y} + \frac{1}{y} = \frac{\alpha}{x + \alpha y} - \frac{1}{y} + \frac{1}{y} = \frac{\alpha}{x + \alpha y}.$$

Hence the equation

$$\int_0^x \frac{1}{x^* + \alpha y} dx^* + \ln y = a \text{ constant } M \qquad\text{(IV)}$$

is an integral of the equation

$$dx + \alpha\, dy = 0, \qquad\text{(V)}$$

where α is a homogeneous expression of degree zero in x and y [Clairaut (1739: 435)].

192. Clairaut (1740a).
193. Euler (1740a).
194. Euler (1734–1735a and 1734–1735b: republished in Euler (1936: 36–56, 57–75). Concerning Euler's proof of the equality of mixed second-order partial differential coefficients [equality (7.3.4.7)], see Fraser (1989: 319–321).
195. Truesdell (1960: 166, 203).
196. L'Académie Royale des Science, Registres des Procès-Verbaux, 31 May 1741, p. 157; 7 June 1741, p. 186.
197. Clairaut (1740b).
198. In a letter to Mairan, dated 15 June 1741, Fontaine said that he would send Mairan his treatise on integral calculus in a sealed envelope (pli cacheté), but Fontaine asked that the parcel not be opened before Clairaut's mémoire was printed. As I mentioned earlier, this letter from Fontaine to Mairan has been sold (see note 177 of Chapter 7). It is reported somewhat later, in the Académie Royale des Sciences, Registres des Procès-Verbaux, 29 November 1741, pp. 462–463, that Mairan received Fontaine's pli cacheté from D'Onsenbray on 17 June 1741 and that D'Alembert and De Gua were appointed to be referees of the contents of the pli in the event that Fontaine requested that the pli be opened.
199. Clairaut (1740b: 295).
200. For a brief history of attempts to prove equality (7.3.4.7) from the early eighteenth century until 1936, see Higgins (1940). Concerning Euler's proof of the equality of mixed second-order partial differential coefficients [equality (7.3.4.7)], see Fraser (1989: 319–321). For a modern demonstration of Leibniz's inversion of differentiation and integration in which the author uses Fubini's theorem, see Flanders (1973: 616). For a modern proof that Leibniz's inversion of differentiation and integration follows from Fubini's theorem and that equality (7.3.4.7) follows from Leibniz's reversal of differentiation and integration, see Seeley (1961).

201. Clairaut (1740b: 297).
202. Ibid., p. 304.
203. More generally, if any *two* of

$$\frac{dL}{dy} = \frac{dM}{dx}, \frac{dL}{dz} = \frac{dN}{dx}, \frac{dM}{dz} = \frac{dN}{dy}$$

hold for an equation $L(x, y, z)\, dx + M(x, y, z)\, dy + N(x, y, z)\, dz = 0$ for which

$$L\left(\frac{dM}{dz} - \frac{dN}{dy}\right) + M\left(\frac{dL}{dz} - \frac{dN}{dx}\right) + N\left(\frac{dL}{dy} - \frac{dM}{dx}\right) = 0$$

is satisfied, then the third follows immediately (when the corresponding factor L, M, or N is $\neq 0$). Hence the differential $L\, dx + M\, dy + N\, dz$ is exact in this case. I state this general result, because I shall have occasion to call attention to it later.

204. Thus Struik (1933: 101) stated that Clairaut proved the sufficiency of the "conditional equation" (7.3.4.37) for the integrability of equation (7.3.4.28) as well as the necessity. René Taton and Adolf P. Juskevic have maintained, however, that neither Clairaut nor Euler rigorously demonstrated the sufficiency of the "conditional equation" (7.3.4.37) for the integrability of equation (7.3.4.28) (see Euler (1741: 87–88, note 1).

205. Lagrange (1772: 355–358). In this *mémoire* Lagrange wrote the general first-order partial "differential equation" in two variables as

$$Z(x, y, z, p, q) = 0,$$

where $z = z(x, y)$ is a solution of the equation, $p = dz/dx$, and $q = dz/dy$. Lagrange assumed that this equation could be solved explicitly for q:

$$q = f(x, y, z, p).$$

He now made use of the "conditional equation"

$$p\frac{dq}{dz} - q\frac{dp}{dz} + \frac{dq}{dx} - \frac{dp}{dy} = 0$$

for the integrability of the first-order "differential" equation of first degree

$$dz - p(x, y, z)\, dx - q(x, y, z)\, dy = 0$$

in three finite, independent variables x, y and z. [These last two equations are easily seen to be the same as Fontaine's two equations (7.3.3.111) and (7.3.3.97) with the notations changed. We shall see in Chapter 9 that equations (7.3.3.97) and (7.3.4.28) are equivalent and that equations (7.3.3.111) and (7.3.4.37) are equivalent.] Using the equation

$$q = f(x, y, z, p),$$

Lagrange substituted $f(x, y, z, p)$ for q in equation

$$p\frac{dq}{dz} - q\frac{dp}{dz} + \frac{dq}{dx} - \frac{dp}{dy} = 0,$$

which resulted in a linear first-order partial "differential" equation for $p = p(x, y, z)$ by treating x, y, and z as three finite, independent variables. Assuming that this equation has a solution $p = P(x, y, z)$, then

$$q = f(x, y, z, p)$$

becomes

$$q = f(x, y, z, p) = f(x, y, z, P(x, y, z)) = Q(x, y, z),$$

where it is clear that $p = P(x, y, z)$ and $q = Q(x, y, z)$ together satisfy the equation

$$p\frac{dq}{dz} - q\frac{dp}{dz} + \frac{dq}{dx} - \frac{dp}{dy} = 0.$$

Lagrange then showed that this last equation is a sufficient condition for the integrability of equation

$$dz - p(x, y, z)dx - q(x, y, z)dy = 0.$$

In the case in question this meant that

$$dz - P(x, y, z)dx - Q(x, y, z)dy = 0$$

is an integrable equation. Moreover, Lagrange's method of proving that equation

$$p\frac{dq}{dz} - q\frac{dp}{dz} + \frac{dq}{dx} - \frac{dp}{dy} = 0$$

is a sufficient condition for the integrability of equation

$$dz - p(x, y, z)dx - q(x, y, z)dy = 0$$

incidentally furnished a method of constructing a solution of equation

$$dz - p(x, y, z)dx - q(x, y, z)dy = 0$$

when the condition for the integrability of this equation was satisfied. In the case in question this meant that Lagrange could construct an equation

$$N(x, y, z) = k,$$

where k is a constant, that is a solution of the equation

$$dz - P(x, y, z)dx - Q(x, y, z)dy = 0,$$

which means that an expression $N(x, y, z)$ exists such that

$$dN(x, y, z) = I(x, y, z)(dz - P(x, y, z)dx - Q(x, y, z)dy)$$

for some integrating factor $I(x, y, z)$. For if the equation

$$N(x, y, z) = k$$

is differentiated (that is, if the differential of the equation is taken), then the equation

$$0 = dk = dN(x, y, z) = I(x, y, z)(dz - P(x, y, z)dx - Q(x, y, z)dy)$$

results, from which the equation

$$dz - P(x, y, z)dx - Q(x, y, z)dy = 0$$

follows by dividing both sides of the preceding equation by $I(x, y, z)$. Now, the equation

$$N(x, y, z) = k,$$

when k is a constant, implicitly determines a function z of x and y: $z = z(x, y)$. This function satisfies the equation

$$dz = P(x, y, z(x, y))dx + Q(x, y, z(x, y))dy,$$

which means in particular that

$$\frac{dz}{dx} = P(x, y, z(x, y))$$

and that

$$\frac{dz}{dy} = Q(x, y, z(x, y)).$$

But then it follows from working backward that

$$\frac{dz}{dy} = Q(x, y, z(x, y)) = f(x, y, z(x, y), P(x, y, z(x, y))) = f\left(x, y, z(x, y), \frac{dz}{dx}\right).$$

In other words, the equation

$$q = f(x, y, z, p)$$

holds, where $p = dz/dx$ and $q = dz/dy$. In this way Lagrange reduced the problem of solving the general first-order partial "differential equation"

$$q = f(x, y, z, p),$$

where $p = dz/dx$ and $q = dz/dy$, to the problem of solving the linear first-order partial "differential equation" for $p = p(x, y, z)$ gotten by using the conditional equation

$$p\frac{dq}{dz} - q\frac{dp}{dz} + \frac{dq}{dx} - \frac{dp}{dy} = 0$$

for the integrability of the first-order "differential" equation of first degree

$$dz - p(x, y, z)\,dx - q(x, y, z)\,dy = 0$$

in three finite, independent variables x, y and z. But to reduce one problem to the other he had to demonstrate that the conditional equation

$$p\frac{dq}{dz} - q\frac{dp}{dz} + \frac{dq}{dx} - \frac{dp}{dy} = 0$$

was indeed a sufficient condition for the integrability of the first-order "differential" equation of first degree

$$dz - p(x, y, z)\,dx - q(x, y, z)\,dy = 0$$

in three finite, independent variables x, y and z.

206. Clairaut (1740b; 320).
207. In fact these conditions give rise to a special case of the general result stated in note 203 of Chapter 7.
208. Clairaut (1740b: 323).
209. Here Clairaut has proven the second assertion that I made in note 176 of Chapter 7 (the converse of the homogeneous function theorem).
210. Clairaut (1739: 433).
211. Clairaut (1740b: 307–308).
212. Clairaut (1739: 433).
213. Ibid., p. 425.
214. Clairaut (1731b: 14, §30).
215. This ordinary "differential" equation with singular solution appears in Clairaut (1734a: 210, 213).
216. Engelsman (1984: 30).
217. I have already cited two examples of Clairaut's ignorance: his "oversights" in 1734, which Fontaine pointed out (see note 32 of Chapter 5); and his ignorance of prior publication of the reversal of differentiation and integration (see note 161 of Chapter 7). In fact, Clairaut's published work abounds in such examples. In his Paris Academy *mémoire* of 1739 on integral calculus, Clairaut applied his method of integration to what he believed to be an intractable first-order ordinary "differential" equation of first degree

$$adx + bdy + cxdx + exdy + fydx + gydy = 0$$

[Clairaut (1739: 431–433)], which he said "a skilled geometer" had tried to solve "in vain" (p. 426). Clairaut again integrated this equation exactly the same way in his second Paris Academy *mémoire*, of 1740, on integral calculus [(1740b: 302–303)]. But Jakob Hermann had already integrated the equation in an article that he wrote which treated this equation and which appears in Volume 2 of the Saint Petersburg Academy's *Commentarii* (1727), published in 1729. Moreover, he did so using a method that was a good deal easier than Clairaut's. Namely, by making linear changes of variables, Hermann simply reduced the equation to a first-order ordinary "differential" equation of first degree in two new variables where the coefficients of the two first-order differentials in the two new variables are both homogeneous expressions of degree 1 in the two new variables, so that the new variables in the new equation can be separated. The author of a French review of Volume 2 of the *Commentarii*, published in the *Journal de Trévoux* for November 1737, refers specifically to Hermann's method of integrating the equation [see the *Journal de Trévoux*, November 1737, pp. 1976–1995, on pp. 1976–1978; or the reproduction in facsimile: *Journal de Trévoux*, Volume 37 (Geneva: Slatkine Reprints, 1968), pp 528–533, on pp. 528–529]. Clairaut had returned to Paris from Lapland late in the summer of 1737, before this issue of the *Journal de Trévoux* was published. Clairaut had apparently even seen Volume 2 of the *Commentarii* itself, as we shall observe in a moment, yet he seems to have missed Hermann's integration. [Actually, Christian Goldbach had accomplished even earlier essentially the same thing that Hermann did, on page 208 of a paper that appears in Volume 1 of the Saint Petersburg Academy's *Commentarii* (1726), published in 1728. Goldbach, however, dealt mainly with an equation of another kind in his paper.] In fact Hermann very likely plagiarized. In a letter to Montmort in 1718 Johann I Bernoulli cited the evidence that Hermann had "borrowed" from him without acknowledgment the linear changes of variables which converted the original first-order ordinary "differential" equation of first degree

$$adx + bdy + cxdx + exdy + fydx + gydy = 0$$

into one whose variables could be separated. [See Montmort (1718a: 3r) and Johann I Bernoulli (1718b: 5v–6r).] In his paper of 1734 which touched off his dispute with Fontaine, Clairaut mentioned papers on "reciprocal trajectories" by Bernoulli, Henry Pemberton, and Euler, published in *Acta Eruditorum* for 1718, 1719, 1720, and in Volume 2 of the Saint Petersburg Academy's *Commentarii* [Clairaut (1734a: 196–197)]. Now, the allusion to the article in the *Acta* for 1720 can only be a reference to the survey of methods of finding orthogonal trajectories which was published in that volume of the *Acta* under Nikolaus II Bernoulli's name. Furthermore, in the paper in Volume 2 of the *Commentarii* which Clairaut cited, Euler took as his starting point the supplements to Nikolaus II Bernoulli's survey, which were published in 1721 in Volume 7 of the supplements to the *Acta*. This was even made perfectly clear in that French review of Volume 2 of the *Commentarii* published in the *Journal de Trévoux* for November 1737, on pp. 1981–1982 of this issue of the journal (or see the reproduction in facsimile: *Journal de Trévoux*, Volume 37 (Geneva: Slatkine Reprints, 1968), p. 530). Finally, Clairaut had praised Johann I Bernoulli's solution of 1730 of the tautochrones in resistant media, at the time that it was read before the Paris Academy [see Clairaut (1731a)]. We recall that in this paper [Johann I Bernoulli (1730c)] Johann I Bernoulli had called attention specifically to one of the supplements to his son Nikolaus's article in the *Acta* for 1720, appearing in Volume 7 of supplements to the *Acta*. And yet despite all of these references, Clairaut remained ignorant of much of the contents of the article in the *Acta* for 1720, as well as the contents of the supplements to that article published in Volume 7 of the supplements to the *Acta*, until late in 1740, when Euler brought this literature to Clairaut's attention in his first letter to Clairaut. In addition, Clairaut only learned in this letter, which was Euler's reply to Clairaut's first letter to him, that the "homogeneous function theorem" in two variables could be found in Proposition 14 of Volume 2 of Euler's *Mechanica*, even though Euler's treatise had been in print since 1736. (At the same time, Euler informed Clairaut that he discussed the "homogeneous function theorem" in an arbitrary number of variables in his work in Volume 7 of the Saint Petersburg Academy's *Commentarii*, a volume that he told Clairaut had just been published.) Clairaut cited Euler as the independent discoverer of what he had called until then "Monsieur Fontaine's theorem." Clairaut did this in a footnote in his Paris Academy *mémoire* of 1740 on integral calculus (1740b: 322). Moreover, in (1731b), in which Clairaut pioneered the study of curves with double curvature, meaning curves skewed in space, Clairaut mentioned in the preface to the work having "learned that the representation of surfaces by means of equations in three variables appears in an article by the 'famous Monsieur Bernoulli' in *Acta Eruditorum*." After a thorough search, however, Julian Lowell Coolidge found nothing in *Acta Eruditorum* by either Jakob or Johann I Bernoulli which had anything to do with such a representation. Instead, Coolidge found that Johann did allude vaguely to Cartesian geometry in three-dimensional space in his correspondence with Leibniz and with the Swedish mathematician Samuel Klingstierna, respectively, in 1715 and 1728, respectively, but nowhere did Coolidge find Johann talking about such things in *Acta Eruditorum* (see Coolidge (1948: 80);

and 1940, 1963: 135). Clairaut must have based his remarks entirely upon hearsay. Finally, Euler's essay on the causes of the tides, for which Euler was awarded a share of the Paris Academy's prize of 1740, includes a certain linear, second-order ordinary "differential" equation and its solution which are essentially the same as ones that played a fundamental part later in Clairaut's *Théorie de la lune*, published in 1752. D'Alembert reprimanded Clairaut for thinking that his treatment of this equation was new and for acting as if he did not know that it had been solved before (see Wilson [1980: 98]). Since Clairaut had been one of the judges of the Paris Academy's contest of 1740, he had read Euler's paper carefully at that time. Consequently D'Alembert had all the more reason to be angry at Clairaut.

218. For example, see Taton (1972: 54).
219. See note 180 of Chapter 7 for an example.
220. Engelsman (1984: 99).
221. Clairaut assumed here that if equation (7.3.5.4) represents a surface, then equation (7.3.5.4) is an integrable equation, because the "conditional equation" (7.3.5.5) for the integrability of equation (7.3.5.4) must hold if equation (7.3.5.4) expresses a surface. In fact, it was only demonstrated much later, by Lagrange, that the conditional equation (7.3.5.5) is a *sufficient* condition for the integrability of equation (7.3.5.4), not just a *necessary* condition for the integrability of equation (7.3.5.4). (See note 205 of Chapter 7.) In the text above I showed that if $\alpha(x,y,p)$ and $\pi(x,y,p)$ are *homogeneous* expressions of degree zero in x,y, and p in equation (7.3.3.97), then equation (7.3.4.24) follows from equation (7.3.3.111) and equation (7.3.4.25), equation (7.3.4.26) follows from equation (7.3.3.111) and equation (7.3.4.27), and that these two results together with a third one of the same kind actually prove that if $\alpha(x,y,p)$ and $\pi(x,y,p)$ are *homogeneous* expressions of degree zero in x,y, and p in equation (7.3.3.97), then the conditional equation (7.3.3.111) is a sufficient condition for the integrability of equation (7.3.3.97).
222. Clairaut (1740b: 308).
223. Simonov (1968: 136–137).
224. Ince (1944: 76, including note*).
225. See Euler (1741: 87–88, note 1) and Taton (1951: 297–299).
226. Clairaut (1740b: 312).
227. To see that the converse is true, apply note 203 of Chapter 7 to the integrable equation

$$dz - A(x,y) - B(x,y)\,dy = 0.$$

The same result would follow if, instead of 1, the coefficient of dz were any expression $F(z)$ of z alone. The reason is the same as that used to deduce (7.3.4.44) from (7.3.4.43'). See as well note 221 of Chapter 7.
228. See Lacroix's annotation in Montucla (1802, 1968: vol. 3, 344).
229. Clairaut (1739: 433).
230. *Histoire de L'Académie Royale des Sciences,* 1725 (Paris: Imprimerie Royale, 1727), p. 48.
231. Truesdell (1960: 150, note 4).
232. Clairaut (1742c: 108).
233. Clairaut (1734b).
234. Fontaine (1734c).
235. Clairaut (1735).
236. Clairaut (1736: 1–3). Clairaut succeeded in criticizing Buffon in the printed Paris Academy *Mémoires* for 1745 as well [see Waff (1976: 163–164)].
237. Clairaut (1739: 425).
238. Ibid., p. 433.
239. Fontaine (1741). Fontaine sent this letter to Mairan about a week after he had dispatched his *pli cacheté* to the Paris Academy (see notes 177 and 198 of Chapter 7). On 13 June 1741, just two or three days before he had the *pli cacheté* delivered to the Paris Academy, Fontaine sent Mairan a letter "concerning his scientific work and in which he claimed the discovery of a theorem in integral calculus which Clairaut took the credit for." This letter has also been sold. The sale is listed in the "Fichier Charavay," 74, FLO-FOR, No. 25237 (192) [*Salle des Manuscrits, Bibliothèque Nationale* (Paris)]. The passage in quotation marks is my translation of a passage in French included in the announcement of the sale of the letter.
240. *L'Académie Royale des Sciences, Registres des Procès-Verbaux,* 1 July 1741, p. 221.
241. Clairaut (1739: 425, note ∗).

242. *L'Académie Royale des Sciences, Registres des Procès-Verbaux*, 29 November 1741, pp. 462–463.
243. D'Alembert and De Gua (1742); and *L'Académie Royale des Sciences, Registres de Procès-Verbaux*, 17 January 1742, p. 14.
244. Fontaine (1764b: 30); and Fontaine (1770: 30).
245. Condorcet (1771: 114). Condorcet was Fontaine's follower in mathematics, and Fontaine was Condorcet's first supporter at the Paris Academy [see Granger (1956: 5 or 1989: 5); and Sergescu (1951: 234). A careful study of the Paris Academy's records of attendance, which appear in the Academy's *Registres des Procès-Verbaux*, reveals Fontaine's pattern of attendance. Here we must take into account the fact that the annual academic vacation ran from the second week of September until about the middle of November, during which time the Academy did not meet. Fontaine missed Academy sessions from 4 August 1745 until 16 November 1746 and from 8 February 1747 until 7 June 1749. His brief return in between the periods of absence can be explained as follows: from 21 January 1747 until 4 February 1747 he read a paper to the Academy on the theory of equations [published as Fontaine (1747)]. Fontaine then missed Academy sessions from 23 August 1749 until 16 January 1754 and from 22 June 1754 until 20 November 1754. Here Fontaine's return in between the periods of absence can be accounted for as follows: from 9 February 1754 until 20 March 1754 he read a paper on the integral calculus to the Academy which was never published in the Academy's *Mémoires*. Fontaine next missed Academy sessions from 22 February 1755 until 19 March 1755 and from 22 March 1755 until 16 June 1756. Here his brief return in between the periods of absence can be explained by his having had to read, on 19 March 1755, a report to the Academy, which he wrote with his former protegé Etienne de Montigny, on a paper submitted to the Academy in which the author treated an isoperimetrical problem. Fontaine then missed Academy sessions from 17 July 1756 until 18 November 1758. His return to the Academy from mid June to mid July of 1756 can be accounted for as follows: on 15 May 1756, Clairaut and D'Alembert read a report to the Academy on some work submitted to the Academy by the Genevan mathematician Louis Necker, who had visited Paris earlier in the year. In his work Necker dealt with some problems in mechanics. Among other things, Necker found tautochrones in a medium, where he assumed various conditions pertaining to resistance. He also found tautochrones in a vacuum, where he assumed that friction exists subject to various conditions. Necker's work favorably impressed both Clairaut and D'Alembert. But Fontaine continued to take pride in the solutions of tautochrones in resistant media which he had found in 1734 [Fontaine (1734b)]. He no doubt returned to Paris in order to learn specifically what Necker had achieved concerning such problems. On 16 June 1756 and 23 June 1756, Necker was, respectively, nominated and elected corresponding member of the Academy. To make a long story short, from 4 August 1745 until 18 November 1758, a period of more than thirteen years, Fontaine missed the majority of the sessions of the Paris Academy. His returns to the Academy during the period can always be related to matters that concerned him personally.
246. Boucharlat (1816: 182). See Condorcet (1771: 112) as well.
247. Boucharlat (1816: 182). Jean-Louis Boucharlat, the author of Fontaine's biographical sketch cited here, was a mathematician educated at the Ecole Polytechnique and the author of several nineteenth-century mathematics texts. His daughter married the French economist Henri-Napoléon Mathon de Fogères, who was mayor of Bourg-Argental, Fontaine's mother's place of origin and the home of the Mathon family. Henri-Napoléon Mathon de Fogères published the 6th edition of Boucharlat's *Eléments de calcul différentiel et de calcul intégral* (Paris: Bachlier, 1852). His grandfather, Joseph Mathon de Fogères, was the brother of the mathematician Jacques Mathon de La Cour, director of the Académie des Beaux Arts de Lyon and co-runner-up, with Euler, of the Paris Academy's contest of 1753, which Daniel Bernoulli won. Also a native of Bourg-Argental, Jacques Mathon de La Cour was undoubtedly Fontaine's closest intimate and confidant. Boucharlat (1820) also wrote a biographical sketch of Jacques Mathon de La Cour. The genealogy that I give here helps explain why Boucharlat knew things about Fontaine that Condorcet did not. These are revealed in the correspondence between Fontaine and Jacques Mathon de La Cour, which Boucharlat quotes from in his two biographical sketches. In these two sketches Boucharlat quotes as well Jacques Mathon de La Cour concerning Fontaine. As a result of his daughter's marriage to a member of the Mathon family, Boucharlat evidently had access to the Mathon family's papers, which seem to have included the correspondence between Fontaine and Jacques Mathon de La Cour.

8. Interlude II

1. Maupertuis (1743).
2. The essays by MacLaurin, Daniel, Bernoulli, and Euler were first published in the *Recueil des pièces qui ont remporté le prix de l'Académie Royale des Sciences en 1740* (Paris, 1741), pp. 1–350. They were republished, with typographical errors corrected, in the Jesuit edition of the *Principia*: Newton [1739–1742: vol. 3 (1742)]. Hereafter I shall refer to these corrected versions: Daniel Bernoulli (1740), MacLaurin (1740a), and Euler (1740b). For an English summary of Bernoulli's essay plus commentary, see Lubbock (1830: 1–16).
3. Daniel Bernoulli (1733: 11v).
4. Hahn (1971: 80) seems to have been the first historian to recognize that seniority largely served as the basis for promotions in the eighteenth-century Paris Academy of Sciences. Rappaport (1981: 237) found that out of some one hundred forty promotions in seven decades, more than three-quarters did occur in order of seniority. She also found that most of the exceptions to the pattern of seniority took place during the early decades of the eighteenth century, where there is evidence that merit played some part in promotions. She finds that the firm adoption of a system of seniority seems to date from about midcentury. Although Hahn (1971: 80) concluded that promotions in the Paris Academy had a basis in seniority, Rappaport (1981: 227, 237) observes that another historian, Jean Torlais, contended that promotions in the Paris Academy were based on merit. It is obviously the failure of the members of the Academy to write the rules that governed promotions in the Academy into the Academy's statutes which has produced this conflict of opinions and caused historians to remain uncertain about the bases of promotion. But evidence abounds that seniority was the de facto basis for promotion in the Academy.
5. Daniel Bernoulli (1734: 415).
6. René Taton, for instance, remarked during a conversation with me that Clairaut addressed no one in correspondence with such familiarity as he did Daniel Bernoulli.
7. Clairaut (1891–1892: 259–286).
8. Daniel Bernoulli (1739: 455).
9. Todhunter (1873: §209, §233) or (1962: §209, §233).
10. Daniel Bernoulli (1740: 161 and the figure on p. 157).
11. Clairaut (1809: Figure 12); and Daniel Bernoulli (1740: figure on p. 145). Instead of Corollary 2 to Proposition 91 of Book I of the *Principia*, which Clairaut, following Newton, used to calculate the magnitude of the attraction at a pole of a homogeneous figure shaped like an ellipsoid of revolution flattened at its poles which attracts according to the universal inverse-square law and whose ellipticity is infinitesimal (see note 6 of Chapter 6), which is what Newton implied that he had used himself to do such calculations, Bernoulli used Corollary 1 to Proposition 90 of Book I of the *Principia* to calculate the magnitude of the attraction at a pole of a homogeneous figure shaped like an ellipsoid of revolution elongated at its poles which attracts according to the universal inverse-square law and whose ellipticity is infinitesimal [Daniel Bernoulli (1740: 144)].
12. Bernoulli (1740: figure on p. 143).
13. Bernoulli (1740: 146).
14. Issues No. 445 and No. 449 of the *Philosophical Transactions*, which include Clairaut's two papers on the theory of the earth's shape, both have "Printed for T. Woodward... Printers to the Royal Society, 1739" written at the end. In October of 1738, James Stirling told Colin MacLaurin that he "barely saw" Clairaut's first paper (1737) "before it was printed." See Stirling (1738: 305).
15. Bernoulli (1740: 139, 155).
16. Ibid., pp. 139, 148.
17. Aiton (1955: 207).
18. Burstyn (1962: 1020).
19. Clairaut (1731b and 1732b).
20. Clairaut (1733a; 1733b, esp. 409–416). Before entering the Paris Academy of Sciences, Clairaut, along with a number of other future members of the Paris Academy of his generation, belonged to the short-lived "Société des Arts," whose members dedicated themselves to the advancement of the applied sciences. However, there is no tangible evidence that this Society influenced Clairaut's mathematics of the early 1730s. Concerning the "Société des Arts," see Hahn (1963).
21. See note 63 of Chapter 5.
22. Euler (1740a: 74). Euler concluded that this was true from the contrapositive of the undemonstra-

ted assertion appearing in Clairaut's second paper on the theory of the earth's shape (1738) published in the *Philosophical Transactions* that if the density decreases from the center to the surface of a figure shaped like an ellipsoid of revolution flattened at its poles which attracts according to the universal inverse-square law, whose ellipticity is infinitesimal, which revolves around its axis of symmetry, and at whose surface the principle of the plumb line holds, then the figure is less flat than the corresponding homogeneous figure shaped like an ellipsoid of revolution flattened at its poles which attracts according to the universal inverse-square law, whose ellipticity is infinitesimal, which revolves around its axis of symmetry, and whose columns from center to surface all balance or weigh the same, meaning the homogeneous figure shaped like an ellipsoid of revolution flattened at its poles which attracts according to the universal inverse-square law, whose ellipticity is infinitesimal, which revolves around its axis of symmetry, whose columns from center to surface all balance or weigh the same, and whose ratio of the magnitude of the centrifugal force per unit of mass to the magnitude of the effective gravity at its equator is the same as the ratio of the magnitude of the centrifugal force per unit of mass to the magnitude of the effective gravity at the equator of the figure whose density decreases from center to surface. Applied to the earth, Clairaut's assertion means that the earth's ellipticity should be less than 1/230 [see equality (6.2.4') and the text accompanied by notes 41 and 42 of Chapter 6].
23. Daniel Bernoulli (1740: 162, 164).
24. In Euler (1741: 87), Euler told Clairaut that he had not yet seen either MacLaurin's or Daniel Bernoulli's essays on the tides and that consequently he was in no position to discuss them yet. Euler added, however, that Bernoulli had mentioned some of his particular views concerning the problem of the tides in their correspondence. For example, Euler said that Bernoulli had told him that the absolute force of the sun was greater than Newton had believed it to be. Euler told Clairaut that, on the contrary, he had found the value of the force to be almost twice as small as Newton's value. Moreover, Euler told Clairaut that Bernoulli believed the ratio of the forces of the sun and moon which move the seas to differ significantly from Newton's value of the ratio, while he, Euler, determined a value of the ratio which agreed with Newton's. Euler said that he could find no mistakes in his own arguments and that he did not know the reasoning that Bernoulli had used to arrive at his particular values. Thus, even if Euler had seen Bernoulli's essay on the tides by February of 1741, which seems practically impossible, he nevertheless would have had reasons not to accept all of the conclusions that Bernoulli had come to in it. Concerning Daniel Bernoulli's method of estimating the ratio of lunar tides to solar tides (in other words, the ratio of the lunar force to the solar force) and his criticisms of Newton's method for determining such a ratio, see Aiton (1955: 216–220) and Wilson (1980: 88–89). In his "Nouvelles Tables astronomique pour calculer la place du soleil" of 1744, Euler *still* showed no signs of being aware of Bernoulli's criticism of Newton's value for the ratio of the mass of the moon to the mass of the earth appearing in Bernoulli's essay on the tides [see Wilson (1980: 88–89)]. Daniel Bernoulli discussed the earth's shape and the variation of its density in a letter to Euler dated 25 June 1740. Euler did the same in his reply to Bernoulli dated (26 September according to an alternative calendar) 15 September 1740. Both Bernoulli and Euler referred in their letters to Clairaut's second paper on the theory of the earth's shape (1738) published in the *Philosophical Transactions*. Euler's letter to Bernoulli appears in Euler and Daniel Bernoulli (1906–1907: 145–153, where p. 146 concerns Clairaut). Summaries of the two letters appear in Euler (1975: p. 29, R. 137 and R. 138). Bernoulli equated the weights of the axes of his stratified figure shaped like an ellipsoid of revolution elongated at its poles which attracts according to the universal inverse-square law, which is attracted by the sun and moon according to the universal inverse-square law, which revolves around the sun and around the center of gravity of the system consisting of the earth and moon, and whose ellipticity is infinitesimal in order to determine *Bb*. Thus he implicitly assumed that his stratified figure was a figure whose columns from center to surface all balance or weigh the same. In Chapter 9 we shall discover that Bernoulli's approach to his particular stratified figure with its particular internal structure has problems.
25. Taton (1975 and 1978: 496–501).
26. Newton (1966: vol. 1, iii). To talk of *the* French translation of the *Principia* is a bit misleading. The extant copies of Du Châtelet's translation vary. Moreover, half of the pages of the edition published in 1966, the edition that I have quoted, are not facsimiles of pages of any known copy, despite the fact that this edition appears to be reproduced in facsimile! These pages were altered during photoreproduction, although the editors do not call the reader's attention to this fact. The

variations, and in some instances discrepancies, in the different extant copies of Du Châtelet's translation are discussed in Cohen (1968). For an account of the way that Du Châtelet's project came about, see Taton (1969). See as well Taton's (1970) review of the edition published in 1966.

27. Newton (1966: vol. 2, 260–261).
28. This is discussed in Greenberg (1986).
29. Clairaut (1738b: 278).
30. Stirling (1738: 305). Clairaut's letter to Stirling, dated 2 October 1738, is Clairaut (1738d).
31. Clairaut (1743: figure on p. 168, figure on p. 170).
32. MacLaurin (1740a: 256–261). See as well MacLaurin (1801: vol. 2, 112–115, §634; 116–117, §636–§638) and Clairaut (1743: 168–174).
33. Clairaut (1741a).
34. MacLaurin (1740b: 337).
35. Clairaut (1741d: 90).
36. Clairaut (1741b: 348).
37. Ibid., p. 349.
38. Clairaut (1741c: 360, and the figure on p. 365).
39. In fact, Clairaut tacitly made a fourth assumption here – namely, that the equation

$$\frac{P' - P}{P} = \frac{10}{4m} - \alpha, \tag{6.2.4}$$

for densities

$$D(r) = fr^p + gr^q + hr^s + ir^t + \text{etc.} \tag{6.2.1}$$

that vary *continuously* (in fact, *differentiably*) from center to surface as a function of r, which Clairaut derived in his second paper on the theory of the earth's shape (1738) published in the *Philosophical Transactions*, holds as well when the density varies discontinuously as a *step function* of r. That is, equation (6.2.4) also holds when the strata have *finite* thicknesses, which the strata do have in MacLaurin's model described in the text. As I mentioned earlier, all of Clairaut's results for solid, stratified figures shaped like ellipsoids of revolution flattened at their poles which attract according to the universal inverse-square law, whose ellipticities are infinitesimal, which revolve around their axes of symmetry, and at whose surfaces the principle of the plumb line holds, which Clairaut derived in his second paper on the theory of the earth's shape (1738) published in the *Philosophical Transactions, do* continue to hold true for stratified figures shaped like ellipsoids of revolution flattened at their poles which attract according to the universal inverse-square law, whose ellipticities are infinitesimal, which revolve around their axes of symmetry, at whose surfaces the principle of the plumb line holds, and whose densities vary as step functions from their centers to their surfaces (see note 23 of Chapter 6).

40. Clairaut (1741b: 348–349).
41. MacLaurin (1740a: 260–261). See as well MacLaurin (1801: vol. 2, 116, §636).

9. Clairaut's mature theory of the earth's shape (1741–1743)

1. Clairaut (1743).
2. Clairaut's correspondence with Euler is published in Euler (1980: 65–246). I thank René Taton for having made this correspondence available to me while it was still in the press, three years before it appeared in 1980. Clairaut's correspondence with MacLaurin is published in MacLaurin (1982).
3. Clairaut (1742a: 127).
4. Clairaut (1742b: 145, the editors' second annotation).
5. Both Todhunter and Truesdell attributed this generalization of the principle of balanced columns to Huygens (see note 8 of Chapter 4), although Huygens never actually used the generalized principle of balanced columns in practice.
6. Clairaut (1743: 3–5, and the figure on p.4).
7. Ibid., pp. 5–7, and the figure on p.6.
8. Ibid., p.2. For a recent history of the principle of solidification, see Casey (1992a).
9. Ibid., p.5.

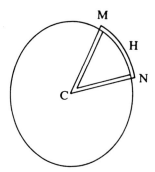

Figure 66. Figure used to show that the principles of balanced columns and the plumb line are equivalent when Clairaut's second new principle of equilibrium holds.

10. Ibid., p.6.
11. Ibid., pp.7–9, and the figure on p.8.
12. Truesdell (1954: XX).
13. If the principle of the plumb line holds at the surface of the homogeneous fluid mass, the effort along channel *MHN* is zero, where *PEpe* is the closed curve at the surface of the homogeneous fluid mass in the plane of channel *MHN* (See Figure 66). If the effort around the closed channel *MCNHM* in the plane of *PEpe* is also zero, then the effort along channel *MCN* is zero. That is, columns *MC* and *NC* weigh the same. Conversely, if columns *MC* and *NC* weigh the same and the effort around the closed channel *MCNHM* is zero, then the effort along channel *MHN* is zero. But *M* and *N* can be chosen to be arbitrarily close to each other. Hence the effective gravity at *M* must be perpendicular to the surface of the homogeneous fluid mass at *M*. (Note that it is assumed once again here that Clairaut's second new principle of equilibrium not only holds for closed plane or skew channels that lie entirely within the homogeneous fluid mass and that are filled with fluid from the homogeneous fluid mass, but that it also holds for closed plane or skew channels filled with fluid from the homogeneous fluid mass parts of which lie *at the surface* of the homogeneous fluid mass.)
14. Clairaut (1743: 28–30, and the figure on p. 29). I say "in general," because sometimes they may be equal. For example, if *ab* is taken to lie along the surface and *cd* is allowed to shrink into the single point *C* at the center, then *ac* and *bd* are two columns from center to surface. But we shall see in a moment that there are forces directed toward fixed, centrally located points, whose magnitudes at points depend upon the angles that the line segments joining the points to the center make with an axis chosen as an axis of reference as well as upon the distances to the points, which *do* permit the principles of balanced columns and the plumb line to hold simultaneously.
15. Ibid., pp. 31–33.
16. Ibid., pp. xxxii–xxxiii.
17. Ibid., pp. 9–12. Truesdell (1954: XX) finds, however, that Clairaut did not even state this result clearly, much less adequately demonstrate it. In Figure 67 *ab* and *αβ* designate two channels filled with homogeneous fluid of the same density. The two channels revolve around an axis *Pp* at the same angular speed. The ends *a* and *α* of the two channels are the same distance from the axis *Pp*, and the ends *b* and *β* of the two channels are also the same distance from the axis *Pp*. Clairaut intended to show that the total efforts produced by centrifugal force along the two channels *ab* and *αβ* filled with homogeneous fluid of the same density are the same. Here I do not expand Clairaut's demonstration of this result, as I have done in many of the demonstrations in the text, but simply reproduce the demonstration as he gave it. Clairaut imagined the channels *ab* and *αβ*, respectively, to be divided into an infinite number of infinitesimal homogeneous fluid right circular cylinders ("*petits cylindres*") *mn* and *uv*, respectively, which are in a one-to-one correspondence with each other. The ends *m* and *u*, respectively, of *mn* and *uv*, respectively, are the same distance from the axis *Pp*, and the ends *n* and *v*, respectively, of *mn* and *uv*, respectively, are also the

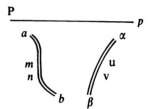

Figure 67. Clairaut's figure used to show that the total efforts produced by centrifugal force along the two plane channels ab and αβ filled with homogeneous fluid of the same density which revolve around the axis Pp at the same angular speed are equal (Clairaut 1743: figure on p. 9).

same distance from the axis Pp. Because mn and uv are infinitesimal, Clairaut assumed that the magnitude of the centrifugal force per unit of mass is constant at all fluid particles that make up mn and that the magnitude of the centrifugal force per unit of mass is constant at all fluid particles that make up uv. Now, since all parts of the two channels ab and $αβ$ revolve around the axis Pp at the same angular speed (or, as Clairaut expressed the matter, they all take the same amount of time to revolve around the axis Pp once, meaning that they all have the same period of revolution), the magnitude of the centrifugal force per unit of mass is directly proportional to the distance from the axis Pp. Hence the magnitude of the centrifugal force per unit of mass at m and the magnitude of the centrifugal force per unit of mass at u are the same, because m and u are the same distance from the axis Pp. Now, the line through m which is perpendicular to the axis Pp indicates the direction of the centrifugal force per unit of mass at m, and the line through u which is perpendicular to the axis Pp indicates the direction of the centrifugal force per unit of mass at u. Clairaut stated that the magnitudes of the components of the centrifugal force per unit of mass at m and at u, respectively, which have the same directions as mn and uv, respectively, are inversely proportional to the lengths of mn and uv "by the theory of inclined planes" [Clairaut (1743: 10)]. By this Clairaut meant the following. Let F stand for the magnitude of the centrifugal force per unit of mass at m and at u; let F_{mn} stand for the magnitude of the component of the centrifugal force per unit of mass at m which is directed toward n; and let F_{uv} stand for the magnitude of the component of the centrifugal force per unit of mass at u which is directed toward v. Then it follows from Figures 68 and 69 that

$$F_{mn} = F\cos\theta$$

and that

$$F_{uv} = F\cos\phi.$$

Hence

$$\frac{F_{mn}}{F_{uv}} = \frac{\cos\theta}{\cos\phi} = \frac{x/mn}{x/uv} = \frac{uv}{mn}. \tag{I}$$

In fact Figures 68 and 69 are not completely specified. (There is no explanation why the two sides of the two right triangles in the two figures which indicate the directions of the centrifugal force per unit of mass at m and at u have the same length x, nor is any proof given that these two sides of the two right triangles in the diagram do indeed have the same length.) Moreover, the diagram cannot be completely specified based only on the information that Clairaut gives, for reasons that I shall explain in a moment. Now, because mn and uv are infinitesimal, Clairaut also assumed that the magnitudes of the components of centrifugal force per unit of mass which are directed toward n are the same at all fluid particles that make up mn and that the magnitudes of the components of centrifugal force per unit of mass which are directed toward v are constant at all fluid particles that make up uv. Now, in the text I show several times in two different ways why the masses of infinitesimal homogeneous fluid right circular cylinders can be thought of as having units of

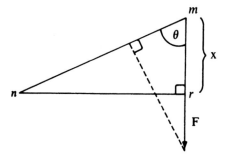

Figure 68. Figure used to illustrate the application of the "theory of inclined planes."

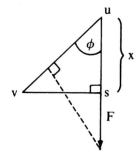

Figure 69. Figure used to illustrate the application of the "theory of inclined planes."

length. But then the effort produced by centrifugal force along *mn* is simply $F_{mn} \times mn$, and the effort produced by centrifugal force along *uv* is simply $F_{uv} \times uv$. But it follows from equation (I) that these two products are the same. That is, the effort produced by centrifugal force along *mn* is the same as the effort produced by centrifugal force along *uv*. Consequently the total effort produced by centrifugal force along channel *ab* is the same as the total effort produced by centrifugal force along channel $\alpha\beta$.

Now, the channels *ab* and $\alpha\beta$ in Figure 67 appear to be plane channels in the plane of the page. But Clairaut evidently meant to state that the same result holds for plane channels *ab* and $\alpha\beta$ which do not both lie in the plane of the page, for plane channels *ab* and $\alpha\beta$ which do lie in the same plane that includes the axis *Pp* but the plane is not the plane of the page, for plane channels *ab* and $\alpha\beta$ which do lie in the same plane but the axis *Pp* does not lie in the plane, and for channels *ab* and $\alpha\beta$ one or both of which are skew channels, where all such channels *ab* and $\alpha\beta$ are filled with homogeneous fluid of the same density, for he claimed to show that it followed from what he had demonstrated that the total effort produced by centrifugal force around any arbitrary closed plane or skew channel *abcd* that is filled with homogeneous fluid and that revolves around an *axis Pp* is equal to zero. Nevertheless, in claiming to show how this statement follows from the preceding result, he made use of a diagram (Figure 70) that appears to represent a closed plane channel *abcd* in the plane of the page, not a closed skew channel, which is filled with homogeneous fluid and which revolves around the axis *Pp*. As I have mentioned several times already, the diagrams that appear in Clairaut's treatise of 1743 often do not help in trying to determine exactly

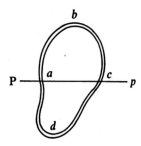

Figure 70. *Figure used to show that the total effort produced by centrifugal force around a closed plane channel filled with homogeneous fluid which revolves around an axis Pp is equal to zero (Clairaut 1743: figure on p. 11).*

what it was that Clairaut was trying to prove. The diagrams frequently only add to the uncertainty and confuse matters even further instead of help clarify them. Thus the demonstrations that appear in Clairaut's treatise of 1743 and the diagrams that accompany them are often ambiguous, which frequently makes them difficult to interpret. If Clairaut stated that the general result in question is true, which it is, it nevertheless seems likely that Clairaut only meant to try to demonstrate this result for the special case where the channels ab and $\alpha\beta$ in Figure 67 appear to be plane channels in the plane of the page and this case alone, for when Clairaut really meant to talk about skew channels filled with homogeneous fluid, he had the means to make this clear in his diagrams. Namely, when he demonstrated results that he specifically meant to apply to skew channels filled with homogeneous fluid, he used diagrams that show skew channels filled with homogeneous fluid, as in the case of the Figure 41.

In other words, he could represent skew channels in three dimensions using diagrams when he wanted to.

And in fact if we assume that $abcd$ designates a closed plane channel in the plane of the page which is filled with homogeneous fluid and which revolves around the axis Pp, then we can complete the incomplete Figures 68 and 69 and give a complete demonstration of the statement that the magnitudes of the components of the centrifugal force per unit of mass at m and at u, respectively, which have the same directions as mn and uv, respectively, in Figure 67 are inversely proportional to the lengths of mn and uv "by the theory of inclined planes" in this particular case. For in this particular case the channels ab and $\alpha\beta$ in the figure can be taken to be plane channels in the plane of the page. Then mn can be thought of as the hypotenuse of an infinitesimal right triangle mrn, where the infinitesimal line segment mr is part of the line through m which is perpendicular to the axis Pp and where the infinitesimal line segment rn is perpendicular to mr. Similarly, uv can be imagined to be the hypotenuse of an infinitesimal right triangle usv, where the infinitesimal line segment us is part of the line through u which is perpendicular to the axis Pp and where the infinitesimal line segment sv is perpendicular to us. Then mr is part of the line that indicates the direction of the centrifugal force per unit of mass at m, and us is part of the line that specifies the direction of the centrifugal force per unit of mass at u. In fact r and s are the same distance from the axis Pp; hence mr and us have the same length. Moreover, rn and sv are parallel to Pp. Once again the magnitude of the centrifugal force per unit of mass at m and the magnitude of the centrifugal force per unit of mass at u will be the same. Then let F stand for the magnitude of the centrifugal force per unit of mass at m and at u; let F_{mn} stand for the magnitude of the component of the centrifugal force per unit of mass at m which is directed toward n; and let F_{uv} stand for the magnitude of the component of the centrifugal force per unit of mass at u which is directed toward v. Now in this particular case, as I just mentioned, mr and us do have the same length, which I designate by x. Then it follows again from Figures 68 and 69 that $F_{mn} = F\cos\theta$ and that $F_{uv} = F\cos\phi$. Hence equation (I) results in the same way that we saw earlier. Clairaut, however, did not give enough information to ensure that the two right triangles in Figures 68 and 69 have sides mr and us that have the same length x, where mr and us, respectively, indicate the directions of the centrifugal force per unit of mass at m and at u, respectively.

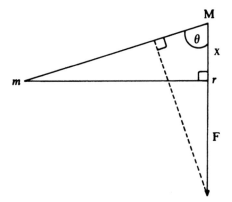

Figure 71. Figure used to illustrate the application of the "theory of inclined planes."

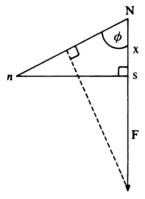

Figure 72. Figure used to illustrate the application of the "theory of inclined planes."

18. Clairaut (1743: 10, 12–16, and the figure on p. 12). As I remarked in the text (see Figure 41), the angle between Mr and the line in the plane of the meridian at the surface of the figure of revolution $AFGB$ which passes through M and which indicates the direction of the attraction at M is the same as the angle between Ns and the line in the plane of the meridian at the surface of the figure of revolution $AFGB$ which passes through N and which indicates the direction of the attraction at N. Moreover, the magnitudes of the attraction at M and at N are equal. Furthermore, as I mentioned in the text, Mr and Ns are parallel to each other and have the same length x and the magnitude of the component of the attraction at M which is directed toward r and the magnitude of the component of the attraction at N which is directed toward s are equal. Then let F stand for the magnitude of the component of the attraction at M which is directed toward r and for the magnitude of the component of the attraction at N which is directed toward s; let F_{Mm} stand for the magnitude of the component of F at M which is directed toward m; let F_{Nn} stand for the magnitude of the component of F at N which is directed toward n; and let x stand for the length of Mr and Ns. Then it follows for essentially the same reasons given in the demonstration in Note 17 of Chapter 9, that is, by applying what Clairaut called "the theory of inclined planes" to the right triangles Mrm and Nsn in Figures 71 and 72, that $F_{Mm} = F \cos \theta$ and that $F_{Nn} = F \cos \phi$.

Hence

$$\frac{F_{Mm}}{F_{Nn}} = \frac{\cos\theta}{\cos\phi} = \frac{x/Mm}{x/Nn} = \frac{Nn}{Mm}. \tag{I}$$

Now, let G stand for the magnitude of the attraction at M and at N; let G_{Mm} stand for the magnitude of the component of the attraction at M which is directed toward m, and let G_{Nn} stand for the magnitude of the component of attraction at N which is directed toward n. Then it seems pretty obvious from the definitions of G_{Mm}, G_{Nn}, F_{Mm}, and F_{Nn} that

$$G_{Mm} = F_{Mm} \quad \text{and} \quad G_{Nn} = F_{Nn}. \tag{II}$$

But it follows from (II) that

$$\frac{G_{Mm}}{G_{Nn}} = \frac{F_{Mm}}{F_{Nn}}. \tag{III}$$

Thus it follows from (I) and (III) that

$$\frac{G_{Mm}}{G_{Nn}} = \frac{Nn}{Mm}. \tag{IV}$$

19. Clairaut (1743: 40–41, and the figure on p. 41).
20. Ibid., pp. 40–43, and the figure on p. 41.
21. Ibid., p. 43.
22. For discussions of the meaning of the term "particles" in the eighteenth century, including evidence that the term did not signify discrete entities at that time, see Truesdell (1954: XLIII, LXXXI; 1984: 444, 512–514).
23. Clairaut (1743: 50–51).
24. Ibid., pp. 35–36, and the figure on p. 36. Truesdell (1954: XXI) observes that P and Q in the differential $Pdy + Qdx$ have force per unit of mass or force per unit of volume as units of measure, since the density was assumed to be uniform. Consequently in the language of the time, $Pdy + Qdx$, where $dy^2 + dx^2 = ds^2$, and

$$\int_\gamma Pdy + Qdx$$

have the same units of measure as "effort" or "weight," because the one-dimensional quantity ds has units of mass. Truesdell (1954: XXI) also says that what Clairaut here calls the "effort" we would now call the "elementary work." Indeed, the work done by the force per unit of mass above in moving a particle whose mass is 1 along a path γ cannot be distinguished in terms of units from the effort produced by the force per unit of mass along γ (where units are concerned), because ds can be thought of as having either units of mass or length, since the density is assumed to be uniform.
25. Huygens (1690: 152, 156) stated that Newton's principle of balanced columns could be generalized to cover *all* channels that join pairs of points at the surface (see note 8 of Chapter 4 of this book). Huygens, however, did not actually *apply* such a generalization himself, nor is it certain that he even could have. Huygens used ratios and proportions in his calculations, but these only work for *straight* channels. Huygens did not have the mathematics needed to calculate "efforts" along *curved plane* or *skew* channels.
26. Clairaut (1743: 37).
27. This result is analogous to the fundamental theorem of the calculus which enables one to evaluate a definite integral by use of an indefinite integral (that is, by use of an antidifferential). Of course, the fundamental theorem of the calculus,

$$d\int_{x_0}^{x} f(x)\,dx = f(x)\,dx,$$

had not yet been rigorously proven by this time. Indeed, the definite integral of a function of one variable had not yet even been rigorously defined nor its existence rigorously established. It goes without saying that the same is true of line integrals of differentials $P(y, x)\,dy + Q(y, x)\,dx$. For the standard way of showing that a differential $P(y, x)\,dy + Q(y, x)\,dx$ whose line integral is path independent is necessarily a complete differential, where y and x are rectangular coordinates; see Ahlfors (1966: 106–107). Widder (1961: 227–230) shows how Green's theorem is now used to establish that if $P(y, x)$ and $Q(y, x)$ are defined in a simply connected domain, where y and x are rectangular coordinates of the points of the simply connected, then the equality

$$\frac{dP}{dx} = \frac{dQ}{dy} \qquad (7.3.4.6''')$$

holds at all points of the simply connected domain if and only if the line integrals

$$\int_\gamma Pdy + Qdx$$

of the complete differential $Pdy + Qdx$ around all closed plane curves γ that lie within the simply connected domain are equal to zero. What is certain is that Clairaut did not establish that if

$$\frac{dP}{dx} = \frac{dQ}{dy}, \qquad (I)$$

then in certain circumstances the path-independent integral

$$\phi(x, y) \equiv \int_{(x_0, y_0)}^{(x, y)} Pdy + Qdx \qquad (II)$$

exists and satisfies

$$d\phi = Pdy + Qdx. \qquad (III)$$

The trouble is that the line integral on the right-hand side of (II) may *not* be path independent even though (I) holds. In other words, when (I) holds, the differential $Pdy + Qdx$ is complete, meaning that an expression $\phi(x, y)$ exists whose total differential is the differential $Pdy + Qdx$. However, the line integral on the right-hand side of (II) will be path independent only if the expression $\phi(x, y)$ is a *single-valued* function of x and y. In his Paris Academy *mémoires* of 1739 and 1740 (1740b) on integral calculus, Clairaut constructed $\phi(x, y)$ that satisfies (III) when (I) holds using algebraic methods that did not involve line integrals.

28. Concerning this trend, see Bos (1974–75) and Engelsman (1984).
29. Clairaut (1743: 39, and the figure on p. 38).
30. Ibid., p. 45, and the figure on p. 44.
31. See note 4 of Chapter 5.
32. Clairaut (1743: 45).
33. Ibid., pp. 46–49. In fact, Clairaut only really proved that $\omega(y, x)$ is constant along a meridian of *any particular* level surface–that is, constant along solutions of the equation $Rdy + Qdx = 0$. It does not follow from his argument that $\omega(y, x)$ is constant *from* meridian *to* meridian of these level surfaces. Of course, $\omega(y, x)$ everywhere constant is obviously sufficient for $\omega(y, x)$ to have the same constant value along a meridian of any particular level surface. But in order for the converse to hold, Clairaut's hypotheses would have to be strengthened somewhat.
34. Ibid., pp. 39, 49–50.
35. Ibid., pp. 79–82, and the figure on p. 79.
36. Ibid., pp. 83–90, and the figure on p. 89.
37. Clairaut's physical argument does not uniquely specify what I have called the "residue." Letting $|LC| \to 0$ in the line integral of $Pdy + Qdx$ along another circular arc instead, say LT in Figure 73, would lead to a different "residue," which when added to the line integral of $Pdy + Qdx$ along the straight channel MC, would not result in a sum equal to the desired integral of $Pdy + Qydx$.
    Clairaut discovered, without realizing it, that the line integral of the differential $Pdy + Qydx$ may not be the integral (meaning the antidifferential) of $Pdy + Qydx$, even though $Pdy + Qydx$ satisfies

$$\frac{dP}{dx} = \frac{d(Qy)}{dy}.$$

This happens when the integral of $Pdy + Qydx$ is not a single-valued function. This occurs in

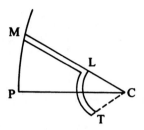

Figure 73. Variation of Figure 47.

differentials and integrals expressed in polar coordinates, because of the fact that the radian angle x is not a single-valued function of position.

Clairaut gave four examples in polar coordinates. In each case, the line integral of $Pdy + Qydx$ around a circle of radius R centered at the origin is not equal to zero:

a.
$$P(y,x) = Y(y), Q(y,x) = \frac{X(x)}{y}, Pdy + Qydx = Y(y)dy + X(x)dx,$$

and

$$\oint Pdy + Qydx = \int_0^{2\pi} X(x)dx \neq 0$$

(unless $X(x) \equiv 0$; $X(x)$ is an even or an odd function of x; etc).

b.
$$P(y,x) = \frac{y}{\sqrt{x^2+y^2}}, Q(y,x) = \frac{x}{y\sqrt{x^2+y^2}}, Pdy + Qydx$$

$$= \frac{y}{\sqrt{x^2+y^2}}dy + \frac{x}{\sqrt{x^2+y^2}}dx,$$

and

$$\oint Pdy + Qydx = \int_0^{2\pi} \frac{xdx}{\sqrt{x^2+R^2}} \neq 0.$$

c. $$P(y,x) = x^2y, Q(y,x) = xy, Pdy + Qydx = x^2ydy + xy^2dx,$$

and

$$\oint Pdy + Qydx = \int_0^{2\pi} xR^2dx \neq 0.$$

d.
$$P(y,x) = \frac{x}{\sqrt{a^2+xy}}, Q(y,x) = \frac{1}{\sqrt{a^2+xy}}, Pdy + Qydx = \frac{x}{\sqrt{a^2+xy}}dy$$

$$+ \frac{y}{\sqrt{a^2+xy}}dx,$$

and

$$\oint Pdy + Qydx = \int_0^{2\pi} \frac{R}{\sqrt{a^2+Rx}}dx \neq 0.$$

Thus in each case, the line integral of $Pdy + Qydx$ around closed plane paths is not always equal to zero, even though $Pdy + Qydx$ is a complete differential [that is, the equality

$$\frac{dP}{dx} = \frac{d(Qy)}{dy}$$

holds].

Clairaut, in effect, stumbled upon this, in discovering that (1) in conjunction with (3) and his mathematical representation of (2) did not lead to the same equation of shape as did (5) together with his mathematical representation of (4) in examples (a) and (b). Clairaut, however, could not accept this finding, because he thought that the completeness of the differential $Pdy + Qydx$ was equivalent to the values of the line integral of $Pdy + Qydx$ equaling zero around all closed plane paths, which is not true. (The former is necessary but not sufficient for the latter.)

Clairaut found that the "paradox" did not arise in examples (c) and (d) and that in these examples he did not need to calculate what I have called the "residue." In fact, examples (c) and (d) illustrate the same difficulties as examples (a) and (b) do. Hence we must examine Clairaut's analysis more carefully. It conceals other problems. In each of the four examples, Clairaut assumed that $P(y, x)$ and the integral (that is, the antidifferential) $\phi(y, x)$ of $Pdy + Qydx$ are well defined at the origin–in particular, defined at the origin as the limits of $P(y, x)$ and $\phi(y, x)$ in letting $|AC| \rightarrow 0$. That is, in each example, Clairaut applied the principle of the plumb line by finding an expression

$$\phi(y, x) \qquad\qquad\qquad (\mathrm{I})$$

that satisfies $d\phi = Pdy + Qydx$, where $\phi(0, 0) = 0$ determines the constant of integration, so that $\phi(0, 0)$ is tacitly assumed to be well defined. Then, in applying the principle of balanced columns, Clairaut in effect calculated the total effort produced by the attraction along the straight channel MC whose radian angle is x as follows

$$\int_0^y P(y, x)\, dy = \int_{(0,0)}^{(y,x)} Pdy + Qydx = \int_0^y d\phi(y, x) = \phi(y, x)\bigg|_0^y = \phi(y, x) - \phi(0, x),$$

where the value of x remains constant. I shall denote $\phi(y, x) - \phi(0, x)$ by (II). In examples (c) and (d), (I) and (II) came out to be the same; in examples (a) and (b) they did not.

Now, neither of the two results (I) and (II) makes any sense if $P(y, x)$ and $\phi(y, x)$ do not approach unique limits, as $(y, x) \rightarrow$ the origin. In example (a), $P(y, x) \equiv Y(y)$, which is independent of x. Hence $P(y, x) \rightarrow$ the origin, no matter how $(y, x) \rightarrow$ the origin. But $\phi(y, x) = \int X(x)dx + \int Y(y)dy$ does not approach a unique value, as $(y, x) \rightarrow$ the origin. For example, if $(y, x) \rightarrow$ the origin along the straight channel MC whose radian angle is x, then $\phi(y, x) \rightarrow (0, x) = \int X(x)dx$, whose value depends upon x. Thus $\phi(0, 0)$ does not exist, and (I) and (II) do not come out to be the same. In example (b)

$$P(y, x) = \frac{y}{\sqrt{x^2 + y^2}} \rightarrow 0,$$

as $y \rightarrow 0$ along any straight channel MC whose radian angle is $x \neq 0$; but along straight channel MC whose radian angle x is 0, $P(y, x) = P(y, 0) = 1 \not\rightarrow 0$, as $y \rightarrow 0$. At the same time, $\phi(y, x) = \sqrt{x^2 + y^2}$; but as $(y, x) \rightarrow$ the origin along the straight channel MC whose radian angle is x, $\phi(y, x) \rightarrow \phi(0, x) = x$, whose value depends upon x. Thus $P(0, 0)$ and $\phi(0, 0)$ are not well defined, and, again, (I) and (II) do not come out to be the same. In example (c), $P(y, x) = x^2 y$, and $P(y, x) \rightarrow 0$, no matter how $(y, x) \rightarrow$ the origin. The same is true of $\phi(y, x) = \frac{1}{2}x^2 y^2$; it too approaches 0, no matter how $(y, x) \rightarrow$ the origin. Hence in this example, $\phi(0, 0)$ is well defined (it equals 0), and $\phi(0, x) = 0$, so that (I) and (II) happen to come out to be the same. Finally, in example (d), $P(y, x) = x/\sqrt{a^2 + xy}$. Along the straight channel MC whose radian angle is $x = 0$, $P(y, x) = P(y, 0) = 0$. Thus $P(y, x) \rightarrow 0$, as $(y, x) \rightarrow$ the origin along this straight channel. But along a straight channel MC whose radian angle is $x \neq 0$, $P(y, x) \rightarrow x/a \neq 0$, as $(y, x) \rightarrow$ the origin along this straight channel. Hence $P(0, 0)$ is not well defined. Nevertheless, $\phi(y, x) = 2\sqrt{a^2 + xy} - 2a \rightarrow 0$, no matter how $(y, x) \rightarrow$ the origin. Thus $\phi(0, 0)$ is well defined (it equals 0),

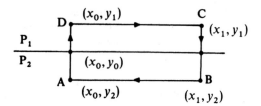

Figure 74. *Figure used to demonstrate Clairaut's fundamental theorem for stratified figures of equilibrium.*

and $\phi(0, x) = 0$, so that and (I) and (II) again happen to come out to be the same. I have gone through the four examples in detail in order to make clear where Clairaut's analysis falls short.

Later D'Alembert discovered that in order that the line integral of the differential $A(y, x)dy + B(y, x)dx$ always equal zero around closed plane paths when $dA/dx = dB/dy$, the attraction (A, B) must have what we now call a single-valued potential [see Truesdell (1954: CXV–CXVI)], which the attractions (P, Qy) do not have.

38. Clairaut (1742c: 103).
39. Clairaut (1743: 94–101, including the figure on p. 98).
40. Clairaut (1731b: 15); and Coolidge (1940, 1963: 135–136).
41. Clairaut (1740b: 308–310, and the figure on p. 309).
42. Ibid., pp. 310–311.
43. Ibid., pp. 311–312.
44. Clairaut (1743: 101).
45. Ibid., pp. 105–128. Clairaut's theory of capillary rise, the first quantitative theory of the phenomenon, was unsuccessful. It failed to account for the rule, then well established, that capillary rise in a tube varies inversely as the tube's diameter. Concerning Clairaut's theory of capillary attraction, see Waff (1976: 75–76); and Bikerman (1978: 103–106).
46. Clairaut (1743: 128–131, and the figure on p. 129).
47. Demonstration of Clairaut's fundamental theorem for stratified figures of equilibrium: Suppose that P_1, P_2 are the densities of the fluid in the two strata to either side of an interface, where the two densities are different. In conformance with Clairaut's configuration of individually homogeneous, finitely thin strata, I shall consequently be assuming that density is discontinuous across the interface. Let $A(x, y)$ denote the magnitude of the component of effective gravity $\vec{F}$ at a point (x, y) which is parallel to the interface. Suppose that $A(x, y)$ is uniformly continuous in (x, y). Let (x_0, y_0) be a point on the interface. $0 = \oint P\vec{F} \cdot d\,\vec{s}$ expresses total zero effort around the closed channel DCBA, or

$$0 = \oint P\vec{F} \cdot d\,\vec{s} = \int_{x_0}^{x_1} P_1 A(x, y_1)dx - \int_{x_0}^{x_1} P_2 A(x, y_2)dx + \int_{AD-BC} P\vec{F} \cdot d\,\vec{s}$$

in other words. This can be rewritten as

$$0 = (P_1 - P_2)\left(\int_{x_0}^{x_1} A(x, y_0)dx + \left[\int_{x_0}^{x_1} A(x, y_1)dx - \int_{x_0}^{x_1} A(x, y_0)dx\right]\right)$$

$$+ P_2\left(\left[\int_{x_0}^{x_1} A(x, y_1)dx - \int_{x_0}^{x_1} A(x, y_0)dx\right] - \left[\int_{x_0}^{x_1} A(x, y_2)dx - \int_{x_0}^{x_1} A(x, y_0)dx\right]\right)$$

$$+ \int_{AD-BC} P\vec{F} \cdot d\,\vec{s}.$$

Now, suppose we let $y_1 \to y_0, y_2 \to y_0$ in this argument. Then

$$\int_{AD} P\vec{F} \cdot d\,\vec{s} \to 0, \quad \int_{BC} P\vec{F} \cdot d\,\vec{s} \to 0, \quad \text{and} \quad \int_{AD-BC} P\vec{F} \cdot d\,\vec{s} \to 0.$$

Meanwhile,

$$\int_{x_0}^{x_1} A(x, y_1)dx - \int_{x_0}^{x_1} A(x, y_0)dx \to 0, \quad \int_{x_0}^{x_1} A(x, y_2)dx - \int_{x_0}^{x_1} A(x, y_0)dx \to 0,$$

by the uniform continuity of $A(x, y)$. So, by letting $y_1 \to y_0$, $y_2 \to y_0$, we obtain

$$0 = (P_1 - P_2)\int_{x_0}^{x_1} A(x, y_0)dx.$$

Moreover, since the preceding argument is independent of the way x_1 is chosen, this is true for all $x_1 > x_0$. Applying the mean value theorem for integrals,

$$0 = (P_1 - P_2)\int_{x_0}^{x_1} A(x, y_0)dx = (P_1 - P_2)A(\hat{x}, y_0)(x_1 - x_0),$$

for some choice of $\hat{x}, x_0 \leqslant \hat{x} \leqslant x_1$, because of the continuity of $A(x, y)$ in x. If we then divide through by $x_1 - x_0$, and let $x_1 - x_0 \to 0$, we obtain $0 =$
$(P_1 - P_2)A(\hat{x}, y_0) \to (P_1 - P_2)A(x_0, y_0)$. That is, $0 = (P_1 - P_2)A(x_0, y_0)$, again by the continuity of $A(x, y)$ in x. Finally, since $P_1 - P_2 \neq 0$, this means $A(x_0, y_0)$ must $= 0$, or that there is no component of effective gravity tangent to the interface. Alternatively, if $A(x_0, y_0)$ were $\neq 0$ – say, $A(x_0 y_0) > 0$ – then $A(x, y_0)$ would also be > 0 for all x in some sufficiently small interval $[x_0, x_1]$, by continuity of $A(x, y)$ in x. Hence $\int_{x_0}^{x_1} A(x, y_0)dx$ would be > 0 for some sufficiently small interval $[x_0, x_1]$, so that $(P_1 - P_2)\int_{x_0}^{x_1} A(x, y_0)dx$ would $\neq 0$ for this same interval as well, contradicting

$$(P_1 - P_2)\int_{x_0}^{x_1} A(x, y_0) = 0$$

for all $x_1 > x_0$, hence for all sufficiently small intervals $[x_0, x_1]$ as well. [Actually, it is enough to assume $A(x, y)$ to be continuous in (x, y) in the preceding argument, for it is enough to consider sufficiently small rectangles $DCBA$, all of which are contained in a closed, bounded neighborhood of (x_0, y_0) on which $A(x, y)$ is continuous, hence uniformly continuous, since a continuous function on a closed, bounded set in the plane is uniformly continuous on that set.]

48. Clairaut (1743: 131–132, and the figure on p. 129).
49. Ibid., pp. 132–134, and the figure on p. 133.
50. Ibid., pp. 134–135.
51. Ibid., pp. 135–137. Had Clairaut realized this in 1738, he might not have come to some of his erroneous conclusions that appear at the end of his second paper on the theory of the earth's shape published in the *Philosophical Transactions*. But there is no real evidence that Clairaut even thought about level surfaces of effective gravity at that time.
52. Ibid., pp. 16–18. and the figures on p. 17 and p. 12. When the attraction is a central force whose magnitude at a point depends solely upon the distance of the point from a fixed center of force C, then if the infinitesimal homogeneous fluid right circular Mm and Nn lie in the same plane, if this plane also includes the fixed center of force C, and if M and N are the same distance from the fixed center of force C, in which case the magnitude of the attraction at M and the magnitude of the attraction at N are the same, which are all conditions that are fulfilled in the problem discussed in the text, it again follows that the magnitude of the component of the attraction at M which is directed toward m and the magnitude of the component of the attraction at N which is directed toward n are inversely proportional to the lengths of the infinitesimal homogeneous fluid right circular cylinders Mm and Nn by applying what Clairaut called "the theory of inclined planes" to the right triangles Mrm and Nsn in Figures 71 and 72. This means the following. Let F stand for the magnitude of the attraction at M and at N; let F_{Mm} stand for the magnitude of the component of the attraction at M which is directed toward m; and let F_{Nn} stand for the magnitude of the component of the attraction at N which is directed toward n. As I explained in the text, the two sides Mr and Ns of the two right triangles Mrm and Nsn in Figures 71 and 72 are both directed toward the fixed center of force C, and, moreover, Mr and Ns have the same length, which I have designated by x in Figures 71 and 72. Then it follows for essentially the same reasons given in the demonstration in note 17 of Chapter 9, that is, by applying what Clairaut

called "the theory of inclined planes" to the right triangles Mrm and Nsn in Figures 71 and 72, that $F_{Mm} = F\cos\theta$ and that $F_{Nn} = F\cos\phi$. Hence

$$\frac{F_{Mm}}{F_{Nn}} = \frac{\cos\theta}{\cos\phi} = \frac{x/Mm}{X/Nn} = \frac{Nn}{Mm}. \tag{I}$$

It is surprising that Clairaut made the error I mentioned in the text, which made his proof of the result in question there circular, because in fact we saw in note 17 of Chapter 9 that Clairaut correctly proved a very similar result, in which "attraction" is replaced by "centrifugal force per unit of mass," by applying in this instance what he called "the theory of inclined planes" to arrive at the result in question, from which it follows, as he also correctly concluded, that the total effort produced by centrifugal force around a closed plane channel filled with homogeneous fluid which lies in a plane that revolves around an axis that lies in the plane is equal to zero.

53. Clairaut (1743: 19–22, including the figure on p. 19); and Maupertuis (1732b: figure on p. 63).
54. Clairaut (1743: 22–23).
55. Ibid., pp. 56–60.
56. Ibid., pp. 60–61.
57. Ibid., pp. 52–55.
58. Ibid., pp. 64–65, and the figure on p. 64. Concerning the meaning of a "particle" of fluid at this time, see note 22 of Chapter 9.
59. Clairaut (1743: 68–69).
60. Ibid., p. 67.
61. Ibid., pp. 65–68.
62. Ibid., pp. 68–69. Concerning the "coefficient lemma for total differentials," see Engelsman (1984: 148–149). For Clairaut's application of the lemma to show that when the magnitudes of the attraction at points in the plane of $PEpe$ are assumed to depend solely upon the distances of the points from the directrix $KLkl$ in that plane, then the relevant differential in two rectangular coordinates turns out to be a complete differential, whatever the shape of the directrix $KLkl$, see Clairaut (1743: 68–69).
63. Clairaut (1743: 69–71, including the figure on p. 69).
64. Ibid., pp. 71–73.
65. Ibid., pp. 74–77.
66. Concerning the meaning of a "particle" of fluid at this time, see note 22 of Chapter 9.
67. Clairaut (1743: 137).
68. Clairaut (1742c: 103).
69. Euler (1742a: 111).
70. Ibid., p. 111.
71. *Journal de Trévoux*, July 1742, pp. 1286–1307, on pp. 1288, 1295; or reproduced in facsimile: *Journal de Trévoux*, 42 (Geneva: Slatkine Reprints, 1968), pp. 327–333, on pp. 328, 330.
72. Clairaut (1742d: 118–119).
73. Euler (1742b: 125).
74. Ibid., p. 125.
75. Truesdell (1960: 176; 1984: 323).
76. Truesdell (1954: XVIII; 1984: 329–330, 365).
77. Truesdell (1954: XVIII). Truesdell notes that Euler realized that this way of handling the problem was defective and that Euler attempted to introduce corrections a posteriori.
78. Aiton (1955: 207, 221–233) says that Euler recognized that the horizontal and not, as Newton had supposed, the vertical components of the disturbing forces are the components that are effective in the generation of the tides. And yet for some reason, in attempting to formulate a dynamical theory of the tides, Euler entirely neglected the horizontal motion of the water, which is the main phenomenon, and confined his attention instead to the relatively insignificant oscillations caused by the vertical components of the disturbing forces. (See as well note 77 of Chapter 9.)
79. Clairaut (1742a: 127–128).
80. Clairaut appears to state the transmissibility of pressure ("Pascal's principle") here. However, Clairaut did not define pressure in his treatise of 1743, and it is not clear that he had a notion of it which was independent of and distinct from weight. Pressure had no part in his theory of figures of equilibrium. That is to say, he applied no such idea in practice. Clairaut used the term

"pression" in developing his theory of figures of equilibrium to mean something that the effective gravity ("la pesanteur") ultimately caused. In other words he used the term pression to mean what he usually called the "effort" or weight ("poids") [for example, see Clairaut (1743: 42, 49, 51, 100, 134)]. [Whether Pascal himself had a true concept of pressure and its properties is unclear. He may have had an idea that was qualitative, but it is doubtful that he had one that was precise or exact; see Truesdell (1954: XII, note 2).]

81. Clairaut seems to make the matter of whether fluid particles are pressed against with equal force in all directions or not depend upon the kind of forces of attraction which act. In this way he appears to make the existence of pressure contingent upon the forces of attract which act, when in fact pressure and its properties are valid for all fluids and for all laws of force. This is another reason for thinking that Clairaut did not have a true notion of pressure which was independent of the forces of attraction which act (see note 80 of Chapter 9).

82. Clairaut (1742a: 128).

83. Forty years ago, Truesdell (1954: XIX–XX, footnote 4) called his reader's attention to this remarkable dialogue between Clairaut and Euler in which the two mathematicians were at cross-purposes. The notion of "unstable equilibrium" has been traced back to Roger Joseph Boscovich's *De litteraria expeditione per Pontificiam ditionem ad dimetiendos duos meridiani gradus et corrigendam mappam*, published in Rome toward the end of 1755, in which Boscovich apparently introduced the idea of "stability" in connection with figures of equilibrium for the first time. In the first chapter of Book 5 of his *Voyage astronomique et géographique*, the enlarged French edition of the Latin work of 1755, published in 1770, Boscovich analyzed solid nucleii surrounded by thin, fluid layers whose densities differed from the densities of the nucleii. He stated that the equilibrium of the fluid layers would be unstable, in the event that they were shaped like elongated ellipsoids. He meant by this that a slight change in the shape of the layer would cause the shape of the layer to deviate further and further from the initial shape. Boscovich doubted that his adversary D'Alembert recognized the difference between "stable" and "unstable" equilibrium, in D'Alembert (1747), in which D'Alembert also investigated solid nucleii surrounded by thin, fluid layers. According to Boscovich, D'Alembert showed his ignorance in declaring in volume 1 of his *Opuscules mathématiques*, published in 1761, that no one until then had ever considered this kind of question in research on the earth's shape. Boscovich viewed the idea applied to figures of equilibrium as having its foundation in his theory of matter — his famous lattices of attracting and repelling "point"-particles. Boscovich first put forward this theory of matter in 1745, and he included it in his *Philosophiae naturalis Theoria*, published in 1758. In this work Boscovich showed that his law of forces allowed particles to exist in configurations that were either stable or unstable, depending on the positions of the particles. [See Markovic (1960: 32).] D'Alembert disputed the statements made by the translator of Boscovich's Latin work of 1755 into French, in which the translator attributed the notion of unstable equilibrium to Boscovich. In volume 6 of his *Opuscules mathématiques*, published in 1773, D'Alembert ascribed the idea of stability of equilibrium to Daniel Bernoulli instead [see Todhunter (1873: §470, §567, and §590); and Pappas (1987a: 132–134)]. In fact, Laplace was the first to demonstrate convincingly using mathematics that a homogeneous fluid figure shaped like an elongated ellipsoid of revolution which attracts in accordance with the universal inverse-square law can never be a figure of equilibrium, much less be a stable figure of equilibrium (see note 40 of Chapter 2).

84. Clairaut (1743: xxxv).

85. Clairaut (1742e).

86. Truesdell (1954: XIX–XX; 1960: 222–223, footnote 1; 1984: 322); and Berteloni Meli (1990: 40). Daniel Bernoulli too developed his equilibrium theory of the tides, in his essay of 1740 [Daniel Bernoulli (1740)], in an accelerated frame of reference. In fact, Bernoulli seems to have had the clearest idea of centrifugal force as a "fictitious" force [see Berteloni Meli (1990: 41–43)].

87. The equations of hydrostatics only emerged for the first time later in the work of Euler. Euler was the first person to introduce into hydrostatics a clear concept of fluid pressure which was independent of weight. In Euler (1755: 227–231), Euler applied Newton's second law of motion expressed in rectangular Cartesian coordinates to hydrostatics conceived in terms of pressure and external forces to derive the equations of hydrostatics [see Truesdell (1954: XLII–XLIII, LXXV–LXXXI; 1960: 250–254)]. This fact serves as one more piece of evidence for doubting that Clairaut truly had the idea of pressure (see notes 80 and 81 of Chapter 9).

88. Clairaut (1743: xxxvii). At one point in some work on hydrodynamics in 1752, Euler claimed to prove that all fluid motions are potential flows, from which it follows that a fluid cannot rotate like a rigid body. Oddly enough, despite having read Clairaut's treatise of 1743, both Euler and D'Alembert failed to recognize that figures of relative equilibrium were counterexamples to their assertions that potential flow is necessary! [See Truesdell (1954: LXXIV, footnote 2).]

89. Clairaut (1760: 392). In fact, D'Alembert did first attempt to deal with this problem in D'Alembert (1747). D'Alembert also discussed the stability of homogeneous figures of equilibrium shaped like flattened ellipsoids of revolution which attract according to the universal inverse-square law in D'Alembert (1773: §27–§31). (See as well note 83 of Chapter 9.) For a history of investigations of the stability of homogeneous figures of equilibrium which attract according to the universal inverse-square law – that is, investigations of the question whether homogeneous, rotating, oscillating fluid masses that attract according to the universal inverse-square law approach figures of equilibrium or not – see Lützen (1984: 120–150).

90. Clairaut (1763; 1764: 285). Daniel Bernoulli tried to explain his theory of a vibrating string to Clairaut in Daniel Bernoulli (1758).

91. Although Euler did not actually employ the terms "stable equilibrium" and "unstable equilibrium" here, he had brought up the ideas that they involve in his *Scientia navalis*, which he completed in 1738 (see note 76 of Chapter 9).

92. Euler (1755: 219–220) [reappearing in Euler (1954: 2–53, on pp. 4–5)].

93. Truesdell (1954: LXXXI); and see note 22 of Chapter 9 as well.

94. This is probably one of the things that Truesdell (1954: XIX–XX, footnote 4) had in mind, when he concluded after summarizing forty years ago this remarkable correspondence between Euler and Clairaut concerning hydrostatics: "that Euler's great work in this field [fluid mechanics] was done entirely after the appearance of Clairaut's book (1743) may be more than accident."

95. For a modern version of Clairaut's problem, whose solution goes beyond Clairaut's first-order analysis to include terms of second order and makes use of subsequent developments in harmonic analysis, including Tesseral Harmonics to handle asymmetry around the axis of rotation, thus variations of effective gravity with longitude for a given latitude, see Kopal (1960). Kopal not only produces a second-order theory starting from Clairaut's theory, but he extends Clairaut's theory to include nonradial oscillations as well.

96. Clairaut (1743: 209–211, and the figure on p. 210).

97. See note 23 of Chapter 6.

98. Clairaut (1743: 217–218). Here I shall sketch Clairaut's demonstration, indicate a few features of it, fill in some missing details, and show how the demonstration basically follows from the way that Clairaut applied the principle of the plumb line in 1738. Whereas Clairaut just stated certain results that he used, I shall derive some of those that he did not prove. Figure 56 is in the plane of a meridian of the figure of revolution. MX is perpendicular to PME at M. MT is perpendicular at M to the ellipse that is geometrically similar to the ellipse KNL that passes through M. And $QM \equiv q$. Clairaut states that

$$CV = 2q\delta, \quad \text{where} \quad \delta \equiv \frac{CE - CP}{CP}. \tag{I}$$

I shall now prove that this is true.

$$\frac{y^2}{PC^2} + \frac{x^2}{CE^2} = 1 \tag{II}$$

is the equation of ellipse PME. From (II) it follows that

$$y = \frac{PC}{CE}\sqrt{CE^2 - x^2}. \tag{III}$$

If we differentiate (III), we find that

$$\frac{dy}{dx} = -\frac{PC}{CE}\frac{x}{\sqrt{CE^2 - x^2}}. \tag{IV}$$

From (IV) it follows that

$$\frac{-1}{dy/dx} = \frac{CE}{PC}\frac{\sqrt{CE^2 - x^2}}{x}. \tag{V}$$

Now, from (V) it follows that

$$\frac{y - QC}{x - QM} = \frac{CE}{PC} \frac{\sqrt{CE^2 - QM^2}}{QM} \tag{VI}$$

is the equation of the line through M which is perpendicular to PME at M. If we set

$$y = 0 \tag{VII}$$

in (VI), we find that

$$-QC = (x - QM) \frac{CE}{PC} \frac{\sqrt{CE^2 - QM^2}}{QM}. \tag{VIII}$$

From (VIII) it follows that

$$x = \frac{QM(CE/PC)(\sqrt{CE^2 - QM^2}/QM) - QC}{(CE/PC)(\sqrt{CE^2 - QM^2}/QM)}, \tag{IX}$$

or that

$$x = \frac{(CE/PC)\sqrt{CE^2 - QM^2} - QC}{(CE/PC)(\sqrt{CE^2 - QM^2}/QM)}. \tag{X}$$

From (X) it follows that

$$\frac{x}{QM} = \frac{(CE/PC)\sqrt{CE^2 - QM^2} - QC}{(CE/PC)\sqrt{CE^2 - QM^2}}. \tag{XI}$$

From (XI) it follows that

$$\frac{x}{QM} = 1 - \frac{QC}{(CE/PC)\sqrt{CE^2 - QM^2}}, \tag{XII}$$

when $y = 0$ in (VI). Now, the point M, whose rectangular coordinates are (QM, QC), is located on the ellipse PME. Hence

$$\frac{QC^2}{PC^2} + \frac{QM^2}{CE^2} = 1. \tag{XIII}$$

From (XIII) it follows that

$$\frac{QC}{PC} = \frac{\sqrt{CE^2 - QM^2}}{CE}. \tag{XIV}$$

From (XII) and (XIV) it follows that

$$\frac{x}{QM} = 1 - \frac{QC}{(CE/PC)(CE(QC/PC))} = 1 - \frac{1}{CE^2/PC^2} = 1 - \frac{PC^2}{CE^2} = \frac{CE^2 - PC^2}{CE^2}. \tag{XV}$$

Now,

$$
\begin{aligned}
\frac{CE^2 - PC^2}{CE^2} &= \left(\frac{CE - PC}{CE}\right)\left(\frac{CE + PC}{CE}\right) = \left(\frac{CE - PC}{CP + (CE - PC)}\right)\left(\frac{CE + PC}{CE}\right) \\
&= \left(\frac{CE - PC}{CP}\right)\left(\frac{1}{1 + ((CE - PC)/CP)}\right)\left(1 + \frac{PC}{CE}\right). \\
&= \left(\frac{CE - PC}{CP}\right)\left(\frac{1}{1 + ((CE - PC)/CP)}\right)\left(1 + \frac{PC}{CP + (CE - PC)}\right) \\
&= \left(\frac{CE - PC}{CP}\right)\left(\frac{1}{1 + ((CE - PC)/CP)}\right)\left(1 + \frac{1}{1 + ((CE - PC)/CP)}\right).
\end{aligned}
\tag{XVI}
$$

But

$$\frac{CE - PC}{CP} \equiv \delta$$

is infinitesimal ($\delta^2 \ll \delta$). Thus to terms of first order in δ, (XVI) can be written as

$$\frac{CE^2 - PC^2}{CE^2} = \delta\left(\frac{1}{1+\delta}\right)\left(1 + \frac{1}{1+\delta}\right) = \delta(1-\delta)(1+1-\delta)$$

(XVII)

$$= \delta(1-\delta)(2-\delta) = 2\,\delta.$$

If we combine (XV) and (XVII), we find that

$$\frac{x}{QM} = 2\,\delta.$$

(XVIII)

Finally, $QM \equiv q$ and x is the value of the abscissa when

$$y = 0$$

(VII)

in (VI). That is, x is the abscissa where line (VI) crosses the equatorial radius CE. Thus $x = CV$, and (XVIII) can be written as

$$\frac{CV}{q} = 2\,\delta.$$

(XIX)

From (XIX) it follows that

$$CV = 2q\delta.$$

(I)

Moreover, we have found that equality (I) only holds true to terms of first order in δ.
In 1738 Clairaut expressed the principle of the plumb line as

$$\frac{\text{Effective Gravity } (M, CX)}{\text{Effective Gravity } (M, CM)} = \frac{|CX|}{|CM|}.$$

(6.2.3)

(See note 32 of Chapter 6.) Clairaut (1743: 216) expressed the same principle as

$$\frac{\text{Effective Gravity } (M, CV)}{\text{Effective Gravity } (M, CM)} = \frac{|CV|}{|CM|}$$

(XX)

instead. I shall now show that (XX) does express the principle of the plumb line.
Since MX is perpendicular to the outer ellipse PME at M, and the effective gravity at M is perpendicular to ellipse PME too, it follows that

$$\text{Effective Gravity at } M = |\text{Effective Gravity at } M| \, \widehat{MX}$$

$$= \frac{|\text{Effective Gravity at } M|}{|MX|} MX.$$

(XXI)

Now, angle MCX is a right angle by construction; hence

$$MX = MC + CX.$$

(XXII)

At the same time, since angle MXC is *almost* a right angle (see note 28 of Chapter 6), it follows that

$$CV \simeq CX + XV$$

(XXIII)

and that

$$CX \simeq CV - XV.$$

(XXIV)

It follows from (XXI), (XXII), and (XXIV) that

$$\text{Effective Gravity at } M = \left(\frac{|\text{Effective Gravity at } M|}{|MX|}\right)|MC|\,\widehat{MC}$$

$$+\left(\frac{|\text{Effective Gravity at } M|}{|MX|}\right)|CV|\,\widehat{CV} \qquad (\text{XXV})$$

$$-\left(\frac{|\text{Effective Gravity at } M|}{|MX|}\right)|XV|\,\widehat{XV}.$$

From (XXV) we conclude that

$$\text{Effective Gravity at } (M, CV) = \left(\frac{|\text{Effective Gravity at } M|}{|MX|}\right)|CV| \qquad (\text{XXVI})$$

and that

$$\text{Effective Gravity at } (M, CM) = \left(\frac{|\text{Effective Gravity at } M|}{|MX|}\right)|MC|. \qquad (\text{XXVII})$$

Dividing the left-hand side of (XXVI) by the left-hand side of (XXVII), and dividing the right-hand side of (XXVI) by the right-hand side of (XXVII), we find that

$$\frac{\text{Effective Gravity } (M, CV)}{\text{Effective Gravity } (M, CM)} = \frac{|CV|}{|CM|}. \qquad (\text{XX})$$

Unlike (6.2.3), however, form (XX) of the principle of the plumb line only holds good to terms of first order, because of (XXIII) and (XXIV). Now, $|PC| = 1$ (normalization). Without loss of generality and for the sake of the argument, let us assume that the figure is flattened, so that $|PC| < |EC|$. Then if we define by

$$|CM| = |PC| + \eta, \qquad (\text{XXVIII})$$

then

$$1 + \eta = |PC| + \eta = |CM| \leqslant |EC|.$$

Hence

$$\eta = \frac{\eta}{1} = \frac{\eta}{|PC|} \leqslant \frac{|EC| - 1}{|PC|} = \frac{|EC| - |PC|}{|PC|} \equiv \delta. \qquad (\text{XXIX})$$

But δ is infinitesimal ($\delta^2 \ll \delta$). Hence it follows from (XXIX) that η is infinitesimal too ($\eta^2 \ll \eta$). Therefore

$$\frac{|CV|}{|CM|} = \frac{|CV|}{1 + \eta} = |CV|(1 - \eta) \qquad (\text{XXX})$$

to terms of first order. But

$$|CV| = 2q\delta \qquad (\text{I})$$

to terms of first order, and

$$q \equiv |QM| \leqslant |EC| = 1 + \delta.$$

Thus

$$|CV| = 2q\delta \leqslant 2(1 + \delta)\delta \simeq 2\delta \qquad (\text{XXXI})$$

to terms of first order. Then if δ is a sufficiently small infinitesimal, 2δ will be infinitesimal too and consequently so will $|CV| = 2q\delta$. Hence $|CV|\eta$ is a term of second order, so that

$$|CV|(1 - \eta) = |CV| \qquad (\text{XXXII})$$

to terms of first order. Then it follows from (XX), (XXX), (XXXII), and (I) that

$$\frac{\text{Effective Gravity } (M, CV)}{\text{Effective Gravity } (M, CM)} = \frac{|CV|}{|CM|} = 2q\delta \qquad (\text{XXXIII})$$

to terms of first order. We note that

$$\frac{\text{Effective Gravity } (M, CV)}{\text{Effective Gravity } (M, CM)} \qquad (\text{XXXIV})$$

must be infinitesimal, since $2q\delta$ is infinitesimal. Now,

$$\text{Effective Gravity}\,(M, CM) = \text{Attraction}\,(M, CM) - \psi, \tag{XXXV}$$

where $\psi \equiv$ the magnitude of the component of the centrifugal force per unit of mass at M in the direction of CM. Then

$$\frac{\text{Effective Gravity}\,(M, CV)}{\text{Attraction}\,(M, CM)} = \frac{\text{Effective Gravity}\,(M, CV)}{\text{Effective Gravity}\,(M, CM) + \psi}$$

$$= \frac{\text{Effective Gravity}\,(M, CV)}{\text{Effective Gravity}\,(M, CM)}\left(\frac{1}{1 + \Phi}\right), \tag{XXXVI}$$

where $\Phi \equiv \dfrac{\psi}{\text{Effective Gravity}\,(M, CM)}$ is infinitesimal $(\Phi^2 \ll \Phi)$.

Then, since (XXXIV) is also infinitesimal,

$$\left(\frac{\text{Effective Gravity}\,(M, CV)}{\text{Effective Gravity}\,(M, CM)}\right)\Phi$$

is a term of second order. Consequently

$$\frac{\text{Effective Gravity}\,(M, CV)}{\text{Effective Gravity}\,(M, CM)}\left(\frac{1}{1 + \Phi}\right) = \frac{\text{Effective Gravity}\,(M, CV)}{\text{Effective Gravity}\,(M, CM)}(1 - \Phi)$$

$$= \frac{\text{Effective Gravity}\,(M, CV)}{\text{Effective Gravity}\,(M, CM)} \tag{XXXVII}$$

to terms of first order. Combining (XXXVI) and (XXXVII), we find that

$$\frac{\text{Effective Gravity}\,(M, CV)}{\text{Effective Gravity}\,(M, CM)} = \frac{\text{Effective Gravity}\,(M, CV)}{\text{Attraction}(M, CM)} \tag{XXXVIII}$$

holds to terms of first order. In other words, in expressing the principle of the plumb line, only the attraction need appear in the denominator. The centrifugal force per unit of mass can be disregarded. That is,

$$\frac{\text{Effective Gravity}\,(M, CV)}{\text{Attraction}\,(M, CM)} = \frac{|CV|}{|CM|} = 2q\delta \tag{XXXIX}$$

holds true to terms of first order. Clairaut then calculated the magnitude of the attraction at the surface of a stratified spherical figure that attracts according to the universal inverse-square law, whose strata have the same densities as the strata of the figure in question, and whose radius equals the equatorial radius of the figure in question. He found the value to be

$$2cA + \frac{2}{3}c - \frac{2}{3}ca^3, \tag{XL}$$

where $c = 2\pi$ and

$$A \equiv \int_{r=0}^{r=a} R(r)\,r\,dr$$

[Clairaut (1743: 214–215)].
Then

$$\text{Attraction}\,(M, CM) = 2cA + \frac{2}{3}c - \frac{2}{3}ca^3 + \gamma, \tag{XLI}$$

where

$$\beta \equiv \frac{\gamma}{\text{Attraction}\,(M, CM)}$$

is infinitesimal $(\beta^2 \ll \beta)$, for the same reasons as those given in note 29 of Chapter 6. But it follows from (XXXIV) and (XXXVIII) that

$$\frac{\text{Effective Gravity }(M, CV)}{\text{Attraction }(M, CM)} \tag{XLII}$$

is also infinitesimal. Hence

$$\left(\frac{\text{Effective Gravity }(M, CV)}{\text{Attraction }(M, CM)}\right)\beta$$

is a term of second order. Consequently

$$\frac{\text{Effective Gravity }(M, CV)}{2\,cA + \frac{2}{3}c - \frac{2}{3}ca^3} = \frac{\text{Effective Gravity }(M, CV)}{\text{Attraction }(M, CM) - \gamma}$$

$$= \frac{\text{Effective Gravity }(M, CV)}{\text{Attraction }(M, CM)}\left(\frac{1}{1 - \beta}\right)$$

$$= \frac{\text{Effective Gravity }(M, CV)}{\text{Attraction }(M, CM)}(1 + \beta) \tag{XLIII}$$

$$= \frac{\text{Effective Gravity }(M, CV)}{\text{Attraction }(M, CM)}$$

to terms of first order. Then it follows from (XXXIX) and (XLIII) that

$$\frac{\text{Effective Gravity }(M, CV)}{2\,cA + \frac{2}{3}c - \frac{2}{3}ca^3} = \frac{|CV|}{|CM|} = 2\,q\delta \tag{XLIV}$$

to terms of first order. This explains why in expressing form (XXXIX) of the principle of the plumb line, Clairaut took the magnitude of the attraction in the denominator to be the magnitude of the attraction at the surface of a stratified spherical figure that attracts according to the universal inverse-square law, whose strata have the same densities as the strata of the figure in question, and whose radius equals the equatorial radius of the figure in question.
Now,

Effective Gravity (M, CV) = the component of the attraction at M in
the direction of CV − the component of the centrifugal force per unit of mass at M in
the direction of CV. (XLV)

Clairaut found that

the component of the attraction at M in the direction of CV

$$\tag{XLVI}$$

$$= \frac{4}{3}cqD + \frac{4}{3}cq\,\delta - \frac{4}{3}cqa^5\alpha,$$

where

$$D \equiv \int_{r=0}^{r=a,\ \rho(a)=\alpha} R(r)\,d(r^5\,\rho(r))$$

[Clairaut (1743: 212–213)]. Since CV is perpendicular to the axis of rotation of the figure, the component of the centrifugal force per unit of mass at M in the direction of CV = the centrifugal force per unit of mass at M. Let $F \equiv$ the magnitude of the centrifugal force per unit of mass at the equator of the figure in question. Let $G \equiv$ the magnitude of the attraction at the equator of the figure in question. Then

$$\phi_* \equiv \frac{F}{G} \tag{XLVII}$$

is infinitesimal ($\phi_*^2 \ll \phi_*$). Let $H \equiv$ the magnitude of the attraction at the equator of the stratified spherical figure whose radius equals the equatorial radius of the figure in question, whose strata have the same densities as the strata of the figure in question, and which attracts according to the universal inverse-square law. Then

$$\zeta \equiv \frac{G - H}{H} \tag{XLVIII}$$

is infinitesimal ($\zeta^2 \ll \zeta$). Hence $\zeta \phi_*$ is a term of second order. Consequently it follows from (XLVII) and (XLVIII) that

$$F = G\phi_* = (H + H\zeta)\phi_* = H(1 + \zeta)\phi_* = H\phi_* \tag{XLIX}$$

to terms of first order.

Moreover, if $J \equiv$ the magnitude of the effective gravity at the equator of the figure in question, and if

$$\phi \equiv \frac{F}{J}, \tag{L}$$

then

$$\phi \equiv \frac{F}{J} = \frac{F}{G - F} = \frac{F}{G}\left(\frac{1}{1 - (F/G)}\right) = \phi_*\left(\frac{1}{1 - \phi_*}\right) = \phi_*(1 + \phi_*) = \phi_* \tag{LI}$$

to terms of first order, since ϕ_* is infinitesimal ($\phi_*^2 \ll \phi_*$). Then from (XLIX) and (LI) it follows that

$$F = H\phi \tag{LII}$$

to terms first order. But

$$H = 2cA + \tfrac{2}{3}c - \tfrac{2}{3}ca^3. \tag{XL}$$

Therefore it follows from (LII) and (XL) that

$$F = (2cA + \tfrac{2}{3}c - \tfrac{2}{3}ca^3)\phi. \tag{LIII}$$

Then

the magnitude of the centrifugal force per unit of mass at M

$$= qF = 2cAq\phi + \tfrac{2}{3}cq\phi - \tfrac{2}{3}ca^3 q\phi \tag{LIV}$$

[Clairaut (1743: 215)]. If (XLV), (XLVI), and (LIV) are combined, then

Effective Gravity (M, CV)

$$= \tfrac{4}{3}cqD + \tfrac{4}{5}cq\delta - \tfrac{4}{3}cqa^5\alpha + 2cAq\phi + \tfrac{2}{3}cq\phi - \tfrac{2}{3}ca^3 q\phi \tag{LV}$$

results. Then it follows from (XLIV) and (LV) that

$$\frac{\tfrac{4}{3}cqD + \tfrac{4}{5}cq\delta - \tfrac{4}{3}cqa^5\alpha + 2cAq\phi + \tfrac{2}{3}cq\phi - \tfrac{2}{3}ca^3 q\phi}{2cA + \tfrac{2}{3}c - \tfrac{2}{3}ca^3} = 2q\delta. \tag{LVI}$$

Equation (LVI) expresses the principle of the plumb line. The qs cancel out, and the equation can be solved for δ:

$$\delta = \frac{6D - 6a^5\alpha + 15A\phi + 5\phi - 5a^3\phi}{30A + 4 - 10a^3} \tag{LVII}$$

[Clairaut (1743: 217)].

99. Clairaut (1743: 220–221). As I say, in deriving this result, Clairaut assumed that $\rho(r) \equiv \rho(a) = \alpha$, for $0 \leqslant r \leqslant a$, as well as assumed that $\alpha = \delta$. In fact, however, if the densities of the solid strata are all assumed to be the same, there should be no need to have to assume that $\rho(r) = \rho(a)$ for $0 \leqslant r < a$. It should be enough to assume that $\rho(a) \equiv \alpha = \delta$ in deducing the result. That is, it should be enough to assume that the solid strata all have the same density, whose value is less than the density of the surrounding homogeneous fluid layer in equilibrium, and that the ellipticity α of the *surface* of the solid part is the same as the ellipticity δ of the outer surface of the homogeneous fluid layer in equilibrium. The reason is that no matter how the ellipticities of strata might vary in the interior of the solid part – that is, no matter how the ellipticities $\rho(r)$ of the strata are chosen for $0 \leqslant r < a$ – the solid part, which is a figure shaped like a flattened

ellipsoid of revolution, comes out homogeneous with an ellipticity equal to $\rho(a) \equiv \alpha$, if the densities of its strata are all assumed to be the same. Hence the solid figure always attracts the same way, no matter how the ellipticities of its strata may be assumed to vary. Hence these variations should not affect the result in question. All that matters is the ellipticity α of the *surface* of the solid figure. In other words, if the strata of the solid figure shaped like a flattened ellipsoid of revolution all have the same densities, it should not matter whether the strata all have the same ellipticity as the surface of the solid figure or not. This figure will be homogeneous however the ellipticities of its strata vary. Hence this figure will attract the same way no matter how the ellipticities of its strata may vary. In fact it can be shown that allowing the ellipticities $\rho(r)$ to vary for $0 \leqslant r < a$ cannot effect the result in question. We can see that this is true by checking the expressions A and D defined in note 98 of Chapter 9.

Let $R(r) \equiv 1 + f$, where f is a constant that can be positive or negative. Let $\rho(r) \equiv \alpha$ for $0 \leqslant r < a$. Clairaut calculated that

$$A \equiv \int_{r=0}^{r=a} R(r)\, rr\, dr = (1+f) \int_{r=0}^{r=a} r^2\, dr = (1+f)\left.\frac{r^3}{3}\right|_{r=0}^{r=a} = (1+f)\frac{a^3}{3} \qquad (I)$$

and that

$$D \equiv \int_{r=0}^{r=a,\,\rho(a)=\alpha} R(r)\, d(r^5\, \rho(r)) = (1+f)\alpha \int_{r=0}^{r=a} d(r^5) = (1+f)\,\alpha r^5 \Big|_{r=0}^{r=a} = (1+f)\alpha a^5 \qquad (II)$$

[Clairaut (1743: 219)]. Now, let $R(r) \equiv 1 + f$ again, but this time assume that $\rho(r) \not\equiv \alpha$ for $0 \leqslant r < a$ and that $\rho(a) = \alpha$. Then

$$A = (1+f)\frac{a^3}{3} \qquad (I)$$

again, and

$$D \equiv \int_{r=0}^{r=a,\,\rho(a)=\alpha} R(r)\, d(r^5\, \rho(r)) = (1+f) \int_{r=0}^{r=a,\,\rho(a)=\alpha} d(r^5\, \rho(r)) = (1+f)r^5\, \rho(r) \Big|_{r=0}^{r=a,\,\rho(a)=\alpha}$$
$$= (1+f)a^5\, \rho(a) = (1+f)a^5\, \alpha. \qquad (III)$$

But (II) and (III) are the same. Hence A and D, respectively, turn out to have the same values, whether $\rho(r) \equiv \alpha$ for $0 \leqslant r \leqslant a$, or $\rho(r) \not\equiv \alpha$ for $0 \leqslant r < a$ and $\rho(a) = \alpha$.

100. Clairaut (1743: 222).
101. Ibid., p. 223.
102. Ibid., pp. 223–224. For some unknown reason, Clairaut erroneously stated that Newton's "theory required that Jupiter have an ellipticity that is less than the ellipticity observed" (p. 223), when in fact Newton's theoretical value for Jupiter's ellipticity was greater than the ellipticity observed, as Clairaut himself had correctly pointed out in his second paper on the theory of the earth's shape published in the *Philosophical Transactions*.
103. Clairaut (1743: 224–225). Again, it is unnecessary to assume that $\rho(r) \equiv \alpha$ for $0 \leqslant r < a$, for the same reasons as the ones that I explained in note 99 of Chapter 9. The result may appear paradoxical at first sight, since we saw in note 40 of Chapter 2 that homogeneous fluid figures shaped like elongated ellipsoids of revolution which attract according to the universal inverse-square law cannot be figures of equilibrium. This was only stated to be true, however, in note 40 of Chapter 2 for *homogeneous* fluid figures shaped like elongated ellipsoids of revolution which attract according to the universal inverse-square law, not for a figure shaped like an elongated ellipsoid of revolution which attracts in accordance with the universal inverse-square law and which is made up of a solid homogeneous core surrounded by a homogeneous fluid layer whose *density differs from the density* of the core.
104. Ibid., pp. 225–226. The equation is

$$10\, A\delta - 2\, D = 5\, A\phi, \qquad (I)$$

where

$$A \equiv \int_{r=0}^{r=1} R(r)\, rr\, dr \qquad (9.4.4)$$

and

$$D \equiv \int_{r=0}^{r=1, \rho(1)=\delta} R(r)\, d(r^5 \rho(r)). \tag{II}$$

Equation (I) is the equation to which equation (LVII) in note 98 of Chapter 9 reduces when $\alpha = \delta$ and $a = 1$. Equation (I) holds provided the densities $R(r)$ vary as a function of r in any way that permits the integrals A and D to exist and not equal zero. Thus, for example, $R(r)$ can be a step function of r. (See notes 23 and 45 of Chapter 6.) Since δ appears as an upper limit of integration of integral (II) that expresses D, D may have to be integrated by parts in order to solve equation (I) for δ for a given function $R(r)$ of r, if no further restrictions are given [say, on the form that $\rho(r)$ has as a function of r]. Let us note that the law of the resultant of attraction not only depends upon $R(r)$ but upon $\rho(r)$ as well. If the principle of the plumb line holds at the surfaces of two rotating stratified figures shaped like flattened ellipsoids of revolution which attract according to the universal inverse-square law, whose ellipticities are infinitesimal, whose strata have the same densities, and the ellipticities of whose strata vary like different functions $\rho(r)$ and $\rho_*(r)$ of r, respectively, where $\delta = \rho(1) = \rho_*(1)$ [in other words, the two figures have the same ellipticity δ, and the function $R(r)$ of r, hence the value of A, is the same for both figures, but $\rho(r)$ is not identically equal to $\rho_*(r)$ for all r, which means that $\rho(r) \neq \rho_*(r)$ for some value of r where $0 \leqslant r < 1$] and where $\rho(r)$ and $\rho_*(r)$ vary in such a way as to make

$$D \equiv \int_{r=0}^{r=1, \rho(1)=\delta} R(r)\, d(r^5 \rho(r)) \tag{II}$$

less than

$$D_* \equiv \int_{r=0}^{r=1, \rho_*(1)=\delta} R(r)\, d(r^5 \rho_*(r)), \tag{II'}$$

then it follows from equation (I) that

$$\phi \neq \phi_*,$$

since δ and A have the same values for both figures. In other words, the ratios of the magnitudes of the centrifugal force per unit mass to the magnitudes of the effective gravity at the equators of the two rotating figures must not be the same, when the principle of the plumb line holds at the surfaces of both figures, even though the two figures have the same ellipticity. For example, as I shall now show, if $0 \leqslant \rho(r) \leqslant \rho_*(r)$ for $0 \leqslant r < 1$, then if $\rho(r) < \rho_*(r)$ for some $0 \leqslant r < 1$, then $D < D_*$.

If $0 \leqslant R(r)$ is nonincreasing, and $0 \leqslant \rho(r) \leqslant \rho_*(r)$, on $[0, 1]$ then

$$\int_{r=0}^{r=1} R\, d(r^5 \rho) \leqslant \int_{r=0}^{r=1} R\, d(r^5 \rho_*).$$

This is easily seen as follows:

$$0 \leqslant R r^5 \rho \Big|_{r=0}^{r=1} = R(1)\rho(1) \leqslant R(1)\rho_*(1) = R r^5 \rho_* \Big|_{r=0}^{r=1},$$

because $0 \leqslant R(1)$ and $0 \leqslant \rho(1) \leqslant \rho_*(1)$. Now,

$$0 \leqslant \int_{r=0}^{r=1} r^5 \rho \left| \frac{dR}{dr} \right| dr \leqslant \int_{r=0}^{r=1} r^5 \rho_* \left| \frac{dR}{dr} \right| dr,$$

because $0 \leqslant \rho(r) \leqslant \rho_*(r)$, on $[0, 1]$. Then, since

$$\frac{dR}{dr} \leqslant 0 \quad \text{on } [0, 1], \quad \frac{dR}{dr} = -\left| \frac{dR}{dr} \right|, \quad \text{so}$$

$$\int_{r=0}^{r=1} r^5 \rho_* \, dR = \int_{r=0}^{r=1} r^5 \rho_* \frac{dR}{dr} \, dr = -\int_{r=0}^{r=1} r^5 \rho_* \left|\frac{dR}{dr}\right| \, dr \leqslant -\int_{r=0}^{r=1} r^5 \rho \left|\frac{dR}{dr}\right| \, dr$$

$$= \int_{r=0}^{r=1} r^5 \rho \frac{dR}{dr} \, dr = \int_{r=0}^{r=1} r^5 \rho \, dR \leqslant 0.$$

Combining

$$0 \leqslant Rr^5 \rho \Big|_{r=0}^{r=1} \leqslant Rr^5 \rho_* \Big|_{r=0}^{r=1}$$

and

$$0 \leqslant -\int_{r=0}^{r=1} r^5 \rho \, dR \leqslant -\int_{r=0}^{r=1} r^5 \rho_* \, dR,$$

we obtain:

$$\int_{r=0}^{r=1} R \, d(r^5 \rho) = Rr^5 \rho \Big|_{r=0}^{r=1} - \int_{r=0}^{r=1} r^5 \rho \, dR \leqslant Rr^5 \rho_* \Big|_{r=0}^{r=1} - \int_{r=0}^{r=1} r^5 \rho_* \, dR = \int_{r=0}^{r=1} R \, d(r^5 \rho_*).$$

So, letting

$$D \equiv \int_{r=0}^{r=1} R \, d(r^5 \rho) \quad \text{and} \quad D_* \equiv \int_{r=0}^{r=1} R \, d(r^5 \rho_*),$$

we have $D \leqslant D_*$. Now, suppose $\rho(1) = \rho_*(1)$ and $D = D_*$. Then

$$0 = \int_{r=0}^{r=1} R \, d(r^5(\rho_* - \rho)) = Rr^5 (\rho_* - \rho) \Big|_{r=0}^{r=1} - \int_{r=0}^{r=1} r^5 (\rho_* - \rho) \, dR$$

$$= \underbrace{R(1)(\rho_*(1) - \rho(1))}_{\substack{\| \\ 0}} - \int_{r=0}^{r=1} r^5 (\rho_* - \rho) \frac{dR}{dr} \, dr = \int_{r=0}^{r=1} r^5 (\rho_* - \rho) \left|\frac{dR}{dr}\right| \, dr.$$

Now, for $0 < r \leqslant 1$, $0 < r^5$, while $0 \leqslant \rho_* - \rho$ and $0 \leqslant |dR/dr|$. Thus we must have $(\rho_* - \rho)|dR/dr| \equiv 0$. For example, if

$$\frac{dR}{dr} \not\leqslant 0$$

(which means R is decreasing everywhere, and is nowhere locally constant), then $\rho_* - \rho \equiv 0$ must be true in this case. That is, $\rho(r) = \rho_*(r)$, for all $0 \leqslant r \leqslant 1$.

Hence if $\rho(r) \not\leqslant \rho_*(r)$ for some $0 \leqslant r < 1$, and

$$\frac{dR}{dr} \not\leqslant 0,$$

then $D \not\leqslant D_*$.

105. Clairaut (1743: 228–229). For Clairaut's equation of 1738 for the ellipticity α, see note 36 of Chapter 6.
106. Ibid., pp. 226–227. To say that Clairaut truly demonstrated this result is to exaggerate. What he claimed to show is that if

$$\rho(r) = \delta\left(\frac{1}{r^2} - u\right), \tag{I}$$

where u is a constant and $0 < u < 1$, then

$$\delta < \frac{5}{4}\phi. \tag{II}$$

In fact, as I shall show in a moment, Clairaut made an error while deriving (II) from (I). Moreover, even if there were no mistake in the proof, the result arrived at would still not establish the condition described by Clairaut, stated in the text. I shall now give a revised and corrected presentation of Clairaut's argument. It leads to a condition like the one that Clairaut expressed verbally, but unfortunately it is one that also turns out not to be very useful. I shall give as well some examples of what Clairaut thought followed as a "corollary" of his derivation of (II) from (I), although in fact it turns out that it is not a corollary. It is this corollary that he actually made use of later in his treatise. As we shall see below, grounds do exist, however, for thinking that the corollary is true, at least in some instances.

Clairaut let $\rho(r) \equiv \delta((1/r^2) - u)$, $0 \leqslant u$. In fact, assuming, as Clairaut did, that the ellipticities $\rho(r)$ are all nonnegative, for $0 \leqslant r \leqslant 1$, then we need $u \leqslant 1$ as well. Otherwise, if $1 < u$, then for $0 < r < 1$ and sufficiently near 1,

$$1 < \frac{1}{r^2} < u, \quad \text{hence} \quad \frac{1}{r^2} - u < 0,$$

so that $\rho(r) = \delta((1/r^2) - u)$ would be < 0 for these r near 1.

Clairaut argued as follows:

$$D \equiv \int_{r=0}^{r=1} Rd(r^5\rho) = \int_{r=0}^{r=1} Rd(r^5\delta((1/r^2) - u)) = \int_{r=0}^{r=1} Rd(\delta r^3 - \delta u r^5)$$

$$= 3\delta \int_{r=0}^{r=1} Rr^2 dr - \delta u \int_{r=0}^{r=1} Rd(r^5), \quad \text{which, integrating by parts,}$$

$$= 3\delta \int_{r=0}^{r=1} Rr^2 dr - \delta u R r^5 \Big|_{r=0}^{r=1} + \delta u \int_{r=0}^{r=1} r^5 dR, \quad \text{or}$$

$$D = 3\delta A - \delta G, \quad \text{where} \quad A \equiv \int_{r=0}^{r=1} Rr^2 dr, \quad \text{and}$$

$$G = u R r^5 \Big|_{r=0}^{r=1} - \int_{r=0}^{r=1} u r^5 dR.$$

Here I assume that $0 \leqslant R(r)$ decreases differentiably on $[0,1]$ (so that dR/dr exists), but the following arguments could be modified to take care of decreasing step functions as well. Let $uH \equiv G$. Since

$$\frac{dR}{dr} \leqslant 0 \quad \text{on } [0,1], \quad r^5 \frac{dR}{dr} \leqslant 0 \quad \text{on } [0,1]$$

as well, so that

$$\int_{r=0}^{r=1} r^5 dR = \int_{r=0}^{r=1} r^5 \frac{dR}{dr} dr \leqslant 0, \quad \text{hence} \quad 0 \leqslant -\int_{r=0}^{r=1} r^5 dR.$$

Moreover,

$$0 \leqslant R(1) = r^5 R \Big|_{r=0}^{r=1}, \quad \text{so} \quad 0 \leqslant R r^5 \Big|_{r=0}^{r=1} - \int_{r=0}^{r=1} r^5 dR \equiv H.$$

If either $0 < R(1)$, or $dR/dr < 0$ for some $0 \leqslant r \leqslant 1$ (in fact, *both* will be true in general), then clearly $0 < H$, so that $uH \equiv G = 0$ iff $u = 0$. Next, Clairaut said that

$$10\delta A - 2D = 5A\phi$$

(see note 104 of Chapter 9). Making substitutions, he obtained

$$10\delta A - 2(3\delta A - \delta G) = 5A\phi, \quad \text{or}$$

$$4\delta A + 2\delta G = 5A\phi, \quad \text{so that}$$

$$\delta = \frac{5A\phi}{4A + 2G} = \frac{5\phi}{4 + ((2G)/A)}.$$

Since A and G are both $\geqslant 0$,

$$\frac{5\phi}{4+((2G)/A)} \leqslant \frac{5}{4}\phi.$$

Clairaut concluded that unless $u = 0$ (that is, $\rho(r) = \delta/r^2$), in which case $G = 0$, the inequality holds.

Clairaut's argument contains flaws. He obviously saw that if $\rho(r) \equiv \delta/r^2$, in which case $\rho(1) = \delta$, then

$$10A\delta - 2D = 5A\phi$$

(see note 104 of Chapter 9), where

$$D \equiv \int_{r=0}^{r=1,\rho(1)=\delta} Rd(r^5\rho) = 3A\delta.$$

So, in this case,

$$10A\delta - 2(3A\delta) = 5A\phi, \quad \text{or} \quad \delta = \tfrac{5}{4}\phi.$$

[In other words if the density $R(r)$ decreases, say, differentiably from center to surface (so that dR/dr exists), then if $\rho(r) = \delta/r^2$, the ellipticity $\rho(1) = \delta$ of the figure will be the same as that for the corresponding homogeneous figure ($\tfrac{5}{4}\phi$), regardless of the particular $R(r)$.] The trouble is, with $\rho(r)$ defined as

$$\delta\left(\frac{1}{r^2} - u\right), \quad 0 < u < 1,$$

δ is not the ellipticity of the figure; rather $\rho(1) = \delta(1-u) \equiv \delta*$ is, and so $10\delta*A - 2D = 5A\phi$, where

$$D \equiv \int_{r=0}^{r=1,\rho(1)=\delta*} Rd(r^5\rho)(= 3\delta A - \delta G)$$

is what is actually true (see note 104 of Chapter 9), and not $10\delta A - 2D = 5A\phi$, which is what Clairaut wrote. That is, $10\,\delta(1-u)\,A - 2D = 5A\phi$ is what is actually true. Clairaut guessed that $\rho(r)$ ought to be such that, if δ is the ellipticity of the figure, $\rho(r)/\delta$ should be bounded away from $1/r^2$, to get the desired inequality $\delta < \tfrac{5}{4}\phi$, and his arguments can be modified to obtain such a result. Namely, with $\rho(r)$ defined as

$$\delta\left(\frac{1}{r^2} - u\right), \quad 0 < u < 1, \quad \rho(1) = \delta(1-u) < \frac{5}{4}\phi$$

is what is desired. But if $\delta \leqslant \tfrac{5}{4}\phi$, then certainly

$$\delta(1-u) < \delta \leqslant \tfrac{5}{4}\phi, \quad \text{if} \quad 0 < u < 1,$$

so that in effect it is indeed enough to establish conditions such that $\delta \leqslant \tfrac{5}{4}\phi$, when $\rho(r) = \delta((1/r^2) - u)$, $0 < u < 1$, thus $\rho(1) = \delta(1-u)[\neq \delta]$ is the ellipticity of the figure. Making substitutions (see note 104 of Chapter 9), we obtain

$$10\delta(1-u)A - 2(3\delta A - \delta G) = 5A\phi, \quad \text{or}$$

$$4\delta A + 2\delta G - 10\delta uA = 5A\phi, \quad \text{so}$$

$$\delta = \frac{5A\phi}{4A + 2G - 10uA} = \frac{5\phi}{4 + ((2G)/A) - 10u}.$$

Now,

$$\delta = \frac{5\phi}{4+((2G)/A) - 10u} \quad \text{is} \quad \leqslant \frac{5}{4}\phi$$

when and only when

$$0 \leqslant \frac{2G}{A} - 10u.$$

That is,

$$0 \leqslant \frac{2(Rr^5 \,|_{r=0}^{r=1} - \int_{r=0}^{r=1} r^5 dR)u}{A} - 10u, \quad \text{or}$$

$$0 \leqslant \left(\frac{2(Rr^5 \,|_{r=0}^{r=1} - \int_{r=0}^{r=1} r^5 dR)}{A} - 10 \right) u.$$

So, when $0 < u$, then

$$\frac{2(Rr^5 \,|_{r=0}^{r=1} - \int_{r=0}^{r=1} r^5 dR)}{A} - 10$$

must be $\geqslant 0$. That is,

$$10A \leqslant 2(Rr^5 \Big|_{r=0}^{r=1} - \int_{r=0}^{r=1} r^5 dR), \quad \text{or}$$

$$0 \leqslant 2(Rr^5 \Big|_{r=0}^{r=1} - \int_{r=0}^{r=1} r^5 dR) - 10 \int_{r=0}^{r=1} Rr^2 dr,$$

if $0 < u$, since $0 \leqslant A = \int_{r=0}^{r=1} Rr^2 dr$.

This imposes another condition that $R(r)$ must satisfy (that is, in addition to $R(r)$ decreasing on $[0,1]$) in order to have $\delta \leqslant \frac{5}{4}\phi$, if $0 < u$. We recall that

$$Rr^5 \Big|_{r=0}^{r=1} - \int_{r=0}^{r=1} r^5 dR \equiv H,$$

so the condition above can be written as $0 \leqslant 2H - 10A$. Moreover, we recall that $0 < H$ in general. So, it is necessary and sufficient that $A \leqslant H/5$. So, for example, if $R(0) < \infty$, then since

$$\int_{r=0}^{r=1} Rr^2 dr \leqslant \int_{r=0}^{r=1} R(0)r^2 dr = R(0)\frac{r^3}{3}\Big|_{r=0}^{r=1} = \frac{R(0)}{3},$$

and

$$Rr^5 \Big|_{r=0}^{r=1} = R(1),$$

then for $A \leqslant H/5$ to be satisfied, it is sufficient (but not necessary) that

$$\frac{R(0)}{3} \leqslant \frac{R(1) - \int_{r=0}^{r=1} r^5 dR}{5}, \quad \text{or}$$

$$5R(0) \leqslant 3R(1) - 3\int_{r=0}^{r=1} r^5 dR,$$

hence

$$0 < \frac{5R(0) - 3R(1)}{3} \leqslant -\int_{r=0}^{r=1} r^5 dR.$$

Thus the preceding arguments will hold, for given $R(1) < R(0)$, provided that $R(r)$ decreases fast enough on $[0,1]$ to make

$$-\int_{r=0}^{r=1} r^5 dR$$

greater than or equal to

$$\frac{5R(0) - 3R(1)}{3}$$

[but of course not so fast as to make

$$-\int_{r=0}^{r=1} r^5 dR$$

infinite – that is, fail to exist].

It is easy to find examples that *don't* work. E.g., $R(r) = r^{-m}, 0 < m$. Here $R(0) = \infty$. For

$$A \equiv \int_{r=0}^{r=1} Rr^2\, dr = \int_{r=0}^{r=1} r^{2-m}\, dr$$

even to exist, we must have $m < 3$. [In such cases,

$$-\int_{r=0}^{r=1} r^5 dR$$

exists as well, because

$$-\int_{r=0}^{r=1} r^5 dR = m\int_{r=0}^{r=1} r^5 r^{-m-1}\, dr = m\int_{r=0}^{r=1} r^{4-m}\, dr,$$

which exists when $0 < m < 3$.]

The condition to be satisfied is: $A \leqslant H/5$. Thus the question is this: Is

$$\int_{r=0}^{r=1} r^{-m} r^2 dr \leqslant \frac{r^{-m} r^5\,\big|_{r=0}^{r=1} - \int_{r=0}^{r=1} r^5(-m r^{-m-1})\, dr}{5},$$

for $0 < m < 3$? That is, is

$$\int_{r=0}^{r=1} r^{2-m}\, dr \leqslant \frac{r^{5-m}\,\big|_{r=0}^{r=1} + m\int_{r=0}^{r=1} r^{4-m}\, dr}{5}, \quad 0 < m < 3?$$

or

$$\frac{r^{3-m}}{3-m}\bigg|_{r=0}^{r=1} \leqslant \frac{1 + m(r^{5-m}/(5-m))\big|_{r=0}^{r=1}}{5}, \quad 0 < m < 3?$$

or

$$\frac{1}{3-m} \leqslant \frac{1+(m/(5-m))}{5} = \frac{5}{5(5-m)} = \frac{1}{5-m}, \quad 0 < m < 3?$$

or

$$5 - m \leqslant 3 - m, \quad 0 < m < 3?$$

or

$$5 \leqslant 3, \quad 0 < m < 3?$$

No!

$R(r) = e^{-r}$ is another example that doesn't work. Here $R(0) < \infty$, and

$$0 < A = \int_{r=0}^{r=1} e^{-r} r^2 dr = -e^{-r} r^2\bigg|_{r=0}^{r=1} + 2\int_{r=0}^{r=1} e^{-r} r\, dr$$

$$= -e^{-r} r^2\bigg|_{r=0}^{r=1} + 2\left(-e^{-r} r\bigg|_{r=0}^{r=1} + \int_{r=0}^{r=1} e^{-r}\, dr\right)$$

$$= -e^{-r} r^2\bigg|_{r=0}^{r=1} + 2\left(-e^{-r} r\bigg|_{r=0}^{r=1} - e^{-r}\bigg|_{r=0}^{r=1}\right)$$

$$= -e^{-1} + 2(-e^{-1} - e^{-1} + 1) = -5e^{-1} + 2 = 2 - \frac{5}{e} > 0.$$

(The last inequality can also be easily seen from:

$$\frac{5}{e} < \frac{5}{2\frac{17}{24}} < \frac{5}{2\frac{1}{2}} = 2.)$$

Moreover,

$$\int_{r=0}^{r=1} e^{-r} r^5 \, dr = -e^{-r} r^5 \Big|_{r=0}^{r=1} + 5 \int_{r=0}^{r=1} e^{-r} r^4 dr$$

$$= -e^{-1} + 5 \left[-e^{-r} r^4 \Big|_{r=0}^{r=1} + 4 \int_{r=0}^{r=1} e^{-r} r^3 \, dr \right]$$

$$= -e^{-1} + 5 \left[-e^{-1} + 4 \left\{ -e^{-r} r^3 \Big|_{r=0}^{r=1} + 3 \int_{r=0}^{r=1} e^{-r} r^2 \, dr \right\} \right]$$

$$= -e^{-1} + 5 [-e^{-1} + 4 \{ -e^{-1} + 3 (2 - 5e^{-1}) \}]$$

$$= -e^{-1} - 5e^{-1} - 20e^{-1} + 60(2 - 5e^{-1})$$

$$= -26e^{-1} + 120 - 300e^{-1} = 120 - 326 \, e^{-1}.$$

Since

$$e^{-r} r^5 \Big|_{r=0}^{r=1} = e^{-1}, \quad H = e^{-1} + 120 - 326 \, e^{-1},$$

and the question is: Is

$$2 - 5e^{-1} \leqslant \frac{e^{-1} + 120 - 326 \, e^{-1}}{5} = \frac{120 - 325 \, e^{-1}}{5} = 24 - 65 \, e^{-1}?$$

That is, is $60 \, e^{-1} \leqslant 22$? Or, is $\frac{60}{22} \leqslant e$? No, because $\frac{60}{22} = 2.727272\ldots$, so that $e = 2.7182818284\ldots < 2.727272\ldots = \frac{60}{22}$.

So, Clairaut's theorem, even when modified to rectify errors, is not very useful. The corrected demonstration entails conditions on decreasing $R(r)$ that, for example, do not hold for densities $R(r) = r^{-m}, 0 < m < 3$, or $R(r) = e^{-r}$. What is more, his argument, even when revised, contains other difficulties. It cannot possibly apply to cases where $\rho(r) \nrightarrow \infty$, as $r \to 0$, because for $0 < u < 1$,

$$\rho(r) \equiv \delta \left(\frac{1}{r^2} - u \right) \to \infty, \quad \text{as } r \to 0.$$

I now consider $\rho(r)$ of the form $\rho(r) \equiv r^n \delta, 0 < n$. Then $\rho(r)$ is increasing on $[0, 1]$, and $\rho(r) \nrightarrow \infty$, as $r \to 0$. We note that

$$\rho(r) \equiv r^n \delta < \frac{\delta}{r^2}, \quad 0 < r < 1, \quad \text{for all } 0 \leqslant n,$$

since $r^{n+2} < 1, 0 < r < 1$, for all $0 \leqslant n$; and so

$$\frac{\rho(r)}{\delta} = r^n < \frac{1}{r^2}, \quad 0 < r < 1. \quad \text{That is, } \frac{\rho(r)}{\delta}$$

is bounded away from $1/r^2$, where $\rho(1) = \delta$ is the ellipticity of the figure in this case. I consider $\rho(r)$ of such a form because the assertion that Clairaut actually puts to use is the following: If $R(r)$ is decreasing on $[0, 1]$, and $\rho(r)$ is increasing on $[0, 1]$, then $\rho(1) = \delta < \frac{5}{4} \phi$. Clairaut took this as a "corollary" of the preceding affirmation, which it is not – first, because Clairaut's demonstration of that affirmation contains errors; second, in the modified demonstration, the ellipticity of the figure is $\delta(1 - u)$, not δ; third, a result involving $\rho(r)$ where $\rho(r) \nrightarrow \infty$, as $r \to 0$, cannot follow from a result for

$$\rho(r) = \delta \left(\frac{1}{r^2} - u \right),$$

which $\to \infty$, as $r \to 0$. I merely illustrate the assertion that Clairaut makes actual use of, via examples.

Let $\rho(r) = \delta r^n$, $0 \leqslant n$. The $\rho(r)$ is increasing on $[0, 1]$, if $0 < n$, and $\rho(1) = \delta$, for $0 \leqslant n$. Now,

$$D \equiv \int_{r=0}^{r=1} Rd(r^5 \rho) = \int_{r=0}^{r=1} Rd(r^5 \delta r^n) = \delta \int_{r=0}^{r=1} Rd(r^{5+n})$$

$$= (5 + n)\delta \int_{r=0}^{r=1} Rr^{5+n-1} \, dr.$$

Then (see note 104 of Chapter 9):

$$A\delta - 2(5 + n)\delta \int_{r=0}^{r=1} Rr^{5+n-1} \, dr = 5A\phi, \quad \text{so}$$

$$\delta = \frac{5A\phi}{10A - 2(5 + n)\int_{r=0}^{r=1} Rr^{5+n-1} \, dr} = \frac{5\phi}{10 - \dfrac{2(5 + n)\int_{r=0}^{r=1} Rr^{5+n-1} \, dr}{A}} < \frac{5}{4}\phi$$

$$\Rightarrow 4 < 10 - \frac{2(5 + n)\int_{r=0}^{r=1} Rr^{4+n} \, dr}{A}$$

$$\Rightarrow 2(5 + n)\int_{r=0}^{r=1} Rr^{4+n} \, dr < 6\int_{r=0}^{r=1} Rr^2 \, dr.$$

Let $R = r^{-m}$, $0 \leqslant m < 3$. Then

$$A \equiv \int_0^1 Rr^2 \, dr$$

exists; and since $4 + n - m > 1$ for $0 \leqslant n$, $0 \leqslant m < 3$,

$$D = (5 + n)\delta \int_{r=0}^{r=1} r^{-m} r^{5+n-1} \, dr = (5 + n)\delta \int_{r=0}^{r=1} r^{4+n-m} \, dr$$

exists too, for $0 \leqslant n$, $0 \leqslant m < 3$. So, the question is: Is

$$2(5 + n)\int_{r=0}^{r=1} r^{4+n-m} \, dr < 6\int_{r=0}^{r=1} r^{2-m} \, dr ?$$

or: Is

$$2(5 + n)\frac{r^{5+n-m}}{5 + n - m}\bigg|_{r=0}^{r=1} < 6\frac{r^{3-m}}{3 - m}\bigg|_{r=0}^{r=1} ?$$

or: Is

$$\frac{2(5 + n)}{5 + n - m} < \frac{6}{3 - m},$$

for all $0 \leqslant n$, $0 \leqslant m < 3$? In other words, is

$$\frac{5 + n}{5 + n - m} < \frac{3}{3 - m}, \quad \text{or}$$

$$\frac{1}{1 - \frac{m}{5+n}} < \frac{1}{1 - \frac{m}{3}}, \quad \text{or} \quad \frac{m}{5 + n} < \frac{m}{3} ?$$

For this to be true, m cannot $= 0$ in $R(r) = r^{-m}$. That is, $R(r)$ cannot be *constant*, but *must decrease* for the result to hold. If $0 < m < 3$, then $m/(5 + n)$ is in fact $< m/3$, for all $0 \leqslant n$. [Note: the fact that the inequality holds for $0 = n$ means that $\rho(1) = \delta < \frac{5}{4}\phi$ even when $\rho(r)$ is not increasing, but is constant ($\equiv \delta$), when $R(r) = r^{-m}$, $0 < m < 3$.] More generally, the necessity of assuming that $R(r)$ decrease on $[0, 1]$ can be seen as follows: Suppose $R(r) = r^m$, $0 \leqslant m$. Then the question becomes: Is

$$(5+n)\int_{r=0}^{r=1} r^{m+4+n}\,dr < 3\int_{r=0}^{r=1} r^{m+2}\,dr, \quad \text{or}$$

$$(5+n)\frac{r^{m+5+n}}{m+5+n}\Big|_{r=0}^{r=1} < 3\frac{r^{m+3}}{m+3}\Big|_{r=0}^{r=1}, \quad \text{or}$$

$$\frac{5+n}{5+n+m} < \frac{3}{3+m}, \quad \text{or}\; \frac{1}{1+(m/(5+n))} < \frac{1}{1+(m/3)}, \quad \text{or}$$

$$\frac{m}{3} < \frac{m}{5+n},$$

which is never true for any $0 \leqslant n$, if $0 \leqslant m$.

If, more generally, $\rho(r) = \delta f(r)$, where $0 < f(r)$ is nondecreasing on $[0, 1]$, and $f(1) = 1$, so $\rho(1) = \delta$, then

$$D \equiv \int_{r=0}^{r=1} R\,d(r^5\rho) = \int_{r=0}^{r=1} R\,d(r^5\delta f(r))$$

$$= \int_{r=0}^{r=1} R\left(\delta 5r^4 f(r) + \delta r^5\frac{df(r)}{dr}\right)dr, \quad \text{so that}$$

$$10\,A\delta - 2\delta\int_{r=0}^{r=1} R\left(5r^4 f(r) + r^5\frac{df(r)}{dr}\right)dr = 5\,A\,\phi, \quad \text{and thus}$$

$$\delta = \frac{5A\phi}{10A - 2\int_{r=0}^{r=1}(5r^4 f(r) + r^5\,df(r)/dr)\,dr}$$

$$= \frac{5\phi}{10 - \dfrac{2\int_{r=0}^{r=1} R\,(5r^4 f(r) + r^5\,df(r)/dr)\,dr}{A}}, \quad \text{which is} < \frac{5}{4}\phi$$

if $\quad 4 < 10 - \dfrac{2\int_{r=0}^{r=1} R(5r^4 f(r) + r^5\,df(r)/dr)\,dr}{A}, \quad \text{or}$

$$2\int_{r=0}^{r=1} R\left(5r^4 f(r) + r^5\frac{df(r)}{dr}\right)dr < 6\int_{r=0}^{r=1} Rr^2\,dr, \quad \text{or}$$

$$\int_{r=0}^{r=1} R\left(5r^4 f(r) + r^5\frac{df(r)}{dr}\right)dr < 3\int_{r=0}^{r=1} Rr^2\,dr.$$

Suppose $0 < f(r)$ is nondecreasing on $[0, 1]$ – hence

$$0 \leqslant \frac{df(r)}{dr} -$$

such that

$$\frac{df(r)}{dr} \leqslant \frac{n}{r}f(r),$$

in which case $f(r) \leqslant r^n$, with equality if

$$\frac{df(r)}{dr} = \frac{n}{r}f(r).$$

That is, if

$$\frac{df(r)}{dr} = \frac{n}{r}f(r),$$

then $f(r) = r^n$; if

$$\frac{df(r)}{dr} < \frac{n}{r} f(r),$$

then $f(r) < r^n$. Then

$$\int_{r=0}^{r=1} R\left(5r^4 f(r) + r^5 \frac{df(r)}{dr}\right) dr$$

$$\leqslant \int_{r=0}^{r=1} R\left(5r^4 f(r) + r^5 \frac{n}{r} f(r)\right) dr$$

$$= (5+n) \int_{r=0}^{r=1} Rr^4 f(r) dr \leqslant (5+n) \int_{r=0}^{r=1} Rr^4 r^n dr$$

$$= (5+n) \int_{r=0}^{r=1} Rr^{n+4} dr.$$

So, in this case, it is, again, enough (though not necessary) for $R(r)$ to satisfy the condition:

$$(5+n) \int_{r=0}^{r=1} Rr^{n+4} dr < 3 \int_{r=0}^{r=1} Rr^2 dr,$$

in addition to decreasing on $[0, 1]$.

I have provided these examples to show that the "corollary" that Clairaut eventually applies, although not demonstrated in his treatise of 1743, probably can be furnished with foundations.

107. Clairaut (1743: 229–232). Clairaut mentioned the third of the three examples described in these pages of his treatise of 1743 in a letter to MacLaurin dated 19 November 1742. See Clairaut (1742f: 377).
108. Clairaut (1743: 247–249).
109. Ibid., pp. 249–250. I emphasize the subtle differences in Clairaut's derivations of equation

$$\frac{P - \Pi}{\Pi} = 2\varepsilon - \delta \tag{9.4.1'}$$

in his second paper on the theory of the earth's shape published in the *Philosophical Transactions* (1738) and in his treatise of 1743 on the earth's shape. In that second paper, Clairaut had determined both the "sin² law" of the variation in the increase in the magnitude of the effective gravity with latitude at the surfaces of the flattened figures he considered in that paper and the expression for the ratio $(P - \Pi)/\Pi$ specifically when the principle of the plumb line was assumed to hold at the surfaces of these figures. In order to simplify the expression for $(P - \Pi)/\Pi$ which he found, Clairaut made use of the equation in note 36 of Chapter 6, an equation that also followed from having assumed that the principle of the plumb line held at the surfaces of the figures that he investigated in his second paper on the theory of the earth's shape published in the *Philosophical Transactions*. This equation and the "sin² law" just mentioned were valid when the ellipticities δ of the figures and the ratios ϕ of the magnitudes of the centrifugal force per unit of mass to the magnitudes of the effective gravity at their equators were infinitesimal. After simplying the expression for $(P - \Pi)/\Pi$, equation (9.4.1') resulted. In his treatise of 1743, however, Clairaut showed instead that the "sin² law" of the variation in the increase in the magnitude of the effective gravity with latitude at the surfaces of the figures that he now considered, which generalized the ones that he had studied in 1738, held *whenever* the ellipticities δ of the figures and the ratios ϕ of the magnitudes of the centrifugal force per unit of mass to the magnitudes of the effective gravity at their equators were infinitesimal (see note 108 of Chapter 9). That is, he demonstrated that the principle of the plumb line *did not* have to hold at the surfaces of the figures in order that the "sin² law" of the variation in the increase in the magnitude of the effective gravity with latitude hold at the surfaces of the figures. Moreover, in his treatise of 1743, Clairaut did not assume that the figures had to be flattened. He showed that the figures could be elongated and that the "sin² law" of the variation in the increase in the magnitude of the effective gravity with latitude could still hold at the surfaces of these figures. He then computed the ratio

$(P - \Pi)/\Pi$ for the figures in question. Next he used the equation $10 A\delta - 2D = 5 A\phi$ [equation (I) in note 104 of Chapter 9], which is valid when ϕ is infinitesimal and the principle of the plumb line holds at the surfaces of the figures. Again, in deriving this equation in his treatise of 1743, he did not assume that the figures had to be flattened; they could be elongated. He used equation $10 A\delta - 2D = 5 A\phi$ to convert the expression for $(P - \Pi)/\Pi$ into $2\varepsilon - \delta$. In this way he arrived at equation (9.4.1') in his treatise of 1743. In transforming the expression for $(P - \Pi)/\Pi$ into $2\varepsilon - \delta$, Clairaut made use of the fact that since $2cA$, $c = 2\pi$, is the magnitude of the attraction at the surface of a stratified spherical figure which attracts according to the universal inverse-square law, whose radius equals the equatorial radius of a stratified figure shaped like an ellipsoid of revolution which attracts in accordance with the universal inverse-square law, whose ellipticity is infinitesimal, and whose strata have the same densities as the strata of this nonspherical figure (in other words, the function $R(r)$ of r that appears in the definition of A in (9.4.4) is the same for both figures), then

$$\frac{P - \Pi}{\Pi} = \frac{P - \Pi}{2cA}$$

holds to terms of first order for this nonspherical figure. Clairaut evidently concluded that this was true for reasons like those that I used to derive equation (LVII) from the principle of the plumb line in note 98 of Chapter 9. Be that as it may, the chains of reasoning that underline Clairaut's derivations of equation (9.4.1') in 1738 and 1743 differ slightly from each other. In using the term "arbitrary" in the text, I mean in the sense that the result stated is true as long as the function $R(r)$ of r is such that the integrals

$$A \equiv \int_{r=0}^{r=1} R(r)rr dr \tag{9.4.4}$$

and

$$D \equiv \int_{r=0}^{r=1,\rho(1)=\delta} R(r)d(r^5\rho(r))$$

exist (see note 104 of Chapter 9) and are not equal to zero (see note 45 of Chapter 6). Thus, for example, $R(r)$ can be a step function of r (see note 23 of Chapter 6). We recall that the law of the resultant of attraction depends upon both the densities $R(r)$ and the ellipticities $\rho(r)$ of the strata. As a result, two rotating stratified figures shaped like ellipsoids of revolution which attract according to the universal inverse-square law, whose ellipticities $\delta = \rho(1) = \rho_*(1)$ are the same, and at whose surfaces the principle of the plumb line holds, can nevertheless have different ratios ϕ and ϕ_* of the magnitudes of the centrifugal force per unit mass to the magnitudes of the effective gravity at their equators (see note 104 of Chapter 9). Consequently, the homogeneous figures of equilibrium shaped like flattened ellipsoids of revolution which attract in accordance with the universal inverse-square and which correspond to each of the two stratified figures will have different infinitesimal ellipticities

$$\varepsilon = \frac{5}{4}\phi \tag{9.4.2}$$

and

$$\varepsilon_* = \frac{5}{4}\phi_*, \tag{9.4.2'}$$

too. Similarly, the relative increases in the magnitudes of the effective gravity

$$\frac{P - \Pi}{\Pi} \quad \text{and} \quad \frac{P_* - \Pi_*}{\Pi_*}$$

from the equator to a pole along the surfaces of the two rotating, stratified figures will differ from each other accordingly. For this reason, equations like

$$\frac{P - \Pi}{\Pi} = 2\varepsilon - \delta \tag{9.4.1'}$$

and

$$\frac{P_* - \Pi_*}{\Pi_*} = 2\varepsilon_* - \delta \qquad (9.4.1'')$$

can both hold at the same time, where δ has the same value in both equations. The specific equation that is valid in any particular problem depends upon the given conditions, namely, the densities $R(r)$ and the ellipticities $\rho(r)$ of the strata.

110. Clairaut (1743: 253). We recall that in 1732–33, Maupertuis expounded the laws of attraction appearing in the second edition of the *Principia* (see note 52 of Chapter 5), while in 1734 he tried to explain the theory of the earth's shape published in the third edition of the *Principia* (see note 51 of Chapter 5).

111. Clairaut (1743: 253–256). Newton's Italian defender Paolo Frisi later criticized Clairaut's judgement. Clairaut replied by translating the passage in question appearing in his treatise of 1743 into English and publishing the translation along with a further clairfication of the mistake that he thought Newton had made in Clairaut (1753: 83–85).

112. Clairaut (1743: 260–262). Clairaut noted that since the value of ε in equation (9.4.1') is $\frac{1}{230}$ in the case of the earth, the equation (9.4.1') that corresponds to the earth is

$$\frac{P - \Pi}{\Pi} = \frac{1}{115} - \delta \qquad (I)$$

if the earth is flattened. Here the earth's infinitesimal ellipticity δ is positive. He next remarked that if the earth's infinitesimal ellipticity were negative instead, therefore had the form $-\delta$ where δ is positive, in which case the earth would be elongated instead of flattened, equation (9.4.1') would then be

$$\frac{P - \Pi}{\Pi} = \frac{1}{115} + \delta \qquad (II)$$

in this case. We note that equation (II) implicitly requires that the principle of the plumb line hold at the surface of the figure shaped like an elongated ellipsoid of revolution which attracts according to the universal inverse-square law, which revolves around its axis of symmetry, and whose infinitesimal ellipticity is $-\delta$, a principle that Clairaut believed is observed to hold at the earth's surface, despite Kuhn's claims to the contrary. Comparing the right-hand sides of equations (I) and (II), Clairaut concluded that "the [relative] decrease in [the magnitude of the] effective gravity toward the equator would be appreciably greater along the [surface of the] elongated spheroid than along the [surface of the] flattened spheroid (pp. 260–261)," which, we recall, is also what Mairan tried to argue in his Paris Academy *mémoire* of 1720. It thus appeared at first sight that the observed relative decrease in the magnitude of the effective gravity from a pole to the equator along the earth's surface might accord better with a figure that is elongated than a figure that is flattened. The trouble, however, was that the relative decrease in the magnitude of the effective gravity from a pole to the equator along the surface of the elongated figure was *too* great. To illustrate that problem, Clairaut took Cassini's value $-\frac{1}{93}$ of the earth's ellipticity, and he substituted it for δ in the right-hand side of equation (II), which became $\frac{1}{115} + \frac{1}{93} = \frac{1}{51}$. But $\frac{1}{51}$ was far larger than the observed relative decrease in the magnitude of the effective gravity from a pole to the equator along the earth's surface. That is, in order that the actual relative decrease be as large as $\frac{1}{51}$, errors in the measurements from which the observed relative decrease was deduced would have had to be made which were much larger than the errors that could reasonably have been expected to be made. [We observe that in order to apply this reasoning to Cassini's figure shaped like an elongated ellipsoid of revolution which attracts according to the universal inverse-square law and whose ellipticity is infinitesimal, the figure *cannot* be assumed to be *homogeneous*. That is, the figure *must* be assumed to be *stratified*. That is, its density varies from it center to its surface. This is true because if the principle of the plumb line holds at all points on the surface of a homogeneous figure shaped like an ellipsoid of revolution which attracts according to the universal inverse-square law, then the principle of balanced columns must hold for the figure as well, because Clairaut's second new principle of equilibrium necessarily holds for such a figure. However, as I indicated in Chapter 2, the principle of balanced columns can never hold for a homogeneous figure shaped like an elongated ellipsoid of revolution which attracts according to the universal inverse-square law. Instead such a homogeneous figure that satisfies the conditions necessary for equilibrium which Clairaut knew must be shaped like a flattened ellipsoid of revolution whose infinitesimal ellipticity ε is determined by the equation

$$\varepsilon = \frac{5}{4}\phi, \tag{9.4.2}$$

where ϕ is the ratio of the magnitude of the centrifugal force per unit of mass to the magnitude of the effective gravity at the equator of the figure. And when ϕ has the earth's value $\frac{1}{288}$, then $\varepsilon = \frac{1}{230}$. In other words, a figure shaped like an elongated ellipsoid of revolution which attracts according to the universal inverse-square law, whose value of ϕ is $\frac{1}{288}$, and which satisifies equation (9.4.1) – that is, equation (II) – above which means that the principle of the plumb line holds at its surface, cannot possibly be homogeneous.] Clairaut's conclusions depend, of course, upon the assumption that the earth attracts according to the universal inverse-square law of attraction, which was not Mairan's hypothesis of attraction.

113. Clairaut (1743: 265–266, and the figure on p. 266).
114. If $\int_\gamma \mathbf{F} \cdot d\mathbf{s} = 0$ around every closed plane or skew channel γ that lies within the interior of a fluid figure and that is filled with fluid from the figure, where $\mathbf{F}$ is the effective gravity, which is true when figures attract according to the universal inverse-square law, as we have seen in the text, then Clairaut's arguments can be suitably modified so as to demonstrate that if the density P varies *continuously* in a heterogeneous fluid figure of equilibrium, then the surfaces of constant density within the heterogeneous fluid figure of equilibrium must be "level surfaces" of the effective gravity $\mathbf{F}$ as well. This I shall now prove using line integrals. I employ line integrals in order to pattern the proof upon the arguments that involve the channels which Clairaut made use of to derive his principles of equilibrium. More precisely, if instead of a stratified fluid figure of equilibrium made up of a finite number of individually homogeneous strata whose thicknesses are all finite, we consider a stratified fluid figure of equilibrium whose density P varies continuously, then the demonstration in note 47 of Chapter 9 can be appropriately modified in such a way as to produce the same result as that demonstration, *provided* $\int_\gamma \mathbf{F} \cdot d\mathbf{s} = 0$ around all closed plane or skew channels γ that lie inside the stratified fluid figure of equilibrium whose density P varies continuously and that are filled with fluid from the stratified fluid figure of equilibrium, where $\mathbf{F}$ is the effective gravity.

In hypothesizing that the density P varies continuously, I assume the following: Let Q stand for a point of a surface of constant density. Suppose that the effective gravity at Q is not perpendicular to this surface at Q. Then the effective gravity at Q must have a component at Q which is tangent to this surface at Q. Consequently one can choose a closed plane curve $HKNQR$ in this surface such that the line that indicates the direction of this component of the effective gravity at Q, which is tangent to this surface at Q, lies in the plane of $HKNQR$ and is tangent to $HKNQR$ at Q. (Such a closed plane curve $HKNQR$ is not uniquely determined.) Let the plane of this closed curve $HKNQR$ be the xy – plane. Then $y = a$ constant is the local equation of this curve. In other words, $y = a$ constant is the local equation of a cross section of the surface of constant density. Then we assume that dP/dy is not equal to zero on any open interval, no matter how small.

I present as well a more algebraic demonstration of this same finding, which Lagrange introduced in 1760–61. In his customary fashion, Lagrange considered the analytic or algebraic notion of the integral of a differential alone in arriving at the result. He made no use of line integrals in his proof. We consider a small region through which neighboring surfaces of constant density pass, whose portions within the region we can take to be parallel to each other. Let $A(x, y) \equiv$ the component of $\mathbf{F}$ at (x, y) which is parallel to these portions of neighboring surfaces of constant density, and let $B(x, y) \equiv$ the component of $\mathbf{F}$ at (x, y) which is normal to these portions of neighboring surfaces of constant density. Then

$$0 = \oint P\,\mathbf{F} \cdot d\mathbf{s} = \int_{x_0}^{x_1} P(y_1)A(x, y_1)dx - \int_{x_0}^{x_1} P(y_0)A(x, y_0)\,dx$$
$$+ \int_{y_0}^{y_1} P(y)B(x_0, y)dy - \int_{y_0}^{y_1} P(y)B(x_1, y)\,dy.$$

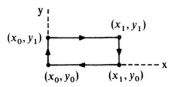

Figure 75. "y = a constant" is the equation of surfaces of constant density P.

To simplify things, I shall assume that all functions are analytic in a region that includes the paths of integration. Then the expression on the right-hand side of the preceding equation can be written as

$$\left[\int_{x_0}^{x_1} \frac{\partial}{\partial y}(P(y)A(x,y))\bigg|_{(x,y_0)} dx\right](y_1 - y_0) + O((y_1 - y_0)^n)$$
$$+ P(y_0)[B(x_0,y_0) - B(x_1,y_0)](y_1 - y_0) + O(y_1 - y_0)^n),$$

where $O((y_1 - y_0)^n)$ denotes terms in $(y_1 - y_0)^n$, where n is $\geqslant 2$. If we now divide through by $y_1 - y_0 \neq 0$, then take the limit as $y_1 - y_0 \to 0$, we obtain

$$0 = \int_{x_0}^{x_1} \frac{\partial}{\partial y}(P(y)A(x,y))\bigg|_{(x,y_0)} dx + P(y_0)[B(x_0,y_0) - B(x_1,y_0)].$$

Now, this equation can be further rewritten as

$$0 = \frac{dP}{dy}\bigg|_{y=y_0} \int_{x_0}^{x_1} A(x,y_0)dx + P(y_0)\int_{x_0}^{x_1} \frac{\partial A}{\partial y}(x,y)\bigg|_{(x,y_0)} dx$$
$$+ P(y_0)[B(x_0,y_0) - B(x_1,y_0)].$$

Again, this can be written as

$$0 = \frac{dP}{dy}\bigg|_{y=y_0}[A(x_0,y_0)(x_1 - x_0) + O((x_1 - x_0)^n)]$$
$$+ P(y_0)\left[\frac{\partial A}{\partial y}(x,y)\bigg|_{(x_0,y_0)}(x_1 - x_0) + O((x_1 - x_0)^n)\right]$$
$$- P(y_0)\left[\frac{\partial B}{\partial x}(x,y)\bigg|_{(x_0,y_0)}(x_1 - x_0) + O((x_1 - x_0)^n)\right],$$

where $O((x_1 - x_0)^n)$ denotes terms in $(x_1 - x_0)^n$, where n is $\geqslant 2$. If we now divide through by $x_1 - x_0 \neq 0$, then take the limit as $x_1 - x_0 \to 0$, we obtain

$$0 = \frac{dP}{dy}\bigg|_{y=y_0} A(x_0,y_0) + P(y_0)\left[\frac{\partial A}{\partial y}(x,y)\bigg|_{(x_0,y_0)} - \frac{\partial B}{\partial x}(x,y)\bigg|_{(x_0,y_0)}\right].$$

In particular, if the effective gravity $\mathbf{F} = (A, B)$ is such that

$$\int_\gamma \mathbf{F} \cdot d\mathbf{s} = 0$$

around all closed plane or skew channels γ that lie inside the stratified fluid figure of equilibrium whose density P varies continuously and that are filled with fluid from the stratified fluid figure of equilibrium, then

$$\frac{\partial A}{\partial y} = \frac{\partial B}{\partial x},$$

and if, as we have assumed, dP/dy is not equal to zero on any open interval, no matter how small, so that

$$\frac{dP}{dy} \neq 0$$

except perhaps at isolated points, then it follows from the last equation that

$$0 = A(x_0, y_0),$$

or that there is no component of the effective gravity which is tangent to a surface of constant density. In other words, the surfaces of constant density are level surfaces of the effective gravity F. This proof involves more than what can be read back into Clairaut's own demonstration of his fundamental theorem for stratified fluid figures of equilibrium made up of finite numbers of unmixed, individually homogeneous strata whose thicknesses are all finite. In D'Alembert (1752), D'Alembert gave an argument to show that Clairaut's theory was not general. Namely, he showed that

$$\int_{\gamma} \mathbf{F} \cdot d\mathbf{s} = 0$$

is *not* true in general around closed plane or skew channels γ that lie inside a heterogeneous fluid figure of equilibrium and that are filled with homogeneous fluid or, for that matter, are filled with heterogeneous fluid from the heterogeneous fluid figure of equilibrium. Moreover, he showed that the surfaces of constant density within a heterogeneous fluid figure of equilibrium whose density P varies continuously need not in general be level surfaces of the effective gravity F. In proving this, D'Alembert introduced

$$\int_{\gamma} P\mathbf{F} \cdot d\mathbf{s} = 0$$

around all closed channels γ that lie inside a heterogeneous fluid figure of equilibrium and that are filled with fluid from the heterogeneous fluid figure of equilibrium as the principle governing the equilibrium of heterogeneous fluid masses. My demonstration shows that in the event that

$$\int_{\gamma} \mathbf{F} \cdot d\mathbf{s} = 0$$

does happen to hold around all closed channels γ that lie inside a heterogeneous fluid figure of equilibrium whose density P varies continuously and that are filled with fluid from the heterogeneous fluid figure of equilibrium, then it turns out that the surfaces of constant density within the heterogeneous fluid figure of equilibrium are all level surfaces of the effective gravity F. It would appear that D'Alembert judged on the face of it that Clairaut did indeed mean to claim that the same laws always govern the equilibrium of both homogeneous and heterogeneous fluid figures in *all* instances, for D'Alembert intended to correct any wrong impression that

$$\int_{\gamma} \mathbf{F} \cdot d\mathbf{s} = 0$$

around closed channels γ that lie inside a fluid mass and that are filled with homogeneous fluid or, for that matter, are filled with fluid from the mass – a result that need not hold in a heterogeneous fluid mass in equilibrium – is the condition for the equilibrium of a heterogeneous fluid mass. Although Clairaut did not exhibit in his treatise of 1743 a mathematical representation of a total effort equal to zero around a closed channel that lies within the interior of a *heterogeneous* fluid figure of equilibrium and that is filled with fluid from the heterogeneous figure, it hardly seems reasonable to think that he could have failed to realize that

$$\int_{\gamma} P\mathbf{F} \cdot d\mathbf{s} = 0$$

around all closed channels γ that lie inside a heterogeneous fluid figure of equilibrium and that are filled with fluid from the heterogeneous fluid figure of equilibrium expresses this condition in general, and in particular when the density P of the figure varies continuously, because (1)

$$\int_\Gamma P\mathbf{F}\cdot d\mathbf{s}$$

obviously expresses the "effort" along a channel Γ that lies inside a heterogeneous fluid mass whose variable density is P and that is filled with fluid from the heterogeneous fluid mass, since $P\mathbf{F}\cdot d\mathbf{s}$ has units of "weight"; (2) if the principle of solidification is applied to a closed channel γ that lies within the interior of a heterogeneous fluid figure of equilibrium and that is filled with fluid from the heterogeneous fluid figure of equilibrium, then

$$\int_\gamma P\mathbf{F}\cdot d\mathbf{s} = 0$$

around the channel is what follows; and (3) the only way to make any sense at all out of Clairaut's own derivation of his fundamental theorem for stratified fluid figures of equilibrium composed of finite numbers of individually homogeneous strata whose thicknesses are all finite, which I derived mathematically in note 47 of Chapter 9, is to assume that

$$\int_\gamma P\mathbf{F}\cdot d\mathbf{s} = 0$$

expresses the total effort equal to zero around closed channels γ that lie within the interiors of such figures and that are filled with fluid from the figures. Indeed, as I explained in Section 2 of Chapter 9, Clairaut's theory of stratified fluid figures of equilibrium composed of finite numbers of individually homogeneous strata whose thicknesses are all finite in fact implicitly involves D'Alembert's condition for equilibrium. I note that Clairaut and D'Alembert were not on speaking terms in 1752, because of their bitter rivalry (see note 143 of Chapter 9). This could have partly induced D'Alembert to call in question the generality of Clairaut's theory of figures of equilibrium at this time. Apart from politics, however, D'Alembert also probably meant to criticize Clairaut's applications of his theorem I mentioned earlier to cases that Clairaut did not appear to cover in his exposition of his theory – namely, to figures whose densities vary *continuously*.

In 1760–61, Lagrange showed by simply algebraically manipulating differentials, which was typical of Lagrange's mathematical style, that if the density P of a heterogeneous fluid figure of equilibrium varies continuously, and if

$$\int_\gamma \mathbf{F}\cdot d\mathbf{s} = 0$$

around all closed plane or skew channels γ that lie within the interior of a heterogeneous fluid figure of equilibrium and that are filled with fluid from the heterogeneous fluid figure of equilibrium, where $\mathbf{F}$ is the effective gravity, which Clairaut had shown to be true when bodies attract according to the universal inverse-square law, then surfaces of constant density within the heterogeneous fluid figure of equilibrium are always level surfaces of the effective gravity $\mathbf{F}$. This Lagrange proved as follows: he argued that if $\vec{F} = (\pi, \bar\omega, \psi)$ represents effective gravity, and if $D(x, y, z) = a$ constant k is a surface of constant density, which appears as

$$\frac{dD}{dx}dx + \frac{dD}{dy}dy + \frac{dD}{dz}dz = 0$$

in differential form, then the conditions

$$\frac{d}{dy}(D\pi) = \frac{d}{dx}(D\bar\omega), \quad \frac{d}{dz}(D\pi) = \frac{d}{dx}(D\psi)$$

for equilibrium (in other words $\oint D\vec{F}\cdot d\vec{s} = 0$) can be used to transform the equation

$$\frac{dD}{dx}dx + \frac{dD}{dy}dy + \frac{dD}{dz}dz = 0$$

for a surface of constant density into the equation

$$\frac{1}{D}\frac{dD}{dx}(\pi dx + \bar{\omega}dy + \psi dz) + \left(\frac{d\bar{\omega}}{dx} - \frac{d\pi}{dy}\right)dy + \left(\frac{d\psi}{dx} - \frac{d\pi}{dz}\right)dz = 0.$$

But this reduces to $\pi dx + \bar{\omega}dy + \psi dz = 0$ precisely when

$$\int_\gamma \mathbf{F}\cdot d\mathbf{s} = 0$$

around all closed plane or skew channels γ that lie inside a heterogeneous fluid figure of equilibrium and that are filled with fluid from the heterogeneous fluid figure of equilibrium, because in that event (1)

$$\frac{d\bar{\omega}}{dx} = \frac{d\pi}{dy} \quad \text{and} \quad \frac{d\psi}{dx} = \frac{d\pi}{dz};$$

and (2) dD/dx is $\neq 0$ (in particular, D is not constant, even locally). But $\pi dx + \bar{\omega}dy + \psi dz = 0$ is the equation of a level surface of the effective gravity $\mathbf{F} = (\pi, \bar{\omega}, \psi)$. [See Lagrange (1760–61: 283–284), which reappears in Lagrange (1867: 454–455).]

115. Clairaut (1743: 273, and the figure on p. 272).
116. Ibid., p. 275.
117. Ibid., p. 276.
118. In fact, Clairaut actually wrote the "differential" equation (9.4.5) as

$$dd\rho = \rho\,dr^2\left(\frac{6}{rr} - \frac{2Rr}{\int_{r=0}^{r=r}Rrr\,dr}\right) - \frac{2Rrr\,dr}{\int_{r=0}^{r=r}Rrr\,dr}d\rho$$

(Ibid., p. 276). I rewrote it in the text by dividing both sides of the equation by dr^2. I did so because the equation written this way looks like a "differential equation." For a detailed discussion of the two kinds of equations and their differences, see Bos (1974–75). What is important is that the "differential" equation, the "differential equation," and the modern differential equation are identical. That is, all three equations express the same conditions. This can be explained as follows: if differential coefficients $\rho^1, \rho^2, \ldots, \rho^n$ are defined by

$$d\rho = \rho^1\,dr, \quad d\rho^1 = \rho^2\,dr, \quad d\rho^2 = \rho^3\,dr, \ldots, d\rho^{n-1} = \rho^n\,dr,$$

then if dr is held constant (that is, r functions as an independent variable), then

$$d^n\rho = \rho^n\,dr^n,$$

or

$$\frac{d^n\rho}{dr^n} = \rho^n,$$

which means that ρ^n as defined must equal what we mean today by the nth derivative of ρ with respect to r. For further discussion of the equivalance stated here, see Bos (1974–75: 31, 72–74; 1986: 113–114); Fraser (1985: 39); and Vygodsky (1975: 326–330).
119. Clairaut (1743: 276). In fact, Clairaut actually wrote the "differential" equation (9.4.7) as:

$$dt + tt\,dr = \frac{rr\,dR}{\int_{r=0}^{r=r}Rrr\,dr} + \frac{6dr}{rr}.$$

I rewrote it in the text by dividing both sides of the equation by dr, in which case it looks like a "differential equation." But the "differential" equation, the "differential equation", and the modern differential equation do not really differ in any essential way (see note 118 of Chapter 9).

120. Clairaut (1764: 285).
121. Clairaut (1743: 276).
122. Kline (1972: 483); Pepe (1981: 72; 1986: 199–201).
123. Clairaut (1743: 276).
124. It was not until 1841 that Liouville demonstrated conclusively that the solution $y = y(x)$ of the Riccati equation

$$\frac{dy}{dx} + ay^2 = x^2,$$

where $a > 0$, cannot be expressed as a finite combination of integrals of elementary functions of x [see Petrovskii (1969: 330); Kasper (1980); and Lützen (1990: 353)]. Moreover, the generalized Riccati equation

$$\frac{dy}{dx} + \psi y^2 + \phi y + \chi = 0,$$

where ψ, ϕ, and χ are functions of x, is not in general even reducible to quadratures $\int f(x)\,dx$, $\int g(y)\,dy$. The equation therefore defines a family of transcendental functions that are essentially distinct from the elementary transcendents [see Ince (1944: 23) and Lützen (1990: 353)]. In general then Riccati equations are neither integrable in finite terms nor even reducible to quadratures. Consequently the solutions of the Riccati equation and the generalized Riccati equation have no finite representations in the sense that I discussed in Chapter 7.
125. Clairaut (1743: 278).
126. See note 106 of Chapter 9.
127. Clairaut (1743: 278). This condition is

$$\delta(q - \tfrac{1}{2} - n) - \delta a(q + \tfrac{1}{2} + n) = \tfrac{5}{2}\phi, \quad \text{where} \quad q = \sqrt{n^2 + 3n + \tfrac{25}{4}}.$$

This condition is derived as follows: The first differentiation of

$$5r^2 \rho \int Rrr\,dr = \int Rd(r^5 \rho) + Fr^5 + \tfrac{5}{2} A\phi r^5 - r^5 \int Rd\rho$$

produces an equation

$$\left(\frac{1}{r^2}\frac{d\rho}{dr} + \frac{2\rho}{r^3}\right)\int Rrr\,dr = F - \int Rd\rho + \tfrac{5}{2} A\phi,$$

which at $r = 1$ becomes

$$\left(\frac{d\rho}{dr}\bigg|_{r=1} + 2\rho(1)\right) A = \tfrac{5}{2} A\phi, \quad \text{or} \quad \frac{d\rho}{dr}\bigg|_{r=1} + 2\rho(1) = \tfrac{5}{2}\phi.$$

In other words the two initial conditions

$$\rho(1), \quad \frac{d\rho}{dr}\bigg|_{r=1}$$

[which happen, in fact, to be sufficient to fix uniquely a solution to the second order ordinary "differential" equation for $\rho(r)$] must satisfy the equation

$$\frac{d\rho}{dr}\bigg|_{r=1} + 2\rho(1) = \tfrac{5}{2}\phi,$$

if the integro-differential equation is to be satisfied as well. In terms of the solution

$$\rho = \delta a r^{-n-5/2-\sqrt{nn+3n+(25/4)}} + \delta r^{-n-5/2+\sqrt{nn+3n+(25/4)}}$$

to the second-order ordinary "differential" equation, this translates into the condition

$$\delta(\sqrt{nn + 3n + \tfrac{25}{4}} - \tfrac{1}{2} - n) - \delta a(\sqrt{nn + 3n + \tfrac{25}{4}} + \tfrac{1}{2} + n) = \tfrac{5}{2}\phi$$

on the constants of integration a, δ. Any solution to the second-order ordinary "differential" equation, whose initial conditions (equivalently, constants of integration) satisfy this relation, cannot fail to be a solution to the equation obtained by the first differentiation of the integro-differential equation as well.

The reason is as follows: Suppose that an expression that has form

$$F(y(r), \frac{d}{dr} y(r), r)$$

satisfies

$$\frac{d}{dr} F(y(r), \frac{d}{dr} y(r), r) = 0$$

when $y(r) = \rho(r)$. Then of course

$$F(\rho(r), \frac{d}{dr} \rho(r), r) = a \text{ constant.}$$

If, moreover,

$$F(\rho(1), \frac{d}{dr} \rho(r)\Big|_{r=1}, 1) = 0$$

then the value of the constant must be zero. That is,

$$F(\rho(r), \frac{d}{dr} \rho(r), r) = 0.$$

128. As Clairaut expressed the matter, every solution of the "differential" equation

$$\frac{d^2\rho}{dr^2} = \frac{6\rho}{rr} - \frac{2R\rho r}{\int_{r=0}^{r=r} Rrrdr} - \frac{2Rrr}{\int_{r=0}^{r=r} Rrrdr} \frac{d\rho}{dr} \tag{9.4.5}$$

which happens to satisfy as well the "differential" equation

$$\left(\frac{1}{r^2}\frac{d\rho}{dr} + \frac{2\rho}{r^3}\right) \int_{r=0}^{r=r} Rrrdr = F - \int_{r=0}^{r=r} Rd\rho + \tfrac{5}{2}A\phi,$$

which is arrived at by differentiating the integro-differential equation

$$5r^2\rho \int_{r=0}^{r=r} Rrrdr = \int_{r=0}^{r=r} Rd(r^5\rho) + Fr^5 + \tfrac{5}{2}A\phi r^5 - r^5 \int_{r=0}^{r=r} Rd\rho \tag{9.4.3}$$

once, cannot fail to satisfy the integro-differential equation (9.4.3) itself too [Clairaut (1743: 275, 278)]. This can be understood as follows: Now, in order to specify $\rho(1)$, $d\rho/dr\big|_{r=1}$ uniquely (or, alternatively, a, δ) the integro-differential equation itself should determine a similar relation between $\rho(1)$ and $d\rho/dr\big|_{r=1}$ (or a and δ). But the problem is this: there is no second, independent condition on $\rho(1)$, $d\rho/dr\big|_{r=1}$ determined in this way, so that $\rho(1)$ can be chosen arbitrarily in effect. Or, to put it in terms of a and δ, either one of a, δ may be chosen arbitrarily, and then the other chosen to satisfy

$$\delta(\sqrt{nn + 3n + \tfrac{25}{4}} - \tfrac{1}{2} - n) - \delta a(\sqrt{nn + 3n + \tfrac{25}{4}} + \tfrac{1}{2} + n) = \tfrac{5}{2}\phi$$

which appears to be the only condition that does need to be satisfied,

This is true in this particular case for the following reason: Choosing a and δ to satisfy

$$\delta(\sqrt{nn + 3n + \tfrac{25}{4}} - \tfrac{1}{2} - n) - \delta a(\sqrt{nn + 3n + \tfrac{25}{4}} + \tfrac{1}{2} + n) = \tfrac{5}{2}\phi,$$

which amounts to choosing

$$\rho(1), \frac{d\rho}{dr}\bigg|_{r=1} \qquad \text{so that} \qquad \frac{d\rho}{dr}\bigg|_{r=1} + 2\rho(1) = \tfrac{5}{2}\phi,$$

would merely assure that the corresponding solution to

$$\frac{d^2\rho}{dr^2} = \frac{6\rho}{r^2} - \frac{2R\rho r}{\int Rr dr} - \frac{2Rrr}{\int Rr dr}\frac{d\rho}{dr}$$

will also satisfy

$$\left(\frac{1}{r^2}\frac{d\rho}{dr} + \frac{2\rho}{r^3}\right)\int Rr dr = F - \int R d\rho + \tfrac{5}{2}A\phi$$

as well.

$$5r^2\rho \int Rr dr - \int Rd(r^5 \rho) - Fr^5 - \tfrac{5}{2}A\phi r^5 + r^5 \int Rd\rho$$

will certainly = a constant in that case, but, in general, $\rho(1), d\rho/dr|_{r=1}$ chosen as above will not be enough to insure that this constant be zero. To guarantee that this will be the case, $\rho(1), d\rho/dr|_{r=1}$ must also be chosen so that

$$5r^2\rho \int Rr dr - \int Rd(r^5 \rho) - Fr^5 - \tfrac{5}{2}A\phi r^5 + r^5 \int Rd\rho \text{ at } r = 1$$

will be zero – in other words, so that

$$5\rho(1)A - \int_{r=0}^{r=1} Rd(r^5 \rho) - \tfrac{5}{2}A\phi = 0.$$

Or, writing this as

$$3\rho(1)A - \int_{r=0}^{r=1} Rd(r^5 \rho) = \tfrac{5}{2}A\phi - 2\rho(1)A,$$

and then making use of the first condition

$$\tfrac{5}{2}\phi - 2\rho(1) = \frac{d\rho}{dr}\bigg|_{r=1},$$

we must have

$$3\rho(1)A - \int_{r=0}^{r=1} Rd(r^5 \rho) = A\frac{d\rho}{dr}\bigg|_{r=1}.$$

That is, the pair of conditions

$$\frac{d\rho}{dr}\bigg|_{r=1} + 2\rho(1) = \tfrac{5}{2}\phi, \quad 3\rho(1)A - A\frac{d\rho}{dr}\bigg|_{r=1} = \int_{r=0}^{r=1} Rd(r^5 \rho)$$

on $\rho(1), d\rho/dr|_{r=1}$ must hold. However, these two equations in $\rho(1), d\rho/dr|_{r=1}$ turn out not to fix $\rho(1), d\rho/dr|_{r=1}$ uniquely for the following reason: If we take the general solution

$$\rho(r) = \delta a r^{-n - 5/2 - \sqrt{nn + 3n + 25/4}} + \delta r^{-n - 5/2 + \sqrt{nn + 3n + 25/4}}$$

to

$$\frac{d^2\rho}{dr^2} = \frac{6\rho}{r^2} - \frac{2R\rho r}{\int Rrrdr} - \frac{2Rrr}{\int Rrrdr}\frac{d\rho}{dr},$$

when $R(r) = r^n$, and then substitute into

$$3\rho(1)A - A\frac{d\rho}{dr}\bigg|_{r=1} = \int_{r=0}^{r=1} Rd(r^5\rho)$$

– or, alternatively, into

$$3\rho(1)A - A\frac{d\rho}{dr}\bigg|_{r=1} = R(r^5\rho)\bigg|_{r=0}^{r=1}$$

$$- \int_{r=0}^{r=1} r^5\rho dR = R(1)\rho(1) - \int_{r=0}^{r=1} r^5\rho\frac{dR}{dr}dr$$

after integrating by parts, which is easier to work with – we obtain an *identity* valid for all a and δ. In other words, when $R(r) = r^n$, *every* solution

$$\rho(r) = \delta a r^{-n-5/2-\sqrt{nn+3n+25/4}} + \delta r^{-n-5/2+\sqrt{nn+3n+25/4}}$$

to the second-order ordinary "differential" equation satisfies

$$3\rho(1)A - A\frac{d\rho}{dr}\bigg|_{r=1} = \int_{r=0}^{r=1} Rd(r^5\rho),$$

so that this condition places no restrictions on a, δ (that is, on $\rho(1)$, $d\rho/dr|_{r=1}$). This is what Clairaut must have meant when he said (p. 278) that any solution to

$$\frac{d^2\rho}{dr^2} = \frac{6\rho}{r^2} - \frac{2R\rho r}{\int Rrrdr} - \frac{2Rrr}{\int Rrrdr}\frac{d\rho}{dr},$$

which happens to satisfy

$$\left(\frac{1}{r^2}\frac{d\rho}{dr} + \frac{2\rho}{r^3}\right)\int Rrrdr = F - \int Rd\rho + \tfrac{5}{2}A\phi$$

as well [that is, for which $\rho(1)$, $d\rho/dr|_{r=1}$ satisfy $d\rho/dr|_{r=1} + 2\rho(1) = \tfrac{5}{2}\phi$] cannot fail to satisfy

$$5r^2\rho\int Rrrdr = \int Rd(r^5\rho) + Fr^5 + \tfrac{5}{2}A\phi r^5 - r^5\int Rd\rho$$

at the same time.

129. Clairaut (1743: 278–279).
130. Ibid., p. 281. To see that the exponent

$$-n - \tfrac{5}{2} - \sqrt{n^2 + 3n + \tfrac{25}{4}} = -(n + \tfrac{5}{2} + \sqrt{n + 3n + \tfrac{25}{4}})$$

in the first term on the right-hand side of expression (9.4.10) is always negative when n is negative, I show that

$$n + \tfrac{5}{2} + \sqrt{n^2 + 3n + \tfrac{25}{4}}$$

is always positive when n is negative. Let $n = -m$, $m > 0$. Then

$$n^2 + 3n + \tfrac{25}{4} = m^2 - 3m + \tfrac{25}{4} > m^2 - 5m + \tfrac{25}{4} = (\tfrac{5}{2} - m)^2 \geqslant 0.$$

Consequently

$$\sqrt{n^2 + 3n + \tfrac{25}{4}} > \sqrt{(\tfrac{5}{2} - m)^2} = |\tfrac{5}{2} - m| = |\tfrac{5}{2} + n| \geqslant -(\tfrac{5}{2} + n).$$

Hence

$$n + \tfrac{5}{2} + \sqrt{n^2 + 3n + \tfrac{25}{4}} > 0.$$

Alternative derivation: Let $n = -m$, $m > 0$. Then

$$n^2 + 3n + \tfrac{25}{4} = m^2 - 3m + \tfrac{25}{4} > m^2 - 5m + \tfrac{25}{4} = n^2 + 5n + \tfrac{25}{4} = (n + \tfrac{5}{2})^2 \geqslant 0.$$

Therefore

$$\sqrt{n^2 + 3n + \tfrac{25}{4}} > \sqrt{(n + \tfrac{5}{2})^2} = |n + \tfrac{5}{2}| \geqslant -(n + \tfrac{5}{2}).$$

Hence

$$n + \tfrac{5}{2} + \sqrt{n^2 + 3n + \tfrac{25}{4}} > 0.$$

To see that the exponent

$$-n - \tfrac{5}{2} + \sqrt{n^2 + 3n + \tfrac{25}{4}}$$

in the second term on the right-hand side of expression (9.4.10) is always positive when n is negative, let $n = -m$, $m > 0$. Then

$$n^2 + 3n + \tfrac{25}{4} = m^2 - 3m + \tfrac{25}{4} > m^2 - 5m + \tfrac{25}{4} = n^2 + 5n + \tfrac{25}{4} = (n + \tfrac{5}{2})^2 \geqslant 0.$$

Therefore

$$\sqrt{n^2 + 3n + \tfrac{25}{4}} > \sqrt{(n + \tfrac{5}{2})^2} = |n + \tfrac{5}{2}| \geqslant n + \tfrac{5}{2}.$$

Hence

$$-n - \tfrac{5}{2} + \sqrt{n^2 + 3n + \tfrac{25}{4}} > 0.$$

Alternative derivation: Let $n = -m$, $m > 0$. Then

$$n^2 + 3n + \tfrac{25}{4} = m^2 - 3m + \tfrac{25}{4} > m^2 - 5m + \tfrac{25}{4} = (\tfrac{5}{2} - m)^2 \geqslant 0.$$

Consequently

$$\sqrt{n^2 + 3n + \tfrac{25}{4}} > \sqrt{(\tfrac{5}{2} - m)^2} = |\tfrac{5}{2} - m| = |\tfrac{5}{2} + n| \geqslant \tfrac{5}{2} + n.$$

Hence

$$-n - \tfrac{5}{2} + \sqrt{n^2 + 3n + \tfrac{25}{4}} > 0.$$

131. Clairaut (1743: 282-283). The corresponding equation

$$\frac{d^2\rho}{dr^2} = \frac{6\rho}{rr} - \frac{2R\rho r}{\int_{r=0}^{r=r} Rr dr} - \frac{2Rr r}{\int_{r=0}^{r=r} Rr r dr} \frac{d\rho}{dr} \qquad (9.4.5)$$

can be written in this instance as

$$\frac{Rrr}{\int Rrr dr} = \frac{6\rho - r^2 d^2\rho/dr^2}{2\rho r - 2r^2 d\rho/dr}.$$

Here the right-hand side

$$\frac{6\rho - r^2 d^2\rho/dr^2}{2\rho r - 2r^2 d\rho/dr} \equiv P(r)$$

is known, since $\rho = \rho(r)$ is given, while the left-hand side has a logarithmic integral (that is, $w(r) \equiv \ln |\int Rrr dr|$ satisfies the first-order "differential" equation $dw/dr = P(r)$) from which it readily follows that

$$R(r) = \frac{P(r)}{rr} c^{\int P(r) dr}$$

(where c satisfies $\ln c = 1$, or $c = e$ in other words).

Clairaut noted that when the ellipticity $\rho(1)$ of a planet and the ratio ϕ of the magnitude of the centrifugal force per unit mass to the magnitude of the effective gravity at the equator of the planet are known from observations, it might appear at first sight to be enough to consider any function $\rho(r)$ of r which satisfies $\lim_{r\to 1}\rho(r) = \rho(1)$, in investigating how the density $R(r)$ might vary as a function of r in the interior of the planet. [While Clairaut did not mention this, the first-order ordinary "differential" equation $dw/dr = P(r)$ does not have unique solutions. One constant of integration must be specified. This means that the expressions $R(r)$ which correspond to solutions of this equation all have the form

$$R(r) = \frac{k P(r)}{rr} e^{\int P(r) dr},$$

where k is a real-valued constant that does not equal zero which must be specified. Clairaut implicitly chose k to equal 1.] However, just as in the preceding problem, the constant ϕ in the integro-differential equation (9.4.3) disappears when this equation is differentiated twice to produce equation (9.4.5). Consequently, Clairaut pointed out, it does not suffice to consider any function $\rho(r)$ of r which simply satisfies $\lim_{r\to 1}\rho(r) = \rho(1)$, in order to investigate how the density $R(r)$ might vary as a function of r in the interior of a planet. The function $\rho(r)$ must also fulfill the condition

$$d\rho/dr\Big|_{r=1} + 2\rho(1) = \tfrac{5}{2}\phi,$$

just as in the preceding problem, in order for the integro-differential equation (9.4.3) to be satisfied at the same time. (See note 127 of Chapter 9.) Therefore, Clairaut concluded, if the ellipticity $\rho(1)$ of a planet and the ratio ϕ of the magnitude of the centrifugal force per unit mass to the magnitude of the effective gravity at the equator of the planet are given by observation, only functions $\rho(r)$ of r which satisfy both $\lim_{r\to 1} = \rho(1)$ and $d\rho/dr|_{r=1} + 2\rho(1) = \tfrac{5}{2}\phi$ can be considered when investigating how the density $R(r)$ might vary as a function of r in the interior of the planet, assuming, of course, that the planet's internal structure obeys the laws of hydrostatics (that is, assuming that the planet is a stratified figure of equilibrium shaped like a flattened ellipsoid of revolution whose ellipticity is infinitesimal, which revolves around its axis of symmetry, and which attracts according to the universal inverse-square law) [Clairaut (1743: 286)].

132. Clairaut (1743: 286–291).
133. Ibid., pp. 291–292.
134. Ibid., p. 155.
135. Ibid., p. 155.
136. Ibid., pp. 229–232. In a letter to MacLaurin dated 19 November 1742, in which he acknowledged receipt of a copy of MacLaurin's *A Treatise of Fluxions*. Clairaut already suggested that his discovery explained why their findings in these cases disagreed with each other. In this letter Clairaut mentioned specifically the third example in the text, also referred to in note 107 of Chapter 9. [See Clairaut (1742f: 376–377).]
137. See the second half of note 130 of Chapter 9. Thus the exponent in the expression (9.4.10′) for $\rho(r)$ is positive, which means that $\rho(r)$ in (9.4.10′) increases on the closed interval $[0, 1]$.
138. Clairaut (1743: 257).
139. Ibid., p. 258.
140. $\rho(r) = 0$, $R(r) =$ a constant $K \neq 0$ within the spherical nucleus; $\rho(r) = 0$, $R(r) = 0$ inside the cavity; and $\rho(r) =$ a constant $C \neq 0$, $R(r) = K$ within the outer layer.
141. Clairaut (1743: 257–259).
142. In D'Alembert (1752: 202), D'Alembert ascribed the notion of a level surface to MacLaurin.
143. In 1751, Clairaut and D'Alembert severed all personal and professional relations with each other, as a result of their intense rivalry [see Hankins (1970: 36)].

144. Clairaut (1742c: 103).
145. Ibid., p. 103.
146. In the letter that he sent Euler in January of 1742, Clairaut wrote the integro-differential equation

$$5r^2z \int_{r=0}^{r=r} Rrrdr = \int_{r=0}^{r=r} Rd(r^5z) + F + \tfrac{5}{2}(A/m) - r^5 \int_{r=0}^{r=r} Rdz$$

[Clairaut (1742c: 103)]. The term $F + \tfrac{5}{2}(A/m)$ needs to be multipled by r^5 to produce an integro-differential equation corresponding to the one that appears in Clairaut's treatise of 1743 (equation (9.4.3)). Moreover, the first-order ordinary "differential" equation that Clairaut wrote in the same letter has the form

$$dt + \frac{ttds}{S(s)} = ds$$

(ibid., p. 103), whereas the correct first-order ordinary "differential" equation, corresponding to the one that appears in the treatise of 1743 (equation (9.4.7)), should have the form

$$dt + ttds = S(s)ds.$$

147. Clairaut (1742g: 135–136). In this letter, Clairaut wrote the equation as

$$5r^2z \int_{r=0}^{r=r} Rrrdr = \int_{r=0}^{r=r} Rd(r^5z) + hr^5 - r^5 \int_{r=0}^{r=r} Rdz, \quad \text{where } h \equiv F + \tfrac{5}{2}(A/m).$$

148. Clairaut (1742h: 138).
149. In his letter to MacLaurin dated 19 November 1742, Clairaut told MacLaurin that a copy of *A Treatise of Fluxions* had arrived "about six weeks ago." [See Clairaut (1742f: 375).]
150. Clairaut (1743: 231–232).
151. Ibid., p. 258.
152. Clairaut (1742f: 377, and the figure on p. 380).
153. Clairaut (1743: 259).
154. Clairaut (1742f: 378).
155. Clairaut (1743: 157–158).
156. Ibid., pp. 188, 197.
157. Ibid., pp. 191–192. The infinite series for the ellipticity δ is:

$$\delta = \tfrac{5}{4}\phi + \tfrac{5}{224}\phi^2 - \tfrac{135}{6272}\phi^3 + \cdots . \tag{I}$$

Now, ϕ is infinitesimal, which means that $\phi^2 \ll \phi$. Hence

$$\tfrac{5}{4}\phi^2 \ll \tfrac{5}{4}\phi. \tag{II}$$

At the same time

$$\tfrac{5}{224}\phi^2 < \frac{5}{4}\phi^2 \tag{III}$$

is also true. If (II) and (III) are combined, then

$$\tfrac{5}{224}\phi^2 < \tfrac{5}{4}\phi^2 \ll \tfrac{5}{4}\phi \tag{IV}$$

results. It follows from (IV) that

$$\tfrac{5}{224}\phi^2 \ll \tfrac{5}{4}\phi. \tag{V}$$

Alternatively, it follows from $\phi^2 \ll \phi$ that

$$\tfrac{5}{224}\phi^2 \ll \tfrac{5}{224}\phi \tag{VI}$$

as well. At the same time,

$$\tfrac{5}{224}\phi < \tfrac{5}{4}\phi \tag{VII}$$

is also true. If (VI) and (VII) are combined, then

$$\tfrac{5}{224}\phi^2 \ll \tfrac{5}{224}\phi < \tfrac{5}{4}\phi \qquad\qquad\text{(VIII)}$$

results. And it follows from (VIII) that

$$\tfrac{5}{224}\phi^2 \ll \tfrac{5}{4}\phi. \qquad\qquad\text{(V)}$$

But (V) means that the second term $\tfrac{5}{224}\phi^2$ in series (I) and all terms that follow $\tfrac{5}{224}\phi^2$ in series (I) can be neglected. Another way to arrive at the same conclusion is as follows: Let $\alpha \equiv \tfrac{5}{4}\phi$, let $\beta \equiv \tfrac{5}{224}\phi^2$, and let $\gamma \equiv \beta/\alpha$. Again, ϕ is infinitesimal means that $\phi^2 \ll \phi$. Then, since $\tfrac{5}{4}$ is a positive number that is not much larger than 1, α is also frequently infinitesimal ($\alpha^2 \ll \alpha$) if ϕ is infinitesimal. Now,

$$\gamma \equiv \frac{\beta}{\alpha} = \tfrac{1}{56}\phi.$$

Then, since $\tfrac{1}{56}\phi < \phi$ and ϕ is infinitesimal, it follows that γ is infinitesimal too ($\gamma^2 \ll \gamma$). But series (I) can be written in the following way:

$$\delta = \alpha + \beta - \tfrac{135}{6272}\phi^3 + \cdots = \alpha + \alpha\gamma - \tfrac{135}{6272}\phi^3 + \cdots. \qquad\qquad\text{(I$'$)}$$

Then, since α and γ are both infinitesimal, $\alpha\gamma$ is a term of second order. Hence $\alpha\gamma$ and all terms that follow $\alpha\gamma$ in series (I$'$) can be discarded.

158. See Greenberg (1988: 230).
159. Clairaut (1743: 218, and the figure on p. 210). Actually, it should be enough simply to set the densities of all of the strata that make up the nucleus equal to the density of the outer layer to produce this result. The reason is that in setting the densities of the strata all equal to the density of the outer layer, a homogeneous figure shaped like a flattened ellipsoid of revolution which attracts according to the universal inverse-square law, whose ellipticity is infinitesimal, and at whose surface the principle of the plumb line holds results, no matter how the ellipticities of the strata may vary. In particular, then, it should not be necessary to have to set $\rho(r) \equiv \alpha$ for $0 \leqslant r \leqslant a$ and $\alpha = \delta$ to produce the result in question. This in fact can easily be demonstrated. The argument is much the same as the one used in note 99 of Chapter 9. If $R(r) \equiv 1$ for $0 \leqslant r \leqslant 1$, $\rho(r) \equiv \alpha$ for $0 \leqslant r \leqslant a$, and $\alpha = \delta$, then

$$A \equiv \int_{r=0}^{r=a} R(r)rr\,dr = \int_{r=0}^{r=a} r^2\,dr = \frac{r^3}{3}\bigg|_{r=0}^{r=a} = \frac{a^3}{3},$$

and

$$D \equiv \int_{r=0}^{r=a,\rho(a)=\alpha} R(r)d(r^5\rho(r)) = \alpha\int_{r=0}^{r=a} d(r^5) = \alpha r^5\bigg|_{r=0}^{r=a} = \alpha a^5 = \delta a^5.$$

Then from equation (LVII) of note 98 of Chapter 9, it follows that

$$\delta = \frac{6a^5\delta - 6a^5\alpha + 15(a^3/3)\phi + 5\phi - 5a^3\phi}{30(a^3/3) + 4 - 10a^3},$$

and since $\alpha = \delta$,

$$\delta = \frac{5a^3\phi + 5\phi - 5a^3\phi}{4} = \frac{5}{4}\phi.$$

At the same time, if $R(r) \equiv 1$ for $0 \leqslant r \leqslant 1$, $\rho(r) \not\equiv \alpha$ for $0 \leqslant r < a$, and $\alpha \neq \delta$, then

$$A \text{ still} = \frac{a^3}{3},$$

and

$$D \equiv \int_{r=0}^{r=a,\rho(a)=\alpha} R(r)d(r^5\rho(r)) = \int_{r=0}^{r=a,\rho(a)=\alpha} d(r^5\rho(r)) = r^5\rho(r)\bigg|_{r=0}^{r=a,\rho(a)=\alpha} = a^5\rho(a) = a^5\alpha.$$

Then from equation (LVII) of note 98 of Chapter 9, it follows that

$$\delta = \frac{6a^5\alpha - 6a^5\alpha + 15(a^3/3)\phi + 5\phi - 5a^3\phi}{30(a^3/3) + 4 - 10a^3} = \frac{5a^3\phi + 5\phi - 5a^3\phi}{4} = \frac{5}{4}\phi.$$

In short, the same result follows from the two different hypotheses.

160. Clairaut (1743: 188–190); and MacLaurin (1801: vol. 2, pp. 116–117, §637).
161. See notes 7 and 8 of Chapter 1.
162. MacLaurin (1801: vol. 2, pp. 127–131, §648–§653).
163. For example, the attraction at a point p situated at a pole or on the equator equals:

$$\int_{r=0}^{r=R} D(r)\,dA(\epsilon(r),r) = \int_{r=0}^{r=R} D(r)\,d\left(\frac{V(\epsilon(r),r)}{V(\epsilon(r),R)}A(\epsilon(r),R)\right),$$

where $\epsilon(r)$ is the ellipticity as a function of the distance r along the polar or equatorial radius of the figure at whose endpoint p is situated; R is the length of this radius; $A(\epsilon(r),r)$ stands for the magnitude of the attraction at p due to a homogeneous figure shaped like an ellipsoid of revolution which has an axis of symmetry in common with the stratified figure, whose radius measured along the radius on which point p of the stratified figure lies has length r, whose ellipticity is $\epsilon(r)$, whose density equals 1, and which attracts according to the universal inverse-square law; $V(\epsilon(r),r)$ denotes the volume of this homogeneous figure; $D(r)$ symbolizes the density of the stratum whose radius measured along the radius on which point p of the stratified figure lies has length r; and $A(\epsilon(r),R)$ stands for the magnitude of the attraction at p due to a homogeneous figure shaped like an ellipsoid of revolution which has an axis of symmetry in common with the stratified figure, which is confocal with the figure whose attraction at p has magnitude $A(\epsilon(r),r)$, whose density is 1, and on whose surface p lies (in other words, whose radius measured along the radius on which point p of the stratified figure lies has length R), whose volume is $V(\epsilon(r),R)$, and which attracts according to the universal inverse-square law. $A(\epsilon(r),R)$ can be found using MacLaurin's theorems for the magnitude of the attraction at a pole or at the equator of a homogeneous figure shaped like an ellipsoid of revolution which attracts according to the universal inverse-square law. Moreover, $V(\epsilon(r),r)$ and $V(\epsilon(r),R)$ can also be determined. Hence the integral

$$\int_{r=0}^{r=R} D(r)\,dA(\epsilon(r),r)$$

can be computed.

164. See MacLaurin (1801: vol. 2, pp. 142–152, §668–§681).
165. Clairaut (1742f: 376).
166. Clairaut (1742i: 140).
167. Ibid., p. 141.
168. Ibid., p. 140.
169. Ibid., p. 141.
170. Ibid., p. 140. Euler's use of tractrices to construct Riccati equations appears in Euler (1736b), which Clairaut cited in Clairaut (1742e: 9). For further discussion of Clairaut's reference in Clairaut (1742e) to Euler's solving Riccati equations by means of tractrices, see Bos (1988: 59–60, note 69).
171. Engelsman (1984: 141, 156).
172. Ibid., p. 192, note 19.
173. Desaguliers (1725b: 241–242; and 1725c: 280).
174. Daniel Bernoulli (1740: 164). See as well note 23 of Chapter 8.
175. D'Alembert (1770: 36 or 1967b: 36).
176. The equation

$$10\delta A - 2D = 5A\phi$$

[equation (I) of note 104 of Chapter 9], which Clairaut used to come to his conclusions, holds whenever certain integrals exist (principally, A and D). Consequently, the densities $R(r)$ of strata need not decrease continuously as a function of r from the center of a figure to its surface. The equation will also hold when density decreases from the center of a figure to its surface as a step function $R(r)$ of r (see note 104 of Chapter 9). In other words, apart from other considerations, Clairaut's conclusions should be valid for a stratified figure shaped like an ellipsoid of revolution which is made up of a finite number of strata whose thicknesses are all finite, whose strata all have surfaces that are ellipsoids of revolution whose ellipticities are infinitesimal and all of which have a common axis of symmetry (the polar axis of the figure), which revolves around its axis of symmetry, at whose outer surface the principle of the plumb line holds, and which attracts according to the universal inverse-square law, as well a for a stratified figure shaped like an ellipsoid of revolution whose density varies continuously from the center of the figure to its surface, whose strata are ellipsoids of revolution whose ellipticities are infinitesimal and all of which have a common axis of symmetry (the polar axis of the figure), which revolves around its axis of symmetry, at whose surface the principle of the plumb line holds, and which attracts in accordance with the universal inverse-square law. As I mentioned earlier, the real problem is that in arriving at his conclusions, Clairaut assumed that the ellipticities $\rho(r)$ vary from the center of a figure to its surface as a function of r that has the form

$$\rho(r) = \delta\left(\frac{1}{r^2} - u\right),$$

where $0 < u \leqslant 1$ is a constant. As we saw, it turns out that it is probably true that if the ellipticities $\rho(r)$ do increase as a function of r from the center of a figure to its surface, when the density $R(r)$ decreases as a function of r from the center of a figure to its surface, either continuously or as a step function, then the ellipticity δ of a stratified figure shaped like a flattened ellipsoid of revolution whose ellipticity is infinitesimal, whose strata are all ellipsoids of revolution or have surfaces that are all ellipsoids of revolution whose ellipticities are infinitesimal and all of which have a common axis of symmetry (the polar axis of the figure), which revolves around its axis of symmetry, at whose surface the principle of the plumb line holds, and which attracts according to the universal inverse-square law must be less than the ellipticity ϵ of the corresponding homogeneous fluid figure of equilibrium shaped like a flattened ellipsoid of revolution whose ellipticity is infinitesimal, which revolves around its axis of symmetry, and which attracts in accordance with the universal inverse-square law. (We recall that

$$\epsilon = \frac{5}{4}\phi, \tag{9.4.2}$$

where ϕ is the ratio of the magnitude of the centrifugal force per unit of mass to the magnitude of the effective gravity at the equators of the two figures.) The trouble, however, is that contrary to Clairaut's way of thinking, the truth of this statement does not follow as a corollary of Clairaut's derivation that involves

$$\rho(r) = \delta\left(\frac{1}{r^2} - u\right)$$

(see note 106 of Chapter 9).
177. See note 45 of Chapter 6.
178. Clairaut (1743: 294–295).
179. Ibid., pp. 294–295.
180. Ibid., pp. 298–299.
181. In fact, Maupertuis himself was apparently not happy with certain measurements made in Lapland [see Nordmann (1966: 94, note 87)]. In 1738, Maupertuis proposed to measure at his own expense an arc of meridian over the ice of Lake Vättern in Sweden, crossed by the 58th parallel of north latitude but nothing came of this project [see Brown (1976: 192, note 90 and Beeson (1992: 122)]. In 1801–03, Jöns Svanberg measured an arc of meridian near the one measured by the members of the Lapland expedition, and this new measurement showed that Maupertuis had overestimated the length of a degree of latitude in Lapland by about 1000 feet [see Berry (1898; 1961: 279) and Beeson (1992: 115–116, note 77)]. In 1743, Clairaut took $\frac{10}{2025}$ to be the value of the relative increase in the magnitude of the effective gravity from the equator to a pole [Clairaut (1743: 145)]. This is the value that the members of the Lapland expedition had

arrived at. Substituting this value for $(P - \Pi)/\Pi$ in Clairaut's equation

$$\frac{P - \Pi}{\Pi} = \frac{1}{115} - \delta$$

for the earth's ellipticity δ (see note 112 of Chapter 9), we find that

$$\frac{1}{202.5} = \frac{1}{115} - \delta,$$

so that

$$\delta = \frac{1}{115} - \frac{1}{202.5} = 0.0037573 = \frac{1}{270},$$

which is a value less than $\frac{1}{230}$, unlike the value $\frac{1}{177}$ determined from the measurements made by the members of the Lapland expedition, and, moreover, a value that is also considerably closer to the current value $\frac{1}{298}$ for the earth's ellipticity than is the value $\frac{1}{177}$.

182. Clairaut (1743: 301).
183. Ibid., pp. 144–151.
184. Huygens (1690: 159). If we let R stand for the equatorial radius and let r denote the polar radius, Huygens took the ellipticity to be

$$\frac{R - r}{R} \qquad\qquad\qquad\qquad (I)$$

[see Huygens (1690: 156, 159)] instead of

$$\frac{R - r}{r}. \qquad\qquad\qquad\qquad (II)$$

But ratios (I) and (II) differ insignificantly from each other when ratio (II) is infinitesimal. That is, in this case ratios (I) and (II) are equal to terms of first order. This we saw in Chapter 8 in analyzing the essay of 1740 on the tides written by Daniel Bernoulli, who, like Huygens, took ratio (I) to be the ellipticity.

185. Maupertuis (1732b: 8).
186. Clairaut (1743: 141–142). Clairaut reasoned geometrically and rather informally in arriving at

$$\epsilon \simeq \tfrac{1}{2}\phi \qquad\qquad\qquad\qquad (9.6.2)$$

[Clairaut (1743: 142–143)], but (9.6.2) can also be derived analytically and more clearly using Clairaut's equation

$$\frac{\pi}{n + 1}(x^2 + y^2)^{(n+1)/2} - \frac{fy^2}{2r} = A$$

[Clairaut (1743: 59)] for the shape of a homogeneous fluid figure of equilibrium shaped like a figure of revolution attracted by a single, fixed center of force situated at the center of the figure of equilibrium on its axis of rotation which is its axis of symmetry, where the magnitude of the attraction at a point whose distance from the fixed center of force is D is πD^n and where the attraction at a point is directed toward the fixed center of force. Here x and y are rectangular coordinates, and f stands for the magnitude of the centrifugal force per unit of mass at a distance r from the axis of rotation of the figure. If a denotes the polar radius of the figure and if we substitute a for x and 0 for y in this equation, then the equality

$$\frac{\pi}{n + 1}a^{n+1} = A, \quad \text{or} \quad \frac{a}{n + 1}(\pi a^n) = A,$$

results, where πa^n is the magnitude of the attraction at a pole of the figure. At the same time, if b denotes the equatorial radius of the figure and if we replace x by 0 and y by b in the same equation, then the equality

$$\frac{\pi}{n+1}b^{n+1} - \frac{fb^2}{2r} = A, \quad \text{or} \quad \frac{b}{n+1}(\pi b^n) - \frac{b}{2}\left(\frac{fb}{r}\right) = A, \quad \text{or} \quad \pi b^n\left[\frac{b}{n+1} - \frac{b}{2}\phi\right] = A,$$

results, where $f(b/r)$ is the magnitude of the centrifugal force per unit mass at the equator of the figure, πb^n is the magnitude of the attraction at the equator of the figure, and ϕ is the ratio of the magnitude of the centrifugal force per unit mass to the magnitude of the attraction at the equator of the figure. If we divide the left-hand side and the right-hand side, respectively, of the second equality by the left-hand side and the right-hand side, respectively, of the first equality, then the equality

$$\left(\frac{b}{a}\right)^n\left[\frac{b}{a} - \left(\frac{n+1}{2}\right)\frac{b}{a}\phi\right] = 1, \quad \text{or} \quad 1 - \left(\frac{n+1}{2}\right)\phi = \left(\frac{b}{a}\right)^{-(n+1)}$$
$$= (1+\epsilon)^{-(n+1)}$$

follows, where ϵ is the ellipticity. Now, this last equality can be rewritten as

$$(1+\epsilon)^{n+1} = \frac{1}{1 - ((n+1)/2)\phi}.$$

When ϕ is infinitesimal ($\phi^2 \ll \phi$), then $((n+1)/2)\phi$ is frequently infinitesimal when the values of n are not too large and ϕ is not too large an infinitesimal or when the values of n are large but ϕ is a very small infinitesimal. In case $((n+1)/2)\phi$ is infinitesimal, then

$$(1+\epsilon)^{n+1} = \frac{1}{1 - ((n+1)/2)\phi} \simeq 1 + \left(\frac{n+1}{2}\right)\phi$$

to terms of first order. Consequently,

$$1 + \epsilon \simeq \left(1 + \left(\frac{n+1}{2}\right)\phi\right)^{1/(n+1)} \simeq 1 + \left(\frac{1}{n+1}\right)\left(\frac{n+1}{2}\right)\phi = 1 + \frac{1}{2}\phi$$

to terms of first order, from which it follows that

$$\epsilon \simeq \frac{1}{2}\phi \tag{9.6.2}$$

to terms of first order, independently of n. Thus, for example, when $\phi = \frac{1}{288}$, then $\epsilon = \frac{1}{576}$. Here we note that the variation of the magnitude of the effective gravity at the surface of a figure from its equator to one of its poles is not a matter at issue. Nor is it a question here in general of the ellipticities of homogeneous figures shaped like flattened *ellipsoids* of revolution whose ellipticities are infinitesimal. [The equation

$$\frac{P - \Pi}{\Pi} = 2\epsilon - \delta \tag{9.4.1'}$$

only holds for a stratified figure shaped like an ellipsoid of revolution whose ellipticity is infinitesimal, whose strata are all ellipsoids of revolution or have surfaces that are all ellipsoids of revolution whose ellipticities are infinitesimal and all of which have a common axis of symmetry (the polar axis of the figure], which revolves around its axis of symmetry, at whose outer surface the principle of the plumb line holds, and which attracts according to the universal inverse-square law (see note 45 of chapter 6). Let us also note the fundamental difference between Clairaut's test of hypothesized central forces of attraction and Bouguer's test of such hypotheses which I discussed in Chapter 4. Bouguer did not believe it possible that such an attraction, if produced by vortices, could ever be entirely independent of directions – that is, independent of the angles that lines that include the center of force and that indicate the directions of attractions at points make with one of these lines chosen as an axis of reference. Finally, let us observe that the fraction $\frac{1}{288}$ is infinitesimal, and the fraction $\frac{1}{289}$ differs from it by terms of second order and higher order. Consequently, the two fractions are equal to terms of first order. The fraction $\frac{1}{576}$ is also infinitesimal, and the fraction $\frac{1}{578}$ differs from it by terms of second order and higher order. Therefore these two fractions are also equal to terms of first order. That is, in both instances, the fractions differ from each other by negligible amounts. (See note 3 of Chapter 1, note 51 of Chapter 5, and note 14 of Chapter 6.) As a result, either $\frac{1}{288}$ or $\frac{1}{289}$ can be taken to be the value of ϕ in (9.6.2), since (9.6.2) itself only holds to terms of first order too (in other words, it only holds when ϕ is infinitesimal). Huygens took the value of ϕ to be $\frac{1}{289}$; Clairaut took the value of ϕ to be $\frac{1}{288}$.

187. Clairaut (1743: 302–303).
188. Ibid., pp. 195–198.
189. Here I rewrite the expression

$$\left(\frac{r}{R}\right)^3 = \left(\frac{\tau^2}{T^2}\right) \tag{I}$$

in note 53 of Chapter 5 for the ratio of the magnitude of the centrifugal force per unit of mass to the magnitude of the attraction at Jupiter's equator, where $r \equiv$ Jupiter's radius, $R \equiv$ the radius of the orbit of Jupiter's fourth satellite, $\tau \equiv$ the period of revolution of Jupiter's fourth satellite, and $T \equiv$ Jupiter's period of rotation as

$$\frac{\tau^2}{h^3\,T^2}, \tag{II$'$}$$

where $h \equiv R/r$, in which case $1 < h$. In deriving the infinite series for Jupiter's finite ellipticity δ which he published in his treatise of 1743, Clairaut mistakenly equated the expression

$$\frac{2c}{3}\left(\frac{\tau^2}{h^3\,T^2}\right) \tag{III}$$

where $c \equiv 2\pi$, with an infinite series for the magnitude of the centrifugal force per unit of mass at Jupiter's equator consisting of terms in δ of all orders. But expression (III) is simply expression (II$'$) multiplied by the constant $(2c)/3$. Hence expression (III) is essentially the *ratio* of the magnitude of the centrifugal force per unit of mass to the magnitude of the attraction at Jupiter's equator, *not* the magnitude of the *centrifugal force per unit of mass* at Jupiter's equator [see Clairaut (1743: 186–187, 196–197, 303)].

190. This can easily be seen as follows:

$$\frac{1}{9.05} = \frac{1}{9.33 - 0.28} = \frac{1}{9.33}\left(\frac{1}{1 - (0.28/9.33)}\right).$$

Now,

$$\frac{0.28}{9.33} = \frac{28}{933} \simeq \frac{1}{33},$$

and $\frac{1}{33}$ is not infinitesimal. Moreover, even if $\frac{1}{33}$ were infinitesimal, in which case

$$\frac{1}{1 - (0.28/9.33)} \simeq 1 + \left(\frac{0.28}{9.33}\right)$$

would hold to terms of first order, it would still be true that

$$\frac{1}{9.33}\left(1 + \left(\frac{0.28}{9.33}\right)\right)$$

does not equal $1/9.33$ to terms of first order. This is true because

$$\frac{1}{9.33} \times \frac{0.28}{9.33} \simeq \frac{1}{308}$$

is obviously not a product of two infinitesimals. Hence $\frac{1}{308}$ is not a term of second order, and thus it cannot be discarded. We recall that Newton's theory of a homogeneous figure shaped like a flattened ellipsoid of revolution which attracts according to the universal inverse-square law, which revolves around its axis of symmetry, and whose columns from center to surface all balance or weigh the same only holds when the ellipticity of the figure is infinitesimal. But Jupiter's ellipticity is finite, not infinitesimal. Consequently Newton's theory *should not* work when it is applied to Jupiter. Here we see evidence that the value $1/9\frac{1}{3}$ of Jupiter's ellipticity which Newton's theory requires does indeed differ by terms of first order from the value $1/9.05$ of Jupiter's ellipticity deduced from a theory of homogeneous figures of equilibrium shaped like flattened ellipsoids of revolution which attract in accordance with the universal inverse-square law, which revolve around their axes of symmetry, and whose ellipticities are finite. In other words, we have proof that Newton's theory *does not* work when it is applied to Jupiter.

191. Clairaut (1743: 304).

192. In the expression

$$\frac{\tau^2}{h^3 T^2} \tag{II$'$}$$

in note 189 of Chapter 9, both Clairaut and Maupertuis took T to have Cassini's value 9 hours 56 minutes = 596 minutes. Clairaut took τ to have Pound's value 24032 minutes, and Maupertuis took τ to have the value 16 days $16\frac{8}{15}$ hours = 24031.98 minutes = 24032 minutes. In short, Clairaut and Maupertuis both used the same values of T and τ in expression (II$'$). Finally, in expression (II$'$) Clairaut took h to have Pound's value 26.63, and Maupertuis took h to have Cassini's value 23. [Compare Maupertuis (1734b: 72) and Clairaut (1743: 197–198).] Then according to Clairaut's data, $\tau^2 = (24032 \text{ minutes})^2 = 5.77537 \times 10^8 \text{ minutes}^2$, $T^2 = (596 \text{ minutes})^2 = 3.55216 \times 10^5 \text{ minutes}^2$, and $h^3 = (26.63)^3 = 1.88848 \times 10^4$. Therefore

$$\frac{\tau^2}{h^3 T^2} = \frac{5.77537 \text{ minutes}^2}{1.88848(3.55216 \text{ minutes}^2)} \times 10^8 \times 10^{-4} \times 10^{-5}$$

$$= \frac{5.77537}{6.70818} \times 10^{-1} = 0.86094 \times 10^{-1} = 0.086094.$$

Clairaut in fact stated that

$$\frac{1}{11.615} = 0.086095$$

is the value of

$$\frac{\tau^2}{h^3 T^2}, \tag{II$'$}$$

which is a value that differs negligibly from the value calculated using his data. Now, as I just mentioned, the value that Maupertuis took h to have in expression (II$'$) was the only one that differed from Clairaut's. According to Maupertuis's data, $h^3 = (23)^3 = 1.2167 \times 10^4$. Consequently

$$\frac{\tau^2}{h^3 T^2} = \frac{5.77537 \text{ minutes}^2}{1.2167(3.55216 \text{ minutes}^2)} \times 10^8 \times 10^{-4} \times 10^{-5}$$

$$= \frac{5.77537}{4.32191} \times 10^{-1} = 1.3363 \times 10^{-1} = 0.13363.$$

Maupertuis in fact declared that $1/7.48 = 0.13369$ is the value of expression (II$'$), which, again, is a value that differs negligibly from the value calculated using his data. We note that no matter which value of h is chosen, h^3 is a much smaller number than either τ^2 or T^2. As a result it follows from the particular form that expression (II$'$) has that expression (II$'$) will be very sensitive to the error made in measuring h. That is, while small variations of large values of τ and T will not cause the value of expression (II′) to change appreciably, variations of the small value of h will cause the value of expression (II′) to change appreciably. [This is clear from the expression

$$d\left(\frac{\tau^2}{h^3 T^2}\right) = \frac{2\tau}{h^3 T^2}\, d\tau - \frac{2\tau^2}{h^3 T^3}\, dT - \frac{3\tau^2}{h^4 T^2}\, dh$$

for the total differential of (II′).] Moreover, the error made in measuring the small ratio h using apparatus, like telescopes, of Maupertuis's and Clairaut's day is doubtlessly much greater than the error made in measuring the fairly lengthy periods τ and T using the same apparatus in conjunction with the devices of that day for measuring time.

193. Indeed, in explaining how he arrived at the value 1/23.23 of Jupiter's ellipticity δ, Clairaut (1743: 303) referred specifically to the equation that I used in note 186 of Chapter 9 and that I showed there *only* reduces to

$$\delta \simeq \frac{1}{2}\phi$$

when $\delta^2 \ll \delta$ and $\phi^2 \ll \phi$, which is not true in the case of Jupiter, whose ellipticity δ is finite.

194. Clairaut (1743: 27–28).

195. Ibid., pp. 139–141. I have already given an example that illustrates that if a body attracts according to the universal inverse-square law, then values of attraction specified beforehand at points at the surface of the body do not uniquely determine how the body's density varies in the body's interior (see note 38 of Chapter 6). This kind of example helps us to understand the significance of an equation like

$$\frac{P - \Pi}{\Pi} = 2\varepsilon - \delta, \tag{9.4.1'}$$

which only requires that the individually homogeneous strata of a figure shaped like an ellipsoid of revolution whose ellipticity is infinitesimal and which attracts according to the universal inverse-square law themselves be ellipsoids of revolution whose ellipticities are infinitesimal and which all have a common axis of symmetry or else be strata whose thicknesses are finite and whose surfaces are ellipsoids of revolution whose ellipticities are infinitesimal and which all have a common axis of symmetry and does not require that the densities of the strata vary in any particular way.

196. Clairaut (1743: xxvi).

197. The Malebranchist movement in the later 1730s is discussed in Brunet (1931, 1970: ch. 5); and in Aiton (1972: ch. 10).

198. Gamaches (1740: xxxvi–xxxviii of the "Préliminaire," and p. 263). See as well the review of this work in the *Journal des Savants,* October 1740, pp. 620–634, especially p. 627.

199. Markovic (1960: 6–7, 9, 13–16, 28–29, 35). Boscovich showed that when the attraction has two components directed toward the two foci of a homogeneous, stationary figure shaped like an elongated ellipsoid of revolution whose axis of symmetry is its major axis, then if the magnitude of the resultant of the attraction is constant, the principle of the plumb line can hold at the surface of the stationary figure, and the "sin² law" of the variation in the increase in the magnitude of the attraction with latitude can also hold at the surface of the immobile figure. In this case the ellipticity of the homogeneous stationary figure shaped like an elongated ellipsoid of revolution . is not infinitesimal but is finite instead (see note 8 of Chapter 1). Various writers have attributed to Boscovich the original idea of the geoid. See, for example, Cubranic (1977–1982; 67–68); and Nikolic (1961: 325). (See as well note 83 of Chapter 9.)

200. *Journal des Savants,* September 1739, pp. 570–571.

201. Some of Clairaut's letters to Boscovich are published in Boscovich (1912: 216–223). Other, unpublished letters from Clairaut to Boscovich are located in the Boscovich Archives, Bancroft Library, University of California, Berkeley. Concerning Boscovich's relations with Clairaut, see as well Bédarida (1977–82: 23); Vidan (1977–82: 198–199); Markovic (1957: 44–45); Pappas (1987a, b, 1991); and Taton (1987). Maupertuis, however, ridiculed Boscovich's *Dissertatio de telluris figurà* in his *Lettre d'un horloger anglois à un astronome de Pékin,* which Maupertuis published anonymously in 1740. Maupertuis possibly treated Boscovich harshly in this work

because Maupertuis may have mistakenly thought that Boscovich's opposed Newtonianism [see Beeson (1985: 222, including note 58)].

202. Newton (1966: vol. 2, p. 266).
203. D'Alembert (1747: 58–62).
204. Newton (1966: vol. 2, pp. 266–271).
205. René Taton told me this during one of our discussions.
206. Clairaut (1750: 4–5).
207. Ibid., p. 26.
208. Clairaut (1750: 7).
209. Clairaut (1751: 243).
210. Clairaut (1750: 54).
211. For discussions of Clairaut's work on lunar theory and other problems in celestial mechanics, see Chandler (1975); Waff (1975; 1976); and Wilson (1980). See as well note 3 of Chapter 1. In fact, if the kinds of spherical bodies described attract according to a universal $1/r^n$ law, where r stands for distance, then such bodies attract like central forces in their exteriors if and only if $n = 2$. However, after Clairaut began to study the problem of the anomalous motion of the moon's apogee, he thought for a time that the universal inverse-square law of attraction might have to be *modified* in order to solve the problem.
212. Clairaut (1750: 5). If Clairaut wrote these words some time in 1749 and not earlier, which seems likely, then he had discovered by the time that he wrote these words that the problem of the anomalous motion of the moon's apogee could be solved *without* having to modify the universal inverse-square law of attraction. [See note 211 of Chapter 9, Waff (1976: 162, 177, 215), and Wilson (1980: 134–145).]

10. Epilogue

1. Truesdell (1987: 186; 1988: 125; 1991: 4–5); and Brown (1991).
2. Kuhn (1970).
3. For example, the failure to demonstrate Euclid's parallel postulate can be considered an anomaly in Euclidean geometry which led to the formulations of non-Euclidean geometries during the early nineteenth century. The diehards among the geometers resisted this revolution until the late nineteenth century, by which time most of them had died. For an account of the creation of non-Euclidean geometries which depicts their discoveries and the reception of their discoveries in such terms, see Pont (1986).
4. Koyré (1968: 221–260).
5. Ibid., pp. 180–184, 233–235.
6. Truesdell (1954: XLI–XLIV; 1960: 250–254); Aiton (1964: 86; 1972: 197); Hankins (1967; 1970: 220–224).
7. Bertoloni Meli (1990).
8. Concerning the importance of Riemannian geometry and the Ricci tensor in Einstein's creation of the general theory of relativity, see Browder (1976: 542) and Stewart (1981: 50–58).
9. The idea that Clairaut never conceived of a line integral until he had to calculate the effort produced by the effective gravity along a curved plane or skew channel filled with homogeneous fluid accords with the thesis argued in Kac (1982: 633–634) and in Truesdell (1991: 4–5).
10. Kuhn (1977: 64–65, note 32).
11. This trend is emphasized throughout Bos (1974–75) and Engelsman (1984).
12. In Brook Taylor's solutions to isoperimetrical problems, published in his *Methodus incrementorum* (1715), Feigenbaum identifies examples of total differential equations in three variables, where partial differential coefficients are written in capital letters and first-order differentials are expressed in Newton's notation for fluxions. Feigenbaum cites Woodhouse, Struik, and Goldstine as sources for her belief that in introducing total differential equations in his own work of 1744 on isoperimetrical problems (*Methodus inveniendi lineas curvas maximi minimi proprietate gaudente*), Euler drew specifically on Taylor's work [see Feigenbaum (1985: 60–63)]. While it is not impossible that Taylor influenced Euler, it seems much more plausible, taking chronology into consideration, that Euler's use of total differential equations in 1744 derived from his correspondence with Clairaut, which began late in 1740, from Clairaut's two Paris Academy *mémoires* of 1739 (1739) and 1740 (1740b) on integral calculus, in which Fontaine's whole-symbol

notation for partial differential coefficients appeared for the first time on a printed page, and from Clairaut's *Théorie de la figure de la terre, tirée des principles de l'hydrostatique* (1743), in which the whole-symbol notation for partial differential coefficients again appears.

13. Euler (1748: 197).
14. D'Alembert (1749).
15. Montucla (1799–1802, 1968: vol. IV, p. 77, Lalande's annotation); repeated in Lalande (1803: 486).
16. The dictum appears in Charles Bossut's *Histoire des mathématiques* (1802), where Bossut explained that "a famous geometer" offered it as a piece of advice. Everyone assumed that the individual referred to in quotation marks was Bossut's good friend D'Alembert. But in his copies of two editions of this work, Bossut wrote "Fontaine" in the margin alongside "a famous geometer" [see Itard (1949: 281)].
17. Lexell (1781). Hankins (1970: 246) qualified this letter as "gossipy," no doubt because of Lexell's numerous negative remarks about *all* of the *other* French scientists whom he met and described in the letter.
18. Delambre (1801?a: 444v).
19. Truesdell (1955: LXXXIV, including note 6).
20. Cousin (1796: vol. 1, p. xii).
21. Delambre (1801?a: 444r; 1801?b: 568r).
22. Paul Charpit, who attended Cousin's lectures at the Collège de France, recounted one of Cousin's "Fontaine stories" to Louis-François-Antoine Arbogast in 1782. See Charpit (1782: 63v–64r). I am grateful to Ivor Grattan-Guinness for having called my attention to this letter.
23. Delambre (1801?a: 444r; 1801?b: 568r).
24. Lalande (1803: 486). Lalande's story is an anecdote and could be apocryphal, because Lalande was not even born until 1732. He became a member of the Paris Academy of Sciences in 1753.
25. Coolidge (1949, 1963: ch. XIV); Truesdell (1960: 242–243); Mayer (1959: 65–104; 1991); Chouillet (1984); Dedeyan (1987: p. III); Dhombres (1985); Kessler (1981); Morin (1987: ch. II); Perkins (1982: ch. 3–4); Spear (1980); and Washner (1987).
26. Diderot (1765: 287, 289).
27. Bibliothèque de l'Institut (Paris), ms 1826: "Table alphabétique des noms des auteurs dont les ouvrages se trouvaient le 4 Septembre 1984 dans la Bibliothèque de l'Académie Royale des Sciences. Par Monsieur Demours Bibliothèquaire de la d. académie."
28. Fontaine (1764b: 3–4 of the "Table des Mémoires contenus dans ce volume").
29. See the second installment of the review of Fontaine's complete works published in the *Journal de Trévoux* (May 1765, 1176–1194, on 1177–1178; quotation cited above), 1182–1184; or reproduced in facsimile; *Journal de Trévoux,* vol. 65 (Geneva: Slatkine Reprints, 1968), pp. 297–301, on 297–298 (quotation cited above), 298–299.
30. Fontaine (1739–40: 17 February 1740, 23r–25v).
31. Condorcet (1771: 109).
32. D'Alembert (1765a). In this draft of the letter, dated 2 May 1765, D'Alembert noted that the *Mercure* for April 1765 ascribed D'Alembert's principle to Fontaine, and D'Alembert reminded the Journal's editors that Euler had falsely attributed Fontaine's conditional equation to him publicly and that he had promptly rectified this mistake publicly. The title of D'Alembert (1765a) suggests that a letter was published in the *Mercure*. In fact such a letter was actually published [D'Alembert (1765d)].
33. See Fontaine (1767); Lagrange (1760–1761a, b). This suggests that Fontaine did not see Lagrange's first work in the calculus of variations until after he published his complete works of 1764. (See note 111 of Chapter 7.)
34. The machinations can be followed in the Lagrange–D'Alembert correspondence for the years 1768–1770, located in Lagrange (1882). The letter from Lagrange to D'Alembert dated 29 May 1767, which is numbered 48, is out of place in the sequence of letters in the Lagrange–D'Alembert correspondence. It should be relocated between letters 55 and 56. The correct date of the letter is probably 29 May 1768.
35. Concerning Fontaine's Paris Academy *mémoire* [Fontaine (1747)], see note 245 of Chapter 7.
36. D'Alembert (1755: 853–854).
37. Lagrange (1769b).
38. D'Alembert (1769: 118, note 10).
39. See note 238 of Chapter 7.

40. Lagrange first visited Paris late in 1763, and he had to prolong his stay through early 1764 because he became ill. In a letter dated 7 January 1764 to Madame Geoffrin, one of the mistresses of the *salons* of the era, D'Alembert praised Lagrange's talent and added that "Monsieur Fontaine, my colleague in the Academy of Sciences, who, as you know, is not one to bestow praise or engage in panegyrics, will say the same thing about Monsieur de Lagrange as I have" [D'Alembert (1764)]. Fontaine's esteem for Lagrange is also evident in a letter Fontaine wrote in 1765 to Jacques Mathon de La Cour, Fontaine's friend from boyhood and the leading mathematician in the Académie des Beaux Arts de Lyon. (Concerning Mathon de La Cour, see note 247 of Chapter 7.) In the letter Fontaine said: "I regard him [Lagrange] as Europe's foremost geometer, and if and when he should no longer be, he will still be one of the most admirable men that I know." [This passage is quoted in Boucharlat (1816: 182).]
41. Levot (1861: 543).
42. Lagrange (1774b: 270, note 2).
43. Lagrange (1774b. See as well Lagrange (1774c: 18).
44. A copy of this volume can be found today in the archives at the Service Historique de la Marine, Château de Vincennes.
45. Jones (1976: 571).
46. Truesdell (1960: 149–150, 170, 258).
47. D'Alembert (1765c: 381).
48. D'Alembert (1771); and Condorcet (1771: 116).

Bibliography

AGNEW, RALPH PALMERL
1962, 1988   *Calculus. Analytic Geometry and Calculus, with Vectors.* New York, San Francisco, Toronto, and London: McGraw-Hill.
AHLFORS, LARS V.
1966   *Complex Analysis. An Introduction to the theory of Analytic Functions of One Complex Variable.* 2nd ed. New York: McGraw-Hill.
AITON, ERIC J.
1955   "The Contributions of Newton, Bernoulli and Euler to the Theory of the Tides," *Annals of Science, 11*: 206–223.
1964   "The Inverse Problem of Central Forces," *Annals of Science, 20*: 81–99.
1972   *The Vortex Theory of Planetary Motions.* London and New York: Macdonald and American Elsevier.
1985   *Leibniz. A Biography.* Bristol and Boston: Adam Hilger.
1986   "The Application of the Infinitesimal Calculus to Some Physical Problems by Leibniz and His Friends," in *Studia Leibnitiana,* Sonderheft 14: *300 Jahre "Nova Methodus" Von G. W. Leibniz (1684–1984).* Stuttgart: Franz Steiner Verlag Wiesbaden GMBH. Pp. 133–143.
1988   "Polygons and Parabolas: Some Problems Concerning the Dynamics of Planetary Orbits," *Centaurus, 31*: 207–221.
1989   "The Contributions of Isaac Newton, Johann Bernoulli and Jakob Hermann to the Inverse Problem of Central Forces," in *Studia Leibnitiana,* Sonderheft 17: *Der Ausbau des Calculus durch Leibniz und die Brüder Bernoulli.* Ed. Heinz-Jürgen Hess and Fritz Nagel. Stuttgart: Franz Steiner Verlag Wiesbaden GMBH. Pp. 48–58.
BEDARIDA, HENRI
1977–82   "Amitiés françaises du père Boscovich," in *Rudjer Boskovic. Annales de l'Institut Français de Zagreb.* Troisième Série/No.3. Pp. 18–39.
BEESON, DAVID
1985   "Lettre d'un horloger anglois à un astronome de Pékin," *Studies on Voltaire and the Eighteenth Century, 230*: 189–222.
1992   *Maupertuis: An Intellectual Biography.* Oxford: The Voltaire Foundation.
BERNHART, ARTHUR
1959   "Curves of General Pursuit," *Scripta Mathematica, 24*: 189–206.
BERNOULLI, DANIEL
1733   Bibliothèque Nationale (Paris), Fonds Français, Nouvelles Acquisitions ms 9186, fols. 10r–11v: letter from Bernoulli to J.-N. Delisle, 15 ... bre.
1734   Letter from Bernoulli to Leonhard Euler, 18 December, in Fuss (1843, 1968: 415).
1739   Letter from Bernoulli to Leonhard Euler, 7 March, in Fuss (1843, 1968: 455).
1740   "Traité sur le flux et reflux de la mer," in Newton (1739–42: vol. 3, 133–246).
1758   "Lettre de Monsieur Daniel Bernoulli, de l'Académie Royale des Sciences, à Monsieur Clairaut de la même Académie, au sujet des nouvelles découvertes faites sur les vibrations des cordes tendues," *Journal des Savants,* March, 157–166.
1968   *Hydrodynamica, by Daniel Bernoulli, and Hydraulics, by Johann Bernoulli.* Trans. Thomas Carmody and Helmut Kobus. New York: Dover.

BERNOULLI, JAKOB
1697   "Solution problematum fraternorum... ," *Acta Eruditorum*, May, pp. 211–217. Republished in Jakob Bernoulli (1744: vol. 2, 768–778).
1744   *Opera Omnia*. 2 vols. Geneva: Cramer and Philibert.

BERNOULLI, JOHANN I
1702a   "Solution d'un problème concernant le calcul intégral, avec quelques abrégés à ce calcul. Le tout extrait d'une de ses lettres écrite de Groningue le 5 Août 1702," *Mémoires de l'Académie Royale des Sciences*, 1702. Paris: Jean Boudot, 1704. Pp. 289–297.
1702b   "Solution d'un problème concernant le calcul intégral, avec quelques abrégés à ce calcul. Le tout extrait d'une de ses lettres écrite de Groningue le 5 Août 1702," *Mémoires de l'Académie Royale des Sciences*, 1702. Paris: Charles-Estienne Hochereau, 1720. Pp. 296–305.
1704   Universitätsbibliothek, Basel, Bernoulli Archives, ms LIa 665, No. 1: letter from Bernoulli to Montmort, 25 October.
1717   Universitätsbibliothek, Basel, Bernoulli Archives, ms LIa 665, No. 8: letter from Bernoulli to Montmort, 10 July.
1718a   Universitätsbibliothek, Basel, Bernoulli Archives, ms LIa 665, No. 11: letter from Bernoulli to Montmort, 21 May.
1718b   Universitätsbibliothek, Basel, Bernoulli Archives, ms LIa 665, No. 12: letter from Bernoulli to Montmort, 29 September.
1721   Universitätsbibliothek, Basel, Bernoulli Archives, ms LIa 669, No. 74: letter from Bernoulli to Varignon, 17 May.
1723   Universitätsbibliothek, Basel, Bernoulli Archives, ms LIa 661, No. 5: letter from Bernoulli to Mairan, 7 September.
1724   "Lettre de Jean Bernoulli [I] à Jean Jacques de Mairan, 30 Septembre, 1724," in *Universitatis Basileensis Luminibus Atque Ornamentis Bernoullianum*. Ed. Eduardo Hagenbach, Julio Piccard, and Ludovicus Sieber. Basel: Johannes Schweighauser, 1874.
1726   "De integrationibus aequationum differentialum, ubi traditur methodi alicujus specimen integrandi sine pravia separatione indeterminatarum," *Commentarii Academiae Scientiarum Imperialis Petropolitanae*, 1726. Saint Petersburg, 1728. Pp. 167–184. Republished in Johann I Bernoulli (1742b: vol. 3, 108–124).
1728   Universitätsbibliothek, Basel, Bernoulli Archives, ms LIa 658, No. 5: letter from Bernoulli to Fontenelle, 4 March.
1730a   Universitätsbibliothek, Basel, Bernoulli Archives, ms LIa 662, No. 3: letter from Bernoulli to Maupertuis, 29 August.
1730b   Universitätsbibliothek, Basel, Bernoulli Archives, ms LIa 661, No. 34: letter from Bernoulli to Mairan, 13 April.
1730c   "Méthode pour trouver les tautochrones, dans des milieux résistants, comme le quarré des vitesses," *Mémoires de l'Academie Royale des Sciences*, 1730. Paris: Imprimerie Royale, 1732 Pp. 78–101.
1731a   Universitätsbibliothek, Basel, Bernoulli Archives, ms LIa 662, No. 8: letter from Bernoulli to Maupertuis, 1 April.
1731b   Universitätsbibliothek, Basel, Bernoulli Archives, ms LIa 662, No. 6: letter from Bernoulli to Maupertuis, 7 January.
1731c   Universitätsbibliothek, Basel, Bernoulli Archives, ms LIa 662, No. 12: letter from Bernoulli to Maupertuis, 12 August.
1734   Universitätsbibliothek, Basel, Bernoulli Archives, ms LIa 662, No. 33: letter from Bernoulli to Maupertuis, 9 March.
1739   Universitätsbibliothek, Basel, Bernoulli Archives, ms LIa 661, No. 48: letter from Bernoulli to Mairan, 7 April.
1742a   "Remarques sur le livre intitulé *Analyse des infiniments petits, comprenant le calcul intégral, dans toute son étendue*, etc. par Monsieur Stone, de la Sociéte Royale de Londres," in Johann I Bernoulli (1742b: vol. 4, 169–192).
1742b   *Opera Omnia*. 4 vols. Lausanne and Geneva: Marci-Michaelis Bosquet.

BERNOULLI, JOHANN I, and VARIGNON, PIERRE
1988, 1992   *Der Briefwechsel von Johann I Bernoulli mit Pierre Varignon.* Part 1: 1692–1702 and
              Part 2: 1703–1722. Ed. P. Costabel and J. Peiffer. Basel: Birkhäuser Verlag.
BERNOULLI, NIKOLAUS I
1716   Universitätsbibliothek, Basel, Bernoulli Archives, ms LIa 21², letter from Bernoulli to
              Montmort, 31 March.
1718a   Universitätsbibliothek, Basel, Bernoulli Archives, ms LIa 21², letter from Bernoulli to
              Montmort, 15 October.
1718b   Universitätsbibliothek, Basel, Bernoulli Archives, ms LIa 21², letter from Bernoulli to
              Montmort, 3 December.
BERNOULLI, NIKOLAUS II
1721   "Exercitatio geometrica de trajectoriis orthogonalibus... Sectio II," *Acta Eruditorum,*
              vol. 7 of Supplements: 303–326.
BERRY, ARTHUR
1898, 1961   *A Short History of Astronomy from Earliest Times through the Ninetheenth Century.*
              New York: Dover.
BERTOLONI MELI, DOMENICO
1990   "The Relativization of Centrifugal Force," *Isis, 81*: 23–43.
BIKERMAN, J. J.
1978   "Capillarity before Laplace: Clairaut, Segner, Monge, Young," *Archive for History of
              Exact Sciences, 18: 103*–122.
BIREMBAUT, ARTHUR
1957   "L'Académie royale des Sciences en 1780 vue par l'astronome suédois Lexell
              (1740–1784)," *Revue d'Histoire des Sciences et de Leurs Applications, 10*: 148–166.
BLAY, MICHEL
1985–86   "L'Introduction du calcul différentiel en dynamique: l'exemple des forces centrales
              dans les Mémoires de Varignon en 1700," *Sciences et Techniques en Perspective,
              10*: 157–190.
1986   "Deux moments de la critique du calcul infinitésimal: Michel Rolle et George Ber-
              keley," *Revue d'Histoire des Sciences, 39*: 223–253.
1988   "Varignon ou la théorie des projectiles 'comprise en une Proposition générale',"
              Annals of Science, 45: 591–618.
1989a   "Dynamique et calcul infinitésimal: les premiers exposés a l'Académie Royale des
              Sciences de Paris," in *La gravitation newtonienne*: Physique et mécanique de Newton à
              Euler (1). Ed. François De Gandt, Christiane Vilain, and Jeanne Peiffer. *Actes des
              Journées d'Histoire et d'Epistémologie,* à l'Observatoire de Meudon (lère livraison) les
              22 et 23 Juin, organisés par le DARC, Observatoire de Paris-Meudon. Pp. 42–77.
1989b   "Du fondement du calcul différentiel au fondement de la science du mouvement dans
              les 'Elémens de la géométrie de l'infini' de Fontenelle," in *Studia Leibnitiana,* Sonder-
              heft 17: *Der Ausbau des Calculus durch Leibniz und die Brüder Bernoulli.* Ed. Heinz-
              Jürgen Hess and Fritz Nagel. Stuttgart: Franz Steiner Verlag Wiesbaden GMBH. Pp.
              99–122.
1989c   "Les *Elémens de la géométrie de l'infini* de Fontenelle," in *Actes du Colloque Fontenelle.*
              Paris: Presses Universitaires de France. Pp. 505–520.
1990   "Note sur la correspondance entre Jean I Bernoulli et Fontenelle," *Corpus, 13*:
              93–100.
1992   *La naissance de la mécanique analytique. La science du mouvement au tournant des
              XVIIe et XVIIIe siècles.* Paris: Presses Universitaires de France.
BOS, H. J. M.
1974–75   "Differentials, Higher-Order Differentials and the Derivative in the Leibnizian Calcu-
              lus," *Archive for History of Exact Sciences, 14*: 1–90.
1986   "Fundamental Concepts of the Leibnizian Calculus," in *Studia Leibnitiana, Sonder-
              heft 14: 300 Jahre "Nova Methodus" Von G. W. Leibniz* (1684–1984). Stuttgart: Franz
              Steiner Verlag Weisbaden GMBH. Pp. 103–118.
1988   "Tractional Motion and the Legitimation of Transcendental Curves," *Centaurus,
              31*: 9–62.

BOSCOVICH, ROGER JOSEPH
1912   "Drugi Ulomak Boskoviceve Korespondencije," ed. Vladimir Varicak, in *Rad Jugoslavenske Akademije Znanosti i Umjetnosti*. Matematicko-Prirodoslovni Razred. Knija 193 (U Zagreb). Pp. 163–338.
BOSS, VALENTIN
1972   *Newton and Russia. The Early Influence, 1698–1796*. Cambridge, Mass.: Harvard Univ. Press.
BOUCHARLAT, JEAN-LOUIS
1816   "Fontaine des Bertins (Alexis)," *Biographie universelle, ancienne et moderne, redigé par une société de Gens de lettres et de sciences*, vol. 15. Paris: L. G.Michaud. Pp. 179–183.
1820   "Mathon de la Cour (Jacques)," *Biographie universelle, ancienne et moderne, redigé par une société de Gens de lettres et de sciences*, vol. 27. Paris: L. G. Michaud. Pp. 454–455.
BOUGUER, PIERRE
1732   "Sur des nouveles courbes auxquelles on peut donner le nom de lignes de poursuite," *Mémoires de l'Académie Royale des Sciences*, 1732. Paris: Imprimerie Royale, 1735. Pp. 1–14.
1733   "Une base qui est exposée au choc d'un fluide étant donnée, trouver l'espèce de conoïde dont il faut la couvrir pour que l'impulsion soit la moindre qu'il est possible," *Mémoires de l'Académie Royale des Sciences*, 1733. Paris: Imprimerie Royale, 1735. Pp. 85–107.
1734   "Comparaison des deux loix que la terre et les autres planètes doivent observer dans la figure que la pesanteur leur fait prendre," *Mémoires de l'Académie Royale des Sciences*, 1734. Paris: Imprimerie Royale, 1736. Pp. 21–40.
BOYER, CARL B.
1946   "The First Calculus Textbooks," *The Mathematics Teacher, 39*: 159–167.
1968   "The New Math of the 1740s in England and France," *XIIe Congrès International d'Histoire des Sciences*, 1968. Vol. 4. Paris: Albert Blanchard, 1971. Pp. 17–23.
BRIGGS, J. MORTON
1974   "Parent, Antoine," *Dictionary of Scientific Biography*, vol. X. Ed. Charles Coulston Gillispie. New York: Charles Scribners Sons, Pp. 319–320.
BROWDER, FELIX E.
1976   "Does Pure Mathematics Have a Relation to the Sciences?," *American Scientist, 64*: 542–549.
BROWN, GARY I.
1991   "The Evolution of the Term 'Mixed Mathematics'," *Jounral of the History of Ideas, 52*: 81–102.
BROWN, HARCOURT
1975   "From London to Lapland: Maupertuis, Johann Bernoulli and *La Terre applatie, 1728–1738*," in *Literature and History in the Age of Ideas*. Ed. Charles G. S. Williams. Columbus: Ohio State Univ. Press. Pp. 69–94.
1976   "From London to Lapland and Berlin," in Brown's *Science and the Human Comedy*. Toronto and Buffalo: Univ. Toronto Press. Pp. 167–206.
BRUNET, PIERRE
1931   *L'Introduction des théories de Newton en France au XVIIIe siècle avant 1738*. Paris: Albert Blanchard.
1952   *La vie et l'oeuvre de Clairaut (1713–1765)*. Paris: Presses Universitaires de France.
1970   *L'Introduction des théories de Newton en France au XVIIIe siècle avant 1738*. Geneva: Slatkine Reprints. This is Brunet (1931) reproduced in facsimile.
BUFFON, GEORGES-LOUIS LECLERC, COMTE DE
1745   "Réflexions sur la loi de l'attraction," *Mémoires de l'Académie Royale des Sciences*, 1745. Paris: Imprimerie Royale, 1749. Pp. 493–500.
1748   "Réflexions sur la loi de l'attraction et sur le mouvement des apsides," *Académie Royale des Sciences, Registres des Procès-Verbaux*, 24 January. Pp. 18–26.
BURSTYN, HAROLD D.
1962   "Theory and Practice in Man's Knowledge of the Tides," *Xe Congrès International d'Histoire des Sciences* (Ithaca, 1962). Vol. 2. Paris: Hermann, 1964. Pp. 1019–1022.

CASEY, JAMES
1992a   "The Principle of Rigidification," *Archive for History of Exact Sciences*, 43: 329–383.
1992b   "Clairaut's Hydrostatics: A Study in Contrast," *American Journal of Physics*, 60: 549–554.

CASSINI, JACQUES
1718   "De la grandeur et de la figure de la terre," *Suite des Mémoires de l'Académie Royale des Sciences, 1718*. Paris: Imprimerie Royale, 1720 [*recte* 1722].

CASSINI, PAOLO
1975   "Maupertuis et Newton," in *Actes de la journée Maupertuis (Créteil, ler décembre 1973)*. Paris: J. Vrin, 1975. Pp. 113–134.

CASTEL, LOUIS-BERTRAND
1722   Review of *Histoire de l'Académie Royale des Sciences, année 1718, avec les Mémoires de Mathématique et de Physique pour la même année*, in *Journal de Trévoux*, June, 989–997.
1726   Library of the Royal Society (London), *Letter Book*, vol. XVIII, fols. 333–337: letter from Castel to Woolhouse.
1728a   *Mathématique universelle, abrégée à l'usage et à la portée de tout le monde*. Paris: Pierre Simon.
1728b   Letter from Castel to Fontenelle, 20 March, in Fontenelle (1766: 156–157).
1728c   Part I of Castel's review of Fontenelle's *Eléments de la géométrie de l'infini* (1727), in *Journal de Trévoux*, July, pp. 1233–1263.
1730   Review of *Histoire de l'Académie Royale des Sciences, avec les Mémoires de Mathématique et de Physique, pour l'année 1725*, in *Journal de Trévoux*, January 1730, 103–122.
1732   Review of Edmund Stone's *The Method of Fluxions, Both Direct and Inverse* (1730), in *Journal de Trévoux*, January 1732, 103–113.
1735   "Discours Préliminaire servant de supplement à la Préface qui est à la tête de *l'Analyse des Infiniments Petits* de Monsieur le Marquis de l'Hôpital," in Stone (1735: iii–xcxx).
1968a   Facsimile of Castel (1730), in *Journal de Trévoux*, vol. 30. Geneva: Slatkine Reprints. Pp. 32–36.
1968b   Facsimile of Castel (1722), in *Journal de Trévoux*, vol. 22. Geneva: Slatkine Reprints. Pp. 250–252.
1968c   Facsimile of Castel (1728c), in *Journal de Trévoux*, vol. 28. Geneva: Slatkine Reprints. Pp. 298–306.
1968d   Facsimile of Castel (1732), in *Journal de Trévoux*, vol. 32. Geneva: Slatkine Reprints. Pp. 32–34.

CHANDLER, PHILIP
1975   "Clairaut's Critique of Newtonian Attraction: Some Insight Into His Philosophy of Science," *Annals of Science*, 32: 369–378.

CHAPIN, SEYMOUR L.
1968   "The Academy of Sciences during the Eighteenth Century: An Astronomical Appraisal," *French Historical Studies*, 5: 371–404.

CHARPIT, PAUL
1782   Bibliothèque Nationale (Paris), Fonds Français, Nouvelles Acquisitions ms 7546, fols 63r–66v: "Lettre autographe No. 17. Charpit à Arbogast, Paris, 20 Novembre."

CHOUILLET, ANNE-MARIE (ed.)
1984   *1984: L'année Diderot*. Supplement to *Dix-Huitième Siècle*, 1985, 17.

CLAIRAUT, ALEXIS-CLAUDE
1728   "Rémarque sur l'Article §52 de *L'Analyse démontrée*," *Journal de Trévoux*, November, pp. 2164–2166.
1731a   Letter from Clairaut to Cramer, 28 July, in Clairaut and Cramer (1955: 208).
1731b   *Recherches sur les courbes à double courbure*. Paris: Nyon, Didot et Quillau.
1732a   Letter from Clairaut to Cramer, 7 March, in Clairaut and Cramer (1955: 212–213).
1732b   "Des épicycloïdes sphériques," *Mémoires de l'Académie Royale des Sciences, 1732*. Paris: Imprimerie Royale, 1735. Pp. 289–294.
1733a   "Sur quelques questions de maximis et minimis," *Mémoires de l'Académie Royale des Sciences, 1733*. Paris: Imprimerie Royale, 1735. Pp. 186–194.

1733b "Détermination géométrique de la perpendiculaire à la méridienne tracée par Monsieur Cassini. Avec plusieurs méthodes d'en tirer la grandeur et la figure de la terre," *Mémoires de l'Académie Royale des Sciences*, 1733. Paris: Imprimerie Royale, 1735. Pp. 406–416.

1734a "Solutions de plusieurs problemes ou il s'agit de trouver des courbes dont la propriété consiste dans une certaine relation entre leurs branches, exprimée par une équation donnée," *Mémoires de l'Académie Royale des Sciences*, 1734. Paris: Imprimerie Royale, 1736. Pp. 196–215.

1734b "Rémarques sur la méthode de Monsieur Fontaine, pour résoudre le problème où il s'agit de trouver une courbe qui touche les côtés d'un angle constant, dont le sommet glisse dans une courbe donnée," *Mémoires de l'Académie Royale des Sciences*, 1734. Paris: Imprimerie Royale, 1736. Pp. 531–537.

1735 "Examen de la réponse de Monsieur Fontaine à mes objections contre sa méthode pour trouver une courbe qui touche continuellement les côtés d'un angle constant, dont le sommet glisse dans une courbe donnée," *Mémoires de l'Académie Royale des Sciences, 1735*. Paris: Imprimerie Royale, 1738. Pp. 577–580.

1736 "Solution de quelques problèmes de dynamique," *Mémoires de l'Académie Royale des Sciences*, 1736. Paris: Imprimerie Royale, 1739. Pp. 1–22.

1737a "Investigationes aliquot, ex quibus probetur terrae figuram, secumdum leges attractionis, in ratione inversa quadrati distantiarum, maxime ad ellipsin accedere debere," *Philosophical Transactions*, vol. 40 (1737–38), No. 445 (January–June 1738): 19–25. This is the published version of Clairaut (1737e).

1737b Library of the Royal Society (London), ms LBC. 23, fols. 252–255: letter from Clairaut to Mortimer, 20 February.

1737c "De l'aberration apparente des étoiles, causée par le mouvement progressif de la lumière," *Mémoires de l'Académie Royale des Sciences, 1737*. Paris: Imprimerie Royale, 1740. Pp. 205–227.

1737d Library of the Royal Society (London), ms LBC. 24, fols. 79–81: letter from Clairaut to Mortimer, 13 December.

1737e Library of the Royal Society (London), ms LBC. 23, fols. 255–263: "Investigationes aliquot, ex quibis probetur terrae figuram, secumdum leges attraction is, in ratione inversa quadrati distantiarum, maxime ad ellipsin accedere debere." This is the manuscript of Clairaut (1737a).

1738a Library of the Royal Society (London), ms LBC. 25, fols. 24–54: "Recherches sur la figure des planètes qui tournent autour de leur axe, en supposant que leur densité varie de centre à la surface. Par Monsieur Clairaut."

1738b "An inquiry concerning the figure of such planets as revolve about an axis, supposing the density continually to vary from the centre towards the surface," trans. John Colson, in *Philosophical Transactions*, vol. 40 (1737–1738), No. 449 (August–September 1738): 277–306. This is the published English translation of Clairaut (1738a).

1738c Library of the Royal Society (London), ms LBC. 25, fols. 22–23: letter from Clairaut to Mortimer, 2 October.

1738d Letter from Clairaut to Stirling, 2 October, in Tweedie (1922: 176–177).

1739 "Recherches générales sur le calcul intégral," *Mémoires de l'Académie Royale des Sciences*, 1739. Paris: Imprimerie Royale, 1741. Pp. 425–436.

1740a Letter from Clairaut to Leonhard Euler, 17 September, in Euler (1980: 68–71).

1740b "Sur l'intégration ou la construction des équations différentielles du premier ordre," *Mémoires de l'Académie Royale des Sciences*, 1740. Paris: Imprimerie Royale, 1742. Pp. 293–323.

1741a Letter from Clairaut to MacLaurin, 10 February, in MacLaurin (1982: 342–343).

1741b Letter from Clairaut to MacLaurin, 10 April, in MacLaurin (1982: 347–354).

1741c Letter from Clairaut to MacLaurin, 18 September, in MacLaurin (1982: 359–370).

1741d Letter from Clairaut to Leonhard Euler, 12 April, in Euler (1980: 88–91).

1742a Letter from Clairaut to Leonhard Euler, 29 May, in Euler (1980: 127–131).

1742b Letter from Clairaut to Leonhard Euler, 28 December, in Euler (1980: 144–145).

1742c Letter from Clairaut to Leonhard Euler, 4 January, in Euler (1980: 102–110).

1742d Letter from Clairaut to Leonhard Euler, 28 March, in Euler (1980: 118–120).

1742e	"Sur quelques principes qui donnent la solution d'un grand nombre de problèmes de dynamique," *Mémoires de l'Académie Royale des Sciences*, 1742. Paris: Imprimerie Royale, 1745. Pp. 1–52.
1742f	Letter from Clairaut to MacLaurin, 19 November, in MacLaurin (1982: 375–380).
1742g	Letter from Clairaut to Leonhard Euler, 25 July, in Euler (1980: 134–136).
1742h	Letter from Clairaut to Leonhard Euler, 14 September, in Euler (1980: 138–139).
1742i	Letter from Clairaut to Leonhard Euler, 3 December, in Euler (1980: 139–143).
1743	*Théorie de la figure de la terre, tirée des principes de l'hydrostatique.* Paris: David Fils.
1750	"Nouvelle théorie de la figure de la terre, où l'on concilie les mésures actuelles avec les principes de la gravitation universelle," *Pièces qui ont remporté le prix de l'Académie Royale des Sciences, Inscriptions, et Belles Lettres de Toulouse, Depuis l'Année 1747, jusqu'en 1750.* Toulouse: François Forest, 1758.
1751	Letter from Clairaut to Cramer, 8 March, in Clairaut (1891–92: 242–243).
1753	"A Translation and Explanation of Some Articles of the Book Intitled, *Théorie de la Figure de la Terre,*" *Philosophical Transactions*, vol. 48, Part I. Pp. 73–85.
1760	Letter from Clairaut to Saverien, in Muller (1760: 391–396).
1763	Letter from Clairaut de Daniel Bernoulli, 27 December, in Clairaut (1891–92: 283–284).
1764	Letter from Clairaut to Daniel Bernoulli, 15 January, in Clairaut (1891–92: 284–286).
1765	*Catalogue des livres de la bibliothèque de Monsieur Clairaut.* Paris: LeClerc.
1809	"Some Investigations, by which It is Proved that the Figure of the Earth Must Approach Very Near to an Ellipsis, According to the Laws of Attraction Inversely as the Square of the Distances," *Philosophical Transactions.... Abridged*, vol. 8 (1735–1743). London: C. & R. Baldwin. Pp. 118–123. This is an English translation of Clairaut (1737a).
1891–92	"Lettere di Alessio Claudio Clairaut," ed. B. Boncompagni, in *Atti Dell'Accademia Pontificia De'Nuovi Lincei*, 45: 233–291.
1968	Facsimile of Clairaut (1728), in *Journal de Trévoux*, vol. 28. Geneva: Slatkine Reprints. Pp. 528–529.

CLAIRAUT, ALEXIS-CLAUDE, and CRAMER, GABRIEL

1955	"Une correspondance inédite entre Clairaut et Cramer," ed. Pierre Speziali, in *Revue d'Histoire des Sciences et de Leurs Applications*, 8: 193–237.

CLAIRAUT, ALEXIS-CLAUDE, and NICOLE, FRANÇOIS

1739	Report on Fontaine's first paper on integrating in finite terms or reducing to quadratures first-order ordinary "differential" equations of first degree using the partial differential calculus. *L'Académie Royale des Sciences, Registres des Procès-Verbaux*, 4 February, Pp. 20r–20v.

CLARINVAL, ANDRE

1957	"Esquisse historique de la courbe de poursuite," *Archives Internationales d'Histoire des Sciences*, 38–41: 25–37.

COHEN, I. BERNARD

1960	*The Birth of a New Physics.* Garden City N.Y.: Doubleday. Science Study Series S10.
1964	"Isaac Newton, Hans Sloane and the Académie Royale des Sciences," in *Mélanges Alexandre Koyré*, vol. I: *L'aventure de la science*. Paris: Hermann. Pp. 61–116.
1968	"The French Translation of Newton's *Philosophiae Naturalis Principia Mathematica* (1756, 1759, 1966)," *Archives Internationales d'Histoire des Sciences*, 21: 261–290.
1978	*Introduction to Newton's Principia.* Cambridge, Mass.: Harvard Univ. Press.
1985	*The Birth of a New Physics.* 2nd ed. New York: W. W. Norton.
1988	"Newton's Third Law and Universal Gravity," in *Newton's Scientific and Philosophical Legacy.* Ed. P. B. Scheurer and G. Debrock. Dordrecht, Boston, and London: Kluwer Academic Publishers. Pp. 25–53.

CONDORCET, MARIE-JEAN-ANTOINE-NICOLAS CARITAT, MARQUIS DE

1771	"Eloge de Monsieur Fontaine," *Histoire de l'Académie Royale des Sciences*, 1771. Paris: Imprimerie Royale, 1774. Pp. 105–130.
1774	"Eloge de Monsieur La Condamine," *Histoire de l'Académie Royale des Sciences*, 1774. Paris: Imprimerie Royale, 1778. Pp. 85–121.

COOLIDGE, JULIAN LOWELL
1940, 1963   A *History of Geometrical Methods*. Oxford: Oxford Univ. Press, 1940; New York: Dover, 1963.
1948   "The Beginnings of Analytic Geometry in Three Dimensions," *The American Mathematical Monthly*, 55: 76–86.
1949, 1963   *The Mathematics of Great Amateurs*. Oxford: Oxford Univ. Press, 1949; New York: Dover, 1963.

CORRY, LEO
1993   "Kuhnian Issues, Scientific Revolutions, and the History of Mathematics." *Studies in History and Philosophy of Science*, 24: 95–117.

COSTABEL, PIERRE
1966   "Pierre Varignon (1654–1722) et la diffusion en France du calcul différentiel et intégral," *Conferences du Palais de la Découverte*, D 108, 4 décembre 1965. Paris: Palais de la Découverte, 1966. Pp. 7–28.
1976   "Varignon, Pierre," *Dictionary of Scientific Biography*, vol. XIII. Ed. Charles Coulston Gillispie. New York: Charles Scribners Sons. Pp. 584–587.
1983   "La question des forces vives," *Cahiers d'Histoire et de Philosophie des Sciences*, 8: 1–170.
1988   "Science positive et forme de la terre au début du XVIIIe siècle," in *La figure de la Terre du XVIIIe siècle á l'ère spatiale*. Ed. Henri Lacombe and Pierre Costabel. Paris: Gauthier-Villars. Pp. 97–114.

COUDER, ANDRE
1961   "Fontenelle, homme des science," in *Journées Fontenelle organisés au centre international de synthèse (Mai, 1957)*, *Revue de Synthèse*, 1961, 82: 43–51.

COUSIN, JACQUES-ANTOINE-JOSEPH
1796   *Traité de calcul différentiel et de calcul intégral*. 2 vols. Paris: Regent et Bernard. This treatise both expands and refines Cousin's earlier work: *Leçons de calcul différentiel et de calcul intégral* (1777).

CUBRANIC, NIKOLA
1977–82   "Ruder Boskovic et la géodésie scientifique," in *Rudjer Boskovic. Annales de l'Institut Français de Zegreb*. Troisième Série/No. 3. Pp. 62–86.

DAINVILLE, FRANÇOIS DE, S. J.
1954   "L'Enseignement des mathématiques dans les Collèges Jésuites de France du XVIe au XVIIIe siècle," *Revue d'histoire des Sciences et de Leurs Applications* 7: 6–21, 109–123.
1964   "L'Enseignement scientifique dans les collèges des Jésuites," in Taton (1964, 1986: 27–65). Republished in Dainville (1978: 355–391).
1978   *L'éducation des jésuites (XVIe–XVIIIe siècles)*. Paris: Les Editions de Minuit.

D'ALEMBERT, JEAN LEROND
1747   *Réflexions sur la cause générale des vents*. Paris: David l'aîné.
1749   "Extrait d'une lettre de Monsieur D'Alembert à Monsieur de Maupertuis du 16 Novembre 1750," *Mémoires de l'Académie Royale des Sciences et des Belles Lettres*, 1749. Berlin: Haude et Spencer, 1751. P. 372.
1752   *Essai d'une nouvelle théorie de la résistance des fluides*. Paris: David l'aîné.
1755   "Equation," in *Encyclopédie, ou dictionnaire raisonné des sciences, des arts et des metiers, par une société de gens de lettres*, vol. 5. Ed. Denis Diderot and Jean LeRond D'Alembert. Paris: Briasson, David, LeBreton, Durand. Pp. 853–854.
1762   Letter from D'Alembert to Lagrange, 15 November, in Lagrange (1882: 7–8).
1764   Letter from D'Alembert to Madame Geoffrin, 7 January, in Henry (ed.) (1886: 131).
1765a   Bibliothèque de l'Institut (Paris), ms 1792, fols. 401 r–403 v: "Avertissement aux Géomètres inseré dans le *Mercure* de Juin 1765 (Monsieur Fontaine était encore vivant, quand cet avertissement a paru: il n'est mort que 6 ans après, en 1771)."
1765b   "Tautochrone," in *Enclypédie, ou dictionnaire raisonné des sciences, des arts et des metiers, per une société de gens de lettres*, vol. 15. Ed. Denis Diderot. Neuchâtel: Samuel Faulche, Pp. 946.
1765c   "Sur les Tautochrones," *Mémoires de l'Académie Royale des Sciences et Belles-Lettres*, 1765. Berlin: Haude et Spencer, 1767. Pp. 381–413.
1765d   Letter from D'Alembert to the *Mercure de France*, 9 May, in the *Mercure de France*, May 1765, pp. 154–158. This is the published version of D'Alembert (1765a).

1768	Letter from D'Alembert to Lagrange, 20 November, in Lagrange (1882:121).

1768   Letter from D'Alembert to Lagrange, 20 November, in Lagrange (1882:121).

1769   "Recherches sur le calcul intégral," *Mémoires de l'Académie Royale des Sciences*, 1769. Paris: Imprimerie Royale, 1772. Pp. 73–146.

1770   *Traitè de l'équilibre et du mouvement des fluides pour servir de suite au Traité de dynamique*. 2nd ed. Paris: Briasson.

1771   Letter from D'Alembert to Lagrange, 6 September, in Lagrange (1882:210).

1773   "Sur la figure de la terre," *Opuscules mathématiques*, vol. 6. Paris. Pp. 47–67.

1967a   "Tautochrone," in *Encyclopédie, ou dictionnaire raisonné des sciences, des arts et des metiers, par une société de gens de lettres*, vol. 15. Ed. Denis Diderot. Stuttgart-Bad Cannstatt: Friederich Frommann Verlag (Günther Holzboog). Pp. 946. This is D'Alembert (1765b) reproduced in facsimile.

1967b   *Traité de l'équilibre et du mouvement des fluides pour servir de suite au Traité de dynamique*. 2nd ed. Bruxelles: Culture et Civilisation. This is D'Alembert (1770) reproduced in facsimile.

D'ALEMBERT, JEAN LEROND, and DE GUA DE MALVES, JEAN-PAUL

1742   Report on Alexis Fontaine's *pli cacheté* submitted to the Paris Academy of Sciences in June of 1741 and opened at the Paris Academy in November of 1741. *L'Académie Royale des Sciences, Registres des Procès-Verbaux*, 17 January, Pp. 14–21.

DEDEYAN, CHARLES

1987   *Diderot et la pensée anglaise*. Florence: Olschki.

DE GANDT, FRANÇOIS

1986   "Le style mathématique des *Principia* de Newton," *Revue d'Histoire des Sciences*, 39:195–222.

1987   "Le problème inverse (Prop. 39–41)," *Revue d'Histoire des Sciences, 40:* 281–309.

DELAMBRE, JEAN-BAPTISTE-JOSEPH

1801?a Bibliothèque de l'Institut (Paris), ms 2041, No. 78, fols. 443r–446v: "Notice sur Cousin, non imprimée." Cousin died on 29 December 1800.

1801?b Bibliothèque de l'Institut (Paris), ms 2041, No. 89, fols. 568r–571r: "Notice sur Cousin, non imprimée." Cousin died on 29 December 1800.

DELISLE, JOSEPH-NICOLAS

1716   Bibliothéque Nationale (Pairs), Fonds Français, ms 9674, fols. 4r–9r: "Nouvelles réflexions sur la figure de la terre communiquées en 1716."

1717a   Paris Observatory, ms B1, 1: J.-N. Delisle's Correspondance, Tome I, No. 55: copy of letter from Delisle to Abbé Teinturier, 7 February.

1717b   Paris Observatory, ms B1, 1: J.-N. Delisle's Correspondance, Tome 1, No. 58: copy of letter from Delisle to Réaumur, 27 April.

1717c   Paris Observatory, ms B1, 1: J.-N. Delisle's Correspondance, Tome I, No. 59: copy of letter from Delisle to Abbé Bignon, 11 June.

1720a   Paris Observatory, ms B1, 1: J.-N. Delisle's Correspondance, Tome I, No. 194: copy of letter from Delisle to the Chevalier de Louville, 8 August.

1720b   Paris Observatory, ms A7$_{7_{65}}$, fols. 65, 4, A–65, 4, C: "Nouvelles réflexions sur la figure de la terre (avril 1720)."

1724a   Paris Observatory, ms B1, 1: J.-N. Delisle's Correspondance, Tome II, No. 125: copy of letter from Delisle to Delisle de la Croyère, 24 August.

1724b   Paris Observatory, ms B1, 2: J.-N. Delisle's Correspondance, Tome II, No. 128: copy of letter from Delisle to Nicasius Grammatici, ? October.

1724c   Paris Observatory, ms B1, 2: J.-N. Delisle's Correspondance, Tome II, No. 130: copy of letter from Delisle to Newton, 21 December.

1724d   Paris Observatory, ms B1, 2: J.-N. Delisle's Correspondance, Tome II, No. 131: copy of letter from Delisle to Halley, 21 December.

1724e   Paris Observatory, ms B1, 2: J.-N. Delisle's Correspondance, Tome II, No. 132: copy of letter from Delisle to DeMoivre, 24 December.

1724f   Paris Observatory, ms B1, 2: J.-N. Delisle's Correspondance, Tome II, No. 133: copy of letter from Delisle to Innys, 21 December.

1725   Paris Observatory, ms B1, 2: J.-N. Delisle's Correspondance, Tome II, No. 146: copy of letter from Delisle to Vallincourt, 12 March.

1734a   Bibliothèque Nationale (Paris), Fonds Français, ms 9674, fols. 25r–26r: letter from Delisle to Fontenelle, 31 July.

1734b   Archives Nationales (Paris), Fonds de la Marine, sous-série 2JJ62:J.-N. Delisle's Correspondence, Tome IV, No. 121: copy of letter from Delisle to Fontenelle, 31 July. This is a copy of Delisle (1734a).

DELORME, SUZANNE
1957   "La *Géométrie de l'infini* et ses commentateurs de Jean Bernoulli Monsieur de Cury," *Revue d'Histoire des Sciences et de Leurs Applications, 10*: 339–359.
1972   "Fontenelle, Bernard le Bouyer (or Bovier) de," *Dictionary of Scientific Biography*, vol. V. Ed. Charles Coulston Gillispie. New York: Charles Scribners Sons. Pp. 57–63.

DESAGULIERS, JOHN THEOPHILIS
1725a   "A Dissertation Concerning the Figure of the Earth," *Philosophical Transactions*, No. 386 (January–February): 201–222.
1725b   "A Dissertation Concerning the Figure of the Earth Continued," *Philosophical Transactions*, No. 387 (March–April): 239–255.
1725c   "A Dissertation Concerning the Figure of the Earth. Part the Second," *Philosophical Transactions*, No. 388 (May–June): 277–304.

DESAUTELS, ALFRED R.
1956   *Les Mémoires de Trevoux et Le Mouvement des Idées au XVIIIe Siécle*. Rome: Institution Historicum S.I.
1971   "Castel, Louis-Bertrand," *Dictionary of Scientific Biography*, vol. III. Ed. Charles Coulston Gillispie. New York: Charles Scribners Sons. Pp. 114–115.

DESGRAVES, LOUIS
1986   *Montesquieu*. Paris: Editions Mazarine, 1986.

DHOMBRES, JEAN
1985   "Quelques rencontres de Diderot avec les mathématiques," in Anne-Marie Chouillet (ed.), *Denis Diderot, 1713–1784. Actes du Colloque international Diderot (Paris-Sèvres-Reims-Langres, 4–11 juillet 1984)*. Paris: Aux Amateurs de Livres, 1985. Pp. 269–280.

DIDEROT, DENIS
1765   "Juin 1765," in *Correspondance Littéraire, Philosophique et Critique Par Grimm, Diderot, Raynal, Meister, etc.*, vol. 6. Ed. Maurice Tourneux. Paris: Garniers Frères, 1878. Pp. 287–290.

DUBOIS, ELFRIEDA T.
1986   "The Exchange of Ideas Between England and France as Reflected in Learned Journals of the Later Seventeenth and Early Eighteenth Centuries," *History of European Ideas, 7*: 33–46.

ENGELSMAN, STEVEN B.
1984   *Families of Curves and the Origins of Partial Differentiation*. Amsterdam, New York, and Oxford: North-Holland.
1986   "Orthogonaltrajektorien im Prioritässtreit zwischen Leibniz und Newton," in *Studia Leibnitiana*, Sonderheft 14: *300 Jahre "Nova Methodus" Von G. W. Leibniz (1684–1984)*. Stuttgart: Franz Steiner Verlag Wiesbaden GMBH. Pp. 144–156.

EULER, LEONHARD
1729   "Curva tautochrona in fluido resistentiam faciente secumdum quadrata celeritatum," *Commentarii Academiae Scientiarum Imperialis Petropolitanae*, vol. 4, 1729. Saint Peterburg, 1735. Pp. 67–89.
(1734–35) a   "De infinitis curvis ejusdem generis. Seu methodus inveniendi aequationes pro infinitis curvis ejusdem generis," *Commentarii Academiae scientiarum Imperialis Petropolitanae*, vol. 7, 1734–35. Saint Petersburg, 1740. Pp. 174–189. Republished in Euler (1936: 36–56).
(1734–35) b   "Additamentum ad dissertationem de infinitis curvis ejusdem generis," *Commentarii Academiae scientiarum Imperialis Petropolitanae*, vol. 7, 1734–35. Saint Petersburg, 1740. Pp. 184–200. Republished in Euler (1936: 57–75).
1736a   *Mechanica sive motus scientia analytice exposita*. 2 vols. Saint Petersburg. Republished in Euler (1912).
1736b   "De constructione aequationum ope motus tractorii allisque ad methodum tangentium inversam pertinentibus," *Commentarii Academiae scientiarum Imperialis Pet-*

ropolitanae, vol. 8, 1736. Saint Petersburg, 1741. Pp. 66–85. Republished in Euler (1936: 83–107).

1740a   Letter from Euler to Clairaut, (30) 19 October, in Euler (1980: 71–78).

1740b   "Inquisitio physica in causam fluxus ac refluxus maris," in Newton (1739–1742: vol. 3. 283–374).

1741   Letter from Euler to Clairaut, (6 March) 24 February, in Euler (1980: 81–88).

1742a   Letter from Euler to Clairaut, January–February, in Euler (1980: 110–117).

1742b   Letter from Euler to Clairaut, April, in Euler (1980: 120–127).

1748   "Rêflexions sur quelques loix générales de la nature qui s'observent dans les effects des forces quelconques," *Mémoires de l'Académie Royale des Sciences et des Belles Lettres,* 1748. Berlin: Haude et Spencer, 1750. Pp. 189–218.

1755   "Principes généraux de l'état d'équilibre des fluides," *Mémoires de l'Académie Royale des Sciences et des Belles Lettres,* 1755. Berlin, 1757. Pp. 217–273.

1756–57   "Principia motus fluidorum," *Novi Commentarii Academiae scientiarum Imperialis Petropolitanae,* vol. 6, 1756–57. Saint Petersburg, 1761. Pp. 271–311.

1768   Letter from Euler to Lagrange, (16) 5 February, in Euler (1980: 461–464).

1912   *Leonhardi Euleri Opera Omnia,* series II, vols. 1 and 2. Ed. Paul Stackel. Leipzig and Berlin: B. G. Teubner. This is a new edition of Euler (1736a).

1936   *Leonhardi Euleri Opera Omnia,* series I, vol. 22. Ed. Henri Dulac. Basel, Leipzig and Zürich: B. G. Teubner and Orell Füssli.

1954   *Leonhardi Euleri Opera Omnia,* series II, vol. 12. Ed. Clifford A. Truesdell. Zürich and Lausanne: Orell Füssli.

1955   *Leonhardi Euleri Opera Omnia,* series II, vol. 13. Ed. Clifford A. Truesdell. Zürich and Lausanne: Orell Füssli.

1975   *Leonhardi Euleri Opera Omnia,* series IVA, vol. 1. Ed. Adolf P. Juskevic, Vladimir L. Smirnov, and Walter Habicht. Basel: Birkhäuser Verlag.

1980   *Leonhardi Euleri Opera Omnia,* series IVA, vol. 5. Ed. A. Juskevic and R. Taton. Basel: Birkhäuser Verlag.

1986   *Leonhardi Euleri Opera Omnia,* series IVA, vol. 6. Ed. P. Costabel, E. Winter, A. T. Grigorian, and A. P. Juskevic. Basel: Birkhäuser Verlag.

EULER, LEONHARD, and BERNOULLI, DANIEL

1906–07   "Der Briefwechsel zwischen Leonhard Euler und Daniel Bernoulli," ed. G. Eneström, *Bibliotheca Mathematica,* 7: 126–156.

FEIGENBAUM, LENORE

1985   "Brook Taylor and the Method of Increments," *Archive for History of Exact Sciences,* 34: 1–140.

1992   "The Fragmentation of the European Mathematical Community," published in Harman and Shapiro (1992: 383–397).

FLANDERS, HARLEY

1973   "Differentiation under the Integral Sign," *American Mathematical Monthly,* 80 (6, Part I): 615–627.

FONTAINE, ALEXIS

1734a   "Problème: Une courbe étant donnée, trouver celle qui serait décrite par le sommet d'un angle dont les côtés toucheraient continuellement la courbe donnée; et réciproquement la courbe qui doit être décrite par le sommet de l'angle, étant donnée, trouver celle qui sera touchée par les côtés," *Mémoires de l'Académie Royale des Sciences,* 1734. Paris: Imprimerie Royale, 1736. Pp. 527–530.

1734b   "Sur les courbes tautochrones," *Mémoires de l'Académie Royale des Sciences,* 1734. Paris: Imprimerie Royale, 1736. Pp. 371–379.

1734c   "Réponse aux rémarques précédentes," *Mémoires de l'Académie Royale des Sciences,* 1734. Paris: Imprimerie Royale, 1736. P. 538.

1735   "Problème: Une courbe étant donnée, trouver celle qui serait décrite par le sommet d'un angle, dont les côtés toucheraient continuellement la courbe donnée, et vice versa, la courbe décrite par le sommet d'un angle étant donnée, trouver la courbe touchée par less côtés," *Académie Royale des Sciences, Registres des Procès-Verbaux,* 21 January, Pp. 17r–19r.

1737   "Manière d'éviter les suites dans certaines problèmes, dont la solution semble d'abord en dépendre," *Académie Royale des Sciences, Registres des Procès-Verbaux*, 6 July, Pp. 139r–140v.

1739–40 "Sur les différentes forces corps," *Académie Royale des Sciences, Registres des Procès-Verbaux*, 2 December 1739, p. 226r; 23 December 1739, p. 232r; 10 February 1740, p. 21r; 13 February 1740, p. 22v; 17 February 1740, pp. 23r–25v.

1741 Acadèmie des Sciences (Paris), Archives Bertrand, Carton I: letter from Fontaine to Mairan, 22 June.

1747 "Sur la résolution des équations," *Mémoires de l'Académie Royale des Sciences*, 1747. Paris: Imprimerie Royale, 1752. Pp. 665–677.

1764a "Nouvelle méthode pour la solution des problèmes de maximis et minimis," in Fontaine (1764b: 1–5).

1764b *Mémoires données à l'Académie Royale des Sciences non imprimés dans leurs temps*. Paris: Imprimerie Royale.

1767 "Addition à la méthode pour la solution des problèmes de maximis & minimis," *Mémoires de l'Académie Royale des Sciences*, 1767. Paris: Imprimerie Royale, 1770. Pp. 588–613.

1770 *Traité de calcul différentiel et intégral*. Paris: Imprimerie Royale. This is the second, unrevised edition of Fontaine (1764b).

FONTENELLE, BERNARD LE BOVIER DE

1731 Brief account of Fontaine's new method for determining curvatures of plane curves, in *Histoire de l'Académie Royale des Sciences*, 1731. Paris: Imprimerie Royale, 1733. P. 54

1732a Review of Maupertuis's *Discours sur les différentes figures des astres* (1732), in *Histoire de l'Académie Royale des Sciences*, 1732. Paris: Imprimerie Royale, 1735. Pp. 85–93.

1732b Summary of Fontaine's work of 1732 on isoperimetrical problems, in *Histoire de l'Académie Royale des Sciences*, 1732. Paris: Imprimerie Royale, 1735. P. 71.

1734 "Sur les figures que les planètes prennent par la pesanteur," *Histoire de l'Académie Royale des Sciences*, 1734. Paris: Imprimerie Royale, 1736. Pp. 83–94.

1766 *Oeuvres de Monsieur Fontenelle*. Nouvelle Edition, vol. XI. Paris: Les Libraires Associés.

FRASER, CRAIG G.

1985 "D'Alembert's Principle: The original Formulation and Application in Jean D'Alembert's *Traité de Dynamique* (1743). Part I," *Centaurus*, 28: 31–61.

1989 "The Calculus as Algebraic Analysis: Some Observations on Mathematical Analysis in the 18th Century," *Archive for History of Exact Sciences, 39*: 317–335.

FUSS, P. H.

1843 *Correspondance mathématique et physique de quelques célèbres géométres du XVIIIe siècle*, vol. 2. Saint Petersburg.

1968 *Correspondance mathématique et physique de quelques célèbres géométres du XVIIIe siècle*, vol. 2. New York and London: Johnson Reprint Corp. This is Fuss (1843) reproduced in facsimile.

GAMACHES, ETIENNE-SIMON, ABBE DE

1740 *Astronomie physique, ou Principes généraux de la nature, appliqués au mécanisme astronomique, et comparés aux principes de la philosophie de Monsieur Newton*. Paris: C.-A. Jombert.

GILLISPIE, CHARLES COULSTON

1958 "Fontenelle and Newton," in *Isaac Newton's Papers and Letters on Natural Philosophy and Related Documents*. Ed. I. Bernard Cohen. Cambridge, Mass.: Harvard Univ. Press. Pp. 427–443.

GOUGH, J. B.

1975 "Réaumur, René-Antoine Ferchault de," *Dictionary of Scientific Biography*, vol. XI. Ed. Charles Coulston Gillispie. New York: Charles Scribners Sons. Pp. 327–335.

GRANDJEAN DE FOUCHY, JEAN-PAUL

1744 "Eloge de M. l'Abbé de Bragelogne," *Histoire de l'Academie Royale des Sciences*, 1744. Paris: Imprimerie Royale, 1748. Pp. 65–70.

1758 "Eloge de Monsieur Bouguer," *Histoire de l'Académie Royale des Sciences*, 1758. Paris: Imprimerie Royale, 1763. Pp. 127–136.

GRANGER, GILLES-GASTON
1956   *La mathématique sociale du Marquis de Condorcet.* Paris: Presses Universitaires de France.
1989   *La mathématique sociale du Marquis de Condorcet.* 2nd ed. Paris: Editions Odile Jacob.
GREENBERG, JOHN L.
1981   "Alexis Fontaine's 'Fluxio-differential Method' and the Origins of the Calculus of Several Variables," *Annals of Science, 38:* 251–290.
1982   "Alexis Fontaine's Integration of Ordinary Differential Equations and the Origins of the Calculus of Several Variables," *Annals of Science, 39:* 1–36.
1983   "Geodesy in Paris in the 1730s and the Paduan Connection," *Historical Studies in the Physical Sciences, 13:* 239–260.
1984a   "Degrees of Longitude and the Earth's Shape: The Diffusion of a Scientific Idea in Paris in the 1730s," *Annals of Science, 41:* 151–158.
1984b   "Alexis Fontaine's Route to the Calculus of Several Variables," *Historia Mathematica, 11:* 22–38.
1985   "The Origins of Partial Differentiation," *Annals of Science, 42:* 421–429.
1986   "Mathematical Physics in Eighteenth-Century France," *Isis, 77:* 59–78.
1988   "Breaking a 'Vicious Circle': Unscrambling A.-C. Clairaut's Iterative Method of 1743," *Historia Mathematica, 15:* 228–239.
GUERLAC, HENRY
1977   "The Newtonianism of Dortous de Mairan," in Guerlac's *Essays and Papers in the History of Modern Science.* Baltimore and London: The Johns Hopkins Univ. Press. Pp. 479–490.
1979   "Some Areas for Further Newtonian Studies," *History of Science, 17:* 75–101.
1981   *Newton on the Continent.* Ithaca and London: Cornell Univ. Press.
HAHN, ROGER
1963   "Applications of Science to Society: the Societies of Arts," *Studies on Voltaire and the Eighteenth Century, 34–37:* 829–836.
1971   *The Anatomy of a Scientific Institution: The Paris Academy of Sciences, 1666–1803.* Berkeley and London: Univ. California Press.
HANKINS, THOMAS L.
1967   "The Reception of Newton's Second Law of Motion in the Eighteenth Century," *Archives Internationales d'Histoire des Sciences, 20:* 43–65.
1970   *Jean D'Alembert. Science and the Enlightenment.* Oxford: Clarendon Press.
1976   Review of *Actes de la journée Maupertuis (Créteil, ler décembre 1973),* in *Isis, 67:* 484.
HARMAN, P. M., and SHAPIRO, ALAN E. (eds).
1992   *An Investigation of Difficult Things: Essays on Newton and the History of the Exact Sciences.* New York: Cambridge Univ. Press.
HEILBRON, JOHN L.
1983   *Physics at the Royal Society during Newton's Presidency.* Los Angeles: William Andrews Clark Memorial Library UCLA.
HENRY, CHARLES
1940   "Sur quelques billets inédits de Lagrange," ed. Charles Henry, *Bullettino di bibliografia e di storia delle scienze matematiche e fisiche, 19:* 129–135.
HIGGINS, THOMAS J.
1940   "A Note on the History of Mixed Partial Derivatives," *Scripta Mathematica, 7:* 59–62.
HUYGENS, CHRISTIAAN
1690   *Traité de la lumière... Avec un discours de la cause de la pesanteur.* Leiden: Pierre Van Der Aa.
INCE, E. L.
1944   *Ordinary Differential Equations.* New York: Dover.
ITARD, JEAN
1949   Review of F. Le Lionnais, *Les Grands courants de la pensée mathématique,* Cahiers du Sud, 1948, in *Revue d'Histoire des Sciences et de Leurs Applications, 2:* 280–281.
JAMES, GLENN, and JAMES, ROBERT C.
1968   *Mathematics Dictionary.* 3rd ed. Princeton N. J.: D. Van Nostrand.

JARDETZKY, WENCESLAS S.
1958   *Theories of Figures of Celestial Bodies.* New York: Interscience Publishers.
JONES, PHILLIP S.
1976   "Vandermonde, Alexandre-Théophile," *Dictionary of Scientific Biography,* vol. XIII. Ed. Charles Coulston Gillispie. New York: Charles Scribners Sons. Pp. 571–572.
KAC, MARK
1982   "Dehydrated elephants revisited," *The American Scientist, 70:* 633–634
KASPER, TONI
1980   "Integration in Finite Terms: The Liouville Theory," *Mathematics Magazine, 53:* 195–201.
KATZ, VICTOR J.
1981   "The History of Differential Forms from Clairaut to Poincaré," *Historia Mathematica, 8:* 161–188.
KELLOGG, OLIVER DIMON
1953   *Foundations of Potential Theory.* New York: Dover.
KESSLER, MICHAEL
1981   "A Puzzle Concerning Diderot's Presentation of Saunderson's *Palpable Arithmetic,"* *Diderot Studies, 20:* 159–173.
KLEINBAUM, ABBY ROSE
1970   Jean Jacques Dortous de Mairan (1678–1771): A Study of an Enlightenment Scientist. Ph.D. Dissertation at Columbia University.
KLINE, MORRIS
1972   *Mathematical Thought from Ancient to Modern Times.* New York: Oxford Univ. Press.
KOENIG, SAMUEL
1737   Letter from Koenig to Maupertuis, 20 September, in Maupertuis (1896: 107–109 or 1971: 107–109).
KOPAL, ZDENEK
1960   *Figures of Equilibrium of Celestial bodies with Emphasis on Problems of Motion of Artificial Satellites.* Madison: Univ. Wisconsin Press.
KOYRE, ALEXANDRE
1968   *Newtonian Studies.* Chicago: Univ. Chicago Press. (Phoenix Books).
KUHN, THOMAS S.
1962   *The Structure of Scientific Revolutions.* Chicago: Univ. Chicago Press.
1970   *The Structure of Scientific Revolutions.* 2nd ed. Chicago: Univ. Chicago Press.
1977   "Mathematical versus Experimental Traditions in the Development of Physical Science," in Thomas S. Kuhn, *The Essential Tension. Selected Studies in Scientific Tradition and Change.* Chicago and London: Univ. Chicago Press. Pp. 31–65.
LA BEAUMELLE, LAURENT ANGLIVIEL DE
1856   *Vie de Maupertuis.* Paris: Le Doyen.
LAFUENTE, ANTONIO
1983   "La mécanica de fluidos y la teoria de la figure de la Tierra entre Newton y Clairaut (1687–1743)," *Dynamis, 3:* 55–89.
LAFUENTE, ANTONIO and DELGADO, ANTONIO J.
1984   *La geometrización de la tierra: Observaciónes y resultados de la expedición geodesica hispano-francesa al virreinato del Peru (1735–1744).* Madrid: Consejo Superior de Investigaciónes Cientificas Instituto "Arnau de Vilanova."
LAFUENTE, ANTONIO, and PESET, JOSE L.
1984   "La question de la figure de la terre. L'agonie d'un débat scientifique au XVIIIe siècle," *Revue d'Histoire des Sciences, 37:* 235–254.
LAGRANGE, JOSEPH-LOUIS
(1760–61)a "Essai d'une nouvelle méthode pour déterminer les maxima et les minima des formules intégrales indéfinies," *Miscellanea philosophico-mathematica societatis privatae Taurinensia,* vol. 2: 173–195.
(1760–61)b "Application de la méthode exposée dans le mémoire précedent à la solution de différents problèmes de dynamique," *Miscellanea philosophico-mathematica societatis privatae Taurinensia,* vol. 2: 196–298.
1769a   Letter from Lagrange to D'Alembert, 15 July, in Lagrange (1882: 139).

1769b	Letter from Lagrange to D'Alembert, 2 August, in Lagrange (1882: 144–145).
1772	"Sur l'intégration des équations à différences partielles du premier ordre," *Nouveaux Mémoires de l'Académie Royale des Sciences et des Belles Lettres*, 1772. Berlin: Chrétien Fréderic Voss, 1774. Pp. 353–372.
1773	"Attraction des sphéroïdes elliptiques," *Nouveaux Mémoires de l'Académie Royale des Sciences et des Belles Lettres*, 1773. Berlin, 1775. Pp. 121–148.
1774a	Letter from Lagrange to Condorcet, 1 October, in Lagrange (1892: 28).
1774b	Letter from Lagrange to J.-J. de Marguerie, 24 February, in Lagrange (1892: 270–271).
1774c	Letter from Lagrange to Condorcet, 24 February, in Lagrange (1892: 17–18).
1790?	Bibliothèque de l'Institut (Paris), ms 903, fols. 67–81: "Sur les tautochrones." The year 1790 appears on fol. 68, which gives a rough idea when Lagrange wrote the manuscript.
1867a	"Essai d'une nouvelle méthode pour déterminer les maxima et les minima des formules intégrales indéfinies," *Oeuvres de Lagrange*, vol. I. Ed. J.-A. Serret. Paris: Gauthier-Villars. Pp. 335–362. This work is a republication of Lagrange [(1760–61)a].
1867b	"Application de la méthode exposée dans le mémoire précédent à la solution de différents problèmes de dynamique," *Oeuvres de Lagrange*, vol. I. Ed. J.-A. Serret. Paris: Gauthier-Villars. Pp. 365–468. This work is a republication of Lagrange [(1760–61)b].
1882	*Oeuvres de Lagrange*, vol. XIII. Ed. J.-A. Serret. Paris: Gauthier-Villars.
1892	*Oeuvres de Lagrange*, vol. XIV. Eds. J.-A. Serret; Gaston Darboux. Paris: Gauthier-Villars.

LALANDE, JEROME DE

1803	*Bibliographie astronomique; avec l'histoire de l'astronomie depuis 1781 jusqu'à 1802.* Paris: Imprimerie de la République.

LAMB, HORACE

1932	*Hydrodynamics.* 6th ed. Cambridge: Cambridge Univ. Press.

LAPLACE, PIERRE-SIMON, MARQUIS DE

1784	*Théorie du mouvement et de la figure elliptique des planètes.* Paris: Imprimerie de P.-D. Pierres.
1829	*Traité de mécanique céleste*, vol. 2. 2nd ed. Paris: Bachlier, Successeur de Mme Ve Courcier.

LEGENDRE, ADRIEN-MARIE

1785	"Recherches sur l'attraction des sphéroïdes homogènes," *Académie Royale des Sciences. Mémoires de Mathématiques et de Physique. Presentés... Par Divers Savants*, vol. 10. Paris. Pp. 411–434.

LENOBLE, ROBERT

1943, 1971	*Mersenne, ou la naissance du mécanisme.* Paris: J. Vrin.

LEVOT, PROSPER

1861	"Marguerie (Jean-Jacques de)," *Biographie universelle, ancienne et moderne*, nouvelle édition, vol. 26. Paris: C. Desplaces; Leipzig: A. Brockhaus. Pp. 543–547.

LEXELL, ANDERS-JOHANN

1781	Letter from Lexell to Johann-Albrecht Euler, 7 January, in Birembaut (1957: 157).

LOUVILLE, JACQUES-EUGENE D'ALLONVILLE, CHEVALIER DE

1720	Paris Observatory, ms B1, 1: J.-N. Delisle's Correspondence, Tome I, No. 195: letter from Louville to Delisle, 13 August.
1722a	"Sur la cycloïde, ou sur une difficulté de statique proposée à l'Académie avec la réfutation d'un écrit par lequel l'auteur pretendait donner le denoüement de cette difficulté," *Académie Royale des Sciences, Registres de Procès-Verbaux*, 22 June. Pp. 161r–176v.
1722b	"Eclaircissement sur une difficulté de statique proposée à l'Académie," *Mémoires de l'Académie Royale des Sciences*, 1722. Paris: Imprimerie Royale, 1724. Pp. 128–142.

LUBBOCK, J. W.

1830	*Account of the Traité sur le flux et réflux de la mer, of Daniel Bernoulli and A Treatise on the Attraction of Ellipsoids.* Pall Mall East and Cambridge: C. Knight and J. J. Deighton.

LÜTZEN, JESPER
1984   "Joseph Liouville's Work on the Figures of Equilibrium of a Rotating Mass of Fluid,"
            Archive for History of Exact Sciences, 30: 113–166.
1990   *Joseph Liouville 1809–1882: Master of Pure and Applied Mathematics.* New
            York: Springer Verlag.
MACLAURIN, COLIN
1740a   "De causa physica fluxus et refluxis maris," in Newton (1739–1742: vol. 3, 247–282).
1740b   Letter from MacLaurin to Stirling, 6 December, in MacLaurin (1982: 336–338).
1801   *A Treatise of Fluxions*, 2nd ed. 2 vols. London: Knight and Compton.
1982   *The Collected Letters of Colin MacLaurin.* Ed. Stella Mills. Nantwich: Shiva.
MAHEU, GILLES
1966   "La vie scientifique au milieu du XVIIIe siècle: Introduction à la publication des
            lettres de Bouguer à Euler," *Revue d'Histoire des Sciences, 19*: 206–224.
MAHONEY, MICHAEL S.
1975   "Saurin, Joseph," *Dictionary of Scientific Biography*, vol. XII. Ed. Charles Coulston
            Gillispie. New York: Charles Scribners Sons Pp. 117–119.
1984a   "On Differential Calculuses," *Isis, 75*: 366–372.
1984 b   "Changing Canons of Mathematical and Physical Intelligibility in the Later 17th
            Century," *Historia Mathematica, 11*: 417–423.
1990   "Infinitesimals and Transcendent Relations: The Mathematics of Motion in the Late
            Seventeenth Century," appearing in *Reappraisals of the Scientific Revolution*. Ed.
            David C. Lindberg and Robert S. Westman. New York: Cambridge Univ. Press.
            Pp. 461–491.
MAIRAN, JEAN-JACQUES DORTOUS DE
1720   "Recherches géomètriques sur la dimunition des dégres terrestres en allant de
            l'équateur vers les pôles. Où l'on examine les conséquences qui en résultent, tant à
            l'égard de la figure de la terre, que de la pesanteur, des corps, et de l'accourcissement
            du pendule," *Mémoires de l'Académie Royale des Science*, 1720. Paris: Imprimerie
            Royale, 1722. Pp. 231–277.
1722   "Lettre de Monsieur de Mairan de l'Académie Royale des Sciences écrite aux Auteurs
            du Journal," *Journal des Savants*, 7 September, p. 570.
1724   Universitätsbibliothek, Basel, Bernoulli Archives, ms LIa 661, No. 9∗: letter from
            Mairan to Johann I Bernoulli, *circa* 8 August.
1736   Universitätsbibliothek, Basel, Bernoulli Archives, ms LIa 661, No. 54∗: letter from
            Mairan to Johann I Bernoulli, 22 March.
1743   Letter from Mairan to Colin MacLaurin, 1 January, in MacLaurin (1982: 381–382).
MANCOSU, PAOLO
1989   "The Metaphysics of the Calculus: A Foundational Debate in the Paris Academy of
            Sciences 1700–1706," *Historia Mathematica, 16*: 224–248.
MARKOVIC ZELJKO
1957   "Le voyage de R. Boskovic en France en 1759/60," in *Rudzer Boskovic. Grada Knjiga*
            II. Zagreb: Jugoslavenska Academija Znanosti I Umjetnosti. Pp. 33–47.
1960   "R. J. Boskovic et la théorie de la figure de la terre," *Conference du Palais de la
            Découverte*, D 77, 5 Novembre 1960. Paris: Palais de la Découverte, 1961. Pp. 5–45.
MARSAK, LEONARD
1959   "Bernard de Fontenelle: The Idea of Science in the French Enlightenment," *Transac-
            tions of the American Philosophical Society, 49*: 3–64.
MARTIN, J.-P
1987   *La figure de la terre: récit de l'éxpedition française en laponie suédoise (1736–1737).*
            Cherbourg: Isoète
MAUPERTUIS, PIERRE-LOUIS MOREAU DE
1730a   "La courbe *Descensus Aquabilis* dans un milieu resistant comme une puissance
            quelconque de la vitesse," *Mémoires de l'Académie Royale des Sciences*, 1730.
            Paris: Imprimerie Royale, 1732. Pp. 233–242.
1730b   Universitätsbibliothek, Basel, Bernoulli Archives, ms LIa 662, No. 5∗: letter from
            Maupertuis to Johann I Bernoulli, late October.
1731a   Universitätsbibliothek, Basel, Bernoulli Archives, ms LIa 662, No. 9∗: letter from
            Maupertuis to Johann I Bernoulli, 11 March.

1731b Universitätsbibliothek, Basel, Bernoulli Archives, ms LIa 662, No. 14∗: letter from Maupertuis to Johann I Bernoulli, 12 September.

1731c "Balistique arithmétique," *Mémoires de l'Académie Royale des Sciences*, 1731. Paris: Imprimerie Royale, 1733. Pp. 297–298.

1731d "Sur la séparation des indéterminées," *Académie Royale des Sciences, Registres des Procés-Verbaux*, 28 April, Pp. 82v–86v.

1731e "Sur la séparation des indéterminées dans les équations différentielles," *Mémoires de l'Académie Royale des Sciences*, 1731. Paris: Imprimerie Royale, 1733. Pp. 103–109.

1731f Universitätsbibliothek, Basel, Bernoulli Archives, ms LIa 662, No. 13∗: letter from Maupertuis to Johann I Bernoulli, 30 July.

1732a "De figuris quas fluida rotata induere possunt problemata duo; cum conjectura de stellis quae aliquando prodeunt vel deficieunt; et de annulo Saturni," *Philosophical Transactions*, No. 422 (January–March): 240–256.

1732b *Discours sur les différentes figures des astres: d'où l'on tire des conjectures sur les étoiles qui paraissent changer de grandeur: et sur l'anneau de Saturne avec une exposition abbrégée des Systèmes de Monsieur Descartes et de Monsieur Newton.* Paris: Imprimerie Royale.

1732c Universitätsbibliothek, Basel, Bernoulli Archives, ms LIa 662, No. 23∗: letter from Maupertuis to Johann I Bernoulli, 4 August.

1732d "Sur les courbes de poursuite," *Mémoires de l'Académie Royale des Sciences*, 1732. Paris: Imprimerie Royale, 1735. Pp. 15–16.

1732e "Sur les loix de l'attraction," *Mémoires de l'Académie Royale des Sciences*, 1732. Paris: Imprimerie Royale, 1735. Pp. 343–362.

1733a "Sur la figure de la terre, et sur les moyens que l'astronomie et la géographie fournissent pour la déterminer," *Mémoires de l'Académie Royale des Sciences*, 1733. Paris: Imprimerie Royale, 1735. Pp. 153–164.

1733b Universitätsbibliothek, Basel, Bernoulli Archives, ms LIa 662, No. 32∗: letter from Maupertuis to Johann I Bernoulli, 21 September.

1733c "Sur les loix de l'attraction," *Académie Royale des Sciences, Registres des Procés-Verbaux*, 28 February, Pp. 40r–51r.

1733d Universitätsbibliothek, Basel, Bernoulli Archives, ms LIa 662, No. 30∗: letter from Maupertuis to Johann I Bernoulli, 31 March.

1734a "Sur la figure des astres," *Académie Royale des Sciences, Registres des Procés-Verbaux*, 28 August. Pp. 237v–259r.

1734b "Sur les figures des corps célestes," *Mémoires de l'Académie Royale des Sciences*, 1734. Paris: Imprimerie Royale, 1736. Pp. 55–100.

1735 "Sur la figure de la terre," *Mémoires de l'Académie Royale des Sciences*, 1735. Paris: Imprimerie Royale, 1738. Pp. 98–105.

1736 "Traité des tractoires," *Académie Royale des Sciences, Registres des Procés-Verbaux*, 24 March. Pp. 52r–62v.

1738a *Examen désintéressé des différents ouvrages qui ont été faits pour déterminer la figure de la terre.* Oldenburg: Theobald Bachmuller.

1738b Letter from Maupertuis to Leonhard Euler, 20 May, in Euler (1986: 38).

1740a "Principes du repos des corps," *Académie Royale des Sciences, Registres des Procés-Verbaux*, 20 February. Pp. 27r–30v.

1740b "Loi du repos des corps," *Mémoires de l'Académie Royale des Sciences*, 1740. Paris: Imprimerie Royale, 1742. Pp. 170–176.

1741 *Examen désintéressé des différents ouvrages qui ont été faits pour déterminer la figure de la terre.* Amsterdam.

1743 Letter from Maupertuis to Johann II Bernoulli, 2 November, quoted in Brown (1976: 194).

1809 "Two Problems Concerning the Figure Assumed by Revolving Fluids; With Conjectures Concerning Stars Which Sometimes Appear and Disappear; and on Saturn's Rings," *Philosophical Transactions... Abridged*, vol. 7. London: C. & R. Baldwin. Pp. 519–528. This is an English translation of Maupertuis (1732a).

1896 *Maupertuis et ses Correspondants.* Ed. Achille LeSueur. Montreuil-sur-Mer.

1971 *Maupertuis et ses Correspondants.* Ed. Achille LeSueur. Geneva: Slatkine Reprints. This is Maupertuis (1896) reproduced in facsimile.

MAUPERTUIS, PIERRE-LOUIS MOREAU DE, and NICOLE, FRANÇOIS
1731   Report on Fontaine's new method of determining curvatures of plane curves, in
            Académie Royale des Sciences, Registres des Procès-Verbaux, 11 July 1731.
            Pp. 168r–168v.
MAYER, JEAN
1959   *Diderot, homme de science.* Rennes: Imprimerie Bretonne.
1991   "Diderot et le calcul des probabilités dans l'*Encyclopédie*," *Revue d'Histoire des
            Sciences*, 44: 375–391.
McCLELLAN III, JAMES E.
1981   "The Académie Royale des Sciences, 1699–1793: A statistical Portrait," *Isis*,
            72: 541–567.
McMILLAN, CYNTHIA ANNE
1970   The concept of the Mathematical Infinite in French Thought 1670–1760. Ph.D.
            Dissertation at the University of Virginia.
MERLEAU-PONTY, JACQUES
1975   Comments made during a round-table discussion, published in *Actes de la journée
            Maupertuis (Çréteil, 1 décembre 1973)*. Paris: J. Vrin, 1975. P. 137.
MONTMORT, PIERRE-REMOND DE
1703a   Universitätsbibliothek, Basel, Bernoulli Archives, ms LIa 665, No. 1∗: letter from
            Montmort to Johann I Bernoulli, 27 February.
1703b   "Solution du problème que Monsieur le Marquis de l'Hôpital à proposé aux
            Géomètres dans le *Journal des Savants* de 1692, p. 598," *Journal des Savants*, 19
            February, Pp. 117–121.
1703c   "Probleme: Rectifier la Cissoïde," *Journal des Savants*, 26 February, Pp. 137–140.
1704   Universitätsbibliothek, Basel, Bernoulli Archives, ms LIa 665, No. 1a∗: letter from
            Montmort to Johann I Bernoulli, 1 September.
1709   Universitätsbibliothek, Basel, Bernoulli Archives, ms LIa 665, No. 4∗: letter from
            Montmort to Johann I Bernoulli, 15 Septermber.
1712   Letter from Montmort to Nikolaus I Bernoulli, 8 June, in Montmort (1713: 355–356
            or 1980: 355–356).
1713   *Essay d'analyse sur les jeux de hazard.* 2nd ed. Paris: Jacques Quillau.
1715a   Universitätsbibliothek, Basel, Bernoulli Archives, ms LIa 22²: letter from Montmort
            to Nikolaus I Bernoulli, 8 June.
1715b   Letter from Montmort to Brook Taylor, 2 September or 6 September, in Taylor
            (1793: 81–83).
1716a   Universitätsbibliothek, Basel, Bernoulli Archives, ms LIa 22²: letter from Montmort
            to Nikolaus I Bernoulli, 28 April.
1716b   Letter from Montmort to Brook Taylor, 31 March, in Taylor (1793: 84–88).
1716c   Letter from Montmort to Brook Taylor, 12 April, in Taylor (1793: 93–98).
1718a   Universitätsbibliothek, Basel, Bernoulli Archives, ms LIa 665, No. 18∗: letter from
            Montmort to Johann I Bernoulli, 26 June.
1718b   Universitätsbibliothek, Basel, Bernoulli Archives, ms LIa 665, No. 19∗: letter from
            Montmort to Johann I Bernoulli, 28 October.
1718c   Universitätsbibliothek, Basel, Bernoulli Archives, ms LIa 665, No. 20∗: letter from
            Montmort to Johann I Bernoulli, 30 November.
1718d   Universitätsbibliothek, Basel, Bernoulli Archives, ms LIa 22²: letter from Montmort
            to Nikolaus I Bernoulli, 8 November.
1718e   Library of St. Johns College, Cambridge: letter from Montmort to Brook Taylor, 5
            August.
1980   *Essay d'analyse sur les jeux de hazard.* 2nd ed. New York: Chelsea. This is Montmort
            (1713) reproduced in facsimile.
MONTUCLA, JEAN-ETIENNE
1799–1802   *Histoire des mathematiques.* 2nd ed. 4 vols. Volumes I and II. Paris: Henri Agasse,
            1799. Volumes III and IV completed by Jérôme de Lalande. Paris: Henri Agasse,
            1802.
1968   *Histoire des mathematiques.* 2nd ed. 4 vols. Paris: Albert Blanchard. This is Montucla
            (1799–1802) reproduced in facsimile.

MORERE, JEAN-EDOUARD
1965 "La photométrie: les sources de l'*Essai d'Optique sur la gradation de la lumière de Pierre Bouguer*, 1729," *Revue d'Histoire des Sciences et de Leurs Applications*, 18: 337–384.

MORIN, ROBERT
1987 *Diderot et l'imagination. Annales Littéraires de l'Université de Besançon*. Paris: Les Belles Lettres.

MOUY, PAUL
1938 "Malebranche et Newton," *Revue de Métaphysique et de Morale*, 45: 411–435.

MULLER, JOHN
1760 *Traité analytique des sections coniques, fluxions et fluentes. Avec un essai sur les quadratures, et un Traité du mouvement*. Paris: Charles-Antoine Jombert.

NEWTON, ISAAC
1739–1742 *Philosophiae naturalis principia mathematica*. 4 vols. Ed. Thomas LeSeur and Franciscus Jacquier. Geneva: Barrillot and Filii.

1966 *Principes mathématiques de la philosophie naturelle. Traduction de la Marquise du Chastellet augmentée des commentaires de Clairaut*. 2 vols. Paris: Albert Blanchard.

1972 *Philosophiae Naturalis Principia Mathematica. The Third Edition (1726) with Variant Readings*. 2 vols. Ed. Alexandre Koyré and I. Bernard Cohen. Cambridge, Mass.: Harvard Univ. Press.

1974 *The Mathematical Papers of Isaac Newton*, vol. 6 (1684–1691). Ed. Derek T. Whiteside. Cambridge: Cambridge Univ. Press.

1977 *The Correspondence of Isaac Newton*, vol. 7 (1718–1727). Ed. A. Rupert Hall and Laura Tilling. Cambridge: Cambridge Univ. Press.

NICOLE, FRANÇOIS
1703 "Problème resolu par Monsieur Nicolle," *Journal des Savants*, 19 March, Pp. 190–192.

1715 "Méthode générale pour déterminer la nature des courbes qui coupent une infinité d'autres courbes données de position, en faisant toûjours un angle constant," *Mémoires de l'Académie Royale des Sciences*, 1715. Paris: Imprimerie Royale, 1717. Pp. 49–61.

1717 "Traité du calcul des différences finies," *Mémoires de l'Académie Royale des Sciences*, 1717. Paris: Imprimerie Royale, 1719. Pp. 7–21.

1718 "Solution d'un problème proposé par Monsieur Bernoulli, professeur de mathématiques à Basle," *Académie Royale des Sciences, Registres des Procès-Verbaux*, 9 April. Pp. 97v–101r.

1725 "Solution nouvelle d'un problème proposé aux géomètres Anglais par feu Monsieur Leibniz, peu de temps avant sa mort," *Mémoires de l'Académie Royale des Sciences*, 1725. Paris: Imprimerie Royale, 1727. Pp. 130–153.

1737 "Usage des suites pour la résolution de plusieurs problémes de la méthode inverse des tangentes," *Mémoires de l'Académie Royale des Sciences*, 1737. Paris: Imprimerie Royale, 1740. Pp. 50 bis–85.

NIDERST, ALAIN
1984 "Fontenelle et la science de son temps," in *Studies on Voltaire and the Eighteenth Century*, 228: 171–178.

1989 *Fontenelle. Actes du Colloque tenu à Rouen du 6 au 10 octobre 1987*. Ed. Alain Niderst. Paris: Presses Universitaire de France, 1989.

1991 *Fontenelle*. Paris: Plon.

NIKOLIC, DJORDJE
1961 "Roger Boscovic et la géodésie moderne," *Archives Internationales d'Histoire des Sciences*, 54–57: 315–335.

PAPPAS, JOHN N.
1986 "Inventaire de la correspondence de d'Alembert," in *Studies on Voltaire and the Eighteenth Century*, 245: 131–276.

1987a "Les Relations Entre Boscovich et D'Alembert," in the *Proceedings of the Bicentennial commemoration of R. G. Boscovich. Milano, September 15–18*. Eds. M. Bossi; P. Tucci. Pp. 121–148.

1987b   "R. J. Boscovich et l'Académie des Sciences de Paris." To appear in the proceedings of
           the symposium *La Rencontre des Cultures et des Domaines du Savoir Chez Rudjer
           Josip Boskovic (1711–1787)*. Ve Symposium International d'études sur l'aire cul-
           turelle croate, held in Paris, December 4th and 5th, 1987. The proceedings of the
           conference will be published by Mirko Drazen Grmek and Henrik Hager in *Croatica
           Parisiensia* (Presses de l'Université de Paris-Sorbonne). Pappas's article will also
           appear in *Revue d'Histoire des Sciences* in 1995.
1991   "Documents inédits sur les relations de Boscovich avec la France," *Physis, 28* (new
           series): 163–198.

PEPE, LUIGI
1981   "Il calcolo infinitesimale in Italia agli inizi del secolo XVIII," *Bollettino di storia delle
           Scienze Matematiche, 1*: 43–101.
1984   "Sulla trattatistica del calcolo infinitesimale in Italia nel secolo XVIII," *Atti del
           Convegno "La Storia delle Matematiche in Italia." Cagliari, 29–30 settembre e 1 ottobre
           1982*. Bologna, 1984. Pp. 145–227.
1986   "Les Mathématiciens Italiens et le Calcul Infinitésimal au début du XVIIIe siècle," in
           Studia Leibnitiana, Sonderheft 14: *300 Jahre "Nova Methodus" von G. W. Leibniz
           (1684–1984)*. Stuttgart: Franz Steiner Verlag Wiesbaden GMBH. Pp. 192–201.

PERKINS, MERLE L.
1982   "Diderot and the Time–Space Continuum: His Philosophy, Aesthetics and Politics,"
           Studies on Voltaire and the Eighteenth Century, 211. Oxford: Voltaire Foundation.

PETROVSKII, I. G.
1969   "Ordinary Differential Equations," in *Mathematics. Its Contents, Methods, and
           Meaning*. 3 vols. Ed. A. D. Aleksandrov, A. N. Kolmogorov, and M. A. Lavrent'ev.
           Trans. S. H. Gould; T. Bartha. Cambridge, Mass.: MIT Press. Vol. 1. Pp. 311–356.

PONT, JEAN-CLAUDE
1986   *L'Aventure des Parallèles. Histoire de la Géométrie Non Euclidienne: Précurseurs et
           Attardés*. Berne: Editions Peter Lang.

POPKIN, RICHARD H.
1979   *The History of Scepticism From Erasmus to Spinoza*. Berkeley, Los Angeles, and
           London: Univ. California Press.

PROTTER, MURRAY H., and MORREY, JR., CHARLES B.
1964   *Modern Mathematical Analysis*. Reading, Mass.: Addison-Wesley.

RAPPAPORT, RHODA
1981   "The Liberties of the Paris Academy of Sciences, 1716–1785," in *The Analytic Spirit.
           Essays in the History of Science in Honor of Henry Guerlac*. Ed. Harry Woolf. Ithaca and
           London: Cornell Univ. Press. Pp. 225–253.

ROBINET, ANDRE
1978   *Oeuvres complètes de Malebranche*, vol. XX: *Documents biographiques et biblio-
           graphiques*. Ed. André Robinet. 2nd ed. Paris: J. Vrin.

ROCHE, DANIEL
1969   "Un savant et sa bibliothèque au XVIIIe siècle. Les livres de Jean-Jacques Dortous de
           Mairan," *Dix-Huitième Siècle, 1*: 47–88.

SALOMON-BAYET, CLAIRE
1975   "Maupertuis et l'Institution," in *Actes de la journée Maupertuis (Créteil, 1 décembre
           1973)*. Paris: J. Vrin, 1975. Pp. 183–201.

SAULMON
1717   Académie des Sciences (Paris), "Dossier Saulmon," letter from Saulmon (quoting
           Guisnée) to Abbé Bignon, 4 November.

SAUNDERS, E. STEWART
1977–78   "The Archives of the Académie des Sciences," *French Historical Studies, 10*: 696–702.

SAURIN, JOSEPH
1721   "Eclaircissement sur une difficulté proposée aux mathématiciens par Monsieur le
           Chevalier de Louville (le 30 avril 1720]," *Académie Royale des Sciences, Registres des
           Procès-Verbaux*, 28 June. Pp. 153r–174v.
1722a   "Eclaircissement sur une difficulté proposée aux mathématiciens par Monsieur le

Chevalier de Louville [le 30 avril 1720]," *Mémoires de l'Académie Royale des Sciences, 1722.* Paris: Imprimerie Royale, 1724. Pp. 70–95.

1722b   *Académie Royale des Sciences, Registres des Procès-Verbaux,* 27 June. Pp. 176v–178r.

SCHIER, DONALD S.

1941   *Louis-Bertrand Castel, Anti-Newtonian Scientist.* Cedar Rapids: Torch Press.

SEELEY, ROBERT T.

1961   "Fubini Implies Leibniz Implies $F_{yx} = F_{xy}$," *American Mathematical Monthly, 68:* 56–57.

SERGESCU, PIERRE

1951   "La contribution de Condorcet à l'Encyclopédie," *Revue d'Histoire des Sciences et de Leurs Applications, 4:* 233–237.

SHAPIRO, ALAN E.

1974–75   "Light, Pressure, and Rectilinear Propagation: Descartes' Celestial Optics and Newton's Hydrostatics," *Studies in History and Philosophy of Science, 5:* 239–296.

SHEA, WILLIAM R.

1988   "The Unfinished Revolution: Johann Bernoulli (1667–1748) and the Debate between the Cartesians and the Newtonians," in William R. Shea (ed.), *Revolutions in Science: Their Meaning and Relevance.* Canton, Mass.: Watson, Science History Publications. Pp. 70–92.

SIMONOV, N. I.

1968   "Sur les recherches d'Euler dans la domaine des équations différentielles," *Revue d'Histoire des Sciences et de Leurs Applications, 21:* 131–156.

SPEAR, FREDERICK A.

1980   "Bibliographie de Diderot: Repertoire analytique international," *Histoire des Idées et Critique Littéraire, 187.* Geneva: Droz. And its supplements.

STEVIN, SIMON

1634   *Oeuvres mathématiques de Simon Stevin.* Ed. Albert Girard. Leiden.

STEWART, IAN

1981   "The Science of Significant Form," *The Mathematical Intelligencer, 3:* 50–58.

STIRLING, JAMES

1738   Letter from Stirling to MacLaurin, 26 October, in MacLaurin (1982: 304–306).

STONE, EDMUND

1735   *Analise des infiniments petits, comprenant le calcul intégral dans toute son étendue... servant de suite aux infiniments petits de Monsieur le Marquis de L'Hôpital,* traduit en François par Monsieur Rondet. Paris: Julien-Michel Gandouin, et Pierre-François Giffart.

STRUIK, DIRK J.

1933   "Outline of a History of Differential Geometry," *Isis, 19:* 92–120.

SYNGE, JOHN L., and GRIFFITH, BYRON A.

1959   *Principles of Mechanics.* 3rd ed. (International Student Edition.) New York, Toronto, London, and Tokyo: McGraw-Hill and Kogakusha.

TATON, RENE

1951   *L'Oeuvre scientifique de Gaspard Monge.* Paris: Presses Universitaires de France.

1958   "Réaumur mathématicien," *Revue d'Histoire des Sciences et de Leurs Applications, 11:* 130–133.

1964, 1986   *Enseignement et diffusion des sciences en France au XVIIIe siècle.* Ed. René Taton. Paris: Hermann.

1969   "Madame du Châtelet, traductrice de Newton," *Archives Internationales d'Histoire des Sciences, 22:* 185–210.

1970   Review of Newton (1966), in *Revue d'Histoire des Sciences, 23:* 175–180.

1972   "Fontaine (Fontaine des Bertins), Alexis," *Dictionary of Scientific Biography,* vol. V. Ed. Charles Coulston Gillispie. New York: Charles Scribners Sons. Pp. 54–55.

1975   Comments made during a round-table discussion, published in *Actes de la journée Maupertuis (Créteil, 1 décembre 1973).* Paris: J. Vrin, 1975. Pp. 137–138.

1978   "Sur la diffusion des théories Newtoniennes en France. Clairaut et le problème de la figure de la terre," *Vistas in Astronomy, 22:* 485–509.

1987   "Les relations entre R. J. Boskovic et Alexis Clairaut." To appear in the same volume
            as Pappas (1987b).
1988   "L'Expédition geodésique de Laponie (avril 1736–août 1737)," in *La figure de la Terre
            du XVIIIe siécle à l'ère spatiale*. Ed. Henri Lacombe and Pierre Costabel.
            Paris: Gauthier-Villars. Pp. 115–138.
TAYLOR, BROOK
1793   *Contemplatio Philosophica: A Posthumous Work*. London: W. Bulmer.
TERRALL, MARY
1992   "Representing the Earth's Shape: The Polemics Surrounding Maupertuis's Expedi-
            tion to Lapland," *Isis, 83*: 218–237.
TODHUNTER, ISAAC
1873   *A History of the Mathematical Theories of Attraction and the Figure of the Earth*.
            London: Macmillan.
1962   *A History of the Mathematical Theories of Attraction and the Figure of the Earth*. New
            York: Dover. This is Todhunter (1873) reproduced in facsimile.
TORLAIS, JEAN
1961   *Un esprit encyclopédique en dehors de "L'Encyclopédie": Réaumur d'après les docu-
            ments inédits*. Paris: Albert Blanchard.
TRUESDELL, CLIFFORD A.
1954   "Rational Fluid Mechanics, 1687–1765," in Euler (1954: IX–CXXV).
1955   "Editor's Introduction," in Euler (1955: IX–CXVIII).
1960   "The Rational Mechanics of Flexible or Elastic Bodies, 1638–1788," *Leonhardi Euleri
            Opera Omnia*, series II, vol. 11 (2). Zürich: Orell Füssli.
1984   *An Idiot's Fugitive Essays on Science. Methods, Criticism, Training, Circumstances*.
            New York, Berlin, Heidelberg, and Tokyo: Springer-Verlag.
1987   Review of *The Higher Calculus: A History of Real and Complex Analysis from Euler to
            Weierstrauss*, by Umberto Bottazzini. Trans. Warren Van Egmond. In the *Bulletin
            (New Series) of the American Mathematical Society, 17*: 186–189.
1988   "History of Mathematics Written for Mathematicians," *Archives Internationales
            d'Histoire des Sciences, 38*: 125–137. This is an expanded version of Truesdell (1987).
1991   "Sophie Germain: Fame Earned by Stubborn Error," *Bollettino di Storia delle Scienze
            Matematiche, 11*(2): 3–24.
TWEEDIE, CHARLES
1922   *James Stirling: A Sketch of His Life and Works along with His Scientific Correspon-
            dence*. Oxford: Clarendon Press.
VAN MAANEN, JAN A.
1987   Facets of Seventeenth-Century Mathematics in the Netherlands. Ph.D. Dissertation,
            Utrecht State Univ. (The Netherlands).
VIDAN, GABRIJELA
1977–82   "Un abbé à partie: le révérend père Boscovich à Paris," in *Rudjer Boskovic. Annales de
            l'Institut Français de Zagreb*. Troisième Série/No. 3. Pp. 184–218.
VOLTAIRE, FRANÇOIS
1954   *Voltaire's Correspondence*, vol. 7. Ed. Theodore Besterman. Geneva: Institut et Musée
            Voltaire.
1969   *The Complete Works of Voltaire*, vol. 88. Ed. Theodore Besterman. Geneva and
            Toronto: Institute et Musée Voltaire and Univ. Toronto Press.
VYGODSKY, M.
1975   *Mathematical Handbook. Higher Mathematics*. Trans. George Yankovsky. Mos-
            cow: Mir.
WAFF, CRAIG B.
1975   "Alexis Clairaut and His Proposed Modification of Newton's Inverse-Square Law of
            Gravitation," in *Avant Avec Après Copernic. La réprésentation de l'Univers et es
            conséquences épistémologiques*. Paris: Albert Blanchard. Pp. 281–288.
1976   Universal Gravitation and the Motion of the Moon's Apogee: The Establishment and
            Reception of Newton's Inverse-Square Law, 1687–1749. Ph.D. Dissertation, The
            Johns Hopkins Univ.

WAHSNER, RENATE
1987   "Das Verhältnis von Mathematik und Physik aus der Sicht von Denis Diderot,"
              NTM, 24: 13–20.
WEINSTOCK, ROBERT
1952, 1974   *Calculus of Variations with Applications to Physics and Engineering.* New York:
              McGraw-Hill, 1952; Dover, 1974.
1984   "Newton's *Principia* and the External Gravitational Field of a Spherically Symmetric
              Mass Distribution," *American Journal of Physics, 52*: 883–890.
WEISS, CHARLES
1825   "Savérien (Alexandre)," *Biographie universelle, ancienne et moderne, redigé par une
              société de Gens de lettres et de sciences,* vol. 40. Paris: L. G. Michaud. Pp. 513–515.
WHITESIDE, DEREK T.
1961   "Patterns of Mathematical Thought in the Later Seventeenth Century," *Archive for
              History of Exact Sciences, 1*: 179–388.
WIDDER, DAVID V.
1961   *Advanced Calculus.* 2nd ed. Englewood Cliffs, N. J.: Prentice-Hall.
WILSON, CURTIS A.
1980   "Perturbations and Solar Tables from Lacaille to Delambre: The Rapprochement of
              Observation and Theory, Part I," *Archive for History of Exact Sciences, 22*: 53–188.
WOODHOUSE, ROBERT
1810   *A Treatise on Isoperimetrical Problems and the Calculus of Variations.* Cambridge.
1964   *A History of the Calculus of Variations in the Eighteenth Century.* New York: Chelsea.
              This is Woodhouse (1810) reproduced in facsimile.

Name index

779